Viral Genome Replication

Craig E. Cameron · Matthias Götte ·
Kevin D. Raney
Editors

Viral Genome Replication

Editors
Craig E. Cameron
Pennsylvania State University
University Park
PA, USA
cec9@psu.edu

Matthias Götte
McGill University
Montreal, Quebec
Canada
matthias.gotte@mcgill.ca

Kevin D. Raney
University of Arkansas for Medical Sciences
Little Rock, AR
USA
raneykevind@uams.edu

ISBN 978-0-387-89425-6 e-ISBN 978-0-387-89456-0
DOI 10.1007/b135974

Library of Congress Control Number: 2009920268

Printed on acid-free paper

springer.com

Preface

Currently, there is no single source that permits comparison of the factors, elements, enzymes and/or mechanisms employed by different classes of viruses for genome replication. As a result, we (and our students) often restrict our focus to our particular system, missing out on the opportunity to define unifying themes in viral genome replication or benefit from the advances in other systems. For example, extraordinary biological and experimental paradigms that have been established over the past 5 years for the DNA replication systems of bacteriophage T4 will likely be of great value to anyone interested in studying a replisome from any virus. These studies could easily go unnoticed by animal RNA and DNA virologists. It is our hope that this monograph will cross-fertilize and invigorate the field, as well as encourage students into this area of research.

The monograph has been divided into eight parts. Chapters appearing in Parts I–VI are intended to compare and contrast the replication and/or transcription processes and corresponding "players" of the indicated family of viruses. We are interested in the sequence of events that lead to production of mRNA and progeny genomes as well as the *cis*-acting elements and *trans*-acting factors and enzymes (viral and cellular) that are required for these processes. Chapters appearing in Part VII are intended to provide a more biochemical and biophysical perspective of the replication and/or transcription process. Chapters appearing in Part VIII are intended to provide a practical perspective on viral replication and its inhibition.

Acknowledgments

The editors are indebted to investigators comprising the genome-replication community for contributing chapters to this monograph. In addition, we appreciate the willingness of so many members of this research community to serve as external reviewers of manuscripts, most of whom preferred remaining anonymous. We are especially grateful to Robert Craigie, Roland Marquet, Luis Menendez-Arias, Nicolas Sluis-Cremer, Rodney Russell and Alice Telesnitsky.

The editors also wish to thank Jamie Arnold for contributions at each stage of the process. Finally, CEC thanks members of his laboratory for proofreading, including Spencer Weeks, Eric Smidansky, Hyung Suk Oh, Ibrahim Moustafa, Maria F. Lodeiro, Cheri A. Lee, Elena Gazina, and Daniel Cordek.

Contents

Contributors

Diego E. Alvarez, PhD Molecular Virology Laboratory, Fundación Instituto Leloir, Av Patricias Agentinas 435, Buenos Aires 1405, Argentina, dalvarez@leloir.org.ar

Graciela Andrei, PhD Laboratory of Virology, Rega Institute for Medical Research, Katholieke Universiteit Leuven, B-3000 Leuven, Belgium, graciela.andrei@rega.kuleuven.ac.be

Todd Appleby, PhD Gilead Sciences, 333 Lakeside Drive, Foster City, CA 94404, USA, Todd.Appleby@gilead.com

Matthew P. Badtke, PhD Department of Molecular, Cellular and Developmental Biology, University of Michigan, Ann Arbor, MI 48109, USA, badtkemp@umich.edu

George A. Belov, PhD Picornavirus Replication Section, Laboratory of Infectious Disease, National Institute of Allergy and Infectious Diseases, National Institutes of Health, Bethesda, MD 20982, USA, gbelov@niaid.nih.gov

Stephen J. Benkovic, PhD Department of Chemistry, Pennsylvania State University, 414 Wartik Laboratory, University Park, PA 16802, USA, sjb1@psu.edu

Kathleen Boyle, PhD Microbiology & Mol Genetics, Medical College of Wisconsin, 8701 Watertown Plank Road, Milwaukee, WI 53226, USA, kboyle@mcw.edu

Erik De Clercq, PhD Laboratory of Virology, Rega Institute for Medical Research, Katholieke Universiteit Leuven, B-3000 Leuven, Belgium, erik.declercq@rega.kuleuven.be

Paul Digard, PhD Department of Pathology, University of Cambridge, Tennis Court Road, Cambridge, CB2 1QP, United Kingdom, pd1@mole.bio.cam.ac.uk

James Dolan, PhD Department of Microbiology and Immunology, Northwestern university Feinberg School of Medicine, 303 E. Chicago Ave., Chicago, IL 60611, USA, j-dolan@northwestern.edu

Ellie Ehrenfeld, PhD Picornavirus Replication Section, Laboratory of Infectious Disease, National Institute of Allergy and Infectious Diseases, National Institutes of Health, Bethesda, MD 20982, USA, eehrenfeld@niaid.nih.gov

Debra Elton, PhD Centre for Preventative Medicine, The Animal Health Trust, Lanwades Park, Kentford, Newmarket, Suffolk, CB8 7UU, United Kingdom, debra.elton@aht.org.uk

Cristina Ferrer-Orta, PhD Institut de Biologia molecular de Barcelona (CSIC), Parc Cientific de Barcelona, Baldiri i Reixac 15, E-08028 Barcelona, Spain, cfocri@ibmb.csic.es

Claudia V. Filomatori, PhD Molecular Virology Laboratory, Fundacion Instituto Leloir, Av Patricias Agentinas 435, Buenos Aires 1405, Argentina, cfilomatori@leloir.org.ar

Román Galetto, PhD Cellectis SA, 102 avenue Gaston Roussel, 93235 Romainville, France, roman.galetto@cellectis.com

Andrea V. Gamarnik, PhD Molecular Virology Laboratory, Fundacion Instituto Leloir, Av Patricias Agentinas 435, Buenos Aires 1405, Argentina, agamarnik@leloir.org.ar

Reuben S. Harris, PhD Department of Biochemistry, Molecular Biology and Biophysics, University of Minnesota, 321 Church Street South East, 6-155 Jackson Hall, Minneapolis, MN 55455, USA, rsh@umn.edu

Fred L. Homa, PhD Department of Molecular Genetics and Biochemistry, University of Pittsburgh, W1256 BSTWR, 200 Lothrop Street, Pittsburgh, PA 15261, USA, flhoma@pitt.edu

Nestor G. Iglesias, PhD Molecular Virology Laboratory, Fundacion Instituto Leloir, Av Patricias Agentinas 435, Buenos Aires 1405, Argentina, giglesias@leloir.org.ar

C. Cheng Kao, PhD Department of Biology, Indiana University, Bloomington, IN 47405, USA, ckao@indiana.edu

Rebecca S. LaRue, PhD Department of Biochemistry, Molecular Biology and Biophysics, 321 Church Street South East, 6-155 Jackson Hall, Minneapolis, MN 55455, USA, larue005@umn.edu

Stuart F.J. Le Grice, PhD HIV Drug Resistance Program, National Cancer Institute – Frederick, Building 535, Room 312, P.O. Box B, Frederick, MD 21702-1201, USA, slegrice@ncifcrf.gov

Jonathon Leis, PhD Department of Microbiology and Immunology, Northwestern university Feinberg School of Medicine, 303 E. Chicago Ave., Chicago, IL 60611, USA, j-leis@northwestern.edu

Brett D. Lindenbach, PhD Section of Microbial Pathogenesis, Yale University School of Medicine, 295 Congress Ave., BCMM 354C, New Haven, CT 06536-0812, USA, brett.lindenbach@yale.edu

Shenping Liu, PhD Exploratory Medicinal Sciences, Pfizer Inc., Eastern Point Rd., Groton, CT 06340, USA, shenping.liu@pfizer.com

David Loakes, PhD Medical Research Council, Laboratory of Molecular Biology, Hills Road, Cambridge, CB2 2QH, United Kingdom, David.Loakes@MRC-LMB.CAM.AC.UK

Bruno Marchand, PhD Department of Molecular Microbiology and Immunology, University of Missouri School of Medicine, 471d Life Sciences Center, 1201 E. Rollins Drive, Columbia, MO 65211-7310, USA, marchandb@missouri.edu

Sarah M. McDonald, PhD Department of Laboratory of Infectious Diseases, National Institute of Allergy and Infectious Diseases, National Institutes of Health, Bethesda, MD 20892, USA, mcdonaldsa@niaid.nih.gov

Jennifer T. Miller, PhD HIV Drug Resistance Program, National Cancer Institute, Frederick, MD 21702, USA, jtmiller@ncifcrf.gov

Peter D. Nagy, PhD Department of Plant Pathology, University of Kentucky, 201F Plant Science Building, Lexington, KY 40546, USA, pdnagy2@uky.edu

Matteo Negroni, PhD Unité de Régulation Enzymatique des Activités Cellulaires, CNRS-URA 2185, Institut Pasteur, Paris, France; Present address: Architecture et reactivité de l'ARN, Université de Strasbourg, CNRS, IBMC, 15 rue René Descartes, 67084, Strasbourg, France, m.negroni@ibmc.u-strasbg.fr

Scott W. Nelson, PhD Department of Biochemistry, Biophysics, and Molecular Biology 1210 Molecular Biology Building Iowa State University Ames, IA 50011, USA, swn@iastate.edu

Smita S. Patel, PhD Department of Biochemistry, Robert Wood Johnson Medical School, 675 Hoes Lane, Piscataway, NJ 08854, USA, patelss@umdnj.edu

John T. Patton, PhD Laboratory of Infectious Diseases, National Institute of Allergy and Infectious Diseases, National Institutes of Health, Bethesda, MD 20892, USA, jpatton@niaid.nih.gov

Aniko V. Paul, PhD Department of Molecular Genetics and Microbiology, SUNY at Stony Brook, Nicolls Road, Stony Brook, NY 11794-5222, USA, apaul@notes.cc.sunysb.edu

Judit Pogany, PhD Plant Pathology, University of Kentucky, 201F Plant Science Building, Lexington, KY 40546, USA, jpoga2@uky.edu

Vaishnavi Rajagopal, PhD Department of Biochemistry, Robert Wood Johnson Medical School, 675 Hoes Lane, Piscataway, NJ 08854, USA, rajagova@umdnj.edu

Stefan G. Sarafianos, PhD Department of Molecular Microbiology and Immunology, University of Missouri School of Medicine, 471d Life Sciences Center, 1201 E. Rollins Drive, Columbia, MO 65211-7310, USA, sarafianoss@missouri.edu

Stanley G. Sawicki, PhD Department of Medical Microbiology and Immunology, University of Toledo, College of Medicine, 3055 Arlington Avenue, Toledo, OH 43614, USA, stanley.sawicki@utoledo.edu

April J. Schumacher, PhD Department of Biochemistry, Molecular Biology and Biophysics, 321 Church Street South East, 6-155 Jackson Hall, Minneapolis, MI 55455, USA, schu1480@umn.edu

I-hung Shih, PhD Gilead Sciences, 333 Lakeside Drive, Foster City, CA 94404, USA, Ihung.Shih@gilead.com

Robert Snoeck, PhD Laboratory of Virology, Rega Institute for Medical Research, Katholieke Universiteit Leuven, B-3000 Leuven, Belgium, Robert.Snoeck@rega.kuleuven.be

Michelle M. Spiering, PhD Department of Chemistry, 104 Chemistry Building, The Pennsylvania State University, University Park, PA 16802, USA, mms36@psu.edu

Mark D. Stenglein, PhD Department of Biochemistry, Molecular Biology and Biophysics, 321 Church Street South East, 6-155 Jackson Hall, Minneapolis, MN 55455, USA, sten0171@umn.edu

John E. Tavis, PhD Molecular Microbiology and Immunology, St Louis University School of Medicine, 1100 South Grand Blvd., St Louis, MO 63104, USA, tavisje@slu.edu

Timothy L. Tellinghuisen, PhD Department of Infectology, The Scripps Research Institute, Scripps Florida, 5353 Parkside Drive, RF-2, Jupiter, FL 33458, USA, tellint@scripps.edu

Laurence Tiley, PhD Department of Veterinary Medicine, University of Cambridge, Madingley Road, Cambridge, CB3 0ES, United Kingdom, lst21@cam.ac.uk

Kathleen Too, PhD Medical Research Council, Laboratory of Molecular Biology, Hills Road, Cambridge, CB2 2QH, United Kingdom, kathleentoo@gmail.com

Juan C. de la Torre, PhD Department of Immunology and Microbial Science, The Scripps Research Institute IMM-6, 10550 N Torrey Pines Road, La Jolla, CA 92037, USA, juanct@scripps.edu

Paula Traktman, PhD Department of Microbiology and Molecular Genetics, Medical College of Wisconsin, 8701 Watertown Plank Road, Milwaukee, WI 53226, USA, ptrakt@mcw.edu

Anne-Mieke Vandamme, PhD Katholieke Universiteit Leuven, Laboratory for Clinical and Epidemiological Virology, AIDS Reference Laboratory, Rega Institute and University Hospitals, Minderbroedersstraat 10, B-3000 Leuven, Belgium, annemie.vandamme@uz.kuleuven.ac.be

Kristel Van Laethem, PhD Katholieke Universiteit Leuven, Rega Institute for Medical Research, Clinical and Epidemiological Virology, Minderbroedersstraat 10, 3000 Leuven, Belgium, kristel.vanlaethem@uz.kuleuven.ac.be

Nuria Verdaguer, PhD Institut de Biologia molecular de Barcelona (CSIC), Parc Cientific de Barcelona, Baldiri i Reixac 15, E-08028 Barcelona, Spain, nvmcri@ibmb.csic.es

Sandra K. Weller, PhD Department of Molecular, Microbial and Structural Biology, University of Connecticut Health Center, 263 Farmington Avenue, Farmington, CT 06030-3205, USA, weller@nso2.uchc.edu

Michaela Wendeler, PhD HIV Drug Resistance Program, National Cancer Institute, Frederick, MD 21702, USA, mwendeler@ncifcrf.gov

Sean P.J. Whelan, PhD Department of Microbiology and Molecular Genetics, Harvard Medical School, 200 Longwood Avenue, Boston, MA 02115, USA, swhelan@hms.harvard.edu

Eckard Wimmer, PhD Molecular Genetics & Microbiology, SUNY at Stony Brook, Nicolls Road, Stony Brook, NY 11794-5222, USA, ewimmer@ms.cc.sunysb.edu

Guanghui Yi, PhD Department of Biology, Indiana University, Bloomington, IN 47405, USA, guanghuiyi@tamu.edu

Weidong Zhong, PhD Gilead Sciences, Inc., 333 Lakeside Drive, Foster City, CA 94404, USA, weidong.zhong@gilead.com

Zhihao Zhuang, PhD Department of Chemistry & Biochemistry, 214A Drake Hall, University of Delaware, Newark, DE 19716, USA, zzhuang@udel.edu

Part I
Genome Replication Strategies

Chapter 1
Model of Picornavirus RNA Replication

Aniko V. Paul, George A. Belov, Ellie Ehrenfeld, and Eckard Wimmer

Introduction

The virus family *Picornaviridae* represents a large number of human and animal pathogens, which can cause a variety of diseases ranging from the benign (common cold) to the serious (poliomyelitis). These small non-enveloped plus-stranded RNA viruses have been grouped into nine genera of which five are well known: Enterovirus, Rhinovirus, Hepatovirus, Cardiovirus, and Aphthovirus. The life cycle of picornaviruses begins with attachment to a susceptible host cell, entry, and the delivery of the RNA genome into the cytoplasm (Semler and Wimmer 2002). The RNA is translated into a large polyprotein, which is processed into functional precursor and mature proteins. The nonstructural proteins of the virus and cellular proteins assemble with the parental RNA to form replication complexes on the surface of membranous vesicles where RNA replication takes place. The progeny RNA are encapsidated prior to being released from the host cell.

The RNA genome of picornaviruses (~7500 nucleotides) contains a long 5′ non-translated region (5′NTR), a single open reading frame, and a short 3′ NTR followed by a poly(A) tail (Fig. 1.1). At the 5′-end the RNA is covalently linked to a tyrosine residue in a small peptide called VPg. Picornaviruses use the same basic steps to replicate their genomes as other plus-strand RNA viruses. First the parental RNA is copied into a complementary minus strand yielding a double-stranded replicative intermediate. The minus strand then serves as the template for the production of progeny plus strands. There is also an important difference, however, between the RNA replication strategy of picornaviruses and of other plus-strand RNA viruses. While most other plus-strand RNA viruses start the synthesis of their RNA strands by de novo initiation, picornaviruses use a uridylylated form of the VPg peptide as primer for the production of both plus- and minus-strand RNAs. The enzyme primarily responsible for RNA synthesis is the RNA-dependent RNA polymerase,

A.V. Paul (✉)
Department of Molecular Genetics and Microbiology, SUNY at Stony Brook,
Nicolls Road, Stony Brook, NY 11794-5222, USA
e-mail: apaul@notes.cc.sunysb.edu

C.E. Cameron et al. (eds.), *Viral Genome Replication*,
DOI 10.1007/b135974_1, © Springer Science+Business Media, LLC 2009

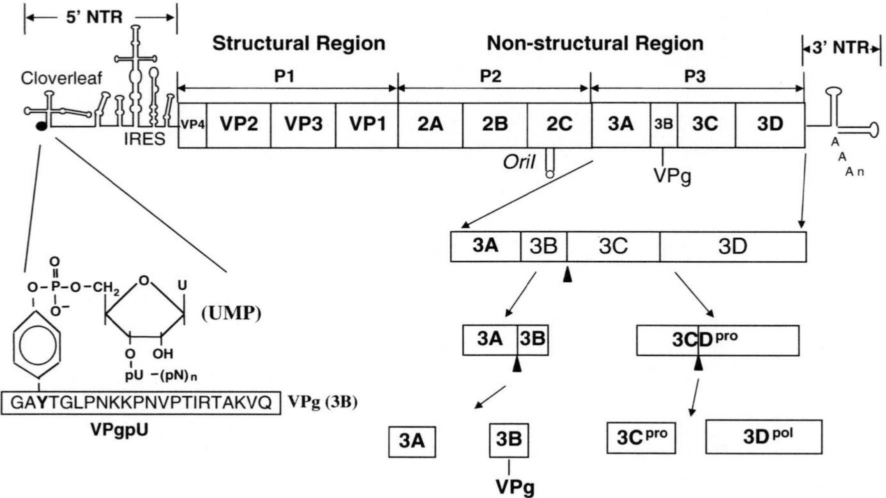

Fig. 1.1 Genomic structure of PV and processing of the P3 domain of the polyprotein. The single-stranded RNA genome of PV is shown with the terminal protein VPg at the 5′-end of the 5′NTR and the 3′NTR with the poly(A) tail. The 5′NTR contains a cloverleaf-like structure and a large IRES element. The attachment site of the 5′-terminal UMP of the RNA to the tyrosine of VPg is shown enlarged. The *oriI* element is located in the coding region of $2C^{ATPase}$. The polyprotein contains structural (P1) and nonstructural (P2 and P3) domains. The *vertical lines* within the polyprotein box represent proteinase cleavage sites. Processing of the P3 domain is shown enlarged.

which requires not only viral but also cellular proteins and *cis*-acting RNA elements to achieve complete replication of the viral RNA genomes.

In this review an attempt will be made to summarize what is known predominantly about the genome replication of poliovirus, the prototype of *Picornaviridae*. Because of the limited scope of this article we will neither be able to discuss in detail the current literature available on all picornavirus RNA replication nor to acknowledge the contribution of every investigator. Principally, progress in five areas have greatly advanced our understanding of poliovirus genome replication during the last 15 years: (i) the development of a de novo cell-free poliovirus replication system, (ii) the elucidation of the mechanism of VPg uridylylation, (iii) the discovery of *cis*-acting genomic RNA structures, (iv) the identification of cellular proteins essential for RNA synthesis, and (v) the characterization of cellular membranous structures involved in genome replication. We suggest that the reader consults previous review articles listed for some early references that could not be accommodated in this article. Finally, we should emphasize that the proposed models of RNA replication are highly speculative and are expected to change as more information accumulates.

Viral and Cellular Factors Involved in Replication

Viral Proteins

The single open reading frame of picornavirus RNAs is translated into a large polyprotein, which is processed by viral proteinases into a variety of precursor and mature proteins (Fig. 1.1). The polyprotein consists of three domains. The P1 domain contains the structural proteins that make up the capsid of the virus while the nonstructural proteins (P2 and P3) are involved in RNA replication and in promoting changes in cellular metabolism. It has been known for a long time that all of the nonstructural proteins of poliovirus have functions in RNA replication. Since picornavirus genomes have a limited coding capacity the virus has adapted to use the genetic information encoded in the RNA multiple times in the form of different precursor and mature proteins. For example, evidence has been presented suggesting that minus-strand RNA synthesis requires large precursors of P2 proteins (P2/P3 or 2BC/P3) (Jurgens and Flanegan 2003).

1. Proteins of the P2 domain. The P2 domain of the polyprotein is processed into a precursor (2BC) and mature proteins ($2A^{pro}$, 2B, and $2C^{ATPase}$) (Leong et al. 2002; Paul 2002; Skern et al. 2002). Protein $2A^{pro}$ is a proteinase in entero- and rhinoviruses whose primary function is to separate the structural and nonstructural domains of the polyprotein but it also has functions in the inhibition of cellular translation and transcription and in RNA replication. The roles of proteins 2B and of its precursor 2BC in RNA replication are not well understood but it is known that they are related to the biochemical and structural changes that occur in the infected cell (Egger et al. 2002; Paul 2002; see below). Expression of 2B in mammalian cells leads to a block of secretory transport, disassembly of the Golgi complex, permeabilization of the plasma membrane, and induction of membrane proliferation and rearrangements. Expression of 2BC results in membrane rearrangements leading to the formation of vesicles. The most conserved protein among picornaviruses is a membrane-bound polypeptide $2C^{ATPase}$ (Leong et al. 2002; Paul 2002). Biochemical and genetic studies have implicated this protein in a variety of functions during the viral life cycle such as uncoating, host cell membrane rearrangements, RNA replication, and encapsidation. The protein contains N- and C-terminal amphipathic helices and RNA-binding domains. There is an N-terminal membrane-binding domain and a cysteine-rich Zn^{++}-binding domain near the C-terminus. In vitro purified $2C^{ATPase}$ exhibits ATPase activity, which is blocked by guanidine hydrochloride, a potent inhibitor of RNA replication in vivo (Pfister et al. 2000 and refs. therein). Although the protein contains conserved motifs typical of helicases so far no helicase activity of the protein has been detected.

2. Proteins of the P3 domain. The proteins derived from the P3 domain are directly involved in RNA replication (Cameron et al. 2002; Leong et al. 2002; Paul 2002). Initial cleavage of the P3 domain yields two relatively stable and very important precursors, 3AB and $3CD^{pro}$. In vitro biochemical studies have shown that the small 3AB protein has multiple functions in RNA replication: (a) 3AB stimulates the polymerization activity of RNA polymerase $3D^{pol}$; (b) 3AB is a non-specific RNA-binding protein, which, however, forms a specific complex with proteinase $3CD^{pro}$ at either the 5′-cloverleaf structure or at the 3′NTR of the viral RNA; (c) 3AB stimulates the autoprocessing of $3CD^{pro}$; (d) the membrane-bound form of 3AB is required for processing by $3CD^{pro}$; (e) 3AB has nucleic acid chaperone and helix destabilizing activities (DeStefano and Titilope 2006). Yeast two-hybrid and biochemical analyses have indicated that 3AB strongly interacts with $3D^{pol}$ and the sequences primarily responsible for this interaction reside in the 3B domain (Y3, K9K10, R17) of the protein (Paul 2002; Paul et al. 2003a). Three amino acids (F377, R379, V391) on the surface of $3D^{pol}$ in a hydrophobic patch were recently identified as binding partners of 3AB (Lyle et al. 2002). Protein 3AB has the propensity to dimerize and form oligomers in solution with both the N-terminal and hydrophobic domain of 3A involved in these interactions (Paul 2002; Strauss et al. 2003). Our recent studies with synthetic membranes suggest that the hydrophobic anchor sequence of 3A forms a mixture of transmembrane and non-transmembrane topographies but adopts only a non-transmembrane configuration in the context of the 3AB protein (Fujita et al. 2007).

Proteolytic processing of 3AB by $3CD^{pro}$ yields 3A and VPg (Leong et al. 2002; Paul 2002). The 3A protein is 87 amino acids long and consists of a soluble cytosolic domain (58 residues), which forms a symmetric dimer (Strauss et al. 2003), a 22-residue long hydrophobic and membrane-binding domain followed by seven additional residues at the C-terminus. The 3A protein inhibits ER to Golgi membrane and secretory protein traffic and induces specific translocation of some ADP Ribosylation Factors (ARF) proteins to membranes (Belov et al. 2005). Studies by the yeast and mammalian two-hybrid systems showed that 3A multimerizes and interacts with $2C^{ATPase}$ and 2B (Teterina et al. 2006; Yin et al. 2007). Mutants resistant to Enviroxime, an antiviral drug that blocks PV RNA replication, map to the 3A sequences supporting a critical role for 3A (or 3AB) in RNA replication.

The VPgs of all picornaviruses are small peptides 21–24 amino acids in length with an absolutely conserved Tyr at position 3. Tyr3 links VPg via a phosphodiester bond to the 5′-terminal UMP of the genome (Fig. 1.1; Wimmer et al. 1993; Paul 2002; Paul et al. 2003a). Entero- and rhinovirus VPgs contain several fully or highly conserved amino acids (Y3, G5, P7, K9, K10, P14, R17), which are required for function in vivo. Interestingly, when two VPgs are introduced in tandem into the PV genome the resulting virus, which has a quasi-infectious growth phenotype, retains only the N-terminal VPg. The replacement of PV VPg with that of HRV14 or HRV16, but not with that of HRV2, results in viable poliovirus (Cheney et al. 2003; Paul et al. 2003a). In contrast to other picornaviruses, foot-and-mouth disease virus (FMDV) encodes in tandem, and uses at random, three distinct VPg peptides (3B1–3B3), which are 23 or 24 amino acids long (Nayak et al. 2005). Each of the VPgs can be uridylylated in vitro although 3B3 is the best substrate for FMDV $3D^{pol}$. Recently two different kinds of structures were proposed for PV VPg. The first structure was predicted by computational modeling and was found to have two antiparallel B strands with the N- and C-termini of the peptide located in close proximity (Tellez et al. 2006) The second structure, determined by NMR, consisted of a large loop (residues 1–14) from which the reactive tyrosine (Y3) projects outward, and of an α-helix (residues 18–21) at the C-terminus (Schein et al. 2006). The amino acids conserved in the VPgs of picornaviruses were located on the same face of the structure as Y3.

The second important precursor of the P3 domain is $3CD^{pro}$, which together with $3C^{pro}$ processes most of the entero- and rhinovirus polyprotein into precursor and mature proteins (Leong et al. 2002; Paul 2002). $3CD^{pro}$ possesses no polymerase activity but it has essential functions in RNA replication as a RNA-binding protein. The RNA-binding domain of the protein is located in $3C^{pro}$ but the $3D^{pol}$ domain of the protein modulates this activity. The crystal structure of PV $3CD^{pro}$ revealed a poorly ordered polypeptide linker between the structurally conserved $3C^{pro}$ and $3D^{pol}$ domains (Marcotte et al. 2007). $3CD^{pro}$ forms several important RNA/protein complexes that are required in RNA replication and these will be discussed later. Studies with the in vitro translation/RNA replication system of Molla et al. (1991) indicated a role for PV $3CD^{pro}$ also in virus maturation, which required both the RNA-binding activity of the $3C^{pro}$ domain and the integrity of interface I in the $3D^{pol}$ domain (Franco et al. 2005).

Processing of the $3CD^{pro}$ precursor yields proteinase $3C^{pro}$ and RNA polymerase $3D^{pol}$. Crystal structures of several picornavirus $3C^{pro}$ proteins (HAV, PV1, HRV14, HRV2) were published and shown to contain a protein fold similar to serine proteinases such as chymotrypsin (Skern et al. 2002). The structure of the PV $3C^{pro}$ protein indicated the formation of dimers and this was confirmed by biochemical experiments (Pathak et al. 2007).

The RNA polymerase $3D^{pol}$ of picornaviruses possesses two major types of synthetic activities in vitro (Cameron et al. 2002; Paul 2002). It elongates RNA or DNA primers on homopolymeric or heteropolymeric RNA templates or catalyzes the covalent attachment of UMP to the hydroxyl group of tyrosine in VPg (Paul et al. 1998). The second reaction requires an RNA template, which can be either poly(A) or an adenylate residue in the *cis*-replicating RNA element *oriI*. The products of the reactions are VPgpU and VPgpUpU, the primers for the synthesis of plus and minus-RNA strands. Crystal structures have been determined for a number of picornavirus RNA polymerases (PV, HRV14, HRV16, HRV1B, and FMDV) and these are discussed by N. Verdaguer and colleagues in another chapter of this book. These structures display a common architecture characteristic of all RNA polymerases, which is that of a right hand with finger, thumb, and palm domains. The purified PV RNA polymerase has been found to exhibit a high level of cooperativity with respect to RNA binding and template usage, suggesting that polymerase/polymerase interactions are important for function. The dimerization/oligomerization of PV $3D^{pol}$ was confirmed by both the yeast and mammalian two-hybrid analysis (Teterina et al. 2006 and refs. therein) and such interactions were also observed in the crystal structure of the protein.

Cellular Proteins

Since plus-strand RNA viruses possess small RNA genomes that encode only a limited number of proteins they seek to supplement their existing synthetic capabilities with cellular proteins (Paul 2002). Several lines of evidence, involving both genetic and biochemical approaches, suggest that this is the case. First, it is known that the replication of RNA viruses is cell-type specific suggesting their dependence on cell-specific factors. Second, a number of host proteins have been identified that interact with viral genomic RNAs or replication proteins and some of these are essential to viral RNA replication.

1. PCBP. Poly(rC)-binding protein 2 (PCBP2), also known as hnRNP E2, or αCP-2, has functions both in the translation and in the replication of PV RNA and possibly also in RNA stability (Paul 2002; Walter et al. 2002). PCBP2 is an RNA-binding protein with a strong preference for poly(rC) sequences. It contains three hnRNP K-homology domains, the first and third of which mediate poly(rC) binding. The protein has been shown to form homodimers and to interact with other hnRNP proteins. For picornavirus RNA containing type I IRES elements, PCBP2 binds to domain IV of the IRES that is essential for translation initiation. In addition, PCBP2 binds to stem-loop B of the 5′-cloverleaf and an adjacent C-rich region in the spacer between the cloverleaf and the IRES (Toyoda et al. 2007). Together with $3CD^{pro}$, this interaction is required for viral RNA synthesis.

2. Sam68. Previous studies using yeast two-hybrid analyses have identified cellular protein Sam68 that interacts with PV $3D^{pol}$ and is relocalized from the nucleus to the cytoplasm upon PV infection (Paul 2002). No function has as yet been assigned to Sam68 in poliovirus replication.

3. Nucleolin. This nuclear protein was found to interact with the 3′NTR of wt PV RNA but not with the RNA of replication-defective mutants (Paul 2002). As with Sam68, no function has as yet been assigned to nucleolin in poliovirus replication.

4. Poly(A)-binding protein (PABP). Herold and Andino (2001) have observed that human PABP interacts in vitro with PV $3CD^{pro}$, PCBP2, and the 3′NTR-poly(A). These observations led to the proposal that the PV genome circularizes via an interaction of PABP, $3CD^{pro}$, and the 5′ cloverleaf on one hand and of PABP and the 3′NTR-poly(A) of the genome on the other.

5. Heterogeneous nuclear ribonucleoprotein C (hnRNP C). This cellular protein that is abundant in the nucleus belongs to a family of RNP motif RNA-binding proteins (Brunner et al. 2005). Using GST-pull down assays it was demonstrated that hnRNPC1 binds to PV $3CD^{pro}$, as well as to the P2 and P3 precursors of the nonstructural proteins. In addition, hnRNPC can be co-immunoprecipitated with PV plus and minus-strand RNA in HeLa extracts suggesting a possible role for hnRNP C in plus-strand RNA synthesis.

6. Reticulon 3. Using yeast two-hybrid analyses, a cellular ER-associated protein, reticulon 3, was recently identified as an interacting partner of enterovirus 71 $2C^{ATPase}$ (Tang et al. 2006). The N-terminal domain of $2C^{ATPase}$, which has both RNA- and membrane-binding activity, was found to interact with reticulon 3. Reduced production of reticulon 3 by RNA interference reduced the synthesis of viral proteins, replicative double-stranded RNA, and plaque formation. Reticulon 3 could also interact with the $2C^{ATPase}$ proteins of PV and CAV16, suggesting that it may be a common factor for the replication of enteroviruses. The function of reticulon 3 was proposed to be to anchor the $2C^{ATPase}$ protein to the membranes but its role needs to be further studied.

7. Other host proteins. The replication of PV in the in vitro translation/replication system and in Xenopus oocytes was found to be dependent on one or more unknown cellular factors. There are numerous other host cell proteins that have been identified through their ability to interact with *cis*-acting RNA elements in the picornavirus genomes (Paul 2002). However, it is not clear that these RNA/protein interactions are biologically important for picornavirus RNA replication.

Cis-*Acting RNA Elements*

The genomes of plus-strand RNA viruses harbor a large amount of genetic information of which much resides in highly structured RNA elements. Most studies in the past concentrated on the role of the 5′NTR and 3′NTR in RNA replication and only recently has the importance of internal *cis*-replicating elements been recognized (Paul 2002).

1. The 5′ cloverleaf (oriL). The 5′-terminal sequences of entero- and rhinovirus RNAs contain a cloverleaf structure (stem-loops A-D) in which the terminal UMP is covalently linked to the hydroxyl group of a tyrosine in the genome-linked protein VPg (Figs. 1.1 and 1.2A). The cloverleaf forms two essential RNP complexes with $3CD^{pro}$ in the presence of either PCBP2 or protein 3AB (Paul 2002). Stem-loop B binds either PCBP or 3AB while a tetra loop in stem-loop D interacts with $3CD^{pro}$ (Rieder et al. 2003). Mutations that disrupt complex formation abolish RNA replication but do not affect translation. Interestingly, not only the C residues in stem-loop B of the cloverleaf are required for PCBP binding and RNA replication but also an adjacent C-rich sequence in the spacer between the cloverleaf and the IRES (Toyoda et al. 2007). Thus, this short segment of spacer sequence is an essential part of the 5′-terminal *cis*-acting element (*oriL*) of the poliovirus genome. The solution structure of a consensus entero- and rhinovirus cloverleaf stem-loop D was determined by NMR and was shown to have an elongated helical stem capped by a UACG tetra loop with a wobble UG closing base pair (Du et al. 2004).

2. The 3′NTR-poly(A) (oriR). The heteropolymeric regions of the 3′NTR in different picornaviruses are very diverse and their functions are unknown although genetic evidence supports their role in RNA replication (Fig. 1.2C; Agol et al. 1999; Paul 2002). A "kissing interaction" between stem-loops X and Y of the PV 3′NTR was found to be important for RNA replication.

The poly(A) tail of picornaviruses is genetically encoded (Wimmer et al. 1993) unlike the poly(A) tails of cellular mRNAs, which are added post-transcriptionally. Efficient RNA replication and infectivity of the viral RNA requires the presence of a poly(A) tail with at least 20 nt (Silvestri et al. 2006). A detailed analysis of the poly(A) tail of CVB3 revealed that while the poly(A) tail is about 80 nt long

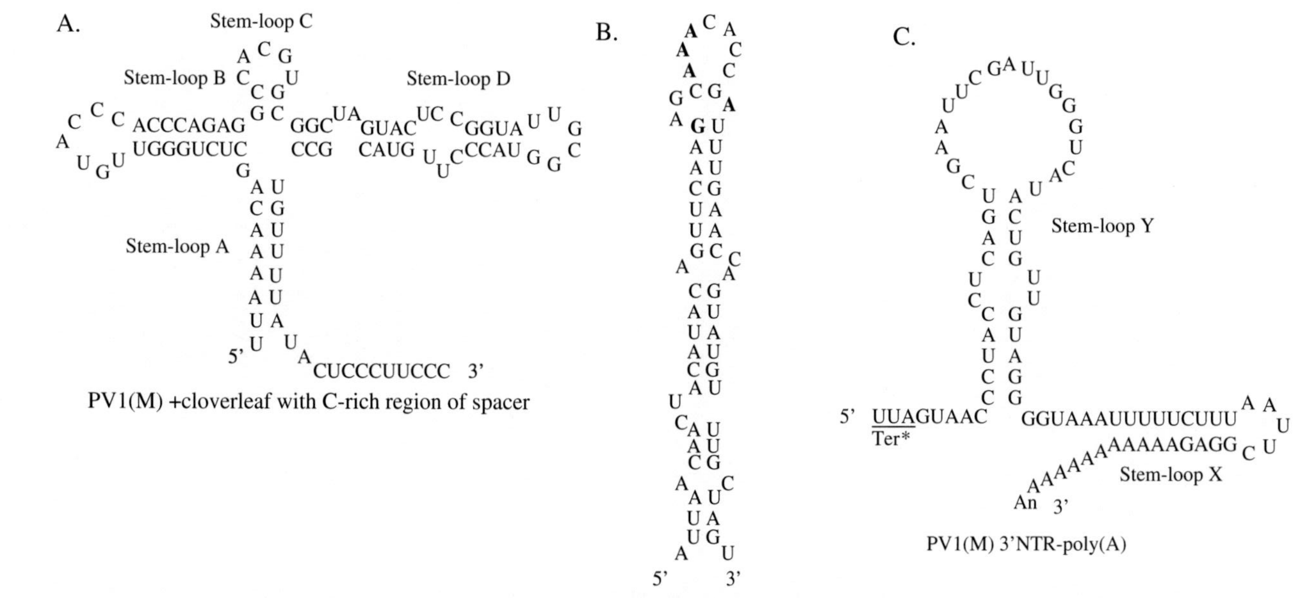

Fig. 1.2 The nucleotide sequence and structure of the PV1 *cis*-replicating elements. (**A**) The 5′-terminal cloverleaf followed by a C-rich region in the spacer between the cloverleaf and the IRES. (**B**) The PV *oriI* [cre(2C)] element and (**C**) the 3′NTR with the poly(A) tail. The conserved entero- rhinoviral cre(2c) sequences are indicated in *bold*.

the complementary poly(U) tract contains only about 20 nts (van Oij et al. 2006a). The 3′NTR controls the length of the poly(A) tail and ensures efficient minus-strand RNA synthesis but apparently it has no effect on poly(U) length.

3. The internal origin of replication (oriI or cre). Analyses of picornaviruses genomes revealed an important *cis*-acting RNA element mapping either to the coding sequences or to the 5′NTR (Fig. 1.2B; Paul 2002). First discovered in the coding sequence of capsid protein VP1 of human rhinovirus 14 (HRV14) (McKnight and Lemon 1998), *oriI* elements have subsequently been identified in $2C^{ATPase}$ of poliovirus and coxsackie virus B3, in $2A^{pro}$ of HRV2, and in the capsid protein VP2 of cardioviruses (for refs. see van Oij et al. 2006b). An exception is the *oriI* of FMDV, which was found to be located in the 5′NTR (Mason et al. 2002). These *oriIs* all consist of a small RNA stem-loop structure made of quite diverse nucleotide sequences. Entero- and rhinovirus *oriIs*, however, contain a conserved motif (Fig. 1.2B; $G_1XXXA_5A_6A_7XXXXXXA_{14}$), which is critically important for function (Yang et al. 2002; Yin et al. 2003). Within this motif, the A_5 residue templates the linkage of both UMPs to VPg by a "slide back" mechanism in a reaction catalyzed by $3D^{pol}$ and stimulated by $3CD^{pro}$ (Fig. 1.3; Paul 2002; Paul et al. 2003b). The products are VPgpU and VPgpUpU, the primers for RNA synthesis. The solution structure of a 33-nt segment the HRV14 *oriI* was recently determined by NMR spectroscopy (Thiviyanathan et al. 2004). It contains a large open loop with 14 nucleotides that derives stability from base-stacking interaction. The two conserved adenylates are oriented to the inside of the loop. Interestingly, the poliovirus *oriI* structure can be moved to different positions within the genome without affecting function (Yin et al. 2003). Recent studies by Crowder and Kirkegaard (2005) have shown that mutants of the PV *oriI* can inhibit PV replication in a *trans*-dominant manner in vivo.

4. The Internal Ribosomal Entry Site (IRES). The poliovirus IRES is located in the 5′NTR between nucleotides 124 and about 630 whose primary function is to promote cap-independent translation (Wimmer et al. 1993; Paul 2002). Numerous

Template 5' $G_1xxxA_5A_6A_7xxxxxxA_{14}$

Protein priming 5' $G_1xxx\mathbf{A_5}A_6A_7xxxxxxA_{14}$
U-VPg

Slide back 5' $G_1xxxA_5\mathbf{A_6}A_7xxxxxxA_{14}$
U-VPg

Elongation 5' $G_1xxx\mathbf{A_5}\mathbf{A_6}A_7xxxxxxA_{14}$
U-U-VPg

Fig. 1.3 The "slide back" mechanism of VPg uridylylation. The first UMP is linked to VPg on the A_5 template nucleotide of the PV1 *oriI*. VPgpU slides back to hybridize with A_6 and the second UMP is templated again by A_5 yielding VPgpUpU. Nucleotides A_5 and A_6 involved in the reaction are shown in *bold*.

genetic studies suggest that the IRES also contains signals for RNA replication in stem-loops II, IV, and V. However, other results are difficult to reconcile with a direct role of the IRES in RNA replication. For example, the IRES of PV1 can be replaced with totally different IRESes from EMCV or HCV but the resulting chimeras have growth properties similar to that of wt poliovirus (Gromeier et al. 1996 and refs. therein). Furthermore, using the in vitro translation/RNA replication system Murray et al. (2004) showed that poliovirus RNA replication was not absolutely dependent on the IRES although the replication of genome length viral RNAs was stimulated by the presence of the IRES in the template RNAs.

5. The cloverleaf at the 3′-end of minus strands. Using 3′-terminal fragments of PV minus-strand RNA, the binding of both cellular and viral ($2C^{ATPase}$, 2BC) proteins derived from virus-infected cell extracts has been demonstrated (Paul 2002). The biological significance of some of these RNA/protein interactions is not yet known. Sharma et al. (2005) recently demonstrated with in vitro translation/RNA replication reactions that the 5′-terminal sequence of stem A in the plus strand, and consequently the 3′-terminal sequence of the minus strand, was required for the efficient plus-strand RNA synthesis.

Membrane Structures

1. Morphological organization of replication complexes. The complexity of the numerous factors that participate in viral RNA synthesis requires that some mechanism exist to topologically coordinate and concentrate the multiple components to function in concert. All positive-strand RNA viruses, including picornaviruses, induce the reorganization of membranes from various sub-cellular organelles (endoplasmic reticulum (ER), Golgi, endosomes, etc.) to form functional scaffolds on which genome replication occurs. In most cases new virus-induced structures are formed that appear by electron microscopy as clusters of heterogeneous sized vesicles concentrated near the nucleus and eventually occupying nearly all the cytoplasm (Fig. 1.4).

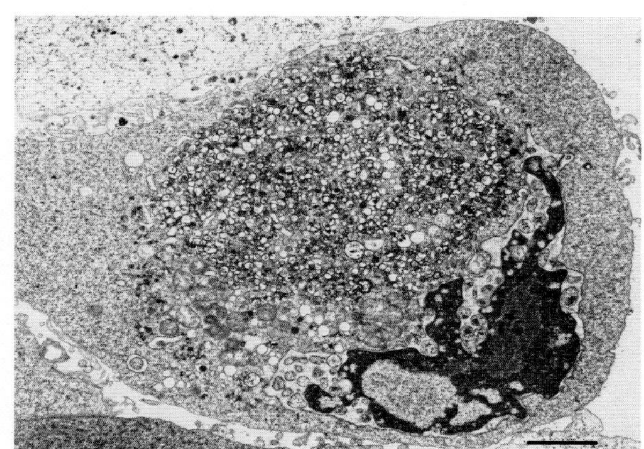

Fig. 1.4 Electron microscopic picture of PV-infected Hep-2 cells. Numerous vesicles can be seen 9 hours post-infection. The bar represents 2 μm. The picture is a gift of K. Bienz and D. Egger. It should be noted that Jackson et al. 2005 have observed some double membrane vesicles in PV-infected cells, which are not apparent on the picture shown here.

The most detailed account of the development of this membrane remodeling has been obtained for poliovirus (Egger et al. 2002; Egger and Bienz 2005). Characteristic vesicles were detected by electron microscopy at 2 h p. i., initially associated with the ER and then clustered in the perinuclear region. Replicating RNAs were located in electron-dense patches in close vicinity to budding vesicles on modified ER and later associated with vesicles. When lysates from infected cells were analyzed by density gradient centrifugation, polymerase activity co-purified with smooth membranes. These replication complexes looked like loosely associated rosettes of membranous vesicles surrounding more dense structures, where actual replication sites were located. When provided with nucleotides and optimal reaction conditions, they could support RNA replication in vitro. It is not yet known how the replication complexes are attached to the membranes but the hydrophobic domains of 3AB, 3A and 2BC, 2B and $2C^{ATPase}$, the latter possibly in conjunction with reticulon 3, are likely to mediate membrane binding.

2. *Viral proteins involved in membrane remodeling.* Expression of all poliovirus nonstructural proteins from non-replicating RNA constructs resulted in membrane rearrangements typical of those found in infected cells (Egger et al. 2002; Egger and Bienz 2005), indicating that viral proteins alone are sufficient to induce characteristic vesicles. Among individual virus proteins that might perform this function, attention was drawn to proteins with intrinsic membrane-targeting properties. Domains in proteins 2B, $2C^{ATPase}$, and 3A and their precursors confer the ability to bind to membranes. Expression of these individual proteins in cells caused intracellular membrane modifications, and when 2BC was co-expressed with 3A, the ultra structure and biochemical properties of the induced vesicles appeared very similar to vesicles found during normal infection. Nevertheless, when cells expressing individual proteins were infected with poliovirus, the pre-formed vesicles were not used in virus replication. This result could mean either that replication vesicles must be formed in *cis*, close to the place of RNA translation, or that vesicles induced by expression of a single viral protein are not the same as those formed when all poliovirus proteins are present. It has been suggested that expression of poliovirus proteins may modify early steps of the secretory pathway (Belov and Ehrenfeld 2007; Egger and Bienz 2005) and/or autophagy (Jackson et al. 2005) but the precise cellular pathways that are utilized in virus-induced membrane remodeling have not yet been elucidated and are currently under investigation in several laboratories.

VPg Uridylylation and RNA Synthesis In Vitro

With Purified Proteins

Purified poliovirus RNA polymerase catalyzes the uridylylation of VPg on a poly(A) template yielding VPgpU and VPgpUpU. These precursors are elongated into VPg-linked poly(U), the 5′-end of minus strands (Paul et al. 1998).

$$VPg + 3D^{pol} + poly(A) + UTP + Mn^{++}(Mg^{++}) \dashrightarrow VPgpU + VPgpUpU \dashrightarrow VPg - poly(U)$$

The enzyme can also use an *oriI* containing PV RNA as template for VPg uridylylation but this reaction requires the stimulatory activity of $3CD^{pro}$ or $3C^{pro}$ (Paul 2002; Pathak et al. 2002).

$$VPg + 3D^{pol} + \textit{oriI}\ RNA + 3CD^{pro} + UTP + Mg^{++}(Mn^{++}) \dashrightarrow VPgpU + VPgpUpU$$

The elongation of the uridylylated VPg precursors in vitro into minus-strand RNA on a PV plus-strand RNA template is very inefficient suggesting that other factors are also required for this process (Paul 2002). In contrast, when PV RNA or another poly(A)-tailed RNA template is incubated with purified $3D^{pol}$ and an oligo(U) primer full-length minus strands can be synthesized.

With Crude Replication Complexes

When crude replication complexes (CRCs) isolated from poliovirus-infected cells are supplied with UTP in vitro they synthesize VPgpU and VPgpUpU in a reaction that is sensitive to the presence of detergents (NP40) (Paul 2002). The uridylylated VPg precursors can be chased into both double- and single-stranded viral RNAs.

With In Vitro Translation/RNA Replication Complexes

As discussed above, dissection and reconstitution of individual steps (partial reactions) that are part of the overall RNA replication mechanism can be performed in vitro with purified components, or analyzed after isolation of replication complexes from infected cells. An additional method for studying viral RNA replication in vitro was developed by Molla et al. (1991) and modified by Barton et al. (2002).

Uridylylation of VPg to form VPgpU and VPgpUpU occurs in the extract in excess of their utilization as primer for RNA chain elongation. Both positive- and negative RNA strands synthesized in vitro are linked to VPg; however, there is some controversy regarding the requirement for *oriI* to serve as template for VPg uridylylation to prime synthesis of negative strands in vitro (see below).

Although uridylylation of VPg can be catalyzed by $3D^{pol}$ in a defined reaction devoid of any membranes (Paul 2002; Nayak et al. 2005), VPg uridylylation formed after translation of poliovirus RNA in HeLa cell extracts was completely eliminated by treatment with non-ionic detergents, suggesting that in vivo this reaction is tightly coupled to the replication complex associated with membranes (Egger et al. 2002; Fogg et al. 2003; Paul 2002). These data, in conjunction with the demonstration that addition of detergent prevented initiation of synthesis of new molecules by replication complexes isolated from infected cells, suggest that the initiation reaction is the membrane-requiring step of viral RNA synthesis. Indeed, addition of even mild detergent abolishes synthesis of poliovirus in the HeLa cell-free extract (Molla et al. 1993). Although membranes are essential for picornavirus RNA replication, their organization into the morphological structures found in infected cells seems to be unnecessary for replication in vitro. Rosettes or vesicle structures typical of poliovirus replication complexes isolated from infected cells were not seen in cell extracts that actively synthesized viral RNA (Fogg et al. 2003).

Proposed Model of Picornavirus RNA Replication

Since virus-infected cells contain both VPgpUpU- and VPg-linked plus- and minus-strand RNAs (Paul 2002), there is little doubt that protein-priming is involved in the initiation of both RNA strands. This hypothesis is supported by the observation that the PV RNA polymerase is strictly primer dependent.

Model of Minus-Strand RNA Synthesis

Prior to minus-strand RNA synthesis translation must be terminated because the ribosomes and the RNA polymerase would have to proceed on the same template but in opposite directions (Paul 2002). It was proposed that the switch from translation to replication occurs when the concentration of $3CD^{pro}$ reaches a critical level. At that time $3CD^{pro}$ interacts with the cloverleaf and sequesters PCBP2 from the IRES thereby shutting off translation and promoting minus-strand RNA synthesis. One problem with this model is that for the most part protein synthesis and RNA replication co-exist in the infected cell (Agol et al. 1999).

Plus-strand RNA viruses initiate negative strand RNA synthesis at the 3′-end of the genome, which is the poly(A) tail in picornavirus RNAs (Agol et al. 1999; Paul 2002). However, the poly(A) tail cannot be the sole determinant of the initiation of negative strand RNA synthesis since the RNA polymerase must discriminate between cellular mRNAs and the viral RNA. For many years it was assumed that the 3′NTR was the only site of recognition in picornavirus RNAs by $3D^{pol}$. This hypothesis was difficult to accept after it was found that the PV 3′NTR can be replaced by the 3′NTR of HRV14 or even deleted and still yield viable virus (Brown et al. 2005). An alternate model was proposed by Herold and Andino (2001) in which the specificity of selection was provided by the viral cloverleaf, which interacted with PCBP2 and $3CD^{pro}$ on the one hand and PABP bound to the poly(A) on the other, thus linking the ends of the viral RNA and effectively circularizing it. This model was based on the observation that all of these *cis*-acting elements and proteins interact in vitro and are required for efficient minus-strand RNA synthesis. In addition, the involvement of a circularized genome in RNA replication is supported by the observation that the 5′ cloverleaf is required in *cis* for minus-strand RNA synthesis (Barton et al. 2001).

Currently two models are being considered to explain the mechanism of VPg-primed negative strand RNA synthesis. According to the first model VPg is

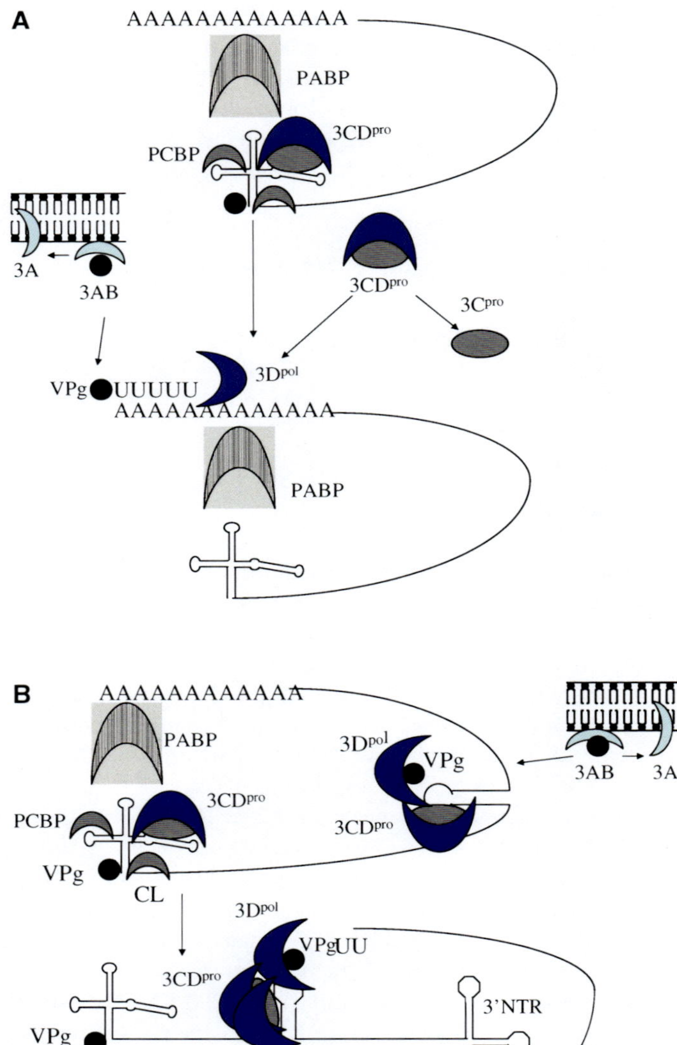

Fig. 1.5 Model of PV minus-strand RNA synthesis. (**A**) VPg is uridylylated on the poly(A) tail and VPgpU is elongated into minus-strand RNA. (**B**) VPg is uridylylated on the *oriI* and VPgpUpU is transferred to the 3′-end of poly(A) before elongation into minus strands. See the text for details of the model.

uridylylated on the poly(A) tail of PV RNA and the resulting VPgpU is immediately elongated into minus strands (Murray and Barton 2003; Morasco et al. 2003). This model is supported by several lines of evidence. First, purified 3D^{pol} catalyzes the uridylylation of VPg in vitro on a poly(A) template yielding VPgpUpU, which is elongated into VPg-linked poly(U) (Paul et al. 1998). Second, the length of the poly(A) tail on PV RNA is an important determinant of minus-strand RNA synthesis both in the in vivo and in the in vitro translation/RNA replication system (van Oij et al. 2006a). Third, mutations in the *oriI* of PV RNA that destroy its structure inhibit viral growth in vivo and VPg uridylylation in vitro translation/RNA replication reactions but have no effect on minus-strand RNA synthesis in the same system (Murray and Barton 2003; Morasco et al. 2003).

In the second model VPgpUpU is made on the PV *oriI* and is subsequently translocated to the 3′-end of the poly(A) tail where it is used as primer for minus-strand RNA synthesis. This model is supported by studies of minus-strand RNA synthesis in the in vitro translation/RNA replication system by point mutants of CVB3 *oriI*. van Oij et al. (2006b) have observed that point mutations in the *oriI* RNA, which do not affect its structure, inhibit both plus and minus-strand RNA synthesis. These investigators proposed that in the in vitro system poly(A) is only used as an alternate template to *oriI* for the uridylylation of VPg when the structure of the *oriI* is disrupted. Under these conditions no RNP complex can form, which would sequester the replication proteins.

Figure 1.5 illustrates both models of minus-strand RNA synthesis in which either the poly(A) tail (A) or the *oriI* (B) is the template for uridylylation of VPg. In each case the first step is the circularization of the genome followed by processing of 3CD^{pro} to yield 3C^{pro} and 3D^{pol}. The RNA polymerase forms a complex with VPg, derived from membrane-bound 3AB, and uridylylates it on the poly(A) tail (A). VPgpUpU is elongated into VPg-linked poly(U) and minus-strand RNA (A). In model B the VPgpUpU made on the *oriI* is translocated to the poly(A) tail where it is elongated into VPg-linked poly(U) and minus-strand RNA. The final product according to both models is a double-stranded replicative form.

Model of Plus-Strand RNA Synthesis

It has been generally accepted that the double-stranded RF structure formed after minus-strand RNA synthesis is a true intermediate in replication (Paul 2002). Therefore, before plus-strand synthesis can begin the end of the RF has to be unwound. It has been proposed that 2C^{ATPase} is responsible for the unwinding of the ends of the duplex molecule because the protein has a conserved helicase motif as well as ATPase activity. However, no helicase activity has been found to be associated with this protein. It is more likely that the unwinding of the end of the RF and the formation of the plus- and minus-strand cloverleaves is facilitated by the binding of a complex of viral and cellular proteins. Since the double-stranded form of picornavirus RNA is infectious it has also been suggested that a cellular helicase is responsible for unwinding the end of the RF.

The in vitro reaction in which VPgpUpU is made on the PV *oriI* with purified protein $3D^{pol}$, $3CD^{pro}$, and synthetic VPg has been thoroughly characterized (Paul 2002). Subsequently, studies with the in vitro translation/RNA replication system have significantly enhanced our understanding of the relationship between VPg uridylylation and RNA replication. First, these studies have provided convincing evidence that the VPgpUpU precursors used for PV plus-strand synthesis are produced on the *oriI* [cre(2C)] RNA (Murray and Barton 2003; Morasco et al. 2003). Second, they showed that the synthesis of VPgpUpU requires membranes (Fogg et al. 2003). Murray and Barton (2003) have proposed that during minus-strand RNA synthesis the circularized genome is disassembled and $3CD^{pro}$ translocates to and enhances the formation of the *oriI* structure where VPg is then uridylylated by $3D^{pol}$. The priming of plus-strand RNA synthesis by VPgpUpU is quite inefficient (Murray and Barton 2003). It is estimated about 500 molecules of VPgpUpU and about 20 plus strands are made for each minus-strand RNA. While the elegant studies using the in vitro translation/replication system have yielded important clues of poliovirus genome replication, their validity in vivo has not been confirmed in all cases.

Figure 1.6 illustrates the proposed model of plus-strand RNA synthesis. Before the synthesis of minus-strand RNA starts or reaches the $2C^{ATPase}$ coding sequences a dimer of $3CD^{pro}$ binds to the upper stem of the *oriI* and destabilizes it (Pathak et al. 2007; Yang et al. 2004; Yin et al. 2003). $3D^{pol}$ is then recruited to the *oriI* by

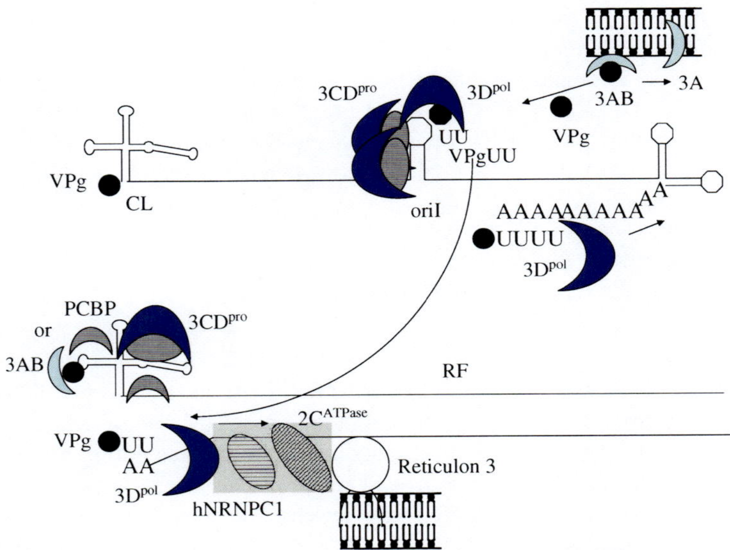

Fig. 1.6 Model of PV plus-strand RNA synthesis. The end of double-stranded RF is unwound by the binding of cellular and viral proteins. VPg is uridylylated on the *oriI* and VPgpUpU is transferred to the 3′-end of minus strands. VPgpUpU primes plus-strand RNA synthesis. See the text for details of the model.

an interaction between the $3C^{pro}$ domain of $3CD^{pro}$ and $3D^{pol}$ (Pathak et al. 2007). VPg, which is derived from the cleavage of 3AB (Liu et al. 2007), interacts with $3D^{pol}$ and is uridylylated. After minus-strand RNA synthesis is completed the end of the RF is unwound and the formation of the minus-strand cloverleaf is enhanced by the binding of hnRNP C and of $2C^{ATPase}$, possibly in a complex with reticulon 3, to the 3′-terminus of minus-strand RNA. In parallel, the formation of the plus-strand cloverleaf is promoted by the interaction of the 5′-terminal plus-strand RNA sequences with PCBP2/$3CD^{pro}$ or 3AB/$3CD^{pro}$ complexes or both. In this context it is interesting to note that 3AB was recently shown to have helix destabilizing activity (DeStefano and Titilope 2006). Once the end of RF is unwound, VPgpUpU is translocated from the *oriI* to the 3′-end of the minus strand. The two 3′-terminal As of minus strand RNA hybridize with the two Us of VPgpUpU, which then leads to the priming of plus-strand RNA synthesis. This model is consistent with the finding that the sequence in stem A of the cloverleaf, and consequently the 3′-terminal sequence in negative strands, is required for efficient initiation of plus-strand RNA synthesis (Sharma et al. 2005).

Some Unanswered Questions About Picornavirus Replication

Despite many years of work on picornavirus RNA replication numerous unanswered questions remain (Agol et al. 1999; Paul 2002). The most important of these questions concerns the viral and cellular factors that are required for the elongation of the uridylylated VPg primers into full-length minus and plus strands. Similarly, nothing is known about the process by which uridylylated VPg is transferred from the *oriI* to the 3′-end of minus strands prior to plus-strand RNA synthesis. Another important question deals with the nature of the true substrate for uridylylation in vivo. In in vitro reactions VPg and 3BC and to a lesser extent 3BCD function as substrates for uridylylation but 3AB does not (Marcotte et al. 2007; Fujita et al. 2007). On the other hand our recent genetic and biochemical studies suggest that in vivo VPg, derived from 3AB, is the substrate of $3D^{pol}$ in the uridylylation reaction (Liu et al. 2007). In this context it should be noted that initially both 3A and VPg have to be delivered to the replication complex in the form of large P3 precursors (Liu et al. 2007; Paul 2002).

Concluding Remarks

During the past 20 years a great deal of information has accumulated on the structure and properties of the viral nonstructural proteins and *cis*-replicating elements. This information was derived from genetic experiments and biochemical studies with purified protein and RNA factors, with the in vitro translation/RNA replication system and with cell-imaging techniques. It is becoming increasingly clear that reconstitution of an in vitro replication complex from purified components will be very difficult, if not impossible. It is more likely that from now on most of the information will come from in vivo experiments or from the in vitro translation/RNA replication system. Hopefully a combination of the different experimental approaches will lead to a better understanding of picornavirus RNA replication in the future.

References

Agol, V. I., Paul, A. V., and Wimmer, E. 1999. Paradoxes of the replication of picornaviral genomes. Virus Res. 62:129–147.

Barton, D. J., O′Donnell, B. J., and Flanegan, J. B. 2001. 5′ cloverleaf in poliovirus RNA is a cis-acting replication element required for negative-strand synthesis. EMBO J. 20:1439–1448.

Barton, D. J., Morasco, B. J., Smerage, L. E., and Flanegan, J. B. 2002. Poliovirus RNA replication and genetic complementation in cell-free reactions. *In* Semler, B. L. and Wimmer, E. (eds), Molecular Biology of Picornaviruses. pp. 461–473. ASM Press, Washington, DC 20036-2904.

Belov, G. A., Fogg, M. H., and Ehrenfeld, E. 2005. Poliovirus proteins induce membrane association of GTPase ADP-ribosylation factor. J. Virol. 79: 7207–7216.

Belov, G., and Ehrenfeld, E. 2007. Involvement of cellular membrane traffic proteins in poliovirus replication. Cell Cycle 6:36–38.

Brown, D. M., Cornell, C. T., Tran, G. P., Nguyen, J. H. C., and Semler, B. L. 2005. An authentic 3′ noncoding region is necessary for efficient poliovirus replication. J. Virol. 79: 11962–11973.

Brunner, J. E., Nguyen, J. H. C., Roehl, H. H., Ho, T. V., Swiderek, K. M., and Semler, B. L. 2005. Functional interaction of heterogeneous nuclear ribonucleoprotein C with poliovirus RNA synthesis initiation complexes. J. Virol. 79:3254–3266.

Cameron, C. E., Gohara, D. W., and Arnold, J. J. 2002. Poliovirus RNA polymerase ($3D^{pol}$): Structure, function and mechanism. *In* Semler, B. L. and Wimmer, E. (eds), Molecular Biology of Picornaviruses. pp. 225–269. ASM Press, Washington, DC 20036-2904.

Cheney, I. W., Naim, S., Shim, J. H., Reinhardt, M., Pai, B, Wu, J. Z., Hong, Z., and Zhong, W. 2003. Viability of poliovirus/rhinovirus VPg chimeric viruses and identification of an amino acid residue in the VPg gene critical for viral RNA replication. J. Virol. 77:7434–7443.

Crowder, S., and Kirkegaard, K. 2005. Trans-dominant inhibition of RNA viral replication can slow the growth of drug-resistant viruses. Nat. Genet. 37:701–709.

DeStefano, J. J., and Titilope, O. 2006. Poliovirus protein 3AB displays nucleic acid chaperone and helix-destabilizing activities. J. Virol. 80:1662–1671.

Du, Z., Yu, J., Ulyanov, N. B., Andino, R., and James, T. L. 2004. Solution structure of a consensus stem-loop D RNA domain that plays important roles in regulating translation and replication in enteroviruses and rhinoviruses. Biochemistry 43:11959–11972.

Egger, D., Gosert, R., and Bienz, K. 2002. Role of cellular structures in viral RNA replication, *In* Semler, B. L. and Wimmer, E. (eds), Molecular Biology of Picornaviruses. pp. 247–255. ASM Press, Washington, DC 20036-2904.

Egger, D., and Bienz, K. 2005. Intracellular location and translocation of silent and active poliovirus replication complexes. J. Gen. Virol. 86:707–718.

Fogg, M. H., Teterina, N. L., and Ehrenfeld, E. 2003. Membrane requirements for uridylylation of the poliovirus VPg protein and viral RNA synthesis in vitro. J. Virol. 77:11408–11416.

Franco, D., Pathak, H. B., Cameron, C. E., Rombaut, B., Wimmer, E., and Paul, A. V. 2005. Stimulation of poliovirus RNA synthesis and virus maturation in a HeLa cell free in vitro translation-RNA replication system by viral protein $3CD^{pro}$. Virol. J. 2:86.

Fujita, K., Krishnakumar, S. S., Franco, D., Paul, A. V., London, E., and Wimmer, E. 2007. Membrane topography of the hydrophobic anchor sequence of poliovirus 3A and 3AB proteins and the functional effect of 3A/3AB membrane association upon RNA replication. Biochemistry 46:5185–5199.

Gromeier, M., Alexander, L., and Wimmer, E. 1996. IRES substitution eliminates neurovirulence in intergeneric poliovirus recombinants. Proc. Natl. Acad. Sci. (USA) 93:2370–2375.

Herold, J., and Andino, R. 2001. Poliovirus RNA replication requires genome circularization through a protein/protein bridge. Mol. Cell 7:581–591.

Jackson, W. T., Giddings, T. H., Taylor, M. P., Mulinyawe, S., Rabonovitch, M., Kopito, R. R., and Kirkegaard, K. 2005. Subversion of cellular autophagosomal machinery by RNA viruses. PLoS Biol. 3:861–871.

Jurgens, C., and Flanegan, J. B. 2003. Initiation of poliovirus negative-strand RNA synthesis requires precursor forms of P2 proteins. J. Virol. 77:1075–1083.

Leong, L. E.-C., Cornell, C. T., and Semler, B. L. 2002. Processing determinants and functions of cleavage products of picornavirus polyproteins. *In* Semler, B. L. and Wimmer, E. (eds), Molecular Biology of Picornaviruses. pp. 187–199. ASM Press, Washington, DC 20036-2904.

Liu, Y., Franco, D., Paul, A. V., and Wimmer, E. 2007. Tyrosine 3 of poliovirus peptide VPg (3B) has an essential function in the context of its precursor protein 3AB. J. Virol. 81:5669–5684.

Lyle, J. M., Clewell, A., Richmond, K., Richards, O. C., Hope, D. A., Schultz, S. C., and Kirkegaard, K. 2002. Similar structural basis for membrane localization and protein priming by an RNA-dependent RNA polymerase. J. Biol. Chem. 277:16324–16331.

Marcotte, L. L., Wass, A. B., Gohara, D. W., Pathak, H. B., Arnold, J. J., Filman, D. J., Cameron, C. E., and Hogle, J. M. 2007. Crystal structure of poliovirus 3CD: Virally-encoded protease and precursor to the RNA-dependent RNA polymerase. J. Virol. 81:3583–3596.

Mason, P. W., Bezborodova, S. V., and Henry, T. M. 2002. Identification and characterization of a cis-acting replication element (cre) adjacent to the IRES of foot-and-mouth disease virus. J. Virol. 76:9686–9694.

McKnight, K. L., and Lemon, S. M. 1998. The rhinovirus type 14 genome contains an internally located RNA structure that is required for viral replication. RNA. 4:1569–1584.

Molla, A., Paul, A. V., and Wimmer, E. 1991. Cell-free de novo synthesis of poliovirus. Science 254:1647–1651.

Molla, A., Paul, A. V., and Wimmer, E. 1993. Effects of temperature and lipophilic agents on poliovirus formation and RNA synthesis in a cell-free system. J. Virol. 67:5932–5938.

Morasco, J. B., Sharma, N., Parilla, J., and Flanegan, J. B. 2003. Poliovirus cre(2C)-dependent synthesis of VPgpUpU is required for positive but not negative-strand RNA synthesis. J. Virol. 77:5136–5144.

Murray, K. E., and Barton, D. J. 2003. Poliovirus cre-dependent VPg uridylylation is required for positive-strand RNA synthesis but not for negative-strand RNA synthesis. J. Virol. 77: 4739–4750.

Murray, K. E., Steil, B. P., Roberts, A. W., and Barton, D. J. 2004. Replication of poliovirus RNA with complete internal ribosome entry site deletions. J. Virol. 78:1393–1402.

Nayak, A., Goodfellow, I. G., and Belsham, G. J. 2005. Factors required for the uridylylation of the foot-and-mouth disease virus 3B1, 3B2, and 3B3 peptides by the RNA-dependent RNA polymerase ($3D^{pol}$) in vitro. J. Virol. 79:7698–7706.

Pathak, H. B., Ghosh, S. K. B., Roberts, A. W., Sharma, S. D., Yoder, J. D., Arnold, J. J., Gohara, D. W., Barton, D. J., Paul, A. V., and Cameron, C. E. 2002. Structure–function relationships of the RNA-dependent RNA polymerase from poliovirus ($3D^{pol}$). A surface of the primary oligomerization domain functions in capsid precursor processing and VPg uridylylation. J. Biol. Chem. 277:31551–31562.

Pathak, H. B., Arnold, J. E., Wiegand, N., Hargittai, M. R. S., and Cameron, C. E. 2007. Picornavirus genome replication: Assembly and organization of the VPg uridylylation ribonucleoprotein complex. J. Biol. Chem. 282:16202–16213.

Paul, A. V., van Boom, J. H., Fillipov, D., and Wimmer, E. 1998. Protein-primed RNA synthesis by purified poliovirus RNA polymerase. Nature 393:280–284.

Paul, A. V. 2002. Possible unifying mechanism of picornavirus genome replication. *In* Semler, B. L., and Wimmer, E. (eds), Molecular Biology of Picornaviruses. pp. 227–246. ASM Press, Washington, DC 20036-2904.

Paul, A. V., Peters, J., Yin, J., van Boom, J., and Wimmer. E. 2003a. Biochemical and genetic studies of the VPg uridylylation reaction catalyzed by the RNA polymerase of poliovirus. J. Virol. 77:891–904.

Paul, A. V., Yin, J., Mugavero, J., Rieder, E., Liu, Y., and Wimmer, E. 2003b. A "slide-back" mechanism for the initiation of protein-primed RNA synthesis by the RNA polymerase of poliovirus. J. Biol. Chem. 278:43951–43960.

Pfister, T, Jones, K. W., and Wimmer, E. 2000. A cysteine-rich motif in poliovirus protein 2C(ATPase) is involved in RNA replication and binds zinc in vitro. J Virol. 74:334–343.

Rieder, E. Xiang, W., Paul, A., and Wimmer, E. 2003. Analysis of the cloverleaf element in a HRV14/poliovirus chimera: Correlation of subdomain D structure, ternary protein complex formation, and virus replication. J. Gen. Virol. 84:2203–2216.

Schein, C. H., Oezguen, N., Volk, D. E., Garimella, R., Paul, A., and Braun, W. 2006. NMR structure of the viral peptide linked to the genome (VPg) of poliovirus. Peptides 27:1676–1684.

Sharma, N., O′Donnell, B. J., and Flanegan, J. B. 2005. 3′-terminal sequence in poliovirus negative-strand templates is the primary cis-acting element required for VPgpUpU-primed positive-strand initiation. J. Virol. 79:3565–3577.

Silvestri, L. S., Parilla, J. M., Morasco, B. J., Ogram, S. A., and Flanegan, J. B. 2006. Relationship between poliovirus negative-strand RNA synthesis and the length of the 3′ poly(A) tail. Virology 345:509–519.

Skern, T., Hampolz, B., Guarne, A., Fita, I., Bergmann, E., Petersen, J., and James, M. N. G. 2002. Structure and function of picornavirus proteinases. *In* Semler, B. L. and Wimmer, E. (eds), Molecular Biology of Picornaviruses. pp. 199–227. ASM Press, Washington, DC 20036-2904.

Strauss, D. M., Glustrom, L. W., and Wuttke, D. S. 2003. Towards an understanding of the poliovirus replication complex: The solution structure of the soluble domain of the poliovirus 3A protein. J. Mol. Biol. 330:225–234.

Tang, W.-F., Yang, S.-Y., Wu, B.-W., Jheng, J.-R., Chen, Y.-L., Shih, C.-H., Lin, K.-H., Lai, H.-C., Tang, P., and Horng, J.-T. 2006. Reticulon 3 binds the 2C protein of enterovirus 71 and is required for viral replication. J. Biol. Chem. 282:5888–5898.

Tellez, A. B., Crowder, S., Spagnolo, J. F., Thompson, A. A., Peersen, O. B., Brutlag, D. L., and Kirkegaard, K. 2006. Nucleotide channel of RNA-dependent RNA polymerase used for intermolecular uridylylation of protein primer. J. Mol. Biol. 357:665–675.

Teterina, N., Levinson, E., Rinaudo, M. S., Egger, D., Bienz, K., Gorbalenya, A. E., and Ehrenfeld, E. 2006. Evidence for functional protein interactions required for poliovirus RNA replication. J. Virol. 80:5327–5337.

Thiviyanathan, V., Yang, Y., Kaluarachchi, K., Rijnbrand, R., Gorenstein, D. G., and Lemon, S. M. 2004. High-resolution structure of picornaviral internal cis-acting RNA replication element (cre). Proc. Natl. Acad. Sci. USA 101:12688–12693.

Toyoda, H., Fujita, K., Paul, A. V., and Wimmer, E. 2007. Replication of poliovirus requires binding of the poly(rC) binding protein to the cloverleaf as well as to the adjacent C-rich spacer sequence between the cloverleaf and the IRES. J. Virol. 81:10017–10028.

van Oij, M. J. M., Polacek, C., Glaudemans, D. H. R. F., Kuijpers, J., van Kuppelveld, F. J. M., Andino, R., Agol, V. I., and Melchers, W. J. G. 2006a. Polyadenylation of genomic RNA and initiation of antigenomic RNA in a positive-strand RNA virus are controlled by the same cis-element. Nucleic Acids Res. 34:2953–2965.

van Oij, M. J. M., Vogt, D. A., Paul, A., Castro, C., Kuijpers, J., van Kuppelveld, F. J. M., Cameron, C. E., Wimmer, E., Andino, R., and Melchers, W. J. G. 2006b. Structural and functional characterization of the coxsackievirus B3 cre(2C): Role of cre(2C) in negative-and positive-strand RNA synthesis. J. Gen. Virol. 87:103–113.

Walter, B. L., Parsely, T. B., Ehrenfeld, E., and Semler, B. L. 2002. Distinct poly(rC) binding protein KH domain determinants for poliovirus translation initiation and viral RNA replication. J. Virol. 76:12008–12022.

Wimmer, E., Hellen, C. U. T., and Cao, X. 1993. Genetics of poliovirus. Annu. Rev. Genet., 27:353–436.

Yang, Y., Rijnbrand, R., McKnight, K. L., Wimmer, E., Paul, A., Martin, A., and Lemon, S. M. 2002. Sequence requirements for viral RNA replication and VPg uridylylation directed by the internal cis-acting replication element (cre) of human rhinovirus type 14. J. Virol. 76: 7485–7494.

Yang, Y., Rijnbrand, R., Watowich, S., and Lemon, S. M. 2004. Genetic evidence for an interaction between a picornaviral cis-acting RNA replication element and 3CD protein. J. Biol. Chem. 279:12659–12667.

Yin, J., Paul, A. V., Wimmer, E., and Rieder, E. 2003. Functional dissection of a poliovirus cis-acting replication element [PV-cre(2C)]: Analysis of single- and dual-cre viral genomes and proteins that bind specifically to PV-cre RNA. J. Virol. 77:5152–5166.

Yin, J., Liu, Y., Wimmer, E., and Paul, A. V. 2007. Complete protein linkage map between the P2 and P3 nonstructural proteins of poliovirus. J. Gen. Virol. 88:2259–2267.

Chapter 2
Coronavirus Genome Replication

Stanley G. Sawicki

Viruses belonging to the family Coronaviridae are unique among RNA viruses because of the unusually large size of their genome, which is of messenger- or positive- or plus-sense. It is ~30,000 bases or 2–3 times larger than the genomes of most other RNA viruses. Coronaviruses belong to the order Nidovirales, the other three families being the Arteriviridae, Toroviridae and Roniviridae. (For a review of classification and evolutionary relatedness of Nidovirales see Gorbalenya et al. 2006.) This grouping is based on the arrangement and relatedness of open reading frames within their genomes and on the presence in infected cells of multiple subgenomic mRNAs that form a 3′-co-terminal, nested set with the genome. Among the Nidovirales, coronaviruses (and toroviruses) are unique in their possession of a helical nucleocapsid, which is unusual for plus-stranded but not minus-stranded RNA viruses; plus-stranded RNA-containing plant viruses in the Closteroviridae and in the Tobamovirus genus also possess helical capsids. Coronaviruses are very successful and have infected many species of animals, including bats, birds (poultry) and mammals, such as humans and livestock. Coronavirus species are classified into three groups, which were based originally on cross-reacting antibodies and more recently on nucleotide sequence relatedness (Gonzalez et al. 2003). There have been several reviews of coronaviruses published recently and the reader is referred to them for more extensive references (Enjuanes et al. 2006; Masters 2006; Pasternak et al. 2006; Sawicki and Sawicki 2005; Sawicki et al. 2007; Ziebuhr 2005).

The genome of coronaviruses is depicted in Fig. 2.1. Its length varies from ~27.5 to ~31 kb among the various species of coronaviruses. The 5′-end is capped although the exact structure of the capped 5′-end has not been determined. The 3′-end is polyadenylated and the genome, as well as subgenomic mRNAs, can be isolated by oligo (dT) chromatography. At the 5′-end there is an untranslated region (5′-UTR) of ~200–500 nucleotides (nts) before the initiation codon for the open reading frame (ORF) that is translated from the genome (ORF1).

S.G. Sawicki (✉)
Department of Medical Microbiology and Immunology, University of Toledo, College of Medicine, 3055 Arlington Avenue, Toledo, OH 43614, USA
e-mail: stanley.sawicki@utoledo.edu

C.E. Cameron et al. (eds.), *Viral Genome Replication*,
DOI 10.1007/b135974_2,

~30,000 bases

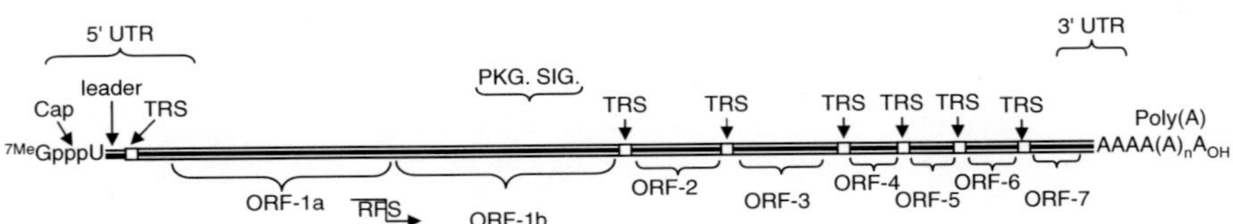

TRS - Transcriptional Regulatory Sequence [UCUAAAC FOR MHV]
RFS - Ribosome Shifting Sequence
ORF - Open Reading Frame
5'-UTR - 5' Untranslated Region (~275 bases)
3'-UTR - 3' Untranslated Region (~80 bases)
Leader - Leader sequence (~65 bases)

Fig. 2.1 Coronavirus genome.

At the 3′-end there is an untranslated region (3′-UTR) of ~250–500 nts after the end of the last ORF and before the poly(A). ORF1 is divided into two large open reading frames (ORF1a and ORF1b); the end of ORF1a overlaps the beginning of ORF1b. A ribosome frame-shifting sequence (RFS) at the end of ORF1a causes the genome to be translated into two unusually long polyproteins, pp1a and pp1ab (see below). After ORF1 there is a series of multiple ORFs, depending on the virus, which are each preceded by a short repeated sequence called the transcription regulating sequence (TRS) immediately upstream of the initiating AUG for that ORF. A TRS is also found about 65 nts from the 5′-end of the genome. The sequence at the 5′ end of the genome, up to this first TRS, is called the leader sequence (Fig. 2.1). The organization of multiple genes was first observed with IBV when its genome was sequenced, which was a feat of manual sequencing skill (Boursnell et al. 1987). After MHV and other coronaviruses were sequenced and shown to have a similar size and organization, equine arteritis virus (EAV, the type member of the Arteriviridae) was sequenced and found to have a similar organization of genes but with half the number of bases as coronaviruses (den Boon et al. 1991). Another distinguishing feature between coronaviruses and arteriviruses is that while coronaviruses have helical nucleocapsids, arteriviruses have the more usual, for plus-stranded RNA viruses, icosahedral-type nucleocapsids. With group 2a coronaviruses, a packaging signal in ORF1b (Chen et al. 2007a) permits the genome, but not the subgenomic mRNA, to be assembled into virions. Some species of coronaviruses package varying amounts of subgenomic mRNAs into virions or membranous structures that have the same density of virions.

The genome replication strategy of coronaviruses, which was originally proposed in 1996 (Sawicki and Sawicki 1995), is depicted in Fig. 2.2. The ORF1 in the genome is translated to form the replicase which can then copy the genome continuously from one end to the other to produce a complementary copy of the genome, i.e., the genomic minus-strand template, that serves in turn to be copied into more

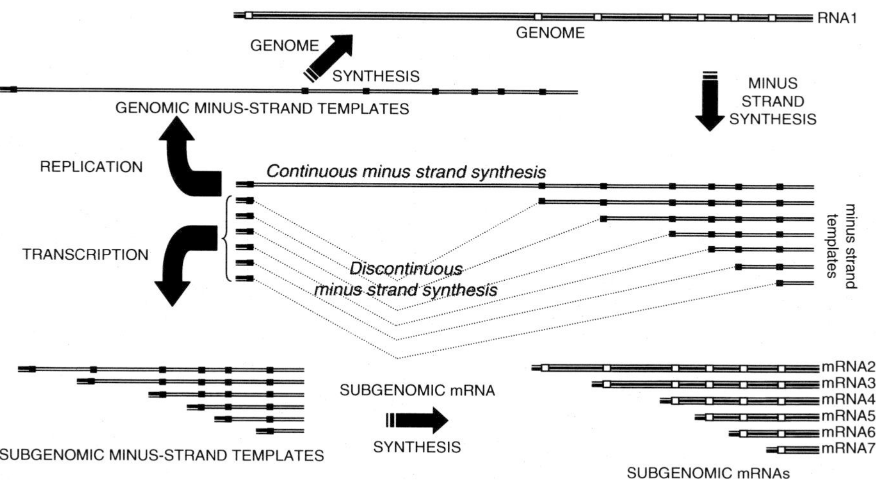

Fig. 2.2 Coronavirus genome replication.

genomes, i.e., genome replication. In addition to making genomic minus-strand templates, the replicase appears to recognize sites at or surrounding the internal TRS, and after copying that internal TRS it then moves discontinuously, or translocates, to the 5′-end of the genome, thereby bypassing a large section of the intervening sequence between any one of the TRS elements and the leader sequence at the 5′-end of the genome. It then continues elongation by copying the leader sequence. Because of this discontinuous event, subgenomic minus-strand templates are produced that also contain a sequence complementary to the leader sequence, i.e., the anti-leader, at the 3′-ends of both genomic and subgenomic minus-strand templates. The subgenomic minus-strand templates, as well as the genomic minus-strand template, would be recognized by the viral transcriptase and copied into subgenomic mRNAs or genomes, respectively. I will refer to the activity of the replication/transcription complex (RTC) that recognizes the genome and synthesizes minus strands as the replicase and the activity of the RTC that recognizes the minus strands (genome-sized as well as subgenome-sized) and synthesizes plus strands as the transcriptase. As discussed below, these are two distinguishable activities of the RTC: The replicase recognizes only the genome as a template and copies it into both genomic and subgenomic minus strands and the transcriptase recognizes both the genomic minus-strand templates and the subgenomic minus-strand templates and copies them into genomes and subgenomic mRNAs, respectively. Because only the genome acts as a template for the production of subgenomic minus-strand templates, a replication signal would be only present on the genome but missing from the subgenomic mRNAs. In contrast, both the genomic and the subgenomic minus strands appear to contain a transcription signal that determines their capacity to serve as templates for plus-strand synthesis.

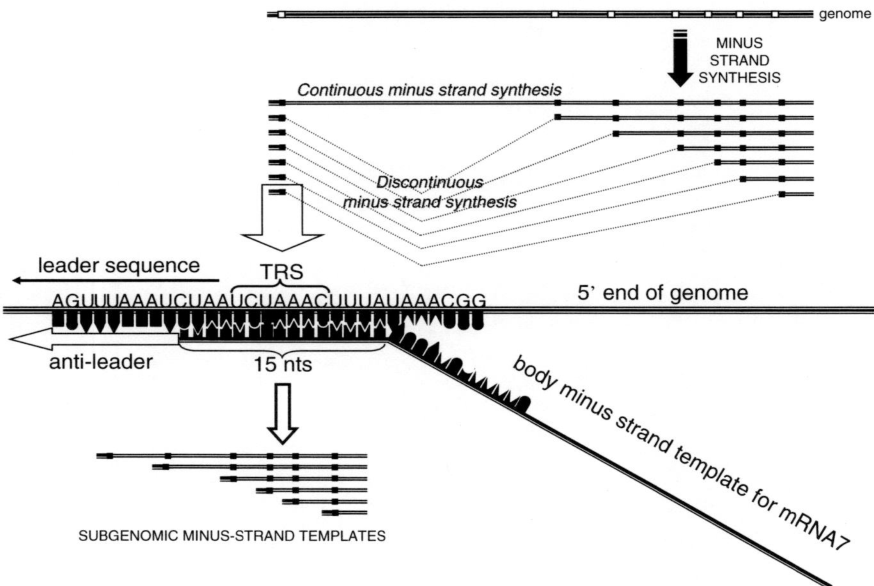

Fig. 2.3 3′-Discontinuous extension of subgenomic minus-strand templates.

Figure 2.3 depicts the key event in the discontinuous synthesis of subgenomic minus-strand templates. The replicase is thought to pause after copying the TRS element and then move with the nascent subgenomic minus strand, which has an anti-TRS at its 3′-end, to the TRS at the end of the leader where it serves to prime and resume elongation before terminating and completing the synthesis of a minus-strand template. Thus, termination of minus-strand synthesis would be the same for genomic as well as subgenomic minus-strand templates. This has been termed facilitated recombination (Brian and Spaan 1997) and creates a subgenomic minus-strand template where the body of the minus strand is joined to the anti-leader at the TRS (actually the complement of the TRS), which results in the subgenomic minus-strand templates all having the same 3′-end as the genomic minus-strand templates. Because they all possess identical 3′- and 5′-ends, all of the minus-strand templates would be equally recognized by the transcriptase. Thus, for coronaviruses to replicate their genome, they need only two activities: One, the replicase that recognizes the genome as a template to make both genomic and subgenomic minus-strand templates and a second, the transcriptase, that recognizes both the genomic and the subgenomic minus-strand templates for the transcription of the viral plus strands. Furthermore, both the genome and all the subgenomic mRNAs have the same 5′-end, which would give each the same ribosome recognition signal. With such a scheme, not only the relative abundance of the different plus strands, but also the relative abundance of the different viral proteins would be determined solely at the level of the minus-strand synthesis. Thus, the crucial determinant or key event in coronavirus genome replication is how the virus determines how much of a particular minus-strand template to produce, i.e., its relative abundance. Each minus-strand template then would be equally susceptible to being copied into a plus-strand RNA because each has the same 5′- and 3′-ends. Furthermore, each plus strand would be equally susceptible to interacting with ribosomes because they all have the same 5′-end sequence and all are polyadenylated, although the genome might be more or less efficiently translated compared to the subgenomic mRNAs because it has a longer 5′-UTR. The initiating AUG on the subgenomic mRNA is very close to the TRS, while on the genome there are ~250 nucleotides between the TRS and the initiating AUG for ORF1. Thus, coronaviruses appeared to have evolved a genome replication strategy that simplifies the problem of coordinating mRNA and protein abundance (gene expression) by focusing on controlling minus-strand template abundance. Thus, the answer to the question "Why do coronaviruses, and also arteriviruses, but not toroviruses or roniviruses possess a leader?" is that they regulate the expression of their genes by controlling minus-strand template abundance. Their regulation of minus-strand template abundance must be considered as a mechanism driving their capacity to have larger RNA genomes and/or many more genes than most other RNA viruses and as responsible for their species diversity.

The genome replication strategy of coronaviruses presented in Fig. 2.2 is based on

(1) Subgenomic mRNA constitutes a 3′-nested set with the genome and they all contain a leader sequence at their 5′-ends; and the leader sequence occurs only once in the genome, also at its 5′-end (Lai et al. 1983; Spaan et al. 1983);
(2) Splicing, i.e., fusion of the 5′- and 3′-sequences of the genome and deletion of the intervening sequences, does not occur (Jacobs et al. 1981; Stern and Sefton 1982);
(3) Subgenomic, in addition to genomic, minus-strand templates are present in infected cells at similar ratios as their corresponding plus strands (Sethna et al. 1989). This corrected the earlier reports that found only genomic minus-strand templates in infected cells (Baric et al. 1983; Lai et al. 1982).
(4) Subgenomic minus-strand templates are present in replication intermediates (RIs) that are actively engaged in plus-strand synthesis (Sawicki and Sawicki 1990),
(5) Replicative form (RF) RNA, i.e., the RNase resistant double-stranded core, with subgenomic minus strands do not arise from replication intermediates (RIs) whose templates were genomic minus strands (Sawicki et al. 2001; Sawicki and Sawicki 1990);
(6) The subgenomic minus strands contained the same anti-leader sequence at their 3′-ends as did the genomic minus strands (Sawicki and Sawicki 1995; Sethna et al. 1991);
(7) Subgenomic mRNA (Brian et al. 1994) or defective interfering (DI) RNA containing only the leader and the TRS at their 5′-end cannot replicate in the presence of helper virus (Makino et al. 1991) but can if they contain at

least ~250 nts of 5′-end of the genome (Brian et al. 1994; Makino et al. 1991; Masters et al. 1994);

(8) RIs containing subgenomic minus-strand templates exist in infected cells and treatment with RNase generate the appropriate RF RNA (Sawicki and Sawicki 1990).

The reader is directed to (Sawicki and Sawicki 2005) for a more detailed account of the history of coronavirus transcription and the other two models proposed for generating subgenomic mRNA by coronaviruses. Eric Snijder and his students and colleagues adopted the discontinuous transcription model (den Boon et al. 1996; van Dinten et al. 1997) to explain EAV genome replication and devised elegant experiments using the infectious clone of EAV and site specific mutations to validate the proposal that it was during minus-strand synthesis that the discontinuous event occurs, whereby nascent minus strands pause at the TRS, relocate and recognize the TRS at the 5′-end of the genome and then act as a primer and complete elongation of the subgenomic minus strands (see Pasternak et al. 2001 for details).

In order to understand how coronaviruses replicate their genome, several questions must be answered: What viral proteins are required for coronavirus genome replication and how exactly do they function? What are the template requirements that specifically permit the viral replicase to recognize the coronavirus genome and copy it into minus-strand templates for genome and subgenomic mRNA? What are the template requirements that specifically permit the transcriptase to recognize the minus-strand templates and copy them into genome and subgenomic mRNA? And, what does the host supply for the replication of the coronavirus genome?

Coronaviruses are typical plus-stranded RNA virus. They do not package a RNA-dependent RNA polymerase in their virions and do not bring this enzyme into the infecting cell. Therefore, they must synthesize such a polymerase by translating its core components from the genome. Figure 2.4 depicts the translational products of ORF1. Two things are striking about the initial polyproteins (pp1a and pp1ab) that are formed. First is their unusually large size (~7,100 amino acids or ~800 KDa) and second is the large number of potential protein products, i.e., 15–16 (called nsp for *n*on*s*tructural *p*roteins and numbered according to their order from the N-terminus to the C-terminus of pp1a and pp1ab), that would be formed after proteolytic processing by either the papain-like cysteine proteases (PL^{PRO}) or the poliovirus 3C-like or coronavirus "main" protease (M^{PRO}) included within pp1a and pp1ab. Sequence analysis of the nonstructural proteins (nsps) predicts that they are associated with at least eight enzymatic activities (Snijder et al. 2003). Bartlam et al. (2007) review the structural proteomics approach to determining the structure–function relationship of the nsp of SARS-CoV, many of which have been crystallized (Cheng et al. 2005; Egloff et al. 2004; Joseph et al. 2006, 2007; Ricagno et al. 2006; Su et al. 2006; Sutton et al. 2004; Yang et al. 2003; Zhai et al. 2005). Some of these activities, e.g., proteinases, RNA-dependent RNA polymerase (RdRp) and helicase (HEL), are common to RNA viruses but others appear to be unique to coronaviruses. Recently, nsp8 was shown to be a second RdRp in addition to nsp12 but one that is less processive and causes the synthesis of complementary

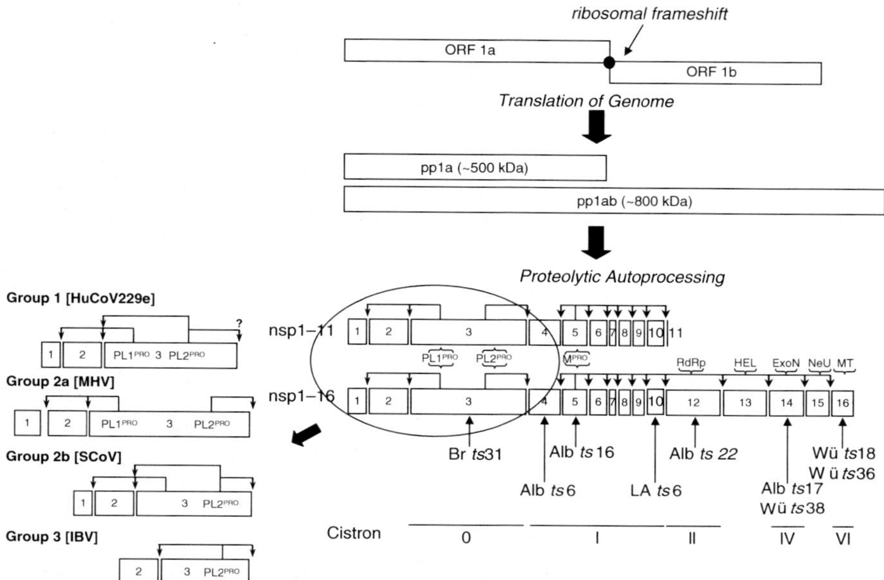

Fig. 2.4 Synthesis and processing of the pp1a and pp1ab polyproteins produced from ORF1 and the location of temperature-sensitive mutants of MHV that do not make viral RNA at 40C and their groupings into cistrons.

oligonucleotides of ~6 residues in a reaction whose fidelity is relatively low. Distant structural homology between the C-terminal domain of nsp8 and the catalytic palm subdomain of RdRps of RNA viruses suggests a common origin of the two coronavirus RdRps, which however may have evolved different sets of catalytic residues (Imbert et al. 2006). Clearly, most of the enzymatic functions associated with coronavirus nsps are concerned with viral RNA synthesis but it should also be noted that some of these activities might have relevance to cellular processes. For example, nsp3 in addition to containing PL^{pro} has been shown to express a deubiquitinating activity and is capable of de-ISGylating protein conjugates (Barretto et al. 2005; Chen et al. 2007b; Ratia et al. 2006), perhaps to subvert cellular processes and facilitate viral replication. Also, the adenosine diphosphate-ribose 1″-phosphatase (ADRP) activity of nsp3, which is not required for coronavirus genome replication (Egloff et al. 2006), may act to influence the levels of cellular ADP-ribose, a key regulatory molecule. Also nsp1, which is probably not essential for genome replication (Graham and Denison 2006; Ziebuhr et al. 2007), is proposed to cause degradation of host mRNA in SARS-CoV infected cells (Kamitani et al. 2006). Thus, it is important to discern those activities or functions that are required to produce viral RNA from those that influence the infected cells to allow viral RNA synthesis and/or to prevent an anti-viral response from foiling genome replication.

If all of the coronavirus proteins were to be assembled into a replicase, it would rival the size and complexity of eukaryotic transcription complexes. Do all of these proteins actually function directly in coronavirus genome replication? Based on sequence analysis, the part of ORF1 starting with PL2^{PRO} at the carboxyl half of nsp3 to the end of nsp16 is highly conserved among coronaviruses, while the sequence from nsp1 to the middle of nsp3 is not highly conserved. Group 3 coronaviruses (Fig. 2.4) exemplified by IBV do not encode an nsp1. Also, reverse genetic experiments showed that nsp1 and nsp2 are not essential for MHV and SARS-CoV genome replication (Deming et al. 2006; Graham et al. 2005; Zust et al. 2007) although recently an RNA stem-loop within nsp1 of group 2a coronavirus might be required for the genome to serve as a template for minus-strand synthesis (Brown et al. 2007). Using classical (forward) genetics or complementation analysis of temperature sensitive (*ts*) mutants (Sawicki et al. 2005; Helen Stokes and Stuart Siddell, personal communication) *ts* mutants that cannot synthesize viral RNA at 39–40ºC (the non-permissive temperature) could be grouped into at least five complementation groups or cistrons 0, I, II, IV and VI. These cistrons were mapped to nsP3, nsp4-10, nsp12, nsp14 and nsp16, respectively. The *ts* mutants tested with causal mutations in nsp4, nsp5 and nsp10 all were found to belong to the same complementation group, i.e., cistron I, suggesting that they are cis-acting. This means that either the polyprotein nsp4/5/6/7/8/9/10 +/– 11 functions in genome replication as the unprocessed polyprotein or nsp4, 5, 6, 7, 8, 9 and 10/11 associate with one another before they are proteolytically processed into individual proteins and thus are not individually diffusible (*trans-active*). Recently, a single nucleotide mutation that caused an arginine to proline substitution in nsp13 (HEL) was found to be lethal for IBV (Fang et al. 2007). Interestingly, this mutation produces the same phenotype of blocking subgenomic mRNA synthesis but allowing genomic RNA synthesis as was found by van Dinten et al. (1997, 2000) for a point mutation in the helicase of EAV. Therefore, it is reasonable to predict that *ts* mutants will be found that have a casual mutation in nsp13 and this may give another cistron, although it is possible that nsp13 will function together with nsp12 or nsp14 and be assigned to cistron II or cistron IV, respectively. A recent report (Eckerle et al. 2006) claimed that the putative active site residues of nsp14 could not be substituted without loss of replication in culture, supporting its essential role. However, whatever functions nsp14 serves appear to be retained by uncleaved or partially processed nsp14, since abolition of either the amino-terminal or carboxy-terminal cleavage site allowed recovery of viable virus. No *ts* mutants with an RNA-negative phenotype and a causal mutation in nsp15 have been found, although single amino acid substitution of its homologue in EAV did result in loss of viral replication (Ivanov et al. 2004), and it would appear that nsp15 probably functions in genome replication, although it might also map to cistron IV or cistron VI. Thus, there might be only five cistrons that encompass replication/transcription functions of pp1a and pp1ab, a result that would argue that certain partially cleaved nonstructural polyproteins are functional in the RTC. At this time it is premature to propose a model for how the viral proteins that are required for coronavirus genome assemble and function in genome replication.

In addition to the nsps that function in viral RNA synthesis, the nucleocapsid protein (N) has been implicated in virus RNA synthesis (Almazan et al. 2004; Bost et al. 2000; Chang and Brian 1996; Shi and Lai 2005; van der Meer et al. 1999) and its expression rescues recombinant coronaviruses from cells transfected with infectious RNA (Almazan et al. 2000; Casais et al. 2001; Coley et al. 2005; Yount et al. 2000, 2003, 2002). According to our model (Sawicki et al. 2007), the subgenomic mRNA expressing N would form almost immediately after the initiation of viral RNA synthesis, in addition to it being present in the infected cell because it would be brought in with the infecting virus. Therefore, it likely does not serve as a replication–transcription switch. It could act as an RNA chaperone, as proposed recently for the N protein of hantaviruses (Mir and Panganiban 2006) and facilitate folding of the genome RNA to permit its copying for the production of a genome-length minus strand. In the case of coronaviruses, such activity could be relevant to, for example, the initiation of minus-strand synthesis or, perhaps, during template switching at the TRS element during discontinuous synthesis. Second, it has not escaped our notice that coronaviruses possess helical nucleocapsids. Thus, similar to many minus-strand virus strategies, its role may be to produce a template that is "configured" to balance the ratio of RTCs engaged in the synthesis of templates either for genome or for subgenomic mRNA. Supporting such a possibility is the observation that replication and transcription from the EAV genome, a virus that has an icosahedral nucleocapsid structure, does not appear to involve N protein function (Molenkamp et al. 2000).

A number of host proteins have been reported to interact with viral RNA (Shi and Lai 2005) but it is unclear what roles these would play in the replication of coronavirus genomes especially since recently it was shown that the region to which these proteins bind can be deleted without preventing the virus from replicating (Goebel et al. 2007). The RTC is associated with double-membrane structures located between the endoplasmic reticulum and the Golgi compartment (Brockway et al. 2003; Gosert et al. 2002; Prentice et al. 2004a,b; Snijder et al. 2006). Transmembrane domains in nsp3, nsp4 and nsp6 are believed to act to anchor the RTC to membranes.

What are the template requirements for the formation of the RTC and for it to make minus-strand templates and genomic and subgenomic mRNA? Using a model analogous to that for picornavirus replication–transcription (Bedard and Semler 2004), the 3′- and 5′-ends of the coronavirus genome may interact, either directly (RNA to RNA) or indirectly (protein to RNA or protein to protein), to form the promoter for minus-strand synthesis. Only genomes containing a 5′-element downstream of the leader would be able to engage the 3′-end to serve as templates for minus-strand synthesis. The subgenome-length mRNAs would be missing the 5′-element (although they would all contain the 3′-element) and this provides an explanation for why they are not able to replicate (Sawicki et al. 2007). Using defective interfering (DI) RNA, it has been proposed that stem-loop (SL) structure within the coding region of nsp1 was required for the replication of the DI-RNA (Brown et al. 2007). Four other SL structures (SLI-IV) located in the 5′-untranslated region (5′-UTR) of the coronavirus genome are implicated in replication and transcription (Brian and Baric 2005). A region of the 5′-UTR, including the 3′-end of the leader, has been postulated to function in joining of the body to the leader during minus-strand synthesis (Wu and Brain 2007; Wu et al. 2006).

Two regions of the 3′-untranslated region (3′-UTR) contain *cis*-acting regulatory elements that play a role in coronavirus RNA synthesis (Brian and Baric 2005). The first region of ~150 nucleotides adjoins the poly(A) stretch and is predicted to form a number of different stem-loop structures. It also contains 55 3′-terminal nucleotides next to the poly(A) that acts as a "minimal promoter" for MHV minus-stand synthesis in a DI-RNA (Lin et al. 1994). The second region contains two stem-loop structures, known as the bulged-stem-loop (BSL) and the hairpin-type pseudoknot (PK). The PK structure involves nucleotides at the base of the BSL structure, which means that the structures are mutually exclusive. It has been proposed that this may represent a form of "molecular switch", related in some, as yet unknown way, to different modes of RNA synthesis (Goebel et al. 2004).

In order for coronavirus to replicate their genome, coronaviruses must create two kinds of machines to synthesize RNA. One recognizes the genome as a template and synthesizes minus strands using both continuous synthesis to make templates for genome synthesis and discontinuous synthesis to make templates for subgenomic mRNA synthesis. The other macromolecular machine makes viral genomes and subgenomic mRNA using the minus-strand templates and continuous transcription. Besides having to recognize different templates and to use or not use discontinuous RNA synthesis, what other differences are there? One is that whereas minus-strand synthesis requires newly made proteins, i.e., minus-strand synthesis is inhibited almost immediately by inhibiting protein synthesis with cycloheximide, plus-strand synthesis continues, in the absence of protein synthesis, for at least 1 hour before decaying (Sawicki and Sawicki 1986). This suggests that only newly made, i.e., nascent, viral proteins function in minus-strand synthesis (replicase) and they are "converted" to plus-strand activity (transcriptase) by the mature RTC. It is possible that there are two independent pathways, one leading to formation of a replicase and one leading to the formation of a transcriptase. Use of temperature-sensitive (*ts*) mutants (Sawicki et al. 2005) supports this notion. Shifting certain *ts* mutants from the permissive (33ºC) to the non-permissive (39ºC), after minus- and plus-strand synthesis had commenced at 33ºC, caused minus-strand synthesis to cease almost immediately while plus strand continued for 1 hour and then declined slowly. Temperature shift caused other mutants to stop plus-strand synthesis.

This is an exciting time both for coronavirologists (and Nidovirologists) who are using forward and reverse genetic and biochemical approaches to unravel the novel discontinuous mechanism of subgenomic minus-strand synthesis and for crystallographers who are probing the domain arrangements and structures of the viral nonstructural proteins, many of which are being found to possess novel folds. Not reviewed in this article are the additional issues of current and future interest that include the mechanism that allows such large RNA genome to avoid error catastrophe and the evolutionary implications of such mechanisms for viral–host interactions in their natural hosts, which include birds and mammals.

References

Almazan, F., C. Galan, and L. Enjuanes. 2004. The nucleoprotein is required for efficient coronavirus genome replication. J Virol **78:**12683–8.

Almazan, F., J. M. Gonzalez, Z. Penzes, A. Izeta, E. Calvo, J. Plana-Duran, and L. Enjuanes. 2000. Engineering the largest RNA virus genome as an infectious bacterial artificial chromosome. Proc Natl Acad Sci U S A **97:**5516–21.

Baric, R. S., S. A. Stohlman, and M. M. Lai. 1983. Characterization of replicative intermediate RNA of mouse hepatitis virus: presence of leader RNA sequences on nascent chains. J Virol **48:**633–40.

Barretto, N., D. Jukneliene, K. Ratia, Z. Chen, A. D. Mesecar, and S. C. Baker. 2005. The papain-like protease of severe acute respiratory syndrome coronavirus has deubiquitinating activity. J Virol **79:**15189–98.

Bartlam, M., Y. Xu, and Z. Rao. 2007. Structural proteomics of the SARS coronavirus: a model response to emerging infectious diseases. J Struct Funct Genomics **8(2–3):**85.

Bedard, K. M., and B. L. Semler. 2004. Regulation of picornavirus gene expression. Microbes Infect **6:**702–13.

Bost, A. G., R. H. Carnahan, X. T. Lu, and M. R. Denison. 2000. Four proteins processed from the replicase gene polyprotein of mouse hepatitis virus colocalize in the cell periphery and adjacent to sites of virion assembly. J Virol **74:**3379–87.

Boursnell, M. E., T. D. Brown, I. J. Foulds, P. F. Green, F. M. Tomley, and M. M. Binns. 1987. Completion of the sequence of the genome of the coronavirus avian infectious bronchitis virus. J Gen Virol **68(Pt 1):**57–77.

Brian, D. A., and R. S. Baric. 2005. Coronavirus genome structure and replication. Curr Top Microbiol Immunol **287:**1–30.

Brian, D. A., R. Y. Chang, M. A. Hofmann, and P. B. Sethna. 1994. Role of subgenomic minus-strand RNA in coronavirus replication. Arch Virol Suppl **9:**173–80.

Brian, D. A., and W. J. M. Spaan. 1997. Recombination and coronavirus defective interfering RNAs. Semin Virol **8:**101–11.

Brockway, S. M., C. T. Clay, X. T. Lu, and M. R. Denison. 2003. Characterization of the expression, intracellular localization, and replication complex association of the putative mouse hepatitis virus RNA-dependent RNA polymerase. J Virol **77:**10515–27.

Brown, C. G., K. S. Nixon, S. D. Senanayake, and D. A. Brian. 2007. An RNA stem-loop within the bovine coronavirus nsp1 coding region is a cis-acting element in defective interfering RNA replication. J Virol **81:**7716–24.

Casais, R., V. Thiel, S. G. Siddell, D. Cavanagh, and P. Britton. 2001. Reverse genetics system for the avian coronavirus infectious bronchitis virus. J Virol **75:**12359–69.

Chang, R. Y., and D. A. Brian. 1996. cis Requirement for N-specific protein sequence in bovine coronavirus defective interfering RNA replication. J Virol **70:**2201–7.

Chen, S. C., E. van den Born, S. H. van den Worm, C. W. Pleij, E. J. Snijder, and R. C. Olsthoorn. 2007a. New structure model for the packaging signal in the genome of group IIa coronaviruses. J Virol **81:**6771–4.

Chen, Z., Y. Wang, K. Ratia, A. D. Mesecar, K. D. Wilkinson, and S. C. Baker. 2007b. Proteolytic processing and deubiquitinating activity of papain-like proteases of human coronavirus NL63. J Virol **81:**6007–18.

Cheng, A., W. Zhang, Y. Xie, W. Jiang, E. Arnold, S. G. Sarafianos, and J. Ding. 2005. Expression, purification, and characterization of SARS coronavirus RNA polymerase. Virology **335:** 165–76.

Coley, S. E., E. Lavi, S. G. Sawicki, L. Fu, B. Schelle, N. Karl, S. G. Siddell, and V. Thiel. 2005. Recombinant mouse hepatitis virus strain A59 from cloned, full-length cDNA replicates to high titers in vitro and is fully pathogenic in vivo. J Virol **79:**3097–106.

Deming, D. J., R. L. Graham, aM. R. Denison, and R. S. Baric. 2006. MHV-A59 ORF1a replicase protein nsp7-nsp10 processing in replication. Adv Exp Med Biol **581:**101–4.

den Boon, J. A., M. F. Kleijnen, W. J. Spaan, and E. J. Snijder. 1996. Equine arteritis virus subgenomic mRNA synthesis: analysis of leader-body junctions and replicative-form RNAs. J Virol **70:**4291–8.

den Boon, J. A., E. J. Snijder, E. D. Chirnside, A. A. de Vries, M. C. Horzinek, and W. J. Spaan. 1991. Equine arteritis virus is not a togavirus but belongs to the coronaviruslike superfamily. J Virol **65:**2910–20.

Eckerle, L. D., S. M. Brockway, S. M. Sperry, X. Lu, and M. R. Denison. 2006. Effects of mutagenesis of murine hepatitis virus nsp1 and nsp14 on replication in culture. Adv Exp Med Biol **581:**55–60.

Egloff, M. P., F. Ferron, V. Campanacci, S. Longhi, C. Rancurel, H. Dutartre, E. J. Snijder, A. E. Gorbalenya, C. Cambillau, and B. Canard. 2004. The severe acute respiratory syndrome-coronavirus replicative protein nsp9 is a single-stranded RNA-binding subunit unique in the RNA virus world. Proc Natl Acad Sci U S A **101:**3792–6.

Egloff, M. P., H. Malet, A. Putics, M. Heinonen, H. Dutartre, A. Frangeul, A. Gruez, V. Campanacci, C. Cambillau, J. Ziebuhr, T. Ahola, and B. Canard. 2006. Structural and functional basis for ADP-ribose and poly(ADP-ribose) binding by viral macro domains. J Virol **80:** 8493–502.

Enjuanes, L., F. Almazan, I. Sola, and S. Zuniga. 2006. Biochemical aspects of coronavirus replication and virus-host interaction. Annu Rev Microbiol **60:**211–30.

Fang, S., B. Chen, F. P. Tay, B. S. Ng, and D. X. Liu. 2007. An arginine-to-proline mutation in a domain with undefined functions within the helicase protein (Nsp13) is lethal to the coronavirus infectious bronchitis virus in cultured cells. Virology **358:**136–47.

Goebel, S. J., B. Hsue, T. F. Dombrowski, and P. S. Masters. 2004. Characterization of the RNA components of a putative molecular switch in the 3′ untranslated region of the murine coronavirus genome. J Virol **78:**669–82.

Goebel, S. J., T. B. Miller, C. J. Bennett, K. A. Bernard, and P. S. Masters. 2007. A hypervariable region within the 3′ cis-acting element of the murine coronavirus genome is nonessential for RNA synthesis but affects pathogenesis. J Virol **81:**1274–87.

Gonzalez, J. M., P. Gomez-Puertas, D. Cavanagh, A. E. Gorbalenya, and L. Enjuanes. 2003. A comparative sequence analysis to revise the current taxonomy of the family Coronaviridae. Arch Virol **148:**2207–35.

Gorbalenya, A. E., L. Enjuanes, J. Ziebuhr, and E. J. Snijder. 2006. Nidovirales: evolving the largest RNA virus genome. Virus Res **117:**17–37.

Gosert, R., A. Kanjanahaluethai, D. Egger, K. Bienz, and S. C. Baker. 2002. RNA replication of mouse hepatitis virus takes place at double-membrane vesicles. J Virol **76:**3697–708.

Graham, R. L., and M. R. Denison. 2006. Replication of murine hepatitis virus is regulated by papain-like proteinase 1 processing of nonstructural proteins 1, 2, and 3. J Virol **80:**11610–20.

Graham, R. L., A. C. Sims, S. M. Brockway, R. S. Baric, and M. R. Denison. 2005. The nsp2 replicase proteins of murine hepatitis virus and severe acute respiratory syndrome coronavirus are dispensable for viral replication. J Virol **79:**13399–411.

Imbert, I., J. C. Guillemot, J. M. Bourhis, C. Bussetta, B. Coutard, M. P. Egloff, F. Ferron, A. E. Gorbalenya, and B. Canard. 2006. A second, non-canonical RNA-dependent RNA polymerase in SARS coronavirus. Embo J **25:**4933–42.

Ivanov, K. A., T. Hertzig, M. Rozanov, S. Bayer, V. Thiel, A. E. Gorbalenya, and J. Ziebuhr. 2004. Major genetic marker of nidoviruses encodes a replicative endoribonuclease. Proc Natl Acad Sci U S A **101:**12694–9.

Jacobs, L., W. J. Spaan, M. C. Horzinek, and B. A. van der Zeijst. 1981. Synthesis of subgenomic mRNA′s of mouse hepatitis virus is initiated independently: evidence from UV transcription mapping. J Virol **39:**401–6.

Joseph, J. S., K. S. Saikatendu, V. Subramanian, B. W. Neuman, A. Brooun, M. Griffith, K. Moy, M. K. Yadav, J. Velasquez, M. J. Buchmeier, R. C. Stevens, and P. Kuhn. 2006. Crystal structure of nonstructural protein 10 from the severe acute respiratory syndrome coronavirus reveals a novel fold with two zinc-binding motifs. J Virol **80:**7894–901.

Joseph, J. S., K. S. Saikatendu, V. Subramanian, B. W. Neuman, M. J. Buchmeier, R. C. Stevens, and P. Kuhn. 2007. Crystal structure of a monomeric form of severe acute respiratory syndrome coronavirus endonuclease nsp15 suggests a role for hexamerization as an allosteric switch. J Virol **81:**6700–8.

Kamitani, W., K. Narayanan, C. Huang, K. Lokugamage, T. Ikegami, N. Ito, H. Kubo, and S. Makino. 2006. Severe acute respiratory syndrome coronavirus nsp1 protein suppresses host gene expression by promoting host mRNA degradation. Proc Natl Acad Sci U S A **103:** 12885–90.

Lai, M. M., C. D. Patton, R. S. Baric, and S. A. Stohlman. 1983. Presence of leader sequences in the mRNA of mouse hepatitis virus. J Virol **46:**1027–33.

Lai, M. M., C. D. Patton, and S. A. Stohlman. 1982. Replication of mouse hepatitis virus: negative-stranded RNA and replicative form RNA are of genome length. J Virol **44:**487–92.

Lin, Y. J., C. L. Liao, and M. M. Lai. 1994. Identification of the cis-acting signal for minus-strand RNA synthesis of a murine coronavirus: implications for the role of minus-strand RNA in RNA replication and transcription. J Virol **68:**8131–40.

Makino, S., M. Joo, and J. K. Makino. 1991. A system for study of coronavirus mRNA synthesis: a regulated, expressed subgenomic defective interfering RNA results from intergenic site insertion. J Virol **65:**6031–41.

Masters, P. S. 2006. The molecular biology of coronaviruses. Adv Virus Res **66:**193–292.

Masters, P. S., C. A. Koetzner, C. A. Kerr, and Y. Heo. 1994. Optimization of targeted RNA recombination and mapping of a novel nucleocapsid gene mutation in the coronavirus mouse hepatitis virus. J Virol **68:**328–37.

Mir, M. A., and A. T. Panganiban. 2006. Characterization of the RNA chaperone activity of hantavirus nucleocapsid protein. J Virol **80:**6276–85.

Molenkamp, R., H. van Tol, B. C. Rozier, Y. van der Meer, W. J. Spaan, and E. J. Snijder. 2000. The arterivirus replicase is the only viral protein required for genome replication and subgenomic mRNA transcription. J Gen Virol **81:**2491–6.

Pasternak, A. O., W. J. Spaan, and E. J. Snijder. 2006. Nidovirus transcription: how to make sense...? J Gen Virol **87:**1403–21.

Pasternak, A. O., E. van den Born, W. J. Spaan, and E. J. Snijder. 2001. Sequence requirements for RNA strand transfer during nidovirus discontinuous subgenomic RNA synthesis. Embo J **20:**7220–8.

Prentice, E., W. G. Jerome, T. Yoshimori, N. Mizushima, and M. R. Denison. 2004a. Coronavirus replication complex formation utilizes components of cellular autophagy. J Biol Chem **279:**10136–41.

Prentice, E., J. McAuliffe, X. Lu, K. Subbarao, and M. R. Denison. 2004b. Identification and characterization of severe acute respiratory syndrome coronavirus replicase proteins. J Virol **78:**9977–86.

Ratia, K., K. S. Saikatendu, B. D. Santarsiero, N. Barretto, S. C. Baker, R. C. Stevens, and A. D. Mesecar. 2006. Severe acute respiratory syndrome coronavirus papain-like protease: structure of a viral deubiquitinating enzyme. Proc Natl Acad Sci U S A **103:**5717–22.

Ricagno, S., M. P. Egloff, R. Ulferts, B. Coutard, D. Nurizzo, V. Campanacci, C. Cambillau, J. Ziebuhr, and B. Canard. 2006. Crystal structure and mechanistic determinants of SARS coronavirus nonstructural protein 15 define an endoribonuclease family. Proc Natl Acad Sci U S A **103:**11892–7.

Sawicki, S. G., and D. L. Sawicki. 1986. Coronavirus minus-strand RNA synthesis and effect of cycloheximide on coronavirus RNA synthesis. J Virol **57:**328–34.

Sawicki, S. G., and D. L. Sawicki. 1990. Coronavirus transcription: subgenomic mouse hepatitis virus replicative intermediates function in RNA synthesis. J Virol **64:**1050–6.

Sawicki, S. G., and D. L. Sawicki. 1995. Coronaviruses use discontinuous extension for synthesis of subgenome- length negative strands. Adv Exp Med Biol **380:**499–506.

Sawicki, S. G., and D. L. Sawicki. 2005. Coronavirus transcription: a perspective. Curr Top Microbiol Immunol **287:**31–55.

Sawicki, S. G., D. L. Sawicki, and S. G. Siddell. 2007. A contemporary view of coronavirus transcription. J Virol **81:**20–9.

Sawicki, S. G., D. L. Sawicki, D. Younker, Y. Meyer, V. Thiel, H. Stokes, and S. G. Siddell. 2005. Functional and genetic analysis of coronavirus replicase-transcriptase proteins. PLoS Pathog **1:**e39.

Sawicki, D., T. Wang, andS. Sawicki. 2001. The RNA tructures engaged in replication and transcription of the A59 strain of ouse hepatitis virus. J Gen Virol **82:**385–96.

Sethna, P. B., M. A. Hofmann, and D. A. Brian. 1991. Minus-strand copies of replicating coronavirus mRNAs contain antileaders. J Virol **65:**320–5.

Sethna, P. B., S. L. Hung, and D. A. Brian. 1989. Coronavirus subgenomic minus-strand RNAs and the potential for mRNA replicons. Proc Natl Acad Sci U S A **86:**5626–30.

Shi, S. T., and M. M. Lai. 2005. Viral and cellular proteins involved in coronavirus replication. Curr Top Microbiol Immunol **287:**95–131.

Snijder, E. J., P. J. Bredenbeek, J. C. Dobbe, V. Thiel, J. Ziebuhr, L. L. Poon, Y. Guan, M. Rozanov, W. J. Spaan, and A. E. Gorbalenya. 2003. Unique and conserved features of genome and proteome of SARS- coronavirus, an early split-off from the coronavirus group 2 lineage. J Mol Biol **331:**991–1004.

Snijder, E. J., Y. van der Meer, J. Zevenhoven-Dobbe, J. J. Onderwater, J. van der Meulen, H. K. Koerten, and A. M. Mommaas. 2006. Ultrastructure and origin of membrane vesicles associated with the severe acute respiratory syndrome coronavirus replication complex. J Virol **80:** 5927–40.

Spaan, W., H. Delius, M. Skinner, J. Armstrong, P. Rottier, S. Smeekens, B. A. van der Zeijst, and S. G. Siddell. 1983. Coronavirus mRNA synthesis involves fusion of non-contiguous sequences. Embo J **2:**1839–44.

Stern, D. F., and B. M. Sefton. 1982. Synthesis of coronavirus mRNAs: kinetics of inactivation of infectious bronchitis virus RNA synthesis by UV light. J Virol **42:**755–9.

Su, D., Z. Lou, F. Sun, Y. Zhai, H. Yang, R. Zhang, A. Joachimiak, X. C. Zhang, M. Bartlam, and Z. Rao. 2006. Dodecamer structure of severe acute respiratory syndrome coronavirus nonstructural protein nsp10. J Virol **80:**7902–8.

Sutton, G., E. Fry, L. Carter, S. Sainsbury, T. Walter, J. Nettleship, N. Berrow, R. Owens, R. Gilbert, A. Davidson, S. Siddell, L. L. Poon, J. Diprose, D. Alderton, M. Walsh, J. M. Grimes, and D. I. Stuart. 2004. The nsp9 replicase protein of SARS-coronavirus, structure and functional insights. Structure **12:**341–53.

van der Meer, Y., E. J. Snijder, J. C. Dobbe, S. Schleich, M. R. Denison, W. J. Spaan, and J. K. Locker. 1999. Localization of mouse hepatitis virus nonstructural proteins and RNA synthesis indicates a role for late endosomes in viral replication. J Virol **73:**7641–57.

van Dinten, L. C., J. A. den Boon, A. L. Wassenaar, W. J. Spaan, and E. J. Snijder. 1997. An infectious arterivirus cDNA clone: identification of a replicase point mutation that abolishes discontinuous mRNA transcription. Proc Natl Acad Sci U S A **94:**991–6.

van Dinten, L. C., H. van Tol, A. E. Gorbalenya, and E. J. Snijder. 2000. The predicted metal-binding region of the arterivirus helicase protein is involved in subgenomic mRNA synthesis, genome replication, and virion biogenesis. J Virol **74:**5213–23.

Wu, H. Y., and D. A. Brian. 2007. 5′-proximal hot spot for an inducible positive-to-negative-strand template switch by coronavirus RNA-dependent RNA polymerase. J Virol **81:**3206–15.

Wu, H. Y., A. Ozdarendeli, and D. A. Brian. 2006. Bovine coronavirus 5′-proximal genomic acceptor hotspot for discontinuous transcription is 65 nucleotides wide. J Virol **80:** 2183–93.

Yang, H., M. Yang, Y. Ding, Y. Liu, Z. Lou, Z. Zhou, L. Sun, L. Mo, S. Ye, H. Pang, G. F. Gao, K. Anand, M. Bartlam, R. Hilgenfeld, and Z. Rao. 2003. The crystal structures of severe acute respiratory syndrome virus main protease and its complex with an inhibitor. Proc Natl Acad Sci U S A **100:**13190–5.

Yount, B., K. M. Curtis, and R. S. Baric. 2000. Strategy for systematic assembly of large RNA and DNA genomes: transmissible gastroenteritis virus model. J Virol **74:**10600–11.

Yount, B., K. M. Curtis, E. A. Fritz, L. E. Hensley, P. B. Jahrling, E. Prentice, M. R. Denison, T. W. Geisbert, and R. S. Baric. 2003. Reverse genetics with a full-length infectious cDNA of severe acute respiratory syndrome coronavirus. Proc Natl Acad Sci U S A **100:**12995–3000.

Yount, B., M. R. Denison, S. R. Weiss, and R. S. Baric. 2002. Systematic assembly of a full-length infectious cDNA of mouse hepatitis virus strain A59. J Virol **76:**11065–78.

Zhai, Y., F. Sun, X. Li, H. Pang, X. Xu, M. Bartlam, and Z. Rao. 2005. Insights into SARS-CoV transcription and replication from the structure of the nsp7-nsp8 hexadecamer. Nat Struct Mol Biol **12:**980–6.

Ziebuhr, J. 2005. The coronavirus replicase. Curr Top Microbiol Immunol **287:**57–94.

Ziebuhr, J., B. Schelle, N. Karl, E. Minskaia, S. Bayer, S. G. Siddell, A. E. Gorbalenya, and V. Thiel. 2007. Human coronavirus 229E papain-like proteases have overlapping specificities but distinct functions in viral replication. J Virol **81:**3922–32.

Zust, R., L. Cervantes-Barragan, T. Kuri, G. Blakqori, F. Weber, B. Ludewig, and V. Thiel. 2007. Coronavirus non-structural protein 1 is a major pathogenicity factor: Implications for the rational design of coronavirus vaccines. PLoS Pathog **3:**e109.

Chapter 3
Flaviviruses

Néstor G. Iglesias, Claudia V. Filomatori, Diego E. Alvarez, and Andrea V. Gamarnik

Abstract Flavivirus genome amplification is a complex process that involves the viral RNA, cellular and viral proteins, and a sophisticated architecture of cellular membranes induced by viral infection. The viral RNA is not just a passive template; it plays an active role acquiring dynamic tertiary structures during viral replication. RNA synthesis is regulated by *cis*-acting elements present at the 5′- and 3′-ends of the viral genome. These elements include complementary sequences that mediate genome cyclization through direct RNA–RNA interactions. Studies from many laboratories have provided compelling evidence supporting the notion that a circular conformation of the viral RNA is essential for flavivirus RNA replication. In addition, an RNA element located within the viral 5′UTR has been found to bind the viral polymerase and promote RNA synthesis. In this chapter, we describe viral proteins and RNA structures involved in flavivirus genome amplification and provide working models that explain the need of long-range RNA–RNA interactions during viral RNA synthesis.

Introduction: The Viral Life Cycle

Flaviviruses comprise one of the three genera within the Flaviviridae family; the other two are the Pestivirus and Hepacivirus. The Flavivirus genus includes, among others, the medically important mosquito borne dengue virus (DENV), yellow fever virus (YFV), West Nile virus (WNV), Japanese encephalitis virus (JEV), and the tick-borne encephalitis virus (TBEV). Flaviviruses are enveloped viruses with a single stranded, ~11 kb, positive-sense RNA genome with a type 1 cap (m7GpppAmp) structure at the 5′-end (Wengler and Gross 1978; Cleaves and Dubin 1979). In contrast to cellular mRNAs, flavivirus genomes are not polyadenylated (Wengler and Gross 1978). The viral RNA encodes a single long open reading frame (ORF)

A.V. Gamarnik (✉)
Molecular Virology Laboratory, Fundacion Instituto Leloir, Av Patricias Argentinas 435, Buenos Aires 1405, Argentina
e-mail: agamarnik@leloir.org.ar

C.E. Cameron et al. (eds.), *Viral Genome Replication*,
DOI 10.1007/b135974_3,

flanked by highly structured 5′ and 3′ untranslated regions (UTRs) of ~100 and 350–700 nucleotides, respectively. As for all positive-stranded RNA viruses, the flavivirus genomic RNA is infectious and serves as mRNA. Translation of the single ORF at the rough ER produces a large polyprotein that is cleaved cotranslationally and posttranslationally into at least 10 proteins. The N-terminal of the polyprotein encodes the three structural proteins (C-prM-E), followed by seven non-structural (NS) proteins (NS1-NS2A-NS2B-NS3-NS4A-NS4B-NS5) (Rice et al. 1985) (Fig. 3.1a). The amino termini of prM, E, NS1, and NS4B are generated upon cleavage by the host signal peptidase in the ER lumen, while the processing of most of the NS proteins and the carboxyl terminus of the C protein is carried out by the viral NS3 serine protease in the cytoplasm of the infected cell (Fig. 3.1b). NS3 requires the cofactor NS2B for protease activity. In addition, NS3 comprises RNA triphosphatase and helicase activities. NS5 is the RNA-dependent RNA polymerase (RdRp), which carries a methyltransferase (MTase) domain in its NH2 terminus.

The viral replication cycle is similar for all flaviviruses (Lindenbach and Rice 2001). The virus enters a host cell via receptor-mediated endocytosis. Upon internalization and acidification of the endosome, fusion of viral and vesicular membranes

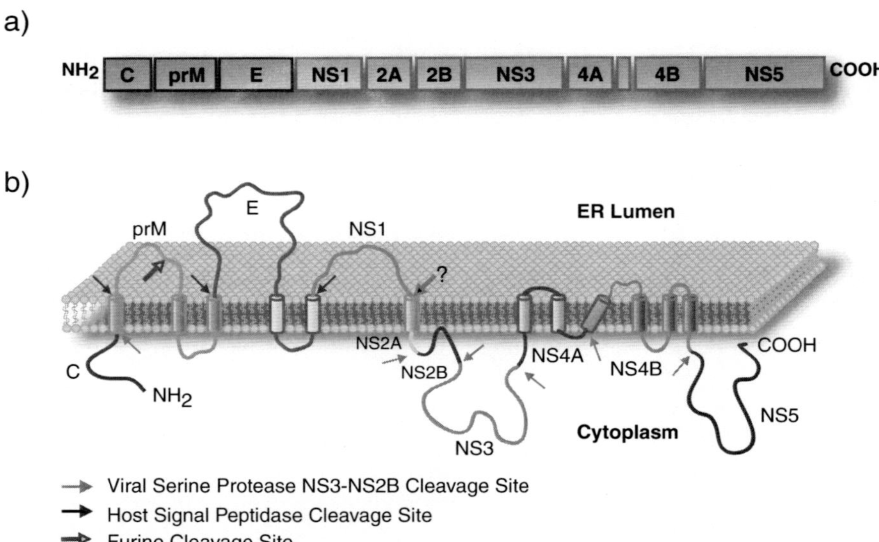

Fig. 3.1 Flavivirus proteins. (**a**) Schematic representation of the viral polyprotein. The three structural proteins C, capsid; prM precursor to membrane protein; E, envelope; and the seven non-structural (NS) proteins NS1-NS2A-NS2B-NS3-NS4A-NS4B-NS5 are shown. (**b**) Membrane topology of the flavivirus polyprotein. The predicted orientation of the viral proteins across the endoplasmic reticulum (ER) membrane is shown. *Trans*-membrane domains are indicated by *cylinders* and *arrows* indicate the cleavage site of specific enzymes. The *question mark* indicates cleavage by an unknown ER enzyme.

allows entry of the nucleocapsid into the cytoplasm. After translation of the viral RNA, virus-induced hypertrophy of intracellular membranes occurs, originating membranous structures in which RNA synthesis takes place (for review see Westaway et al. 2003). Based on *trans*-complementation studies, genome packaging appears to be coupled to RNA replication (Khromykh et al. 2001b). Nascent virus particles pass through the Golgi apparatus, where prM is cleaved by furin and virion maturation occurs. Finally, the viral progeny is exocytosed via secretory vesicles.

Flavivirus Replication Complexes

Flavivirus RNA replication occurs in close association with cellular membranes in so-called viral replication complexes (RCs). Replication begins with the synthesis of a negative-strand RNA, which serves as a template for the synthesis of additional positive-strand genomic RNA. The enzymatic reaction is catalyzed by the RdRp activity of the viral NS5 protein, in association with the viral protease/helicase NS3, other viral NS proteins, and presumably host factors. RNA synthesis is asymmetric, leading to a 10- to 100-fold excess of positive over negative strands (Cleaves et al. 1981; Uchil and Satchidanandam 2003b). Negative strands continue to accumulate throughout the infection and have been isolated exclusively in double-stranded forms. Three species of viral RNA can be metabolically labeled: a ribonuclease-resistant double-stranded RNA (dsRNA) called replicative form (RF); a form partially resistant to ribonucleases, likely composed by RNAs with complementary nascent elongating strands, known as replicative intermediates (RI); and the genomic vRNA that is fully sensitive to ribonucleases. The three RNA forms have also been described for in vitro RNA polymerase reactions using infected cell extracts (Grun and Brinton 1986; Chu and Westaway 1987; You and Padmanabhan 1999; Uchil and Satchidanandam 2003b).

Different lines of evidence revealed that RNA replication appears to be confined to discrete foci, mainly in the perinuclear region (Ng et al. 1983; Westaway et al. 1997; Mackenzie et al. 1999). Data on the composition of the RCs in flavivirus-infected cells were obtained by confocal and electron microscopy together with co-immunoprecipitations using specific antibodies to different NS proteins and to dsRNA. The results indicated that proteins NS1, NS2A, NS3, NS4A, NS5, and for some viruses NS4B co-localize with dsRNA (Mackenzie et al. 1996; Westaway et al. 1997; Mackenzie et al. 1998; Miller et al. 2006). Interestingly, the DENV and YFV NS5 RNA polymerases were found predominantly in the nucleus, showing weak staining in the perinuclear region that co-localized with the RCs (Buckley et al. 1992; Kapoor et al. 1995; Miller 2006; Uchil et al. 2006). This observation is in agreement with the finding that a small amount of NS5 is involved in active RNA replication (Grun and Brinton 1987; Chu and Westaway 1992; Uchil and Satchidanandam 2003a). Furthermore, active RCs were also found in the nucleus of infected cells (Uchil et al. 2006).

Membranes have been suggested to play a structural and organizational role in flavivirus replication, possibly offering a suitable microenvironment for viral

RNA synthesis and viral morphogenesis. The membranous structures found in flavivirus-infected cells seem to originate from different cellular organelles (for review see Mackenzie 2005). Convoluted membranes (CM) and paracrystalline structures (PC) are the putative sites of viral polyprotein processing, whereas proliferating ER and vesicles of 70–100 nm in diameter, found as vesicle packets (VP) enclosed by an outer membrane, may represent the sites of viral RNA replication. The association of the replicative forms of the viral RNA with the VP has been shown by electron microscopy and confirmed by biochemical analysis (Mackenzie et al. 1996; Grief et al. 1997; Westaway et al. 1997; Uchil and Satchidanandam 2003a). Biochemical studies with flavivirus-infected cells and cell extracts active for RNA synthesis were used to probe the architecture of the RC (Uchil and Satchidanandam 2003a). Treatment of the extracts with nucleases in the presence or absence of different detergents suggested that the three viral RNA species (RF, RI, and vRNA) reside in a membrane enclosed nuclease-resistant compartment. It was proposed that the RF resides within the inner membrane of a double membranous structure, whereas the nascent genomic RNA was extruded from the vesicles but retained inside the outer bounding membrane of the VP (Uchil and Satchidanandam 2003a).

It has been shown that protein NS4A is required for induction of the membranous structures CM/PC by different flaviviruses. Cleavage of a 2 K domain at the C-terminus of NS4A by the viral serine protease leads to a large accumulation of intracellular membranes in DENV-infected cells (Miller et al. 2007) and relocalization of matured NS4A to Golgi membranes in WNV-infected cells (Roosendaal et al. 2006). These results suggest that proteolytic processing of NS4A could regulate the membrane rearrangements observed upon infection. A role for NS1 during RNA replication was proposed. Mutation of the N-glycosylation sites of NS1 led to a dramatic defect in RNA replication (Muylaert et al. 1996). In addition, deletions within YFV NS1 resulted in viruses with defects at early stages of RNA replication, presumably during minus-strand synthesis (Lindenbach and Rice 1997). Interestingly, this defect was suppressed by a mutation in NS4A, providing genetic evidence for NS1–NS4A interaction in RNA replication (Lindenbach and Rice 1999). Although NS1, NS2A, and NS4B of different flaviviruses have been implicated in RNA synthesis, their precise roles in this process remain unclear.

Multifunctional Viral Proteins Involved in Flavivirus RNA Replication

NS5 is the largest and the most conserved of the flavivirus proteins. It contains an N-terminal S-adenosyl-methionine (SAM)-dependent methyl transferase domain (MTase) and a C-terminal RdRp domain. Recently, the structure of the NS5 C-terminal domain of WNV and DENV revealed a classical RdRp-fold bearing palm, thumb, and fingers motifs (Malet et al. 2007; Yap et al. 2007). The presence of a priming loop found in these structures is consistent with a primer-independent (de novo) mechanism of initiation of RNA synthesis proposed for flaviviruses. The NS5 from WNV, DENV, and YFV were shown to possess guanine N7 and ribose 2′-O MTase activities involved in formation of the 5′ cap (Egloff et al. 2002; Ray et al. 2006; Dong et al. 2007; Zhou et al. 2007). The crystal structures of WNV and DENV MTase domain showed a single binding site for the methyl donor SAM (Egloff et al. 2002; Zhou et al. 2007). In addition, a positively charged surface adjacent to the SAM binding site was proposed to be the recognition site for the capped-RNA substrate. Deletions or mutations within the MTase domain were shown to be lethal for the replication of WNV and DENV (Khromykh et al. 1998, 1999b; Ray et al. 2006). In addition, mutagenesis of the Kunjin polymerase active site motif confirmed that it is essential for viral replication and that polymerase activity could be supplied in *trans* from a Kunjin replicon. However, *trans*-complementation in cells expressing only NS5 was found to be inefficient, suggesting that co-translational expression of additional NS proteins may be required for the RdRp to associate with other replicase components (Khromykh et al. 1999a).

The multifunctional NS3 protein bears a protease, helicase, nucleotide triphosphatase (NTPase), and 5′RNA triphosphatase (RTPase) activities. Its N-terminal 180 amino acids comprise the serine protease, and its C-terminal region has conserved domains found in the DEXH family of NTPase/RNA helicases. RNA-stimulated NTPase and RNA unwinding activities have been characterized in DENV, WNV, YFV, and JEV (Warrener et al. 1993; Li et al. 1999; Utama et al. 2000; Borowski et al. 2001; Benarroch et al. 2004; Yon et al. 2005). Crystal structures have been recently reported for the helicase of DENV and YFV (Wu et al. 2005; Xu et al. 2005). A study with a DENV infectious clone showed that an active helicase was essential for virus viability (Matusan et al. 2001). Although helicases have been implicated in the replication of flaviviruses genomes, their precise role in RNA synthesis remains unknown. Possible functions include unwinding dsRNA intermediates that arise during RNA amplification, destabilizing secondary structures of the RNA to increase polymerase processivity, or participating in RNA recruitment at specific subcellular locations.

Two-hybrid systems and co-immunoprecipitation studies using infected cells and recombinant proteins demonstrated that the helicase domain of NS3 of DENV binds NS5 (Kapoor et al. 1995; Johansson et al. 2001). This interaction occurred in the absence of other viral proteins but was dependent on the NS5 phosphorylation state (Kapoor et al. 1995). NS5 has been detected in both the cytoplasm and the nucleus. Only a hyperphosphorylated form of NS5, which was unable to interact with NS3, has been detected in the nucleus of DENV-infected cells (Kapoor et al. 1995; Brooks et al. 2002). NS5 contains two functional nuclear localization sequences (NLSs) and binding to importin β1 was demonstrated (Johansson et al. 2001; Brooks et al. 2002). Whether the nuclear localization of NS5 plays a role in the viral replication cycle or is part of a mechanism used by the virus to alter a cell function is currently unknown. The N-terminal region of NS5 that was shown to be required to bind NS3 was also found to bind importin β1, suggesting a competition between NS3 and importin β1 that may play a role in controlling the subcellular localization of NS5.

According to the crystal structure of NS5, the putative binding site for NS3 would be near the entrance of the RNA template tunnel, consistent with the proposed role of its activity in RNA unwinding during RNA synthesis (Malet et al. 2007). In addition, NS5 was reported to stimulate the NTPase and RTPase activities of NS3 (Yon et al. 2005), suggesting a functional interaction between these two viral proteins.

5′UTR Elements and Promoter Signals for RNA Synthesis

The 5′UTRs of flavivirus RNAs are relatively short and almost complete sequence conservation was observed among different strains of the same virus. In contrast, sequence comparisons between different flaviviruses, such as DENV, WNV, YFV, and SLEV, showed little conservation (Brinton and Dispoto 1988). Interestingly, the predicted secondary structure of 5′UTRs of different flaviviruses is very similar. These structures consist of a large stem-loop with a side stem-loop (named Stem-Loop A, SLA, Fig. 3.2). The conservation of this secondary structure in unrelated flaviviruses was taken as evidence of its possible importance in viral replication. In most cases, a second short stem-loop named SLB, which includes or ends at the translation initiator codon, could be formed downstream of SLA. The structure of SLA and the sequence upstream of the translation initiator AUG were found to be essential *cis*-acting elements for viral RNA synthesis (Cahour et al. 1995; Filomatori et al. 2006; Kofler et al. 2006). In addition, a conserved stem-loop structure present just downstream of the initiator AUG of DENV type 2 (DENV2) RNA was reported to act as a regulatory element important for start codon selection in translation initiation (Clyde and Harris 2006).

Cahuer and co-workers reported the first evidence of functional 5′UTR elements in flavivirus replication in vivo (Cahour et al. 1995). In this study, deletions from 5 to 25 nucleotides were incorporated throughout the 5′UTR by mutagenesis of an infectious DNA copy of DENV4. The dominant effect of the deletions appeared to be at the level of RNA synthesis and many of the mutations were found to be lethal. More recently, a similar approach was used to study the role of SLA structure during DENV2 replication (Filomatori et al. 2006). In this case, site-directed mutagenesis was done to (a) open the bottom of the stem of SLA structure, (b) reconstitute the stem with sequences that differed from the wild type, and (c) substitute nucleotides at the top and side loops of SLA. Alteration of the stem of SLA was found to be lethal, while reconstitution of the stem yielded an infectious RNA with a phenotype similar to the parental virus. In addition, infectious viruses could not be recovered after transfection of RNAs carrying substitutions at the top loop of SLA. In this case, revertant viruses with single substitutions, partially recovering the wild type sequence, were rescued in cell culture. In the same study, using a DENV2 replicon system that allows discrimination between translation and RNA synthesis, it was shown that the SLA structure was essential for RNA amplification, whereas no crucial role of SLA was connected with translation of the input RNA.

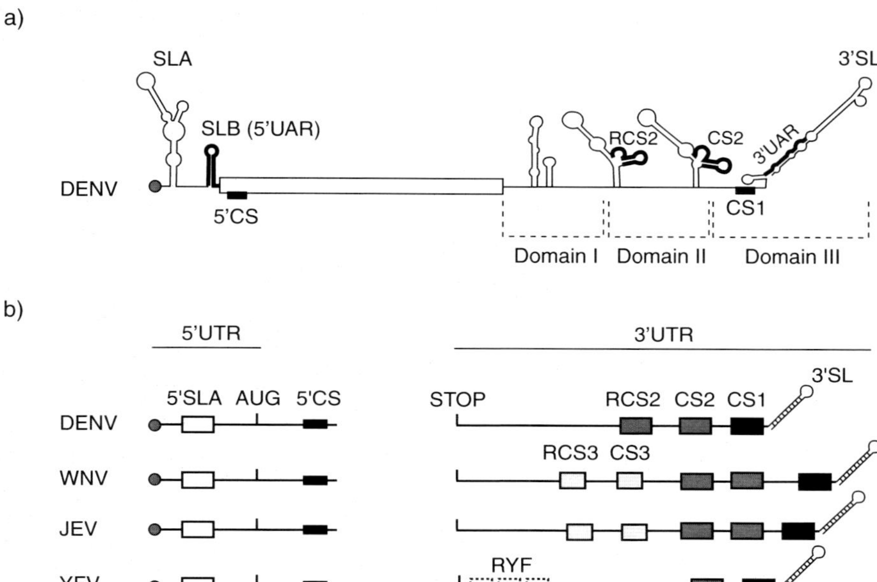

Fig. 3.2 Conserved 5′ and 3′UTR RNA structures of mosquito-borne flaviviruses. (**a**) Schematic representation of RNA elements found at the 5′ and 3′UTRs of DENV genome. The predicted secondary structures of defined domains are indicated: at the 5′-end, stem-loop A (SLA), stem-loop B (SLB, containing 5′UAR), and 5′CS are indicated; at the 3′-end, domains I, II, and III are shown. In addition, location of the 3′ stem-loop (3′SL) and the conserved sequence RCS2, CS2, and CS1 are indicated. (**b**) RNA elements of 5′ and 3′UTRs of dengue virus (DENV), West Nile virus (WNV), Japanese encephalitis virus (JEV), and yellow fever virus (YFV) are schematically shown. At the 5′-end the 5′SLA, the translation initiator codon AUG, and the cyclization sequence 5′CS are indicated. The presence and relative location of the different conserved RNA elements at the 3′UTRs, CS1, CS2, RCS2, CS3, RCS3, and the yellow fever tandem repeats (RYF) are indicated by *boxes* in different *colors*.

RNA–protein interaction studies have been used to demonstrate direct binding of DENV polymerase to RNA molecules carrying the SLA structure. Moreover, using an in vitro assay to measure RdRp activity of DENV2 NS5, it was shown that SLA was a critical determinant for template specificity (Filomatori et al. 2006). The polymerase was able to initiate de novo and copy RNA templates bearing the SLA, while it was very inefficient in copying viral or non-viral RNA templates lacking this structure. A remarkable correlation between the requirement of DENV SLA sequence/structure for viral replication in transfected cells and the need of this element for in vitro polymerase activity was observed (Fig. 3.3). Based on in vivo and in vitro results, it was proposed that the SLA functions as the promoter for DENV negative-strand RNA synthesis. Further studies are necessary to extrapolate these observations to other flaviviruses.

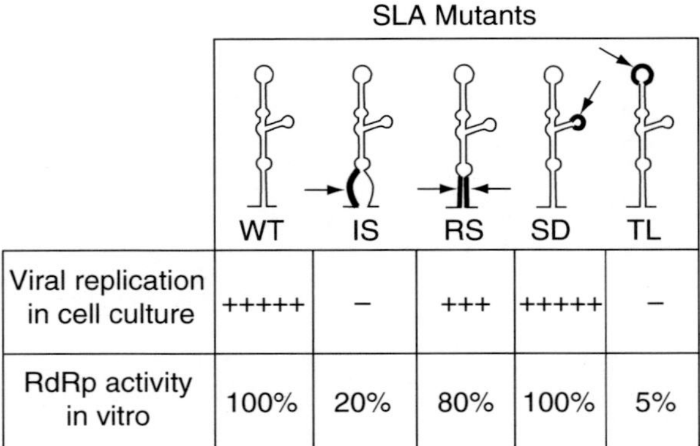

	WT	IS	RS	SD	TL
Viral replication in cell culture	+++++	–	+++	+++++	–
RdRp activity in vitro	100%	20%	80%	100%	5%

Fig. 3.3 Structure–function correlation of DENV stem-loop A (SLA) mutants. On the *top*, schematic representation of RNA mutants: interrupted stem (IS), reconstituted stem (RS), side loop mutant (SD), and top loop mutant (TL). The level of DENV replication in BHK cells transfected with full-length viral RNAs carrying the respective mutation is compared with the viral RdRp activity observed in vitro using RNA templates of 160 nucleotides carrying the same mutations (Filomatori et al. 2006).

Cis-Acting Elements at the 3′UTRs

The 3′UTR sequences exhibit great variability among different flaviviruses; however, several conserved features and conserved secondary structures have been elucidated (Wallner et al. 1995; Proutski et al. 1997; Rauscher et al. 1997; Shurtleff et al. 2001; Thurner et al. 2004; Gritsun and Gould 2007). The 3′UTRs are between 350 and 700 nucleotides long and can be divided into three domains based on sequence/structure conservation (Fig. 3.2a). Domain I is the region immediately following the stop codon that is hypervariable and contains deletions and insertions in most flaviviruses. YF contains unique tandem repeats within domain I, known as RYF (Bryant et al. 2005) (Fig. 3.2b). Domain II is a region of moderate conservation comprising several hairpin motifs, including a characteristic dumbell (DB) structure with a conserved sequence named CS2 motif, present in all mosquito-borne flaviviruses (Olsthoorn and Bol 2001; Gritsun and Gould 2006; Romero et al. 2006). This DB structure is repeated in tandem in members of the DEN and JE subgroups, containing a repeated conserved sequence (RCS2) motif (Fig. 3.2b). Domain III is the most conserved region of flavivirus 3′UTRs, bearing a terminal stable stem-loop structure (3′SL). The presence of the 3′SL has been supported by secondary structure predictions, co-variation analysis, and biochemical probing. An essential role of the 3′SL in flavivirus replication has been extensively documented (for review see Markoff 2003). Upstream of the 3′SL there is a highly conserved sequence named CS1 motif (Hahn et al. 1987).

Work from many laboratories allowed to define several essential elements within domain III of the 3′UTR: (i) a pentanucleotide sequence CAGAC mostly present in a loop of the 3′SL (Wengler and Castle 1986; Khromykh et al. 2003; Tilgner and

Shi 2004; Elghonemy et al. 2005); (ii) a region within CS1 that contains a complementary sequence to a region present within the coding sequence of the capsid protein (Hahn et al. 1987; Men et al. 1996; Corver et al. 2003); (iii) the 3′ terminal nucleotides of the 3′SL, including the last CU_{OH} conserved in all mosquito and tick-borne flaviviruses (Khromykh et al. 2003; Tilgner and Shi 2004); (iv) a region within the stem of the 3′SL that contains a complementary sequence to a region present upstream of the translation initiator AUG at the viral 5′UTR (named UAR in DENV and CSA in TBEV) (Alvarez et al. 2005b; Kofler et al. 2006); and (v) specific bulges within the 3′SL structure (Yu and Markoff 2005). Deletions or mutations within any of these *cis*-acting RNA elements abolish viral replication. In contrast, RNA structures within domains I and II are considered dispensable for flavivirus replication. However, these structures are believed to serve as replication enhancers because deletion mutants within domains I and II exhibit decreased viral RNA synthesis and attenuation (Men et al. 1996; Mandl et al. 1998; Bredenbeek et al. 2003; Lo et al. 2003; Alvarez et al. 2005a).

Inverted Complementary Sequences in Flavivirus RNAs

Flavivirus genomes possess inverted complementary sequences at the ends of the RNA (Fig. 3.4), similar to those observed in the negative-strand RNA bunya-, arena-, and orthomyxoviruses. These inverted complementary sequences have been suggested to allow the ends of the genome to associate through base pairing, leading to circular conformations of the RNA (panhandle-like structures).

Two pairs of inverted complementary sequences can be found at the ends of mosquito- and tick-borne flavivirus genomes (Fig. 3.4). In all mosquito-borne flaviviruses (MBF), there is a region within CS1 that complements perfectly with a sequence located in the coding region of the C protein. This pair of complementary sequences is known as cyclization sequence 5′–3′CS (Hahn et al. 1987) (Fig. 3.4). The second pair of inverted complementary sequences was first noticed using folding prediction algorithms on flavivirus RNAs (Hahn et al. 1987; Khromykh et al. 2001a; Thurner et al. 2004). A sequence located just upstream of the translation initiator AUG at the 5′UTR was found to be complementary to a region present within the stem of the 3′SL. This pair of complementary sequences is known as cyclization sequence 5′–3′UAR (the name stands for *u*pstream *A*UG *r*egion) (Alvarez et al. 2005b) (Fig. 3.4). Alignment of MBF sequences indicates high conservation of 5′–3′CS, whereas less sequence conservation was observed within 5′–3′UAR (Fig. 3.4). Two pairs of complementary sequences, CSA and CSB, were proposed as possible cyclization elements in the case of tick-borne flaviviruses (TBF) (Mandl et al. 1993; Khromykh et al. 2001a). The 5′CSA is located upstream of the initiator AUG and is complementary to the 3′CSA located within the stem of the 3′SL, which is reminiscent of the location of the MBF 5′–3′UAR. The 5′–3′CSB sequences are found in similar locations as the MBF 5′–3′CS.

Requirement of 5′–3′CS base pairing was first analyzed in RdRp reactions using DENV-infected cell extracts and exogenous RNA templates including the 5′- and

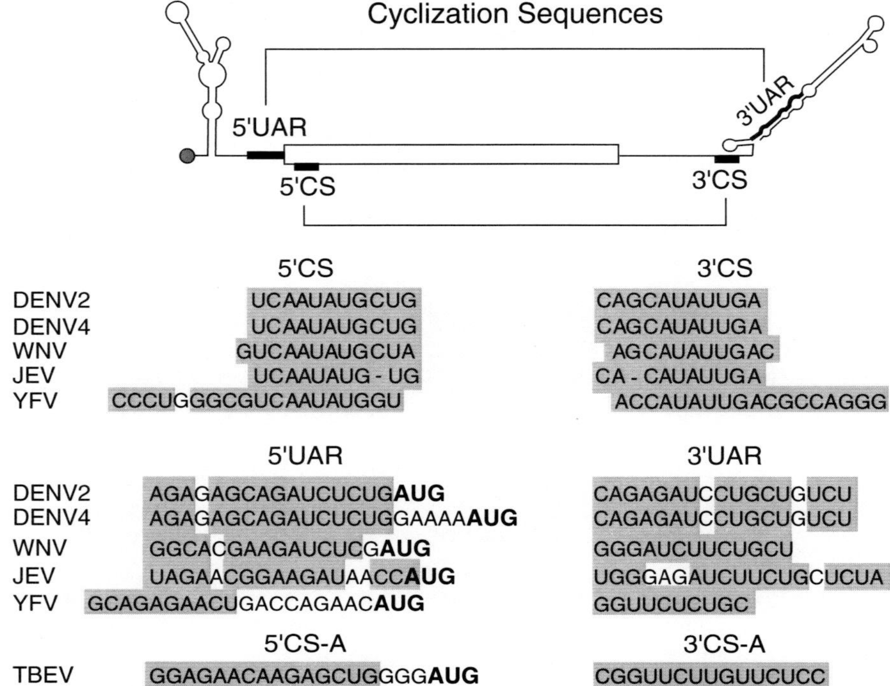

Fig. 3.4 Sequence and location of flavivirus cyclization sequences. On the *top*, a schematic representation shows the location of 5′–3′UAR and 5′–3′CS regions of mosquito-borne flaviviruses. The *bottom* panels show the nucleotide sequences of the complementary regions 5′–3′CS and 5′–3′UAR of dengue virus type 2 and 4 (DENV2, Genebank number U87412 and DENV4 Genebank number M14931), West Nile virus (WNV, Genebank number M12294), Japanese encephalitis virus (JEV, Genebank number NC001437), and yellow fever virus (YFV, Genebank number NC002031); and the 5′–3′CSA of tick-borne encephalitis virus (TBEV, Genebank number U27495). The *gray boxes* denote the inverted complementary sequences. In *boldface* the translation initiator AUG is indicated.

3′-end viral sequences. In this study, it was shown that 5′–3′CS complementarity was necessary for polymerase activity (You and Padmanabhan 1999). Subsequent studies specifically addressed the requirement of sequence complementarity in vivo using a Kunjin virus replicon system (Khromykh et al. 2001a). In agreement with the in vitro study, specific mutations in the 5′CS or 3′CS abolished RNA amplification while reconstitution of potential base pairings with foreign sequences restored replicon replication. More recent studies confirmed the requirement of CS complementarity for RNA amplification of DEN and WN viruses (Lo et al. 2003; Alvarez et al. 2005a).

The specific role of 5′–3′UAR complementarity was addressed using infectious DENV2 and DENV replicons. These studies indicated that mismatches within 5′–3′UAR did not alter translation of the viral RNA but greatly decreased RNA synthesis, leading in some cases to undetectable levels of viral replication. Compensatory mutations that restored 5′–3′UAR base pairing rescued RNA synthesis. In addition, the mutants with the compensatory changes within UAR were shown to replicate less efficiently than the parental virus, suggesting that 5′ and 3′UAR sequences could play additional roles during viral replication (Alvarez et al. 2005b; Alvarez et al. 2006).

The requirement of the putative cyclization sequences CSA and CSB in TBEV was recently investigated using a replicon system (Kofler et al. 2006). This work provided clear evidence that complementarity of 5′–3′CSA, which is analogous in location to the 5′–3′UAR cyclization elements in DENV, was essential for TBEV replication. Interestingly, no crucial function was connected with the CSB elements, suggesting that only one pair of the two putative complementary sequences would be required to mediate 5′–3′ interactions in TBF genomes. In summary, there is compelling evidence indicating that sequence complementarity between the ends of flavivirus genomes is essential for viral RNA synthesis.

Cyclization of the Viral Genome

Direct interaction between two RNA molecules carrying the 5′ and 3′ terminal sequences of a flavivirus genome was first observed using psoralen/UV cross-linking, and a role of 5′–3′CS complementarity for the interaction was proposed (You et al. 2001). More recently, an electrophoretic mobility shift assay was employed to study the formation of RNA–RNA complexes with molecules carrying the terminal DENV sequences (Alvarez et al. 2005b). In this work, the specific contribution of both cyclization elements, 5′–3′CS and 5′–3′UAR, was demonstrated by mutagenesis analysis. Single mismatches within the complementary sequences were shown to increase the apparent dissociation constants of specific RNA–RNA complexes.

The first direct evidence of long-range RNA–RNA interactions between the ends of a flavivirus RNA was obtained by visualization of individual molecules using atomic force microscopy (AFM) (Alvarez et al. 2005b). Because single-stranded RNA molecules acquire compact tertiary structures that preclude visualization of intramolecular contacts, the RNAs used in that analysis were hybridized with antisense RNA molecules to generate elongated double-stranded segments (Fig. 3.5). This strategy allowed visualization of long-range RNA–RNA contacts at single-stranded regions. Using AFM in air, a model RNA molecule of 2 kb carrying the 5′ and 3′ terminal sequences of DENV2 as well as single molecules of the full-length genomic RNA were visualized in linear and circular conformations, whereas control molecules with deletions of 3′CS and 3′UAR were only observed in linear forms. Cyclization of DENV RNA was observed in the absence of proteins (Fig. 3.5). However, it is possible that binding of cellular or viral proteins to the ends of the RNA enhances or disrupts genome cyclization.

Interaction of the viral protein NS5 with capped- and uncapped-RNA molecules corresponding to 5′-end ~200 nucleotides of WNV and DENV genomes was

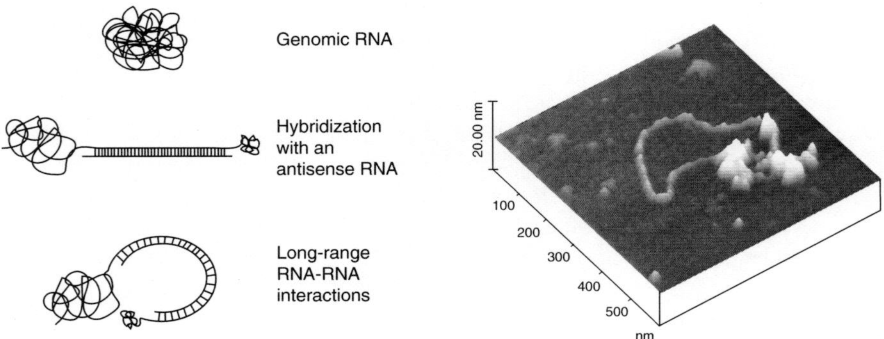

Fig. 3.5 Cyclization of DENV genomic RNA. On the *left*, a schematic representation of the viral genome in a compact conformation and the strategy used to visualize RNA–RNA contacts hybridizing the central part of the molecule with an antisense RNA molecule. On the *right*, a representative image of a single molecule of DENV genomic RNA obtained by tapping mode atomic force microscopy in circular conformation is shown. The 10.7 kb RNA molecule was hybridized with an antisense RNA of 3.3 kb resulting in a linear double-stranded region with single-stranded overhangs of 6970 and 451 nucleotides at the 5′- and 3′-ends, respectively (Alvarez et al. 2005b).

recently demonstrated (Filomatori et al. 2006; Dong et al. 2007). Deletion analysis of 5′-end sequences of DENV RNA indicated that the SLA structure was essential while 5′CS and 5′UAR sequences were dispensable for NS5 binding. Interestingly, interaction of DENV NS5 with an RNA–RNA complex formed between the 5′- and 3′- ends of the viral genome was observed, suggesting that the viral polymerase recognizes SLA even in the context of interacting 5′- and 3′-end viral sequences. Moreover, binding of NS2A, NS3, and NS5 to the 3′UTR of different flaviviruses has been also reported (Chen et al. 1997; Mackenzie et al. 1998). Whether binding of these viral proteins or other *trans*-acting factors modulates long-range RNA–RNA interactions in the context of the viral genome remains to be defined.

Proteins Interacting with the Viral RNA

Several reports have shown that defined RNA elements present at the 3′UTRs of flavivirus genomes differentially enhanced viral replication in distinct host cells. For instance, specific nucleotides of the bottom long stem of WNV and DENV 3′SL greatly enhanced replication competence in mosquito cells but had no effect on replication in mammalian cells (Zeng et al. 1998; Yu and Markoff 2005). In addition, it was found that deletion of the variable region encompassing domain I of DENV 3′UTR, reduced viral replication in mammalian cells without altering replication in mosquito cells (Alvarez et al. 2005a). These and other observations suggest that host cell specific factors bind the viral 3′UTRs. Significant effort has been made to identify host proteins that interact with the viral RNA. The eukaryotic elongation factor 1 alpha (EF-1 α) was identified to bind the 3′SL of WNV and

DENV (Blackwell and Brinton 1997; De Nova-Ocampo et al. 2002). Binding of EF-1 α to the viral RNA was mapped by footprinting analysis. This study defined a main binding site in the middle of the 3′SL (Blackwell and Brinton 1997). In addition, the human La autoantigen and the human PTB were found to interact with the 3′UTR of genomic DENV4 (De Nova-Ocampo et al. 2002; Garcia-Montalvo et al. 2004). In the case of JEV, a 36 kDa protein MOV34 was found to bind the 3′SL (Ta and Vrati 2000).

Search for proteins that bind the negative-strand RNA of flaviviruses has also been pursued. Four host proteins that bind specifically to the 3′ terminal sequences of the negative strand of WNV RNA were detected using BHK cell extracts (Shi et al. 1996). Purification of one of these proteins revealed to be TIAR, an RNA-binding protein containing three RNA recognition motifs (Li et al. 2002). Furthermore, a related protein TIA-1 was also shown to bind the same RNA. Interestingly, the growth of WNV was inhibited in a TIAR knockout cell line, indicating the functional importance of this protein. In addition, the human La autoantigen, calreticulin, and the protein disulfide isomerase were shown to interact in vitro with the 3′-end of the negative strand of DENV4 (Yocupicio-Monroy et al. 2003). Although several host proteins have been identified that bind the viral RNA and a functional role was proposed for some of them, the participation of these proteins during flavivirus RNA replication remains to be defined.

A Model for Minus-Strand RNA Synthesis

Filomatori and co-workers demonstrated that RNA molecules of ~160 nucleotides carrying the DENV SLA structure were efficient templates for in vitro RNA polymerase activity, whereas longer RNA molecules of ~2000 nucleotides, carrying the SLA in the same location were inefficient templates (Filomatori et al. 2006). In contrast, when the two pairs of cyclization sequences (5′–3′CS and 5′–3′UAR) were introduced at the ends of the RNA, polymerase activity became independent on the length of the template. Thus, it was hypothesized that the distance between the promoter SLA and the 3′-end of the template was critical for in vitro polymerase activity and that long-range RNA–RNA interactions would bring the 3′-end of the molecule near the SLA. This idea was consistent with the previous work of Padmanabhan and collaborators (Ackermann and Padmanabhan 2001; You et al. 2001).

A model for DENV minus-strand RNA synthesis was proposed (Fig. 3.6). In this model, the viral NS5 protein binds the promoter SLA at the 5′-end of the RNA, ~11 kb away from the initiation site. Cyclization of the viral genome through long-range RNA–RNA interactions could place the 3′-end of the RNA near the polymerase-SLA complex, allowing initiation of RNA synthesis. Therefore, only molecules in circular conformation would be competent templates for minus-strand RNA synthesis. In addition, it is possible that interaction of the polymerase with specific nucleotides within the SLA could induce conformational changes in the protein, facilitating the recognition of the 3′-end of the template. According to the

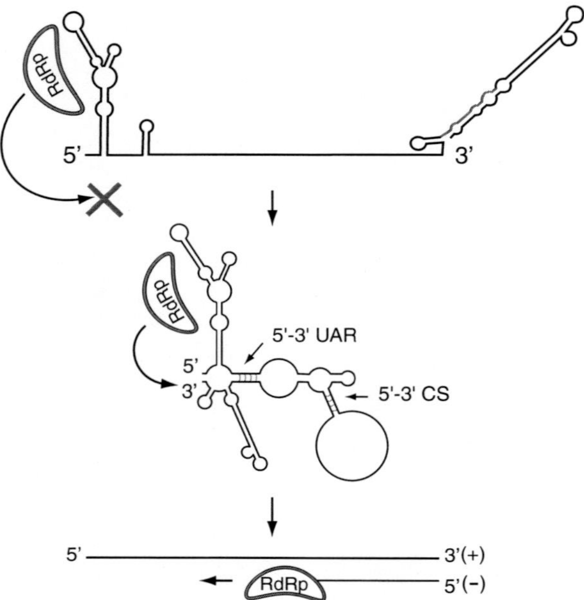

Fig. 3.6 Model for DENV minus-strand RNA synthesis. The viral RNA-dependent RNA polymerase (RdRp) binds the promoter SLA at the 5′-end of the RNA. The genome in a linear conformation does not permit the RdRp to initiate minus-strand RNA synthesis, however, cyclization of the RNA through 5′–3′CS and 5′–3′UAR contacts allows the RdRp to reach the 3′ initiation site.

crystal structure of DENV RdRp the template channel has dimensions that would only permit access to a ssRNA chain (Yap et al. 2007). Therefore, it is likely that the 3′SL structure unwinds before entering the template channel of the enzyme during initiation of minus-strand RNA synthesis. This process could be aided by the helicase activity of NS3 or by cellular *trans*-acting factors interacting with the viral 3′SL. Alternatively, base pairings between 5′ and 3′UAR of MBF or between 5′ and 3′CSA of TBF genomes, which were predicted to open the bottom half stem of the 3′SL, could release the 3′-end nucleotides, rendering the structural changes around the 3′SL presumably necessary for the initiation process.

In summary, the model proposes a core promoter at the 5′-end of the genome and long-range RNA–RNA interactions as essential elements for initiation of RNA synthesis. Although the 3′SL structure has been shown to be essential for flavivirus RNA synthesis, the molecular details by which this element participates during the process remain unclear. While it is not surprising to find a core promoter for RNA synthesis at the 3′-end of a viral genome, it is intriguing why certain plus-strand RNA viruses would have promoters or enhancer elements for RNA replication at the 5′-end of the RNA. In this case, the requirement of genome cyclization may provide advantages for viral replication such as control mechanisms to amplify only full-length templates or coordination of translation, RNA synthesis, and RNA packaging by overlapping signals involved in these processes. Further analysis of the RNA conformations required in each viral process will help to clarify the molecular details by which flaviviruses replicate their genomes.

Perspectives

Much has been learned in the last years about flavivirus RNA replication. A model for minus-strand RNA synthesis has been proposed that explains the need of genome cyclization, and roles of viral proteins and RNA *cis*-acting elements have been uncovered. An important aspect in flavivirus RNA replication that remains undefined is the mechanism of positive-strand RNA amplification. It has been proposed that the RF form is the template for genome amplification. We can speculate that the SLA could also serve as the promoter for positive-strand RNA synthesis. In this case, the SLA would have to work in *trans*, transferring the polymerase to the 3′-end of the negative strand, to initiate positive-strand RNA synthesis. A similar strategy has been previously proposed for the cloverleaf structure present at the 5′-end of poliovirus RNA (Andino et al. 1990). For DENV, a *trans*-initiation activity of SLA in vitro has been observed (Ackermann and Padmanabhan 2001; You et al. 2001; Filomatori et al. 2006), but not experimental evidence in vivo has yet been provided. Otherwise, the negative-strand RNA could carry its own promoter element either at the 5′- or the 3′-end of the molecule to facilitate positive-strand RNA amplification. In vivo experiments that allow discrimination between negative- and positive-strand RNA synthesis will be necessary to identify *cis*- and *trans*-acting factors specifically involved in each of these processes.

Formation of the cap at the 5′-end of the viral RNA requires the MTase activity of NS5. Because the cap structure precedes the SLA, it is possible that binding of NS5 to the promoter element is also involved in cap methylation. Consistent with this idea, recent studies using a recombinant WNV MTase have reported a requirement of SLA sequences for in vitro cap RNA methylation. The challenge will be to define how both enzymatic activities of NS5 are coordinated during flavivirus RNA synthesis.

Another intriguing question is: How is the vRNA released from the VP to the cytosol for translation and RNA packaging? The close association between the VP and the CM/PC reveals a level of organization that might allow the vRNA to be transported to the sites of protein synthesis or RNA packaging. Additional studies are needed to understand how the viral RNA is recruited to specific places of the infected cell for each step of the viral life cycle.

Finally, different RNA viruses use long-range RNA–RNA interactions as a strategy to allow *cis*-acting regulatory elements such as enhancers, promoters, and silencers to act from long distances (Pogany et al. 2003; Ray and White 2003). In addition, the dynamics of RNA tertiary structures allow modulation of specific functions of RNA–RNA contacts by *trans*-acting factors. In the future, we will have to uncover the functional significance and the underlying connections of these common viral strategies.

References

Ackermann, M. and Padmanabhan, R. 2001. De novo synthesis of RNA by the dengue virus RNA-dependent RNA polymerase exhibits temperature dependence at the initiation but not elongation phase. J Biol Chem 276: 39926–37.

Alvarez, D. E., De Lella Ezcurra, A. L., Fucito, S. and Gamarnik, A. V. 2005a. Role of RNA structures present at the 3′UTR of dengue virus on translation, RNA synthesis, and viral replication. Virology 339: 200–12.

Alvarez, D. E., Lodeiro, M. F., Filomatori, C. V., Fucito, S., Mondotte, J. A. and Gamarnik, A. V. 2006. Structural and functional analysis of dengue virus RNA. Novartis Found Symp 277: 120–32.

Alvarez, D. E., Lodeiro, M. F., Luduena, S. J., Pietrasanta, L. I. and Gamarnik, A. V. 2005b. Long-range RNA–RNA interactions circularize the dengue virus genome. J Virol 79: 6631–43.

Andino, R., Rieckhof, G. E. and Baltimore, D. 1990. A functional ribonucleoprotein complex forms around the 5′ end of poliovirus RNA. Cell 63: 369–80.

Benarroch, D., Selisko, B., Locatelli, G. A., Maga, G., Romette, J. L. and Canard, B. 2004. The RNA helicase, nucleotide 5′-triphosphatase, and RNA 5′-triphosphatase activities of dengue virus protein NS3 are Mg2+-dependent and require a functional Walker B motif in the helicase catalytic core. Virology 328: 208–18.

Blackwell, J. L. and Brinton, M. A. 1997. Translation elongation factor-1 alpha interacts with the 3′ stem-loop region of West Nile virus genomic RNA. J Virol 71: 6433–44.

Borowski, P., Niebuhr, A., Mueller, O., Bretner, M., Felczak, K., Kulikowski, T. and Schmitz, H. 2001. Purification and characterization of West Nile virus nucleoside triphosphatase (NTPase)/helicase: evidence for dissociation of the NTPase and helicase activities of the enzyme. J Virol 75: 3220–9.

Bredenbeek, P. J., Kooi, E. A., Lindenbach, B., Huijkman, N., Rice, C. M. and Spaan, W. J. 2003. A stable full-length yellow fever virus cDNA clone and the role of conserved RNA elements in flavivirus replication. J Gen Virol 84: 1261–8.

Brinton, M. A. and Dispoto, J. H. 1988. Sequence and secondary structure analysis of the 5′-terminal region of flavivirus genome RNA. Virology 162: 290–9.

Brooks, A. J., Johansson, M., John, A. V., Xu, Y., Jans, D. A. and Vasudevan, S. G. 2002. The interdomain region of dengue NS5 protein that binds to the viral helicase NS3 contains independently functional importin beta 1 and importin alpha/beta-recognized nuclear localization signals. J Biol Chem 277: 36399–407.

Bryant, J. E., Vasconcelos, P. F., Rijnbrand, R. C., Mutebi, J. P., Higgs, S. and Barrett, A. D. 2005. Size heterogeneity in the 3′ noncoding region of South American isolates of yellow fever virus. J Virol 79: 3807–21.

Buckley, A., Gaidamovich, S., Turchinskaya, A. and Gould, E. A. 1992. Monoclonal antibodies identify the NS5 yellow fever virus non-structural protein in the nuclei of infected cells. J Gen Virol 73 (Pt 5): 1125–30.

Cahour, A., Pletnev, A., Vazielle-Falcoz, M., Rosen, L. and Lai, C. J. 1995. Growth-restricted dengue virus mutants containing deletions in the 5′ noncoding region of the RNA genome. Virology 207: 68–76.

Chen, C. J., Kuo, M. D., Chien, L. J., Hsu, S. L., Wang, Y. M. and Lin, J. H. 1997. RNA-protein interactions: involvement of NS3, NS5, and 3′ noncoding regions of Japanese encephalitis virus genomic RNA. J Virol 71: 3466–73.

Chu, P. W. and Westaway, E. G. 1987. Characterization of Kunjin virus RNA-dependent RNA polymerase: reinitiation of synthesis in vitro. Virology 157: 330–7.

Chu, P. W. and Westaway, E. G. 1992. Molecular and ultrastructural analysis of heavy membrane fractions associated with the replication of Kunjin virus RNA. Arch Virol 125: 177–91.

Cleaves, G. R. and Dubin, D. T. 1979. Methylation status of intracellular dengue type 2 40 S RNA. Virology 96: 159–65.

Cleaves, G. R., Ryan, T. E. and Schlesinger, R. W. 1981. Identification and characterization of type 2 dengue virus replicative intermediate and replicative form RNAs. Virology 111: 73–83.

Clyde, K. and Harris, E. 2006. RNA secondary structure in the coding region of dengue virus type 2 directs translation start codon selection and is required for viral replication. J Virol 80: 2170–82.

Corver, J., Lenches, E., Smith, K., Robison, R. A., Sando, T., Strauss, E. G. and Strauss, J. H. 2003. Fine mapping of a cis-acting sequence element in yellow fever virus RNA that is required for RNA replication and cyclization. J Virol 77: 2265–70.

De Nova-Ocampo, M., Villegas-Sepulveda, N. and del Angel, R. M. 2002. Translation elongation factor-1alpha, La, and PTB interact with the 3′ untranslated region of dengue 4 virus RNA. Virology 295: 337–47.

Dong, H., Ray, D., Ren, S., Zhang, B., Puig-Basagoiti, F., Takagi, Y., Ho, C. K., Li, H. and Shi, P. Y. 2007. Distinct RNA elements confer specificity to flavivirus RNA cap methylation events. J Virol 81: 4412–21.

Egloff, M. P., Benarroch, D., Selisko, B., Romette, J. L. and Canard, B. 2002. An RNA cap (nucleoside-2′-O-)-methyltransferase in the flavivirus RNA polymerase NS5: crystal structure and functional characterization. EMBO J 21: 2757–68.

Elghonemy, S., Davis, W. G. and Brinton, M. A. 2005. The majority of the nucleotides in the top loop of the genomic 3′ terminal stem loop structure are cis-acting in a West Nile virus infectious clone. Virology 331: 238–46.

Filomatori, C. V., Lodeiro, M. F., Alvarez, D. E., Samsa, M. M., Pietrasanta, L. and Gamarnik, A. V. 2006. A 5′ RNA element promotes dengue virus RNA synthesis on a circular genome. Genes Dev 20: 2238–49.

Garcia-Montalvo, B. M., Medina, F. and del Angel, R. M. 2004. La protein binds to NS5 and NS3 and to the 5′ and 3′ ends of Dengue 4 virus RNA. Virus Res 102: 141–50.

Grief, C., Galler, R., Cortes, L. M. and Barth, O. M. 1997. Intracellular localisation of dengue-2 RNA in mosquito cell culture using electron microscopic in situ hybridisation. Arch Virol 142: 2347–57.

Gritsun, T. S. and Gould, E. A. 2006. Direct repeats in the 3′ untranslated regions of mosquito-borne flaviviruses: possible implications for virus transmission. J Gen Virol 87: 3297–305.

Gritsun, T. S. and Gould, E. A. 2007. Direct repeats in the flavivirus 3′ untranslated region; a strategy for survival in the environment? Virology 358: 258–65.

Grun, J. B. and Brinton, M. A. 1986. Characterization of West Nile virus RNA-dependent RNA polymerase and cellular terminal adenylyl and uridylyl transferases in cell-free extracts. J Virol 60: 1113–24.

Grun, J. B. and Brinton, M. A. 1987. Dissociation of NS5 from cell fractions containing West Nile virus-specific polymerase activity. J Virol 61: 3641–4.

Hahn, C. S., Hahn, Y. S., Rice, C. M., Lee, E., Dalgarno, L., Strauss, E. G. and Strauss, J. H. 1987. Conserved elements in the 3′ untranslated region of flavivirus RNAs and potential cyclization sequences. J Mol Biol 198: 33–41.

Johansson, M., Brooks, A. J., Jans, D. A. and Vasudevan, S. G. 2001. A small region of the dengue virus-encoded RNA-dependent RNA polymerase, NS5, confers interaction with both the nuclear transport receptor importin-beta and the viral helicase, NS3. J Gen Virol 82: 735–45.

Kapoor, M., Zhang, L., Ramachandra, M., Kusukawa, J., Ebner, K. E. and Padmanabhan, R. 1995. Association between NS3 and NS5 proteins of dengue virus type 2 in the putative RNA replicase is linked to differential phosphorylation of NS5. J Biol Chem 270: 19100–6.

Khromykh, A. A., Kenney, M. T. and Westaway, E. G. 1998. trans-Complementation of flavivirus RNA polymerase gene NS5 by using Kunjin virus replicon-expressing BHK cells. J Virol 72: 7270–9.

Khromykh, A. A., Kondratieva, N., Sgro, J. Y., Palmenberg, A. and Westaway, E. G. 2003. Significance in replication of the terminal nucleotides of the flavivirus genome. J Virol 77: 10623–9.

Khromykh, A. A., Meka, H., Guyatt, K. J. and Westaway, E. G. 2001a. Essential role of cyclization sequences in flavivirus RNA replication. J Virol 75: 6719–28.

Khromykh, A. A., Sedlak, P. L., Guyatt, K. J., Hall, R. A. and Westaway, E. G. 1999. Efficient trans-complementation of the flavivirus kunjin NS5 protein but not of the NS1 protein requires its coexpression with other components of the viral replicase. J Virol 73: 10272–80.

Khromykh, A. A., Sedlak, P.L., and Westaway, E. G. 1999b. trans-Complementation analysis of the flavivirus Kunjin NS5 gene reveals an essential role for translation of its N-terminal half in RNA replication. J Virol 73: 9247–55.

Khromykh, A. A., Varnavski, A. N., Sedlak, P. L. and Westaway, E. G. 2001b. Coupling between replication and packaging of flavivirus RNA: evidence derived from the use of DNA-based full-length cDNA clones of Kunjin virus. J Virol 75: 4633–40.

Kofler, R. M., Hoenninger, V. M., Thurner, C. and Mandl, C. W. 2006. Functional analysis of the tick-borne encephalitis virus cyclization elements indicates major differences between mosquito-borne and tick-borne flaviviruses. J Virol 80: 4099–113.

Li, H., Clum, S., You, S., Ebner, K. E. and Padmanabhan, R. 1999. The serine protease and RNA-stimulated nucleoside triphosphatase and RNA helicase functional domains of dengue virus type 2 NS3 converge within a region of 20 amino acids. J Virol 73: 3108–16.

Li, W., Li, Y., Kedersha, N., Anderson, P., Emara, M., Swiderek, K. M., Moreno, G. T. and Brinton, M. A. 2002. Cell proteins TIA-1 and TIAR interact with the 3′ stem-loop of the West Nile virus complementary minus-strand RNA and facilitate virus replication. J Virol 76: 11989–2000.

Lindenbach, B. D. and Rice, C. M. 1997. trans-Complementation of yellow fever virus NS1 reveals a role in early RNA replication. J Virol 71: 9608–17.

Lindenbach, B. D. and Rice, C. M. 1999. Genetic interaction of flavivirus nonstructural proteins NS1 and NS4A as a determinant of replicase function. J Virol 73: 4611–21.

Lindenbach, B. and Rice, C. 2001. Flaviviridae: the viruses and their replication. In: Fields Virology, ed. D. M. Knipe and P. M. Howley, pp. 991–1041. Philadelphia: Lippincott-Raven.

Lo, M. K., Tilgner, M., Bernard, K. A. and Shi, P. Y. 2003. Functional analysis of mosquito-borne flavivirus conserved sequence elements within 3′ untranslated region of West Nile virus by use of a reporting replicon that differentiates between viral translation and RNA replication. J Virol 77: 10004–14.

Mackenzie, J. 2005. Wrapping things up about virus RNA replication. Traffic 6: 967–77.

Mackenzie, J. M., Jones, M. K. and Westaway, E. G. 1999. Markers for trans-Golgi membranes and the intermediate compartment localize to induced membranes with distinct replication functions in flavivirus-infected cells. J Virol 73: 9555–67.

Mackenzie, J. M., Jones, M. K. and Young, P. R. 1996. Immunolocalization of the dengue virus nonstructural glycoprotein NS1 suggests a role in viral RNA replication. Virology 220: 232–40.

Mackenzie, J. M., Khromykh, A. A., Jones, M. K. and Westaway, E. G. 1998. Subcellular localization and some biochemical properties of the flavivirus Kunjin nonstructural proteins NS2A and NS4A. Virology 245: 203–15.

Malet, H., Egloff, M. P., Selisko, B., Butcher, R. E., Wright, P. J., Roberts, M., Gruez, A., Sulzenbacher, G., Vonrhein, C., Bricogne, G., Mackenzie, J. M., Khromykh, A. A., Davidson, A. D. and Canard, B. 2007. Crystal structure of the RNA polymerase domain of the West Nile virus non-structural protein 5. J Biol Chem 282: 10678–89.

Mandl, C. W., Holzmann, H., Kunz, C. and Heinz, F. X. 1993. Complete genomic sequence of Powassan virus: evaluation of genetic elements in tick-borne versus mosquito-borne flaviviruses. Virology 194: 173–84.

Mandl, C. W., Holzmann, H., Meixner, T., Rauscher, S., Stadler, P. F., Allison, S. L. and Heinz, F. X. 1998. Spontaneous and engineered deletions in the 3′ noncoding region of tick-borne encephalitis virus: construction of highly attenuated mutants of a flavivirus. J Virol 72: 2132–40.

Markoff, L. 2003. 5′ and 3′ NCRs in Flavivirus RNA, Adv Virus Res 59: 177–228.

Matusan, A. E., Pryor, M. J., Davidson, A. D. and Wright, P. J. 2001. Mutagenesis of the dengue virus type 2 NS3 protein within and outside helicase motifs: effects on enzyme activity and virus replication. J Virol 75: 9633–43.

Men, R., Bray, M., Clark, D., Chanock, R. M. and Lai, C. J. 1996. Dengue type 4 virus mutants containing deletions in the 3′ noncoding region of the RNA genome: analysis of growth restriction in cell culture and altered viremia pattern and immunogenicity in rhesus monkeys. J Virol 70: 3930–7.

Miller, S., Kastner, S., Krijnse-Locker, J., Buhler, S. and Bartenschlager, R. 2007. The non-structural protein 4A of dengue virus is an integral membrane protein inducing membrane alterations in a 2 K-regulated manner. J Biol Chem 282: 8873–82.

Miller, S., Sparacio, S. and Bartenschlager, R. 2006. Subcellular localization and membrane topology of the dengue virus type 2 Non-structural protein 4B. J Biol Chem 281: 8854–63.

Muylaert, I. R., Chambers, T. J., Galler, R. and Rice, C. M. 1996. Mutagenesis of the N-linked glycosylation sites of the yellow fever virus NS1 protein: effects on virus replication and mouse neurovirulence. Virology 222: 159–68.

Ng, M. L., Pedersen, J. S., Toh, B. H. and Westaway, E. G. 1983. Immunofluorescent sites in vero cells infected with the flavivirus Kunjin. Arch Virol 78: 177–90.

Olsthoorn, R. C. and Bol, J. F. 2001. Sequence comparison and secondary structure analysis of the 3′ noncoding region of flavivirus genomes reveals multiple pseudoknots. RNA 7: 1370–7.

Pogany, J., Fabian, M. R., White, K. A. and Nagy, P. D. 2003. A replication silencer element in a plus-strand RNA virus. EMBO J 22: 5602–11.

Proutski, V., Gould, E. A. and Holmes, E. C. 1997. Secondary structure of the 3′ untranslated region of flaviviruses: similarities and differences. Nucleic Acids Res 25: 1194–202.

Rauscher, S., Flamm, C., Mandl, C. W., Heinz, F. X. and Stadler, P. F. 1997. Secondary structure of the 3′-noncoding region of flavivirus genomes: comparative analysis of base pairing probabilities. RNA 3: 779–91.

Ray, D., Shah, A., Tilgner, M., Guo, Y., Zhao, Y., Dong, H., Deas, T. S., Zhou, Y., Li, H. and Shi, P. Y. 2006. West Nile virus 5′-cap structure is formed by sequential guanine N-7 and ribose 2′-O methylations by nonstructural protein 5. J Virol 80: 8362–70.

Ray, D. and White, K. A. 2003. An internally located RNA hairpin enhances replication of Tomato bushy stunt virus RNAs. J Virol 77: 245–57.

Rice, C. M., Lenches, E. M., Eddy, S. R., Shin, S. J., Sheets, R. L. and Strauss, J. H. 1985. Nucleotide sequence of yellow fever virus: implications for flavivirus gene expression and evolution. Science 229: 726–33.

Romero, T. A., Tumban, E., Jun, J., Lott, W. B. and Hanley, K. A. 2006. Secondary structure of dengue virus type 4 3′ untranslated region: impact of deletion and substitution mutations. J Gen Virol 87: 3291–6.

Roosendaal, J., Westaway, E. G., Khromykh, A. and Mackenzie, J. M. 2006. Regulated cleavages at the West Nile virus NS4A-2 K-NS4B junctions play a major role in rearranging cytoplasmic membranes and Golgi trafficking of the NS4A protein. J Virol 80: 4623–32.

Shi, P. Y., Li, W. and Brinton, M. A. 1996. Cell proteins bind specifically to West Nile virus minus-strand 3′ stem-loop RNA. J Virol 70: 6278–87.

Shurtleff, A. C., Beasley, D. W., Chen, J. J., Ni, H., Suderman, M. T., Wang, H., Xu, R., Wang, E., Weaver, S. C., Watts, D. M., Russell, K. L. and Barrett, A. D. 2001. Genetic variation in the 3′ non-coding region of dengue viruses. Virology 281: 75–87.

Ta, M. and Vrati, S. 2000. Mov34 protein from mouse brain interacts with the 3′ noncoding region of Japanese encephalitis virus. J Virol 74: 5108–15.

Thurner, C., Witwer, C., Hofacker, I. L. and Stadler, P. F. 2004. Conserved RNA secondary structures in Flaviviridae genomes. J Gen Virol 85: 1113–24.

Tilgner, M. and Shi, P. Y. 2004. Structure and function of the 3′ terminal six nucleotides of the west nile virus genome in viral replication. J Virol 78: 8159–71.

Uchil, P. D., Kumar, A. V. and Satchidanandam, V. 2006. Nuclear localization of flavivirus RNA synthesis in infected cells. J Virol 80: 5451–64.

Uchil, P. D. and Satchidanandam, V. 2003a. Architecture of the flaviviral replication complex. Protease, nuclease, and detergents reveal encasement within double-layered membrane compartments. J Biol Chem 278: 24388–98.

Uchil, P. D. and Satchidanandam, V. 2003b. Characterization of RNA synthesis, replication mechanism, and in vitro RNA-dependent RNA polymerase activity of Japanese encephalitis virus. Virology 307: 358–71.

Utama, A., Shimizu, H., Morikawa, S., Hasebe, F., Morita, K., Igarashi, A., Hatsu, M., Takamizawa, K. and Miyamura, T. 2000. Identification and characterization of the RNA helicase activity of Japanese encephalitis virus NS3 protein. FEBS Lett 465: 74–8.

Wallner, G., Mandl, C. W., Kunz, C. and Heinz, F. X. 1995. The flavivirus 3′-noncoding region: extensive size heterogeneity independent of evolutionary relationships among strains of tick-borne encephalitis virus. Virology 213: 169–78.

Warrener, P., Tamura, J. K. and Collett, M. S. 1993. RNA-stimulated NTPase activity associated with yellow fever virus NS3 protein expressed in bacteria. J Virol 67: 989–96.

Wengler, G. and Castle, E. 1986. Analysis of structural properties which possibly are characteristic for the 3′-terminal sequence of the genome RNA of flaviviruses. J Gen Virol 67 (Pt 6): 1183–8.

Wengler, G. and Gross, H. J. 1978. Studies on virus-specific nucleic acids synthesized in vertebrate and mosquito cells infected with flaviviruses. Virology 89: 423–37.

Westaway, E. G., Mackenzie, J. M., Kenney, M. T., Jones, M. K. and Khromykh, A. A. 1997. Ultrastructure of Kunjin virus-infected cells: colocalization of NS1 and NS3 with double-stranded RNA, and of NS2B with NS3, in virus-induced membrane structures. J Virol 71: 6650–61.

Westaway, E. G., Mackenzie, J. M. and Khromykh, A. A. 2003. Kunjin RNA replication and applications of Kunjin replicons. Adv Virus Res 59: 99–140.

Wu, J., Bera, A. K., Kuhn, R. J. and Smith, J. L. 2005. Structure of the Flavivirus helicase: implications for catalytic activity, protein interactions, and proteolytic processing. J Virol 79: 10268–77.

Xu, T., Sampath, A., Chao, A., Wen, D., Nanao, M., Chene, P., Vasudevan, S. G. and Lescar, J. 2005. Structure of the dengue virus helicase/nucleoside triphosphatase catalytic domain at a resolution of 2.4 A. J Virol 79: 10278–88.

Yap, T. L., Xu, T., Chen, Y. L., Malet, H., Egloff, M. P., Canard, B., Vasudevan, S. G. and Lescar, J. 2007. Crystal structure of the dengue virus RNA-dependent RNA polymerase catalytic domain at 1.85-Angstrom resolution. J Virol 81: 4753–65.

Yocupicio-Monroy, R. M., Medina, F., Reyes-del Valle, J. and del Angel, R. M. 2003. Cellular proteins from human monocytes bind to dengue 4 virus minus-strand 3′ untranslated region RNA. J Virol 77: 3067–76.

Yon, C., Teramoto, T., Mueller, N., Phelan, J., Ganesh, V. K., Murthy, K. H. and Padmanabhan, R. 2005. Modulation of the nucleoside triphosphatase/RNA helicase and 5′-RNA triphosphatase activities of dengue virus type 2 nonstructural protein 3 (NS3) by interaction with NS5, the RNA-dependent RNA polymerase. J Biol Chem 280: 27412–9.

You, S., Falgout, B., Markoff, L. and Padmanabhan, R. 2001. In vitro RNA synthesis from exogenous dengue viral RNA templates requires long range interactions between 5′- and 3′-terminal regions that influence RNA structure. J Biol Chem 276: 15581–91.

You, S. and Padmanabhan, R. 1999. A novel in vitro replication system for dengue virus. Initiation of RNA synthesis at the 3′-end of exogenous viral RNA templates requires 5′- and 3′-terminal complementary sequence motifs of the viral RNA. J Biol Chem 274: 33714–22.

Yu, L. and Markoff, L. 2005. The topology of bulges in the long stem of the flavivirus 3′ stem-loop is a major determinant of RNA replication competence. J Virol 79: 2309–24.

Zeng, L., Falgout, B. and Markoff, L. 1998. Identification of specific nucleotide sequences within the conserved 3′-SL in the dengue type 2 virus genome required for replication. J Virol 72: 7510–22.

Zhou, Y., Ray, D., Zhao, Y., Dong, H., Ren, S., Li, Z., Guo, Y., Bernard, K. A., Shi, P. Y. and Li, H. 2007. Structure and function of flavivirus NS5 methyltransferase. J Virol 81: 3891–903.

Chapter 4
Hepatitis C Virus Genome Replication

Brett D. Lindenbach and Timothy L. Tellinghuisen

Introduction

Over 170 million people are infected with hepatitis C virus (HCV), a major cause of acute and chronic liver disease that can lead to cirrhosis and hepatocellular carcinoma (Alter and Seeff 2000). The success of this virus is largely due to its ability to cause persistent infections – often lasting for decades – in over 70% of infected individuals. Thus, HCV infection leads to a dynamic interplay between viral replication, host antiviral responses, and viral countermeasures to evade those responses. Understanding these processes will be crucial for devising effective strategies to combat this virus and alleviate the human suffering it exacts. In this chapter we review the current understanding of HCV genome replication, emphasizing the role of viral and host factors in this process. Where applicable, we will draw comparisons to other viruses within this volume. Nevertheless, due to space limitations this review is not meant to be comprehensive, and we apologize in advance to authors whose work could not be cited.

Overview of the HCV Life Cycle

HCV is an enveloped, positive-strand RNA virus classified within the family *Flaviviridae*. The life cycle of HCV (Fig. 4.1) therefore shares overall similarity to the flaviviruses (Chapter 3) and other positive-strand RNA viruses. Upon infection of a host cell, HCV particles are taken up by receptor-mediated endocytosis and trafficked to endosomes, where the low pH of this compartment induces fusion of the viral envelope and bounding endosomal membrane. The nucleocapsid is then uncoated to release the viral genome into the cytoplasm (step 1), where it can be directly translated to produce the viral structural and non-structural (NS) proteins (step 2). Viral NS proteins and host factors assemble into a cytoplasmic,

B.D. Lindenbach (✉)
Section of Microbial Pathogenesis, Yale University School of Medicine, 295 Congress Ave., BCMM 354C, New Haven, CT 06536-0812, USA
e-mail: brett.lindenbach@yale.edu

C.E. Cameron et al. (eds.), *Viral Genome Replication*,
DOI 10.1007/b135974_4, © Springer Science+Business Media, LLC 2009

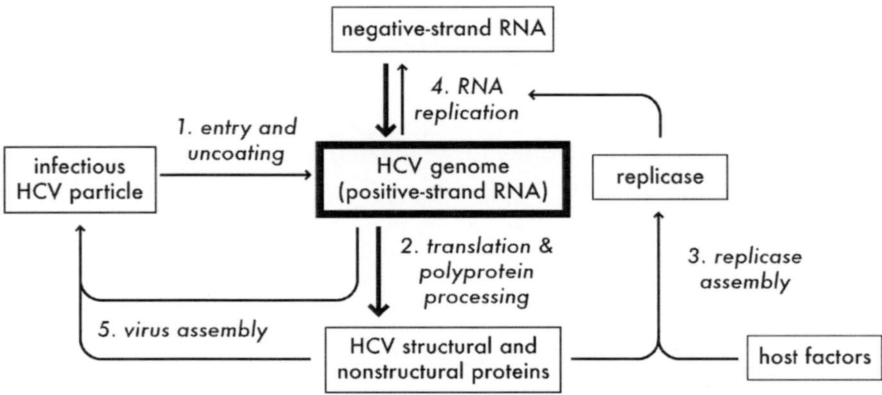

Fig. 4.1 Overview of the HCV life cycle.

membrane-bound RNA replicase (step 3), which then recruits the HCV genome out of translation and into replication (step 4). After RNA synthesis, new viral genomes can be recycled back into translation and replication or packaged by viral structural proteins into nascent viral particles (step 5).

Until recently, only limited aspects of the HCV life cycle could be studied because efficient viral culture systems did not exist. Once the viral genome was fully sequenced, infectious cDNA clones were constructed and shown to initiate replication upon intrahepatic inoculation into chimpanzees (Kolykhalov et al. 1997; Yanagi et al. 1997). While these reverse genetic systems were functional in vivo, they were obviously limited by the ethical and practical issues of primate research and did not permit viral replication to be studied in cell culture. The first broadly useful system for studying HCV RNA replication came when Lohmann et al. (1999) engineered bicistronic "subgenomic" replicons to express the selectable marker gene Neo, and selected for rare HCV replication events after transfecting this RNA replicon into the human hepatoma line, Huh-7 (Nakabayashi et al. 1982). Further growth of Neo-resistant cells selected for mutant replicons with increased RNA replication (Blight et al. 2000; Lohmann et al. 2001, 2003). Thus, cell culture-adapted replicons allowed the intracellular aspects of the viral life cycle to be studied and provided much-needed cell-based assays to screen for HCV-specific antivirals. More recently, Takaji Wakita and other investigators showed that the HCV strain JFH-1 was capable of producing infectious virus in cell culture (Lindenbach et al. 2005; Wakita et al. 2005; Zhong et al. 2005), and additional HCV cell culture systems have recently become available (reviewed in (Tellinghuisen et al. 2007)). Thus, the tools to study the complete life cycle of HCV are now in hand.

HCV Genomes

The HCV genome is a monopartite, single-stranded RNA, 9.6 kb in length (Choo et al. 1991). Unlike most cellular mRNAs, the HCV genome lacks a 5′ cap and does not encode a 3′ polyadenosine tail (Tanaka et al. 1995; Kolykhalov et al. 1996).

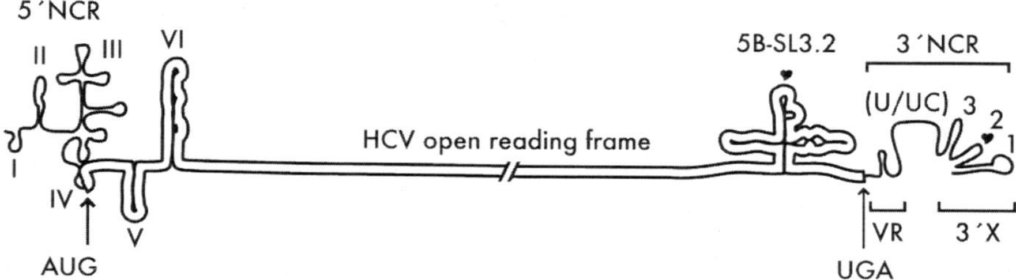

Fig. 4.2 Features of the HCV genome. This map of the HCV genome highlights important secondary structures within the HCV genomic RNA, as described in the text. The large open reading frame is indicated as an open bar. Loop regions involved in the kissing interaction between 5BSL-3.2 and 3′X SL2 are indicated by hearts. AUG, start codon; UGA, stop codon; VR, variable region.

Instead, the HCV genome encodes several *cis*-acting RNA elements (CREs) that regulate genome translation, RNA replication, and most likely, packaging. Known CREs include sequences and secondary structures within the 5′ and 3′ noncoding regions (NCRs) and the large open reading frame (Fig. 4.2). Unlike the picornaviruses (Chapter 1), HCV does not appear to encode a viral genome-linked protein.

As for all positive-strand RNA viruses, which encode polymerases that lack proof reading activity, HCV exhibits a high degree of genetic variability. Given the size of the HCV genome and an estimated mutation rate of $\approx 10^{-4}$ misincorporations/nt (Crotty et al. 2001), one can calculate that mutants will quickly accumulate and predominate in HCV populations of even modest size. It has been estimated that a chronically infected person makes 10^{12} virions/day (Perelson et al. 2005). Thus, the sequence diversity present even within a single person is huge. This swarm of genetically related viruses, with no real "wild-type" master sequence, is often referred to as a "quasi-species". On the global scale, with over 10^8 infected people, the amount of sequence diversity available to HCV is astronomical. Of course evolutionary fitness varies greatly among these populations. As a result, HCV has evolved into six metastable genotypes, which differ by more than 30% at the nucleotide level (Simmonds et al. 2005).

The HCV 5′ NCR is 341 nt in length, well conserved, and highly structured (Fig. 4.2); it has at least two major functions in the viral life cycle. First, it encodes an internal ribosome entry site (IRES) that allows for cap-independent translation of the viral genome, described below. Second, the 5′ NCR includes one or more overlapping CREs necessary for genome replication. While efficient RNA replication requires nearly the entire 5′ NCR, the minimal 5′ replication element is encoded within the first 120 nt, which includes stem-loops I and II (Friebe et al. 2001; Kim et al. 2002b). Intriguingly, the liver-specific cellular micro RNA (miRNA)-122 binds to the unstructured spacer between these stem-loops, and this interaction is required for an early step in HCV replication (Jopling et al. 2005). Furthermore, by making intergenotypic recombinants with reduced rates of positive- and negative-strand synthesis, Binder and colleagues have confirmed that the 5′ NCR (or more likely, its reverse complement in the negative strand) plays an important role in positive-strand synthesis (Binder et al. 2007b).

The 3′ NCR consists of three subdomains: (1) a short (≈40 nt) variable region located immediately downstream of the termination codon; (2) a polyuridine/polypyrimidine (U/UC) tract of variable length; and (3) a highly conserved, 98 nt region termed the 3′X domain (Fig. 4.2) (Tanaka et al. 1995; Kolykhalov et al. 1996). The poly (U/UC) tract and 3′X domain are required for RNA replication, while the variable region seems to influence replication efficiency (Yanagi et al. 1999; Kolykhalov et al. 2000; Friebe and Bartenschlager 2002; Yi and Lemon 2003). The minimal poly (U/UC) tract appears to be between 26 and 52 nt in length and requires an uninterrupted homopolyuridine tract, suggesting that it is recognized by a *trans*-acting factor (Friebe and Bartenschlager 2002; Yi and Lemon 2003; You and Rice 2007). The 3′X domain consists of a highly stable 3′ terminal stem-loop, 3′ SL1, and two metastable stem-loops, 3′ SL2 and 3′ SL3 (Tanaka et al. 1996; Blight and Rice 1997). Again, by using intergenotypic chimeras Binder and colleagues confirmed that the 3′X domain serves as an important CRE for negative-strand synthesis (Binder et al. 2007b).

Given that the large open reading frame encompasses over 94% of the HCV genome, it is no surprise that HCV coding region also encodes CREs. Sequences with dual overlapping functions place tight constraints on codon usage, which can be used to help identify such internal CREs (reviewed in (Branch et al. 2005)). As an example, anomalous codon usage was originally thought to reflect a selective pressure to retain an alternative reading frame within the core gene (reviewed in (Branch et al. 2005)). Subsequent genetic analysis determined that this alternative reading frame was dispensable for virus replication, but identified an important CRE, stem-loop VI, embedded within the core gene (Fig. 4.2) (McMullan et al. 2007). While this region is not required for RNA replication in the context of subgenomic RNA replicons, full-length HCV genomes containing mutations in this structure are highly attenuated in vitro and in vivo, and selectively recover the wild-type sequence. Interestingly, one leg of stem-loop VI can base pair with the 5′ NCR, overlapping the miRNA-122 binding site, and thereby down-regulate IRES-mediated translation (Kim et al. 2003).

Embedded within the NS5B gene is another important CRE, 5BSL3.2 (Fig. 4.2) (Lee et al. 2004; You et al. 2004). A key feature of 5BSL3.2 is that one of the loop regions can base pair with the loop region of SL2 in the 3′X region of the genome (Friebe et al. 2005). Mutations in either loop region destroy HCV RNA replication, while compensatory mutations restore it (Friebe et al. 2005; You and Rice 2007). It is not yet known how this long-distance "kissing" interaction regulates HCV RNA replication, but it is interesting to note that base pairing between the coding region and both HCV NCRs is functionally important. This may reflect a general strategy to maintain genome integrity.

In addition to several CREs, the HCV genome exhibits a number of interesting features, such as a low frequency of UA and UU dinucleotides, which may reflect a selection for viral genomes that are poor substrates for the interferon-inducible RNase L (Han and Barton 2002). The HCV genome is also predicted to contain an unusually high rate of internal base pairing, a feature that correlates with a high rate of persistent infection among positive-strand RNA viruses (Simmonds et al. 2004). Thus, these higher-order structures may help to circumvent innate antiviral defenses.

Translation and Polyprotein Processing

As mentioned, the HCV 5′ NCR encodes an IRES that directs cap-independent translation of the viral genome (recently reviewed in (Fraser and Doudna 2007)). The minimal HCV IRES is encoded by sequences within stem-loops II through IV (Fig. 4.2). The central part of this structure, stem-loop III, can directly bind the ribosomal 40S subunit, positioning the start codon within the ribosomal P site (Pestova et al. 1998). Importantly, this interaction appears to bypass the need for canonical eukaryotic translation initiation factors (eIFs), and functionally resembles the mechanism of prokaryotic initiation via the Shine-Delgarno sequence. Structural studies indicate that stem-loops II and III induce a conformational change in the 40S subunit, allowing the RNA-binding cleft to open (Spahn et al. 2001). Subsequently, the IRES-40S complex recruits eIF3 and the eIF2•GTP•Met-tRNA ternary complex, forming a 48S intermediate complex (Ji et al. 2004; Otto and Puglisi 2004). Interestingly, stem-loop IIIb of the HCV IRES functionally and structurally mimics the binding of the 5′ cap binding complex eIF4F (Siridechadilok et al. 2005). Following GTP hydrolysis, these initiation factors are released and the ribosomal 60S subunit is recruited into an IRES-80S complex, which is then capable of initiating protein synthesis (Ji et al. 2004; Otto and Puglisi 2004).

There has been some debate about how downstream sequences influence HCV IRES activity, and it now appears that secondary structures are disfavored in the proximal region of the core gene (Rijnbrand et al. 2001). Although originally controversial (reviewed in (Tellinghuisen et al. 2007)), it is also now clear that the HCV 3′ NCR can enhance HCV IRES-mediated translation in hepatic cells, perhaps by promoting ribosomal recycling analogous to the function of the polyA tail in cellular mRNAs (Ito et al. 1998; McCaffrey et al. 2002; Bradrick et al. 2006; Song et al. 2006). In addition, a number of cellular factors have been shown to influence HCV IRES activity in *trans* including the La autoantigen, which may assist in recruiting the 40S ribosomal subunit to the AUG start codon (reviewed in (Lindenbach et al. 2007b)).

Translation of the HCV genome produces a large polyprotein that is cotranslationally and post-translationally cleaved by cellular and viral proteases into at least ten discrete products (Fig. 4.3). These include signal peptidase cleavage at the core/E1, E1/E2, E2/p7, and p7/NS2 junctions. In addition, the mature form of core is generated via an intramembrane cleavage of the C-terminal anchor by signal peptide peptidase. The remaining polyprotein processing steps are catalyzed by two HCV-encoded proteases. The C-terminal domain of NS2 encodes a cysteine protease that is responsible for cleaving the NS2/3 junction (Grakoui et al. 1993; Hijikata et al. 1993). The crystal structure of this domain revealed that it forms an unusual homodimeric protease with twin composite active sites (Lorenz et al. 2006).

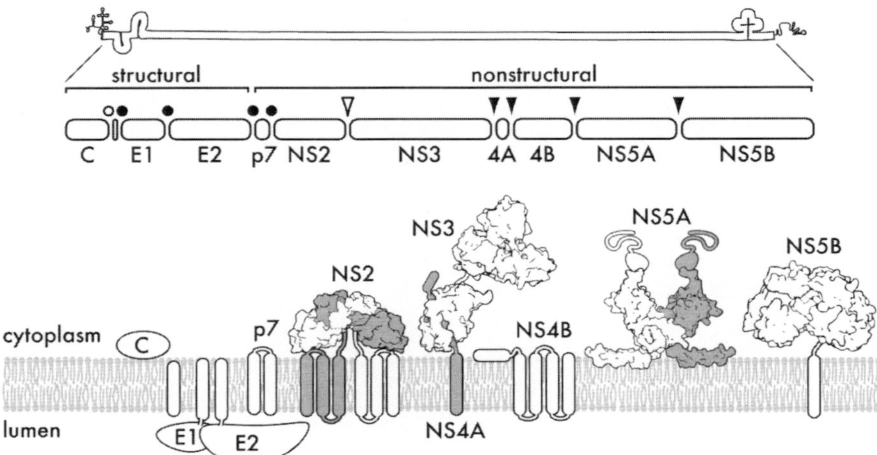

Fig. 4.3 Features of the HCV proteins. The top illustration indicates the order of the HCV gene products as they are translated in the polyprotein. Open bullet, signal peptide peptidase cleavage; closed bullet, signal peptidase cleavage; open arrowhead, NS2-3 cysteine protease cleavage; closed arrowhead, NS3-4A serine protease cleavage. The bottom illustration indicates the topology of HCV proteins. Where available, atomic coordinates were used to render the illustration to approximate scale. Since NS2 and NS5A were crystallized as dimers, one monomer of each is colored gray.

As described below, all downstream cleavages are mediated by the NS3-4A serine protease.

The core protein and E1 and E2 glycoproteins are structural components of HCV virions, while p7 and NS2 proteins appear to be involved in virus assembly (Jones et al. 2007; Steinmann et al. 2007). The remaining NS proteins are responsible for modulating the intracellular aspects of the HCV life cycle, including RNA replication. Since the core, E1, E2, p7, and NS2 are dispensable for the replication of subgenomic replicons (Lohmann et al. 1999), we therefore turn our attention to the NS proteins involved in RNA replication.

Replicase Components

NS3-4A

NS3 (70 kDa) is a key component of the HCV replicase, encoding an N-terminal serine protease domain and a C-terminal RNA helicase/nucleoside triphosphatase (NTPase) domain. The activities of both enzymes are essential for viral replication. The serine protease domain of NS3 can cleave NS3/4A in *cis* and then interact with NS4A, which contributes one ß-strand to the chymotrypsin-like fold and activates the serine protease activity (reviewed in (Penin et al. 2004b)). In turn, NS4A anchors the NS3-4A complex to cellular membranes via an N-terminal membrane anchor. The NS3-4A serine protease is responsible for cleaving the viral polyprotein at the NS4A/B, NS4B/5A, and NS5A/5B junctions. In addition, NS3-4A can cleave the cellular proteins IPS-1 and TRIF, which normally transduce signals to activate gene expression in response to viral infection (Li et al. 2005a, b; Meylan et al. 2005). By cleaving these substrates, NS3-4A helps to circumvent cellular antiviral defenses, as described below.

NS3-4A is a member of the superfamily 2 RNA helicases, which use the energy from ATP hydrolysis to power double-stranded RNA unwinding. Recent enzymatic studies have revealed that NS3-4A unwinds 18-bp segments via several discrete 3 bp steps, each of which uses a spring-loaded mechanism to coordinate ATP hydrolysis with smaller, 1 bp advances along the substrate (Serebrov and Pyle 2004; Dumont et al. 2006; Myong et al. 2007). Although the NS3 helicase domain is functional on its own, full helicase activity requires full-length NS3 and NS4A, which contribute to substrate binding (Pang et al. 2002; Beran et al. 2007). Furthermore, NS3-4A helicase appears to function as a dimer or other higher-order multimer (Serebrov and Pyle 2004; Mackintosh et al. 2006). NS3-4A preferentially binds polyuridine, which stimulates NTPase activity in vitro (Suzich et al. 1993; Kanai et al. 1995); it is interesting to speculate that the poly (U/UC) may perform a similar function in vivo, targeting NS3-4A helicase activity to the 3′ NCR. Despite these details, the precise function of the NS3-4A helicase during viral replication remains unknown. One important clue is that the NS3-4A helicase is genetically linked to NS5A, NS5B, and the 5′ NCR for efficient positive-strand synthesis (Binder et al. 2007b). Thus, perhaps the HCV helicase is responsible for fraying the double-stranded product of negative-strand synthesis, revealing a CRE that directs positive-strand synthesis.

At 54-aa, NS4A (8 kDa) is the smallest NS protein. As mentioned, NS4A serves to anchor the NS3-4A complex to cellular membranes, contributes to the folding of the serine protease, and stimulates RNA helicase activity. In addition, NS4A plays an important albeit unclear role in NS5A hyperphosphorylation (described below) (Kaneko et al. 1994; Koch and Bartenschlager 1999; Lindenbach et al. 2007a). Mutagenesis of the C-terminal acidic region in NS4A revealed that the efficiency of RNA replication correlates with NS4A's ability to mediate NS5A phosphorylation (Lindenbach et al. 2007a). Some of these NS4A-mediated replication defects were suppressed by second-site changes in NS3, indicating additional functional interactions between these two proteins.

NS4B

NS4B is a small (27 kDa) hydrophobic integral membrane protein that co-translationally associates with ER membranes via an internal signal sequence and at least four central transmembrane spanning helices (Hügle et al. 2001; Lundin et al. 2003; Elazar et al. 2004). Despite the relative hydrophobicity of the protein, the bulk of NS4B appears to be on the cytoplasmic face of the ER membrane, particularly regions at the N- and C-termini, although the topology of the N-terminus remains an area of debate, and may involve other HCV non-structural proteins, such as NS5A (Lundin et al. 2003; Elazar et al. 2004; Lundin et al. 2006). NS4B may also be palmitoylated, which can apparently facilitate its oligomerization (Yu et al. 2006). For some HCV strains, the central cytoplasmic loop of NS4B apparently encodes a nucleotide-binding motif and an in vitro GTPase activity has been demonstrated for this protein (Einav et al. 2004). However the relevance of these observations remains unclear, as at least one cell culture-adaptive mutation disrupts this motif yet leads to increased RNA replication (Bartenschlager et al. 2004).

Several studies have implicated NS4B in RNA replication. A number of cell culture-adaptive mutations have been mapped to NS4B (reviewed in (Bartenschlager et al. 2004)), and allelic variation within NS4B correlates with RNA replication efficiency (Blight 2007). One specific role for NS4B may be to serve as a scaffold for viral replicase assembly. Overexpression of NS4B induces the rearrangement of cellular membranes into structures that resemble the sites of RNA replication (Egger et al. 2002; Gosert et al. 2003) and can trigger ER stress leading to an unfolded protein response (Zheng et al. 2005). NS4B has recently been shown to interact with Rab5, a regulator of membrane fusion, as well as other components of the early endosomal compartment (Stone et al. 2007). NS4B also appears to regulate the activity of sterol regulatory element binding proteins (SREBPs), major regulators of lipid metabolism, perhaps for the purpose of membrane synthesis and reorganization (Lundin et al. 2006).

NS5A

NS5A is a large (56–58 kDa), hydrophilic, RNA-binding phosphoprotein of unknown function. Yet recent biochemical and genetic experiments have provided new insights into this enigmatic protein. NS5A is peripherally anchored to intracellular membranes via an N-terminal amphipathic helix (Brass et al. 2002; Penin et al. 2004a; Sapay et al. 2006; Brass et al. 2007). Deletion or alteration of this helix leads to a diffuse cytoplasmic localization of NS5A and is lethal for RNA replication (Elazar et al. 2003; Penin et al. 2004a). Furthermore, other amphipathic viral membrane anchors cannot substitute for the NS5A membrane anchor, suggesting that this region likely interacts with other HCV replicase components (Lee et al. 2006; Teterina et al. 2006).

Limited proteolysis of purified NS5A suggests that the remainder of NS5A folds into three domains separated by two flexible, low complexity sequence blocks (Tellinghuisen et al. 2004). Domain I contains four conserved cysteines that coordinate a single zinc atom (Tellinghuisen et al. 2004). These residues are essential for RNA replication, presumably via their structural role in metal ion coordination. The structure of domain I has been determined by X-ray crystallography, revealing a novel protein fold for NS5A (Tellinghuisen et al. 2005). One obvious feature of domain I is a large, basic groove formed by the interface of monomers in the dimeric

NS5A structure. It is tempting to speculate that this groove might represent the site of RNA binding to NS5A, although this remains to be experimentally determined (Huang et al. 2005; Tellinghuisen et al. 2005). Domain I also contains a large, conserved region on its surface that includes residues involved in NS5B binding and inhibiting polymerase activity (Shirota et al. 2002). NS5A domains II and III are poorly conserved and not well characterized. Domain II appears to contain some alpha helical content but attempts at determining the structure of this domain indicate that it may be natively unfolded (Liang et al. 2006, 2007). Domain III appears even more plastic than domain II, as this region can tolerate large deletions and insertions without disrupting RNA replication (Moradpour et al. 2004b; Appel et al. 2005b; Liu et al. 2006; McCormick et al. 2006).

NS5A clearly has an important role in HCV RNA replication. First, a number of cell culture-adaptive mutations that greatly enhance RNA replication have been mapped to NS5A (Blight et al. 2000; Lohmann et al. 2001). Conversely, RNA replication is ablated by a number of mutations in NS5A (Elazar et al. 2003; Penin et al. 2004a; Tellinghuisen et al. 2004; Appel et al. 2005b). Furthermore, NS5A colocalizes with other replicase components at the site of active RNA synthesis (Moradpour et al. 2004b). NS5A is the only HCV NS protein that can be complemented in *trans*, suggesting that this protein may be a dynamic component of the replicase that can enter and exit the replicase throughout the HCV life cycle (Appel et al. 2005a; Tong and Malcolm 2006).

One of the most striking aspects of NS5A is the correlation between NS5A phosphorylation and RNA replication. NS5A exists in basally phosphorylated (56 kDa) and hyperphosphorylated (58 kDa) forms, based on their mobility in SDS-PAGE (Kaneko et al. 1994; Tanji et al. 1995). Soon after cell culture-adaptive mutations were discovered, it was apparent that NS5A mutations that dramatically increase RNA replication also tend to decrease NS5A hyperphosphorylation, suggesting that these phosphorylation events may regulate the level of replication (Blight et al. 2000). The relevant basal and hyperphosphorylation acceptor sites have not been fully defined, but the predominant sites of phosphorylation are likely to be serine residues (Tanji et al. 1995; Reed et al. 1997; Reed and Rice 1999; Katze et al. 2000; Appel et al. 2005b). Nonetheless, accumulating evidence suggests that basal phosphorylation primarily targets residues in domains II and III, whereas hyperphosphorylation sites cluster in domain I and the low complexity linker between domains I and II. Residues implicated in basal phosphorylation are not required for RNA replication (Appel et al. 2005b). Hyperphosphorylation of NS5A requires the expression of NS3, NS4A, NS4B, and NS5A in *cis* (Koch and Bartenschlager 1999; Neddermann et al. 1999). While the mechanisms linking NS5A hyperphosphorylation and RNA replication remain to be determined, one important clue is that the NS5A phosphoforms differentially interact with hVAP-A, a SNARE-like vesicle sorting protein that has been implicated in HCV replication (Evans et al. 2004; Randall et al. 2007).

Perhaps the most compelling data in the area of NS5A phosphorylation and replication comes from the study of NS5A kinases. One large-scale compound library screen for small molecules that decrease the hyperphosphorylation of NS5A turned up molecules that allow RNA replication to proceed efficiently even in the absence of adaptive mutations (Neddermann et al. 2004). Some of these compounds likely block casein kinase 1α (CKIα), and genetic silencing of CKIα produces results similar to treatment of cells with these inhibitors (Quintavalle et al. 2006). Phosphorylation of NS5A by CKIα requires prior phosphorylation of nearby residues in NS5A, presumably by one or more additional kinases, indicating that NS5A phosphorylation is likely to be a cascade of events (Quintavalle et al. 2007). Indeed, a number of other potential NS5A kinases have been identified, including AKT, casein kinase II, p70s6K, MEK, and MKK1 (see (Huang et al. 2007) for a recent review). Perhaps the best-characterized potential NS5A kinase activity, after that of CKIα, is casein kinase II (CKII). A number of biochemical experiments have implicated CKII, or a related CMGC kinase family member, as a kinase that phosphorylates multiple sites in NS5A domain III (Reed et al. 1997; Kim et al. 1999; Huang et al. 2004). Despite progress in identifying NS5A kinases, more work is needed to understand the network of NS5A phosphorylation events and their role in regulating HCV genome replication.

NS5A has been reported to interact with a large constellation of host proteins involved in innate immunity, signaling, apoptosis, and lipid trafficking (Macdonald and Harris 2004). Many of these interactions, although interesting, have no demonstrated effect on HCV biology. For instance, a recent siRNA screen targeting a number of published NS5A-interaction partners confirmed only a few essential interactions (Randall et al. 2007). Nevertheless, some NS5A-interacting host factors do have an essential role in HCV genome replication. Three recent examples include FBL-2, FKBP8, and TBC120. The interaction of NS5A with the geranylgeranylated F-box protein FBL-2 is required for replication of a genotype 1b replicon, and small molecule inhibitors of geranylgeranylation or siRNA silencing of FBL-2 inhibit(s) this replicon (Wang et al. 2005). FBL-2, like other F-box proteins, is believed to target proteins for degradation, although its specific substrates remain to be identified. NS5A also interacts with FKBP8, an immunophilin that shares similarity with the cyclophilin family of peptidyl-prolyl *cis–trans* isomerases (PPIs), although FKBP8 appears to lack PPI activity (Okamoto et al. 2006). Interestingly, FKBP8 binds to NS5A as a trimeric complex with the chaperone HSP90 to modulate HCV RNA replication. The significance of these interactions for HCV protein folding or post-translational modification is not understood. TBC120 is an NS5A-binding host protein that is essential for HCV RNA replication (Sklan et al. 2007a, b). This protein appears to be similar to the Rab GTPase-activating proteins, and like VAP-A, may play a role in membrane trafficking and reorganization during the HCV life cycle.

NS5B

NS5B (68 kDa) is a central component of the HCV replicase, the RNA-dependent RNA polymerase (RdRp) that synthesizes all viral RNAs. NS5B was initially predicted to function as an RNA polymerase based on the presence of the conserved

GDD motif common to the active site of other polymerases (Choo et al. 1989). Mutation of this GDD motif abolishes infectivity of HCV transcripts in chimpanzees and blocks RNA replication in cell culture (Lohmann et al. 1999; Kolykhalov et al. 2000). NS5B is a hydrophilic protein that associates with ER-derived membranes via a C-terminal hydrophobic tail-anchor that can post-translationally insert into membranes (Schmidt-Mende et al. 2001; Ivashkina et al. 2002). The NS5B tail-anchor is required for HCV RNA replication, and its removal leads to nuclear localization of NS5B (Moradpour et al. 2004a), although it can be functionally replaced by a similar tail-anchor sequence from a poliovirus protein (Lee et al. 2006). Nevertheless, removal of this sequence allows the expression and purification of soluble NS5B that retains polymerase activity, which has allowed extensive structural and enzymatic analyses on NS5B (Lohmann et al. 1997; Ferrari et al. 1999).

A number of crystal structures of the soluble, tail-anchor deleted form of NS5B have been generated (Ago et al. 1999; Bressanelli et al. 1999; Lesburg et al. 1999; Bressanelli et al. 2002; O′Farrell et al. 2003). These structures have been reviewed elsewhere in great detail (De Francesco et al. 2003). The overall fold of NS5B is similar to that of other single chain polymerases, with a classic right-hand topology containing distinct palm, finger, and thumb domains. Like other polymerases, the palm domain of NS5B contains the residues responsible for catalysis, nucleotide binding, and RNA template coordination. Unlike other polymerases, extensive interactions exist between the finger and thumb domains in NS5B, resulting in a fully enclosed, preformed active site capable of binding nucleotides without further conformational changes. It is thought that this closed form of NS5B may represent the structure of the polymerase during strand initiation, and that further conformational changes are required for elongation. More recent structural efforts have captured an open form of the polymerase in which thumb domain movements disrupt contact with the finger domain, which may represent a processive form of the polymerase (Biswal et al. 2005). Another unique feature of the NS5B polymerase is the presence of a ß-hairpin loop near the active site, which may position the 3′-end of the template in the proper orientation relative to the active site (Hong et al. 2001; O′Farrell et al. 2003; Ranjith-Kumar et al. 2003; Kim et al. 2005). Once the template is properly positioned, the ß-hairpin may be displaced from the active site to allow the large double-stranded RNA product to exit the active site region. Another unusual feature of NS5B is the presence of a GTP-binding allosteric regulatory site in the thumb domain (Bressanelli et al. 2002). A structure of NS5B complexed with non-nucleoside inhibitors suggests the importance of the region of the thumb subdomain near this allosteric site in conformational changes required for the transition of NS5B from the initiation state to an elongation state (O′Farrell et al. 2003). The residues comprising the GTP-binding site are not required for the in vitro polymerase activity of NS5B but are essential for RNA replication in the replicon system (Ranjith-Kumar et al. 2003; Cai et al. 2005). Another unusual regulatory feature observed in the structure of NS5B is a long C-terminal loop that encircles the thumb domain and inserts in the region of the active site (Lévêque et al. 2003). This loop decreases the RNA-binding and polymerase activities of NS5B in vitro (Lévêque et al. 2003; Ranjith-Kumar et al. 2003). It is clear NS5B possesses many features that regulate its activity, what remains to be understood is how these features function dynamically in the context of replication.

The enzymatic activity of NS5B has been extensively studied (see (Lohmann et al. 2000) for review). Although NS5B is capable of extending annealed RNA and DNA primers or self-primed "copy back" templates (Behrens et al. 1996; Lohmann et al. 1997; Al et al. 1998; Yamashita et al. 1998), NS5B most likely uses primer-independent "de novo" initiation during authentic RNA replication (Luo et al. 2000; Zhong et al. 2000). A crystal structure of NS5B with bound nucleotides strongly supports this model, as this structure is similar to the de novo initiation complex of the bacteriophage phi 6 polymerase (Bressanelli et al. 2002). Furthermore, the aforementioned ß-hairpin and C-terminal regulatory loop likely favor de novo initiation over copy back and primer extension activities by excluding double-stranded templates (Cheney et al. 2002; Lévêque et al. 2003; Ranjith-Kumar et al. 2003). NS5B prefers to initiate with a purine residue templated by a free 3′-end, but surprisingly, initiation can also occur on circular RNA templates, indicating that a free 3′-end is not absolutely required for de novo initiation (Kao et al. 2000; Ranjith-Kumar and Kao 2006). Early work with NS5B indicated that it is also capable of adding non-templated residues to the 3′-end of templates via a terminal transferase-like activity (Behrens et al. 1996; Ranjith-Kumar et al. 2001; Shim et al. 2002). These findings are controversial, as disparity exists in observing this activity among different research groups, and it has been suggested that terminal transferase activity may be a copurifying enzyme from the expression host (Lohmann et al. 1997; Yamashita et al. 1998; Oh et al. 1999; Kashiwagi et al. 2002). Nevertheless, one group reported that NS5B terminal transferase activity is dependent on active site residues within NS5B (Ranjith-Kumar et al. 2001).

In addition to the complexities of regulation that became apparent from NS5B structures, additional levels of regulation of NS5B exist. NS5B forms oligomers and exhibits cooperativity in RNA synthesis (Wang et al. 2002), suggesting that polymerase activity is regulated by homotypic intermolecular interactions. Similar findings have been made for the poliovirus RdRP (Chapter 1). In addition NS5B polymerase activity is modulated by a number of HCV replicase proteins, including NS3, a positive stimulator, and NS4A and NS5A, both negative regulators (Piccininni et al. 2002; Shirota et al. 2002). Clearly, the activity of NS5B in the viral replicase might be quite different from that observed in vitro using purified NS5B. Additionally, a number of host cell proteins have been shown to interact with and modify the activity of NS5B. NS5B can be phosphorylated by the cellular kinase PRK2, and this modification increases HCV RNA replication (Kim et al. 2004). In addition, NS5B, like NS5A, interacts with vesicle sorting proteins like hVAP-A and B. The interaction of NS5B with hVAP-B appears to increase polymerase stability, and therefore RNA replication (Tu et al. 1999; Gao et al. 2004). Perhaps the most exciting observation in this area in recent years is the interaction of NS5B with cyclophilins, another class of PPIs. This interaction was uncovered when cyclosporin A (CsA) was found to suppress HCV RNA replication in a dose-dependent manner (Watashi et al. 2003; Nakagawa et al. 2004). It was later found that CsA disrupts the interaction of NS5B with cyclophilin B, which is required for efficient recruitment and replication of HCV RNA (Watashi et al. 2005). HCV replication is reduced by knockdown of cyclophilin B, or by the non-immunosuppressive CsA derivative, DEBI0-025 (Nakagawa et al. 2005; Paeshuyse et al. 2006). Thus, the immunosuppressive activity of CsA is not required for inhibition of replication. A number of HCV mutants resistant to CsA have been selected, and mapping studies suggest both NS5B and NS5A may play a role in the inhibitory activities of CsA (Fernandes et al. 2007; Robida et al. 2007). CsA is not likely to become a widely used HCV antiviral in the immediate future as this drug is only highly effective in inhibiting genotype 1b replicons (Ishii et al. 2006).

Membrane Alterations

Like all positive-strand RNA viruses, HCV genome replication is membrane associated. HCV replication occurs within a dense cluster of perinuclear vesicles often referred to as the "membranous web" (Fig. 4.4A) (Egger et al. 2002). The source of these membranes is likely to be the ER or a closely related compartment (reviewed in (Bartenschlager et al. 2004)) and can be induced by expressing NS4B (Egger et al. 2002). The membranous web has been positively identified as the site of HCV synthesis RNA through metabolic labeling of nascent RNAs and colocalization of viral replicase components (Gosert et al. 2003; Moradpour et al. 2004b). Biochemical analysis of membrane fractions from HCV replicon-bearing cells indicate that RdRP activity is protected from nuclease and protease digestion within a detergent-sensitive compartment (Miyanari et al. 2003; Aizaki et al. 2004; Yang et al. 2004; Quinkert et al. 2005). Thus, it is thought that HCV replication occurs within these vesicles (Fig. 4.4B). This model is in agreement with studies on numerous other positive-strand RNA viruses. Recent biophysical studies on the Flock House virus

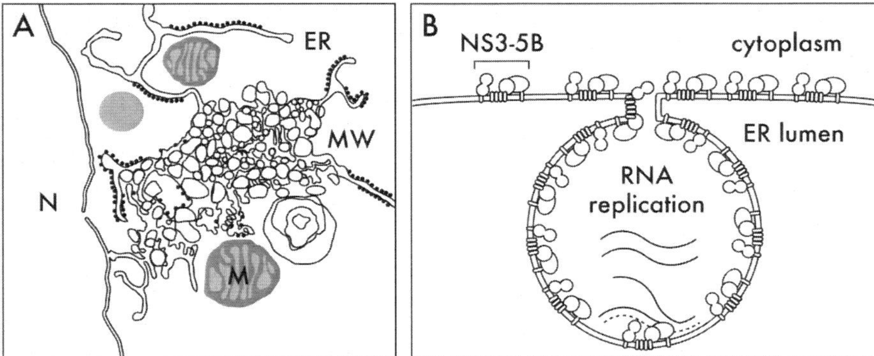

Fig. 4.4 Membrane interactions of HCV. A. An illustration of HCV-induced membrane rearrangements, as interpreted by the authors from an electron micrograph published by Gosert et al. (2003). N, nucleus; ER, endoplasmic reticulum; MW, membranous web; M, mitochondrion. **B.** A model for the HCV replicase within a spherule.

replicase indicated that these "spherules" are invaginations lined with viral replicase proteins, contain a single negative-strand intermediate and only a few positive strands, and retain communication with the cytosol via a thin neck (Kopek et al. 2007). It has been further suggested that these structures bear structural and functional homology to incompletely budded retrovirus particles or icosahedral, double-stranded RNA virus particles (Schwartz et al. 2002).

HCV replication induces genes involved in lipid metabolism, including ATP citrate lyase and acetyl-CoA synthetase (Su et al. 2002; Kapadia and Chisari 2005). Thus, membrane proliferation is likely to be required for membranous web formation and replicase assembly. Indeed, HCV replication is stimulated by saturated and monounsaturated fatty acids and inhibited by polyunsaturated fatty acids or inhibitors of lipid synthesis (Kapadia and Chisari 2005). The role of cholesterol in this process remains obscured by the fact that for genotype 1b replicons, RNA replication is dependent on the mevalonate biosynthetic pathway, most likely for the geranylgeranylation of the NS5A-interacting protein FBL2 (Ye et al. 2003; Kapadia and Chisari 2005; Wang et al. 2005). Nevertheless, extraction of cellular cholesterol with methyl-ß-cyclodextrin – a crude method to be sure – had only a modest effect on HCV genome replication (Aizaki et al. 2004; Kapadia et al. 2007). Thus, the role of cholesterol in HCV replicase function remains unclear, and it would be interesting to reassess this in the context of a replicon that is not reliant on FBL2.

Mechanisms of RNA Replication

As for all positive-strand viruses, the flow of genetic information is relatively straightforward: the positive-strand genomic RNA is used to make a negative-strand RNA intermediate, which then serves as a template for synthesizing new positive strands (Fig. 4.5). However this model is deceptively simple, as there are multiple levels of regulation controlling this process, and the mechanisms of HCV RNA replication are only beginning to be understood. The complexity of intracellular events associated with HCV infection is staggering, with the viral genome serving as an mRNA for translation of viral proteins, as a template for RNA replication, and as carrier of genetic information within progeny virions. Clearly, the trafficking

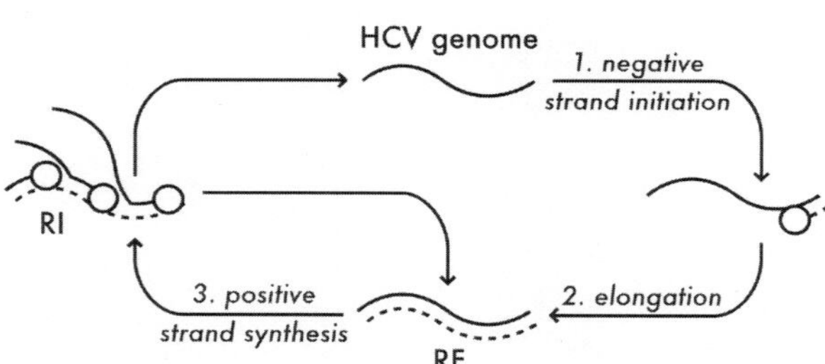

Fig. 4.5 HCV RNA replication cycle.

of viral RNA between these processes must be regulated to avoid conflicts. For instance, it seems unlikely that the viral genome can be simultaneously used as a substrate for translation, with ribosomes moving down the genome in the 5′ to 3′ direction, and as a template for negative-strand synthesis, with the viral replicase copying the RNA in the 3′ to 5′ direction. Thus, it is clear that all of the steps in the viral life cycle are highly coordinated.

By simple and elegant methods of quantitation, Quinkert et al. (2005) determined that each HCV replicon-bearing cell contains about 1,000 positive-strand RNAs, 100 negative strands, and about 1,000,000 copies of each viral protein. Thus, the viral genome serves as a template for translation far more often than as a template for RNA replication. An excess of viral structural proteins makes sense, as the formation of nascent virus particles will likely require at least 180 copies of each structural protein for each positive-strand RNA that is packaged (assuming a T=3 particle). Given that they are derived from a single polyprotein precursor, HCV NS proteins must be generated in a roughly equimolar amount as the viral structural proteins. The function of excess NS proteins is not yet clear, as only a small fraction of them are sequestered within the membrane-bound replicase at any given time (Miyanari et al. 2003; Quinkert et al. 2005). As described below, there is increasing evidence that some of the HCV NS proteins have gained additional, non-replicative activities such as to manipulate the innate antiviral response and cell-signaling pathways.

As alluded to earlier, the HCV genome must be recruited out of translation and into a membrane-bound replicase. The signals controlling this transition are not yet fully understood, although a number of important clues have recently emerged. Based on what is known about the picornaviruses (Chapter 1), this switch likely involves the 5′ and 3′ NCRs, which function in both translation and replication. As mentioned previously, HCV RNA replication requires miR-122 binding to the 5′ NCR (Jopling et al. 2005). This is particularly interesting given that (1) miRNAs can reduce translation of cellular mRNAs by sequestering them within specialized cytoplasmic processing "P" bodies (reviewed in (Parker and Sheth 2007)); and (2) P-bodies appear to recruit the genome of brome mosaic virus, another positive-strand RNA virus (Chapter 5), out of translation and into replication (Beckham et al. 2007). Thus, it is tempting to speculate that the binding of miRNA-122 to the HCV genome may regulate the switch from genome translation to RNA replication. In line with this model, cellular mRNAs targeted by miRNA-122 are translationally silenced and targeted to P-bodies; this silencing is derepressed by the cellular protein HuR (Bhattacharyya et al. 2006). Interestingly, HuR binds to the positive- and negative-stranded forms of the HCV genome, and is required for the replication cycle of HCV (Spångberg et al. 2000; Randall et al. 2007). It will be interesting indeed to see whether miRNA-122 and HuR are relevant to the utilization of HCV genomes in translation vs. replication. Another seemingly important clue is that the polypyrimidine tract binding protein (PTB) binds to both the 5′ NCR and core-coding region of the HCV RNA, where it modulates translation from the viral IRES, and to the 3′ NCR where it may suppress RNA replication

(Tsuchihara et al. 1997; Ito and Lai 1999; Anwar et al. 2000; Tischendorf et al. 2004). In addition, it is interesting to note that the polycytidine-binding protein 2 (PCBP-2) binds to the HCV 5′ NCR (Spångberg and Schwartz 1999; Fukushi et al. 2001); during poliovirus replication, this protein interacts with the viral 5′ NCR and RdRP to control the switch between translation and replication (Chapter 1).

Once HCV RNA has been recruited into the replicase, RNA synthesis presumably begins (Fig. 4.5). Aside from what has been learned regarding strand initiation and elongation with purified NS5B polymerase in vitro, little is known about this process. Membrane extracts from HCV replicon-bearing cells have been used to study replicase-associated RdRP activity, but these reactions only appear to involve strand elongation and not initiation on new templates (Ali et al. 2002; Hardy et al. 2003; Lai et al. 2003; Aizaki et al. 2004; Yang et al. 2004; Quinkert et al. 2005). Since all of the viral genetic material must be copied during each replication cycle, it is thought that negative-strand synthesis must begin via a primer-independent de novo initiation event. As mentioned earlier, the 3′-end of the HCV genome is folded in a stable stem-loop structure. When the authentic 3′-end of the HCV genome is used as a template for de novo initiation in an in vitro reaction, only internal initiation products (i.e., 5′ truncated negative strands) are generated (Kao et al. 2000; Oh et al. 2000; Sun et al. 2000; Kim et al. 2002a). However the addition of a few unpaired nts to the 3′-end allows template-length negative strands to be synthesized (Oh et al. 2000). Thus, perhaps NS3-4A RNA helicase is needed to unwind the 3′X SL1 to allow authentic initiation. Alternatively, it has been suggested that the NS5B terminal transferase activity provides these unpaired 3′-ends (Ranjith-Kumar et al. 2001). It remains unclear how these extensions would be subsequently resolved.

Once the negative-strand RNA is synthesized it remains associated with the positive-strand RNA, either in partially double-stranded replicative intermediates (RI) or fully double-stranded replicate forms (RF) (Fig. 4.5) (Ali et al. 2002). The negative-strand RNA then serves as a template to direct the synthesis of multiple positive strands, leading to asymmetry in RNA synthesis, with approximately ten positive strands generated for each minus strand (Lanford et al. 1995; Lohmann et al. 1999; Miyanari et al. 2003; Aizaki et al. 2004; Ranjith-Kumar et al. 2004). How this asymmetry is regulated remains unknown, but likely involves CREs, the composition of the viral replicase, or the differential processivity of the replicase on different templates. Recent studies with chimeric replicons have suggested that genotype-specific contacts required for efficient negative-strand synthesis are made between the 3′X tail and the NS5B polymerase, whereas genotype-specific positive-strand synthesis likely utilizes a CRE on the 3′-end of the negative-strand RNA (i.e., the reverse complement of the 5′ NCR) and requires NS3, NS5A, and NS5B (Binder et al. 2007b). These data provide the first evidence of differential requirements of replicase proteins for replication at the 5′- and 3′-end of the genome, and as such, may be important clues for the regulation of strand synthesis asymmetry.

Cellular Response to Infection

As with all viruses that are capable of establishing persistent infections, HCV must face the innate antiviral response of the host cell. It is thought that cells recognize and respond to the unusual features of viral genomes, such as double-stranded RNAs or RNAs lacking a 5′-methylated cap (Hornung et al. 2006: Pichlmair, 2006 #1699). Identifying how HCV manages to deal with this response has been an area of intense investigation for many years.

Much of the early work in this area was focused on the NS5A protein and its role in interferon resistance via interaction with PKR, a double-stranded RNA sensor (see (Tan and Katze 2001) for review). A region in NS5A, termed the interferon sensitivity-determining region (ISDR), was found to possess a high mutation rate in clinical samples, which weakly correlated with sensitivity to IFN therapy (Enomoto et al. 1995, 1996). Furthermore, the NS5A ISDR was shown to bind PKR and inhibit the IFN-induced activity of PKR on downstream targets, most notably eIF2α, thereby counteracting the antiviral effects of PKR (Gale et al. 1997, 1998a, b). Yet the significance of the ISDR is an area of debate, with some groups showing no relationship between ISDR and IFN response in patients (Pawlotsky 1999). Furthermore, deletion or mutation of the ISDR does not affect the IFN sensitivity of HCV replicons, suggesting this sequence does not play a direct role in IFN response in cell culture (Blight et al. 2000; Liu et al. 2006). In addition, NS5A has been shown to induce interleukin-8, which antagonizes the antiviral effects of IFN (Polyak et al. 2001). Although none of these reported NS5A activities seem to be the major mechanism by which HCV escapes antiviral defenses, it is possible that they act synergistically in the context of an authentic infection and modulate IFN sensitivity. Indeed, the ability of NS5A to manipulate many host-signaling pathways and interact with a diverse range of host-signaling molecules, few of which have a dramatic effect on virus replication, may collectively represent clues toward the overall manipulation of the host cell by HCV (see (Macdonald and Harris 2004) for review).

More recently the focus has shifted to the role of the NS3-4A complex in circumventing innate antiviral defenses. NS3-4A antagonizes at least two key innate antiviral defenses by short-circuiting the transduction of the viral RNA-sensing signals of the retinoic acid inducible gene-I (RIG-I) helicase and the Toll-like receptor-3 (TLR3) systems (see (Johnson and Gale 2006) for review). In the manipulation of both pathways, the NS3-4A protease cleaves key adapter proteins that are required for effective signal transduction to IRF-3 and NF-κB, important transcription factors for the innate immune response. NS3-4A manipulates the TLR3 pathway by cleaving an adapter molecule, TRIF, required for the activation of IRF-3 and NF-κB in response to extracellular forms of double-stranded RNA (Ferreon et al. 2005; Li et al. 2005a). In the case of the RIG-I pathway, the NS3-4A protease cleaves a CARD domain-containing adapter protein designated IPS-1 (also known as Cardif, MAVS, and VISA) (Li et al. 2005b; Meylan et al. 2005). IPS-1 is an outer mitochondrial protein responsible for transducing signals from both RIG-I and another viral RNA-sensing helicase, MDA5, to IRF-3 and NF-κB. Cleavage of IPS-1 blocks the induction of gene transcription in response to cytoplasmic RNAs, allowing the virus to escape from this potent antiviral pathway (Breiman et al. 2005; Foy et al. 2005; Karayiannis 2005). Further evidence that RIG-I limits HCV replication is the observation that the Huh-7.5 hepatoma cell line, a clone of Huh-7 cells that is highly permissive for HCV RNA replication, expresses a dominant negative form of RIG-I (Sumpter et al. 2005). Nevertheless, the relative importance of the RIG-I pathway in mediating Huh-7.5 permissiveness has been questioned by at least one other group (Binder et al. 2007a), so it seems likely that the picture is not yet complete.

References

Ago, H., Adachi, T., Yoshida, A., Yamamoto, M., Habuka, N., Yatsunami, K. and Miyano, M. 1999. Crystal structure of the RNA-dependent RNA polymerase of hepatitis C virus. Structure Fold Des 7: 1417–26.

Aizaki, H., Lee, K. J., Sung, V. M., Ishiko, H. and Lai, M. M. 2004. Characterization of the hepatitis C virus RNA replication complex associated with lipid rafts. Virology 324: 450–61.

Al, R. H., Xie, Y., Wang, Y. and Hagedorn, C. H. 1998. Expression of recombinant hepatitis C virus non-structural protein 5B in Escherichia coli. Virus Res 53: 141–9.

Ali, N., Tardif, K. D. and Siddiqui, A. 2002. Cell-free replication of the hepatitis C virus subgenomic replicon. J Virol 76: 12001–7.

Alter, H. J. and Seeff, L. B. 2000. Recovery, persistence, and sequelae in hepatitis C virus infection: a perspective on long-term outcome. Semin Liver Dis 20: 17–35.

Anwar, A., Ali, N., Tanveer, R. and Siddiqui, A. 2000. Demonstration of functional requirement of polypyrimidine tract-binding protein by SELEX RNA during hepatitis C virus internal ribosome entry site-mediated translation initiation. J Biol Chem 275: 34231–5.

Appel, N., Herian, U. and Bartenschlager, R. 2005a. Efficient rescue of hepatitis C virus RNA replication by trans-complementation with nonstructural protein 5A. J Virol 79: 896–909.

Appel, N., Pietschmann, T. and Bartenschlager, R. 2005b. Mutational analysis of hepatitis C virus nonstructural protein 5A: potential role of differential phosphorylation in RNA replication and identification of a genetically flexible domain. J Virol 79: 3187–94.

Bartenschlager, R., Frese, M. and Pietschmann, T. 2004. Novel insights into hepatitis C virus replication and persistence. Adv Virus Res 63: 71–180.

Beckham, C. J., Light, H. R., Nissan, T. A., Ahlquist, P., Parker, R. and Noueiry, A. 2007. Interactions between brome mosaic virus RNAs and cytoplasmic processing bodies. J Virol 81: 9759–68.

Behrens, S. E., Tomei, L. and De Francesco, R. 1996. Identification and properties of the RNA-dependent RNA polymerase of hepatitis C virus. EMBO J 15: 12–22.

Beran, R. K., Serebrov, V. and Pyle, A. M. 2007. The serine protease domain of hepatitis C viral NS3 activates RNA helicase activity by promoting the binding of RNA substrate. J Biol Chem 282: 34913–20.

Bhattacharyya, S. N., Habermacher, R., Martine, U., Closs, E. I. and Filipowicz, W. 2006. Stress-induced reversal of microRNA repression and mRNA P-body localization in human cells. Cold Spring Harb Symp Quant Biol 71: 513–21.

Binder, M., Kochs, G., Bartenschlager, R. and Lohmann, V. 2007a. Hepatitis C virus escape from the interferon regulatory factor 3 pathway by a passive and active evasion strategy. Hepatology 46: 1365–74.

Binder, M., Quinkert, D., Bochkarova, O., Klein, R., Kezmic, N., Bartenschlager, R. and Lohmann, V. 2007b. Identification of determinants involved in initiation of hepatitis C virus RNA synthesis by using intergenotypic replicase chimeras. J Virol 81: 5270–83.

Biswal, B. K., Cherney, M. M., Wang, M., Chan, L., Yannopoulos, C. G., Bilimoria, D., Nicolas, O., Bedard, J. and James, M. N. 2005. Crystal structures of the RNA-dependent RNA polymerase genotype 2a of hepatitis C virus reveal two conformations and suggest mechanisms of inhibition by non-nucleoside inhibitors. J Biol Chem 280: 18202–10.

Blight, K. J. 2007. Allelic variation in the hepatitis C virus NS4B protein dramatically influences RNA replication. J Virol 81: 5724–36.

Blight, K. J., Kolykhalov, A. A. and Rice, C. M. 2000. Efficient initiation of HCV RNA replication in cell culture. Science 290: 1972–4.

Blight, K. J. and Rice, C. M. 1997. Secondary structure determination of the conserved 98-base sequence at the 3′ terminus of hepatitis C virus genome RNA. J Virol 71: 7345–52.

Bradrick, S. S., Walters, R. W. and Gromeier, M. 2006. The hepatitis C virus 3′-untranslated region or a poly(A) tract promote efficient translation subsequent to the initiation phase. Nucleic Acids Res 34: 1293–303.

Branch, A. D., Stump, D. D., Gutierrez, J. A., Eng, F. and Walewski, J. L. 2005. The hepatitis C virus alternate reading frame (ARF) and its family of novel products: the alternate reading frame protein/F-protein, the double-frameshift protein, and others. Semin Liver Dis 25: 105–17.

Brass, V., Bieck, E., Montserret, R., Wolk, B., Hellings, J. A., Blum, H. E., Penin, F. and Moradpour, D. 2002. An amino-terminal amphipathic alpha-helix mediates membrane association of the hepatitis C virus nonstructural protein 5A. J Biol Chem 277: 8130–9.

Brass, V., Pal, Z., Sapay, N., Deleage, G., Blum, H. E., Penin, F. and Moradpour, D. 2007. Conserved determinants for membrane association of nonstructural protein 5A from hepatitis C virus and related viruses. J Virol 81: 2745–57.

Breiman, A., Grandvaux, N., Lin, R., Ottone, C., Akira, S., Yoneyama, M., Fujita, T., Hiscott, J. and Meurs, E. F. 2005. Inhibition of RIG-I-dependent signaling to the interferon pathway during hepatitis C virus expression and restoration of signaling by IKKepsilon. J Virol 79: 3969–78.

Bressanelli, S., Tomei, L., Rey, F. A. and De Francesco, R. 2002. Structural analysis of the hepatitis C virus RNA polymerase in complex with ribonucleotides. J Virol 76: 3482–92.

Bressanelli, S., Tomei, L., Roussel, A., Incitti, I., Vitale, R. L., Mathieu, M., De Francesco, R. and Rey, F. A: 1999. Crystal structure of the RNA-dependent RNA polymerase of hepatitis C virus. Proc Natl Acad Sci USA 96: 13034–9.

Cai, Z., Yi, M., Zhang, C. and Luo, G. 2005. Mutagenesis analysis of the rGTP-specific binding site of hepatitis C virus RNA-dependent RNA polymerase. J Virol 79: 11607–17.

Cheney, I. W., Naim, S., Lai, V. C., Dempsey, S., Bellows, D., Walker, M. P., Shim, J. H., Horscroft, N., Hong, Z. and Zhong, W. 2002. Mutations in NS5B polymerase of hepatitis C virus: impacts on in vitro enzymatic activity and viral RNA replication in the subgenomic replicon cell culture. Virology 297: 298–306.

Choo, Q.-L., Kuo, G., Weiner, A. J., Overby, L. R., Bradley, D. W. and Houghton, M: 1989. Isolation of a cDNA clone derived from a blood-borne non-A, non-B viral hepatitis genome. Science 244: 359–62.

Choo, Q.-L., Richman, K. H., Han, J. H., Berger, K., Lee, C., Dong, C., Gallegos, C., Coit, D., Medina-Selby, A., Barr, P. J., Weiner, A. J., Bradley, D. W., Kuo, G. and Houghton, M: 1991. Genetic organization and diversity of the hepatitis C virus. Proc Natl Acad Sci USA 88: 2451–5.

Crotty, S., Cameron, C. E. and Andino, R. 2001. RNA virus error catastrophe: direct molecular test by using ribavirin. Proc Natl Acad Sci USA 98: 6895–900.

De Francesco, R., Tomei, L., Altamura, S., Summa, V. and Migliaccio, G. 2003. Approaching a new era for hepatitis C virus therapy: inhibitors of the NS3-4A serine protease and the NS5B RNA-dependent RNA polymerase. Antiviral Res 58: 1–16.

Dumont, S., Cheng, W., Serebrov, V., Beran, R. K., Tinoco, I., Jr., Pyle, A. M. and Bustamante, C. 2006. RNA translocation and unwinding mechanism of HCV NS3 helicase and its coordination by ATP. Nature 439: 105–8.

Egger, D., Wolk, B., Gosert, R., Bianchi, L., Blum, H. E., Moradpour, D. and Bienz, K. 2002. Expression of hepatitis C virus proteins induces distinct membrane alterations including a candidate viral replication complex. J Virol 76: 5974–84.

Einav, S., Elazar, M., Danieli, T. and Glenn, J. S. 2004. A nucleotide binding motif in hepatitis C virus (HCV) NS4B mediates HCV RNA replication. J Virol 78: 11288–95.

Elazar, M., Cheong, K. H., Liu, P., Greenberg, H. B., Rice, C. M. and Glenn, J. S. 2003. Amphipathic helix-dependent localization of NS5A mediates hepatitis C virus RNA replication. J Virol 77: 6055–61.

Elazar, M., Liu, P., Rice, C. M. and Glenn, J. S. 2004. An N-terminal amphipathic helix in hepatitis C virus (HCV) NS4B mediates membrane association, correct localization of replication complex proteins, and HCV RNA replication. J Virol 78: 11393–400.

Enomoto, N., Sakuma, I., Asahina, Y., Kurosaki, M., Murakami, T., Yamamoto, C., Izumi, N., Marumo, F. and Sato, C: 1995. Comparison of full-length sequences of interferon-sensitive and resistant hepatitis C virus 1b. Sensitivity to interferon is conferred by amino acid substitutions in the NS5A region. J Clin Invest 96: 224–30.

Enomoto, N., Sakuma, I., Asahina, Y., Kurosaki, M., Murakami, T., Yamamoto, C., Ogura, Y., Izumi, N., Marumo, F. and Sato, C: 1996. Mutations in the nonstructural protein 5A gene and response to interferon in patients with chronic hepatitis C virus 1b infection. N Engl J Med 334: 77–81.

Evans, M. J., Rice, C. M. and Goff, S. P. 2004. Phosphorylation of hepatitis C virus nonstructural protein 5A modulates its protein interactions and viral RNA replication. Proc Natl Acad Sci USA 101: 13038–43.

Fernandes, F., Poole, D. S., Hoover, S., Middleton, R., Andrei, A. C., Gerstner, J. and Striker, R. 2007. Sensitivity of hepatitis C virus to cyclosporine A depends on nonstructural proteins NS5A and NS5B. Hepatology 46: 1026–33.

Ferrari, E., Wright-Minogue, J., Fang, J. W., Baroudy, B. M., Lau, J. Y. and Hong, Z: 1999. Characterization of soluble hepatitis C virus RNA-dependent RNA polymerase expressed in Escherichia coli. J Virol 73: 1649–54.

Ferreon, J. C., Ferreon, A. C., Li, K. and Lemon, S. M. 2005. Molecular determinants of TRIF proteolysis mediated by the hepatitis C virus NS3/4A protease. J Biol Chem 280: 20483–92.

Foy, E., Li, K., Sumpter, R., Jr., Loo, Y. M., Johnson, C. L., Wang, C., Fish, P. M., Yoneyama, M., Fujita, T., Lemon, S. M. and Gale, M., Jr. 2005. Control of antiviral defenses through hepatitis C virus disruption of retinoic acid-inducible gene-I signaling. Proc Natl Acad Sci USA 102: 2986–91.

Fraser, C. S. and Doudna, J. A. 2007. Structural and mechanistic insights into hepatitis C viral translation initiation. Nat Rev Microbiol 5: 29–38.

Friebe, P. and Bartenschlager, R. 2002. Genetic analysis of sequences in the 3′ nontranslated region of hepatitis C virus that are important for RNA replication. J Virol 76: 5326–38.

Friebe, P., Boudet, J., Simorre, J. P. and Bartenschlager, R. 2005. Kissing-loop interaction in the 3′ end of the hepatitis C virus genome essential for RNA replication. J Virol 79: 380–92.

Friebe, P., Lohmann, V., Krieger, N. and Bartenschlager, R. 2001. Sequences in the 5′ nontranslated region of hepatitis C virus required for RNA replication. J Virol 75: 12047–57.

Fukushi, S., Okada, M., Kageyama, T., Hoshino, F. B., Nagai, K. and Katayama, K. 2001. Interaction of poly(rC)-binding protein 2 with the 5′-terminal stem loop of the hepatitis C-virus genome. Virus Res 73: 67–79.

Gale, M., Jr., Blakely, C. M., Kwieciszewski, B., Tan, S. L., Dossett, M., Tang, N. M., Korth, M. J., Polyak, S. J., Gretch, D. R. and Katze, M. G: 1998a. Control of PKR protein kinase by hepatitis C virus nonstructural 5A protein: molecular mechanisms of kinase regulation. Mol Cell Biol 18: 5208–18.

Gale, M. J., Jr., Korth, M. J. and Katze, M. G: 1998b. Repression of the PKR protein kinase by the hepatitis C virus NS5A protein: a potential mechanism of interferon resistance. Clin Diagn Virol 10: 157–62.

Gale, M. J., Jr., Korth, M. J., Tang, N. M., Tan, S. L., Hopkins, D. A., Dever, T. E., Polyak, S. J., Gretch, D. R. and Katze, M. G: 1997. Evidence that hepatitis C virus resistance to interferon is mediated through repression of the PKR protein kinase by the nonstructural 5A protein. Virology 230: 217–27.

Gao, L., Aizaki, H., He, J. W. and Lai, M. M. 2004. Interactions between viral nonstructural proteins and host protein hVAP-33 mediate the formation of hepatitis C virus RNA replication complex on lipid raft. J Virol 78: 3480–8.

Gosert, R., Egger, D., Lohmann, V., Bartenschlager, R., Blum, H. E., Bienz, K. and Moradpour, D. 2003. Identification of the hepatitis C virus RNA replication complex in Huh-7 cells harboring subgenomic replicons. J Virol 77: 5487–92.

Grakoui, A., McCourt, D. W., Wychowski, C., Feinstone, S. M. and Rice, C. M: 1993. A second hepatitis C virus-encoded proteinase. Proc Natl Acad Sci USA 90: 10583–7.

Han, J. Q. and Barton, D. J. 2002. Activation and evasion of the antiviral 2′-5′ oligoadenylate synthetase/ribonuclease L pathway by hepatitis C virus mRNA. RNA 8: 512–25.

Hardy, R. W., Marcotrigiano, J., Blight, K. J., Majors, J. E. and Rice, C. M. 2003. Hepatitis C virus RNA synthesis in a cell-free system isolated from replicon-containing hepatoma cells. J Virol 77: 2029–37.

Hijikata, M., Mizushima, H., Akagi, T., Mori, S., Kakiuchi, N., Kato, N., Tanaka, T., Kimura, K. and Shimotohno, K: 1993. Two distinct proteinase activities required for the processing of a putative nonstructural precursor protein of hepatitis C virus. J Virol 67: 4665–75.

Hong, Z., Cameron, C. E., Walker, M. P., Castro, C., Yao, N., Lau, J. Y. and Zhong, W. 2001. A novel mechanism to ensure terminal initiation by hepatitis C virus NS5B polymerase. Virology 285: 6–11.

Hornung, V., Ellegast, J., Kim, S., Brzozka, K., Jung, A., Kato, H., Poeck, H., Akira, S., Conzelmann, K. K., Schlee, M., Endres, S. and Hartmann, G. 2006. 5′-Triphosphate RNA is the ligand for RIG-I. Science 314: 994–7.

Huang, L., Hwang, J., Sharma, S. D., Hargittai, M. R., Chen, Y., Arnold, J. J., Raney, K. D. and Cameron, C. E. 2005. Hepatitis C virus nonstructural protein 5A (NS5A) is an RNA-binding protein. J Biol Chem 280: 36417–28.

Huang, L., Sineva, E. V., Hargittai, M. R., Sharma, S. D., Suthar, M., Raney, K. D. and Cameron, C. E. 2004. Purification and characterization of hepatitis C virus non-structural protein 5A expressed in Escherichia coli. Protein Expr Purif 37: 144–53.

Huang, Y., Staschke, K., De Francesco, R. and Tan, S. L. 2007. Phosphorylation of hepatitis C virus NS5A nonstructural protein: a new paradigm for phosphorylation-dependent viral RNA replication? Virology 364: 1–9.

Hügle, T., Fehrmann, F., Bieck, E., Kohara, M., Krausslich, H. G., Rice, C. M., Blum, H. E. and Moradpour, D. 2001. The hepatitis C virus nonstructural protein 4B is an integral endoplasmic reticulum membrane protein. Virology 284: 70–81.

Ishii, N., Watashi, K., Hishiki, T., Goto, K., Inoue, D., Hijikata, M., Wakita, T., Kato, N. and Shimotohno, K. 2006. Diverse effects of cyclosporine on hepatitis C virus strain replication. J Virol 80: 4510–20.

Ito, T. and Lai, M. M: 1999. An internal polypyrimidine-tract-binding protein-binding site in the hepatitis C virus RNA attenuates translation, which is relieved by the 3′-untranslated sequence. Virology 254: 288–96.

Ito, T., Tahara, S. M. and Lai, M. M: 1998. The 3′-untranslated region of hepatitis C virus RNA enhances translation from an internal ribosomal entry site. J Virol 72: 8789–96.

Ivashkina, N., Wolk, B., Lohmann, V., Bartenschlager, R., Blum, H. E., Penin, F. and Moradpour, D. 2002. The hepatitis C virus RNA-dependent RNA polymerase membrane insertion sequence is a transmembrane segment. J Virol 76: 13088–93.

Ji, H., Fraser, C. S., Yu, Y., Leary, J. and Doudna, J. A. 2004. Coordinated assembly of human translation initiation complexes by the hepatitis C virus internal ribosome entry site RNA. Proc Natl Acad Sci USA 101: 16990–5.

Johnson, C. L. and Gale, M., Jr. 2006. CARD games between virus and host get a new player. Trends Immunol 27: 1–4.

Jones, C. T., Murray, C. L., Eastman, D. K., Tassello, J. and Rice, C. M. 2007. Hepatitis C virus p7 and NS2 proteins are essential for production of infectious virus. J Virol 81: 8374–83.

Jopling, C. L., Yi, M., Lancaster, A. M., Lemon, S. M. and Sarnow, P. 2005. Modulation of hepatitis C virus RNA abundance by a liver-specific MicroRNA. Science 309: 1577–81.

Kanai, A., Tanabe, K. and Kohara, M: 1995. Poly(U) binding activity of hepatitis C virus NS3 protein, a putative RNA helicase. FEBS Lett 376: 221–4.

Kaneko, T., Tanji, Y., Satoh, S., Hijikata, M., Asabe, S., Kimura, K. and Shimotohno, K: 1994. Production of two phosphoproteins from the NS5A region of the hepatitis C viral genome. Biochem Biophys Res Commun 205: 320–6.

Kao, C. C., Yang, X., Kline, A., Wang, Q. M., Barket, D. and Heinz, B. A. 2000. Template requirements for RNA synthesis by a recombinant hepatitis C virus RNA-dependent RNA polymerase. J Virol 74: 11121–8.

Kapadia, S. B., Barth, H., Baumert, T., McKeating, J. A. and Chisari, F. V. 2007. Initiation of hepatitis C virus infection is dependent on cholesterol and cooperativity between CD81 and scavenger receptor B type I. J Virol 81: 374–83.

Kapadia, S. B. and Chisari, F. V. 2005. Hepatitis C virus RNA replication is regulated by host geranylgeranylation and fatty acids. Proc Natl Acad Sci USA 102: 2561–6.

Karayiannis, P. 2005. The hepatitis C virus NS3/4A protease complex interferes with pathways of the innate immune response. J Hepatol 43: 743–5.

Kashiwagi, T., Hara, K., Kohara, M., Kohara, K., Iwahashi, J., Hamada, N., Yoshino, H. and Toyoda, T. 2002. Kinetic analysis of C-terminally truncated RNA-dependent RNA polymerase of hepatitis C virus. Biochem Biophys Res Commun 290: 1188–94.

Katze, M. G., Kwieciszewski, B., Goodlett, D. R., Blakely, C. M., Neddermann, P., Tan, S. L. and Aebersold, R. 2000. Ser(2194) is a highly conserved major phosphorylation site of the hepatitis C virus nonstructural protein NS5A. Virology 278: 501–13.

Kim, M., Kim, H., Cho, S. P. and Min, M. K. 2002a. Template requirements for de novo RNA synthesis by hepatitis C virus nonstructural protein 5B polymerase on the viral X RNA. J Virol 76: 6944–56.

Kim, S. J., Kim, J. H., Kim, Y. G., Lim, H. S. and Oh, J. W. 2004. Protein kinase C-related kinase 2 regulates hepatitis C virus RNA polymerase function by phosphorylation. J Biol Chem 279: 50031–41.

Kim, Y. K., Kim, C. S., Lee, S. H. and Jang, S. K. 2002b. Domains I and II in the 5′ nontranslated region of the HCV genome are required for RNA replication. Biochem Biophys Res Commun 290: 105–12.

Kim, J., Lee, D. and Choe, J: 1999. Hepatitis C virus NS5A protein is phosphorylated by casein kinase II. Biochem Biophys Res Commun 257: 777–81.

Kim, Y. K., Lee, S. H., Kim, C. S., Seol, S. K. and Jang, S. K. 2003. Long-range RNA-RNA interaction between the 5′ nontranslated region and the core-coding sequences of hepatitis C virus modulates the IRES-dependent translation. RNA 9: 599–606.

Kim, Y. C., Russell, W. K., Ranjith-Kumar, C. T., Thomson, M., Russell, D. H. and Kao, C. C. 2005. Functional analysis of RNA binding by the hepatitis C virus RNA-dependent RNA polymerase. J Biol Chem 280: 38011–9.

Koch, J. O. and Bartenschlager, R: 1999. Modulation of hepatitis C virus NS5A hyperphosphorylation by nonstructural proteins NS3, NS4A, and NS4B. J Virol 73: 7138–46.

Kolykhalov, A. A., Agapov, E. V., Blight, K. J., Mihalik, K., Feinstone, S. M. and Rice, C. M: 1997. Transmission of hepatitis C by intrahepatic inoculation with transcribed RNA. Science 277: 570–4.

Kolykhalov, A. A., Feinstone, S. M. and Rice, C. M: 1996. Identification of a highly conserved sequence element at the 3′ terminus of hepatitis C virus genome RNA. J Virol 70: 3363–71.

Kolykhalov, A. A., Mihalik, K., Feinstone, S. M. and Rice, C. M. 2000. Hepatitis C virus-encoded enzymatic activities and conserved RNA elements in the 3′ nontranslated region are essential for virus replication in vivo. J Virol 74: 2046–51.

Kopek, B. G., Perkins, G., Miller, D. J., Ellisman, M. H. and Ahlquist, P. 2007. Three-dimensional analysis of a viral RNA replication complex reveals a virus-induced mini-organelle. PLoS Biol 5: e220.

Lai, V. C., Dempsey, S., Lau, J. Y., Hong, Z. and Zhong, W. 2003. In vitro RNA replication directed by replicase complexes isolated from the subgenomic replicon cells of hepatitis C virus. J Virol 77: 2295–300.

Lanford, R. E., Chavez, D., Chisari, F. V. and Sureau, C: 1995. Lack of detection of negative-strand hepatitis C virus RNA in peripheral blood mononuclear cells and other extrahepatic tissues by the highly strand-specific rTth reverse transcriptase PCR. J Virol 69: 8079–83.

Lee, H., Liu, Y., Mejia, E., Paul, A. V. and Wimmer, E. 2006. The C-terminal hydrophobic domain of hepatitis C virus RNA polymerase NS5B can be replaced with a heterologous domain of poliovirus protein 3A. J Virol 80: 11343–54.

Lee, H., Shin, H., Wimmer, E. and Paul, A. V. 2004. cis-acting RNA signals in the NS5B C-terminal coding sequence of the hepatitis C virus genome. J Virol 78: 10865–77.

Lesburg, C. A., Cable, M. B., Ferrari, E., Hong, Z., Mannarino, A. F. and Weber, P. C: 1999. Crystal structure of the RNA-dependent RNA polymerase from hepatitis C virus reveals a fully encircled active site. Nat Struct Biol 6: 937–43.

Lévêque, V. J., Johnson, R. B., Parsons, S., Ren, J., Xie, C., Zhang, F. and Wang, Q. M. 2003. Identification of a C-terminal regulatory motif in hepatitis C virus RNA-dependent RNA polymerase: structural and biochemical analysis. J Virol 77: 9020–8.

Li, K., Foy, E., Ferreon, J. C., Nakamura, M., Ferreon, A. C., Ikeda, M., Ray, S. C., Gale, M., Jr. and Lemon, S. M. 2005a. Immune evasion by hepatitis C virus NS3/4A protease-mediated cleavage of the Toll-like receptor 3 adaptor protein TRIF. Proc Natl Acad Sci USA 102: 2992–7.

Li, X. D., Sun, L., Seth, R. B., Pineda, G. and Chen, Z. J. 2005b. Hepatitis C virus protease NS3/4A cleaves mitochondrial antiviral signaling protein off the mitochondria to evade innate immunity. Proc Natl Acad Sci USA 102: 17717–22.

Liang, Y., Kang, C. B. and Yoon, H. S. 2006. Molecular and structural characterization of the domain 2 of hepatitis C virus non-structural protein 5A. Mol Cells 22: 13–20.

Liang, Y., Ye, H., Kang, C. B. and Yoon, H. S. 2007. Domain 2 of nonstructural protein 5A (NS5A) of hepatitis C virus is natively unfolded. Biochemistry 46: 11550–8.

Lindenbach, B. D., Evans, M. J., Syder, A. J., Wolk, B., Tellinghuisen, T. L., Liu, C. C., Maruyama, T., Hynes, R. O., Burton, D. R., McKeating, J. A. and Rice, C. M. 2005. Complete replication of hepatitis C virus in cell culture. Science 309: 623–6.

Lindenbach, B. D., Pragai, B. M., Montserret, R., Beran, R. K., Pyle, A. M., Penin, F. and Rice, C. M. 2007a. The C terminus of hepatitis C virus NS4A encodes an electrostatic switch that regulates NS5A hyperphosphorylation and viral replication. J Virol 81: 8905–18.

Liu, S., Ansari, I. H., Das, S. C. and Pattnaik, A. K. 2006. Insertion and deletion analyses identify regions of non-structural protein 5A of Hepatitis C virus that are dispensable for viral genome replication. J Gen Virol 87: 323–7.

Lohmann, V., Hoffmann, S., Herian, U., Penin, F. and Bartenschlager, R. 2003. Viral and cellular determinants of hepatitis C virus RNA replication in cell culture. J. Virol. 77: 3007–19.

Lohmann, V., Korner, F., Dobierzewska, A. and Bartenschlager, R. 2001. Mutations in hepatitis C virus RNAs conferring cell culture adaptation. J Virol 75: 1437–49.

Lohmann, V., Korner, F., Herian, U. and Bartenschlager, R: 1997. Biochemical properties of hepatitis C virus NS5B RNA-dependent RNA polymerase and identification of amino acid sequence motifs essential for enzymatic activity. J Virol 71: 8416–28.

Lohmann, V., Korner, F., Koch, J. O., Herian, U., Theilmann, L. and Bartenschlager, R: 1999. Replication of subgenomic hepatitis C virus RNAs in a hepatoma cell line. Science 285: 110–3.

Lohmann, V., Roos, A., Korner, F., Koch, J. O. and Bartenschlager, R. 2000. Biochemical and structural analysis of the NS5B RNA-dependent RNA polymerase of the hepatitis C virus. J Viral Hepat 7: 167–74.

Lorenz, I. C., Marcotrigiano, J., Dentzer, T. G. and Rice, C. M. 2006. Structure of the catalytic domain of the hepatitis C virus NS2-3 protease. Nature 442: 831–5.

Lundin, M., Lindstrom, H., Gronwall, C. and Persson, M. A. 2006. Dual topology of the processed hepatitis C virus protein NS4B is influenced by the NS5A protein. J Gen Virol 87: 3263–72.

Lundin, M., Monne, M., Widell, A., Von Heijne, G. and Persson, M. A. 2003. Topology of the membrane-associated hepatitis C virus protein NS4B. J Virol 77: 5428–38.

Luo, G., Hamatake, R. K., Mathis, D. M., Racela, J., Rigat, K. L., Lemm, J. and Colonno, R. J. 2000. De novo initiation of RNA synthesis by the RNA-dependent RNA polymerase (NS5B) of hepatitis C virus. J Virol 74: 851–63.

Macdonald, A. and Harris, M. 2004. Hepatitis C virus NS5A: tales of a promiscuous protein. J Gen Virol 85: 2485–502.

Mackintosh, S. G., Lu, J. Z., Jordan, J. B., Harrison, M. K., Sikora, B., Sharma, S. D., Cameron, C. E., Raney, K. D. and Sakon, J. 2006. Structural and biological identification of residues on the surface of NS3 helicase required for optimal replication of the hepatitis C virus. J Biol Chem 281: 3528–35.

McCaffrey, A. P., Ohashi, K., Meuse, L., Shen, S., Lancaster, A. M., Lukavsky, P. J., Sarnow, P. and Kay, M. A. 2002. Determinants of hepatitis C translational initiation in vitro, in cultured cells and mice. Mol Ther 5: 676–84.

McCormick, C. J., Maucourant, S., Griffin, S., Rowlands, D. J. and Harris, M. 2006. Tagging of NS5A expressed from a functional hepatitis C virus replicon. J Gen Virol 87: 635–40.

McMullan, L. K., Grakoui, A., Evans, M. J., Mihalik, K., Puig, M., Branch, A. D., Feinstone, S. M. and Rice, C. M. 2007. Evidence for a functional RNA element in the hepatitis C virus core gene. Proc Natl Acad Sci USA 104: 2879–84.

Meylan, E., Curran, J., Hofmann, K., Moradpour, D., Binder, M., Bartenschlager, R. and Tschopp, J. 2005. Cardif is an adaptor protein in the RIG-I antiviral pathway and is targeted by hepatitis C virus. Nature 437: 1167–72.

Miyanari, Y., Hijikata, M., Yamaji, M., Hosaka, M., Takahashi, H. and Shimotohno, K. 2003. Hepatitis C virus non-structural proteins in the probable membranous compartment function in viral genome replication. J Biol Chem 278: 50301–8.

Moradpour, D., Brass, V., Bieck, E., Friebe, P., Gosert, R., Blum, H. E., Bartenschlager, R., Penin, F. and Lohmann, V. 2004a. Membrane association of the RNA-dependent RNA polymerase is essential for hepatitis C virus RNA replication. J Virol 78: 13278–84.

Moradpour, D., Evans, M. J., Gosert, R., Yuan, Z., Blum, H. E., Goff, S. P., Lindenbach, B. D. and Rice, C. M. 2004b. Insertion of green fluorescent protein into nonstructural protein 5A allows direct visualization of functional hepatitis C virus replication complexes. J Virol 78: 7400–9.

Myong, S., Bruno, M. M., Pyle, A. M. and Ha, T. 2007. Spring-loaded mechanism of DNA unwinding by hepatitis C virus NS3 helicase. Science 317: 513–6.

Nakabayashi, H., Taketa, K., Miyano, K., Yamane, T. and Sato, J: 1982. Growth of human hepatoma cells lines with differentiated functions in chemically defined medium. Cancer Res 42: 3858–63.

Nakagawa, M., Sakamoto, N., Enomoto, N., Tanabe, Y., Kanazawa, N., Koyama, T., Kurosaki, M., Maekawa, S., Yamashiro, T., Chen, C. H., Itsui, Y., Kakinuma, S. and Watanabe, M. 2004. Specific inhibition of hepatitis C virus replication by cyclosporin A. Biochem Biophys Res Commun 313: 42–7.

Nakagawa, M., Sakamoto, N., Tanabe, Y., Koyama, T., Itsui, Y., Takeda, Y., Chen, C. H., Kakinuma, S., Oooka, S., Maekawa, S., Enomoto, N. and Watanabe, M. 2005. Suppression of hepatitis C virus replication by cyclosporin a is mediated by blockade of cyclophilins. Gastroenterology 129: 1031–41.

Neddermann, P., Clementi, A. and De Francesco, R: 1999. Hyperphosphorylation of the hepatitis C virus NS5A protein requires an active NS3 protease, NS4A, NS4B, and NS5A encoded on the same polyprotein. J Virol 73: 9984–91.

Neddermann, P., Quintavalle, M., Di Pietro, C., Clementi, A., Cerretani, M., Altamura, S., Bartholomew, L. and De Francesco, R. 2004. Reduction of hepatitis C virus NS5A hyperphosphorylation by selective inhibition of cellular kinases activates viral RNA replication in cell culture. J Virol 78: 13306–14.

O'Farrell, D., Trowbridge, R., Rowlands, D. and Jäger, J. 2003. Substrate complexes of hepatitis C virus RNA polymerase (HC-J4): structural evidence for nucleotide import and de-novo initiation. J Mol Biol 326: 1025–35.

Oh, J. W., Ito, T. and Lai, M. M: 1999. A recombinant hepatitis C virus RNA-dependent RNA polymerase capable of copying the full-length viral RNA. J Virol 73: 7694–702.

Oh, J. W., Sheu, G. T. and Lai, M. M. 2000. Template requirement and initiation site selection by hepatitis C virus polymerase on a minimal viral RNA template. J Biol Chem 275: 17710–7.

Okamoto, T., Nishimura, Y., Ichimura, T., Suzuki, K., Miyamura, T., Suzuki, T., Moriishi, K. and Matsuura, Y. 2006. Hepatitis C virus RNA replication is regulated by FKBP8 and Hsp90. Embo J 25: 5015–25.

Otto, G. A. and Puglisi, J. D. 2004. The pathway of HCV IRES-mediated translation initiation. Cell 119: 369–80.

Paeshuyse, J., Kaul, A., De Clercq, E., Rosenwirth, B., Dumont, J. M., Scalfaro, P., Bartenschlager, R. and Neyts, J. 2006. The non-immunosuppressive cyclosporin DEBIO-025 is a potent inhibitor of hepatitis C virus replication in vitro. Hepatology 43: 761–70.

Pang, P. S., Jankowsky, E., Planet, P. J. and Pyle, A. M. 2002. The hepatitis C viral NS3 protein is a processive DNA helicase with cofactor enhanced RNA unwinding. EMBO J 21: 1168–76.

Parker, R. and Sheth, U. 2007. P bodies and the control of mRNA translation and degradation. Mol Cell 25: 635–46.

Pawlotsky, J. M: 1999. Hepatitis C virus (HCV) NS5A protein: role in HCV replication and resistance to interferon-alpha. J Viral Hepat 6 Suppl 1: 47–8.

Penin, F., Brass, V., Appel, N., Ramboarina, S., Montserret, R., Ficheux, D., Blum, H. E., Bartenschlager, R. and Moradpour, D. 2004a. Structure and function of the membrane anchor domain of hepatitis C virus nonstructural protein 5A. J Biol Chem 279: 40835–43.

Penin, F., Dubuisson, J., Rey, F. A., Moradpour, D. and Pawlotsky, J. M. 2004b. Structural biology of hepatitis C virus. Hepatology 39: 5–19.

Perelson, A. S., Herrmann, E., Micol, F. and Zeuzem, S. 2005. New kinetic models for the hepatitis C virus. Hepatology 42: 749–54.

Pestova, T. V., Shatsky, I. N., Fletcher, S. P., Jackson, R. J. and Hellen, C. U: 1998. A prokaryotic-like mode of cytoplasmic eukaryotic ribosome binding to the initiation codon during internal translation initiation of hepatitis C and classical swine fever virus RNAs. Genes Dev 12: 67–83.

Piccininni, S., Varaklioti, A., Nardelli, M., Dave, B., Raney, K. D. and McCarthy, J. E. 2002. Modulation of the hepatitis C virus RNA-dependent RNA polymerase activity by the non-structural (NS) 3 helicase and the NS4B membrane protein. J Biol Chem 277: 45670–9.

Polyak, S. J., Khabar, K. S., Paschal, D. M., Ezelle, H. J., Duverlie, G., Barber, G. N., Levy, D. E., Mukaida, N. and Gretch, D. R. 2001. Hepatitis C virus nonstructural 5A protein induces interleukin-8, leading to partial inhibition of the interferon-induced antiviral response. J Virol 75: 6095–106.

Quinkert, D., Bartenschlager, R. and Lohmann, V. 2005. Quantitative analysis of the hepatitis C virus replication complex. J Virol 79: 13594–605.

Quintavalle, M., Sambucini, S., Di Pietro, C., De Francesco, R. and Neddermann, P. 2006. The alpha isoform of protein kinase CKI is responsible for hepatitis C virus NS5A hyperphosphorylation. J Virol 80: 11305–12.

Quintavalle, M., Sambucini, S., Summa, V., Orsatti, L., Talamo, F., De Francesco, R. and Neddermann, P. 2007. Hepatitis C virus NS5A is a direct substrate of casein kinase I-alpha, a cellular kinase identified by inhibitor affinity chromatography using specific NS5A hyperphosphorylation inhibitors. J Biol Chem 282: 5536–44.

Randall, G., Panis, M., Cooper, J. D., Tellinghuisen, T. L., Sukhodolets, K. E., Pfeffer, S., Landthaler, M., Landgraf, P., Kan, S., Lindenbach, B. D., Chien, M., Weir, D. B., Russo, J. J., Ju, J., Brownstein, M. J., Sheridan, R., Sander, C., Zavolan, M., Tuschl, T. and Rice, C. M.

2007. Cellular cofactors affecting hepatitis C virus infection and replication. Proc Natl Acad Sci USA 104: 12884–9.

Ranjith-Kumar, C. T., Gajewski, J., Gutshall, L., Maley, D., Sarisky, R. T. and Kao, C. C. 2001. Terminal nucleotidyl transferase activity of recombinant Flaviviridae RNA-dependent RNA polymerases: implication for viral RNA synthesis. J Virol 75: 8615–23.

Ranjith-Kumar, C. T., Gutshall, L., Sarisky, R. T. and Kao, C. C. 2003. Multiple interactions within the hepatitis C virus RNA polymerase repress primer-dependent RNA synthesis. J Mol Biol 330: 675–85.

Ranjith-Kumar, C. T. and Kao, C. C. 2006. Recombinant viral RdRps can initiate RNA synthesis from circular templates. RNA 12: 303–12.

Ranjith-Kumar, C. T., Sarisky, R. T., Gutshall, L., Thomson, M. and Kao, C. C. 2004. De novo initiation pocket mutations have multiple effects on hepatitis C virus RNA-dependent RNA polymerase activities. J Virol 78: 12207–17.

Reed, K. E. and Rice, C. M: 1999. Identification of the major phosphorylation site of the hepatitis C virus H strain NS5A protein as serine 2321. J Biol Chem 274: 28011–8.

Reed, K. E., Xu, J. and Rice, C. M: 1997. Phosphorylation of the hepatitis C virus NS5A protein in vitro and in vivo: properties of the NS5A-associated kinase. J Virol 71: 7187–97.

Rijnbrand, R., Bredenbeek, P. J., Haasnoot, P. C., Kieft, J. S., Spaan, W. J. and Lemon, S. M. 2001. The influence of downstream protein-coding sequence on internal ribosome entry on hepatitis C virus and other flavivirus RNAs. Rna 7: 585–97.

Robida, J. M., Nelson, H. B., Liu, Z. and Tang, H. 2007. Characterization of hepatitis C virus subgenomic replicon resistance to cyclosporine in vitro. J Virol 81: 5829–40.

Sapay, N., Montserret, R., Chipot, C., Brass, V., Moradpour, D., Deleage, G. and Penin, F. 2006. NMR structure and molecular dynamics of the in-plane membrane anchor of nonstructural protein 5A from bovine viral diarrhea virus. Biochemistry 45: 2221–33.

Schmidt-Mende, J., Bieck, E., Hugle, T., Penin, F., Rice, C. M., Blum, H. E. and Moradpour, D. 2001. Determinants for membrane association of the hepatitis C virus RNA-dependent RNA polymerase. J Biol Chem 276: 44052–63.

Schwartz, M., Chen, J., Janda, M., Sullivan, M., den Boon, J. and Ahlquist, P. 2002. A positive-strand RNA virus replication complex parallels form and function of retrovirus capsids. Mol Cell 9: 505–14.

Serebrov, V. and Pyle, A. M. 2004. Periodic cycles of RNA unwinding and pausing by hepatitis C virus NS3 helicase. Nature 430: 476–80.

Shim, J. H., Larson, G., Wu, J. Z. and Hong, Z. 2002. Selection of 3′-template bases and initiating nucleotides by hepatitis C virus NS5B RNA-dependent RNA polymerase. J Virol 76: 7030–9.

Shirota, Y., Luo, H., Qin, W., Kaneko, S., Yamashita, T., Kobayashi, K. and Murakami, S. 2002. Hepatitis C virus (HCV) NS5A binds RNA-dependent RNA polymerase (RdRP) NS5B and modulates RNA-dependent RNA polymerase activity. J Biol Chem 277: 11149–55.

Simmonds, P., Bukh, J., Combet, C., Deleage, G., Enomoto, N., Feinstone, S., Halfon, P., Inchauspe, G., Kuiken, C., Maertens, G., Mizokami, M., Murphy, D. G., Okamoto, H., Pawlotsky, J. M., Penin, F., Sablon, E., Shin, I. T., Stuyver, L. J., Thiel, H. J., Viazov, S., Weiner, A. J. and Widell, A. 2005. Consensus proposals for a unified system of nomenclature of hepatitis C virus genotypes. Hepatology 42: 962–73.

Simmonds, P., Tuplin, A. and Evans, D. J. 2004. Detection of genome-scale ordered RNA structure (GORS) in genomes of positive-stranded RNA viruses: Implications for virus evolution and host persistence. RNA 10: 1337–51.

Siridechadilok, B., Fraser, C. S., Hall, R. J., Doudna, J. A. and Nogales, E. 2005. Structural roles for human translation factor eIF3 in initiation of protein synthesis. Science 310: 1513–5.

Sklan, E. H., Serrano, R. L., Einav, S., Pfeffer, S. R., Lambright, D. G. and Glenn, J. S. 2007a. TBC1D20 is a RAB1 GAP that mediates HCV replication. J Biol Chem 282: 36354–61.

Sklan, E. H., Staschke, K., Oakes, T. M., Elazar, M., Winters, M., Aroeti, B., Danieli, T. and Glenn, J. S. 2007b. A Rab-GAP TBC domain protein binds hepatitis C virus NS5A and mediates viral replication. J Virol 81: 11096–105.

Song, Y., Friebe, P., Tzima, E., Junemann, C., Bartenschlager, R. and Niepmann, M. 2006. The hepatitis C virus RNA 3′-untranslated region strongly enhances translation directed by the internal ribosome entry site. J Virol 80: 11579–88.

Spahn, C. M., Kieft, J. S., Grassucci, R. A., Penczek, P. A., Zhou, K., Doudna, J. A. and Frank, J. 2001. Hepatitis C virus IRES RNA-induced changes in the conformation of the 40 s ribosomal subunit. Science 291: 1959–62.

Spångberg, K. and Schwartz, S: 1999. Poly(C)-binding protein interacts with the hepatitis C virus 5′ untranslated region. J Gen Virol 80 (Pt 6): 1371–6.

Spångberg, K., Wiklund, L. and Schwartz, S. 2000. HuR, a protein implicated in oncogene and growth factor mRNA decay, binds to the 3′ ends of hepatitis C virus RNA of both polarities. Virology 274: 378–90.

Steinmann, E., Penin, F., Kallis, S., Patel, A. H., Bartenschlager, R. and Pietschmann, T. 2007. Hepatitis C Virus p7 Protein Is Crucial for Assembly and Release of Infectious Virions. PLoS Pathog 3: e103.

Stone, M., Jia, S., Heo, W. D., Meyer, T. and Konan, K. V. 2007. Participation of rab5, an early endosome protein, in hepatitis C virus RNA replication machinery. J Virol 81: 4551–63.

Su, A. I., Pezacki, J. P., Wodicka, L., Brideau, A. D., Supekova, L., Thimme, R., Wieland, S., Bukh, J., Purcell, R. H., Schultz, P. G. and Chisari, F. V. 2002. Genomic analysis of the host response to hepatitis C virus infection. Proc Natl Acad Sci USA 99: 15669–74.

Sumpter, R., Jr., Loo, Y. M., Foy, E., Li, K., Yoneyama, M., Fujita, T., Lemon, S. M. and Gale, M., Jr. 2005. Regulating intracellular antiviral defense and permissiveness to hepatitis C virus RNA replication through a cellular RNA helicase, RIG-I. J Virol 79: 2689–99.

Sun, X. L., Johnson, R. B., Hockman, M. A. and Wang, Q. M. 2000. De novo RNA synthesis catalyzed by HCV RNA-dependent RNA polymerase. Biochem Biophys Res Commun 268: 798–803.

Suzich, J. A., Tamura, J. K., Palmer-Hill, F., Warrener, P., Grakoui, A., Rice, C. M., Feinstone, S. M. and Collett, M. S: 1993. Hepatitis C virus NS3 protein polynucleotide-stimulated nucleoside triphosphatase and comparison with the related pestivirus and flavivirus enzymes. J Virol 67: 6152–8.

Tan, S. L. and Katze, M. G. 2001. How hepatitis C virus counteracts the interferon response: the jury is still out on NS5A. Virology 284: 1–12.

Tanaka, T., Kato, N., Cho, M. J. and Shimotohno, K: 1995. A novel sequence found at the 3′ terminus of hepatitis C virus genome. Biochem Biophys Res Commun 215: 744–9.

Tanaka, T., Kato, N., Cho, M. J., Sugiyama, K. and Shimotohno, K: 1996. Structure of the 3′ terminus of the hepatitis C virus genome. J Virol 70: 3307–12.

Tanji, Y., Kaneko, T., Satoh, S. and Shimotohno, K: 1995. Phosphorylation of hepatitis C virus-encoded nonstructural protein NS5A. J Virol 69: 3980–6.

Tellinghuisen, T. L., Evans, M. J., von Hahn, T., You, S. and Rice, C. M. 2007. Studying hepatitis C virus: making the best of a bad virus. J Virol 81: 8853–67.

Tellinghuisen, T. L., Marcotrigiano, J., Gorbalenya, A. E. and Rice, C. M. 2004. The NS5A protein of hepatitis C virus is a zinc metalloprotein. J Biol Chem 279: 48576–87.

Tellinghuisen, T. L., Marcotrigiano, J. and Rice, C. M. 2005. Structure of the zinc-binding domain of an essential component of the hepatitis C virus replicase. Nature 435: 374–9.

Teterina, N. L., Gorbalenya, A. E., Egger, D., Bienz, K., Rinaudo, M. S. and Ehrenfeld, E. 2006. Testing the modularity of the N-terminal amphipathic helix conserved in picornavirus 2C proteins and hepatitis C NS5A protein. Virology 344: 453–67.

Tischendorf, J. J., Beger, C., Korf, M., Manns, M. P. and Kruger, M. 2004. Polypyrimidine tract-binding protein (PTB) inhibits Hepatitis C virus internal ribosome entry site (HCV IRES)-mediated translation, but does not affect HCV replication. Arch Virol 149: 1955–70.

Tong, X. and Malcolm, B. A. 2006. Trans-complementation of HCV replication by non-structural protein 5A. Virus Res 115: 122–30.

Tsuchihara, K., Tanaka, T., Hijikata, M., Kuge, S., Toyoda, H., Nomoto, A., Yamamoto, N. and Shimotohno, K: 1997. Specific interaction of polypyrimidine tract-binding protein with the extreme 3′-terminal structure of the hepatitis C virus genome, the 3′X. J Virol 71: 6720–6.

Tu, H., Gao, L., Shi, S. T., Taylor, D. R., Yang, T., Mircheff, A. K., Wen, Y., Gorbalenya, A. E., Hwang, S. B. and Lai, M. M: 1999. Hepatitis C virus RNA polymerase and NS5A complex with a SNARE-like protein. Virology 263: 30–41.

Wakita, T., Pietschmann, T., Kato, T., Date, T., Miyamoto, M., Zhao, Z., Murthy, K., Habermann, A., Krausslich, H. G., Mizokami, M., Bartenschlager, R. and Liang, T. J. 2005. Production of infectious hepatitis C virus in tissue culture from a cloned viral genome. Nat Med 11: 791–6.

Wang, C., Gale, M., Jr., Keller, B. C., Huang, H., Brown, M. S., Goldstein, J. L. and Ye, J. 2005. Identification of FBL2 as a geranylgeranylated cellular protein required for hepatitis C virus RNA replication. Mol Cell 18: 425–34.

Wang, Q. M., Hockman, M. A., Staschke, K., Johnson, R. B., Case, K. A., Lu, J., Parsons, S., Zhang, F., Rathnachalam, R., Kirkegaard, K. and Colacino, J. M. 2002. Oligomerization and cooperative RNA synthesis activity of hepatitis C virus RNA-dependent RNA polymerase. J Virol 76: 3865–72.

Watashi, K., Hijikata, M., Hosaka, M., Yamaji, M. and Shimotohno, K. 2003. Cyclosporin A suppresses replication of hepatitis C virus genome in cultured hepatocytes. Hepatology 38: 1282–8.

Watashi, K., Ishii, N., Hijikata, M., Inoue, D., Murata, T., Miyanari, Y. and Shimotohno, K. 2005. Cyclophilin B is a functional regulator of hepatitis C virus RNA polymerase. Mol Cell 19: 111–22.

Yamashita, T., Kaneko, S., Shirota, Y., Qin, W., Nomura, T., Kobayashi, K. and Murakami, S: 1998. RNA-dependent RNA polymerase activity of the soluble recombinant hepatitis C virus NS5B protein truncated at the C-terminal region. J Biol Chem 273: 15479–86.

Yanagi, M., Purcell, R. H., Emerson, S. U. and Bukh, J: 1997. Transcripts from a single full-length cDNA clone of hepatitis C virus are infectious when directly transfected into the liver of a chimpanzee. Proc Natl Acad Sci USA 94: 8738–43.

Yanagi, M., St Claire, M., Emerson, S. U., Purcell, R. H. and Bukh, J: 1999. In vivo analysis of the 3′ untranslated region of the hepatitis C virus after in vitro mutagenesis of an infectious cDNA clone. Proc Natl Acad Sci USA 96: 2291–5.

Yang, G., Pevear, D. C., Collett, M. S., Chunduru, S., Young, D. C., Benetatos, C. and Jordan, R. 2004. Newly synthesized hepatitis C virus replicon RNA is protected from nuclease activity by a protease-sensitive factor(s). J Virol 78: 10202–5.

Ye, J., Wang, C., Sumpter, R., Jr., Brown, M. S., Goldstein, J. L. and Gale, M., Jr. 2003. Disruption of hepatitis C virus RNA replication through inhibition of host protein geranylgeranylation. Proc Natl Acad Sci USA 100: 15865–70.

Yi, M. and Lemon, S. M. 2003. 3′ nontranslated RNA signals required for replication of hepatitis C virus RNA. J Virol 77: 3557–68.

You, S. and Rice, C. M. 2007. 3′ RNA elements in hepatitis C virus replication: Kissing partners and long poly (U). J Virol 82: 184–95.

You, S., Stump, D. D., Branch, A. D. and Rice, C. M. 2004. A cis-acting replication element in the sequence encoding the NS5B RNA-dependent RNA polymerase is required for hepatitis C virus RNA replication. J Virol 78: 1352–66.

Yu, G. Y., Lee, K. J., Gao, L. and Lai, M. M. 2006. Palmitoylation and polymerization of hepatitis C virus NS4B protein. J Virol 80: 6013–23.

Zheng, Y., Gao, B., Ye, L., Kong, L., Jing, W., Yang, X., Wu, Z. and Ye, L. 2005. Hepatitis C virus non-structural protein NS4B can modulate an unfolded protein response. J Microbiol 43: 529–36.

Zhong, J., Gastaminza, P., Cheng, G., Kapadia, S., Kato, T., Burton, D. R., Wieland, S. F., Uprichard, S. L., Wakita, T. and Chisari, F. V. 2005. Robust hepatitis C virus infection in vitro. Proc Natl Acad Sci USA 102: 9294–9.

Zhong, W., Uss, A. S., Ferrari, E., Lau, J. Y. and Hong, Z. 2000. De novo initiation of RNA synthesis by hepatitis C virus nonstructural protein 5B polymerase. J Virol 74: 2017–22.

Chapter 5
Brome Mosaic Virus RNA Replication and Transcription

Guanghui Yi and C. Cheng Kao

Introduction

Brome mosaic virus (BMV) was first isolated in 1942 from bromegrass (*Bromus inermis*) and has since been documented to infect several monocot and dicot species studied in the laboratory. The impact of BMV, however, is not so much as a plant pathogen, but as a model for in-depth studies of the infection process of positive-stranded RNA viruses. As such, BMV is responsible for several firsts. (1) BMV was among the first to be translated using a cell-free system (Shih and Kaesberg, 1973), allowing studies of cap-dependent translation. (2) The BMV genome was one of the first RNA viruses for which the entire sequence was determined (Ahlquist et al., 1981, 1984a). (3) BMV was the first plant virus to be regenerated from transcripts derived from infectious cDNAs (Ahlquist and Janda, 1984; Ahlquist et al., 1984b). (4) The BMV replicase could be produced from membranes of infected plants and accept exogenously supplied transcripts for RNA synthesis, enabling the dissection of the mechanism of viral RNA synthesis (Hardy et al., 1979). (5) Recombinant BMV proteins were first demonstrated to direct replication and transcription of BMV RNA replicons in *Saccharomyces cerevisiae* (Janda and Ahlquist, 1993), allowing in-depth probing of the requirements of the host.

With regard to RNA replication, the topic of this chapter, extensive effort has been focused on the characterization of the *cis*-acting sequences, the identification of the host proteins, the assembly of the replicase, and the mechanism of RNA-dependent RNA synthesis. After an introduction in the basics of BMV molecular biology, this chapter will emphasize progress in these areas.

G. Yi (✉)
Department of Biology, Indian University, Bloomington, IN 47405, USA
e-mail: guanghuiyi@tamu.edu

C.E. Cameron et al. (eds.), *Viral Genome Replication*,
DOI 10.1007/b135974_5,

BMV RNAs and Replication Proteins

BMV belongs to Bromoviridae family, member of the alphavirus-like superfamily of animal- and plant-infecting viruses. The BMV genome is composed of three mRNA-sense RNAs that are encapsidated separately, named RNA1, RNA2, and RNA3 (Fig. 5.1). RNA1 and RNA2 are monocistronic and encode replication-associated proteins 1a and 2a, respectively. RNA3 is dicistronic and encodes the movement protein 3a and the capsid protein. The capsid protein is translated from a subgenome-length RNA named RNA4 (Fig. 5.1B).

The 5′ untranslated regions (UTRs) of BMV RNA1 and RNA2 share the conserved motif of the TψC loop of tRNA, named box B, which is also found in the RNAs of other bromoviruses (Dzianott and Bujarski, 1991; Romero et al., 1991; Ahlquist, 1992). In BMV RNA3, the box B is located in the intercistronic regulatory region between the movement protein and the capsid coding sequence. In addition, the intercistronic region contains an 18–21 nucleotide oligo(A) tract, a replication enhancer and subgenomic promoter (Ahlquist et al., 1981; French and Ahlquist, 1987; Adkins et al., 1997; Fig. 5.1B). The 3′UTRs of all three BMV RNAs lack a poly(A) tail, but contain a highly conserved and tRNA-like structure, as determined by enzymatic probing and computer modeling (Perret et al., 1989; Rietveld et al., 1983; Felden et al., 1994, 1996).

The replication scheme used by BMV is fairly typical for a plus-strand RNA virus (Fig. 5.1C). After entry into the cell and translation from the BMV RNAs, the

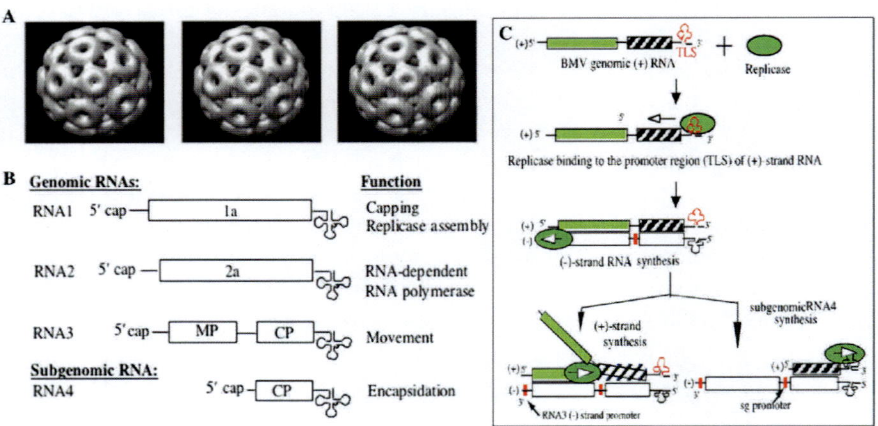

Fig. 5.1 Basic information on BMV. (**A**) Structures of the BMV virion; each virion packages one genomic RNA. The images are reconstructed from negative-stained BMV particles using the EMAN software (Sun et al., 2007). (**B**) Schematics of the BMV genomic RNAs, the subgenomic RNA4, and the functions encoded by each RNA. The cloverleaf denotes a tRNA-like structure present at the 3′ terminus of all BMV positive-strand RNAs. (**C**) Schematic of the BMV replication and transcription mechanism. The locations of the core promoters are in *red*. The BMV replicase is shown as a *green ball* and its direction of movement is denoted with an *arrow*.

1a and 2a protiens and unidentified cellular factors form the BMV replicase. The complex assembles in plant cellular membranes and will bind specifically to the BMV genomic plus-strand RNAs to first synthesize a complementary minus-strand RNA, which then serves as the template for genomic plus-strand RNA synthesis (Fig. 5.1C). The minus-strand RNA3 can also direct subgenomic RNA4 synthesis. RNA4 thus provides a simple means to allow RNA3 to be a dicistronic RNA without breaking rules for cap-dependent translation of both the movement and capsid proteins. Many of the requirements for BMV RNA replication and transcription have been elucidated.

1a and 2a are central players in BMV RNA synthesis. Cells transfected with only RNA1 and RNA2 can replicate in the absence of RNA3 or RNA4 transcription (Gopinath et al., 2005; Annamalai and Rao, 2005; French and Ahlquist, 1987). Thus, the movement and the capsid proteins encoded by BMV RNA3 are not essential parts of the BMV replicase. However, the normal accumulation of molar excesses of BMV plus-strand RNAs relative to the minus-strand RNAs is not observed in the absence of RNA3 (Marsh et al., 1991; Gopinath et al., 2005).

BMV 1a is a multifunctional protein. The N-terminal portion of 1a has a predicted secondary structure that is highly similar to known DNA and RNA methyltransferases, while the C-terminal portion contains helicase-like motifs (Ahlquist et al., 1985; Ahola and Ahlquist, 1999; O'Reilly et al., 1998). The N-terminal 516 residues have been demonstrated to have RNA capping-associated activities, including the ability to form a covalent complex and methylate a guanine nucleotide at the N7 position (Ahola and Ahlquist, 1999; Kong et al., 1999). RNA helicase activity of 1a has not been demonstrated biochemically. However, the C-terminal 424–961 amino acids of 1a, including the NTPase/hel domain, has ATPase activity (Wang et al., 2005). Mutations in both parts of the 1a protein have led to defects in BMV RNA replication.

The BMV 2a protein is an RNA-dependent RNA polymerase. 2a has a large central domain that contains all the hallmark motifs of an RNA-dependent RNA polymerase, flanked by less conserved N- and C-termini (O'Reilly and Kao, 1998; Traynor et al., 1991). The N-terminal domain of 2a interacts with the 1a protein while the C-terminal domain is dispensable for replication in protoplasts (Traynor et al., 1991). Recombinant 2a protein has recently been demonstrated to direct RNA synthesis in vitro (Wierzchoslawski and Bujarski, 2006). As with most recombinant viral RdRps, the most robust activity of 2a is in extending from a primed template. This does not mean that primer extension is the preferred mechanism of initiation by 2a, however. In fact, positive-strand BMV RNAs are capped, and the substrate for the capping reaction is a de novo initiated RNA (Kao et al., 2001).

There is abundant evidence that the BMV 1a and 2a proteins function by forming a complex. A complex is observed when 1a and 2a are expressed in rabbit reticulocytes (Kao et al., 1992; Kao and Ahlquist, 1992) and in yeast in the form of the two-hybrid assay (O'Reilly et al., 1997). 1a and 2a also co-purified with enzymatically active BMV replicase (Quadt et al., 1988) and are co-localized in the endoplasmic reticulum of plant cells, the site of BMV RNA synthesis (Restrepo-Hartwig and Ahlquist, 1996, 1999).

The interaction between 1a and 2a is species specific, as determined by examining the replication of homologous and heterologous combinations of BMV 1a and 2a and the orthologs from the closely related bromovirus, cowpea chlorotic mottle virus (Dinant et al., 1993; O'Reilly et al., 1995, 1997). Attempts to overexpress the 1a protein from the 35S Cauliflower Mosaic virus promoter can lead to the inhibition of both 1a and 2a expression in *Nicotiana benthamiana*. The inhibition acts through box B in the 5′ untranslated region of BMV RNA1 and RNA2. RNA3 is not regulated by 1a since it lacks a box B in its 5′UTR (Gopinath et al., 2005; Yi et al., 2007). These results suggest that there are mechanisms in place for BMV to regulate the translation of the replication proteins and that unregulated production may be detrimental to efficient BMV infection, perhaps through innate host responses.

Properties of the BMV Replicase

Viral nucleic acids are targets for the host innate defense systems (Meylan and Tshopp, 2006). Hence, it is likely that the process of viral RNA replication will include a number of mechanisms that are in place to prevent recognition by the host. For example, double-stranded RNA viruses, minus-sense RNA viruses, and retroviruses replicate within some form of the viral particles, the site where replication-associated enzymes have been cached (Jayaram et al., 2004). However, positive-stranded RNA viruses, due to the need for translation to precede replication, must expose their RNA. To accommodate this exposure as well to protect the viral genome, RNAs are redirected to cellular membranes once the replication proteins are available.

Indeed, for all the positive-strand RNA viruses, RNA replication is associated with intracellular membranes (Schaad et al., 1997). BMV RNA replication occurs on the perinuclear region of the endoplasmic reticulum (ER), both in its natural plant host and in the surrogate host *S. cerevisiae* (Restrepo-Hartwig and Ahlquist, 1996, 1999). The replication protein 1a is the primary viral protein determinant for the subcellular localization of the BMV replication complex (Fig. 5.2). 1a can localize to the cytoplasmic face of ER membranes in the absence of other viral factors and induces spherules serving as replication compartments sequestering viral positive-strand RNA templates in a nuclease-resistant, detergent-susceptible state (Restrepo-Hartwig and Ahlquist, 1999; Schwartz et al., 2002). Membrane flotation gradient analysis with wild type 1a and deletion mutants showed that the sequences in the N-terminal RNA capping domain of 1a mediate membrane association (den boon et al., 2001). In the absence of 1a, 2a is present in the cytoplasm in diffused distribution or in punctate spots that are not apparently associated with cytoplasmic organelles. Thus 1a appears to bring 2a into the membrane-associated replicase complex. Indeed, the interaction between the N-terminus of 2a and the C-terminal helicase-like domain of 1a leads to the formation of double-membrane layers (Chen and Ahlquist, 2000). The expression level of 2a polymerase can also modulate 1a-induced membrane rearrangements (Schwartz et al., 2004). Using monoclonal antibodies raised against the 1a and 2a proteins, it was shown that the region between

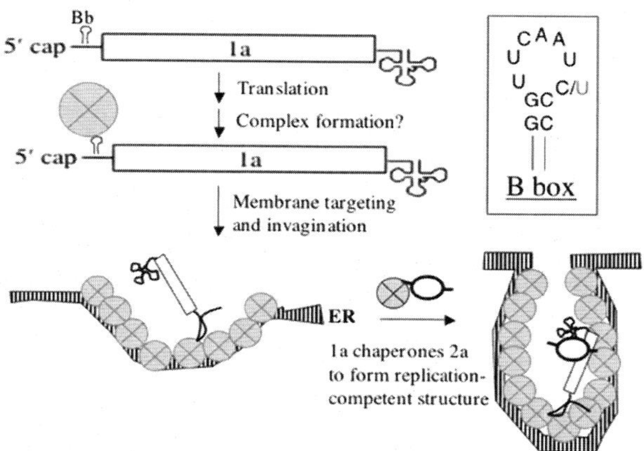

Fig. 5.2 A schematic for the assembly of the BMV replicase. The 1a protein is represented as a sectored *circle*. This complex could involve 1a–1a interaction and/or binding to cellular proteins. The 2a protein is shown as an open oval with thick lines denoting the N- and C-terminal domains. The inset shows the sequence of the Box B, which is identical in RNA1 and RNA2. For RNA3, there is a conserved change of the seventh residue in the loop from a C to a U.

the N-terminal methyltransferase domain and the C-terminal helicase-like domain of 1a and the N-terminus region of 2a protein are exposed on the surface of the solubilized replicase complex. (Dohi et al., 2002).

The capping domain and NTPase/helicase-like domain of 1a contribute to RNA templates recruitment in the formation of the BMV replicase. Mutations in the capping enzyme active site cause defects in template recruitment, negative-strand RNA synthesis (Ahola et al., 2000). Mutations in the helicase motifs in the 1a protein severely inhibited RNA replication and reduced the stability of RNA3, although they did not affect 1a accumulation, localization to perinuclear ER membranes, or recruitment of 2a polymerase (Wang et al., 2005).

The replicase must specifically recognize the viral RNAs. This interaction has been partially elucidated using the yeast system. The 1a replication protein has been demonstrated to recognize the BMV genomic RNA2 and RNA3 through the box B, then recruit the RNAs from translation to replication. RNA1 is likely recognized in a manner similar to RNA2, as they share an identical box B RNA in their 5′UTRs. In the absence of 2a, 1a can induce the association of RNA2 or RNA3 with cellular membrane by the intercistronic sequence (Janda and Ahlquist, 1998; Sullivan and Ahlquist, 1999; Schwartz et al., 2002). The RNA2 5′UTR was sufficient to confer 1a-induced membrane association although sequences in the N-terminal region of the 2a open reading frame could enhance 1a responsiveness (Chen et al., 2001). In addition to recruitment of the RNA presumably through binding the box B, 1a can also recruit RNA2 through its interaction with the N-terminal portion of 2a, presumably when the 2a is being translated from RNA2 (Chen et al., 2003).

The viral RNA is an active participant in the assembly of replicase complex. In yeast, coexpression of BMV RNA3 was required for functional BMV RNA-dependent RNA polymerase activity (Quadt et al., 1995). Deletion analysis showed that the tRNA-like 3′UTR and the intercistronic region are minimally required for in vivo formation of functional RNA-dependent RNA polymerase. This RNA may contribute by recruiting essential host factors to participate in replicase assembly (Quadt et al., 1995). The mechanism of replicase assembly in BMV RNA1 and RNA2 may differ from that of RNA3, since they lack the intercistronic region. Once assembled, the functional BMV replicase does not need to contain RNA, as the RNA is not present in biochemically active preparations of the BMV replicase (Sun et al., 1996), suggesting that once the replicase assembles, a functional complex can be maintained through protein–protein or protein–membrane interaction.

In yeast, the 1a protein can significantly increase the stability of BMV RNA3 by binding to the intercistronic region of RNA3 (Janda and Ahlquist, 1998; Sullivan and Ahlquist, 1999). This requirement is not observed in *N. benthamiana* or barley protoplasts (Gopinath et al., 2005), indicating either that the replicases formed in yeast and plants have distinct properties or that the host degradation pathways have different access to the RNA in plants and in yeast. While this may seem counter-intuitive at first glance, an emerging theme in virus replication is that the virus is far more adaptable than one may expect. For example, flock house virus, a positive-stranded virus of insect cells, was found to replicate perfectly well in two distinct membrane locations when the membrane targeting signal sequence was altered (Miller et al., 2003).

BMV RNA Motifs and the Modes of RNA Replication and Transcription

Efficient viral RNA synthesis requires specific and coordinated interactions between the template RNA and the viral replicase, a membrane-associated complex of viral replication proteins and host-encoded factors (Kao et al., 2001; Lai, 1998). The recognition is likely to be quite complex because an RNA virus not only will express different classes of RNAs [i.e., genomic plus-strand RNA, genomic minus-strand RNA, and possibly subgenomic RNA(s)], but will do so at regulated levels and times (Buck, 1996).

Using a combination of approaches, including genetic analysis in plant protoplasts and a template-specific BMV replicase that can be extracted from BMV-infected plants, the sequences and motifs that can efficiently direct BMV RNA synthesis have been identified (Choi et al., 2004). The replicase-binding sequences are called ‘core promoters’ since they can bind the BMV replicase and direct the initiation of RNA synthesis. As is the case with core promoters for transcription from DNA templates, the viral core promoters direct a basal level of RNA synthesis that can be modulated by positive- and negative-acting sequences (Lai, 1998).

Genomic Minus-Strand Promoter

The promoter for minus-strand RNA synthesis is within the tRNA-like 3′ sequence (Dreher and Hall, 1988a,b). These secondary structures interact with each other to

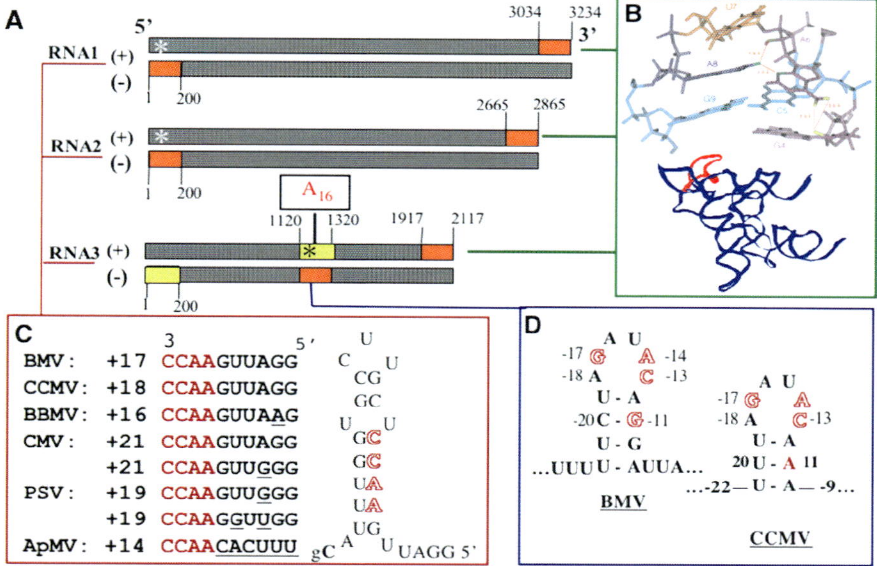

Fig. 5.3 A summary of the locations and relevant features of the three classes of BMV core promoters. (**A**) The locations of the replicase-binding RNA sequences within positive- and negative-strand BMV RNAs. (**B**) Structure of the clamped adenine motif of BMV RNA that directs the initiation of BMV minus-strand RNA synthesis. The lower structure denotes the approximate structure of the tRNA-like structure with SLC in *red*. The upper structure was determined by NMR and shows the essential features that contribute to the formation of a clamped adenine motif. (**C**) The core promoter for genomic plus-strand RNA synthesis. A conserved sequence complementary to the Box B that contains a CCAA motif is *highlighted*. The conservation of the CCAA sequence in related members of the Bromoviridae is listed along with the NMR-derived secondary structure of the RNA that can direct genomic plus-strand RNA synthesis. The nontemplated nucleotide required for genomic plus-strand RNA synthesis is in a lower case "g". (**D**) A summary of the essential residues for replicase binding in the BMV subgenomic promoter. The critical residues are in *outlined letters*. The sequence can fold into a quasi-stable hairpin in both BMV and the related virus, CCMV (cowpea chlorotic mottle virus).

contribute to mimicry of the tRNA-like tertiary structure needed for aminoacylation of the 3′ termini of BMV and CMV RNAs (Fig. 5.3B; Felden et al., 1993; Giege' 1996). A notable exception to the tRNA-like tertiary structure is a complex stem-loop named SLC. Mutations in SLC can severely reduce BMV and CMV replication in protoplasts (Dreher and Hall, 1988a; Rao and Hall, 1993) and SLC was sufficient to interact with the BMV replicase in vitro in the absence of the remainder of the tRNA-like sequence. SLC fused to the 3′ terminal 8 nt of the 3′-terminus of the tRNA-like sequence resulted in an RNA that could direct RNA synthesis in vitro (Fig. 5.3; Chapman and Kao, 1999). Mutations to SLC+8 that decreased BMV RNA replication in vivo have parallel effects on RNA synthesis in vitro (Chapman and Kao, 1999).

The solution structure of the BMV SLC was determined by NMR spectroscopy and found to be composed of two stems, separated by a flexible internal bulge

(Fig. 5.3B). The bulged portion of the RNA was dynamic (Kim et al., 2000). Related viruses also possess a bulged sequence, suggesting that a dynamic motif here in the bulge may be preferred for viral RNA replication. The terminal stem contains a tri-nucleotide loop (5′AUA3′) that is specifically required for interaction with the replicase (Kim et al., 2000). The tri-loop was found to fold into a highly ordered structure called a clamped adenine motif (CAM), with the 5′-most adenine of the tri-loop projected into the solution, primarily due to the base stacking interactions between the 3′-most adenine and the stem-closing C–G base pair (red nucleotides, Fig. 5.3B; Kim et al., 2000). A network of electrostatic interactions also stabilizes the solution-exposed 5′-adenine. Variations of the terminal loop nucleotides that were unable to form a CAM failed to direct efficient RNA synthesis by the BMV replicase (Kim and Tinoco, 2001).

A change of the 3′-most adenine of the tri-loop (5′AUA3′) to a guanine (5′ AUG 3′) resulted in wild-type levels of RNA synthesis. This change should disrupt the normal CAM. However, when the solution structure of an RNA containing the 5′AU<u>G</u>3′ tri-loop (the mutated nucleotide is underlined) was solved using NMR, it was found to form a dramatically altered structure that still retained a solution-exposed and clamped adenine (Kim and Kao, 2001). These studies reveal the features in the RNA core promoter required for recognition by the BMV replicase for minus-strand RNA synthesis in vitro and in vivo.

Genomic Plus-Strand Promoter

BMV genomic plus-strand RNA synthesis in vitro required an adjacent stem-loop with a short single-stranded sequence with nontemplated 3′ nucleotide (Fig. 5.3C; Sivakumaran and Kao, 1999; Sivakumaran et al., 1999). The replication of the related CMV satellite RNA also requires a nontemplated nucleotide (Wu and Kaper, 1994). Nontemplated nucleotide addition is a common property of cellular and viral polymerases (Kumar et al., 2001; Siegel et al., 1997). Therefore, the addition to BMV minus-strand RNA might be by the BMV replicase or a cellular enzyme. Furthermore, since the initiation of minus-strand RNA synthesis occurs from the penultimate nucleotide (Miller et al., 1986; Sun et al., 1996), the requirement may reflect a structural requirement for the BMV polymerase. In the ternary structure of the RdRp from bacteriophage $\phi 6$, a nontemplated nucleotide was required to allow proper contact between the active site and the initiation nucleotide (Bamford et al., 2005).

In addition to the 3′ initiation cytidylate, a highly conserved ca. 9 nt sequence called the cB box exists at the 5′-end of the BMV genomic core promoter from RNA1 and RNA2 (Fig. 5.3C). The cB box is found in RNA1 and RNA2 of all of the Bromoviridae except for most ilarviruses and the alfalfa mosaic virus (Sivakumaran and Kao, 2000). The cB box is complementary to the Box B that is required for replicase assembly (French and Ahlquist, 1987; Marsh and Hall, 1987; Pogue and Hall, 1992; Chen et al., 2003). In positive-strand BMV RNAs, box B is usually positioned upstream of the protein-coding sequence and interacts with the 1a protein in a way that increases the stability of the RNA (Sullivan and Ahlquist, 1999). The cB box appears to be required in a position-specific manner; moving it one nucleotide closer to the 3′ initiation site severely reduced RNA synthesis (Sivakumaran et al., 2000). However, moving the cB box one nucleotide 5′ of its original position was less detrimental to RNA synthesis (Sivakumaran et al., 2000), indicating that some flexibility in the RNA can be used to correctly position box B relative to the replicase subunits.

The synthesis of RNA3 from BMV genomic RNA3 may be different than for other genomic RNAs. For BMV RNA3, the cB box exists in the intercistronic region, over a kilobase from the initiation site for genomic plus-strand RNA synthesis. Since RNA3 is replicated in trans by the viral replicase, it is possible that RNAs that do not encode a subunit of the replicase have different replicase recognition requirements than RNAs that can be translated to provide a subunit of the replicase.

Subgenomic Promoter

A 20 nt 3′ of the initiation cytidylate for BMV RNA4 is sufficient for an accurate initiation of RNA synthesis in vitro (Adkins et al., 1997). Additional sequences 3′ of the core promoter does affect RNA synthesis (Adkins et al., 1997; French and Ahlquist, 1988; Marsh et al., 1988), but does not influence the selection of the initiation site. Single-nucleotide changes identified that positions −11, −13, −14, and −17 relative to the +1 initiation cytidylate were required for efficient RNA synthesis (Fig. 5.3D). While some other positions within this 20 nt sequence also contributed to the level of RNA synthesis, changes at these four positions decreased RNA synthesis by up to 10-fold (Siegel et al., 1997) (Fig. 5.3D). These results suggest that the BMV subgenomic core promoter may be recognized in an RNA sequence-dependent manner, a mechanism similar to the recognition of DNA promoters by DNA-dependent RNA polymerases (Adkins et al., 1998). Consistent with this, RNAs containing nucleotide analogues at these four positions, some of which should retain normal base pairing potential of these nucleotides, significantly decreased RNA synthesis (Siegel et al., 1998). Moieties in these nucleotides of the BMV core promoter are specifically required for RNA synthesis in vitro.

Haasnoot and colleagues proposed that a stable secondary structure exist in the core promoter (Haasnoot et al., 2000). This structure is not required for RNA synthesis by the BMV replicase in vitro, but mutations that prevented stem-loop formation did reduce BMV subgenomic RNA synthesis in protoplasts (Sivakumaran et al., 2004). The stem and specific nucleotide within presumably will bind the replicase to direct the recognition of the initiation cytidylate. Whether the subgenomic core promoter is recognized after the synthesis of the full-length minus-strand RNA or can be recognized during minus-strand RNA3 synthesis is unknown.

The characterization of the three classes of core promoters from both BMV and CMV revealed some similarities and differences in replicase-promoter interaction. In general, each core promoter contains three features that are, to different extents, required for RNA synthesis. First, there is a specificity determinant (SLC for genomic minus-strand initiation, the cB box for genomic plus-strand initiation of RNA1 and RNA2, the nucleotides upstream of the initiation site for the subgenomic promoter). It is not known whether viral or cellular subunits within the replicase interact with the specificity determinants. A second required element is the initiation site, which includes the initiation cytidylate and a few neighboring nucleotides. The viral RdRp subunit within the replicase must recognize this site, since the RdRp polymerizes nucleotides. In addition to the RdRp–initiation site interaction, the base pairing between the initiation cytidylate and the substrate GTP could contribute to specificity for initiation. Specificity recognition of the initiation complex was observed in the ternary crystal structure of the recombinant RdRp from bacteriophage $\phi 6$ (Butcher et al., 2001) and with a functional analysis of the template used by the recombinant RdRps from members of the Flaviviridae (Kim et al., 2000). A third requirement is the template sequence immediately following the initiation cytidylate, which can apparently alter the level of synthesis, perhaps by regulating the efficiency of the replicase transition from initiation to elongation. Plus-strand RNA viruses in the Bromoviridae generally have at least three nucleotides after the initiation cytidylate that will weakly base pair with the nascent RNA. Templates for minus-strand synthesis do not follow this trend (Table 5.1), perhaps reflecting a role in regulating the level of RNA produced.

Despite the three general requirements, there is some fluidity in each of the requirements, because several changes in each core promoter can be tolerated. Even a change of the initiation cytidylate to a uridylate can result in RNA synthesis at about 5% of wild type. This fluidity suggests that extensive molecular communication occurs between the RNA and the replicase to allow some adjustments by conformational changes (induced fit) in the interactions that lead to productive synthesis (Williamson, 2000). An induced fit mechanism provides the best explanation for the recognition of some variants of the BMV subgenomic promoters by the BMV replicase (Stawicki and Kao, 1999) and the cross-recognition of some core promoters between the replicases of BMV, CMV, and Cowpea chlorotic mottle virus (Adkins and Kao, 1998; Chen et al., 2000; Sivakumaran et al., 2000).

Replication Mechanism

The recognition of the core promoters is only the first step in successful BMV RNA replication. This process was studied in detail using the BMV replicase. Overall, the process of BMV RNA synthesis can be divided into several biochemically defined steps, consisting of initiation, abortive initiation, template commitment, elongative RNA synthesis, and termination (Fig. 5.4).

Initiation by the BMV replicase is perhaps the most distinct aspect of viral RNA replication since the nature of the linear viral templates requires that initiation take place at or near the 3′ terminus of the template. For the BMV RNA synthesis, a cytidylate penultimate to the 3′ nucleotide is preferred. The cytidylate is recognized in a sequence-specific manner and is paired with the initiation GTP or GTPi. The recognition of the GTPi has additional requirements in comparison to

Table 5.1 Host factor implicated in BMV replication

Host factors	Function	Effects on BMV replication	References
eIF3	Translation elongation factor	Stimulate (–)-Strand RNA synthesis	Quadt et al. (1993)
DED1	Translation factor	Translation of 2a RNA polymerase	Noueiry et al. (2000)
Lsm1	mRNA turnover	Recruiting RNA from translation to replication	Diez et al. (2000), Noueiry et al. (2003)
YDJ1	Chaperone	(–)-Strand RNA synthesis and assembly replicase complex	Tomita et al. (2003)
OLE1	Fatty acid desaturase	RNA replication	Lee et al. (2001)
Pus4	Pseudouridine synthase	Multiple effects on translation, replication, and systemic infection	Zhu et al. (2007)
App1	Intracellular trafficking	RNA replication and systemic infection	Zhu et al. (2007)
Lsm1p-7p Pat1p/Dhh1p	Decapping cellular RNA and RNA turnover	Recruit RNA for replication	Nouseiry et al. (2003), Mas et al. (2006)
mab1-1, *mab2-1* and *mab3-1*		RNA3 and subgenomic RNA4 replication	Ishikawa et al. (1997)
SIF2 and ~100 host factors in total		Regulate BMV RNA replication and translation	Kushner et al. (2003)

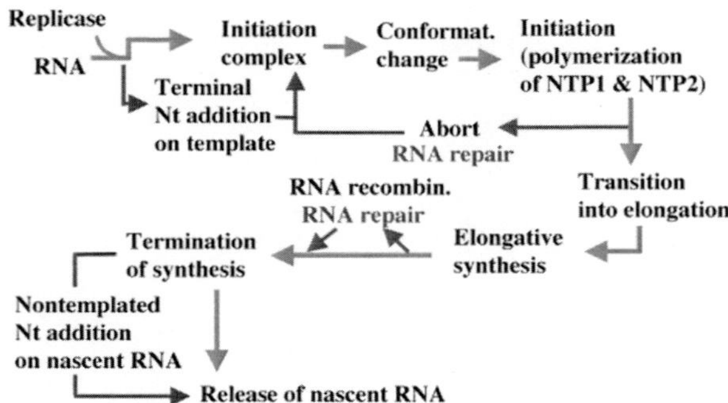

Fig. 5.4 A summary of the steps in the RNA synthesis by BMV replicase. The sequence of events proceeds from the *upper right* according to the direction of the *arrows*.

the recognition of GTP late in RNA synthesis. For example, the GTPi is required at ~15-fold higher concentration for initiation than for elongation (Sun et al., 1996). The GTPi can also be replaced with oligonucleotide primers as long as the primers can maintain base pairing to the initiation cytidylate in the template (Kao and Sun, 1996). Similar requirements are found for several RdRps from the Flaviviridae family (Ranjith-Kumar et al., 2003), demonstrating that this is a basic property of the initiation process by RNA virus polymerases. The presence of the template and the GTPi will increase the stability of the binding by the replicase, preventing inhibition by template mimics such as heparin (Sun and Kao, 1997a,b). In fact, the stability is increased stepwise with the number of nucleotides in the initiating RNA, suggesting that the polymerase will undergo a series of transitions that lead to a productively synthesizing complex.

The transition from initiation and commitment to the template by the BMV replicase is marked by the formation of abortive initiation products (Fig. 5.4). Abortive initiation products are formed by the replicase and released before the replicase transitions to productive elongative synthesis (Carpousis and Gralla, 1980). They are typically present at molar excesses of the full-length products and, for the BMV replicase, range from 2 to 12 nt in length. Interestingly, the abundance of the abortive products decreases after 8 nt, suggesting that the ternary complex is committing to elongation at or shortly after the synthesis of a nascent RNA of 8 nt. Abortive products could have additional roles in the repair of the ends of the RNA (perhaps giving an advantage to viruses with multi-partite genomes that share common 3′ sequences). Rapid repair of short deletions in the 3′-end of the BMV genomic RNAs was observed (Hema et al., 2005).

Much less is known about the regulation of elongation and termination of BMV RNA synthesis. A basic residue in the template will trap the BMV ternary complex depending on its location. When present within the first ten nucleotides from the 3′ termini, the replicase can reinitiate. However, when present later in the template, the replicase is unable to reinitiate (Picard et al., 2005). With regard to termination, the sequence from −4 to −2 of the position from the very 3′ terminus of the template could regulate the proper termination of nascent RNA synthesis; nucleotides that allow stronger base pairing tend to promote the synthesis of a full-length RNA while nucleotides that have weaker base pairing tend to decrease the proportion of full-length RNA synthesis and increase premature termination (Tayon et al., 2001). These steps in BMV RNA synthesis are of interest relative to the mechanism of transcription by DNA-dependent RNA polymerases. A comparison of the requirements for the steps is in Adkins et al. (1998).

Host Factors That Affect BMV Replication

Successful viral replication requires proper interaction between viral and cellular factors. Early in vitro experiments have shown that translation elongation factor EF-1A could bind to the 3′tRNA-like structure of BMV RNA1, although its function in RNA replication remains unclear (Bastin and Hall, 1976). The translational factor eIF3 was co-purified with BMV RNA replicase (Table 5.1; Quadt et al., 1993).

The identification of host factors that regulate BMV replication was accelerated by the ability of BMV to replicate in *S. cerevisiae* (Janda and Ahlquist, 1993; Noueiry and Ahlquist, 2003; Sullivan and Ahlquist, 1997). An initial screen using the mutant yeast strains has identified a number of host factors involved in cellular RNA degradation and fatty acid metabolism, such as Lsm1p-7p and OLE1 that are involved directly or indirectly in regulation of BMV replication complex assembly, template recruitment for replication, or the process of RNA synthesis. A summary of the host factors is presented in Table 5.1. Several are involved in the specific steps of BMV replication. For example, mutation of host gene Lsm1p, which encodes a protein involved in mRNA turnover and other processes, resulted in defects in an early template selection step of BMV RNA replication (Diez et al., 2000). In addition to Lsm1p, all tested components of the Lsm1p-7p/Pat1p/Dhh1p decapping activator complex, which functions in deadenylation-dependent decapping of cellular mRNAs, were required for BMV RNA recruitment (Mas et al., 2006). Additional factors have been identified to be involved in BMV RNA replication, but their exact contributions to BMV-specific processes remains to be determined.

Yeast proteome chips have also been used to identify the host proteins that could bind to the specific BMV RNA (Zhu et al., 2007). Among the ones identified to bind the BMV core promoter for minus-strand RNA synthesis are Pus4, a pseudouridylate synthase, and App1, which is associated with the actin patch. Overexpression of Pus4 and App1 resulted in the inhibition of BMV virion assembly. In all of these cases, it is important to recognize that, while it is informative to see what possibly could interact with BMV, the plant host factors homologous to the yeast proteins should be characterized in order to study an evolved interaction.

Relationship Between Replication, Encapsidation, and Translation

For BMV genomic and subgenomic RNA, viral RNAs are serving as templates for translation and replication as well as encapsidation. Thus these processes may be related and coordinated during virus infection. Coupling packaging and replication

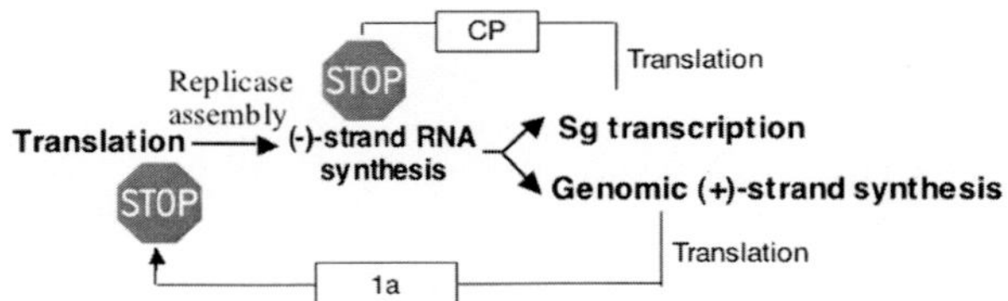

Fig. 5.5 Crosstalk between BMV RNA replication and processes required for BMV infection. The process of BMV replication is shown in *bold* and the regulatory roles of 1a or the capsid protein are denoted by *arrows*.

has been reported for some positive-strand RNA viruses, such as Poliovirus (Nugent et al., 1999), Kunjin virus (Khromykh et al., 2001), Flock House virus (Venter et al., 2005), and Venezuelan equine encephalitis virus (Vovkova et al., 2006). Recently, Annamalai and Rao (2006) demonstrated that efficient packaging of subgenomic RNA4 was functionally coupled to translation of coat protein from replication-derived mRNA, both in vitro assembly assay and *Agrobacterium*-mediated transient in vivo expression system. Packaging of RNA by the BMV CP was nonspecific in the absence of replication, while induction of viral replication increased the specificity of RNA packaging (Annamalai and Rao, 2006). Since 1a recruiting of RNA2 to replication complex required high-efficiency translation of the N-terminal half of the RNA template (Chen et al. 2003), replication and translation might be coupled since recruitment of RNA template to the replication complex is a major step in RNA replication. Recently, our lab found that efficient BMV genomic RNA1 replication required the translation of encoded protein 1a *in cis*, indicating that replication and translation of genomic RNA1 is functionally coupled (in preparation). Coupling replication and translation has been observed for poliovirus (Novak and Kirkegaard, 1994), mouse hepatitis virus (de Groot et al., 1992), and turnip yellow mosaic virus (Weiland and Dreher, 1993). The linkage among these processes may favor the viral replicase to efficiently differentiate viral RNA template from cellular RNA. A schematic for the crosstalks between different BMV processes is summarized in Fig. 5.5.

It is likely that the intersection between viral RNA replication and other processes required for infection (translation, RNA recombination, encapsidation, host innate responses) will provide fertile grounds for future research.

Acknowledgments We thank the NSF (MCB0641362 to C. Kao) for funding this research.

References

Adkins, S., and Kao, C. 1998. Subgenomic RNA promoters dictate the mode of recognition by bromoviral RNA-dependent RNA polymerases. Virology 252:1–8.

Adkins, S., Siegel, R., Sun, J., and Kao, C. 1997. Minimal templates directing accurate initiation of subgenomic RNA synthesis in vitro by the brome mosaic virus RNA-dependent RNA polymerase. RNA 3:634–647.

Adkins, S., Stawicki, S., Faurote, G., Siegel, R., and Kao, C. 1998. Mechanistic analysis of RNA synthesis RNA-dependent RNA polymerase from two promoters reveals similarities to DNA-dependent RNA polymerase. RNA 4:455–470.

Ahlquist, P. 1992. Bromovirus RNA replication and transcription. Curr. Opin. Genet. Dev. 2:71–76.

Ahlquist, P., and Janda, M. 1984. cDNA cloning and in vitro transcription of the complete brome mosaic virus genome. Mol. Cell. Biol. 4:2876–2882.

Ahlquist, P., Dasgupta, R., and Kaesberg, P. 1984a. Nucleotide sequence of brome mosaic virus genome and its implication for virus replication. J. Mol. Biol. 172:369–383.

Ahlquist, P., French, R., Janda, M., and Loesch-Fries, L.S. 1984b. Multicomponent RNA plant virus infection derived from cloned viral cDNA. Proc. Natl Acad. Sci. USA 81: 7066–7070.

Ahlquist, P., Luckow, V., and Kaesberg, P. 1981. Complete nucleotide sequence of brome mosaic virus RNA3. J. Mol. Biol. 153:23–38.

Ahlquist, P., Strauss, E.G., Rice, C.M., Strauss, J.H., Haseloff, J., and Zimmern, D. 1985. Sindbis virus proteins nsP1 and nsP2 contain homology to nonstructural proteins from several RNA plant viruses. J. Virol. 53:536–542.

Ahola, T., and Ahlquist, P. 1999. Putative RNA capping activities encoded by brome mosaic virus: methylation and covalent binding of guanylate by replicase protein 1a. J. Virol. 73:10061–10069.

Ahola, T., den Boon, J.A., and Ahlquist, P. 2000. Helicase and capping enzyme active site mutations in brome mosaic virus protein 1a cause defects in template recruitment, negative-strand RNA synthesis, and viral RNA capping. J. Virol. 74:8803–8811.

Annamalai, P., and Rao, A.L. 2005. Replication-independent expression of genome components and capsid protein of brome mosaic virus in planta: a functional role for viral replicase in RNA packaging. Virology 338:96–111.

Annamalai, P., and Rao, A.L. 2006. Packaging of brome mosaic virus subgenomic RNA is functionally coupled to replication-dependent transcription and translation of coat protein. J. Virol. 80:10096–10108.

Bamford, D.H., Grimes, T.M., and Stuart, D.I. 2005. What does structure tell us about virus evolution. Curr. Opin. Struct. Biol. 15:655–663.

Bastin, M., and Hall, T.C. 1976. Interaction of elongation factor 1a with aminoacylated brome mosaic RNA. J. Virol. 20:117–122.

Buck, K.W. 1996. Comparison of the replication of positive-stranded viruses of plants and animals. Adv. Virus Res. 47:159–251.

Butcher, S.J., Grimes, J.M., Makeyev, E.V., Bamford, D.H., and Stuart, D.I. 2001. A mechanism for initiating RNA-dependent RNA polymerization. Nature 410:235–240.

Carpousis, A.J., and Gralla, J.D. 1980. Cycling of ribonucleic acid polymerase oligonucleotides during initiation in vitro at the lac UV5 promoter. Biochemistry 19:3245–3253.

Chapman, M., and Kao, C. 1999. A minimal RNA promoter for minus-strand RNA synthesis by the brome mosaic virus polymerase complex. J. Mol. Biol. 286:709–720.

Chen, J., and Ahlquist, P. 2000. Brome mosaic virus polymerase-like protein 2a is directed to the endoplasmic reticulum by helicase-like viral protein 1a. J. Virol. 74:4310–4318.

Chen, M.H., Roossinck, M.J., and Kao, C.C. 2000. Efficient and specific initation of subgenomic RNA synthesis by cucumber mosaic virus replicase in vitro requires an upstream RNA stem-loop. J. Virol. 74:11201–11209.

Chen, J., Noueiry, A., and Ahlquist, P. 2001. Brome mosaic virus Protein 1a recruits viral RNA2 to RNA replication through a 5′ proximal RNA2 signal. J. Virol. 75:3207–3219.

Chen, J., Noueiry, A., and Ahlquist, P. 2003. An alternate pathway for recruiting template RNA to the brome mosaic virus RNA replication complex. J. Virol. 77:2568–2577.

Choi, S.K., Hema, M., Gopinath, K., Santos, J., and Kao, C. 2004. Replicase-binding sites on plus- and minus-strand brome mosaic virus RNAs and their roles in RNA replication in plant cells. J. Virol. 78:13420–13429.

de Groot, R.J., van der Most, R.G., and Spaan, W.J. 1992. The fitness of defective interfering murine coronavirus DI-a and its derivatives is decreased by nonsense and frameshift mutations. J. Virol. 66:5898–5905.

den Boon, J.A., Chen, J., and Ahlquist, P. 2001. Identification of sequences in Brome mosaic virus replicase protein 1a that mediate association with endoplasmic reticulum membranes. J. Virol. 75:12370–12381.

Diez, J., Ishikawa, M., Kaido, M., and Ahlquist, P. 2000. Identification and characterization of a host protein required for efficient template selection in viral RNA replication. Proc. Natl Acad. Sci. USA 97:3913–3918.

Dinant, S., Janda, M., Kroner, P., and Ahlquist, P. 1993. Bromovirus RNA replication and transcription require compatibility between the polymerase- and helicase-like viral RNA synthesis proteins. J. Virol. 67:7181–7189.

Dohi, K., Mise, K., Furusawa, I., and OKuno, T. 2002. RNA-dependent RNA polymerase complex of brome mosaic virus: analysis of the molecular structure with monoclonal antibodies. J. Gen. Virol. 83:2879–2890.

Dreher, T.W., and Hall, T.C. 1988a. Mutational analysis of the sequence and structural requirements in brome mosaic virus RNA for minus strand promoter activity. J. Mol. Biol. 201:31–40.

Dreher, T.W., and Hall, T.C. 1988b. Mutational analysis of the tRNA mimicry of brome mosaic virus RNA. Sequence and structural requirements for aminoacylation and 3′-adenylation. J. Mol. Biol. 201:41–55.

Dzianott, A.M., and Bujarski, J.J. 1991. The nucleotide sequence and genome organization of the RNA-1 segment in two bromoviruses: broad bean mottle virus and cowpea chlorotic mottle virus. Virology 185:553–562.

Felden, B., Florentz, C., Giege, R., and Westhof, E. 1994. Solution structure of the 3′-end of brome mosaic virus genomic RNAs. Conformational mimicry with canonical tRNAs. J. Mol. Biol. 235:508–531.

Felden, B., Florentz, C., Westhof, E., and Giege, R. 1993. Non-canonical substrates of aminoacyl-tRNA synthetases: the tRNA-like structure of brome mosaic virus genomic RNA. Biochimie 75:1143–1157.

Felden, B., Florentz, C., Westhof, E., and Giege, R. 1996. Usefulness of functional and structural solution data for the modeling of tRNA-like structures. Pharm. Acta. Helv. 71:3–9.

French, R., and Ahlquist, P. 1987. Intercistronic as well as terminal sequence are required for efficient amplification of brome mosaic virus RNA3. J. Virol. 61:1457–1465.

French, R., and Ahlquist, P. 1988. Characterization and engineering of sequences controlling in vivo synthesis of brome mosaic virus subgenomic RNA. J. Virol. 62:2411–2420.

Giege, R. 1996. Interplay of tRNA-like structures from plant viral RNAs with partners of the translation and replication machineries. Proc. Natl Acad. Sci. USA 93:12078–12081.

Gopinath, K., Dragnea, B., and Kao, C. 2005. Interaction between Brome mosaic virus proteins and RNAs: effects on RNA replication, protein expression, and RNA stability. J. Virol. 79:14222–14234.

Haasnoot, P.C., Brederode, F.T., Olsthoorn, R.C., and Bol, J.F. 2000. A conserved hairpin structure in Alfamovirus and Bromovirus subgenomic promoters is required for efficient RNA synthesis in vitro. RNA 6:708–716.

Hardy, S.F., German, T.L., Loesch-Fries, L.S., and Hall, T.C. 1979. Highly active template-specific RNA-dependent RNA polymerase from barley leaves infected with brome mosaic virus. Proc. Natl Acad. Sci. USA 76:4956–4960.

Hema, M., Gopinath, K., and Kao, C. 2005. Distinct requirements for the repairs of the 3′ terminal sequence of Brome Mosaic Virus: a special role for RNA1. J. Virol. 79:1417–1427.

Ishikawa, M., Diez, J., Restrepo-Hartwig, M., and Ahlquist, P. 1997. Yeast mutations in multiple complementation groups inhibit brome mosaic virus RNA replication and transcription and perturb regulated expression of the viral polymerase-like gene. Proc. Natl Acad. Sci. USA 94:13810–13815.

Janda, M., and Ahlquist, P. 1993. RNA-dependent replication, transcription, and persistence of brome mosaic virus RNA replicons in *Saccharomyces cerevisiae*. Cell 72:961–970.

Janda, M., and Ahlquist, P. 1998. Brome mosaic virus RNA replication protein 1a dramatically increases in vivo stability but not translation of viral genomic RNA3. Proc. Natl Acad. Sci. USA 95:2227–2232.

Jayaram, H., Estes, M.K., and Prasad, B.V. 2004. Emerging themes in rotavirus cell entry, genome organization, transcription and replication. Virus Res. 101:67–81.

Kao, C.C., and Ahlquist, P. 1992. Identification of the domains required for direct interaction of the helicase-like and polymerase-like RNA replication proteins of brome mosaic virus. J. Virol. 66:7293–7302.

Kao, C., and Sun, J. 1996. The RNA-dependent RNA polymerase of a (+)-strand RNA virus uses oligoribonucleotide primers to initiate (–)-strand RNA synthesis. J. Virol. 70:6826–6830.

Kao, C.C., Singh, P., and Ecker, D.J. 2001. De novo initiation of viral RNA-dependent RNA synthesis. Virology 287:251–260.

Kao, C., Quadt, R., Hershberger, R., and Ahlquist, P. 1992. Brome mosaic virus RNA replication proteins 1a and 2a form a complex in vitro. J. Virol. 66:6322–6329.

Khromykh, A.A., Varnavski, A.N., Sedlak, P.L., and Westaway, E.G. 2001. Coupling between replication and packaging of flavivirus RNA: evidence derived from the use of DNA-based full-length cDNA clones of Kunjin virus. J. Virol. 75:4633–4640.

Kim, C.H., and Kao, C.C. 2001. A mutant viral RNA promoter with an altered conformation retains efficient recognition by a viral RNA replicase through a solution-exposed adenine. RNA 7:1476–1485.

Kim, C.H., and Tinoco, I. Jr. 2001. Structural and thermodynamic studies on mutant RNA motifs that impair the specificity between a viral replicase and its promoter. J. Mol. Biol. 307: 827–839.

Kim, C.H., Kao, C.C, and Tinoco, I. Jr. 2000. RNA motifs that determine specificity between a viral replicase and its promoter. Nat. Struct. Biol. 7:415–423.

Kim, M.J., Zhong, W., Hong, Z., and Kao, C.C. 2000. Template nucleotide moieties required for de novo initiation of RNA synthesis by a recombinant viral RNA-dependent RNA polymerase. J. Virol. 74:10312–10322.

Kong, F., Sivakumaran, K., and Kao, C.C. 1999. The N-terminal half of the brome mosaic virus 1a protein as RNA capping-associated activities: specificity for GTP S-adenosylmethionine. Virology 259:200–210.

Kumar, J.K., Tabor, S., and Richardson, C.C. 2001. Role of the C-terminal residue of the DNA polymerase of bacteriophage T7. J. Biol. Chem. 276:34905–34912.

Kushner, D.B., Lindenbach, B.D., Grdzelishvili, V.Z., Noueiry, A.O., Paul, S.M., and Ahlquist, P. 2003. Systematic, genome-wide identification of host genes affecting replication of a positive-strand RNA virus. Proc. Natl Acad. Sci. USA 100:15764–15769.

Lai, M.M.C. 1998. Cellular factors in the transcription and replication of viral RNA genomes: a parallel to DNA-dependent RNA transcription. Virology 244:1–12.

Lee, W.M., Ishikawa, M., and Ahlquist, P. 2001. Mutation of host delta9 fatty acid desaturase inhibits brome mosaic virus RNA replication between template recognition and RNA synthesis. J. Virol. 75:2097–2106.

Marsh, L.E., and Hall, T.C. 1987. Evidence implicating a tRNA heritage for the promoters of positive-strand RNA synthesis in brome mosaic virus and related viruses. Cold Spring Harb. Symp. Quant. Biol. 52:331–341.

Marsh, L.E., Dreher, T.W., and Hall, T.C. 1988. Mutational analysis of the core and modulator sequences of the BMV RNA3 subgenomic promoter. Nucleic Acids Res. 16:981–995.

Marsh, L., Huntley, C.C., Pogue, G.P., Connell, J., and Hall, T.C. 1991. Regulation of (+):(–)-strand asymmetry in replication of brome mosaic virus RNA. Virology 182:76–83.

Mas, A., Alves-Rodrigues, I., Noueiry, A., Ahlquist, P., and Diez, J. 2006. Host deadenylation-dependent mRNA decapping factors are required for a key step in brome mosaic virus RNA replication. J. Virol. 80:246–251.

Meylan, E., and Tschopp, J. 2006. Toll-like receptors and RNA helicases: two parallel ways to trigger antiviral responses. Mol. Cell 22:561–569.

Miller, W.A., Bujarski, J.J., Dreher, T.W., and Hall, T.C. 1986. Minus-strand initiation by brome mosaic virus replicase within the 3′tRNA-like structure of native and modified RNA templates. J. Mol. Biol. 187:537–546.

Miller, D.J., Schwartz, M.D., Dye, B.T., and Ahlquist, P. 2003. Engineered retargeting of viral RNA replication complexes to an alternative intracellular membrane. J. Virol. 77:12193–12202.

Noueiry, A.O., and Ahlquist, P. 2003. Brome mosaic virus RNA replication: revealing the role of the host in RNA virus replication. Annu. Rev. Phytopathol. 41:77–98.

Noueiry, A.O., Chen, J., and Ahlquist, P. 2000. A mutant allele of essential, general translation initiation factor DED1 selectively inhibits translation of a viral mRNA. Proc. Natl Acad. Sci. USA 97:12985–12990.

Noueiry, A.O., Diez, J., Falk, S.P., Chen, J., and Ahlquist, P. 2003. Yeast Lsm1p-7p/Pat1p deadenylation-dependent mRNA-decapping factors are required for brome mosaic virus genomic RNA translation. Mol. Cell. Biol. 23:4096–4106.

Novak, J.E., and Kirkegaard, K. 1994. Coupling between genome translation and replication in an RNA virus. Genes Dev. 14:1726–1737.

Nugent, C.I., Johnson, K.L., Sarnow, P., and Kirkegaard, K. 1999. Functional coupling between replication and packaging of poliovirus replicon RNA. J. Virol. 73:427–435.

O'Reilly, E., and Kao, C. 1998. Analysis of the structure and function of viral RNA-dependent RNA polymerases as guided by computer-assisted structure predictions and the known structures of polymerases. Virology 252:287–303.

O'Reilly, E., Paul, J., and Kao, C. 1997. Analysis of the interaction of viral RNA replication proteins by using the yeast two-hybrid assay. J. Virol. 71:7526–7532.

O'Reilly, E., Tang, N., Ahlquist, P., and Kao, C. 1995. Biochemical and genetic analyses of the interaction between the helicase-like and polymerase-like proteins of the brome mosaic virus. J. Virol. 214:59–71.

O'Reilly, E., Wang, Z., French, R., and Kao, C. 1998. Interaction between the structural domains of the RNA replication proteins of plant-infecting RNA viruses. J. Virol. 72:7160–7169.

Perret, V., Florentz, C., Dreher, T., and Giege, R. 1989. Structural analogies between the 3'tRNA-like structure of brome mosaic virus RNA and yeast tRNATyr revealed by protection studies with yeast tyrosyl-tRNA synthetase. Eur. J. Biochem. 185:331–339.

Picard, D., Kao, C., and Hudak, K.A. 2005. Pokeweed antiviral protein inhibits brome mosaic virus replication in plant cells. J. Biol. Chem. 280:20069–20075.

Pogue, G., and Hall, T.C. 1992. The requirement for a 5′ stem-loop structure in brome mosaic virus replication supports a new model for viral positive strand RNA synthesis. J. Virol. 66:674–684.

Quadt, R., Ishikawa, M., Janda, M., and Ahlquist, P. 1995. Formation of brome mosaic virus RNA-dependent RNA polymerase in yeast requires coexpression of viral proteins and viral RNA. Proc. Natl Acad. Sci. USA 92:4892–4896.

Quadt, R., Kao, C.C., Browning, K.S., Hershberger, R.P., and Ahlquist, P. 1993. Characterization of a host protein associated with brome mosaic virus RNA-dependent RNA polymerase. Proc. Natl Acad. Sci. USA 90:1498–1502.

Quadt, R., Verbeek, H.J., and Jaspars, E.M. 1988. Involvement of a nonstructural protein in the RNA synthesis of brome mosaic virus. Virology 165:256–261.

Ranjith-Kumar, C.T., Santos, J.L., Gutshall, L.L., Johnston, V.K., Lin-Goerke, J., Kim, M.J., Porter, D.J., Maley, D., Greenwood, C., Earnshw, D.L., Baker, A., Gu, B., Silverman, C., Sarisky, R.T., and Kao, C. 2003. Enzymatic activities of the GB virus-B RNA-dependent RNA polymerase. Virology 312:270–280.

Rao, A.L., and Hall, T.C. 1993. Recombination and polymerase error facilitate restoration of infectivity in brome mosaic virus. J. Virol. 67:969–979.

Restrepo-Hartwig, M., and Ahlquist, P. 1996. Brome mosaic virus helicase- and polymerase-like proteins colocalize on the endoplasmic reticulum at sites of viral RNA synthesis. J. Virol. 70:8908–8916.

Restrepo-Hartwig, M., and Ahlquist, P. 1999. Brome mosaic virus RNA replication proteins 1a and 2a colocalize and 1a independently localizes on the yeast endoplasmic reticulum. J. Virol. 73:10303–10309.

Rietveld, K., Pleij, C.W., and Bosch, L. 1983. Three-dimensional models of the tRNA-like 3' termini of some plant viral RNAs. EMBO J. 2:1079–1085.

Romero, J., Dzianott, A.M., and Bujarski, J.J. 1992. The nucleotide sequence and genome organization of the RNA2 and RNA3 segments in broad bean mottle virus. Virolgy 187: 671–681.

Schaad, M.C., Jensen, P.E., and Carrington, J.C. 1997. Formation of plant RNA virus replication complexes on membranes: role of an endoplasmic reticulum-targeted viral protein. EMBO J. 16:4049–4059.

Schwartz, M., Chen, J., Janda, M., Sullivan, M., den Boon, J., and Ahlquist, P. 2002. A positive-strand RNA virus replication complex parallels form and function of retrovirus capsids. Mol. Cell. 9:505–514.

Schwartz, M., Chen, J., Lee, W.M., Janda, M., and Ahlquist, P. 2004. Alternate, virus-induced membrane rearrangements support positive-strand RNA virus genome replication. Proc. Natl Acad. Sci. USA, 101:11263–11268.

Shih, D.S., and Kaesberg, P. 1973. Translation of brome mosaic virus ribonucleic acid in a cell free system derived from wheat embryo. Proc. Natl Acad. Sci. USA 70:1799–1803.

Siegel, R., Adkins, S., and Kao, C. 1997. Sequence-specific recognition of an RNA promoter by a viral RNA polymerase. Proc. Natl Acad. Sci. USA 94:11238–11243.

Siegel, R., Bellon, L., Beigelman, L., and Kao, C. 1998. Identification of the base moieties in an RNA promoter specifically recognized by a viral RNA-dependent RNA polymerase. Proc. Natl Acad. Sci. USA 95:11613–11618.

Sivakumaran, K., and Kao, C. 1999. Initiation of genomic (+)-strand RNA synthesis from DNA and RNA templates by a viral RNA-dependent RNA polymerase. J. Virol. 73:6415–6423.

Sivakumaran, S., and Kao, C. 2000. Initiation of brome mosaic virus genomic RNA synthesis requires a correctly positioned box B. Molecular Plant Pathology 1(6):331–340.

Sivakumaran, K., Bao, Y., Roossinck, M.J., and Kao, C.C. 2000. Recognition of the core RNA promoter for minus-strand RNA synthesis by the replicase of Brome mosaic virus and Cucumber mosaic virus. J. Virol. 74:10323–10321.

Sivakumaran, K., Choi, S.K., Hema, M., and Kao, C.C. 2004. Requirements for brome mosaic virus subgenomic RNA synthesis in vivo and replicase-core promoter interactions in vitro. J. Virol. 78:6091–6101.

Sivakumaran, K., Kim, C.H., Tayon, R., and Kao, C. 1999. RNA sequence and structural determinants for the recognition and efficiency of RNA synthesis by a viral RNA replicase. J. Mol. Biol. 294:667–682.

Stawicki, S., and Kao, C. 1999. Spatial requirements for promoter recognition by a viral RNA-dependent RNA polymerase. J. Virol. 73:198–204.

Sullivan, M., and Ahlquist, P. 1997. Cis-acting signals in bromovirus RNA replication and gene expression: networking with viral proteins and host factors. Semin. Virol. 8:221–230.

Sullivan, M.L., and Ahlquist, P. 1999. A brome mosaic virus intergenic RNA3 replication signal functions with viral replication protein 1a to dramatically stabilize RNA in vivo. J. Virol. 73:2622–2632.

Sun, J., and Kao, C. 1997a. Characterization of RNA products associated with or aborted by a viral RNA-dependent RNA polymerase. Virology 236:348–353.

Sun, J., and Kao, C. 1997b. RNA synthesis by the brome mosaic virus RNA-dependent RNA polymerase: transition from initiation to elongation. Virology 233:63–73.

Sun, J., Adkins, S., Faurote, G., and Kao, C. 1996. Initiation of (–)-strand RNA synthesis catalyzed by the brome mosaic virus RNA-dependent RNA polymerase: synthesis of oligonucleotides. Virology 226:1–12.

Sun, J., DuFort, C., Daniel, M.C., Murali, A., Chen, C., Gopinath, K., Stein, B., De M Rotell, V.M., Holzenburg, A., and Kao, C.C. 2007. Core-controlled polymorphism in virus-like particles. Proc. Natl Acad. Sci. USA 104:1354–1359.

Tayon, R. Jr., Kim, M.-J., and Kao, C. 2001. Factors regulating the completion of RNA synthesis by viral RNA replicases. Nucleic Acids Res. 29:3583–3594.

Tomita, Y., Mizuno, T., Diez, J., Naito, S., Ahlquist, P., and Ishikawa, M. 2003. Mutation of host DnaJ homolog inhibits brome mosaic virus negative-strand RNA synthesis. J. Virol. 77:2990–2997

Traynor, P., Young, B.M., and Ahlquist, P. 1991. Deletion analysis of brome mosaic virus 2a protein: effects on RNA replication and systemic spread. J. Virol. 65:2807–2815.

Venter, P.A., Krishna, N.K., and Schneemann, A. 2005. Capsid protein synthesis from replicating RNA directs specific packaging of the genome of a multipartite, positive-strand RNA virus. J. Virol. 79:6239–6248.

Vovkova, E., Gorchakov, R., and Frolov, I. 2006. The efficient packaging of Venezuelan equine encephalitis virus-specific RNAs into viral particles is determined by nsp1-3 synthesis. Virology 344:315–327.

Wang, X., Lee, W.M., Watanabe, T., Schwartz, M., Janda, M., and Ahlquist, P. 2005. Brome mosaic virus 1a nucleoside triphosphatase/helicase domain plays crucial roles in recruiting RNA replication templates. J. Virol. 79:13747–13758.

Weiland, J.J., and Dreher, T.W. 1993. Cis-preferential replication of the turnip yellow mosaic virus RNA genome. Proc. Natl. Acad. Sci. USA 90:6095–6099.

Wierzchoslawski, R., and Bujarski, J.J. 2006. Efficient in vitro system of homologous recombination in brome mosaic bromovirus. J. Virol. 80:6182–6187.

Williamson, J.R. 2000. Induced fit in RNA-protein recognition. Nat. Struct. Biol. 7:834–837.

Wu, G., and Kaper, J.M. 1994. Requirement of 3′-terminal guanosine in (–)-stranded RNA for in vitro replication of cucumber mosaic virus satellite RNA by viral RNA-dependent RNA polymerase. J. Mol. Biol. 238:655–657.

Yi, G., Gopinath, K., and Kao, C.C. 2007. Selective repression of translation by the brome mosaic virus 1a RNA replication protein. J. Virol. 81:1601–1609.

Zhu, J., Gopinath, K., Murali, A., Yi, G., Hayward, S.D., Zhu, H., and Kao, C. 2007. RNA-binding proteins that inhibit RNA virus infection. Proc. Natl Acad. Sci. USA 104:3129–3134.

Chapter 6
Retroviruses

Román Galetto and Matteo Negroni

Abbreviations HIV, Human immunodeficiency virus; HTLV, Human T-lymphotropic virus; MMTV, Mouse mammary tumor virus; MLV, Murine leukemia viruses; MoMLV, Moloney murine leukemia virus; ORF, Open reading frame; SIV, Simian immunodeficiency virus

Morphology and Taxonomy

Retroviruses are a large group of enveloped RNA viruses infecting vertebrates. The viral particles are spherical and acquire their envelope during budding from the infected cell. The lipid bilayer therefore contains cellular proteins as well as the viral envelope glycoproteins. These glycoproteins are constituted by a transmembrane subunit (TM) associated to the surface protein (SU), present on the virion. Underneath the membrane is a spherical shell constituted by the matrix (MA) protein. Internally is the viral capsid, whose shape varies in different viruses, constituted by the CA protein. This core contains the retroviral enzymes (the reverse transcriptase, RT, the integrase, IN, and the protease, PR), together with the genomic RNA, coated by the nucleocapsid protein (NC).

Retroviruses have been historically divided into four groups on the basis of morphological criteria, through visualization of the virion core by electron microscopy. A-type viruses form characteristic intracellular structures (spheres with an electron-lucent centre and an electron-dense shield), while B- and C-type viruses contain a round inner core located eccentrically or in the middle of the particle, respectively. D-type viruses, in contrast, contain a distinctive cylindrical core. More recently, a new classification based on phylogenetic analysis grouped all retroviruses in seven genera within the Retroviridae family. According to this classification,

M. Negroni (✉)
Unité de Régulation Enzymatique des Activités Cellulaires, CNRS-URA 2185, Institut Pasteur, Paris, France; Present address: Architecture et reactivité de l'ARN, Université de Strasbourg, CNRS, IBMC, 15 rue René Descartes, 67084, Strasbourg, France
e-mail: m.negroni@ibmc.u-strasbg.fr

C.E. Cameron et al. (eds.), *Viral Genome Replication*,
DOI 10.1007/b135974_6,

Table 6.1 Retroviruses genera

Simple retroviruses	
Alpharetrovirus	Rous sarcoma virus (RSV)
	Avian leukosis virus (ALV)
Betaretrovirus	Mouse mammary tumor virus (MMTV)
	Mason-Pfizer monkey virus (M-PMV)
Gammaretrovirus	Murine leukemia viruses (MLV)
	Feline leukemia virus (FeLV)
Complex retroviruses	
Deltaretrovirus	Human T-lymphotropic virus 1 and 2 (HTLV-1, -2)
	Bovine leukemia virus (BLV)
Epsilonretrovirus	Walleye dermal sarcoma virus
	Walleye epidermal hyperplasia virus 1
Lentivirus	Human immunodeficiency virus 1 and 2 (HIV-1, -2)
	Simian immunodeficiency virus (SIV)
Spumavirus	Human foamy virus
	Simian foamy virus

The seven genera of retroviruses are listed, divided into single or complex retroviruses according to the proteins encoded by their genome (either Gag, Pol and Env, for simple retroviruses, or these same proteins plus regulatory proteins for complex retroviruses). Two representative members of each genus are cited as examples.

the alpha, beta and gammaretroviruses are *simple* retroviruses, while deltaretroviruses, epsilonretroviruses, lentiviruses and spumaviruses are considered *complex* (Table 6.1). Simple viruses contain only three main genes (*gag*, *pol* and *env*, see below), whereas complex viruses also encode other small proteins with regulatory functions. In addition, certain genetics elements such as endogenous retroviruses and retrotransposons are closely related to retroviruses. These elements, regarded indeed as defective forms of retroviruses, can spread within the cellular genome by reverse transcription in a similar way as retroviral replication but are (generally for endogenous retroviruses, and always, for retrotransposons) not capable of an extracellular phase. They are thereby normally not horizontally transmissible.

The retroviral genome is constituted by a dimer of two copies of a single-stranded RNA molecule of positive polarity. Indeed, a unique feature of the Retroviridae family is to comprise the only viruses that can be considered as "diploid". The genomic RNA is generated by the host transcriptional machinery and, therefore, is capped in 5′, and has a poly(A) tail at the 3′-end, as any cellular RNA. The two molecules are linked by non-covalent interactions near their 5′-end. The size of each monomer varies from 7 to 13 kb, depending on the virus, and each molecule contains three major coding domains: *gag*, for group-specific antigens; *pol*, for polymerase; and *env*, for envelope. These genes form precursor polyproteins that are then processed after viral assembly, yielding the structural proteins and the enzymes present in the mature infectious particles. Accessory genes (present only in complex retroviruses) are essentially located downstream from *pol* (Fig. 6.1), and their products mostly regulate transcription of viral DNA, splicing and transport of RNA, among other

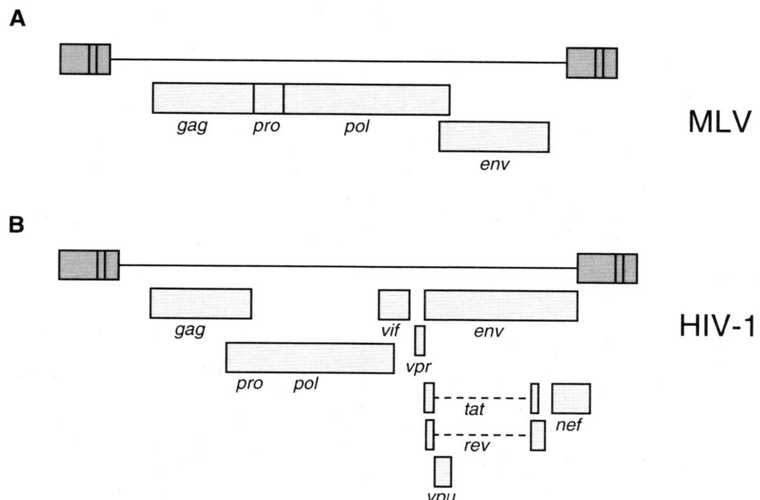

Fig. 6.1 Organization of the genomes of a simple (**panel A**) and of a complex (**panel B**) retrovirus. **Panel A** shows the arrangements in the genome of MLV. In **panel B**, the genetic organization of HIV-1 is represented. ORFs are shown as *white boxes*. ORFs on the same reading frame are on the same line, while those in different frames are represented in different lines. *Dashed lines* indicate spliced introns. *Grey boxes* represent the LTRs.

specific functions. As an example, the Tax and Tat proteins, from HTLV and HIV respectively, have a role in transcriptional activation of the viral promoter; the Rex and Rev proteins, from the same viruses, have a role in nuclear export of full-length and single spliced viral RNAs; the Vif protein of HIV has the capacity to increase infectivity by blocking the action of inhibitory proteins present in certain cell types.

Overview of the Life Cycle

The replication cycle of retroviruses is constituted by a series of steps that, after transferring genetic information from RNA to DNA molecules, leads to the establishment of a persistent infection subsequent to integration of the proviral DNA into the host cells (Fig. 6.2). To initiate infection, retroviruses interact with specific receptors on the surface of target cells by means of the envelope proteins present on the outside of the viral membrane. After the initial binding, the envelope proteins are subjected to conformational changes that lead to the fusion of viral and cell membranes and the subsequent release of the viral core into the cytoplasm. The early events following penetration are, to date, poorly understood. While in some cases it is proposed that uncoating of the viral core occurs after internalization, there are other indications suggesting that this capsid can remain intact, allowing reverse transcription to occur in this confined environment (for a review see Nisole and Saib 2004). Upon penetration of the viral capsid in the cytoplasm of

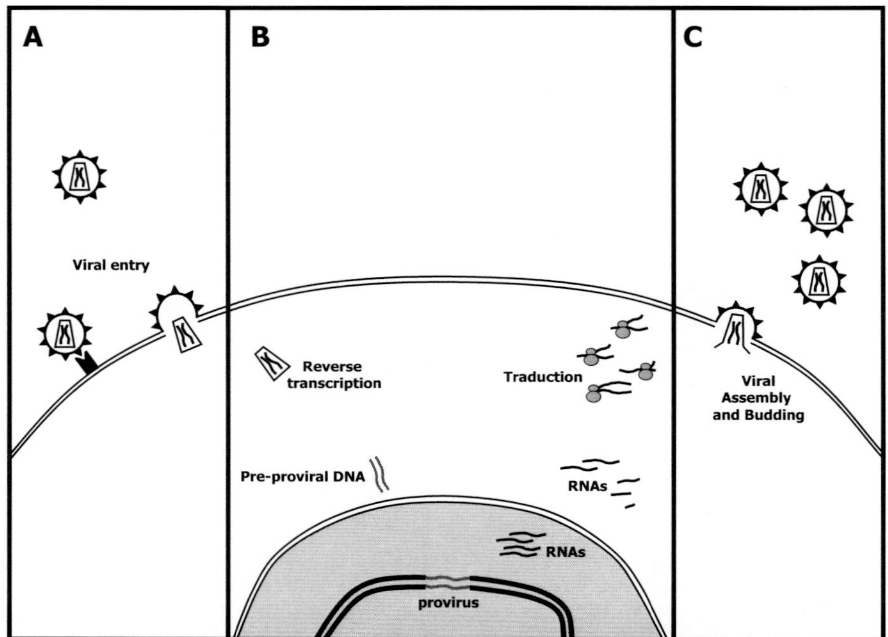

Fig. 6.2 Outline of the infectious cycle of retroviruses. **Panel A**: First steps of the early phase of the replication cycle. Attachment of the virus and binding to the membrane receptors present in susceptible host cells occurs through the Env viral proteins. This interaction leads to conformational changes in the Env proteins that facilitate fusion of cellular and viral membranes, delivering the viral capsid into the cytoplasm. **Panel B**: Reverse transcription takes place in the cytoplasm and yields a double-stranded DNA molecule that integrates in the host genome, generating a provirus. The late stages of the life cycle include the expression of viral RNA from the provirus. Some of these RNAs are spliced and exported to the cytoplasm together with unspliced RNAs. The unspliced RNA serves both as genomic RNA and for synthesis of the Gag and Gag-Pol polyproteins. Spliced forms are used to make Env and the regulatory proteins in complex retroviruses. **Panel C**: Assembly of the viral proteins and encapsidation of the genomic RNAs are represented, leading to the formation of the viral progeny.

the target cell, reverse transcription (detailed in the next section) begins and leads to the generation of double-stranded DNA from the single-stranded RNA genome. This characteristic step has given the name to this family of viruses. The resulting DNA molecule must then enter the nucleus in order to integrate in the host genome, giving rise to a "provirus" that will be permanently established in the host genome. The late phase of the replication cycle, which mostly relies on the cellular machinery, takes place after this integration step. The viral RNA is expressed from its promoter, located in the U3 region, and transcriptional regulation is controlled by viral as well as host transcription factors. In all cases a full-length transcript is generated, corresponding to the full-length viral genomic RNA. However, during the early phase of expression of the proviral DNA, the transcript is processed to give rise to a series of sub-genomic forms that are used for translation. Later, the balance between full-length RNAs and spliced forms is turned in favour of the farmer (late phase) and leads to export from the nucleus of the genomic RNA that will be packaged into the nascent viral particles.

Organization of Retroviral Genomic RNAs

In addition to the sequences coding for enzymatic, structural and regulatory proteins, retroviral genomic RNAs contain a series of sequences that have important functions in different steps of the life cycle. The 5′ portion of the genome contains an untranslated region (5′ UTR) that includes various sequences required for viral replication. This region contains, at its 5′-end, the R sequence (for *repeated*, since another copy of this sequence is present at the 3′-end). R is essential for translocation of the nascent DNA from the 5′- to the 3′-end of the genome during synthesis of the (–) DNA strand (see below). Following the 5′ R lies the U5 region (for unique 5′ sequence) that includes one of the trinucleotide sequences ATT, required for proviral integration. Downstream is the primer-binding site (PBS), a 18 nt long sequence where the cellular tRNA anneals to initiate reverse transcription. The region following the PBS is constituted by the dimer linkage structure (DLS) that contains the sequences used for dimerization and packaging of the RNA in the viral particle. Adjacent to the DLS, on its 3′-end, are located the sequences coding for the viral proteins. These genes are then followed by a short sequence rich in purine residues (polypurine tract, PPT), required for the initiation of the (+) DNA strand. The PPT is followed by the U3 (for unique 3′ sequence), which contains the other ATT sequence required for integration, as well as the regulatory elements necessary for transcription of the integrated provirus. Behind this sequence lies the 3′ copy of R, followed by the poly(A) tail. In several lentiviruses an additional PPT sequence is found in a central position of the genome (central PPT, cPPT), and it is used to prime second strand synthesis.

Genome Replication: From Single-Stranded Genomic RNA to Double-Stranded DNA

From tRNA to (–) DNA Strong Stop Strand Transfer

Beginning of the viral replication cycle can be fixed as the moment when, after binding to the cellular receptor and fusion of cellular and viral membranes, the retroviral core enters the cytoplasm of the target cell. DNA synthesis is set off by the availability of nucleotides, within the cytoplasm of the host cell, to which the viral core is permeable. This allows the reverse transcriptase to begin polymerization of the (–) DNA strand. Nucleotide addition is primed by a tRNA molecule, partially unwound and annealed to the 18 nt of the PBS on the genomic RNA (Fig. 6.3A). The specific tRNA used varies in different retroviruses and is carried along by the viral particle from the previously infected cell. Most retroviruses contain a pool of

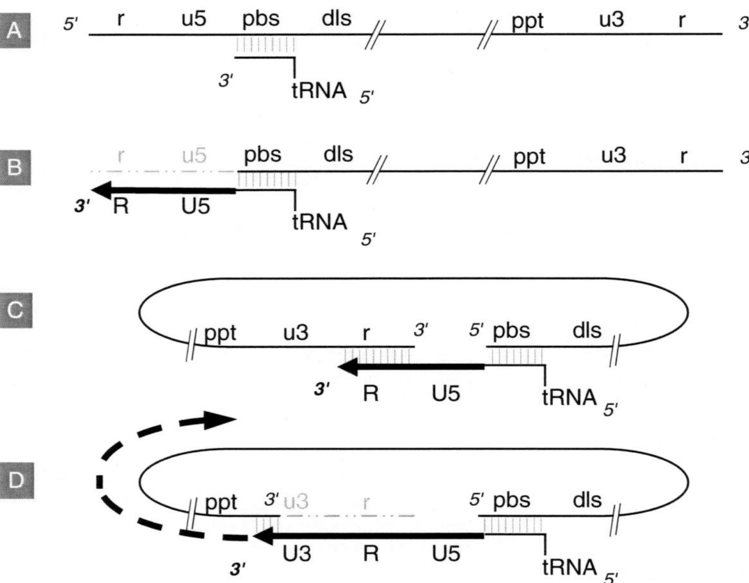

Fig. 6.3 Beginning of reverse transcription and minus DNA strong stop strand transfer. *Black thin lines*: RNA; *black thick lines*: DNA; *gray dotted lines*: RNA degraded by the RT-encoded RNase H activity; *gray vertical lines* indicate the presence of a region where the nucleic acids are annealed; PBS: primer-binding site; DLS: dimer linkage structure, a region involved in RNA dimerization and packaging; PPT: polypurine tract. Regions on the RNA are indicated by lowercase, on the DNA by uppercase. **Panel A**: Outline of the structure of the genomic RNA with tRNA partially unwound and annealed at the primer-binding site. The polarity of the RNA molecules is given. **Panel B**: Synthesis primed using the 3′-OH of the tRNA proceeds through the U5 and R regions and is stopped at the end of the template. The *arrowhead* gives the direction of synthesis and the polarity of the nascent DNA is given in *bold*. In the figure is shown the situation where transfer occurs only after achievement of a complete copy of the tRNA sequence (considered the most frequent situation). **Panel C**: The R sequence at the 3′ of the genomic RNA anneals onto its newly synthesized complementary sequence on the DNA, generating a circular intermediate. **Panel D**: This circularization makes the 3′-OH of the nascent DNA available for continuing DNA synthesis across the U3 region, first, and the rest of the genome then (*dotted thick line*).

tRNAs enriched in the specific tRNA used for priming reverse transcription through an interaction between the tRNA and the Pol domain of the Gag-Pol polyprotein (Khorchid et al. 2000; Peters and Hu 1980). Correct placement of the tRNA on the genomic RNA and its partial unwinding are instead assisted by the NC domain of the Gag precursor (Cen et al. 2000). Selection for the appropriate tRNA is very accurate since, for most viruses, the canonical tRNA cannot be replaced by an alternative one. This strict requirement is due not only to the presence on the genomic RNA of a sequence complementary to that of the 18 nt at the 3′-end of the specific tRNA, but also to the occurrence of additional interactions between the genomic RNA and different regions of the tRNA. The regions involved in these interactions vary with the virus considered.

The tRNA-PBS complex is recognized by the RT and used to prime DNA synthesis. The structural requirements for the specific recognition of this complex have been well characterized for HIV. The specificity of the interactions between the genomic RNA and the tRNA is required, in this case, to avoid steric clashes between the RT and the nucleic acids (Isel et al. 1999). Initiation of reverse transcription using the tRNA-PBS complex has been shown to be functionally distinct from further DNA synthesis when the nascent DNA is used as a primer (elongation step). Both efficient beginning of reverse transcription and transition to an elongation complex require the presence of the modified nucleosides of the $tRNA_3^{Lys}$, the one used by HIV, further underlining the specificity of the complex used for beginning reverse transcription in vivo (Isel et al. 1996).

Once escaped from this initiation complex, the RT travels a few hundreds of nts before reaching the 5′-end of the genomic RNA, where synthesis is abruptly interrupted (Fig. 6.3B). To achieve copying of the remaining part of the genomic RNA (around 95% of the full size) synthesis must be transferred at the 3′-end of the molecule. This process, known as minus DNA strong stop strand transfer, is made possible essentially by two factors: the presence of a repeated sequence (R) at the ends of the genome, and the presence of an RNase H activity in the RT that degrades the genomic RNA once this is part of the heteroduplex formed with the nascent DNA strand. This degradation does not need to be coupled to DNA synthesis, since successful transfer can be observed in viruses that carry an RNase H^- mutant RT, if complemented by mutants of RT that possess a functional RNase H domain but are deprived of a polymerase activity (Telesnitsky and Goff 1993). Degradation of the template RNA can therefore be performed in a polymerase-independent manner and leads to the progressive degradation of the residual R sequence till when the shortened heteroduplex becomes instable and melts. This generates a single-stranded DNA region carrying the sequence complementary to the R sequence that can anneal to the R sequence located at the 3′-end of the genome (Fig. 6.3B). Once annealing has occurred (Fig. 6.3C), DNA synthesis can be resumed and proceed across the internal regions of the genomic RNA (Fig. 6.3D).

Minus DNA strong stop strand transfer has been studied in great detail both in infected cells as well as in reconstituted in vitro systems. Central to this process is the sequence R. This sequence provides homology between the 5′- and the 3′-end of the genome but also contains important signals for viral replication. For instance, in HIV, R contains two stable hairpins, the transactivation response element (TAR) and the polyadenylation signal, involved in transcription and RNA processing. The size of R varies with the virus considered, spanning from the 15 nt of the MMTV to the 247 of the HTLV-2. The analysis, in cell culture, of the effects of shortening R or its replacement by a heterologous sequence, is far from being straightforward, essentially due to the multiple functions of R that make difficult to set apart effects on template switching, from those on reverse transcription, transcription and RNA processing. Despite these difficulties, through combining results obtained after infection of cells in culture with those obtained from in vitro reconstituted reverse transcription reactions using purified components, it has been possible to address the issue of the role of R in strand transfer.

Deletions of R have been shown to affect strand transfer to different degrees in different viruses, and depending on the portion of R deleted. Overall, it appears that reducing the size of R does not necessarily impair significantly the process. For MLV, reducing the size from the 69 nt of the full-length sequence to 12 nt retained comparable infectious titres in single cycle infection assays (Dang and Hu 2001). The same was true when the 3′ R sequence of HIV was reduced to 30 nt (from the 96 of the wild-type sequence) (Berkhout et al. 1995). For SIV, the reduction of infectivity was strongly dependent on the position of the deletion (Brandt et al. 2006).

Replacing R with heterologous sequences has indicated that the role of R in strand transfer goes beyond that of providing a mere support for transferring DNA synthesis from one end of the genomic RNA to the other. In MLV, replacement of R by non-viral sequences reduced viral titre approximately by a fivefold factor (Cheslock et al. 2000). Extensive in vitro reconstituted template switching assays have allowed identifying structural motifs that favour transfer by permitting long distance interactions between portions of the genomic RNA, between the genomic RNA and the tRNA or between the nascent DNA and the genomic RNA (Berkhout et al. 2001; Brule et al. 2000). Finally, the presence of hairpins as the TAR and the polyadenylation hairpins for HIV has been suggested to enhance strand transfer by favouring template switching from internal positions of R (Moumen et al. 2001). Indeed, even if the majority of template switching events seem to occur once synthesis has reached the 5′-end of the genomic RNA, the occurrence of a basal level of strand transfer from internal positions has been described for HIV, SIV and MoMLV (Klaver and Berkhout 1994; Kulpa et al. 1997; Lobel and Goff 1985; Ramsey and Panganiban 1993). It has been proposed that such internal transfer events could allow the virus to bypass the problem of addition of non-templated residues that occurs when DNA synthesis stalls at the 5′-end of the genome. These residues can, in fact, hamper resumption of DNA synthesis after transfer of the nascent DNA at the 3′-end of the genome, if they do not match the sequence present in that position.

In conclusion, the whole transfer process is driven by a delicate equilibrium between the stability of the nucleic acids reactants at different steps of the process that regulate timing and efficiency of melting and annealing of these components. The viral RNA chaperon protein NC, present on the viral RNA during reverse transcription, has been demonstrated to modulate most of the steps that lead to successful strand transfer (Levin et al. 2005).

Synthesis of (–) DNA Across the Genome

RTs are quite slow polymerases. Based on in vitro assays using purified reverse transcriptases, the rates of nucleotide incorporation appeared to vary dramatically, according to the conditions employed. Data from endogenous reverse transcription reactions with various retroviruses were instead more consistent, with estimates that ranged between 0.5 and 6 nt per second (Boone and Skalka 1981; Kung et al. 1981; O'Brien et al. 1994; Rothenberg and Baltimore 1977). Results coherent with these estimates were also obtained using purified HIV-1 or MoMLV RTs (2–0.5 nt/s) when the generation of long DNA molecules was monitored (Negroni and Buc 1999) or when measuring reverse transcription rates in HIV-1 infected cells in culture (Thomas et al. 2007).

A hallmark of synthesis of the (–) DNA strand in most retroviruses is the frequent occurrence of template switching between the two copies of genomic RNA from internal positions of the genome. The requirement for the RT to successfully carry out the mandatory strand transfer during (–) strong stop strand transfer (previous section) has probably selected enzymes that are prone to switch template even when copying internal positions of the RNA. This process, known as copy choice, has been intensively investigated during the last decades. The importance of recombination in the evolution of this group of viruses has been known for a long time, with its involvement in important issues as the uptake of exogenous genes to generate the oncoretroviruses, or the generation of viruses with new characteristics through recombination with endogenous retroviruses. The detection of such events was, however, not indicative of how frequently recombination occurs during the retroviral infectious cycle. It was only with the development of systems that allowed estimating the frequency of generation of template switching events during a single infectious cycle that such an estimate could be obtained. It is now known that reverse transcription generates a chimeric DNA at a frequency ranging between 1 and 3×10^{-4} per nucleotide reverse transcribed, depending on the retrovirus considered (Negroni and Buc 2001).

The involvement of template switching in the replication strategy itself was postulated decades ago with the proposal of the process of *forced copy choice* (Coffin 1979). Based on the observation that fragmented RNA was frequently recovered from retroviral genomic RNA preparations, it was proposed that recombination provides a means for achieving full-length reverse transcription through template switching, whenever reverse transcription encounters a break (Fig. 6.4A). In vitro evidence has subsequently shown that template switching could be obtained with purified components in reconstituted reverse transcription reactions even in the absence of RNA breaks (*copy choice*). Pausing of DNA synthesis has been proposed, in conceptual analogy to the presence of breaks during reverse transcription, to promote copy choice (Fig. 6.4B). The role of stalling of reverse transcription would be to increase the extent of degradation of the template RNA by the RNase H activity, thereby increasing the probability of annealing between the nascent DNA and the "acceptor" RNA (the one onto which DNA synthesis is transferred, whereas the one where synthesis was started is defined as the "donor" RNA) (DeStefano et al. 1994, 1992). Another important parameter for copy choice has been shown to be constituted by the presence of RNA hairpins (Fig. 6.4C). The importance of these hairpins seems crucial on the acceptor RNA (Moumen et al. 2003), and the existence of a window of optimal stability for these structures to efficiently promote copy choice has been observed in cell culture (Galetto et al. 2004). The role of these structures would be to favour the exchange of the template annealed at the 5′-end of the nascent DNA, from the donor onto the acceptor template. The presence of the stem of the hairpin on the acceptor RNA would provide a double-stranded structure

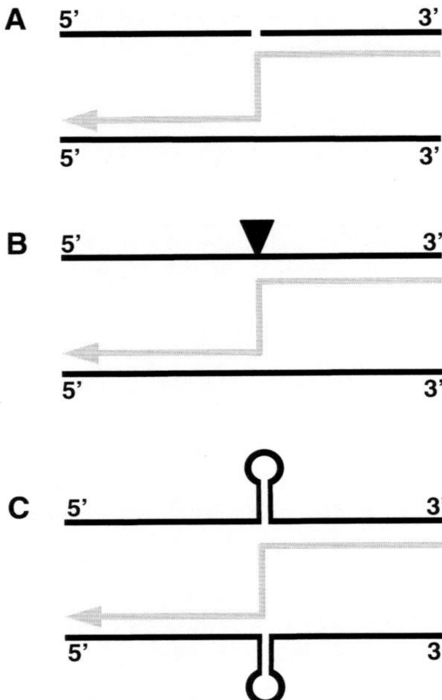

Fig. 6.4 Recombination during synthesis of minus DNA strand (copy choice). Retroviral genomic RNA molecules are represented as *black lines*, and the newly synthesized DNA molecules by *grey arrows*. The *arrow* indicates the direction of synthesis. **Panel A**: Forced copy-choice model for recombination: breaks on the genomic RNA force the reverse transcriptase to transfer synthesis onto to the second genomic RNA molecule. **Panel B**: Pause-induced copy-choice model: stalling of the reverse transcriptase at a pause site (*black triangle*) would induce jumping of the reverse transcriptase. **Panel C**: RNA secondary structure-induced copy-choice model. The *hairpin* indicates a secondary structure on the genomic RNA. Even if this is represented in the acceptor and the donor RNA, its presence in the acceptor molecule appears more important to promote recombination.

that allows strand exchange with the heteroduplex constituted by the nascent DNA and the donor RNA, following a mechanism reminiscent of branch migration in Holiday junctions (Galetto et al. 2006). As for (–) DNA strong stop strand transfer, also copy choice is modulated by the NC protein (for a review see Levin et al. 2005), as judged by experiments in reconstituted in vitro systems.

Frequent template switching has obvious consequences for viral population dynamics and evolution since, if the two RNAs diverge, the generation of a DNA by copying partially each genomic RNA results in genetic recombination. Copackaging of two different RNAs depends on the probability of coinfection of the same cell by at least two different viruses and on the probability of dimerization of their respective genomic RNAs. The frequency at which such events are fulfilled is expected to vary in different retroviruses. For instance, for MoMLV, even when two different genomic RNAs are produced in the same cell, the formation of homozygous viruses is more likely than that of heterozygous particles (Onafuwa et al. 2003). In this case, although template switching occurs at rates comparable to those of HIV, the lower abundance of heterozygous viruses leads to lower rates of genetic recombination for MoMLV than for HIV (Onafuwa et al. 2003). High recombination rates could serve to combine positive mutations, to remove deleterious mutations generated during reverse transcription or, most likely, at both aims with a relative importance of one aspect or the other that might change with the retrovirus considered (also see Chapter 30 on these issues). An implication of the occurrence of frequent recombination on the retroviral replication strategy concerns whether only one or both genomic RNAs are converted into DNA, an issue for which no conclusive evidence has been provided yet. Indeed, even if both RNAs could, in principle, be fully copied in each viral particle, the high frequency of template switching from one genomic RNA to the other suggests that only one genomic DNA molecule is generated. This, at least for retroviruses yielding a high recombination rate.

Synthesis of the (+) DNA Strand and Completion of (–) DNA Strand

Conversion of the genomic RNA into double-stranded DNA is achieved by synthesis of the second (or *plus*) DNA strand by the viral reverse transcriptase itself that, in addition to its RNA-dependent DNA polymerase activity, also possesses a DNA-dependent DNA polymerase activity. Plus DNA synthesis begins while first strand synthesis is still ongoing. As mentioned above, during synthesis of the (–) DNA strand, the genomic RNA is degraded by the RT-encoded RNase H activity, which can act as an exonuclease as well as an endonuclease. Degradation of the genomic RNA is most likely not complete, although the extent at which fragments of the genomic RNA resist to cleavage and remain temporarily associated to the (–) DNA strand has not been conclusively established. These fragments can be used to initiate reverse transcription at multiple sites. Accordingly, several foci where second strand synthesis is started have been identified (Klarmann et al. 1997; Miller et al. 1995; Thomas et al. 2007), suggesting that synthesis of the second DNA strand can be a discontinuous process. Regardless the occurrence of multiple priming, full-length synthesis of the (+) DNA strand always results from extension of the PPT sequence, located immediately 5′ of the U3 sequence (Fig. 6.5A), which is temporarily resistant to cleavage by the RNase H. The mechanisms leading to selection of the correct primer for (+) DNA-strand synthesis are detailed in Chapter 19. Synthesis primed at the PPT will then proceed across the U3, R and U5 sequence (Fig. 6.5B), and will continue, unwinding the RNA duplex on the PBS sequence, along the tRNA until it will be stopped by the presence of a modified nucleotide in the tRNA, immediately downstream of the region complementary to the PBS sequence (Fig. 6.5C). Modified bases are present at this position in all tRNA used as primers by retroviruses (Marquet et al. 1995). The copied portion of the tRNA is then unwound, either intact or after partial cleavage by the RNase H activity (Omer and Faras 1982; Schultz et al. 1995; Smith et al. 1997), becoming single stranded. At the same time, the counterpart of the PBS sequence on the genomic RNA is now available to serve as template for the achievement of (–) DNA-strand synthesis (Fig. 6.5D). As a result, the PBS

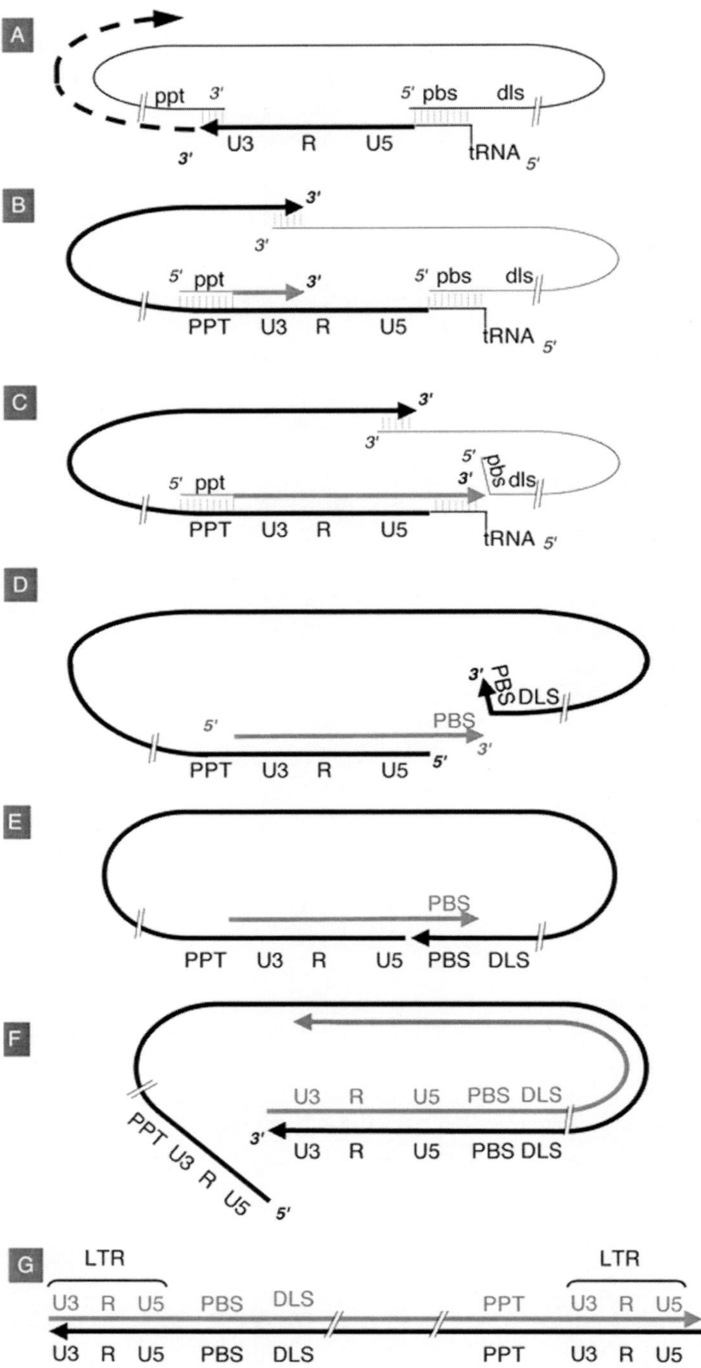

Fig. 6.5 (continued)

sequence has been copied into DNA on the tRNA as well as on the genomic RNA, generating complementary DNA sequences. Annealing between the two resulting sequences is therefore possible, generating a circular DNA intermediate (Fig. 6.5E). This process constitutes the (+) DNA strand transfer. Synthesis of (+) DNA can then proceed through the internal parts of the genome (Fig. 6.5F). Concomitantly, synthesis of (–) DNA proceeds, by strand displacement synthesis along the U5, R and finally the U3 region (Fig. 6.5F). After copying this sequence, synthesis is stopped by the absence of template, since the RNA PPT sequence has eventually been degraded by the RNase H (panel D). As for (–) DNA strong stop strand transfer, also this complex series of unwinding and annealing events is assisted all the way through by the RNA chaperone protein NC (for an exhaustive review on this topic, see Levin et al. 2005). Finally this process results in the generation of a complete double-stranded pre-proviral DNA, competent for integration into the host's genome (Fig. 6.5F).

In several lentiviruses, an additional second DNA synthesis begins from another cleavage-resistant purine-rich sequence, located near the middle of the RNA genome, the cPPT. This concomitant DNA synthesis leads to the generation of a discontinuous (+) DNA strand that is constituted by two parts, partially overlapped. The structure including the overlapped part is called central FLAP and it has been suggested to be involved in transport of genomic DNA from the cytoplasm into the nucleus (Zennou et al. 2000).

Integration

Once the double-stranded final product of reverse transcription is generated, pre-proviral DNA must be integrated in the genome of the infected cell, a process mediated by the retroviral integrase. In order for the pre-proviral DNA to gain access

Fig. 6.5 Plus-strand DNA strand transfer and completion of reverse transcription. Symbols and abbreviations are as in Fig. 6.3. **Panel A** presents the situation given in **panel D** of Fig. 6.3, without drawing the degraded RNA (*dotted grey line* in Fig. 6.3). **Panel B**: Synthesis of (–) DNA proceeds and concomitantly synthesis of (+) DNA (*grey thick line*) begins using as primer the polypurine tract (ppt), which has not been degraded by the RNase H activity of RT. For clarity, pairing of (+) and (–) DNA strands is indicated simply by drawing the two strands running parallel. **Panel C**: Synthesis of both DNA strands progress, and synthesis of plus-DNA strand (*grey*) proceeds along the first 18 nt of the tRNA before being arrested by the presence of a modified nucleotide on the tRNA (see text), and displacing the genomic RNA on the PBS sequence. **Panel D**: Minus-DNA synthesis proceeds across the DLS sequence and copies the PBS region of the genomic RNA. Copied RNA sequences are degraded (for simplicity also the PPT is indicated to be degraded at this moment). **Panel E**: Complementary PBS sequences on the two DNA strands pair. For clarity, starting from this panel the polarity of the DNA strands is simply given by the direction of the *arrows* (5′–3′). **Panel F**: Both syntheses progress, (–) DNA synthesis (*black*) displacing the strand ahead, which is the tail of the same (–) DNA strand. **Panel G**: (+) DNA strand synthesis is completed, generating the full-length structure of pre-proviral DNA ready for integration (the portions constituting the LTRs are indicated).

to the cellular chromosomes, certain retroviruses require disruption of the nuclear membrane or cell-cycle progression of their target cells (reviewed in Yamashita and Emerman 2006). Some simple retroviruses, as gammaretroviruses, need the infected cell to undergo mitosis, likely because they necessitate the nuclear membrane to be disrupted, while alpharetroviruses integrate in resting cells but *require* cell-cycle progression for replication. In the case of lentiviruses, the preintegration complexes can successfully enter the nucleus of non-dividing cells through an active transport of the viral DNA across an intact nuclear membrane, likely mediated by cellular and viral factors. Both ends of the DNA molecules generated after reverse transcription are constituted by the LTRs, which contain the sequences recognized by the integrase. Integration occurs in two steps: in a first instance, removal of two nucleotides from the 3' extremities of the pre-proviral DNA occurs, while in the second step, the 3′OH groups attack the target DNA and, through a strand transfer process, these are ligated to the nicked host DNA (see Chapter 21 for a comprehensive description of the integration process). This process generates a short duplication of sequences flanking the proviral DNA, whose size will depend on the distance at which the attacks were done by the integrase on the host DNA. This distance varies from 4 to 6 nt depending on the retrovirus (for a review see Lewinski and Bushman 2005).

Integration in the DNA of the infected cell does not occur at specific sequences. However, preferences for integration into actively transcribed regions have been reported (Schroder et al. 2002; Wu et al. 2003), as well as weak, but statistically significant preferences for symmetric target sequences (Holman and Coffin 2005; Wu et al. 2005). Nevertheless, different viruses show preferences for different sequences (Mitchell et al. 2004).

Replication of the Viral Genome: From Double-Stranded DNA to RNA

Transcription and RNA Processing

Once the provirus is established, the cis-acting elements that regulate viral transcription, contained in the U3 region of the provirus, lead to synthesis of full-length genomic RNA from the 5′ LTR region. Although most retroviral promoters are efficient enough to constitutively yield high levels of RNA, this transcriptional efficiency will depend on the infected cell type, the availability of the required transcription factors, as well as on the site of integration in the cellular genome (Feinstein et al. 1982). The U3 region contains the core promoter as well as enhancer sequences. The promoter harbours a TATA box and a CCAAT box, which bind TFIIB and CEBP respectively, as any eukaryotic promoter. Various retroviruses also use also enhancers containing recognition motifs for ubiquitous transcription factors, such as Sp1, Nuclear Factor 1 (NF1) or the Ets family of factors; while, others more specific, either tissue-specific or ligand-dependent activators, can modulate transcription in certain retroviruses, as is the case of the glucocorticoid receptors by MMTV (Archer et al. 1992), or NF-kB in the case of HIV (Nabel and Baltimore 1987). On the other hand, certain cell types can produce proteins that have a negative effect on transcription mediated by the LTR. These phenomena have been mostly described for MoMLV, where a protein from embryonic cells can bind to the LTR and inhibit initiation of transcription (Tsukiyama et al. 1990), or a stem-cell factor that binds to the tRNA recognition site (Loh et al. 1990; Petersen et al. 1991).

The genome of complex retroviruses encode viral transcriptional activators, like the protein Tax for HTLV or Tat for HIV which are, by far, the better characterized among retroviruses. In the case of Tax, it will *trans*-activate the RNA transcription by binding specific sequences in the LTR, in conjunction with cellular proteins, and creating a positive feedback loop that will boost transcription (Bex and Gaynor 1998; Yoshida 1994). The Tat protein from HIV binds, instead, to a hairpin structure in the 5′-end of the nascent RNA, creating a complex with host proteins that increases the capability of the RNA polymerase to elongate across the whole provirus (Dingwall et al. 1990; Sharp and Marciniak 1989).

After transcription reaches the 3′ LTR, it can continue on the flanking sequences of the host genome. Nevertheless, the transcripts are cleaved and polyadenylated at the R-U5 border of the LTR, giving rise to a stable unspliced RNA molecule that is exported to the cytoplasm to serve as genomic RNA for the viral progeny and to be used for translation of the Gag and Gag-Pol polyproteins. All the same, a portion of the RNA pool is spliced before being exported from the nucleus, generating subgenomic-sized messenger RNAs. In simple retroviruses a single spliced mRNA encodes the Env glycoprotein, while complex retroviruses yield multiple spliced mRNAs encoding the Env glycoprotein and a variety of auxiliary proteins (Fig. 6.1).

The number of multiple spliced mRNAs and the complexity of the splicing patterns stand for a highly regulated process. Normally, cellular nascent RNAs are processed before nuclear export, in a way that only mature mRNAs will be present in the cytoplasm to lead gene expression. This contrasts the need of retroviruses to avoid the uncontrolled complete splicing of their RNAs. To circumvent this problem, retroviruses use splice sites that do not match consensus splice and donor sequences, therefore lacking efficiency in promoting splicing, and allowing the appearance of all the subgenomic population of mRNAs, that will differ in the number of splicing events and the choice of alternative splice sites (Katz and Skalka 1990). As mentioned before, in the late phases of the infectious cycle, also a completely unspliced RNA must be exported from the nucleus: the genomic RNA. This problem has been intensively studied particularly for HIV, for which export is obtained through the interaction of a highly structured sequence (the Rev responsive element) with the viral protein Rev (Malim et al. 1989), which mediates the interaction with cellular proteins involved in the nuclear export pathway (Strebel 2003).

Translation

Once in the cytoplasm, translation takes place generating precursor proteins that will be processed during and after viral assembly to generate the mature infectious particles. One characteristic of retroviruses is that multiple proteins are encoded by the same ORF, which constitutes an advantage for viral assembly since polyproteins are targeted at the same moment to the site of assembly, also ensuring that the different proteins will be present at proper ratios after processing.

The Gag and Pol proteins are translated from the full-length viral RNA. The Gag polyprotein comprises the MA, CA and NC proteins from N to C terminus. The N terminus of the protein is hydrophobic in order to drive this protein to the cell membrane. In some retroviruses, the N terminus is hydrophobic enough, but in most retroviruses a myristyl residue is attached to augment the hydrophobicity of the protein. The PR (protease) is expressed in different ways in different retroviruses. In some cases it is expressed alone, from a single ORF, while in others it can be fused either to the 3′ terminus of Gag or, in most cases, to the 5′ extremity of Pol, that yields the RT and IN proteins.

Since the amount of proteins with catalytic functions required by the virus is much less than the amount of structural proteins, retroviruses have developed a process by which they modulate the relative abundance of the different proteins. In fact, the full-length RNA will yield either the Gag polyprotein or an even larger polyprotein containing Gag-Pol. The latter polyprotein results from a process of translational read-through that takes place in almost 10% of translations, in which the stop codon for the Gag protein (present, i.e. between Gag and Pol) is misread as a sense codon, allowing translation to proceed and leading to the synthesis of Pol fused to Gag (Yoshinaka et al. 1985). In other cases, when the proteins are in different ORFs, a process of translational frameshifting, in which the ribosomes slips one nucleotide backwards, is used to solve this problem (Jacks and Varmus 1985). These polyproteins are then processed to yield individual proteins (see below).

The *env* gene is instead expressed from a different subgenomic RNA. The protein is directed to the rough endoplasmic reticulum by a hydrophobic signal peptide, which is then removed by a cellular protease, and the protein is heavily glycosylated (Einfeld 1996). It folds and oligomerizes in the endoplasmic reticulum, prior to being imported to the Golgi apparatus, where it is cleaved by cellular furin proteases to form the SU and TM subunits that remain together through non-covalent bonds. The proteins are then transported to the cell membrane for recruitment during budding.

Late Phases of the Infectious Cycle

The Gag precursor protein drives assembly of the retroviral particle, since its presence is sufficient to generate virus-like particles that resemble immature ones, except for the absence of genomic RNA and Env proteins. Most retrovirus assemble at the plasma membrane, though for others assembly takes place in the cytoplasm and they are then transported to the plasma membrane where they acquire the envelope during budding. Three domains of the Gag precursor mediate assembly: The M domain (for membrane-binding) is located in the MA region, at the N-terminus of Gag, and it is required for those retroviruses that assemble at the cell membrane. The I domain (for interaction) is present in the CA and NC portions of the precursor and is necessary for Gag–Gag interactions. The L, or late domain, is necessary for efficient release of viral particles from the cell, as well as for interaction with cellular partners involved in protein sorting. Together with the Gag precursor, Gag-Pol polyproteins will also be incorporated in viral particles, but 10–20 times less abundantly than the Gag precursor. The genomic RNA, which dimerizes prior to encapsidation, is incorporated into the viral particle through interaction between the NC portion of the Gag precursor and the packaging signal contained in the genomic RNA (Paillart et al. 2004). The host tRNA primer, needed to prime first strand DNA synthesis, is also packaged at this stage.

During and after budding of the viral particles from the surface of the infected cells, the Gag and Gag-Pol precursor proteins are cleaved by the viral protease to release the individual proteins present in the infectious virus (Vogt 1996). Precursors are not cleaved until they are assembled, since PR is a homodimer and requires assembly for activation. This maturation process leads to morphological changes in the viral core, which, after being released from the cells, appears as a more dense structure, detached from the viral envelope. The mature particle is thereby generated and can begin a new infectious cycle.

References

Archer TK, Lefebvre P, Wolford RG and Hager GL. (1992) Transcription factor loading on the MMTV promoter: a bimodal mechanism for promoter activation. *Science*, **255**, 1573–6.

Berkhout B, van Wamel J and Klaver B. (1995) Requirements for DNA strand transfer during reverse transcription in mutant HIV-1 virions. *J Mol Biol*, **252**, 59–69.

Berkhout B, Vastenhouw NL, Klasens BI and Huthoff H. (2001) Structural features in the HIV-1 repeat region facilitate strand transfer during reverse transcription. *RNA*, **7**, 1097–114.

Bex F and Gaynor RB. (1998) Regulation of gene expression by HTLV-I Tax protein. *Methods*, **16**, 83–94.

Boone LR and Skalka AM. (1981) Viral DNA synthesized in vitro by avian retrovirus particles permeabilized with melittin. I. Kinetics of synthesis and size of minus- and plus-strand transcripts. *J Virol*, **37**, 109–16.

Brandt S, Grunwald T, Lucke S, Stang A and Uberla K. (2006) Functional replacement of the R region of simian immunodeficiency virus-based vectors by heterologous elements. *J Gen Virol*, **87**, 2297–307.

Brule F, Bec G, Keith G, Le Grice SF, Roques BP, Ehresmann B, Ehresmann C and Marquet R. (2000) In vitro evidence for the interaction of $tRNA_3^{Lys}$ with U3 during the first strand transfer of HIV-1 reverse transcription. *Nucleic Acids Res*, **28**, 634–40.

Cen S, Khorchid A, Gabor J, Rong L, Wainberg MA and Kleiman L. (2000) Roles of Pr55(gag) and NCp7 in $tRNA_3^{Lys}$ genomic placement and the initiation step of reverse transcription in human immunodeficiency virus type 1. *J Virol*, **74**, 10796–800.

Cheslock SR, Anderson JA, Hwang CK, Pathak VK and Hu WS. (2000) Utilization of nonviral sequences for minus-strand DNA transfer and gene reconstitution during retroviral replication. *J Virol*, **74**, 9571–9.

Coffin JM. (1979) Structure, replication, and recombination of retrovirus genomes: some unifying hypotheses. *J Gen Virol*, **42**, 1–26.

Dang Q and Hu WS. (2001) Effects of homology length in the repeat region on minus-strand DNA transfer and retroviral replication. *J Virol*, **75**, 809–20.

DeStefano JJ, Bambara RA and Fay PJ. (1994) The mechanism of human immunodeficiency virus reverse transcriptase-catalyzed strand transfer from internal regions of heteropolymeric RNA templates. *J Biol Chem*, **269**, 161–8.

DeStefano JJ, Mallaber LM, Rodriguez-Rodriguez L, Fay PJ and Bambara RA. (1992) Requirements for strand transfer between internal regions of heteropolymer templates by human immunodeficiency virus reverse transcriptase. *J Virol*, **66**, 6370–8.

Dingwall C, Ernberg I, Gait MJ, Green SM, Heaphy S, Karn J, Lowe AD, Singh M and Skinner MA. (1990) HIV-1 tat protein stimulates transcription by binding to a U-rich bulge in the stem of the TAR RNA structure. *EMBO J*, **9**, 4145–53.

Einfeld D. (1996) Maturation and assembly of retroviral glycoproteins. *Curr Top Microbiol Immunol*, **214**, 133–76.

Feinstein SC, Ross SR and Yamamoto KR. (1982) Chromosomal position effects determine transcriptional potential of integrated mammary tumor virus DNA. *J Mol Biol*, **156**, 549–65.

Galetto R, Giacomoni V, Veron M and Negroni M. (2006) Dissection of a circumscribed recombination hot spot in HIV-1 after a single infectious cycle. *J Biol Chem*, **281**, 2711–20.

Galetto R, Moumen A, Giacomoni V, Veron M, Charneau P and Negroni M. (2004) The structure of HIV-1 genomic RNA in the gp120 gene determines a recombination hot spot in vivo. *J Biol Chem*, **279**, 36625–32.

Holman AG and Coffin JM. (2005) Symmetrical base preferences surrounding HIV-1, avian sarcoma/leukosis virus, and murine leukemia virus integration sites. *Proc Natl Acad Sci U S A*, **102**, 6103–7.

Isel C, Lanchy JM, Le Grice SF, Ehresmann C, Ehresmann B and Marquet R. (1996) Specific initiation and switch to elongation of human immunodeficiency virus type 1 reverse transcription require the post-transcriptional modifications of primer $tRNA_3^{Lys}$. *EMBO J*, **15**, 917–24.

Isel C, Westhof E, Massire C, Le Grice SF, Ehresmann B, Ehresmann C and Marquet R. (1999) Structural basis for the specificity of the initiation of HIV-1 reverse transcription. *EMBO J*, **18**, 1038–48.

Jacks T and Varmus HE. (1985) Expression of the Rous sarcoma virus pol gene by ribosomal frameshifting. *Science*, **230**, 1237–42.

Katz RA and Skalka AM. (1990) Control of retroviral RNA splicing through maintenance of suboptimal processing signals. *Mol Cell Biol*, **10**, 696–704.

Khorchid A, Javanbakht H, Wise S, Halwani R, Parniak MA, Wainberg MA and Kleiman L. (2000) Sequences within Pr160gag-pol affecting the selective packaging of primer $tRNA_3^{Lys}$ into HIV-1. *J Mol Biol*, **299**, 17–26.

Klarmann GJ, Yu H, Chen X, Dougherty JP and Preston BD. (1997) Discontinuous plus-strand DNA synthesis in human immunodeficiency virus type 1-infected cells and in a partially reconstituted cell-free system. *J Virol*, **71**, 9259–69.

Klaver B and Berkhout B. (1994) Premature strand transfer by the HIV-1 reverse transcriptase during strong-stop DNA synthesis. *Nucleic Acids Res*, **22**, 137–44.

Kulpa D, Topping R and Telesnitsky A. (1997) Determination of the site of first strand transfer during Moloney murine leukemia virus reverse transcription and identification of strand transfer-associated reverse transcriptase errors. *EMBO J*, **16**, 856–65.

Kung HJ, Fung YK, Majors JE, Bishop JM and Varmus HE. (1981) Synthesis of plus strands of retroviral DNA in cells infected with avian sarcoma virus and mouse mammary tumor virus. *J Virol*, **37**, 127–38.

Levin JG, Guo J, Rouzina I and Musier-Forsyth K. (2005) Nucleic acid chaperone activity of HIV-1 nucleocapsid protein: critical role in reverse transcription and molecular mechanism. *Prog Nucleic Acid Res Mol Biol*, **80**, 217–86.

Lewinski MK and Bushman FD. (2005) Retroviral DNA integration – mechanism and consequences. *Adv Genet*, **55**, 147–81.

Lobel LI and Goff SP. (1985) Reverse transcription of retroviral genomes: mutations in the terminal repeat sequences. *J Virol*, **53**, 447–55.

Loh TP, Sievert LL and Scott RW. (1990) Evidence for a stem cell-specific repressor of Moloney murine leukemia virus expression in embryonal carcinoma cells. *Mol Cell Biol*, **10**, 4045–57.

Malim MH, Hauber J, Le SY, Maizel JV and Cullen BR. (1989) The HIV-1 rev trans-activator acts through a structured target sequence to activate nuclear export of unspliced viral mRNA. *Nature*, **338**, 254–7.

Marquet R, Isel C, Ehresmann C and Ehresmann B. (1995) tRNAs as primer of reverse transcriptases. *Biochimie*, **77**, 113–24.

Miller MD, Wang B and Bushman FD. (1995) Human immunodeficiency virus type 1 preintegration complexes containing discontinuous plus strands are competent to integrate in vitro. *J Virol*, **69**, 3938–44.

Mitchell RS, Beitzel BF, Schroder AR, Shinn P, Chen H, Berry CC, Ecker JR and Bushman FD. (2004) Retroviral DNA integration: ASLV, HIV, and MLV show distinct target site preferences. *PLoS Biol*, **2**, E234.

Moumen A, Polomack L, Roques B, Buc H and Negroni M. (2001) The HIV-1 repeated sequence R as a robust hot-spot for copy-choice recombination. *Nucleic Acids Res*, **29**, 3814–21.

Moumen A, Polomack L, Unge T, Veron M, Buc H and Negroni M. (2003) Evidence for a mechanism of recombination during reverse transcription dependent on the structure of the acceptor RNA. *J Biol Chem*, **278**, 15973–82.

Nabel G and Baltimore D. (1987) An inducible transcription factor activates expression of human immunodeficiency virus in T cells. *Nature*, **326**, 711–3.

Negroni M and Buc H. (1999) Recombination during reverse transcription: an evaluation of the role of the nucleocapsid protein. *J Mol Biol*, **286**, 15–31.

Negroni M and Buc H. (2001) Mechanisms of retroviral recombination. *Annu Rev Genet*, **35**, 275–302.

Nisole S and Saib A. (2004) Early steps of retrovirus replicative cycle. *Retrovirology*, **1**, 9.

O'Brien WA, Namazi A, Kalhor H, Mao SH, Zack JA and Chen IS. (1994) Kinetics of human immunodeficiency virus type 1 reverse transcription in blood mononuclear phagocytes are slowed by limitations of nucleotide precursors. *J Virol*, **68**, 1258–63.

Omer CA and Faras AJ. (1982) Mechanism of release of the avian retrovirus $tRNA^{Trp}$ primer molecule from viral DNA by ribonuclease H during reverse transcription. *Cell*, **30**, 797–805.

Onafuwa A, An W, Robson ND and Telesnitsky A. (2003) Human immunodeficiency virus type 1 genetic recombination is more frequent than that of Moloney murine leukemia virus despite similar template switching rates. *J Virol*, **77**, 4577–87.

Paillart JC, Shehu-Xhilaga M, Marquet R and Mak J. (2004) Dimerization of retroviral RNA genomes: an inseparable pair. *Nat Rev Microbiol*, **2**, 461–72.

Peters GG and Hu J. (1980) Reverse transcriptase as the major determinant for selective packaging of tRNA's into Avian sarcoma virus particles. *J Virol*, **36**, 692–700.

Petersen R, Kempler G and Barklis E. (1991) A stem cell-specific silencer in the primer-binding site of a retrovirus. *Mol Cell Biol*, **11**, 1214–21.

Ramsey CA and Panganiban AT. (1993) Replication of the retroviral terminal repeat sequence during in vivo reverse transcription. *J Virol*, **67**, 4114–21.

Rothenberg E and Baltimore D. (1977) Increased length of DNA made by virions of murine leukemia virus at limiting magnesium ion concentration. *J Virol*, **21**, 168–78.

Schroder AR, Shinn P, Chen H, Berry C, Ecker JR and Bushman F. (2002) HIV-1 integration in the human genome favors active genes and local hotspots. *Cell*, **110**, 521–9.

Schultz SJ, Whiting SH and Champoux JJ. (1995) Cleavage specificities of Moloney murine leukemia virus RNase H implicated in the second strand transfer during reverse transcription. *J Biol Chem*, **270**, 24135–45.

Sharp PA and Marciniak RA. (1989) HIV TAR: an RNA enhancer? *Cell*, **59**, 229–30.

Smith CM, Potts WB, 3rd, Smith JS and Roth MJ. (1997) RNase H cleavage of $tRNA^{Pro}$ mediated by M-MuLV and HIV-1 reverse transcriptases. *Virology*, **229**, 437–46.

Strebel K. (2003) Virus-host interactions: role of HIV proteins Vif, Tat, and Rev. *Aids*, **17 Suppl 4**, S25–34.

Telesnitsky A and Goff SP. (1993) Two defective forms of reverse transcriptase can complement to restore retroviral infectivity. *EMBO J*, **12**, 4433–8.

Thomas DC, Voronin YA, Nikolenko GN, Chen J, Hu WS and Pathak VK. (2007) Determination of the ex vivo rates of human immunodeficiency virus type 1 reverse transcription by using novel strand-specific amplification analysis. *J Virol*, **81**, 4798–807.

Tsukiyama T, Niwa O and Yokoro K. (1990) Characterization of the negative regulatory element of the 5′ noncoding region of Moloney murine leukemia virus in mouse embryonal carcinoma cells. *Virology*, **177**, 772–6.

Vogt VM. (1996) Proteolytic processing and particle maturation. *Curr Top Microbiol Immunol*, **214**, 95–131.

Wu X, Li Y, Crise B and Burgess SM. (2003) Transcription start regions in the human genome are favored targets for MLV integration. *Science*, **300**, 1749–51.

Wu X, Li Y, Crise B, Burgess SM and Munroe DJ. (2005) Weak palindromic consensus sequences are a common feature found at the integration target sites of many retroviruses. *J Virol*, **79**, 5211–4.

Yamashita M and Emerman M (2006) Retroviral infection of non-dividing cells: old and new perspectives. Virology, **344**, 88–93.

Yoshida M. (1994) Mechanism of transcriptional activation of viral and cellular genes by oncogenic protein of HTLV-1. *Leukemia*, **8 Suppl 1**, S51–3.

Yoshinaka Y, Katoh I, Copeland TD and Oroszlan S. (1985) Murine leukemia virus protease is encoded by the gag-pol gene and is synthesized through suppression of an amber termination codon. *Proc Natl Acad Sci U S A*, **82**, 1618–22.

Zennou V, Petit C, Guetard D, Nerhbass U, Montagnier L and Charneau P. (2000) HIV-1 genome nuclear import is mediated by a central DNA flap. *Cell*, **101**, 173–85.

Chapter 7
Hepadnaviral Genomic Replication

John E. Tavis and Matthew P. Badtke

The Hepadnaviruses

Hepadnaviruses are small, enveloped, hepatotropic DNA viruses. Despite having a DNA genome in the mature virions, they are replicated by reverse transcription of an RNA intermediate (for a comprehensive review see Seeger et al., 2007). Hepadnaviruses infect humans [Hepatitis B virus (HBV)], some primates (wooly monkeys and chimpanzees), rodents (squirrels and woodchucks), and birds (ducks, geese, and storks) (Sprengel et al., 1988; Schoedel et al., 1989; Lanford et al., 1998; Prassolov et al., 2003). The human pathogen HBV chronically infects 350–400 million people world-wide and kills approximately 1 million people each year (Ganem and Prince, 2004). Significant differences exist among the hepadnaviruses, but they are all highly hepatotropic, establish chronic infections, follow the same replication cycle, and have nearly identical genetic organizations. The hepadnaviral genomic replication mechanism has been investigated primarily with HBV and Duck Hepatitis B virus (DHBV).

Hepadnaviral Genomic Structure

Hepadnaviral virion-derived DNA is circular and partially double stranded (Fig. 7.1). The genome is very small, ~3200 bp for the mammalian viruses and ~3000 bp for the avian viruses. The minus-polarity DNA strand is slightly longer than the length of the viral genome, and its 5′ end is covalently attached to the viral reverse transcriptase (P). The plus-polarity DNA strand has a short, capped RNA oligomer at its 5′ end and circularizes the viral DNA by annealing to both the 3′ and the 5′ ends of the minus-polarity DNA strand. The plus-polarity DNA is

J.E. Tavis (✉)
Molecular Microbiology and Immunology, St Louis University School of Medicine, 1100 South Grand Blvd. St Louis, MO 63104, USA
e-mail: tavisje@slu.edu

C.E. Cameron et al. (eds.), *Viral Genome Replication*,
DOI 10.1007/b135974_7,

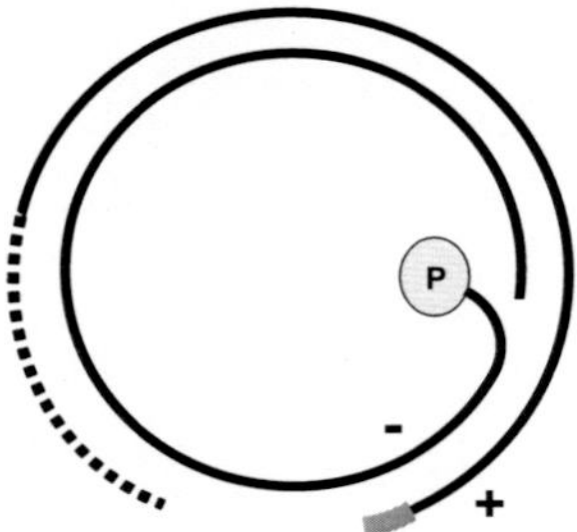

Fig. 7.1 The hepadnaviral genome. DNAs are represented as *black lines*, and their polarities are indicated by + or –. The *dashed* segment of the plus-polarity DNA indicates the variable position of its 3′ end. P covalently bound to the 5′ end of the minus-polarity DNA is represented as an *oval*, and the capped RNA oligomer at the 5′ end of the plus-polarity DNA is in *grey*.

shorter than the full length of the genome, leaving a single-stranded region in the DNA. In the mammalian hepadnaviruses, the plus-strand DNA extends on average about half the length of the genome, whereas in the avian viruses the plus-polarity DNA is nearly full length (Summers et al., 1975; Lien et al., 1987). This unusual structure is a direct result of the reverse transcription mechanism that produces the DNA.

Viral Replication Cycle

Viral replication (Fig. 7.2; reviewed in Seeger et al., 2007) starts with binding of the virus to hepatocytes through an unknown receptor. Fusion of the viral envelope with a cellular membrane releases the subviral core particle into the cytoplasm (Step 1), and then the core particle is transported by a poorly understood process to the nucleus, where it releases the viral DNA (Step 2). The DNA is repaired to a covalently closed circular episome (cccDNA) (Step 3), which is the template for transcription (Step 4). Three or four viral mRNAs are synthesized and transported to the cytoplasm, where they are translated to produce five to seven viral proteins (Step 5) (message and protein numbers depend on the viral species). One of the largest RNAs (the pregenomic RNA, or pgRNA) is encapsidated as a nucleoprotein complex together with P into nascent core particles whose icosahedral shells are composed of the single viral capsid protein, C (Step 6). Reverse transcription occurs in the cytoplasm within subviral core particles, with synthesis of the minus-polarity DNA (Step 7) followed by production of plus-polarity DNA (Step 8). Mature core particles are then either transported back into the nucleus to maintain the pool of transcriptional templates (Step 9) or they bud through post-endoplasmic reticulum/pre-Golgi membranes to acquire the envelope (Step 10). The mature viruses are then secreted from the cell non-cytolytically (Step 11).

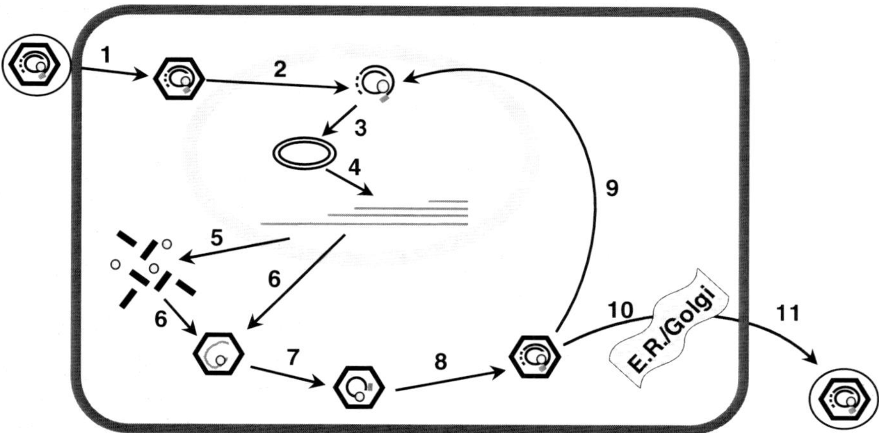

Fig. 7.2 The hepadnaviral replication cycle. The *large rectangle* represents a hepatocyte, the *large oval* is the nucleus, and the "ER/Golgi" shape indicates the post-endoplasmic reticulum/pre-Golgi compartment. RNAs are shown as *grey lines* and DNAs are shown as *black lines*. The *small grey oval* represents P, the *hexagon* is the core particle, and the *circle* surrounding the hexagon is the viral envelope.

Repair of the cccDNA and Transcription of the pgRNA

Once the virion-derived DNA is delivered into the nucleus it is repaired to fully double-stranded cccDNA. The DNA polymerase activity of P is probably not needed for cccDNA formation because ablating DHBV DNA polymerase activity with phosphonoformic acid has no effect on cccDNA accumulation in primary duck hepatocytes (Mason et al., 1987), and nucleoside analog inhibitors only partially reduce formation of HBV cccDNA following infection of permissive *Tupaia* primary hepatocytes (Kock et al., 2003). These data imply that the gap in the plus-polarity DNA is filled by cellular DNA polymerases, but it remains possible that P may contribute to other reactions during repair of the incoming viral DNA to cccDNA.

The cccDNA is the template for viral transcription by host RNA polymerase II (Rall et al., 1983). Three or four sets of 3′ co-terminal mRNAs are produced by transcription from three or four promoters. None of the hepadnaviral RNAs contain classic introns, although spliced forms have been observed in cultured cells and infected liver (Su et al., 1989; Obert et al., 1996). The pgRNA (Fig. 7.3, line 1) is one of the genomic-length RNAs. It plays a key role in reverse transcription because it is the mRNA for both viral proteins required for genomic replication (C and P), and it is also the template for reverse transcription. The pgRNA is terminally redundant due to read-through of the single viral poly-adenylation site during the first pass of the RNA polymerase around the circular cccDNA template, followed by recognition of the same poly-adenylation signal the during second pass (Russnak and Ganem, 1990). The pgRNA contains several elements important for viral replication, especially the direct repeats DR1 and DR2 and a stem loop called ε.

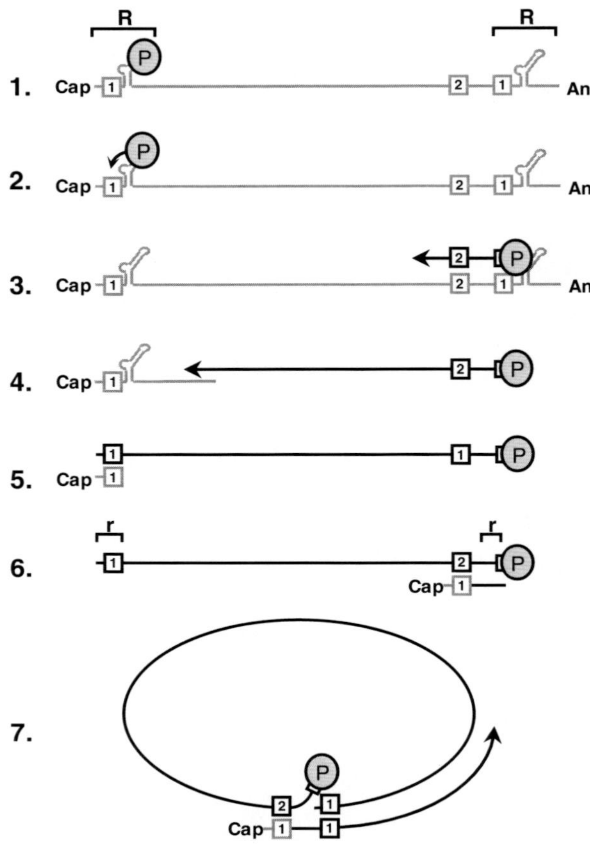

Fig. 7.3 Hepadnaviral reverse transcription. The pgRNA is shown as a *grey line*, Cap indicates the 5′ cap on the pgRNA, and An is the polyA tail. DNAs are shown as *black lines*, and *arrowheads* indicate their growing 3′ ends. R represents the large terminal redundancy in the pgRNA, r is the small terminal redundancy in the minus-polarity DNA, and *boxes* labeled 1 or 2 represent DR1 and DR2. P is represented as a *grey oval*. The nucleic acids are shown fully extended for clarity.

Encapsidation

Hepadnaviral reverse transcription occurs exclusively within nascent cytoplasmic core particles. Core particles are formed in a two-step process, in which P first binds to ε at the 5′ end of the pgRNA, and then the icosahedral capsid polymerizes around the P:pgRNA complex. Neither the P nor the pgRNA enters capsids if this complex does not form (Hirsch et al., 1990; Bartenschlager et al., 1990; Junker-Niepmann et al., 1990). Binding of P to the pgRNA appears to be co-translational because pgRNA molecules that encode P are preferentially packaged in competition experiments employing wild-type and P-deficient pgRNAs (Hirsch et al., 1990). ε is sufficient to direct encapsidation of the mammalian pgRNAs; however, encapsidation of the avian pgRNAs requires a second signal located internally in the pgRNA (Hirsch et al., 1991; Calvert and Summers, 1994).

Binding of P to ε is not a simple binary reaction, but rather requires the active participation of cellular chaperones including HSP90, HSP23, and perhaps others in an ATP-dependent reaction (Hu and Seeger, 1996; Hu et al., 1997). These results are reminiscent of the role of chaperones in binding to steroid receptors and holding them in a binding-competent state (Jakob and Buchner, 1994; Bohen et al., 1995; Kimura et al., 1995). Chaperones also appear to be involved in binding of HBV P to the pgRNA (Park and Jung, 2001; Park et al., 2002a, b).

Minus-Polarity DNA Synthesis

Hepadnaviral reverse transcription was first identified by the seminal studies of Summers and Mason with DHBV (Summers and Mason, 1982). Reverse transcription is primed by P itself, resulting in covalent linkage of the product DNA to P. Priming employs a tyrosine in the terminal protein domain (Y96 in DHBV or Y63 in HBV) (Zoulim and Seeger, 1994; Weber et al., 1994; Lanford et al., 1997). The covalent linkage of the 5′ end of the DNA to P persists throughout reverse transcription and is responsible for the solubility in phenol that is characteristic of hepadnaviral DNAs (Gerlich and Robinson, 1980).

A bulge in the 5′ copy of ε acts as the template for the priming reaction, but DNA synthesis arrests after 3–4 nt (Fig. 7.3, Step 2) (Wang and Seeger, 1993; Tavis et al., 1994; Nassal and Rieger, 1996), yielding a very short protein-linked nascent minus-strand DNA. This nascent DNA is transferred to the 3′ copy of a 12 nt repeat element, direct repeat 1 (DR1) (Fig. 7.3, Step 3) in the first of the three strand-transfer reactions of reverse transcription (Wang and Seeger, 1993; Tavis et al., 1994; Nassal and Rieger, 1996). Homology between the nascent DNA and DR1 contributes to the transfer of the nascent minus-strand DNA to DR1, but the homology is too small to guide transfer to a unique position, so protein contacts and/or higher-order template structures must assist in directing the transfer (Tavis and Ganem, 1995; Loeb and Tian, 1995). Once annealed to the 3′ copy of DR1, the nascent minus-strand DNA is extended to its full length, terminating when P reaches the 5′ end of the pgRNA. Concomitantly with DNA synthesis, the template pgRNA is degraded by the viral RNAseH (Fig. 7.3, Step 4) (Summers and Mason, 1982).

Positive-Polarity DNA Synthesis

Upon completion of the full-length minus-polarity DNA, RNAseH digestion leaves a capped RNA oligomer annealed to the short terminal duplication at the 3′ end of the newly synthesized minus-strand DNA ("r", Fig. 7.3, Step 5). This RNA oligomer is then transferred to a second copy of the direct repeat (DR2) (Fig. 7.3, Step 6) (Seeger et al., 1986; Staprans et al., 1991), where it primes synthesis of positive-polarity DNA. The 5′ end of the plus-polarity DNA remains attached to the RNA primer throughout reverse transcription (Seeger et al., 1986). Transfer of the RNA oligomer fails about 10% of the time, and in this case, in situ priming of the plus-strand DNA from the untransferred primer produces a duplex linear copy of the genome (Staprans et al., 1991). Although these duplex linear genomes are normally dead-end products, they can produce progeny viruses with low efficiency through a complex recombination-mediated process termed illegitimate replication (Wang and Summers, 1995). Following successful transfer of the RNA oligomer, plus-polarity DNA synthesis proceeds a short distance to the 5′ end of the minus-polarity DNA (Fig. 7.3, Step 6). Finally, the third strand transfer circularizes the genome by transferring the growing 3′ end of the plus-strand DNA from the 5′ to the 3′ end of the minus-strand DNA template, employing homology between the 3′ end of the plus-strand DNA and the short terminal redundancy of the minus-strand DNA (Fig. 7.3, Step 7) (Molnar-Kimber et al., 1984). The plus-strand DNA is then extended a variable length along the minus-polarity DNA, where it terminates prematurely, leaving a single-stranded gap of variable length in the progeny viral DNA (Summers et al., 1975; Lien et al., 1987). This partially double-stranded DNA is the mature genome found within extracellular hepadnaviral virions.

Cis- and Trans-Acting Factors in Reverse Transcription

The three strand transfers of reverse transcription must be directed by protein activities somewhat analogous to those involved in DNA recombination because the short nucleotide homologies between the strands are insufficient to provide adequate sequence specificity to the transfers, and at least the plus-strand primer transfer is energetically unfavorable because there is a net loss of base pairing following transfer. Because P is the only viral enzyme involved in reverse transcription, it is assumed that P promotes the transfers. However, the only data supporting this assumption is provided by a deletion of 10 amino acids near the amino-terminus of DHBV P, which has little effect on synthesis of the first few nucleotides of DNA, but greatly diminishes the first strand transfer and subsequent DNA elongation (Gong et al., 2000).

The cis-acting nucleic acid structures contributing to reverse transcription are better understood than are the trans-acting factors. The first such structures to be identified were the direct repeats DR1 and DR2 and the terminal redundancies in the pgRNA and the minus-polarity DNA (Seeger et al., 1986). These structures contribute to reverse transcription by providing the regions of homology needed to guide the strand transfers. ε was initially identified as the primary element of the pgRNA encapsidation signal, and its role as the origin of reverse transcription was subsequently discovered. Although the nucleic acids in Fig. 7.3 are shown fully extended to illustrate reverse transcription as simply as possible, the template strands appear to adopt discrete secondary structures that promote the strand transfers (Fig. 7.4). Base pairing in the pgRNA between a cis-acting element termed

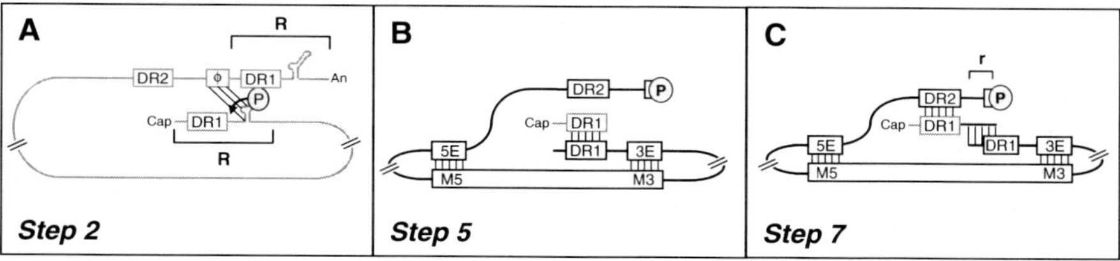

Fig. 7.4 Higher-order nucleic acid structures that guide the strand transfers. Reverse transcription occurs on highly ordered nucleic acid structures that promote the strand transfers. The symbols and step numbers are the same as in Fig. 7.3; *thin lines* between nucleic acids indicate base pairing. (**A**) The juxtaposition of the 5′ copy of ε and the 3′ copy of DR1 through base pairing between ϕ and ε that is believed to promote the minus-polarity DNA strand transfer. (**B**) The structure of the minus-strand DNA that promotes the plus-strand DNA primer transfer from DR1 to DR2. M3 and M5 are the 3′ and 5′ portions of the M element. (**C**) The structure of the DNA immediately after the second strand transfer in plus-polarity DNA synthesis that circularizes the DNA. Adapted from Tang and McLachlan (2002) and Liu et al. (2003).

ϕ (or β) which lies between DR2 and the 3′ copy of DR1 and the 5′ half of ε (Fig. 7.4A) contributes to efficient minus-strand DNA synthesis. Base pairing between ϕ and ε would align the nascent minus-polarity DNA and the 3′ copy of DR1 (Tang and McLachlan, 2002; Shin et al., 2004) and would also destabilize ε; both of these events would aid the first strand transfer. The DHBV plus-polarity template switches require at least three cis-acting sequences in the minus-polarity DNA termed 5E, M, and 3E. 5E and 3E bind to the 5′ (M5) and 3′ (M3) portions of M, and this binding is essential to guide both plus-strand primer translocation and genome circularization (compare Fig. 7.3, Steps 5 and 7 with Fig. 7.4B and C) (Mueller-Hill and Loeb, 2002; Liu et al., 2003). In addition, a small DNA hairpin near the 3′ end of the minus-polarity DNA overlapping the 5′ end of DR1 contributes to efficient plus-polarity primer translocation (Habig and Loeb, 2002). The cis-acting signals in HBV are less well characterized, but HBV appears to contain sequences analogous to 5E, M, and 3E, plus additional cis-acting sequences that contribute efficient synthesis of mature viral DNA (Liu et al., 2004; Lee et al., 2004).

Participation of the Capsid in Reverse Transcription

Reverse transcription occurs within subviral core particles, and mutations to the carboxy-terminal region of DHBV and HBV C inhibit extension of plus-strand DNA and can impede plus-polarity primer translocation (Schlicht et al., 1989; Yu and Summers, 1991; Nassal, 1992). C in immature capsids is phosphorylated at its C-terminus, whereas C in secreted capsids containing mature DNA is fully dephosphorylated (Schlicht et al., 1989; Perlman et al., 2005). Recent data with DHBV indicate that sequential phosphorylation–dephosphorylation of C may regulate DNA synthesis because minus-polarity DNA synthesis requires negative charges on the C-terminus of C, whereas plus-polarity DNA synthesis is inhibited by negative charges on C (Basagoudanavar et al., 2007). In addition to modulating plus-polarity DNA synthesis, blocking phosphorylation of HBV C can inhibit encapsidation of the pgRNA (Gazina et al., 2000) and lead to reverse transcription of spliced RNAs that were aberrantly encapsidated (Kock et al., 2004).

Fidelity of HBV Reverse Transcription

P lacks a proof-reading 3′ to 5′ exonuclease activity and hence hepadnaviral genomic replication is error prone, as is the case for all reverse-transcribing elements. Technical limitations in producing purified P coupled with its inability to accept exogenous primer templates (Radziwill et al., 1988) have prevented direct measurements of P's intrinsic error rate. However, mutations accumulate in the HBV genome at a rate of approximately 4×10^{-5} substitutions per site per year during natural infections (Fares and Holmes, 2002). This mutation fixation rate is about two orders of magnitude lower than similar estimates for Human Immunodeficiency Virus (HIV) ($2.7–6.7 \times 10^{-3}$) (Leitner and Albert, 1999), presumably due to stricter selection pressures stemming from the unusually dense packing of genetic information in the HBV genome: Every nucleotide in the HBV genome encodes protein, over half of the sequences encode two proteins simultaneously in overlapping reading frames, and there are a large number of promoters and cis-acting sequences that are essential for mRNA production and reverse transcription. Therefore, HBV appears to have fewer degrees of freedom to tolerate mutations than do most other viruses.

Structure of P

No three-dimensional structural data exist for P, but homology alignments and mutational analyses have revealed that P contains four domains (Fig. 7.5; Radziwill et al., 1990; Chang et al., 1990). The terminal protein domain (TP) contains the tyrosine residue that primes DNA synthesis and covalently links P to the viral

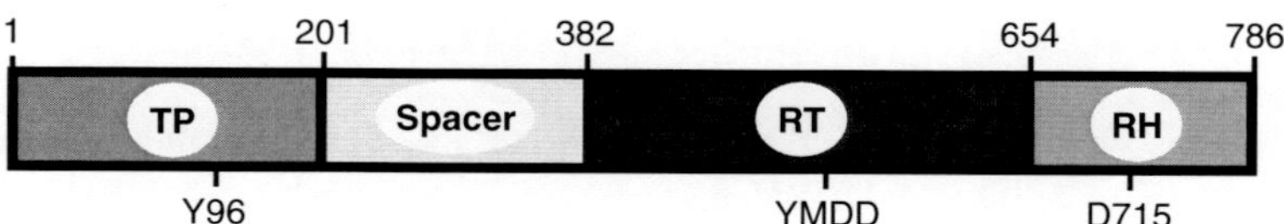

Fig. 7.5 Domain structure of P. The amino acid numbers for the domain boundaries are shown for DHBV P; they are approximate and are based on homology alignments with the other hepadnaviral P proteins and with retroviral reverse transcriptases. TP is the terminal protein domain, RT represents the reverse transcriptase domain, RH indicates the RNAseH domain, Y96 is the tyrosine that forms the covalent linkage to the minus-polarity DNA, YMDD represents amino acids 511–514 that form a key motif of the reverse transcriptase active site, and D715 is a key residue of the RNAseH active site.

DNA (Y96 in DHBV, Y63 in HBV) (Zoulim and Seeger, 1994; Weber et al., 1994; Lanford et al., 1997). The spacer domain has no known function other than to link the TP domain to the rest of the molecule. The reverse transcriptase (RT) and RNAseH (RH) domains contain the two known enzymatic active sites (Radziwill et al., 1990; Chang et al., 1990), and these domains share weak homology with the RT and RH domains of reverse transcriptases from the retroviruses and other retro-elements (Schodel et al., 1988; Poch et al., 1989; McClure, 1993; Li et al., 1995). The three-dimensional structure of part of the HBV RT domain has been modeled based on the HIV RT domain (Das et al., 2001). This model appears to be largely accurate at an atomic level because the mechanism of mutations that provide resistance to nucleoside analog drugs can be productively interpreted from it (reviewed in Shaw et al., 2006). The size and multimeric state of P in viral particles have been difficult to discern due to its low abundance in core particles and its covalent linkage to the product DNA. We recently found that DHBV P acts as a full-length monomer both in vitro and in vivo (Zhang and Tavis, 2006). Therefore, a single P monomer is encapsidated into each nascent core particle, primes DNA synthesis, synthesizes minus- and plus-polarity DNA, and participates in the three strand transfers of DNA synthesis, with all steps after DNA priming performed while P is covalently coupled to its product DNA.

In vitro reconstitution studies with recombinant DHBV P expressed in *Escherichia coli* revealed that production of active P requires addition of the molecular chaperones HSP90, HSP70, HSP40, HSP23, and HOP (Hu and Seeger, 1996; Hu and Anselmo, 2000; Beck and Nassal, 2001; Hu et al., 2002). At least HSP90 and HSP23 remain bound to DHBV P and are encapsidated with the P:pgRNA ribonucleoprotein complex (Hu and Seeger, 1996; Hu et al., 1997). However, the chaperones do not appear to be actively involved in subsequent steps of reverse transcription because inhibiting HSP90 function with geldanamycin does not block reverse transcription when it is added to RNA-containing capsids (Hu et al., 1997). Chaperones are also involved in producing active HBV P (Park and Jung, 2001; Park et al., 2002a, b; Hu et al., 2004). Therefore, P functions as part of a macromolecular complex containing cellular chaperones and viral nucleic acids.

P Is Structurally Dynamic

P must be structurally dynamic to catalyze the complex reverse transcription pathway while covalently bound to its product DNA. Data supporting a dynamic nature for P are provided by the observation that binding of P to ε induces a conformational alteration in the RT domain of P which is essential for encapsidation of the P:pgRNA complex and for enzymatic activation of P (Tavis and Ganem, 1996; Tavis et al., 1998). Presumably, the chaperones bound to P promote this structural alteration. Because ε is the principal cis-acting element of the viral encapsidation signal, the discovery that ε activates P revealed an unexpected parallel in the metabolism of the hepadnaviral and retroviral reverse transcriptases. Most retroviral reverse transcriptases are synthesized as inactive *gag-pol* fusion proteins that are

Fig. 7.6 Dynamic maturation of P during reverse transcription. P_i is the inactive primary translation product, P_p is P in the priming mode that is competent for encapsidation and priming after the structural alteration induced by binding to ε, and P_e is P in the elongation mode that is competent for RNA- and DNA-dependent DNA synthesis following the first strand transfer. The reverse reactions are shown with *dashed arrows* because they either have not yet been demonstrated or are highly disfavored.

encapsidated through polymerization of *gag*, and then are activated by proteolytic cleavage from *gag*. Enzymatic activation of the hepadnaviral P is also coupled to encapsidation, but the virus has used the binding between the P and the encapsidation signal, ε, to trigger enzymatic maturation.

A second observation indicating that P is dynamic is that the DNA polymerase active site of P must accommodate the TP domain and ε in the priming reaction, and later it must accept the elongating DNAs hybridized to either RNA or DNA. A change in the polymerase active site following the first strand transfer is evident biochemically because DNA synthesis prior to strand transfer is resistant to phosphonoformic acid (PFA), whereas later DNA synthesis is sensitive to PFA (Wang and Seeger, 1992; Stashcke and Colacino, 1994).

These data can be summarized in a maturation pathway for P (Fig. 7.6). P is translated as an inactive enzyme (P_i), and chaperone-mediated binding to ε on the pgRNA yields a P_i:ε complex. P then undergoes a structural alteration to the priming mode (P_p) and the P_p:ε complex is encapsidated. Following initiation of DNA synthesis and the first strand transfer, P undergoes its final known alteration and enters the elongation mode (P_e). P_e synthesizes the large majority of the viral genome, and it appears to be the form found in secreted virions.

DNA Synthesis Regulates Envelopment

Hepadnaviral virions exported from the cell contain only mature viral genomes, whereas the immature cores within cells contain genomes at all stages of synthesis (Mason et al., 1982; Weiser et al., 1983). This observation led to the hypothesis that DNA synthesis induces a signal that is passed to the exterior of the core particle to trigger binding to the surface glycoproteins and encapsidation (Summers and Mason, 1982). Envelopment of mutant DHBV and HBV viruses lacking RNAseH activity requires extensive minus-strand DNA synthesis (Gerelsaikhan et al., 1996; Wei et al., 1996), but envelopment of wild-type DHBV requires both minus-strand and plus-strand DNA syntheses (Yu and Summers, 1991; Perlman and Hu, 2003). Therefore, accumulation of double-stranded nucleic acids in cores may trigger the envelopment signal. The envelopment signal itself may involve dephosphorylation of C because intracellular capsids contain phosphorylated C, but C in extracellular virions is fully dephosphorylated (Schlicht et al., 1989; Perlman et al., 2005). The possibility that double-stranded nucleic acids induce a dephosphorylation-dependent change in the conformation of C needed for envelopment is consistent with the ability of certain mutations in HBV C to permit envelopment of capsids containing immature viral DNAs (Yuan et al., 1999).

Outstanding Questions in Hepadnaviral Reverse Transcription

Through a large number of studies conducted since Summers and Mason defined the hepadnaviral reverse transcription pathway (Summers and Mason, 1982), we now understand hepadnaviral genomic replication in considerable molecular detail, especially the nucleic acid intermediates involved. This knowledge has contributed to development of four nucleoside analog drugs that can successfully control viral replication in many individuals. However, development of HBV strains resistant to the nucleoside analogs is a major problem that limits the duration of effective control of viral replication in most patients (Shaw et al., 2006). Consequently, a better understanding of reverse transcription is needed to develop additional antiviral compounds that could be used in combination with the existing nucleoside analogs to forestall development of resistance. To achieve this goal, we must improve our understanding of the enzymology of reverse transcription. Key advances that are needed would be to understand at a molecular level the three strand-transfer reactions, the role that the molecular chaperones play during reverse transcription, and the contribution of the capsid structure to reverse transcription. Most importantly, we need to obtain atomic-resolution structures for each of the various conformations adopted by P during reverse transcription.

References

Bartenschlager, R., Junker-Niepmann, M., Schaller, H., 1990. The P gene product of hepatitis B virus is required as a structural component for genomic RNA encapsidation. J. Virol. 64, 5324–5332.

Basagoudanavar, S. H., Perlman, D. H., Hu, J., 2007. Regulation of hepadnavirus reverse transcription by dynamic nucleocapsid phosphorylation. J. Virol. 81, 1641–1649.

Beck, J., Nassal, M., 2001. Reconstitution of a functional duck hepatitis B virus replication initiation complex from separate reverse transcriptase domains expressed in *Escherichia coli*. J Virol. 75, 7410–7419.

Bohen, S. P., Kralli, A., Yamamoto, K. R., 1995. Hold 'em and Fold 'em: chaperones and signal transduction. Science 268, 1303–1304.

Calvert, J., Summers, J., 1994. Two regions of an avian hepadnavirus RNA pregenome are required in cis for encapsidation. J. Virol. 68, 2084–2090.

Chang, L. J., Hirsch, R. C., Ganem, D., Varmus, H. E., 1990. Effects of insertional and point mutations on the functions of the duck hepatitis B virus polymerase. J. Virol. 64, 5553–5558.

Das, K., Xiong, X., Yang, H., Westland, C. E., Gibbs, C. S., Sarafianos, S. G., Arnold, E., 2001. Molecular modeling and biochemical characterization reveal the mechanism of hepatitis B virus polymerase resistance to lamivudine (3TC) and emtricitabine (FTC). J. Virol. 75, 4771–4779.

Fares, M. A., Holmes, E. C., 2002. A revised evolutionary history of hepatitis B virus (HBV). J. Mol. Evol. 54, 807–814.

Ganem, D., Prince, A. M., 2004. Hepatitis B virus infection – natural history and clinical consequences. N. Engl. J. Med. 350, 1118–1129.

Gazina, E. V., Fielding, J. E., Lin, B., Anderson, D. A., 2000. Core protein phosphorylation modulates pregenomic RNA encapsidation to different extents in human and duck hepatitis B viruses. J. Virol. 74, 4721–4728.

Gerelsaikhan, T., Tavis, J. E., Bruss, V., 1996. Hepatitis B virus nucleocapsid envelopment does not occur without genomic DNA synthesis. J. Virol. 70, 4269–4274.

Gerlich, W. H., Robinson, W. S., 1980. Hepatitis B virus contains protein attached to the 5′ terminus of its complete DNA strand. Cell 21, 801–809.

Gong, Y., Yao, E., Stevens, M., Tavis, J. E., 2000. Evidence that the first strand-transfer reaction of duck hepatitis B virus reverse transcription requires the polymerase and that strand transfer is not needed for the switch of the polymerase to the elongation mode of DNA synthesis. J. Gen. Virol. 81, 2059–2065.

Habig, J. W., Loeb, D. D., 2002. Small DNA hairpin negatively regulates in situ priming during duck hepatitis B virus reverse transcription. J. Virol. 76, 980–989.

Hirsch, R. C., Lavine, J. E., Chang, L. J., Varmus, H. E., Ganem, D., 1990. Polymerase gene products of hepatitis B viruses are required for genomic RNA packaging as well as for reverse transcription. Nature (London) 344, 552–555.

Hirsch, R. C., Loeb, D. D., Pollack, J. R., Ganem, D., 1991. cis-Acting sequences required for encapsidation of duck hepatitis B virus pregenomic RNA. J. Virol. 65, 3309–3316.

Hu, J., Anselmo, D., 2000. In vitro reconstitution of a functional duck hepatitis B virus reverse transcriptase: posttranslational activation by HSP90. J. Virol. 74, 11447–11455.

Hu, J., Flores, D., Toft, D., Wang, X., Nguyen, D., 2004. Requirement of heat shock protein 90 for human hepatitis B virus reverse transcriptase function. J Virol. 78, 13122–13131.

Hu, J., Seeger, C., 1996. Hsp90 is required for the activity of a hepatitis B virus reverse transcriptase. Proc. Natl. Acad. Sci. USA 93, 1060–1064.

Hu, J., Toft, D., Anselmo, D., Wang, X., 2002. In vitro reconstitution of functional hepadnavirus reverse transcriptase with cellular chaperone proteins. J. Virol. 76, 269–279.

Hu, J., Toft, D. O., Seeger, C., 1997. Hepadnavirus assembly and reverse transcription require a multi-component chaperone complex which is incorporated into nucleocapsids. EMBO J. 16, 59–68.

Jakob, U., Buchner, J., 1994. Assisting spontaneity: the role of Hsp90 and small Hsps as molecular chaperones. Trends Biol. Sci. 19, 205–211.

Junker-Niepmann, M., Bartenschlager, R., Schaller, H., 1990. A short cis-acting sequence is required for hepatitis B virus pregenome encapsidation and sufficient for packaging of foreign RNA. EMBO J. 9, 3389–3396.

Kimura, Y., Yahara, I., Lindquist, S., 1995. Role of the protein chaperone YDJ1 in establishing Hsp90-mediated signal transduction pathways. Science 268, 1362–1365.

Kock, J., Baumert, T. F., Delaney, W. E., Blum, H. E., von Weizsacker, F., 2003. Inhibitory effect of adefovir and lamivudine on the initiation of hepatitis B virus infection in primary tupaia hepatocytes. Hepatology 38, 1410–1418.

Kock, J., Nassal, M., Deres, K., Blum, H. E., von Weizšacker, F., 2004. Hepatitis B virus nucleocapsids formed by carboxy-terminally mutated core proteins contain spliced viral genomes but lack full-size DNA. J. Virol. 78, 13812–13818.

Lanford, R. E., Chavez, D., Brasky, K. M., Burns III, R. B., Rico-Hesse, R., 1998. Isolation of a hepadnavirus from the woolly monkey, a New World primate. Proc. Natl. Acad. Sci. USA 95, 5757–5761.

Lanford, R. E., Notvall, L., Lee, H., Beams, B., 1997. Transcomplementation of nucleotide priming and reverse transcription between independently expressed TP and RT domains of the hepatitis B virus reverse transcriptase. J. Virol. 71, 2996–3004.

Lee, J., Shin, M. K., Lee, H. J., Yoon, G., Ryu, W. S., 2004. Three novel cis-acting elements required for efficient plus-strand DNA synthesis of the hepatitis B virus genome. J. Virol. 78, 7455–7464.

Leitner, T., Albert, J., 1999. The molecular clock of HIV-1 unveiled through analysis of a known transmission history. Proc. Natl. Acad. Sci. USA 96, 10752–10757.

Li, M. D., Bronson, D. L., Lemke, T. D., Faras, A. J., 1995. Phylogenetic analyses of 55 retroelements on the basis of the nucleotide and product amino acid sequences of the pol gene. Mol. Biol. Evol. 12, 657–670.

Lien, J. M., Petcu, D. J., Aldrich, C. E., Mason, W. S., 1987. Initiation and termination of duck hepatitis B virus DNA synthesis during virus maturation. J. Virol. 61, 3832–3840.

Liu, N., Ji, L., Maguire, M. L., Loeb, D. D., 2004. cis-Acting sequences that contribute to the synthesis of relaxed-circular DNA of human hepatitis B virus. J. Virol. 78, 642–649.

Liu, N., Tian, R., Loeb, D. D., 2003. Base pairing among three cis-acting sequences contributes to template switching during hepadnavirus reverse transcription. Proc. Natl. Acad. Sci. USA 100, 1984–1989.

Loeb, D. D., Tian, R., 1995. Transfer of the minus strand of DNA during hepadnavirus replication is not invariable but prefers a specific location. J. Virol. 69, 6886–6891.

Mason, W. S., Aldrich, C., Summers, J., Taylor, J. M., 1982. Asymmetric replication of duck hepatitis B virus DNA in liver cells: free minus strand DNA. Proc. Natl. Acad. Sci. USA 79, 3997–4001.

Mason, W. S., Lien, J. M., Petcu, D. J., Coates, L., London, W. T., O'Connell, A., Aldrich, C., Custer, R. P., 1987. In vivo and In vitro studies on duck hepatitis B virus replication. In: W. S. Robinson, K. Koike, H. Well (Eds.), Hepadna Viruses. Alan R. Liss, New York, pp. 3–16.

McClure, M. A., 1993. Evolutionary history of reverse transcriptase. In: A. M. Skalka (Ed.), Reverse Transcriptase. Cold Spring Harbor Laboratory Press, Cold Spring Harbor, pp. 425–444.

Molnar-Kimber, K. L., Summers, J. W., Mason, W. S., 1984. Mapping of the cohesive overlap of duck hepatitis B virus DNA and of the site of initiation of reverse transcription. J. Virol. 51, 181–191.

Mueller-Hill, K., Loeb, D. D., 2002. cis-Acting sequences 5E, M, and 3E interact to contribute to primer translocation and circularization during reverse transcription of avian hepadnavirus DNA. J. Virol. 76, 4260–4266.

Nassal, M., 1992. The arginine-rich domain of the hepatitis B virus core protein is required for pregenome encapsidation and productive viral positive-strand DNA synthesis but not for virus assembly. J. Virol. 66, 4107–4116.

Nassal, M., Rieger, A., 1996. A bulged region of the hepatitis B virus RNA encapsidation signal contains the replication origin for discontinuous first-strand DNA synthesis. J. Virol. 70, 2764–2773.

Obert, S., Zachmann-Brand, B., Deindl, E., Tucker, W., Bartenschlager, R., Schaller, H., 1996. A splice hepadnavirus RNA that is essential for virus replication. EMBO J. 15, 2565–2574.

Park, S. G., Jung, G. H., 2001. Human hepatitis B virus polymerase interacts with the molecular chaperonin Hsp60. J. Virol. 75, 6962–6968.

Park, S. G., Lim, S. O., Jung, G. H., 2002a. Binding site analysis of human HBV Pol for molecular chaperonin, Hsp60. Virology 298, 116–123.

Park, S. G., Rho, J. K., Jung, G., 2002b. Hsp90 makes the human HBV Pol competent for in vitro priming rather than maintaining the human HBV Pol/pregenomic RNA complex. Arch. Biochem. Biophys. 401, 99–107.

Perlman, D., Hu, J. M., 2003. Duck hepatitis B virus virion secretion requires a double-stranded DNA genome. J. Virol. 77, 2287–2294.

Perlman, D. H., Berg, E. A., O′connor, P. B., Costello, C. E., Hu, J., 2005. Reverse transcription-associated dephosphorylation of hepadnavirus nucleocapsids. Proc. Natl. Acad. Sci. USA 102, 9020–9025.

Poch, O., Sauvaget, I., Delarue, M., Tordo, N., 1989. Identification of four conserved motifs among the RNA-dependent polymerase encoding elements. EMBO J. 8, 3867–3874.

Prassolov, A., Hohenberg, H., Kalinina, T., Schneider, C., Cova, L., Krone, O., Frolich, K., Will, H., Sirma, H., 2003. New hepatitis B virus of cranes that has an unexpected broad host range. J. Virol. 77, 1964–1976.

Radziwill, G., Tucker, W., Schaller, H., 1990. Mutational analysis of the hepatitis B virus P gene product: domain structure and RNase H activity. J. Virol. 64, 613–620.

Radziwill, G., Zentgraf, H., Schaller, H., Bosch, V., 1988. The duck hepatitis B virus DNA polymerase is tightly associated with the viral core structure and unable to switch to an exogenous template. Virology 163, 123–132.

Rall, L. B., Standring, D. N., Laub, O., Rutter, W. J., 1983. Transcription of hepatitis B virus by RNA polymerase II. Mol. Cell Biol. 3, 1766–1773.

Russnak, R., Ganem, D., 1990. Sequences 5′ to the polyadenylation signal mediate differential poly(A) site use in hepatitis B viruses. Genes Dev. 4, 764–776.

Schlicht, H. J., Bartenschlager, R., Schaller, H., 1989. The duck hepatitis B virus core protein contains a highly phosphorylated C terminus that is essential for replication but not for RNA packaging. J. Virol. 63, 2995–3000.

Schodel, F., Weimer, T., Will, H., Sprengel, R., 1988. Amino acid sequence similarity between retroviral and *E. coli* RNase H and hepadnaviral gene products. AIDS Res. Human Retroviruses 4, 9–11.

Schoedel, F., Sprengel, R., Weimer, T., Fernholz, D., Schneider, R., Will, H., 1989. Animal hepatitis B viruses. Adv. Viral Oncol. 8, 73–102.

Seeger, C., Ganem, D., Varmus, H. E., 1986. Biochemical and genetic evidence for the hepatitis B virus replication strategy. Science 232, 477–484.

Seeger, C., Zoulim, F., Mason, W. S., 2007. Hepadnaviruses. In: D. M. Knipe, P. Howley, D. E. Griffin, R. A. Lamb, M. A. Martin, B. Roizman, S. E. Straus (Eds.), Fields Virology. Lippincott Williams & Wilkins, Philadelphia, pp. 2977–3029.

Shaw, T., Bartholomeusz, A., Locarnini, S., 2006. HBV drug resistance: mechanisms, detection and interpretation. J. Hepatol. 44, 593–606.

Shin, M. K., Lee, J., Ryu, W. S., 2004. A novel cis-acting element facilitates minus-strand DNA synthesis during reverse transcription of the hepatitis B virus genome. J. Virol. 78, 6252–6262.

Sprengel, R., Kaleta, E. F., Will, H., 1988. Isolation and characterization of a hepatitis B virus endemic in herons. J. Virol. 62, 3832–3839.

Staprans, S., Loeb, D. D., Ganem, D., 1991. Mutations affecting hepadnavirus plus-strand DNA synthesis dissociate primer cleavage from translocation and reveal the origin of linear viral DNA. J. Virol. 65, 1255–1262.

Stashcke, K. A., Colacino, J. M., 1994. Priming of duck hepatitis B virus reverse transcription in vitro: premature termination of primer DNA induced by the 5′-triphosphate of fialuridine. J. Virol. 68, 8265–8269.

Su, T. S., Lai, C. J., Huang, J. L., Lin, L. H., Yauk, Y. K., Chang, C. M., Lo, S. J., Han, S. H., 1989. Hepatitis B virus transcript produced by RNA splicing. J. Virol. 63, 4011–4018.

Summers, J., Mason, W. S., 1982. Replication of the genome of a hepatitis B-like virus by reverse transcription of an RNA intermediate. Cell 29, 403–415.

Summers, J., O′Connell, A., Millman, I., 1975. Genome of hepatitis B virus: restriction enzyme cleavage and structure of DNA extracted from Dane particles. Proc. Natl. Acad. Sci. USA 72, 4597–4601.

Tang, H., McLachlan, A., 2002. A pregenomic RNA sequence adjacent to DR1 and complementary to epsilon influences hepatitis B virus replication efficiency. Virology 303, 199–210.

Tavis, J. E., Ganem, D., 1995. RNA sequences controlling the initiation and transfer of duck hepatitis B virus minus-strand DNA. J. Virol. 69, 4283–4291.

Tavis, J. E., Ganem, D., 1996. Evidence for the activation of the hepatitis B virus polymerase by binding of its RNA template. J. Virol. 70, 5741–5750.

Tavis, J. E., Massey, B., Gong, Y., 1998. The duck hepatitis B virus polymerase is activated by its RNA packaging signal, epsilon. J. Virol. 72, 5789–5796.

Tavis, J. E., Perri, S., Ganem, D., 1994. Hepadnavirus reverse transcription initiates within the stem-loop of the RNA packaging signal and employs a novel strand transfer. J. Virol. 68, 3536–3543.

Wang, G. H., Seeger, C., 1992. The reverse transcriptase of hepatitis B virus acts as a protein primer for viral DNA synthesis. Cell 71, 663–670.

Wang, G. H., Seeger, C., 1993. Novel mechanism for reverse transcription in hepatitis B viruses. J. Virol. 67, 6507–6512.

Wang, W., Summers, J., 1995. Illegitimate replication of linear hepadnavirus DNA through nonhomolgous recombination. J. Virol. 69, 4029–4036.

Weber, M., Bronsema, V., Bartos, H., Bosserhoff, A., Bartenschlager, R., Schaller, H., 1994. Hepadnavirus P protein utilizes a tyrosine residue in the TP domain to prime reverse transcription. J. Virol. 68, 2994–2999.

Wei, Y., Tavis, J. E., Ganem, D., 1996. Relationship between viral DNA synthesis and virion envelopment in hepatitis B viruses. J. Virol. 70, 6455–6458.

Weiser, B., Ganem, D., Seeger, C., Varmus, H. E., 1983. Closed circular viral DNA and asymmetrical heterogeneous forms in livers from animals infected with ground squirrel hepatitis virus. J. Virol. 48, 1–9.

Yu, M., Summers, J., 1991. A domain of the hepadnavirus capsid protein is specifically required for DNA maturation and virus assembly. J. Virol. 65, 2511–2517.

Yuan, T. T., Sahu, G. K., Whitehead, W. E., Greenberg, R., Shih, C., 1999. The mechanism of an immature secretion phenotype of a highly frequent naturally occurring missense mutation at codon 97 of human hepatitis B virus core antigen. J. Virol. 73, 5731–5740.

Zhang, Z., Tavis, J. E., 2006. The duck hepatitis B virus reverse transcriptase functions as a full-length monomer. J. Biol. Chem. 281, 35794–35801.

Zoulim, F., Seeger, C., 1994. Reverse transcription in hepatitis B viruses is primed by a tyrosine residue of the polymerase. J. Virol. 68, 6–13.

Chapter 8
Rhabdoviruses

Sean P.J. Whelan

Introduction

Vesicular stomatitis virus (VSV) has long served as paradigm for understanding the strategies of viral genome replication of the non-segmented negative-strand (NNS) RNA viruses or order *Mononegavirales*. Although the primary focus of this chapter will be to summarize our current knowledge based upon work performed with VSV-Indiana or VSIV it should be emphasized that there are important distinctions between different *Rhabdoviridae* family members, notably some of the plant viruses exhibit a nuclear phase to their replication cycle. In addition, while the general principles from studying the cis-acting signals and trans-acting factors derived from study of VSIV hold true for other viruses, the nucleotide (nt) sequence of the signals themselves are distinct and the protein–protein interactions may be accomplished by distinct residues. The general strategy of gene expression is exploited by other families of viruses within the *Mononegavirales*, notably the *Paramyxoviridae*, *Filoviridae*, and *Bornaviridae*. Readers are referred to other insightful recent reviews for a more detailed discussion of the *Paramyxoviridae* (Cowton et al., 2006; Kolakofsky et al., 2004). In this chapter, I will summarize current knowledge of how VSIV gene expression is controlled following the release of the transcription competent viral core into the host cell. Particular emphasis is placed on the cis-acting signals and the trans-acting factors that control gene expression. A model is presented that summarizes our current understanding of gene expression. Viral entry and exit are fascinating topics in which much recent progress has been made. However, these topics are beyond the scope of this present review.

S.P.J. Whelan (✉)
Department of Microbiology and Molecular Genetics, Harvard Medical School,
200 Longwood Avenue, Boston, MA 02115, USA
e-mail: swhelan@hms.harvard.edu

C.E. Cameron et al. (eds.), *Viral Genome Replication*,
DOI 10.1007/b135974_8, © Springer Science+Business Media, LLC 2009

Classification

The family *Rhabdoviridae* includes viruses that infect a broad range of animal and plant hosts. Currently six genera are recognized within the family, but it should be emphasized that the majority of rhabdoviruses have not been assigned to a specific genus (Table 8.1).

Table 8.1 Classification of *Rhabdoviridae*

Genus	Type species	Other viruses
Vesiculovirus	Vesicular stomatitis Indiana virus (VSIV)	28
Lyssavirus	Rabies virus (RABV)	7
Ephemerovirus	Bovine ephemeral fever virus (BEFV)	5
Novirhabdovirus	Infectious hematopoietic necrosis virus (IHNV)	6
Cytorhabdovirus	Lettuce necrotic yellows virus (LNYV)	8
Nucleorhabdovirus	Potato yellow dwarf virus (PYDV)	6
Unassigned		117

General Overview of the Viral Replication Cycle

Infection of the cell is initiated by the delivery of a ribonucleoprotein (RNP) core of the virus. For VSIV this comprises 11,161-nucleotides (nt) of genomic RNA completely encapsidated by the viral nucleocapsid protein (N) and associated with the RNA-dependent RNA polymerase (RdRP). The viral components of the RdRp are 241 kDa large subunit (L) and an accessory phosphoprotein (P). This RNP core effectively reprograms the cell such that by 8 hours in baby hamster kidney (BHK-21) cells, a single cell infected by a single infectious particle has yielded 10,000 infectious progeny. A simplified diagram representing the virus and the replication cycle is shown in Fig. 8.1. Briefly, following attachment uncoating and release of the RNP core into the cytoplasm of the cell RNA synthesis commences. In the absence of viral protein synthesis primary transcription occurs in which the polymerase makes a 47-nt leader RNA (Le+) from the 3′ end of the genome and 5 capped and polyadenylated mRNAs encoding the N, P, M, G, and L proteins. Viral protein synthesis is essential for synthesis of the complementary antigenomic RNA, to provide a source of soluble N protein, termed N^0 necessary to drive the encapsidation of the nascent RNA chain (Patton et al., 1984). The antigenomic RNA can serve as template for synthesis of the 45-nt minus sense leader RNA (Le-) and also as template for synthesis of full-length complementary genomic RNA. The newly synthesized genomic RNA can feedback into the system to synthesize additional mRNAs in a process termed secondary transcription.

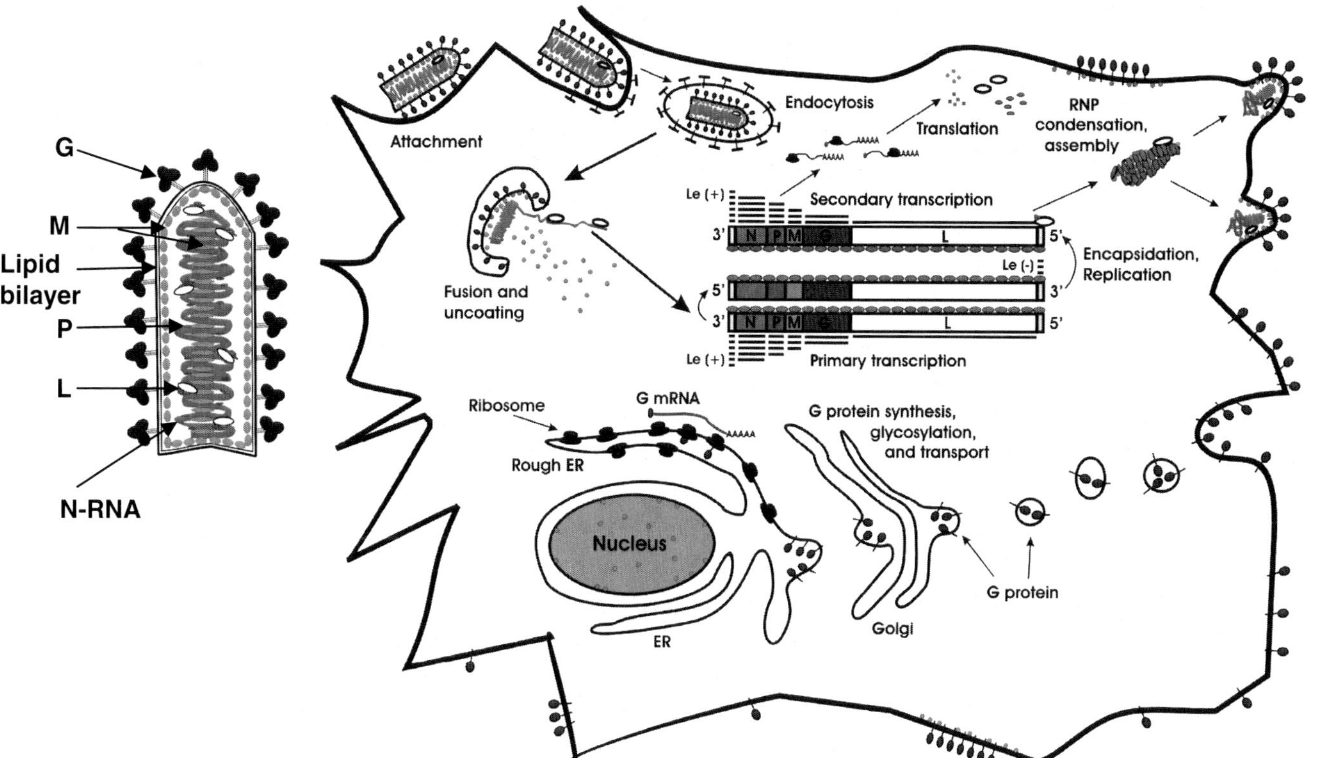

Fig. 8.1 Schematic of VSV particles and viral replication. A diagram of a viral particle is shown, *highlighting* the viral proteins and the genomic N-RNA template. The viral replication cycle is depicted at the *right*. Courtesy of David Cureton, Harvard Medical School.

The Regulatory Regions of the Genome

A schematic of the VSV genome highlighting the regions that regulate polymerase behavior is shown in Fig. 8.2. The boundaries of the viral *N*, *P*, *M*, *G*, and *L* genes are defined by the conserved gene-start (GS) and gene-end (GE) sequences, 3′-UUGUCNNUAG-5′ and 3′-AUACUUUUUUU-5′, respectively. These conserved elements together with the non-transcribed intergenic region (IGR) 3′-G/CA-5′ govern polymerase behavior at each of the gene junctions. The five genes linear arranged 3′ N-P-M-G-L 5′ are flanked by the genomic termini referred to as the 3′ leader (Le) and 5′ trailer (Tr) regions. The short 50-nt Le and 59-nt Tr regions serve as the promoters or "master regulators" of polymerase behavior. Although not every position within the conserved elements or the termini are essential for controlling polymerase activity, these regions are rich with signals that can sometimes serve distinct functions in the template and nascent RNA strands. This remarkable genetic economy results in these control elements amounting to 224 nucleotides or 2% of the viral genome. These signals serve to recruit polymerase, promote the initiation of RNA synthesis, 5′ and 3′ modification of mRNA, termination of RNA synthesis, and the encapsidation of the genomic and antigenomic replication products.

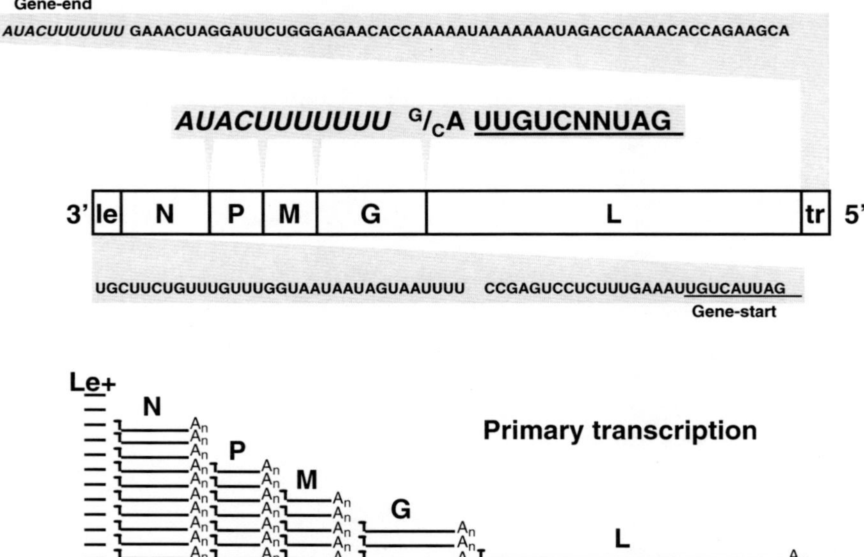

Fig. 8.2 Regulatory regions of the VSIV genome. The viral genome is shown 3′–5′ the leader region, N-P-M-G-L genes and the trailer region. The sequence of the regulatory regions is shown 3′–5′. The gene-start element is highlighted in *green* and the gene-end in *red*. The products of primary transcription are also shown.

The Viral Proteins Required for Gene Expression

The genomic RNA template does not exist as naked RNA, rather it is always found encapsidated by N protein. This encapsidated RNA is completely resistant to ribonuclease attack. It is in this encapsidated form that the RNA-dependent RNA polymerase (RdRP) recognizes the RNA. Although not yet formally demonstrated it seems likely that polymerase must either displace N or substantially remodel the protein in order to access the bases of the RNA and copy the template. The viral components of the RdRP are the P and L proteins. Host cell proteins have been reported to be required for polymerase function although their exact roles in gene expression have not been defined. Viral protein synthesis is essential for genome replication and is necessary to provide N to encapsidate the nascent RNA strand (Patton et al., 1984). Evidence indicates that the polymerase that replicates the genomic RNA is distinct from that responsible for mRNA synthesis and comprises N-P-L, whereas for mRNA synthesis the polymerase comprises L-P-EF1α-Hsp60 and the host guanylyltransferase (Qanungo et al., 2004). Although the viral matrix (M) protein plays a role in downregulating viral transcription and shutting of the host, it is not essential for viral RNA synthesis and will not be considered here. Similarly, the viral attachment glycoprotein (G) plays no direct role in RNA synthesis.

Nucleocapsid Protein

N protein comprises 422 amino acids and has a molecular weight of 48 kDa. N completely encapsidates the negative-sense genomic and positive-sense antigenomic RNAs. Encapsidation is such that the RNA is resistant to cleavage by ribonuclease. The three-dimensional x-ray crystal structure of an N-RNA complex has been solved for both VSIV (Green et al., 2006) and Rabies virus (Albertini et al., 2006). This revealed that N has a bilobed structure with 9-nt of RNA sequestered between the lobes. Neither the genome of VSIV nor the RV is exact multiples of 9, suggesting that some RNA is not encapsidated or that some N protein is not associated with RNA. A striking feature of these structures is that each monomer of N has long arm-like protrusions which in the N–RNA complex mediate association with adjacent subunits holding the structure together. The overall structure indicates that N protein must be either transiently displaced or substantially remodeled during copying of the RNA genome by the polymerase. The mechanism by which this occurs is not understood. For VSIV the residues of N that bind the phosphate backbone of the RNA are R143, R146, K155, K286, R317, and R408. Residue K155 is not conserved among *Rhabdoviridae* but the remaining residues are conserved consistent with their role in RNA binding. How genomic and antigenomic RNAs become encapsidated is not well understood. N protein binding to RNA must be sequence independent, although the nucleation of encapsidation may be favored on RNA of specific sequence. Encapsidation requires N to be in a soluble form, termed N^0, which depends on association with the phopshoprotein P (La Ferla and Peluso, 1989). The residues of N involved in interacting with P are not well understood, although a two hybrid study with the N and P genes of VSNJV implicates the C-terminal 5 amino acids of N. Further, the regions of N that may play a role in recruiting polymerase to the N–RNA template are not well defined. However, trypsin cleavage of the N-RNA template of Rabies virus releases a fragment from the C terminus of N and inhibits binding of the polymerase complex.

Phosphoprotein

P plays at least two roles in infection by serving as an essential polymerase cofactor and by maintaining N protein in a soluble form (P-N^0) that is necessary for RNA encapsidation. For VSIV, P is a highly phosphorylated acidic protein of 265 amino acids. P comprises three distinct domains, an acidic N-terminal domain (domain I) of 150 amino acids that contains three major phosphorylation sites at S60, T62, and S64. Phosphorylation at these residues by host casein kinase II (Barik and Banerjee, 1992) promotes oligomerization of P and assembly of a transcription competent polymerase complex (Gao and Lenard, 1995). The N terminal 123 amino acids of P were shown to bind efficiently to L protein (Emerson and Schubert, 1987). Amino acid changes that prevent phosphorylation at positions 60, 62, and 64 of P are detrimental to mRNA synthesis in reconstituted systems (Pattnaik et al., 1997; Spadafora et al., 1996) and attempts to rescue recombinant virus in which all three sites are altered have been unsuccessful (Das and Pattnaik, 2004). However, overexpression of P that cannot be phosphorylated at these positions partially overcomes the defect in mRNA synthesis (Spadafora et al., 1996). Using a two-hybrid assay, an interaction between VSNJV N and P was mapped to domain I of P, which is presumed to represent the P–N^0 complex (Takacs et al., 1993).

Domain I is separated from domain II by a highly variable hinge region comprising amino acid residues 150–210. Genetic and biochemical evidence indicates that amino acid residues 201–220 of this hinge region play an important role in both mRNA synthesis and genome replication (Das and Pattnaik, 2005). The entire hinge region does not appear to be critical for virus replication as amino acids 150–191 can be deleted and an attenuated virus recovered (Das and Pattnaik, 2005). Deletions within this region suggested that residues 161–210 are required for oligomerization of P (Chen et al., 2006). Consistent with this, a synthetic peptide corresponding to residues 191–210 served as an inhibitor of transcription. The crystal structure of a proteinase K-resistant fragment that spans a portion of domain I and the hinge region (residues 107–177) of P-protein was solved (Ding et al., 2006). This region of P comprised two beta sheets separated by an alpha helical region and was shown to form a dimer. This was suggested to represent the oligomerization domain of P which mediates tetramerization by interactions between dimers (Ding et al., 2006). For Sendai virus, a paramyxovirus, the P protein oligomerization domain was mapped and the crystal structure solved as an unusual tetrameric-coiled coil. Each monomer comprises three short N terminal helical regions followed by an extended C terminal helical region (Tarbouriech et al., 2000). This structure is thus quite distinct from that solved for the proteinase K resistant fragment of VSV P

(Ding et al., 2006). Given the genetic evidence described above it seems likely that this region of P may not represent the full oligomerization domain. Domain II of P comprises residues 210–244 and contains additional phosphorylation sites at S226, S227, and S233. The kinase responsible is unknown, but phosphorylation at these residues is important for RNA replication (Hwang et al., 1999). Domain III is basic and comprises the C-terminal 21 amino acids. The C terminus of P is required to bind the N–RNA template.

Large Polymerase Protein

The large (L) polymerase protein is a multifunctional protein comprising 2109 amino acids. This protein assembles into a complex comprising P–L and possibly host factors that engage the N–RNA template. In addition to catalyzing ribonucleotide polymerization, L is responsible for the cotranscriptional modification of the 5′ and 3′ termini of the viral mRNAs so that they are capped, methylated, and polyadenylated. Thus, L functions to (i) bind template through its interaction with the P protein, although L must gain direct access to the RNA bases during synthesis; (ii) polymerize NTP's to RNA; (iii) generate the mRNA cap structure, a reaction that involves conversion of the 5′ pppRNA to the 5′ pRNA and its transfer onto GDP derived from GTP; (iv) guanine-N-7 methylate the mRNA cap structure; (v) 2′-O ribose methylate the mRNA cap structure; (vi) polyadenylate the mRNA; (vii) replicate the genomic and antigenomic RNA. To date, three-dimensional structures are not available for any portion of the L protein. However, genetic and biochemical evidence has mapped several of the functions within L.

Amino acid sequence alignments of NNS RNA virus L proteins led to the identification of six regions of sequence conservation separated by regions of no or low sequence homology. These regions are thought to represent the functional domains of L protein (Fig. 8.3). Conserved region I (CRI) has been implicated in playing a role in interacting with the P protein and the P–N^0 complex during encapsidation of the nascent RNA chain during replication. This is based on observations with SeV and has not been directly examined with either VSIV or RABV (Chandrika et al., 1995). CRII contains a conserved charged motif that was suggested to play a role in template binding and clustered charged to alanine substitutions in this region of the SeV L protein inhibit RNA synthesis (Smallwood et al., 1999). CRIII contains clearly identifiable motifs found in all polymerases, and alterations to a universally conserved aspartic acid residue (D714A) eliminate polymerase activity in reconstructed RNA synthesis assays (Sleat and Banerjee, 1993). CRIV is poorly characterized, although clusters of charged alanine substitutions differentially affect replication and transcription functions of the SeV L protein (Feller et al., 2000). Positions G1154, T1157, H1227, and R1228 are present within CRV of L and are universally conserved among the NNS RNA virus L proteins. Substitutions at these positions inhibit mRNA cap formation (Li et al., 2007b) and suggest that this region of L may function as the polyribonucleotidyltransferase (Ogino and Banerjee, 2007). CRVI of L protein has been shown to function as the mRNA cap methylating enzyme.

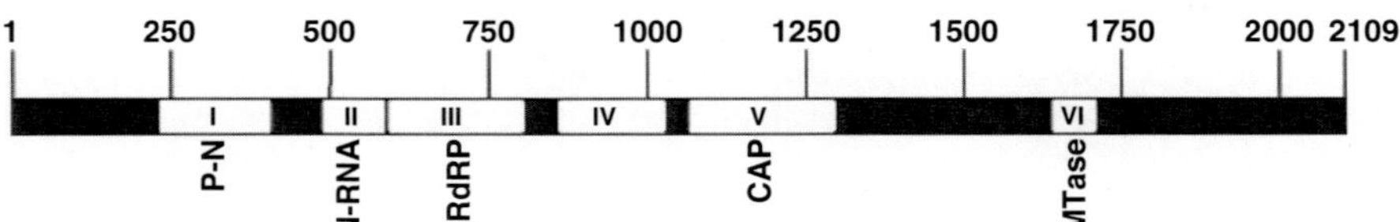

Fig. 8.3 Functions of the large polymerase protein. A diagram of the VSV L protein is shown with the conserved regions I–VI *highlighted*. For VSV genetic and biochemical evidence supports the assignment of the RdRP, capping and methylase activities. Interaction with the N-RNA template and the P–N complex is extrapolated from studies with Sendai virus.

Specifically amino acid substitutions to K1651, D1762, K1795, and E1833 (Li et al., 2005) and a presumed SAM-binding site provided by G1670, G1672, and G1675 (Li et al., 2006) disrupt mRNA cap methylation at both the guanine N-7 and the ribose 2′-O positions of the cap structure. In addition a host range mutant of VSV that is defective in mRNA cap methylation contains amino acid substitutions in CRVI of L (Grdzelishvili et al., 2005). Furthermore, substitutions in CRII and CRVI have been associated with viral resistance to methylase inhibitors (Li et al., 2007a) suggesting that these residues can influence methylation. Amino acids 1638–1673 of VSIV L protein are required for interaction with P protein based on a coimmunprecipitation assay (Canter and Perrault, 1996). This is also consistent with studies on RABV L protein that show the C terminal 566 amino acids of L protein interact with P (Chenik et al., 1998).

Host Factors Associated with the Polymerase Complex

The viral components of the polymerase complex are P and L, which are required to recognize the N-RNA template (Emerson and Wagner, 1972; Emerson and Yu, 1975). Several host factors have been identified that copurify with the polymerase and are proposed to be functionally relevant for virus replication (Das et al., 1998; Gupta et al., 2002). Specifically, the putative transcriptase comprises P, L, EF-1α, Hsp70, and the host RNA guanylyltransferase, whereas the putative replicase comprises N, P, and L proteins (Qanungo et al., 2004). While this report provides evidence for two functionally distinct populations of polymerase, the significance of the host factors in these complexes is not well understood. Evidence is lacking regarding their functional significance, as the sites of interaction with the polymerase proteins have not been defined and the consequences of disruption of these interactions have not been examined.

Initiation of RNA Synthesis

Ultraviolet mapping studies showed that the transcription of the five viral genes was sequential. Low dose UV radiation induces the formation of covalent dimers between adjacent pyrimidine dimers in nucleic acids. Such pyrimidine dimers act as blocks to polymerase during RNA synthesis and thus lead to chain termination. Exposure of VSV to UV radiation allowed the determination of the relative sensitivity of each viral gene. The only gene whose UV target size was approximately equivalent to its physical size was that encoding the N protein. The target size of each of the other genes was substantially greater than their physical size and corresponded nicely to their physical size plus that of the genes that preceded them on the template. This mapped the viral gene order as N-P-M-G-L and demonstrated that transcription occurred by polymerase entering the genome at a single 3′ proximal site (Abraham and Banerjee, 1976; Ball and White, 1976).

How then does polymerase gain access to the N gene? To address this question the VSV transcription reaction was reconstituted in vitro from isolated N-RNA template and polymerase purified from virus. When these components were incubated together with ATP and CTP, a dinucleotide product pppApC was isolated that corresponds to polymerase initiating synthesis at position 1 of the viral genome. Under these partial reaction conditions, initiation at the N gene-start site would be anticipated to yield pppApApC. Such a product was only observed if the reconstituted reactions had been preincubated with all NTP's (Emerson, 1982). This finding lent strong support to the idea that in order to initiate synthesis at the N gene-start site, the polymerase must first initiate synthesis at the 3′ end of the genome. Presumably then, polymerase would terminate at the end of the leader region to release a free leader RNA prior to its initiation at the N gene-start. Consistent with this idea, leader RNA is typically produced in molar excess of N mRNA during in vitro transcription reactions. Combined with the observation that N protein can encapsidate the leader RNA (Blumberg et al., 1981) these findings led to the long held model that polymerase initiates at the 3′ end of the genome to sequentially synthesize the leader RNA and the 5 mRNAs. In the presence of N protein, termination at the end of the leader was suppressed leading to the polymerase reading through the leader-N gene junction to yield a full-length antigenomic RNA. Although this model was consistent with and offered an explanation for much available data, a number of observations were incompatible with this model. Specifically, this model demanded that leader RNA always be synthesized in at least equimolar quantity to N mRNA. Remarkably, for a VSV mutant termed polR1 which has a single amino-acid change in the template-associated N protein, N mRNA is synthesized in an approximately twofold molar excess of leader RNA. This is incompatible with a requirement for 3′ initiation prior to N synthesis and instead supports internal initiation at the N gene-start (Chuang and Perrault, 1997). Second, the availability of reverse genetics permitted application of UV mapping to engineered templates and allowed measurement of the effects of changing the UV target size of the leader region on short genes inserted between the leader and the N gene. In these experiments, it was shown that the UV target size of the leader region could be altered by changing its base composition, such that increasing the number of adjacent uracil bases increased the sensitivity of the gene to UV radiation and decreasing the number of adjacent uracil bases decreased the sensitivity of the gene to UV radiation. Remarkably, changing the sensitivity of the leader gene in this way had no affect on the UV sensitivity of a short 60-nt gene inserted between leader and N as measured in infected cells but did so in vitro (Whelan and Wertz, 2002). Third, polymerase purified from infected cells could be isolated as two functionally distinct pools. In transcription reactions reconstituted with ATP, CTP and GTP these two polymerase fractions made different products. A fraction referred to as the putative replicase generated an 18-nt product. This corresponds to polymerase pausing at the first site for UTP incorporation in the leader region of the genome and indicates 3′-end initiation. A second pool of polymerase, the putative transcriptase, generated a 5-nt product. This would correspond to polymerase pausing at the first site for UTP incorporation following initiation at the N gene-start site (Qanungo et al., 2004). These three studies thus provide independent evidence that the VSV polymerase can initiate internally at the N gene-start site in a manner that does not depend on the prior transcription of the leader RNA.

How polymerase loads on to the template is not understood. The crystal structure of an N-RNA oligomer although not authentic viral sequence likely reflects the authentic template structure (Green et al., 2006). This structure indicates that the N protein must be remodeled or displaced during copying of the RNA chain in order for the polymerase active site to gain access to the bases. How this is achieved is uncertain. One possibility is that the polymerase loads onto to the template at the extreme 3′ end and sequentially remodels the N-RNA complex perhaps through the equivalent of an interaction between disordered regions of N and P as has been proposed for the paramyxoviruses (Kolakofsky et al., 2004). This would permit the polymerase to gain access to the RNA bases. Although displacement or remodeling of N sequentially from the 3′ end is an attractive idea, polymerase could also engage the N-RNA template and introduce a localized distortion such that it gains direct access to the N gene start. It will be of significant interest to determine the crystal structure of the polymerase active site to see if the active site is a closed channel through which RNA must be threaded or whether the active site is not enclosed which would perhaps facilitate internal loading of polymerase on the N-RNA template.

Cis-Acting Sequences Within the Leader Region

However the polymerase gains access to the template to initiate RNA synthesis, the 50-nt leader region serves as a master regulator that controls viral mRNA synthesis and genome replication. Changes engineered into this region of the genome can impact RNA encapsidation by N protein, recruitment of the polymerase to template, leader synthesis, mRNA synthesis, and genome replication. Because this region of the genome is so richly packed with control sequences, dissecting their individual functions has been problematic. In part this has been due to the lack of tractable systems to separately study these processes. However, significant information has been gained from studying the role of the 3′ leader region in control of mRNA synthesis and genome replication (Li and Pattnaik, 1997, 1999; Pattnaik et al., 1995; Wertz et al., 1994; Whelan and Wertz, 1999).

A clue as to functional dissection of the leader region can be obtained by comparing it to the complement of the genomic trailer region, which must also contain essential *cis*-acting signals for RNA encapsidation, polymerase recruitment, synthesis of a short triphosphorylated leader RNA, and replication of the genome. As can be seen from the sequence (Fig. 8.2) the highest regions of sequence conservation are at the 3′ terminus of both the leader and the trailer complement. This likely reflects a shared property of both these promoters in controlling polymerase recruitment, initiation, or encapsidation of the nascent RNA. Positions 19–28 and 34–46 show less conservation and switching these regions between the leader and the trailer complement promoters diminishes mRNA synthesis consistent with these elements serving as a determinant of the promoter for mRNA synthesis. However, the elements in the complement of the trailer region have been described as a replication enhancer element, although this element can be replaced with the sequences from the leader region. The exact details of how this region serves as the promoter for mRNA synthesis have thus not been determined.

The Gene-Start Sequence

For VSIV the conserved GS sequence comprises 10 nucleotides: 3′-UUGUCNNUAG-5′. The role of this sequence element in governing polymerase behavior has been well studied. Initial work used a VSV genomic analog encoding the M and G genes and reported the effect of changes to the conserved elements of the GS. These studies defined a consensus sequence 3′-UYGNNNNNNN-5′ that was required for initiation of mRNA synthesis and efficient G protein expression in cells (Stillman and Whitt, 1997). The conclusion of this study was elegantly revised through the use of an engineered two-segment virus system, in which one segment comprised Le-N-P-M-L-Tr and the other segment comprised Le-G-GFP-CAT-Tr. Using this system, the effects of modifying positions 1–3 of the GS that controlled CAT expression were evaluated. VSV packages its polymerase into the viral particles, and disruption of purified particles with detergent and incubation with NTP's permits RNA synthesis in vitro. This permitted a re-evaluation of how GS1-3 mutations affected gene expression. In contrast to the earlier reported requirement for initiation of 3′-UYG-5′, products were detected for each of the GS mutants in vitro with almost wild-type levels of initiation being detected for U1C and the U2 substitutions. Consistent with the earlier study, transcripts from these mutants were not detected in infected cells suggesting that they may be unstable. Further analysis showed that transcripts synthesized by these mutants were truncated ranging from 40 to 200 nt and were not reactive with an antibody raised against 2,2,7 trimethylguanosine. This suggested that these short RNAs lacked an mRNA cap structure and/or were not methylated indicating that the conserved GS sequence was required not only for mRNA initiation but also to ensure correct cap formation and/or polymerase processivity. This work led to a model in which cap formation was required for the processivity of polymerase (Stillman and Whitt, 1999).

The importance of the GS sequence in initiation and cap formation was also addressed in the context of a 60-nt non-essential gene inserted between the leader and the N genes of the viral genome. Each of the eight conserved positions of the GS sequence were individually altered to each of the other positions and the effects on RNA synthesis examined using detergent activated virus in vitro. With the exception of mutant U8G, products were detected from the +60 gene for each virus. The cap status was evaluated by determining the sensitivity of the RNA to exonuclease digestion and by mapping the 5′ end of the RNA following treatment with the cap-cleaving enzyme tobacco acid pyrophosphatase (TAP). These experiments showed that many single nt changes to the GS inhibit mRNA cap formation, although the most significant effect was seen when the GS deviated from the consensus UYG (Wang et al., 2007).

The above studies provided strong support that the capping apparatus of VSV required a specific sequence provided by the GS, but they could not discriminate whether this was provided by the GS site itself or was present in the nascent RNA chain. Using recombinant VSV polymerase, it was demonstrated that a 5-nt RNA could be capped by L protein in the presence of GTP (Ogino and Banerjee, 2007). In contrast a 5-nt RNA that corresponded to the 5′ end of the VSV leader RNA could not be capped in *trans*. This confirmed that specific signals were required for cap formation and defined that these must be present in the nascent RNA. Thus, the GS element serves as a critical signal in the template that signals polymerase initiation, and its counterpart the 5′ end of the nascent RNA serves as a critical signal for recognition by the capping machinery.

mRNA Cap Formation

Although rhabdoviruses generate a cap structure that is indistinguishable to that on host mRNA's the route by which they achieve this is distinct. Conventional mRNA cap formation requires four reactions. First, the triphosphate end of the mRNA is trimmed by an RNA triphosphatase to yield a diphosphate acceptor ppNpN. An RNA guanylyltransferase then transfers GMP derived from GTP through a 5′–5′ linkage to form GpppNpN. This capped RNA is then sequentially methylated. A guanine-N-7 (G-N-7) MTase transfers a methyl group from *S*-adenosyl-L-methionine (SAM) onto the N7 position of the cap structure to yield 7mGpppNpN. A ribose 2′-O (2′-O) methylase can then transfer a second methyl group using SAM as donor to yield 7mGpppNmpN. With the exception of the RNA triphosphatase reaction which comes in two flavors, a metal-dependent enzyme or a cysteine phosphatase-dependent enzyme, the mechanism by which each step of mRNA cap formation occurs is well conserved among eukaryotes. For NNS RNA viruses, the 5′ end of the pppApApCpApG mRNA is capped by an unusual enzymatic activity, in which the 5′ end of the mRNA is transferred onto GDP derived from GTP through a covalent L-monophosphate RNA intermediate (Abraham et al., 1975; Ogino and Banerjee, 2007). This polyribonucleotidyltransferase activity contrasts with all other known mRNA capping reactions that are catalyzed by an RNA triphosphatase and RNA guanylyltransferase (Furuichi and Shatkin, 2000). The resulting GpppApApCpApG cap structure is then methylated at both G-N-7 and 2′-O positions to yield 7mGpppAmpApCpApG (Moyer et al., 1975). Failure to add the mRNA cap structure leads to frequent intragenic termination, whereas failure to methylate the cap does not lead to such a termination effect. Rather inhibition of methylation can be associated with the synthesis of a giant polyadenylate (Rose et al., 1977). Transcription of mRNA occurs at an approximate rate 3–4 nt s^{-1} but with significant pausing (2.5–5.7 min) in close proximity to the gene junctions (Iverson and Rose, 1981).

The Gene-End Sequence

For VSIV the conserved GE sequence comprises 11 nucleotides: 3′-AUACUUUUUUU-5′. Like the GS element, the role of this sequence in control of polymerase behavior has been extensively studied. This sequence element contains essential signals that signal reiterative synthesis on the U tract to generate the polyadenylate tail and the process of mRNA termination. Consistent with the start-stop mechanism of sequential viral mRNA synthesis, termination of an upstream gene is an essential prerequisite for polymerase to initiate transcription of the downstream gene. For VSIV, the minimal transcription termination signal comprises the GE and the first nucleotide of the IGR (Barr et al., 1997a, b; Hwang et al., 1998; Stillman and Whitt, 1997). This signal is highly efficient in VSIV such that polymerase ignores this signal to generate a read-through RNA approximately 1–3% of the time (Iverson and Rose, 1981). Changes to the 3′-AUAC-5′ sequence diminished the ability of the GE to signal termination, with the C being completely intolerant of alteration (Barr et al., 1997a; Hwang et al., 1998). Similarly, shortening the U-tract by a single nt or inserting a non-U residue within the tract abolished termination (Barr et al., 1997a; Hwang et al., 1998). In contrast, increasing the U tract to 14 residues had a modest effect on termination but diminished transcription of the downstream gene (Barr et al., 1997a). The finding that alterations to the U-tract-inhibited termination as well as reiterative copying of the U tract suggested that either reiterative transcription or synthesis of polyadenylate was critical for termination (Barr et al., 1997a). Replacement of the U7 tract with an A7 tract was tolerated for reiterative copying, but not for termination indicating that U7 in the template or poly A in the nascent mRNA was critical for termination. Further analysis of the AUAC element revealed that the AU rich nature of positions 1–3 rather than its precise sequence was critical for termination (Barr and Wertz, 2001). This suggested that weak base pairing probably between the template and the nascent strand may control the initiation of reiterative transcription on the U7 tract. Precedent for this is provided by *Escherichia coli* RNAP which has been shown to backtrack in the presence of weak base pairing between the template and the nascent strands (Nudler et al., 1997). Perhaps the AU element destabilizes the hybrid between nascent RNA and template to facilitate the initiation of polymerase stuttering on the U7 tract. This process of reiterative transcription to generate the poly A tail is similar to the process of cotranscriptional editing of the P genes of the paramyxoviridae where a slippery sequence leads to the programmed insertion of 1–6 G residues to permit access to alternate open reading frames in the mRNA (Hausmann et al., 1999).

Although the GE of VSV seems to be optimal for signaling termination, in some contexts the GE is completely ignored by the polymerase (Whelan et al., 2000). Specifically, when the GE is located shortly after a GS element, polymerase fails to respond, suggesting that termination depends on a modification either to polymerase or to the nascent mRNA, during an early stage of synthesis (Whelan et al., 2000). This also explains how some *Rhabdoviridae* such as Sigma virus can have overlapping genes in which the GS sequence of the downstream gene is prior to the GE of the upstream gene. Provided that the distance between the GS and the GE is short, the polymerase will not terminate in response to this signal.

In addition to its central role in mRNA termination, the GE also plays a role in transcription of a downstream mRNA. For VSV the GE was duplicated such that an upstream copy was used for termination and polyadenylation of the first mRNA rendering the second copy of the GE amenable for analysis. Using this system, the U7 tract was shown to be required for efficient synthesis of the downstream mRNA. Thus, the GE sequence is multifunctional, providing signals for termination and polyadenylation of the upstream mRNA and also signals for efficient transcription of the downstream gene (Hinzman et al., 2002).

The Intergenic Region

The IGR is 3′-GA-5′, except at the P/M junction where it is 3′-CA-5′. However, this difference does not seem to be significant for viral replication (Ball et al., 1999). The IGR was changed to each of the 15 other possible combinations which showed that first nt affects termination, whereas the second nt primarily affects initiation at the downstream gene (Barr et al., 1997b; Stillman and Whitt, 1997, 1998). More drastic changes to the IGR that altered sequence length are consistent with this general principle, suggesting that its role is primarily to separate the U7 tract from the GS.

Viral Protein Synthesis

The viral mRNA's are efficiently translated by the host translation machinery, while cellular translation is inhibited. How viral mRNAs compete with cellular mRNA for translation is not well understood given that their 5′ cap and 3′ poly A tails are essentially indistinguishable. Two models were postulated for the translational advantage. The first model suggests that a specific attribute of viral mRNA's facilitates their efficient translation, however, no such feature has been defined and attempts to test the role of the termini of a viral mRNA have suggested that it is not the sequence of the RNA rather the act of transcription from the viral genome (Whitlow et al., 2006). The second model postulates that the host translational machinery is sequestered in infected cells so that the viral mRNAs gain access to these components. Viral infection results in alterations to the host translational machinery, notable among which is the dephosphorylation of the eIF-4E binding protein 4E-BP1 (Connor and Lyles, 2002). Eukaryotic initiation factor 4E is a critical component of the host translation machinery being essential for recognition of the G-N-7 methylated mRNA cap structure. Dephosphorylation of the 4E-BP1 leads to association with eIF-4E, effectively sequestering it away from translation. This raises the question of how the viral capped and methylated mRNAs are translated and additional work will be required to understand this.

RNA Encapsidation and Genome Replication

Encapsidation of the genomic and antigenomic RNA's occurs concomitant with their synthesis. Although N protein will bind to other RNA's including the viral mRNAs and the leader RNAs they are not the preferred template for encapsidation in that significant quantities of leader RNA are not encapsidated. A specific sequence that is required for N protein encapsidation has not been defined, but sequence alignments led to initial speculation that an A residue located at every third position in the leader RNA was required for this process (Blumberg et al., 1983). Ongoing encapsidation is necessary to drive genome replication. How this is achieved is not certain, but evidence suggests that the RNA replicase is distinct to the transcriptase comprising N-P-L (Qanungo et al., 2004). During replicative synthesis the polymerase initiates synthesis at the 3′ end of the genome and ignores all the signals that control mRNA synthesis.

Current Model for Viral Gene Expression

Available data support the following model for viral gene expression. The viral polymerase, comprising P and L and possibly host proteins, is recruited to the template. By virtue of the interaction of P and N, the N-RNA is transiently remodeled to permit access of the L protein to the bases of the RNA. The polymerase complex diffuses or scans along the template until it encounters the promoter element which serves to direct initiation of RNA synthesis at the first gene-start signal. For mRNA synthesis, this promoter is provided by sequence elements within the leader region and the conserved elements of the N GS. The polymerase initiates synthesis of a triphosphate RNA pppApApCpApGpNpNpApUpC corresponding to the beginning of the N mRNA. By analogy with transcription in other systems, the initiation phase results in the synthesis and release of many short transcripts. At some point, the polymerase escapes the promoter entering an early elongation phase of RNA synthesis. By analogy to other transcription systems, the polymerase at this stage is no longer undergoing abortive synthesis. However, for VSV the polymerase it is not yet fully processive and terminates frequently during copying of the gene, possibly in response to pause sites that may mimic aspects of the termination signal. This would account for the reported short transcripts in the size range of >12 nt found during transcription reactions in vitro. At some point prior to a nascent RNA chain of 100 nt the polymerase adds the mRNA cap structure to the mRNA and transitions to a fully processive elongation complex. Precisely where this occurs during RNA synthesis is not clear, but the shortest transcript described in vitro that contains an mRNA cap structure at the 5′ terminus is reportedly 37 nt (Piwnica-Worms and Keene, 1983). There is no evidence that directly links the process of 5′ processing and polymerase processivity, but genetic disruption of the *cis*-acting signals that are required for 5′ cap formation or of the regions of the polymerase that control cap formation results in frequent intragenic termination. Following addition of the mRNA cap structure, intragenic termination appears to be rare, such that the polymerase now terminates only in response to an intact authentic gene-end sequence AUACUUUUUUU. Here the AUAC element serves to signal to polymerase to reiteratively copy the U7 tract and generate the polyadenylate tail. The AU rich nature of this signal was critical for the polyadenylation, and this indicated that a hybrid between the nascent RNA and the template may play a central role in regulating polymerase slippage. Indeed it was suggested that the influence of this tetranucleotide was most likely at the onset of the slippage reaction. Following termination, the polymerase can then resume its scanning or diffusion along the template until it encounters the next start sequence. Following the onset of viral protein synthesis, genome replication can begin. Available evidence suggests that a polymerase complex that is distinct from the transcriptase initiates synthesis at the 3′ end of the genome. During replication, the nascent RNA is encapsidated by soluble N protein that is brought to the site of RNA synthesis as a N^0–P complex. Under these conditions the polymerase ignores all the regulatory elements that control mRNA synthesis and produces the full-length antigenome.

Future Perspectives

Vesicular stomatitis virus will continue to serve as an important model system for understanding gene expression in NNS RNA viruses. The future promises to improve our understanding how polymerase gains access to the bases during RNA synthesis, how polymerase activity is regulated between mRNA synthesis and genome replication, and understanding how viral mRNAs are efficiently translated. Structural information regarding the polymerase complex will likely prove invaluable in further understanding this large multifunctional protein and how its different activities in mRNA synthesis including cap formation, reiterative transcription to generate the poly A tail, and genome replication are controlled. Such information might also lead to the rational design of inhibitors that could prove invaluable in battling the significant pathogens in this order of viruses.

References

Abraham, G. and Banerjee, A.K. (1976). *Proc Natl Acad Sci U S A*, **73**, 1504–1508.
Abraham, G., Rhodes, D.P. and Banerjee, A.K. (1975). *Cell*, **5**, 51–58.
Albertini, A.A., Wernimont, A.K., Muziol, T., Ravelli, R.B., Clapier, C.R., Schoehn, G., Weissenhorn, W. and Ruigrok, R.W. (2006). *Science*, **313**, 360–363.
Ball, L.A., Pringle, C.R., Flanagan, B., Perepelitsa, V.P. and Wertz, G.W. (1999). *J Virol*, **73**, 4705–4712.
Ball, L.A. and White, C.N. (1976). *Proc Natl Acad Sci U S A*, **73**, 442–446.
Barik, S. and Banerjee, A.K. (1992). *Proc Natl Acad Sci U S A*, **89**, 6570–6574.
Barr, J.N. and Wertz, G.W. (2001). *J Virol*, **75**, 6901–6913.
Barr, J.N., Whelan, S.P. and Wertz, G.W. (1997a). *J Virol*, **71**, 8718–8725.
Barr, J.N., Whelan, S.P. and Wertz, G.W. (1997b). *J Virol*, **71**, 1794–1801.
Blumberg, B.M., Giorgi, C. and Kolakofsky, D. (1983). *Cell*, **32**, 559–567.
Blumberg, B.M., Leppert, M. and Kolakofsky, D. (1981). *Cell*, **23**, 837–845.
Canter, D.M. and Perrault, J. (1996). *Virology*, **219**, 376–386.
Chandrika, R., Horikami, S.M., Smallwood, S. and Moyer, S.A. (1995). *Virology*, **213**, 352–363.
Chen, M., Ogino, T. and Banerjee, A.K. (2006). *J Virol*, **80**, 9511–9518.
Chenik, M., Schnell, M., Conzelmann, K.K. and Blondel, D. (1998). *J Virol*, **72**, 1925–1930.
Chuang, J.L. and Perrault, J. (1997). *J Virol*, **71**, 1466–1475.
Connor, J.H. and Lyles, D.S. (2002). *J Virol*, **76**, 10177–10187.
Cowton, V.M., McGivern, D.R. and Fearns, R. (2006). *J Gen Virol*, **87**, 1805–1821.
Das, S.C. and Pattnaik, A.K. (2004). *J Virol*, **78**, 6420–6430.
Das, S.C. and Pattnaik, A.K. (2005). *J Virol*, **79**, 8101–8112.
Das, T., Mathur, M., Gupta, A.K., Janssen, G.M. and Banerjee, A.K. (1998). *Proc Natl Acad Sci U S A*, **95**, 1449–1454.
Ding, H., Green, T.J., Lu, S. and Luo, M. (2006). *J Virol*, **80**, 2808–2814.
Emerson, S.U. (1982). *Cell*, **31**, 635–642.
Emerson, S.U. and Schubert, M. (1987). *Proc Natl Acad Sci U S A*, **84**, 5655–5659.
Emerson, S.U. and Wagner, R.R. (1972). *J Virol*, **10**, 297–309.
Emerson, S.U. and Yu, Y. (1975). *J Virol*, **15**, 1348–1356.
Feller, J.A., Smallwood, S., Horikami, S.M. and Moyer, S.A. (2000). *Virology*, **269**, 426–439.
Furuichi, Y. and Shatkin, A.J. (2000). *Adv Virus Res*, **55**, 135–184.
Gao, Y. and Lenard, J. (1995). *EMBO J*, **14**, 1240–1247.
Grdzelishvili, V.Z., Smallwood, S., Tower, D., Hall, R.L., Hunt, D.M. and Moyer, S.A. (2005). *J Virol*, **79**, 7327–7337.
Green, T.J., Zhang, X., Wertz, G.W. and Luo, M. (2006). *Science*, **313**, 357–360.
Gupta, A.K., Mathur, M. and Banerjee, A.K. (2002). *Biochem Biophys Res Commun*, **293**, 264–268.
Hausmann, S., Garcin, D., Delenda, C. and Kolakofsky, D. (1999). *J Virol*, **73**, 5568–5576.
Hinzman, E.E., Barr, J.N. and Wertz, G.W. (2002). *J Virol*, **76**, 7632–7641.
Hwang, L.N., Englund, N., Das, T., Banerjee, A.K. and Pattnaik, A.K. (1999). *J Virol*, **73**, 5613–5620.
Hwang, L.N., Englund, N. and Pattnaik, A.K. (1998). *J Virol*, **72**, 1805–1813.
Iverson, L.E. and Rose, J.K. (1981). *Cell*, **23**, 477–484.
Kolakofsky, D., Le Mercier, P., Iseni, F. and Garcin, D. (2004). *Virology*, **318**, 463–473.
La Ferla, F.M. and Peluso, R.W. (1989). *J Virol*, **63**, 3852–3857.
Li, J., Chorba, J.S. and Whelan, S.P. (2007a). *J Virol*, **81**, 4104–4115.
Li, J., Fontaine-Rodriguez, E.C. and Whelan, S.P. (2005). *J Virol*, **79**, 13373–13384.
Li, J., Rahmeh, A., Morrelli, M. and Whelan, S.P. (2008). *J Virol*, **82**, 775–784.
Li, J., Wang, J.T. and Whelan, S.P. (2006). *Proc Natl Acad Sci U S A*, **103**, 8493–8498.
Li, T. and Pattnaik, A.K. (1997). *Virology*, **232**, 248–259.

Li, T. and Pattnaik, A.K. (1999). *J Virol*, **73**, 444–452.
Moyer, S.A., Abraham, G., Adler, R. and Banerjee, A.K. (1975). *Cell*, **5**, 59–67.
Nudler, E., Mustaev, A., Lukhtanov, E. and Goldfarb, A. (1997). *Cell*, **89**, 33–41.
Ogino, T. and Banerjee, A.K. (2007). *Mol Cell*, **25**, 85–97.
Pattnaik, A.K., Ball, L.A., LeGrone, A. and Wertz, G.W. (1995). *Virology*, **206**, 760–764.
Pattnaik, A.K., Hwang, L., Li, T., Englund, N., Mathur, M., Das, T. and Banerjee, A.K. (1997). *J Virol*, **71**, 8167–8175.
Patton, J.T., Davis, N.L. and Wertz, G.W. (1984). *J Virol*, **49**, 303–309.
Piwnica-Worms, H. and Keene, J.D. (1983). *Virology*, **125**, 206–218.
Qanungo, K.R., Shaji, D., Mathur, M. and Banerjee, A.K. (2004). *Proc Natl Acad Sci U S A*, **101**, 5952–5957.
Rose, J.K., Lodish, H.F. and Brock, M.L. (1977). *J Virol*, **21**, 683–693.
Sleat, D.E. and Banerjee, A.K. (1993). *J Virol*, **67**, 1334–1339.
Smallwood, S., Easson, C.D., Feller, J.A., Horikami, S.M. and Moyer, S.A. (1999). *Virology*, **262**, 375–383.
Spadafora, D., Canter, D.M., Jackson, R.L. and Perrault, J. (1996). *J Virol*, **70**, 4538–4548.
Stillman, E.A. and Whitt, M.A. (1997). *J Virol*, **71**, 2127–2137.
Stillman, E.A. and Whitt, M.A. (1998). *J Virol*, **72**, 5565–5572.
Stillman, E.A. and Whitt, M.A. (1999). *J Virol*, **73**, 7199–7209.
Takacs, A.M., Das, T. and Banerjee, A.K. (1993). *Proc Natl Acad Sci U S A*, **90**, 10375–10379.
Tarbouriech, N., Curran, J., Ruigrok, R.W. and Burmeister, W.P. (2000). *Nat Struct Biol*, **7**, 777–781.
Wang, J.T., McElvain, L.E. and Whelan, S.P. (2007). *J Virol*, **81**, 11499–11506.
Wertz, G.W., Whelan, S., LeGrone, A. and Ball, L.A. (1994). *Proc Natl Acad Sci U S A*, **91**, 8587–8591.
Whelan, S.P., Barr, J.N. and Wertz, G.W. (2000). *J Virol*, **74**, 8268–8276.
Whelan, S.P. and Wertz, G.W. (1999). *J Virol*, **73**, 297–306.
Whelan, S.P. and Wertz, G.W. (2002). *Proc Natl Acad Sci U S A*, **99**, 9178–9183.
Whitlow, Z.W., Connor, J.H. and Lyles, D.S. (2006). *J Virol*, **80**, 11733–11742.

Chapter 9
Orthomyxovirus Genome Transcription and Replication

Paul Digard, Laurence Tiley, and Debra Elton

Viruses from the Orthomyxoviridae family infect a wide range of vertebrate hosts. Five genera are currently recognised: *Influenza A, B* and *C*, *Thogotovirus* (of which the homonymic viruses are the type species) and *Isavirus* (type species infectious salmon anaemia virus). Other so far unclassified and uncharacterised members of the family also exist (Da Silva et al., 2005). These viruses are typified by possessing a single-stranded, negative-sense RNA genome that is split into segments (between 6 and 8 depending on the genus), by replicating their genome in the nucleus and by an unusual mechanism for producing capped mRNAs (discussed in detail later); shared traits that make it likely they are descended from a common ancestor. Influenza A virus is by far the most important in terms of significance to human health and accordingly is the best studied. The World Health Organization estimates that between 5 and 15% of the population in temperate countries are infected each year, resulting in up to 500,000 deaths worldwide. In addition, periodic introduction of antigenically novel viruses from avian reservoirs into the human population causes worldwide pandemics with substantially higher attack rates and mortality levels. The current possibility that a highly pathogenic avian H5N1 strain of influenza A will make this species jump is of great concern (Peiris et al., 2007). Ongoing research has reinforced longstanding concepts that the viral RNA polymerase is an important determinant of host range and pathogenicity (Almond, 1977; Subbarao et al., 1993; Hatta et al., 2001; Tumpey et al., 2005), further justifying research into this area. This chapter provides an overview of orthomyxovirus RNA synthesis, focussing in detail on areas where recent progress has changed the consensus view of the molecular mechanisms involved. For in-depth treatments of other aspects, the reader is referred to other reviews (Amorim and Digard, 2006; Elton et al., 2006; Engelhardt and Fodor, 2006; Ortin and Parra, 2006).

P. Digard (✉)
Department of Pathology, University of Cambridge, Tennis Court Road, Cambridge, CB2 1QP, United Kingdom
e-mail: pd1@mole.bio.cam.ac.uk

C.E. Cameron et al. (eds.), *Viral Genome Replication*,
DOI 10.1007/b135974_9,

Overview of Orthomyxovirus RNA Synthesis

Each orthomyxovirus segment contains conserved sequences at the 3′ and 5′ termini that share partial sequence complementarity (Skehel and Hay, 1978; Robertson, 1979; Desselberger et al., 1980; Staunton et al., 1989) and can base-pair to form a panhandle structure (Hsu et al., 1987; shown schematically in Fig. 9.1a). The genome functions as ribonucleoproteins (RNPs), where each vRNA segment is separately encapsidated by the nucleoprotein (NP) and associated with one copy of the viral RNA-dependent RNA polymerase (Fig. 9.2a). NP plays an essential role in maintaining the structure of the RNPs. The viral polymerase complex is a heterotrimer composed of two basic proteins, PB1 and PB2, and the more acidic PA (Horisberger, 1980; Detjen et al., 1987; Honda et al., 1990).

Unlike most negative-sense RNA viruses, transcription of orthomyxovirus genomes takes place in the nucleus of infected cells (Herz et al., 1981; Siebler et al., 1996). During the infectious cycle, two types of positive-sense RNA molecules are transcribed from vRNA (Fig. 9.1b). Synthesis of capped and polyadenylated messenger RNAs (mRNAs) is primed by short capped oligonucleotides of around 10–12 nucleotides, which are scavenged from host cell pre-mRNAs by the combined cap-binding and endonuclease activities of the PB2 and PB1 components of the polymerase complex, respectively (Plotch et al., 1981; Li et al., 1998). Influenza mRNAs therefore contain non-templated host cell-derived sequences at their 5′ ends (Fig. 9.1a). Transcription terminates 15–17 nt before the 5′ end of the vRNA segment and the mRNA is polyadenylated by a process of stuttering on a poly (U) tract (Fig. 9.1a; Hay et al., 1977a; Robertson et al., 1981; Luo et al., 1991; Poon et al., 1998). In contrast, synthesis of the positive-sense cRNA involves unprimed

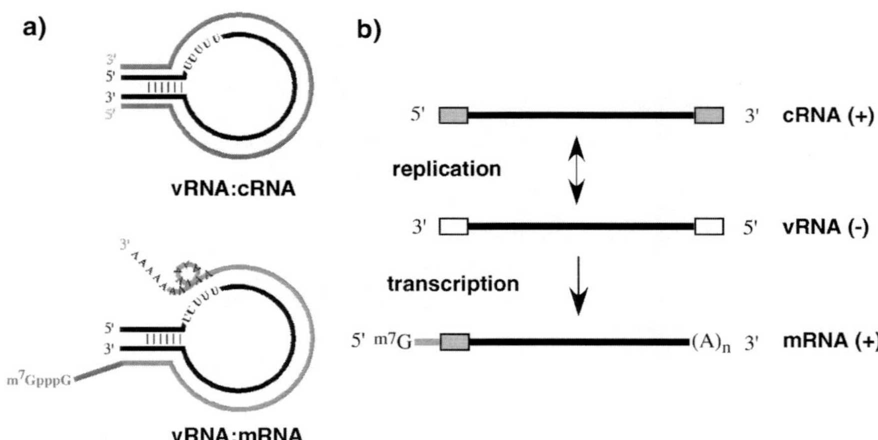

Fig. 9.1 Overview of influenza virus RNA synthesis. (**a**) Schematic depiction of the complementarity between m, c- and vRNA. *Black lines* indicate vRNA, *grey lines* positive-sense RNA as labelled. (**b**) Schematic depiction of the synthetic relationship between influenza virus RNA species. *Boxes* represent the conserved terminal promoter regions.

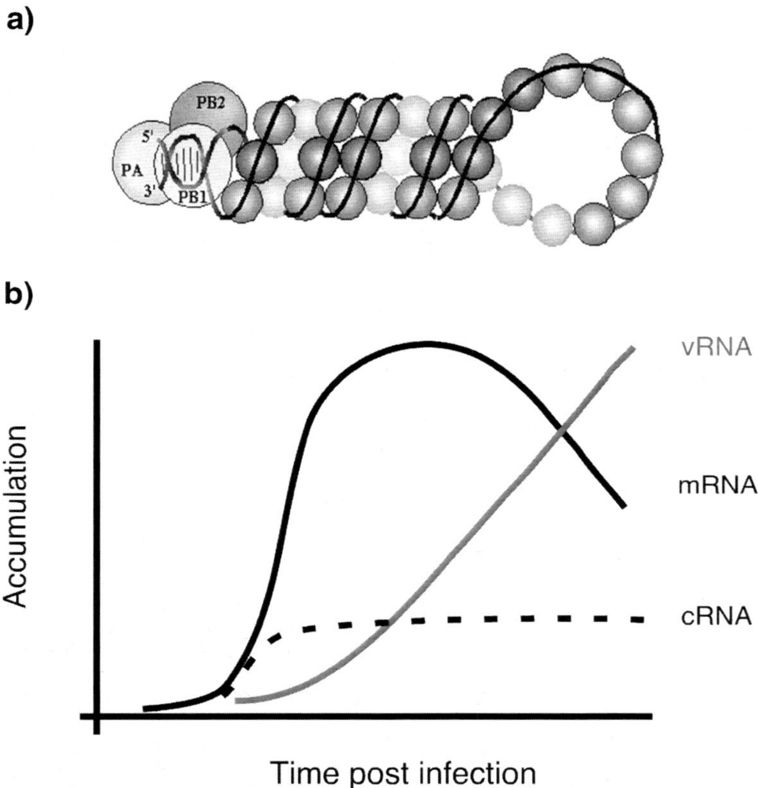

Fig. 9.2 (**a**) Cartoon depiction of an influenza virus RNP. *Small spheres* represent NP, *large spheres* the polymerase subunits and the *black line* genomic RNA. (**b**) Schematic depiction of the temporal regulation of viral RNA accumulation.

initiation (Hay et al., 1982) and read-through of the poly (U) tract to produce unit length copies of the vRNA template (Fig. 9.1a). These cRNAs exist as low abundance RNPs and are the replicative intermediates for synthesis of new copies of vRNA (Fig. 9.1b), required for production of progeny virions. Production of these three species of viral RNA is temporally regulated in infected cells (Fig. 9.2b). Synthesis of mRNA occurs first, catalysed by the input vRNPs and independent of viral protein synthesis (Hay et al., 1977b; Taylor et al., 1977; Barrett et al., 1979). Maximum rates of mRNA synthesis occur around 2.5 h post-infection so the peak amounts of viral mRNA occur relatively early in infection after which levels decline (Hay et al., 1977b; Barrett et al., 1979; Mark et al., 1979; Smith and Hay, 1982; Shapiro et al., 1987; Mullin et al., 2004). Normally, cRNAs can only be detected after mRNA synthesis, consistent with their dependence upon viral protein synthesis (Hay et al., 1977b; Barrett et al., 1979), but their maximal rate of synthesis occurs before that of viral mRNA (Barrett et al., 1979; Shapiro et al.,

1987). Synthesis of vRNA follows cRNA and continues to increase even after synthesis of the other classes of RNA declines (Hay et al., 1977b; Barrett et al., 1979; Shapiro et al., 1987; Mullin et al., 2004). Differential expression of the viral gene products is achieved both transcriptionally and post-transcriptionally. Expression of the polymerase genes remains relatively low throughout infection and this correlates with the low abundance of their mRNAs (Hay et al., 1977b; Smith and Hay, 1982; Enami et al., 1985). NP and NS1 expression predominates at early times post-infection while synthesis of the major virion structural proteins M1 and HA lags until later times. It has been proposed that this reflects the kinetics of individual mRNA synthesis and that this is in turn regulated by differential synthesis of the vRNA templates from which the mRNAs are transcribed (Smith and Hay, 1982; Shapiro et al., 1987; reviewed by Elton et al., 2006). However, there is also the possibility of post-transcriptional regulation occurring via delayed nuclear export of the 'late gene' mRNAs (Hatada et al., 1989; Amorim et al., 2007). A further layer of post-transcriptional regulation undoubtedly occurs for the NS1/NS2 and M1/M2 genes which are expressed through differential splicing of primary transcripts and here the reader is referred to a prior review (Ortin and Parra, 2006).

Structure and Assembly of Viral RNPs

Early electron microscopy studies of virion RNPs showed helical ribbon structures with a terminal loop (Pons et al., 1969; Jennings et al., 1983; shown in cartoon form in Fig. 9.2a). The polymerase is present at one end of the structure, as shown by immunogold labelling (Murti et al., 1988). The polymerase maintains the association of the 5′ and 3′ ends of the RNA (Klumpp et al., 1997) forming the closed structure in the RNP. Artificial complexes generated in vitro with NP and RNA are structurally and biochemically similar to natural RNPs (Yamanaka et al., 1990). Furthermore, the helical form of the RNP is maintained when the RNA is replaced by negatively charged polymers, suggesting that NP determines RNP organisation rather than the viral RNA (Pons et al., 1969). This is supported by electron microscopy of purified RNA-free NP extracted from RNPs, which showed structures morphologically indistinguishable from intact RNPs (Ruigrok and Baudin, 1995). Deletion mutagenesis of NP suggests that two separate regions within the protein are capable of association with the full-length protein, of which a C-terminal sequence is more important for NP–NP oligomerisation (Elton et al., 1999a).

RNPs are flexible entities with variable length, depending on the RNA segment they contain, and are poor subjects for detailed structural analyses. However, much smaller recombinant RNPs have been generated by in vivo amplification in cells expressing the viral polymerase, NP and a model vRNA (Ortega et al., 2000). These more uniform populations of RNPs are amenable to analysis by electron microscopy and image processing and this system has provided valuable low-resolution information on the structure and organisation of the RNP. Initial studies revealed the presence of circular, elliptic or coiled particles, depending on the length of the genomic RNA included (Ortega et al., 2000). From the length of the viral RNA present in these recombinant RNPs and the number of NP monomers observed, it could be calculated that around 24–25 nt are bound per NP molecule. Further analysis of one such recombinant RNP containing nine NP monomers showed a circular structure containing one copy of the polymerase complex (Martin-Benito et al., 2001). Two of the NP monomers are associated with the polymerase complex through non-identical contacts, which might reflect the NP-PB1 and NP-PB2 interactions identified biochemically (Biswas et al., 1998; Medcalf et al., 1999). The NP monomers have a banana-like structure with one main NP–NP contact. Viral RNA is included in these recombinant mini-RNPs and its termini are presumably bound by the polymerase complex, but the resolution of the reconstruction is so far insufficient to permit its localisation (Martin-Benito et al., 2001). Recently, a high-resolution crystal structure has been reported for NP (Ye et al., 2006). The protein crystallised as a trimer in which monomer interactions were primarily mediated by a flexible C-terminal loop, consistent with mutational data examining NP oligomerisation (Elton et al., 1999a). Monomers contain head and body domains made up of non-contiguous sequences arranged into a crescent shape reminiscent of the low-resolution EM pictures. The groove between domains is lined with multiple basic residues and therefore seems a plausible RNA-binding site (Ye et al., 2006). However, individual mutation of many of these residues did not affect NP–RNA interactions (Elton et al., 1999b), perhaps suggesting a degree of redundancy or conformational flexibility in how NP binds RNA, analogous to that proposed for how the vesicular stomatitis virus nucleoprotein binds RNA (Green et al., 2006).

Although there is no detailed structural data available for protein–RNA interactions within the RNP, there are solved structures for the terminal promoter elements of v- and cRNA. Consistent with early sequencing studies noting partial complementarity between the conserved 5′ and 3′ ends of the viral genomic segments (Skehel and Hay, 1978; Robertson, 1979; Desselberger et al., 1980), NMR analysis of short synthetic RNAs shows partial duplexes interrupted by bulge regions (Bae et al., 2001; Lee et al., 2003a; Park et al., 2003). However, not all aspects of the structure of these naked RNA 'panhandles' are compatible with a large body of mutagenesis data probing the function of the promoter elements, suggesting that perhaps an alternative conformation is adopted when the polymerase binds to the genome termini. For an in-depth discussion of this topic the reader is referred to other recent reviews (Elton et al., 2006; Ortin and Parra, 2006).

The enzyme responsible for RNA synthesis in the RNP is the virus polymerase complex, a heterotrimer in which the PB1 subunit constitutes the core to which both PB2 and PA subunits are bound (Digard et al., 1989). Several laboratories have examined the regions of these subunits involved in complex formation (reviewed in Elton et al., 2006). These studies suggest an N-terminal to C-terminal tandem arrangement of the subunits in the order PA-PB1-PB2, but with a further degree of interlinking between PB1 and PB2. However, the first three-dimensional EM reconstruction models reported for the polymerase present in recombinant RNPs show a compact, roughly globular structure in which the location of individual subunits is not apparent (Martin-Benito et al., 2001; Area et al., 2004). The position of specific domains of PB1, PB2 and PA proteins within the polymerase was determined by imaging of RNP-monoclonal antibody complexes or tagged RNPs (Area et al., 2004). Both the N-terminal region of PB2 and the C-terminus of PB1 are close to the areas of the polymerase that contact the adjacent NP monomers in the RNP, in agreement with the reports of in vitro interactions (Biswas et al., 1998; Medcalf et al., 1999). On the other hand, the C-terminal region of PA is opposite to the NP-polymerase contacts. The three-dimensional model reported for the polymerase corresponds to the enzyme present in a mature RNP, which can be activated for transcription in vitro and can be rescued into infectious virus in vivo. It represents the enzyme present in virion RNPs, poised for transcription but still not activated in the absence of a capped primer or nucleotides. Much biochemical evidence indicates that transcriptional activation of the polymerase involves allosteric cross-talk between the various RNAs and subunits (reviewed by Elton et al., 2006) and the latest EM imaging study from the Ortin laboratory provides elegant structural confirmation of this. Image processing analysis of non-RNP-associated polymerase complexes revealed a more open but still globular complex in which several regions (in particular density thought to be PB2) showed conformational changes when compared to RNP-associated polymerase (Torreira et al., 2007).

Limited amounts of other structural data are available for the influenza A virus polymerase complex. Partial proteolysis and functional mapping experiments suggest that the N-terminal 200 or so amino acids of PA form a discrete domain (Sanz-Ezquerro et al., 1996; Hara et al., 2006). Circular dichroism and structural prediction analysis suggests that the C-terminal 75 amino acids of PB1 (with PB2-binding function) form an α-helical domain (Poole et al., submitted for publication). In addition, the first high-resolution structural information for the polymerase has just been reported. NMR and crystal structures of the C-terminal 80 amino acids of PB2 show a compact α–β domain that contains a nuclear localisation signal (NLS) (Tarendeau et al., 2007).

It has long been known that the polymerase is a heterotrimer that exists even when not bound to an RNP (Braam et al., 1983; Detjen et al., 1987) and that PB1 forms the backbone of the complex (Digard et al., 1989). However, much recent work has concerned the mechanism by which the polymerase trimer is assembled, in large part prompted by further elucidation of how the P proteins are trafficked to the nucleus. NLSs have been identified in each of the individual P proteins (Nath and Nayak, 1990; Mukaigawa and Nayak, 1991; Nieto et al., 1994; Tarendeau et al., 2007). However, more recent analysis has confirmed an earlier suggestion that PB1 and PA are not efficiently imported into the nucleus unless in the form of a heterodimer (Nieto et al., 1992; Fodor and Smith, 2004). Further work identified the cellular Ran binding protein 5 (RanBP5, also known as importin ß3) as a binding partner of PB1 or PB1-PA but not the full trimer (Deng et al., 2006a; Mayer et al., 2007). SiRNA-mediated depletion of RanBP5 reduced nuclear import of PB1 and PA, suggesting that this interaction is indeed functionally important for nuclear trafficking of part of the polymerase complex (Deng et al., 2006a). PB2 undergoes efficient nuclear import in the absence of other influenza virus proteins and this is likely mediated by interactions with the canonical cellular import machinery. A bipartite NLS has been identified in the C-terminus of the protein and the structure of this region co-crystallised in complex with importin α5 has been solved (Tarendeau et al., 2007). PB2 has also been shown to interact with cellular hsp90 and to relocalise it to the nucleus (Momose et al., 2002; Naito et al., 2007). As with the PB1-PA dimer and RanBP5, the interaction of PB2 and hsp90 is lost on formation of a full PB1-PB2-PA trimer (Naito et al., 2007). Overall, these data suggest that the trimeric polymerase complex may only form in the nucleus after separate import of individual or subcomplexes of the P proteins along with accessory cellular proteins that perform import and/or chaperone functions (Deng et al., 2006a; Naito et al., 2007). Consistent with this hypothesis, a functional polymerase complex could be assembled in vitro by the addition of separately expressed PB2 to a PB1-PA dimer (Deng et al., 2005). Whether the subcomplexes of P proteins have any functional significance prior to their assembly into a trimer remains controversial (reviewed by Elton et al., 2006).

Mechanism of Viral mRNA Synthesis

Initiation

The influenza virus polymerase complex is essentially inactive for any of its enzymatic functions in the absence of viral RNA. A popular model for the mechanism of mRNA synthesis involves sequential binding of the polymerase to the 5′ and 3′ termini of vRNA, with each interaction causing allosteric changes to the proteins that result in activation of the cap-binding, endonuclease and nucleotide polymerisation functions in a regulated manner. In this model, the polymerase complex binds to the 5′ end of a vRNA segment (Fodor et al., 1994; Tiley et al., 1994), primarily through PB1–RNA interactions (Li et al., 1998; Gonzalez and Ortin, 1999a). This induces a conformational change in the polymerase that activates the cap-binding activity of PB2 (Cianci et al., 1995; Li et al., 1998) and allows the polymerase to bind a cellular pre-mRNA. The 3′ end of the vRNA template then enters the complex through a combination of protein–RNA interactions (again, primarily through PB1; Li et al., 1998; Gonzalez and Ortin, 1999b) and base-pairing between the 5′ and 3′ sequences. This event stimulates endonuclease activity (Cianci et al., 1995; Hagen et al., 1995; Li et al., 1998) and the cellular 5′ cap structure, together with 9–15 nucleotides, is endonucleolytically cleaved by PB1 (Plotch et al., 1981; Li et al., 1998). The 3′ end of the truncated mRNA is then used to prime transcription initiation by PB1 (Braam et al., 1983; Biswas and Nayak, 1994; Asano et al., 1995), with addition of a guanosine residue directed by the second residue of the vRNA template (Plotch et al., 1981). The nascent viral mRNA chain is then elongated by sequential addition of ribonucleotides, as directed by the vRNA template. Characterisation of the polymerase's Km for ATP suggests a transition from an initiation mode to a processive transcription mode between positions 4 and 5 (Klumpp et al., 1998). PB2 releases the cap structure after the first 11–15 nucleotides have been added (Braam et al.,

1983) but the mechanism that triggers this is not known. However, there is evidence that prior to this, binding of the cap structure to PB2 increases the transcriptional activity of PB1 (Penn and Mahy, 1984; Kawakami et al., 1985). Insofar as is known, other orthomyxoviruses (as well as the bunya, arena and tenuiviruses) possess similar 'cap-snatching' mechanisms for transcription initiation. However, thogotovirus generates a shorter, more homogeneous cellular-derived primer containing only the cap structure and one additional nucleotide (Albo et al., 1996; Weber et al., 1996).

The key features of the sequential addition model for transcription initiation are that the polymerase assembles on the 5′ end of vRNA, cap-binding is activated, the polymerase then binds to the 3′ end of vRNA and this results in endonuclease activation. Although much support for the sequential model has been generated, recent studies suggest that this sequence of events is not obligatory. Using an assay where vRNA was added to recombinant polymerase either as a pre-annealed duplex of 5′ and 3′ vRNA or as sequential components, it was shown that the sequence of assembly had a marked effect on the stability of cap-binding but not endonuclease activity (Lee et al., 2003b). Polymerase bound to pre-annealed vRNA template showed high-level capped primer binding and endonuclease activity which resulted in enhanced levels of mRNA transcription activity, compared to that from polymerase bound initially to just 5′ vRNA. However, the low levels of capped-RNA substrate bound by polymerase associated with only the 5′ end of vRNA were cleaved efficiently, indicating that the 3′ end was not required for activation of the endonuclease, and that the original enhancement of endonuclease activity attributed to the presence of the 3′ end actually resulted from increased levels of cap-binding (Lee et al., 2003b). In addition, much of the early in vitro data on influenza virus transcription was gained using rabbit ß-globin mRNA or latterly, a synthetic capped RNA made by in vitro transcription. Although these mRNAs are good substrates for the endonuclease, they are primarily cleaved after a G residue (Plotch et al., 1981) and are not used efficiently as primers for transcription initiation (Rao et al., 2003). However, in vivo the polymerase complex shows a preference for cleavage of host mRNAs after an A, or in around 20% of the available cloned sequences (for influenza A virus), CA residues (reviewed in Elton et al., 2006). Furthermore, when a capped substrate with a CA sequence upstream of the cleavage site was used as a primer the 3′ end of vRNA was not required for full endonuclease activity (Rao et al., 2003). Overall therefore, while many of the details of the sequential model of influenza virus mRNA transcription are intact, some aspects are still not fully defined.

Polyadenylation

Processive synthesis of mRNA halts at a stretch of 5–7 uridine residues ~17 nt from the 5′ end of the vRNA template (Robertson et al., 1981), adjacent to the base-paired region of the panhandle structure (Hsu et al., 1987; Fig. 9.1a). Polymerase stuttering at this site to reiteratively copy the U(5–7) track produces a poly A tail ranging from 60 to 350 residues for mRNA isolated from virus-infected cells (Plotch and Krug,

1977), and up to 120–150 in vitro (Perales et al., 1996; Pritlove et al., 1998). Direct evidence for this being a non-processive, template-directed process rather than a poly A polymerase activity came from a study where the U track was mutated to A_6 leading to synthesis of positive-sense capped RNAs with poly (U) tails (Poon et al., 1999). Initially it was proposed that base-pairing of the vRNA panhandle was a physical block to polymerase processivity, resulting in stuttering on the adjacent U track (Robertson, 1979; Luo et al., 1991). The discovery that the polymerase binds tightly to the 5′ end of vRNA suggested the hypothesis that the polymerase itself prevented processive transcription through the poly (U) stretch by remaining bound to the 5′ end of its template (Fodor et al., 1994; Tiley et al., 1994). In this model, the continued association of the polymerase with the 5′ end of vRNA creates a loop of untranscribed template that becomes progressively shorter until the polymerase is arrested with its active site over the poly (U) stretch immediately adjacent to its 5′ binding site. This steric block forces the polymerase to stutter and polyadenylate the transcript. Experimental evidence supporting this hypothesis comes from the observation that nucleotides required for polymerase binding to the 5′ end of vRNA (Fodor et al., 1994; Tiley et al., 1994) are also essential for polyadenylation (Poon et al., 1998; Pritlove et al., 1998).

It is generally believed that there is a mechanism to couple the mode of initiation by the polymerase complex to that of termination. Hay et al. (1982) observed that most full-length transcripts of the vRNA templates were uncapped, while in vivo and in vitro studies have found that most polyadenylated viral RNAs have host sequences at their 5′ ends (Shaw and Lamb, 1984; Vreede and Brownlee, 2007). In support of a mechanism to couple initiation and termination, transcripts initiated in vitro with a capped primer are also polyadenylated, even in the presence of free NP (Beaton and Krug, 1986). The finding that binding of the polymerase to duplex-form genome termini promotes high levels of cap-primed transcription initiation suggests a mechanism for achieving this coupling, as it is reasonable to suppose that any interaction of the RNA termini in the absence of the polymerase is more likely to happen in *cis* than in *trans* (Lee et al., 2003b). However, this hypothesis is yet to be tested in vivo.

Interactions Between Viral and Cellular Transcription Machinery

One facet of influenza virus transcription where there has been significant recent progress concerns the cell biological aspects of the interaction between host and viral transcription machinery. The point in the cellular transcription cycle at which viral RNPs capture host cell cap structures has not been defined as in theory it could occur at any point before the cellular mRNA is exported to the cytoplasm. The pre-mRNA cap structure normally becomes associated with the cap-binding complex (CBC) soon after transcription initiation (Howe, 2002). The CBC remains bound to the cap up until nuclear export of the mature mRNA so it is plausible that the CBC and the influenza polymerase compete for the mRNA cap. Recent work has shown that the influenza polymerase interacts with the host RNA Pol II (Engelhardt et al., 2005; Mayer et al., 2007) mediated through the C-terminal repeat domain (CTD) of RNA Pol II with a preference for the form phosphorylated on serine 5 (Engelhardt et al., 2005). Serine 5 phosphorylation occurs during initiation of the Pol II transcriptional cycle and is thought to activate the cellular cap synthesis complex (Howe, 2002). Potentially therefore, the influenza polymerase targets Pol II at the initiation stage to compete for newly synthesised cap structures for use as primers (Engelhardt et al., 2005).

A physical association between host and viral transcription machinery may also serve to direct nuclear export of viral mRNAs. The maturation of a cellular mRNA leading up to and including its export to the cytoplasm is thought to be coupled to RNA Pol II transcription through a suite of accessory proteins that are loaded onto the nascent transcript in a sequential fashion (Howe, 2002). Excluding the first dozen or so nucleotides captured from host mRNAs, influenza virus mRNAs are made by the viral polymerase but still need to access cellular machinery for nuclear export. The pathways and mechanisms are currently poorly defined but it has been observed that certain viral transcripts (notably those encoding HA and M1) are retained in the nucleus by drugs that affect the phosphorylation of the RNA Pol II CTD (Vogel et al., 1994; Amorim et al., 2007). The latter study used a variety of chemically and mechanistically distinct Pol II inhibitors to infer that the block to nuclear export of the viral mRNAs was reversible and depended on Pol II transcription (Amorim et al., 2007), a result consistent with recent work regarding the nuclear export of microinjected cellular mRNAs (Tokunaga et al., 2006). Two recent studies concluded that influenza infection somehow downregulates Pol II transcription (Chan et al., 2006; Rodriguez et al., 2007). This occurs late in infection when viral mRNA synthesis is diminishing and thus is consistent with a role for RNA Pol II in viral mRNA expression.

Mechanism of Genome Replication

Synthesis of cRNA

The process of genome replication is less well characterised in comparison to that of mRNA synthesis. Incoming vRNPs are the templates for synthesis of cRNAs which are then used to make more vRNA (Fig. 9.1b). cRNA constitutes only 5–10% of the total plus-sense viral RNA present in infected cells (Hay et al., 1977a; Barrett et al., 1979; Herz et al., 1981; Mullin et al., 2004). Viral mRNAs cannot serve as replicative intermediates because of the host-derived sequences at their 5′ ends and because they are truncated at their 3′ ends when polyadenylated (Fig. 9.1a). cRNAs are uncapped, 5′ triphosphorylated (Hay et al., 1982), full-length copies that are not polyadenylated. They cannot be generated through the endonucleolytic processing of a cap-primed intermediate, as this would leave a monophosphate terminus (Olsen et al., 1996). To generate a full-length copy, the polymerase must read through the polyadenylation signal towards the 5′ end of the vRNA template. Unlike viral mRNA, cRNA is encapsidated by NP to form RNP structures, in much the same way as vRNA (Pons, 1971; Dalton et al., 2006). Thus, cRNA synthesis is mechanistically distinct from viral transcription.

Early studies indicated that RNPs from purified influenza virions are able to synthesise mRNA but not cRNA in vitro (Plotch and Krug, 1977; Skorko et al., 1991), indicating a requirement for other factors besides transcriptionally active RNPs. However, a recent study re-examining this question reached the opposite conclusion and found that virion RNPs were fully competent with no extra factors necessary (Vreede and Brownlee, 2007). Differing methodologies for the detection of cRNA may underlie the discrepancy.

Unquestionably, the same RNPs introduced into a cell by infection act as templates for both mRNA and cRNA synthesis. Nuclear extracts prepared from normal cells infected with influenza virus supported the synthesis of both types of positive-sense RNA (Beaton and Krug, 1984; del Rio et al., 1985; Beaton and Krug, 1986; Takeuchi et al., 1987; Shapiro and Krug, 1988). Early experiments showed that cRNA accumulation was dependent upon synthesis of viral and/or cellular proteins whereas mRNA synthesis was not (Hay et al., 1977b). Isolated RNP complexes recovered from the infected nuclear extracts by centrifugation lost the ability to make cRNA, but this could be restored by addition of the supernatant fraction unless it was immuno-depleted of NP (Beaton and Krug, 1986; Shapiro and Krug, 1988). This early data led to the concept of a 'switch' mechanism operating in infected cells to divert a minor fraction of polymerase activity from transcription to cRNA synthesis. Various hypotheses concerning how such a control mechanism might operate, mostly centred around NP, but also concerning the viral polymerase or putative cellular factors have been proposed. The reader is referred to other reviews where this work is considered in detail (Elton et al., 2006; Ortin and Parra, 2006). Instead, this chapter will focus on recent data suggesting alternative models for how influenza A virus replicates its genome.

The Stabilisation Model for cRNA Synthesis

NP is the prime candidate for a regulatory factor in the active 'switching' hypothesis, from the evidence described above and by analogy with non-segmented negative-sense viruses, where the intracellular concentration of the equivalent N protein is thought to regulate the balance between transcription and replication (Blumberg et al., 1981; Arnheiter et al., 1985). However, experimental manipulation of NP levels showed a slight negative rather than any positive correlation with levels of genome replication in cells (Mullin et al., 2004). In addition, NP is not necessary in vitro for the polymerase to initiate unprimed (replication mode) RNA synthesis (Lee et al., 2002; Deng et al., 2006b; Vreede and Brownlee, 2007) although it may function as a processivity factor (Beaton and Krug, 1986; Shapiro and Krug, 1988). A recent key study has shown that cRNA synthesis can occur in the absence of protein synthesis if a supply of pre-existing polymerase is available (Vreede et al.,

2004). This is dependent on the promoter-binding activity but not on the catalytic activity of the pre-expressed polymerase as a polymerisation defective mutant PB1 can fulfil this role whereas RNA-binding mutants cannot. Pre-existing NP was neither necessary nor sufficient to permit cRNA accumulation. However, its presence substantially increased the levels of cRNA accumulation. Thus a new model for the first step in genome replication proposes that cRNA synthesis is an intrinsic property of negative-sense RNPs but its accumulation is dependent on stabilisation resulting from polymerase binding (and enhanced by NP) rather than by an active switch mechanism (Vreede et al., 2004). Because cRNA molecules are not capped or polyadenylated they are quickly degraded by cellular nucleases unless there is a source of viral RNP polypeptides, particularly the polymerase components, present to encapsidate and stabilise them (Fig. 9.3). Paradoxically, a PB1-PA dimer possessing high levels of promoter-binding activity (Lee et al., 2002; Deng et al., 2005) does not suffice for this purpose (Vreede et al., 2004). This current model suggests that the first event in influenza A virus genome replication is not actively regulated, but is instead a stochastic process based on the probability of the viral polymerase initiating cap-primed or unprimed transcription. Nevertheless, evidence for regulated cRNA synthesis at some level remains. For instance, maximum rates of synthesis and levels of cRNA accumulation are reached early in infection and are not substantially amplified by the subsequent rise in vRNA levels (Barrett et al., 1979; Shapiro et al., 1987; Mullin et al., 2004; Dalton et al., 2006), despite the large pool of non-RNP-associated polymerase that remains in the nucleus until the end of the

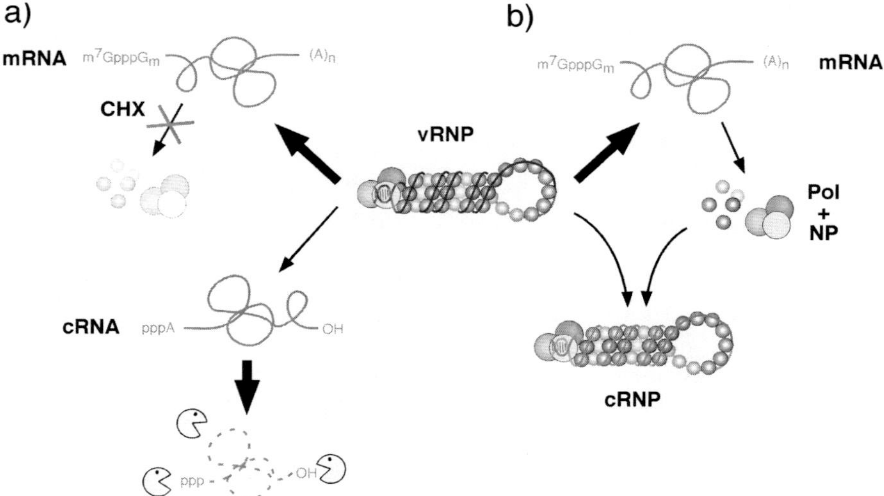

Fig. 9.3 Stabilisation model for influenza virus cRNA synthesis. (**a**) When protein synthesis is blocked by cycloheximide (CHX), RNPs synthesise mRNA and cRNA but the unencapsidated cRNA is quickly degraded. (**b**) In untreated cells, viral mRNAs are translated to produce new polymerase and NP which co-transcriptionally encapsidates nascent cRNA strands and protects them from degradation. After Vreede et al. (2004).

viral lifecycle (Detjen et al., 1987; Carrasco et al., 2004). This perhaps implies that only input (and not newly replicated) vRNA templates are used for cRNA synthesis, but this remains to be determined.

The Mechanism of vRNA Synthesis

The synthesis of vRNA from a cRNA template can be viewed as a simpler process than the transcription or replication of positive-sense RNA, since it is the only type of RNA transcribed from a cRNA template. Like cRNA, initiation of vRNA synthesis is unprimed and the products have 5′ triphosphorylated ends (Young and Content, 1971; Hay et al., 1982). Nevertheless, recent work suggests a significant difference in their modes of transcription initiation. Based on the results of experiments examining the precise origins of the first 2–3 nucleotides polymerised on WT and mutant cRNA and vRNA templates (Deng et al., 2006c) concluded that initiation with ATP occurs at the very 3′ end of the vRNA template and leads to synthesis of pppApG that is subsequently elongated to a full-length cRNA transcript (Fig. 9.4a). vRNA synthesis also initiates with ATP to produce a pppApG dinucleotide, but surprisingly, this is templated by positions 4 and 5 of the 3′ cRNA promoter. The pppApG dinucleotide is then postulated to translocate back to the very 3′ end of the template and there act as a primer for initiation of a nascent full-length vRNA molecule (Fig. 9.4b). Theoretically, the internally templated ApG could also be released by the polymerase to prime transcription in *trans* on other cRNA or vRNA templates (Deng et al., 2006c). Ironically, ApG dinucleotides have been used by influenza scientists as a tool to stimulate in vitro transcription by the viral polymerase for more than 30 years (McGeoch and Kitron, 1975).

Differential activation of the polymerase complex has been observed with the vRNA and cRNA promoters. This may be due to their binding to different sequences within the PB1 subunit, although there is some disagreement on the PB1 sequences involved (Li et al., 1998; Jung and Brownlee, 2006). Binding to the 5′ end of vRNA or cRNA stimulates cap-binding activity of the polymerase (Cianci et al., 1995), which may increase overall levels of transcription through allosteric upregulation of PB1 activity (Penn and Mahy, 1984; Kawakami et al., 1985). However, only binding to vRNA templates triggers cap-primed transcription activity of the polymerase. Until recently, this was thought to be due to a failure of cRNA templates to activate the endonuclease activity of the complex (Cianci et al., 1995; Honda et al., 2001), but as discussed earlier, this may only hold true for certain cap-donor RNAs. Addition of a CA cleavage site-containing cap donor to a reconstituted polymerase complex bound to 5′cRNA stimulated endonuclease activity to levels approaching those achieved with 5′vRNA. However, as these products are not subsequently extended (Rao et al., 2003), the synthesis of non-functional capped vRNA is prevented. Although the cRNA template does not contain a full polyadenylation signal (as it lacks the poly U tract), the polymerase still binds to the 5′ arm of cRNA (Tiley et al., 1994; Cianci et al., 1995; Gonzalez and Ortin, 1999b) and evidence indicates that the cRNA promoter can exist as a panhandle (Elton et al., 2006; Ortin and Parra,

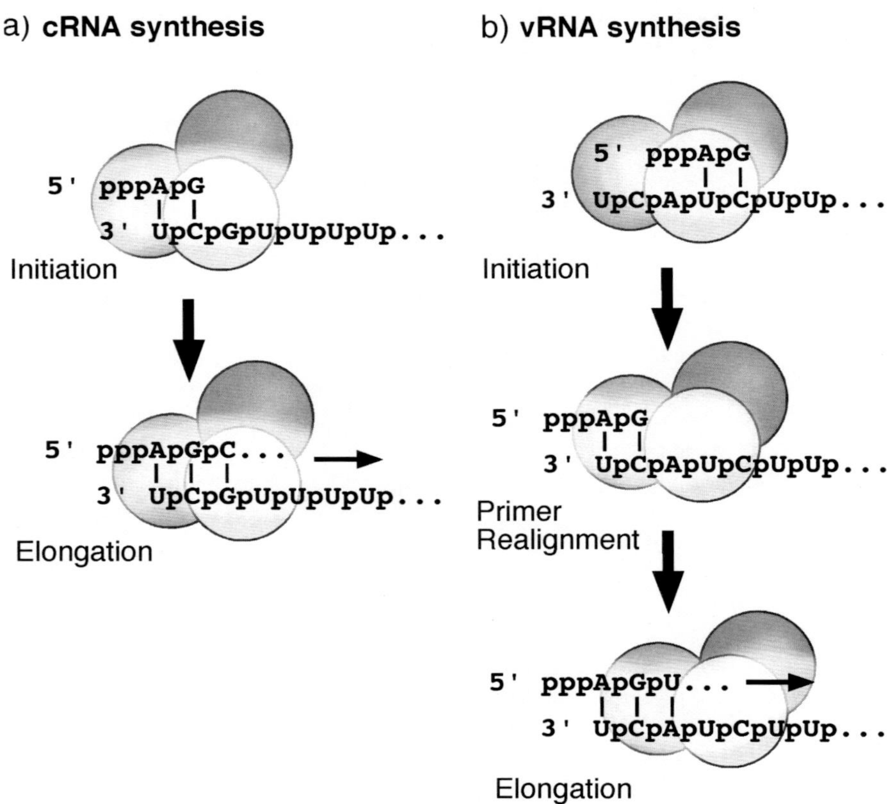

Fig. 9.4 Initiation modes for viral genome replication. (**a**) cRNA synthesis initiates by synthesis of an ApG dinucleotide templated by the 3′ end of vRNA that is then processively extended. (**b**) For vRNA synthesis, ApG is synthesised internally using residues 4 and 5 of the cRNA template, then translocated back to the 3′ end (primer realignment) to prime processive elongation. After Deng et al. (2006).

2006). This raises the question of how the steric block proposed for vRNA-directed polyadenylation is avoided in the case of cRNA to allow synthesis of a full-length vRNA transcript. Estimates of the dissociation constants for the interaction of PB1 with the 5′ ends of vRNA and cRNA are similar (Gonzalez and Ortin, 1999a, b). Nevertheless, the overall interaction of the polymerase with the cRNA promoter is more labile than with the vRNA promoter and significantly more temperature sensitive (Dalton et al., 2006).

The genetics of vRNA synthesis are similar to that of cRNA, with early experiments on *ts* mutants providing evidence that both PA and NP are important (Elton et al., 2006). However, the two polarities of genome replication are separable, since mutants have been isolated that can synthesise positive-sense RNA but appear to be specifically deficient for synthesis of vRNA (Thierry and Danos, 1982; Markushin and Ghendon, 1984) or vice versa (Mena et al., 1999). In addition, mutations in the NS1 gene show a partial deficiency in the accumulation of vRNA, but not in that of cRNA, suggesting that NS1 acts as a cofactor in the second step of viral RNA replication (Falcon et al., 2004). This may be related to the observed association of NS1 with viral RNPs (Marion et al., 1997). Analysis of in vitro transcription reactions carried out with infected cell extracts has shown that, as with cRNA synthesis, a supply of non-RNP-associated NP is required to support vRNA synthesis (Shapiro and Krug, 1988). However, it is notable that prior expression of the polymerase and NP in cells before infection in the presence of a cycloheximide block does not support vRNA synthesis even though cRNA is made (Vreede et al., 2004).

Conclusions

Orthomyxoviral RNA synthesis has been a topic of continual research for over 40 years now and yet despite this, even after significant recent advances, much remains to be discovered. The molecular mechanisms of viral mRNA transcription are still incomplete, especially with regard to the cell biology of the process. New models have been formulated for how influenza virus replicates its genome but these require further testing. The longstanding lack of structural information on the viral RNA synthesis machinery is beginning to be rectified but there is still much work to be done. These remain worthwhile areas of study and may eventually lead to the design of novel antivirals targeted against an enzymatic complex that occupies the coding capacity of well over half of the influenza virus genome.

Acknowledgments Research in the authors' laboratories is supported by grants from the Wellcome Trust, MRC, BBSRC and the Horse Race Betting Levy Board. We thank Eva Kreysa for helpful criticism.

References

Albo, C., Martin, J. and Portela, A. 1996. J Virol 70: 9013–9017.
Almond, J. W. 1977. Nature 270: 617–618.
Amorim, M. J. and Digard, P. 2006. Vaccine 24: 6651–6655.
Amorim, M. J., Read, E. K., Dalton, R. M., Medcalf, L. and Digard, P. 2007. Traffic 8: 1–11.
Area, E., Martin-Benito, J., Gastaminza, P., Torreira, E., Valpuesta, J. M., Carrascosa, J. L. and Ortin, J. 2004. Proc Natl Acad Sci U S A 101: 308–313.
Arnheiter, H., Davis, N. L., Wertz, G., Schubert, M. and Lazzarini, R. A. 1985. Cell 41: 259–267.
Asano, Y., Mizumoto, K., Maruyama, T. and Ishihama, A. 1995. J Biochem (Tokyo) 117: 677–682.
Bae, S. H., Cheong, H. K., Lee, J. H., Cheong, C., Kainosho, M. and Choi, B. S. 2001. Proc Natl Acad Sci U S A 98: 10602–10607.
Barrett, T., Wolstenholme, A. J. and Mahy, B. W. 1979. Virology 98: 211–225.
Beaton, A. R. and Krug, R. M. 1984. Proc Natl Acad Sci U S A 81: 4682–4686.
Beaton, A. R. and Krug, R. M. 1986. Proc Natl Acad Sci U S A 83: 6282–6286.
Biswas, S. K., Boutz, P. L. and Nayak, D. P. 1998. J Virol 72: 5493–5501.
Biswas, S. K. and Nayak, D. P. 1994. J Virol 68: 1819–1826.
Blumberg, B. M., Leppert, M. and Kolakofsky, D. 1981. Cell 23: 837–845.

Braam, J., Ulmanen, I. and Krug, R. M. 1983. Cell 34: 609–618.

Carrasco, M., Amorim, M. J. and Digard, P. 2004. Traffic 5: 979–992.

Chan, A. Y., Vreede, F. T., Smith, M., Engelhardt, O. G. and Fodor, E. 2006. Virology 351: 210–217.

Cianci, C., Tiley, L. and Krystal, M. 1995. J Virol 69: 3995–3999.

Da Silva, E. V., Da Rosa, A. P., Nunes, M. R., Diniz, J. A., Tesh, R. B., Cruz, A. C., Vieira, C. M. and Vasconcelos, P. F. 2005. Am J Trop Med Hyg 73: 1050–1058.

Dalton, R. M., Mullin, A. E., Amorim, M. J., Medcalf, E., Tiley, L. S. and Digard, P. 2006. Virol J 3: 58.

del Rio, L., Martinez, C., Domingo, E. and Ortin, J. 1985. Embo J 4: 243–247.

Deng, T., Engelhardt, O. G., Thomas, B., Akoulitchev, A. V., Brownlee, G. G. and Fodor, E. 2006a. J Virol 80: 11911–11919.

Deng, T., Sharps, J. L. and Brownlee, G. G. 2006b. J Gen Virol 87: 3373–3377.

Deng, T., Sharps, J., Fodor, E. and Brownlee, G. G. 2005. J Virol 79: 8669–8674.

Deng, T., Vreede, F. T. and Brownlee, G. G. 2006c. J Virol 80: 2337–2348.

Desselberger, U., Racaniello, V. R., Zazra, J. J. and Palese, P. 1980. Gene 8: 315–328.

Detjen, B. M., St Angelo, C., Katze, M. G. and Krug, R. M. 1987. J Virol 61: 16–22.

Digard, P., Blok, V. C. and Inglis, S. C. 1989. Virology 171: 162–169.

Elton, D., Digard, P., Tiley, L. and Ortin, J. (2006). Structure and function of the influenza virus RNP. *Influenza virology; current topics*. Wymondham, Caister Academic Press: 1–36.

Elton, D., Medcalf, E., Bishop, K. and Digard, P. 1999a. Virology 260: 190–200.

Elton, D., Medcalf, L., Bishop, K., Harrison, D. and Digard, P. 1999b. J Virol 73: 7357–7367.

Enami, M., Fukuda, R. and Ishihama, A. 1985. Virology 142: 68–77.

Engelhardt, O. G. and Fodor, E. 2006. Rev Med Virol 16: 329–345.

Engelhardt, O. G., Smith, M. and Fodor, E. 2005. J Virol 79: 5812–5818.

Falcon, A. M., Marion, R. M., Zurcher, T., Gomez, P., Portela, A., Nieto, A. and Ortin, J. 2004. J Virol 78: 3880–3888.

Fodor, E., Pritlove, D. C. and Brownlee, G. G. 1994. J Virol 68: 4092–4096.

Fodor, E. and Smith, M. 2004. J Virol 78: 9144–9153.

Gonzalez, S. and Ortin, J. 1999a. J Virol 73: 631–637.

Gonzalez, S. and Ortin, J. 1999b. Embo J 18: 3767–3775.

Green, T. J., Zhang, X., Wertz, G. W. and Luo, M. 2006. Science 313: 357–360.

Hagen, M., Tiley, L., Chung, T. D. and Krystal, M. 1995. J Gen Virol 76 (Pt 3): 603–611.

Hara, K., Schmidt, F. I., Crow, M. and Brownlee, G. G. 2006. J Virol 80: 7789–7798.

Hatada, E., Hasegawa, M., Mukaigawa, J., Shimizu, K. and Fukuda, R. 1989. J Biochem (Tokyo) 105: 537–546.

Hatta, M., Gao, P., Halfmann, P. and Kawaoka, Y. 2001. Science 293: 1840–1842.

Hay, A. J., Abraham, G., Skehel, J. J., Smith, J. C. and Fellner, P. 1977a. Nucleic Acids Res 4: 4197–4209.

Hay, A. J., Lomniczi, B., Bellamy, A. R. and Skehel, J. J. 1977b. Virology 83: 337–355.

Hay, A. J., Skehel, J. J. and McCauley, J. 1982. Virology 116: 517–522.

Herz, C., Stavnezer, E., Krug, R. and Gurney, T., Jr. 1981. Cell 26: 391–400.

Honda, A., Endo, A., Mizumoto, K. and Ishihama, A. 2001. J Biol Chem 276: 31179–31185.

Honda, A., Mukaigawa, J., Yokoiyama, A., Kato, A., Ueda, S., Nagata, K., Krystal, M., Nayak, D. P. and Ishihama, A. 1990. J Biochem (Tokyo) 107: 624–628.

Horisberger, M. A. 1980. Virology 107: 302–305.

Howe, K. J. 2002. Biochim Biophys Acta 1577: 308–324.

Hsu, M. T., Parvin, J. D., Gupta, S., Krystal, M. and Palese, P. 1987. Proc Natl Acad Sci U S A 84: 8140–8144.

Jennings, P. A., Finch, J. T., Winter, G. and Robertson, J. S. 1983. Cell 34: 619–627.

Jung, T. E. and Brownlee, G. G. 2006. J Gen Virol 87: 679–688.

Kawakami, K., Mizumoto, K., Ishihama, A., Shinozaki-Yamaguchi, K. and Miura, K. 1985. J Biochem (Tokyo) 97: 655–661.

Klumpp, K., Ford, M. J. and Ruigrok, R. W. 1998. J Gen Virol 79 (Pt 5): 1033–1045.
Klumpp, K., Ruigrok, R. W. and Baudin, F. 1997. Embo J 16: 1248–1257.
Lee, M. K., Bae, S. H., Park, C. J., Cheong, H. K., Cheong, C. and Choi, B. S. 2003a. Nucleic Acids Res 31: 1216–1223.
Lee, M. T., Bishop, K., Medcalf, L., Elton, D., Digard, P. and Tiley, L. 2002. Nucleic Acids Res 30: 429–438.
Lee, M. T., Klumpp, K., Digard, P. and Tiley, L. 2003b. Nucleic Acids Res 31: 1624–1632.
Li, M. L., Ramirez, B. C. and Krug, R. M. 1998. Embo J 17: 5844–5852.
Luo, G. X., Luytjes, W., Enami, M. and Palese, P. 1991. J Virol 65: 2861–2867.
Marion, R. M., Zurcher, T., de la Luna, S. and Ortin, J. 1997. J Gen Virol 78 (Pt 10): 2447–2451.
Mark, G. E., Taylor, J. M., Broni, B. and Krug, R. M. 1979. J Virol 29: 744–752.
Markushin, S. G. and Ghendon, Y. Z. 1984. J Gen Virol 65 (Pt 3): 559–575.
Martin-Benito, J., Area, E., Ortega, J., Llorca, O., Valpuesta, J. M., Carrascosa, J. L. and Ortin, J. 2001. EMBO Rep 2: 313–317.
Mayer, D., Molawi, K., Martinez-Sobrido, L., Ghanem, A., Thomas, S., Baginsky, S., Grossmann, J., Garcia-Sastre, A. and Schwemmle, M. 2007. J Proteome Res 6: 672–682.
McGeoch, D. and Kitron, N. 1975. J Virol 15: 686–695.
Medcalf, L., Poole, E., Elton, D. and Digard, P. 1999. J Virol 73: 7349–7356.
Mena, I., Jambrina, E., Albo, C., Perales, B., Ortin, J., Arrese, M., Vallejo, D. and Portela, A. 1999. J Virol 73: 1186–1194.
Momose, F., Naito, T., Yano, K., Sugimoto, S., Morikawa, Y. and Nagata, K. 2002. J Biol Chem 277: 45306–45314.
Mukaigawa, J. and Nayak, D. P. 1991. J Virol 65: 245–253.
Mullin, A. E., Dalton, R. M., Amorim, M. J., Elton, D. and Digard, P. 2004. J Gen Virol 85: 3689–3698.
Murti, K. G., Webster, R. G. and Jones, I. M. 1988. Virology 164: 562–566.
Naito, T., Momose, F., Kawaguchi, A. and Nagata, K. 2007. J Virol 81: 1339–1349.
Nath, S. T. and Nayak, D. P. 1990. Mol Cell Biol 10: 4139–4145.
Nieto, A., de la Luna, S., Barcena, J., Portela, A. and Ortin, J. 1994. J Gen Virol 75 (Pt 1): 29–36.
Nieto, A., de la Luna, S., Barcena, J., Portela, A., Valcarcel, J., Melero, J. A. and Ortin, J. 1992. Virus Res 24: 65–75.
Olsen, D. B., Benseler, F., Cole, J. L., Stahlhut, M. W., Dempski, R. E., Darke, P. L. and Kuo, L. C. 1996. J Biol Chem 271: 7435–7439.
Ortega, J., Martin-Benito, J., Zurcher, T., Valpuesta, J. M., Carrascosa, J. L. and Ortin, J. 2000. J Virol 74: 156–163.
Ortin, J. and Parra, F. 2006. Annu Rev Microbiol 60: 305–326.
Park, C. J., Bae, S. H., Lee, M. K., Varani, G. and Choi, B. S. 2003. Nucleic Acids Res 31: 2824–2832.
Peiris, J. S., de Jong, M. D. and Guan, Y. 2007. Clin Microbiol Rev 20: 243–267.
Penn, C. R. and Mahy, B. W. 1984. Virus Res 1: 1–13.
Perales, B., de la Luna, S., Palacios, I. and Ortin, J. 1996. J Virol 70: 1678–1686.
Plotch, S. J., Bouloy, M., Ulmanen, I. and Krug, R. M. 1981. Cell 23: 847–858.
Plotch, S. J. and Krug, R. M. 1977. J Virol 21: 24–34.
Pons, M. W. 1971. Virology 46: 149–160.
Pons, M. W., Schulze, I. T., Hirst, G. K. and Hauser, R. 1969. Virology 39: 250–259.
Poon, L. L., Pritlove, D. C., Fodor, E. and Brownlee, G. G. 1999. J Virol 73: 3473–3476.
Poon, L. L., Pritlove, D. C., Sharps, J. and Brownlee, G. G. 1998. J Virol 72: 8214–8219.
Pritlove, D. C., Poon, L. L., Fodor, E., Sharps, J. and Brownlee, G. G. 1998. J Virol 72: 1280–1286.
Rao, P., Yuan, W. and Krug, R. M. 2003. Embo J 22: 1188–1198.
Robertson, J. S. 1979. Nucleic Acids Res 6: 3745–3757.
Robertson, J. S., Schubert, M. and Lazzarini, R. A. 1981. J Virol 38: 157–163.
Rodriguez, A., Perez-Gonzalez, A. and Nieto, A. 2007. J Virol 81: 5315–5324.
Ruigrok, R. W. and Baudin, F. 1995. J Gen Virol 76 (Pt 4): 1009–1014.

Sanz-Ezquerro, J. J., Zurcher, T., de la Luna, S., Ortin, J. and Nieto, A. 1996. J Virol 70: 1905–1911.
Shapiro, G. I., Gurney, T., Jr. and Krug, R. M. 1987. J Virol 61: 764–773.
Shapiro, G. I. and Krug, R. M. 1988. J Virol 62: 2285–2290.
Shaw, M. W. and Lamb, R. A. 1984. Virus Res 1: 455–467.
Siebler, J., Haller, O. and Kochs, G. 1996. Arch Virol 141: 1587–1594.
Skehel, J. J. and Hay, A. J. 1978. Nucleic Acids Res 5: 1207–1219.
Skorko, R., Summers, D. F. and Galarza, J. M. 1991. Virology 180: 668–677.
Smith, G. L. and Hay, A. J. 1982. Virology 118: 96–108.
Staunton, D., Nuttall, P. A. and Bishop, D. H. 1989. J Gen Virol 70 (Pt 10): 2811–2817.
Subbarao, E. K., London, W. and Murphy, B. R. 1993. J Virol 67: 1761–1764.
Takeuchi, K., Nagata, K. and Ishihama, A. 1987. J Biochem (Tokyo) 101: 837–845.
Tarendeau, F., Boudet, J., Guilligay, D., Mas, P. J., Bougault, C. M., Boulo, S., Baudin, F., Ruigrok, R. W., Daigle, N., Ellenberg, J., Cusack, S., Simorre, J. P. and Hart, D. J. 2007. Nat Struct Mol Biol 14: 229–233.
Taylor, J. M., Illmensee, R., Litwin, S., Herring, L., Broni, B. and Krug, R. M. 1977. J Virol 21: 530–540.
Thierry, F. and Danos, O. 1982. Nucleic Acids Res 10: 2925–2938.
Tiley, L. S., Hagen, M., Matthews, J. T. and Krystal, M. 1994. J Virol 68: 5108–5116.
Tokunaga, K., Shibuya, T., Ishihama, Y., Tadakuma, H., Ide, M., Yoshida, M., Funatsu, T., Ohshima, Y. and Tani, T. 2006. Genes Cells 11: 305–317.
Torreira, E., Schoehn, G., Fernandez, Y., Jorba, N., Ruigrok, R. W., Cusack, S., Ortin, J. and Llorca, O. 2007. Nucleic Acids Res 35: 3774–3783.
Tumpey, T. M., Basler, C. F., Aguilar, P. V., Zeng, H., Solorzano, A., Swayne, D. E., Cox, N. J., Katz, J. M., Taubenberger, J. K., Palese, P. and Garcia-Sastre, A. 2005. Science 310: 77–80.
Vogel, U., Kunerl, M. and Scholtissek, C. 1994. Virology 198: 227–233.
Vreede, F. T. and Brownlee, G. G. 2007. J Virol 81: 2196–2204.
Vreede, F. T., Jung, T. E. and Brownlee, G. G. 2004. J Virol 78: 9568–9572.
Weber, F., Haller, O. and Kochs, G. 1996. J Virol 70: 8361–8367.
Yamanaka, K., Ishihama, A. and Nagata, K. 1990. J Biol Chem 265: 11151–11155.
Ye, Q., Krug, R. M. and Tao, Y. J. 2006. Nature 444: 1078–1082.
Young, R. J. and Content, J. 1971. Nat New Biol 230: 140–142.

Chapter 10
Arenaviruses: Genome Replication Strategies

Juan C. de la Torre

Introduction

Arenaviruses constitute one of the most widespread and diverse groups of rodent-borne viruses, which merit significant attention both as tractable model systems to study acute and persistent viral infections (Oldstone, 2002; Zinkernagel, 2002) and as clinically important human pathogens including several causative agents of severe hemorrhagic fever (HF), chiefly Lassa fever virus (LASV) (Geisbert and Jahrling, 2004; McCormick and Fisher-Hoch, 2002; Peters, 2002). Moreover, evidence indicates that the prototypic Arenavirus lymphocytic choriomeningitis virus (LCMV) is a neglected human pathogen of clinical significance, especially in cases of congenital infection (Barton and Mets, 1999, 2001; Barton et al., 2002; Jahrling and Peters, 1992; Mets et al., 2000; Peters, 2006). No licensed anti-arenavirus vaccines are available, and current anti-arenavirus therapies are limited to the use of the nucleoside analog ribavirin, which is only partially effective and often associated with significant secondary effects including anemia and birth defects. Therefore, it is important to develop novel antiviral strategies to combat arenaviruses. This task will benefit from an improved knowledge about the arenavirus molecular biology, which is the focus of this chapter.

The inability to genetically manipulate the Arenavirus genome has hampered studies aimed at understanding its molecular and cell biology, as well as the role played by each viral gene product in virus–host interactions during both acute and persistent infections and associated diseases. The recent development of reverse genetics systems for several arenaviruses including LCMV (Flatz et al., 2006; Lee et al., 2000; Sanchez and de la Torre, 2006), as well as LASV (Hass et al., 2004) and Tacaribe virus (TCRV) (Lopez et al., 2001), has provided investigators with a novel and powerful approach for the investigation of the *cis*-acting sequences and *trans*-acting factors that control arenavirus replication and gene expression, as well

J.C. de la Torre (✉)
Department of Immunology and Microbial Science, The Scripps Research Institute IMM-6, 10550 N Torrey Pines Road, La Jolla, CA 92037, USA
e-mail: juanct@scripps.edu

C.E. Cameron et al. (eds.), *Viral Genome Replication*,
DOI 10.1007/b135974_10,

as assembly and budding. Moreover, the ability to rescue recombinant LCM viruses from cloned cDNAs with predetermined specific mutations and analyze their phenotypic expression in its natural host, the mouse, has created unique opportunities to investigate arenavirus–host interactions that influence a variable infection outcome ranging from virus control and clearance by the host defenses to long-term chronic infection associated with subclinical disease, and severe acute disease including HF. Likewise these new developments are facilitating a detailed understanding of the arenavirus molecular biology. This new knowledge, in turn, is opening new avenues for the development and evaluation of novel antiviral strategies to combat arenaviral infections.

Arenaviruses Important as Both Model Systems and Human Pathogens

LCMV as a Model to Study Virus–Host Interactions Associated with Both Acute and Chronic Viral Infections

The LCMV system provides a primary workhorse in the fields of immunology and viral pathogenesis and has contributed to the development of multiple key concepts in both disciplines including (1) antiviral tolerance (Oldstone, 2002; Zinkernagel, 2002), (2) virus-induced immunosuppression (Oldstone, 2002; Zinkernagel, 2002), and (3) the ability of noncytolytic persistent riboviruses to avoid elimination by the host immune responses and to induce disease by interfering with specialized functions of infected cells, revealing a new way by which viruses do harm in the absence of the classic hallmarks of cytolysis and inflammation (de la Torre and Oldstone, 1996; Oldstone, 2002). Moreover, LCMV represents an excellent model to unravel the mechanisms of virally induced meningitis and foster the development of novel interventions to ameliorate the symptoms and consequences of this clinically important pathogenic process.

Arenaviruses and the Diseases That They Cause

Arenaviruses cause chronic infections of rodents with a worldwide distribution (Buchmeier et al., 2001). Asymptomatically infected animals move freely in their natural habitat and may invade human habitation. Humans are infected through mucosal exposure to aerosols, or by direct contact of abrade skin with infectious materials.

Several arenaviruses cause severe HF disease in humans and pose a serious public health problem (Buchmeier et al., 2001; McCormick and Fisher-Hoch, 2002; Peters, 2002). In recent years, increased traveling to and from endemic regions has led to the importation of LASV into non-endemic regions including the United States (Freedman and Woodall, 1999; Holmes et al., 1990; Isaacson, 2001). On the other hand, it is worth noting that compelling evidence indicates that LCMV is a neglected human pathogen of clinical significance, especially in cases of congenital infection leading to hydrocephalus, mental retardation, and chorioretinitis in infants (Barton et al., 2002; Jahrling and Peters, 1992; Mets et al., 2000). In addition, LCMV poses special threat to immuno-compromised individuals, as illustrated by recent cases of transplant-associated infections by LCMV with a fatal outcome in the United States (Fischer et al., 2006; Peters, 2006).

Arenavirus Genome Organization

Arenaviruses are enveloped viruses with a bi-segmented negative single-stranded RNA genome and a life cycle restricted to the cell cytoplasm (Buchmeier et al., 2007; Meyer et al., 2002). Individual arenaviruses exhibit some variability in the lengths of the two genomic RNA segments, L (ca. 7.2 kb) and S (ca. 3.5 kb), but their overall organization is well conserved across the virus family. As with other negative strand (NS) RNA viruses, arenaviruses are characterized by a lack of infectivity of their purified genome RNA species and the presence of a virion-associated RNA-dependent-RNA polymerase (RdRp). However, the Arenavirus coding strategy has unique features compared to prototypical NS RNA viruses. Each Arenavirus genome segment uses an ambisense coding strategy to direct the synthesis of two polypeptides in opposite orientation, separated by a non-coding intergenic region (IGR) with a predicted folding of a stable hairpin structure (Buchmeier et al., 2007; Meyer et al., 2002). The S RNA encodes the viral glycoprotein precursor, GPC (ca. 75 kDa), and the nucleoprotein, NP (ca. 63 kDa), whereas the L RNA encodes the viral RNA-dependent-RNA polymerase (RdRp, or L polymerase) (ca. 200 kDa) and a small RING finger protein Z (ca. 11 kDa). The NP and L coding regions are transcribed into a genomic complementary mRNA, whereas the GPC and Z coding regions are not translated directly from genomic RNA, but rather from genomic sense mRNAs that are transcribed using as templates the corresponding antigenome RNA species, which also function as replicative intermediates (Fig. 10.1). The term ambisense refers to this situation in which regions located down- and upstream of the IGR of the S and L genome RNA species are of negative and positive sense, respectively. However, genomic S and L RNAs cannot function as mRNAs and be directly translated into GPC and Z proteins, respectively.

Virions contain the L and S genomic RNAs as helical nucleocapsid structures that are organized into circular configurations, with lengths ranging from 400 to 1300 nm (Young and Howard, 1983). The L and S genomic RNA species are not present in equimolar amounts within virions (L:S ratios ~1:2), and low levels of both L and S antigenomic RNA species are also present within virions. In addition, it has been documented that host ribosomes can be incorporated into virions, but the biological implications of this remain to be determined (Buchmeier et al., 2007; Muller et al., 1983). Likewise, significant levels of the Z mRNA appear to be incorporated into virions (Salvato et al., 1992), but it is unknown whether this reflects a functional requirement or an imprecise encapsidation process.

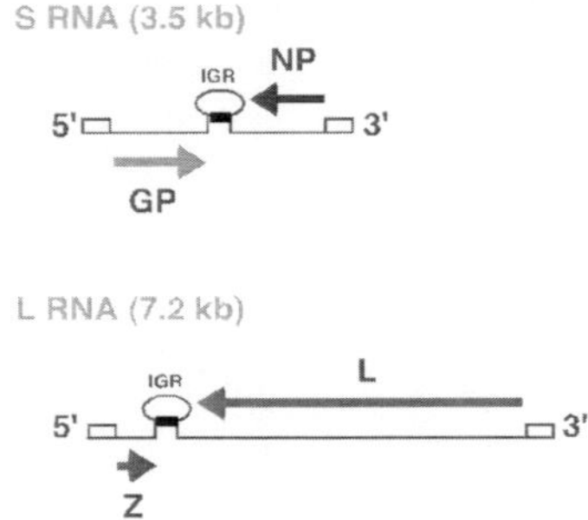

Fig. 10.1 LCMV genome organization.

Terminal Nucleotide Sequences

Arenaviruses exhibit high degree of sequence conservation at the 3′-end of the L and S RNA segments (17 out of 19 nucleotides [nt] are identical), suggesting that this conserved terminal sequence element constitutes the virus promoter for polymerase entry. Arenaviruses, similar to other NS RNA viruses, also exhibit complementarity between the 5′- and 3′-ends of their genomes and antigenomes. The almost exact inverted complement of the 3 ′-end 19 nt is found at the 5′-termini of genomes and antigenomes of arenaviruses. Thus, the 5′- and 3′-ends of both L and S genome segments are predicted to form panhandle structures. This prediction is supported by electron microscopy data showing the existence of circular ribonucleoprotein (RNP) complexes within arenavirus virion particles. This terminal complementarity may reflect the presence at the 5′-ends of *cis*-acting signals sequences that provide a nucleation site for RNA encapsidation, required to generate the nucleocapsid (NC) templates recognized by the virus polymerase. Terminal complementarity may also be a consequence of strong similarities between the genome and antigenome promoters used by the virus polymerases. This terminal complementarity has been proposed to favor the formation of both intra- and inter-molecular L and S duplexes that might be part of the replication initiation complex (Salvato, 1993b). For several arenaviruses, an additional non-templated G residue has been detected on the 5′-end of their genome RNAs (Garcin and Kolakofsky, 1992; Raju et al., 1990).

Intergenic Regions

Arenavirus IGR are predicted to fold into a stable hairpin structure. Transcription termination of the S-derived NP and GP occurs at multiple sites within the predicted stem of the IGR, suggesting that a structural motif rather than a sequence-specific signal promotes the release of the Arenavirus polymerase from the template RNA.

Some arenaviruses including LCMV contain one single predicted stem loop within the S IGR, whereas the S IGR of others, e.g., TCRV, is predicted to contain two distinct IGRs downstream to the translation termination codons from NP and GPC (Buchmeier et al., 2007; Meyer et al., 2002).

Sequence Heterogeneity

The family Arenaviridae comprises two distinct complexes: the LCMV-Lassa complex, which includes the Old World arenaviruses, and the Tacaribe virus (TCRV) complex, which includes all known New World arenaviruses. Early sequence analysis of laboratory-adapted arenaviruses revealed a significant degree of genetic stability with amino acid sequence homologies of 90%–95% among different strains of the prototypic arenavirus lymphocytic choriomeningitis virus (LCMV), whereas significant higher levels of genetic diversity (37%–56%) were observed for homologous proteins of different arenaviruses species (Buchmeier et al., 2007; Southern, 1996). More recent genetic studies on arenavirus field isolates, including Lassa (Bowen et al., 2000), Junin (Garcia et al., 2000), Guanarito (GTO) (Weaver et al., 2000), Pirital (PIR) (Fulhorst et al., 1999), and Whitewater Arroyo (Fulhorst et al., 2001), have revealed a high degree of genetic variation among geographical and temporal isolates of the same virus species. Notably, a remarkably high level of genetic divergence (26% and 16% at the nt and amino acid level, respectively) has been documented among PIR isolates within very small geographic regions (Weaver et al., 2001). The substantial degree of inter- and intra-species genetic variation among arenaviruses appears to have important biological correlates, as suggested by the significant variation in biological properties observed among LCMV strains (Sevilla et al., 2002). Thus, dramatic phenotypic differences have been documented among genetically very closely related LCMV isolates exhibiting only a few amino acid differences in their proteins.

Arenavirus Proteins

The NP is the most abundant viral polypeptide in both infected cells and virions. NP is the main structural element of the viral RNP and plays an essential role in viral RNA synthesis. We have shown that NP exhibits also an IFN counteracting activity (Martinez-Sobrido et al., 2006). The viral glycoprotein precursor GPC is posttranslationally proteolytically processed by the S1P cellular protease to yield the two mature virion glycoproteins GP-1 (40–46 kDa) and GP-2 (35 kDa) (Beyer et al., 2003; Pinschewer et al., 2003b). GPC contains a 58-amino-acid signal peptide (SSP) that is expressed as a stable polypeptide in infected cells that remains stably associated to the GP complex (GPcx). Besides its role in targeting the nascent polypeptide to the endoplasmic reticulum, this SSP likely serves additional roles in the trafficking and function of the viral envelope glycoproteins (Eichler et al.,

2003a, 2004, 2003b; Froeschke et al., 2003; York et al., 2004). GP-1 mediates virus interaction with host cell surface receptors and is held in place at the top of the spike by ionic interactions with the N-terminus of the transmembrane GP-2 (Buchmeier et al., 2001; Neuman et al., 2005).

The Arenavirus L protein has the characteristic sequence motifs conserved among the RdRp (L proteins), of negative strand (NS) RNA viruses (Salvato et al., 1989; Sanchez and de la Torre, 2005; Tordo et al., 1992; Vieth et al., 2004). Detailed sequence analysis and secondary structure predictions have been documented for the LFV L polymerase (Vieth et al., 2004). These studies identified several regions of strong alpha-helical content and a putative coiled-coil domain at the N-terminus, whose functional roles remain to be determined.

The arenavirus RING finger protein Z has no homologue among other known NS RNA viruses. Z is a structural component of the virion (Salvato et al., 1992). In LCMV-infected cells Z has been shown to interact with several cellular proteins including the promyelocytic leukemia (PML) protein (Borden et al., 1998) and the eukaryote translation initiation factor 4E (eIF4E) (Campbell Dwyer et al., 2000; Kentsis et al., 2001), which have been proposed to contribute to the noncytolytic nature of LCMV infection and repression of cap-dependent translation, respectively. Biochemical studies suggested that Z might be the arenavirus counterpart of the matrix (M) protein found in other negative strand RNA viruses (Salvato, 1993a; Salvato et al., 1992). Consistent with this proposal, more recent evidence has shown that Z is the driving force of arenavirus budding (Perez et al., 2003; Strecker et al., 2003; Urata et al., 2006). The expression of Z during the progression from early to late phases of the LCMV life cycle appears to be highly regulated, and thereby Z might play different roles during the life cycle of LCMV.

Arenavirus Life Cycle

Cell Attachment and Entry

Consistent with a broad host range and cell type tropism, a highly conserved and widely expressed cell surface protein, alpha-dystroglycan (aDG) has been identified as a main receptor for LCMV, LASV and Clade C NW arenaviruses (Kunz et al., 2002). However, several other arenaviruses appear to use an alternative receptor (Kunz et al., 2004; Spiropoulou et al., 2002), and recently transferring receptor 1 was identified as a cellular receptor used for entry of the NW HF arenaviruses Machupo and Junin (Radoshitzky et al., 2007). Upon receptor binding, arenavirus virions are internalized by uncoated vesicles and released into the cytoplasm by a pH-dependent membrane fusion step that is mediated by GP-2 (Borrow and Oldstone, 1994; Di Simone and Buchmeier, 1995; Di Simone et al., 1994). This fusion event is mediated by GP-2, which is structurally similar to the fusion active membrane proximal portions of the GP of other enveloped viruses (Gallaher et al., 2001).

RNA Replication and Transcription

The fusion between viral and cellular membranes releases the viral RNP into the cytoplasm, which is ensued by viral RNA synthesis. LCMV mRNAs have extra non-templated nt and a cap structure at their 5′-ends, but the origin of both the cap and 5′-non-templated nt extensions remains to be determined. Transcription termination of subgenomic non-polyadenylated viral mRNAs was mapped to multiple sites within the distal side of the IGR (Meyer and Southern, 1994; Tortorici et al., 2001), suggesting that the IGR acts as a bona fide transcription termination signal for the virus polymerase. The basic steps of Arenavirus RNA replication and gene transcription are illustrated in Fig. 10.2 using the S segment as example; the same scheme applies also to the L segment. NP and L are transcribed into a genomic complementary mRNA, whereas the GPC and Z are not translated directly from genomic RNA, but rather from genomic sense mRNAs that are transcribed using as templates the corresponding antigenome RNA species, which also function as replicative intermediates.

Viral *Trans*-Acting Factors Required for RNA Replication and Transcription

Reverse genetic studies using an LCMV minigenome rescue system identified NP and L as the minimal viral *trans*-acting factors required for efficient RNA synthesis mediated by the virus polymerase (Lee et al., 2000). Similar findings have been now documented for LASV (Hass et al., 2004) and the New World Arenavirus TCRV (Lopez et al., 2001). Notably, both genetic and biochemical evidence have indicated

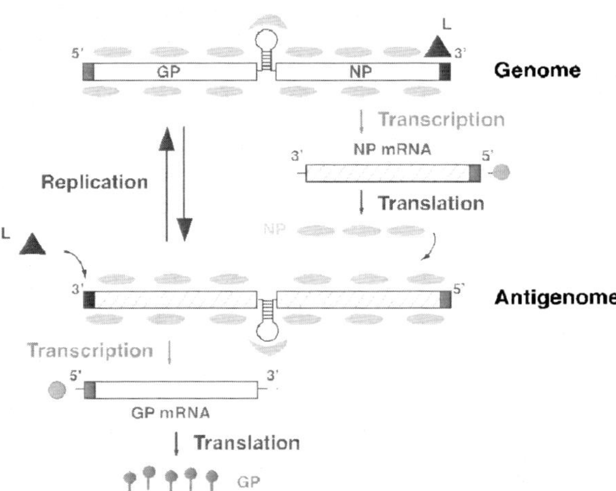

Fig. 10.2 LCMV replication and transcription.

that oligomerization of L is required for the activity of the LCMV L polymerase (Sanchez and de la Torre, 2005), a finding similar to that previously documented for the paramyxoviruses Sendai (Smallwood et al., 2002) and parainfluenza virus 3 (PIV3) (Smallwood and Moyer, 2004).

Cis-Acting Signals Involved in the Regulation of LCMV RNA Synthesis

Sequence Specificity and Structure Define the Functional Genome Promoter

All arenavirus genomes examined to date have a highly conserved sequence element at their 3′-termini (17/19 nt are identical), and the inverted complement of this sequence is found at the genome 5′-termini. Terminal complementarity in L and S RNAs predicts the formation of a conserved and thermodynamically stable panhandle structure that, similar to influenza- and bunya-viruses, was proposed to contribute to the control of RNA synthesis.

Mutation-function analysis of the genome 3′/5′-termini in the control of viral RNA synthesis using a transcription and replication competent LCMV minigenome (MG) rescue system (Perez and de la Torre, 2002) revealed the minimal LCMV genomic promoter to be contained within the 3′-terminal 19 nt. Moreover, deletions and nt substitutions within the MG 5′-end that disrupted terminal complementarity abolished also genome promoter activity. Notably, compensatory mutations that restore paring between the 3′- and 5′-termini did not result in restoration of genome promoter activity. Likewise, these studies did not identify mutations within the promoter sequences that affected independently either RNA replication or transcription. A detailed mutation-function analysis of the virus genome promoter has been documented for LASV (Hass et al., 2006), revealing that the LASV genome promoter also regulates transcription and replication in a coordinated manner. These studies also showed that the LASV genome promoter is composed of two functional elements, a sequence-specific region from residues 1 to 12 and a variable complementary region from residues 13 to 19. The first region appears to interact with the replication complex mainly via base-specific interactions, while in the second region solely base pairing between 3′ and 5′ promoter ends is important for promoter function.

These findings support the view that arenavirus genome promoters regulate transcription and replication in a coordinated manner and that both sequence specificity within the 3′-terminal 19 nt and the integrity of the predicted panhandle structure are required for the activity of the genome promoter. Moreover, the two biosynthetic processes, RNA replication and transcription, directed by the virus polymerase complex appeared to be coordinated by the same *cis*-acting regulatory sequences. Initiation of RNA synthesis by the arenavirus polymerase has been proposed to employ a prime and realign mechanism, which would account for the presence of a non-templated G at the 5′-ends of the arenavirus genomic and antigenomic RNAs (Garcin and Kolakofsky, 1990, 1992). Results obtained using MGs with a variety of 5′-end sequences have provided evidence in support of this model (Perez and de la Torre, 2002).

The IGR Present Within Each Arenavirus Genome Segment is a Bona Fide Transcription Termination Signal, but Plays also a Critical Role in Assembly of Infectious Particles

Arenavirus mRNAs have extra non-templated nucleotides (nt) and a cap structure at their 5′-ends, but the origin of both the cap and 5′-non-templated nt extensions remains to be determined. The 3′-termini of the subgenomic non-polyadenylated viral mRNAs have been mapped to multiple sites within the distal side of the IGR (Meyer et al., 2002; Southern, 1996). All arenavirus IGR sequences are predicted to fold into single or double stem-loop structures (Buchmeier et al., 2001; Meyer et al., 2002; Southern, 1996), suggesting a structure-dependent transcription termination mechanism reminiscent of rho-independent termination in prokaryotes (Yarnell and Roberts, 1999).

Studies using the LCMV MG rescue system where a variety of RNA analogues of the S genome segment, containing or not an IGR, served as a template for synthesis of full-length anti-MG (aMG) replicate and subgenomic size mRNA species for reporter gene expression showed that a MG without IGR was amplified by the virus polymerase with equal efficiency but subgenomic mRNA species were undetectable (Pinschewer et al., 2005). Intriguingly, however, reporter gene expression from IGR-deficient aMG CAT-sense RNA of genomic length was found to be only about 5-fold less efficient than from subgenomic CAT mRNA derived from an IGR-containing MG, but at least 100-fold more efficient than a T7 RNA polymerase transcript with the same sequence. These results validated the IGR as a bona fide transcription termination signal, but revealed also that in the absence of IGR-mediated transcription termination, a fraction of full-length aMG RNA behaves as bona fide mRNA. Conceptually similar findings have been also documented for the NW arena TCRV (Lopez and Franze-Fernandez, 2007). Likewise studies using a TCRV MG rescue system demonstrated that the transcription termination signal provided by the IGR is structure, but not sequence, dependent (Lopez and Franze-Fernandez, 2007).

Unexpectedly, LCMV MGs without IGR were dramatically impaired in their ability to passage reporter gene activity via infectious VLP (Pinschewer et al., 2005), suggesting that in addition to its role in the control of RNA synthesis, the arenavirus IGR plays a role in virus assembly or budding, or both, required for the efficient virus propagation. Whether this role of the IGR depends on sequence specificity or structure, or both, remains to be determined.

Intracellular levels of NP determine levels of viral RNA synthesis but do not regulate the balance between RNA replication and transcription. For arenaviruses it has been proposed, and widely accepted, that intracellular NP levels modulate the balance between RNA replication and transcription. Intracellular NP levels increase during the course of the infection and unfold secondary RNA structures within the IGR. This results in attenuation of structure-dependent transcription termination at the IGR, which promotes replication of genome and antigenome RNA species. However, more recent studies using the LCMV MG rescue system revealed that both RNA replication and transcription were equally enhanced by incrementally increasing amounts of NP up to levels in the range of LCMV-infected cells (Pinschewer et al., 2003a). These data, similar to those described for the paramyxovirus RSV (Fearns et al., 1997), are consistent with a central role for NP in transcription and RNA replication of the LCMV genome, but they do not support a central role of NP levels in balancing the two biosynthetic processes.

Role of the Z Protein in the Control of Arenavirus RNA Synthesis

Z Exhibits a Dose-Dependent Inhibitory Effect on RNA Replication and Transcription of the LCMV MG

Z was not required for intracellular transcription and replication of an LCMV MG, but rather Z exhibited a dose-dependent inhibitory effect on both transcription and replication of LCMV MG (Cornu and de la Torre, 2001, 2002; Cornu et al., 2004; Lee et al., 2000) (see below). Similar findings have been also reported for TV (Lopez et al., 2001) and LASV (Hass et al., 2004).

Mutation-function studies identified regions and specific amino acid residues within Z contributing to its inhibitory activity on RNA synthesis mediated by the LCMV polymerase (Cornu and de la Torre, 2002). Serial deletion mutants of the N- and C-termini of Z showed that the N-terminus (residues 1–16) and C-terminus (residues 79–90) do not contribute to the Z inhibitory activity. Moreover, results from the use of chimera proteins between Z and Xenopus Neuralized, a nonviral RING finger protein, indicated that the structural integrity of the Z ring domain (RD) was required but not sufficient for the inhibitory activity of Z. Likewise, a highly conserved tryptophan (W) residue located at position 36 in ARM-Z, next to the second conserved cysteine (C) of the Z RD, had a major contribution to the Z inhibitory activity. The inhibitory activity of Z on virus RNA synthesis appeared to be related to the degree of genetic proximity between Z and the viral *trans*-acting factors L and NP. Thus, Z proteins from different LCMV strains had similar inhibitory activities on the expression of LCMV MG, whereas the Z protein of the genetically more distantly related TCRV had about 10-fold lower inhibitory activity on LCMV MG expression (Cornu and de la Torre, 2002).

Homotypic viral interference can be readily demonstrated with several arenaviruses including LCMV (Welsh and Pfau, 1972). This phenomenon is not strictly strain specific as illustrated by the existence of interference among different pairs of LCMV strains. Heterotypic interference between arenaviruses has been occasionally reported, and its degree appears to correlate with the genetic relationship of the viruses (Damonte et al., 1983; Welsh and Pfau, 1972). It is therefore plausible that increased expression of Z protein during the virus life cycle might contribute to block replication of an additional infection by a genetically closely related arenavirus. Superinfection exclusion could influence arenavirus evolution and contribute to explain the observed population partitioning in the field, resulting in the maintenance of independent evolutionary lineages of the same strain within a small geographic range.

Cells Expressing Z Become Highly Resistant to Virus Infection Due to a Blockade in Virus RNA Synthesis

Cells transduced with recombinant, replication-deficient adenoviruses expressing Z (rAd-Z) from either LCMV or LASV became highly resistant to infection with LCMV or LASV, respectively (Cornu et al., 2004), whereas cells transduced with a control rAd expressing GFP remained fully susceptible to both LCMV and LASV. This resistance was specific as the rAd-Z transduced cells remained fully susceptible to measles virus (MV) (Cornu et al., 2004). These findings indicated that the Z-mediated inhibitory activity operates also during the course of the natural cycle of virus infection, and it is not a property observed only in the context of a MG system. Cells transduced with rAd-Z remained susceptible to infection with a recombinant VSV where the LCMV G substituted for the VSV G (Cornu et al., 2004), suggesting that Z-mediated resistance to infection was not due to a blockade of virus entry, but rather to a strong inhibitory effect of Z on LCMV RNA replication and transcription.

The TCRV Z protein was reported to interact with the virus L polymerase, and this interaction was proposed to be responsible for TCRV Z-mediated inhibition of RNA replication and expression of a TCRV MG (Jacamo et al., 2003). Intriguingly, for LCMV the use of either biochemical (Co-IP) or genetic (mammalian-TH) approaches have failed to provide evidence of a Z–L interaction. This could be due to intrinsic differences between the biology of TCRV and LCMV, or differences in the two experimental systems. Further studies would be necessary to elucidate the mechanisms by which Z exerts its inhibitory activity on RNA synthesis by the virus polymerase, as well as to determine the biological implications of this Z activity. The recent observation that the activity of the LCMV L polymerase requires an L–L interaction (Sanchez and de la Torre, 2005) raises the possibility that Z might interfere with L–L interaction and thereby affect the virus polymerase activity. An alternative way whereby Z could mediate inhibition of viral RNA synthesis stems from our finding that Z interacts with N. As with the M proteins of several other NS RNA viruses, Z–NP interaction could inhibit the biosynthetic activity of the virus RNP.

Assembly and Budding

GP and Z are Required for the Generation of Infectious VLP

Production of LCMV occurs by budding at the surface of infected cells. For most enveloped NS RNA viruses, this process is assumed to depend on the interaction between the RNP core and the virus-encoded transmembrane glycoproteins (GP), which is mediated by the matrix (M) protein. Arenaviruses do not code for an obvious counterpart of M, but early cross-linking studies showed complex formation between NP and Z, suggesting a possible role of Z in virion morphogenesis (Salvato, 1993b; Salvato et al., 1992). Studies using the LCMV MG rescue system showed that generation of infectious arenavirus-like particles (VLP) required both Z and GP (Lee et al., 2002). Importantly, the correct processing of GPC was strictly required for the generation of either infectious VLPs (Lee et al., 2002) or retroviral pseudotyped particles (Beyer et al., 2003; Pinschewer et al., 2003b). Moreover, correct processing of GPC necessitates the structural integrity of GP-2 cytoplasmic tail (Kunz et al., 2003).

Z is the Driving Force of Arenavirus Budding

The requirement of GP for the generation of infectious VLP was expected due to its role in receptor recognition and virus entry (Kunz et al., 2002), whereas the need for Z indicated a role of this protein in virus assembly or budding, or both. This, in turn, suggested that consistent with earlier biochemical data (Salvato, 1993b; Salvato et al., 1992) and recent ultrastructural data on arenavirus virions determined by cryo-electron microscopy (Neuman et al., 2005), Z could be the Arenavirus functional counterpart of the M proteins that mediate budding in other NS RNA viruses. Studies using reverse genetics approaches to examine the requirement of LCMV proteins for efficient cell release of VLP containing bona fide viral nucleocapsids (NC) revealed that the production of MG RNA-containing NC was not impaired in the absence of GP, but dramatically diminished in the absence of Z (Perez et al., 2003), indicating that Z was playing a central role in LCMV budding.

Z has Features of Bona Fide Budding Proteins and Contains Canonical Late (L) Domain Motifs That are Functionally Active in Promoting Z-Mediated Budding

Consistent with its role as the driving force of arenavirus budding, Z exhibited self-budding activity in the absence of other viral proteins (Perez et al., 2003; Strecker et al., 2003; Urata et al., 2006). A feature characteristic of viral budding proteins is the flexibility of their L domains; one L domain substitutes for another in promoting virion release (Freed, 2002). Z also exhibited this feature as determined by the budding properties of Z-Gag chimeric proteins where Z was fused to an RSV Gag protein that lacked both its membrane targeting and binding signal (M domain) and L domain (Perez et al., 2003).

Consistent with their features of bona fide budding proteins, arenavirus Z contains canonical late (L) domain motifs similar to those present in Gag and M proteins of several viruses (Freed, 2002). The Z protein of LCMV contains a single PPPY motif, whereas the Z protein of the highly pathogenic arenavirus LFV possesses both PTAP and PPPY motifs separated by nine amino acids (Perez et al., 2003; Strecker et al., 2003; Urata et al., 2006). Mutation-function studies confirmed that these L domain motifs present in Z mediated the budding activity of Z (Perez et al., 2003; Strecker et al., 2003). Ebola VP40 protein contains overlapping PTAP and PPXY L domains, but each one of them was found to be sufficient to promote efficient VP40-mediated budding (Licata et al., 2003). In contrast, in the case of

LASV Z both L domains were found to be required for efficient Z-mediated budding (Perez et al., 2003).

Myristoylation of Z is Required for its Budding Activity

Z is devoid of hydrophobic transmembrane domains, but it accumulates near to the inner surface of the plasma membrane and is strongly membrane associated. All known arenavirus Z proteins contain a glycine (G) at position 2 and nearby K and R residues characteristic of a myristoylation motif. Metabolic labeling showed incorporation of [^{3}H]myristic acid by wild-type, but not G2A mutant; Z protein and the mutation G2A abrogated Z-mediated budding without affecting viral RNA replication and transcription (Perez et al., 2004). Likewise, treatment with the myristoylation inhibitor 2-hydroxymyristic acid (2-OHM) inhibited Z-mediated budding, abrogated formation of virus-like particles, and caused a dramatic reduction in virus production in LCMV-infected cells (Perez et al., 2004). Moreover, addition to the N-terminus of Z(G2A) of the myristoylation domain of the tyrosine protein kinase Src restored budding activity in the Z(G2A)G2A (Perez et al., 2004). These findings and similar ones described by others (Strecker et al., 2003, 2006) have also been documented. These findings indicate that myristoylation of Z plays a key role in Arenavirus budding. Similar findings have been also documented for LASV Z.

Z–GP Interact

Based on the roles played by Z and GP in the arenavirus life cycle, it would be predicted that Z and the GP should interact in a manner required for the formation of mature infectious virion particles. Accordingly, recent evidence has shown the subcellular co-localization and biochemical association of Z and GP (Capul et al., 2007). Notably, neither the RING domain nor the L domains were required for this Z–GP interaction (Capul et al., 2007). In contrast myristoylation of Z played a critical role in Z–GP interaction as determined by the failure of a G2A mutant of Z to interact with GP (Capul et al., 2007). These results may reflect that accumulation of Z at certain membranes within the cell might be a limiting factor for its association with GP.

Production of Infectious LCMV from Cloned cDNAs

The ability to generate predetermined specific mutations within the LCMV genome, and analyze their phenotypic expression in appropriate cell culture systems and the virus natural host, the mouse, has represented a major step forward for the elucidation of the molecular and cellular mechanisms underlying LCMV–host interactions, including the bases of LCMV persistence and associated disease. In addition, the procedures developed for the rescue of rLCMV should allow for the rescue of LFV and other HF arenaviruses, which may accelerate the development and fine-tuning of live-attenuated arenavirus vaccines (Lukashevich et al., 2005), and facilitate their safe production for use in endemic areas where they are urgently needed (Geisbert and Jahrling, 2004; Geisbert et al., 2005; McCormick and Fisher-Hoch, 2002).

Rescue of Infectious rLCMV from Cloned cDNAs

Prior to the successful rescue of LCMV entirely from cloned cDNAs, a helper virus-based system was documented that allowed for the rescue of LCMV carrying a recombinant S segment (rS) (Pinschewer et al., 2003b). This system was based on intracellular reconstitution of a recombinant LCMV S (rS) where the glycoprotein of vesicular stomatitis virus (VSVG) was substituted for the glycoprotein of LCMV and produced intracellularly from cDNA under control of a polymerase I promotor. Coexpression of the LCMV proteins NP and L allowed expression of VSVG from rS, and infection of transfected cells with wild-type (wt) LCMV resulted in reassortment of the L segment of wt LCMV with the rS at low frequency. Selection of the rLCMV over the LCMVwt used as helper virus was facilitated by the use of a cell line (SRD-12B) deficient in the S1P protease. The rationale for this approach was based on the fact that LCMV infectivity, but not that of rLCMV/VSVG, requires correct processing of LCMV-GPC by the cellular protease S1P.

The rLCMV/VSVG provided investigators with a powerful tool to rescue rLCM viruses containing engineered rS segments. For this, cells instructed to express the rS RNP of interest are infected with rLCMV/VSVG, and the virus progeny is subjected to selection with a neutralizing antibody to VSV G to eliminate the helper rLCMV/VSVG. This approach however required several rounds of selection and was limited to the rescue of LCMV carrying recombinant S segments. These limitations were circumvented with the development of reverse genetics system to allow for the rescue of infectious LCMV entirely from cloned cDNAs, without the need of using a helper virus. Both a T7 RNA polymerase (T7RP) (Sanchez and de la Torre, 2006) and RNA polymerase I (pol-I) (Flatz et al., 2006) systems have been developed to direct intracellular synthesis to recombinant L and S genome, or antigenome, RNA species. Both the T7RP and pol-I-based rescue systems used pol-II-based expression plasmids to provide the viral *trans*-acting factors L and NP. Both systems exhibited similar efficiencies. The pol-I-based system offers the advantages that (1) the generation of the correct 3′-end of the Sag and Lag RNA species does not depend on the efficiencies of self-cleavable ribozymes and (2) there is no need for a plasmid expressing T7RP. On the other hand this system has the limitation of the species specificity of the pol-I promoters, which determines the need of generating different vectors for efficient intracellular synthesis of virus genome RNA species in cell types from different species.

Production of rLCMV was readily detected 48 h after transfection, which was followed by a rapid increase in virus production reaching titers of 10^7 PFU/ml. These rLCMV exhibited growth and biological properties predicted for LCMV. Notably, similar rescue efficiencies were obtained using genome of antigenome L and S expressing plasmids (Sanchez and de la Torre, 2006), indicating that annealing between viral mRNAs and genome, or antigenome, RNA species does not pose a significant problem for the rescue of arenaviruses.

Use of rLCMV to Address Biological Questions

The rLCMV/VSVG has been used to examine a variety of biological questions, which illustrate the tremendous impetus for arenavirus research derived from the ability to generate rLCMV from cloned cDNAs (Bergthaler et al., 2006; Merkler et al., 2006; Pinschewer et al., 2004). The development and use of reverse genetics approaches have revolutionized the analysis of the *cis*-acting signals and *trans*-acting proteins required for RNA replication, transcription, maturation, and budding of other negative strand RNA viruses (Conzelmann, 2004; Kawaoka, 2004; Neumann et al., 2002). These approaches are now applicable to arenaviruses and will permit to dissect the role, and underlying mechanisms, of each virus gene product to the each of the steps of the arenavirus life cycle.

References

Barton, L. L., and Mets, M. B. (1999). Lymphocytic choriomeningitis virus: pediatric pathogen and fetal teratogen. *Pediatr Infect Dis J* **18**(6), 540–1.

Barton, L. L., and Mets, M. B. (2001). Congenital lymphocytic choriomeningitis virus infection: decade of rediscovery. *Clin Infect Dis* **33**(3), 370–4.

Barton, L. L., Mets, M. B., and Beauchamp, C. L. (2002). Lymphocytic choriomeningitis virus: emerging fetal teratogen. *Am J Obstet Gynecol* **187**(6), 1715–6.

Bergthaler, A., Gerber, N. U., Merkler, D., Horvath, E., de la Torre, J. C., and Pinschewer, D. D. (2006). Envelope exchange for the generation of live-attenuated arenavirus vaccines. *PLoS Pathog* **2**(6), e51.

Beyer, W. R., Popplau, D., Garten, W., von Laer, D., and Lenz, O. (2003). Endoproteolytic processing of the lymphocytic choriomeningitis virus glycoprotein by the subtilase SKI-1/S1P. *J Virol* **77**(5), 2866–72.

Borden, K. L., Campbell Dwyer, E. J., and Salvato, M. S. (1998). An arenavirus RING (zinc-binding) protein binds the oncoprotein promyelocyte leukemia protein (PML) and relocates PML nuclear bodies to the cytoplasm. *J Virol* **72**(1), 758–66.

Borrow, P., and Oldstone, M. B. (1994). Mechanism of lymphocytic choriomeningitis virus entry into cells. *Virology* **198**(1), 1–9.

Bowen, M. D., Rollin, P. E., Ksiazek, T. G., Hustad, H. L., Bausch, D. G., Demby, A. H., Bajani, M. D., Peters, C. J., and Nichol, S. T. (2000). Genetic diversity among Lassa virus strains. *J Virol* **74**(15), 6992–7004.

Buchmeier, M. J., Bowen, M. D., and Peters, C. J. (2001). Arenaviridae: the virus and their replication. 4th ed. *In* Fields Virology (D. M. Knipe, and P. M. Howley, Eds.), Vol. 2, pp. 1635–1668. Lippincott Williams & Wilkins, Philadelphia.

Buchmeier, M. J., Peters, C. J., and de la Torre, J. C. (2007). Arenaviridae: the viruses and their replication. 5th ed. *In* Fields Virology (D. M. Knipe, and P. M. Holey, Eds.), Vol. 2, pp. 1792–1827. Lippincott Williams & Wilkins, Philadelphia.

Campbell Dwyer, E. J., Lai, H., MacDonald, R. C., Salvato, M. S., and Borden, K. L. (2000). The lymphocytic choriomeningitis virus RING protein Z associates with eukaryotic initiation factor 4E and selectively represses translation in a RING-dependent manner. *J Virol* **74**(7), 3293–300.

Capul, A. A., Perez, M., Burke, E., Kunz, S., Buchmeier, M. J., de la Torre, J.C (2007). Arenavirus Z–GP association requires Z myristoylation but not functional RING or L domains. *J Virol* **81**(17), 9451–9460.

Conzelmann, K. K. (2004). Reverse genetics of mononegavirales. *In* Curr. Top. Microbiol. Immunol. (Y. Kawaoka, Ed.), Vol. 283, pp. 1–41.

Cornu, T. I., and de la Torre, J. C. (2001). RING finger Z protein of lymphocytic choriomeningitis virus (LCMV) inhibits transcription and RNA replication of an LCMV S-segment minigenome. *J Virol* **75**(19), 9415–26.

Cornu, T. I., and de la Torre, J. C. (2002). Characterization of the arenavirus RING finger Z protein regions required for Z-mediated inhibition of viral RNA synthesis. *J Virol* **76**(13), 6678–88.

Cornu, T. I., Feldmann, H., and de la Torre, J. C. (2004). Cells expressing the RING finger Z protein are resistant to arenavirus infection. *J Virol* **78**(6), 2979–83.

Damonte, E. B., Mersich, S. E., and Coto, C. E. (1983). Response of cells persistently infected with arenaviruses to superinfection with homotypic and heterotypic viruses. *Virology* **129**(2), 474–8.

de la Torre, J. C., and Oldstone, M. B. A. (1996). The anatomy of viral persistence: mechanisms of persistence and associated disease. *Adv Virus Res* **46**, 311–43.

Di Simone, C., and Buchmeier, M. J. (1995). Kinetics and pH dependence of acid-induced structural changes in the lymphocytic choriomeningitis virus glycoprotein complex. *Virology* **209**(1), 3–9.

Di Simone, C., Zandonatti, M. A., and Buchmeier, M. J. (1994). Acidic pH triggers LCMV membrane fusion activity and conformational change in the glycoprotein spike. *Virology* **198**(2), 455–65.

Eichler, R., Lenz, O., Strecker, T., Eickmann, M., Klenk, H. D., and Garten, W. (2003a). Identification of Lassa virus glycoprotein signal peptide as a trans-acting maturation factor. *EMBO Rep* **4**(11), 1084–8.

Eichler, R., Lenz, O., Strecker, T., Eickmann, M., Klenk, H. D., and Garten, W. (2004). Lassa virus glycoprotein signal peptide displays a novel topology with an extended endoplasmic reticulum luminal region. *J Biol Chem* **279**(13), 12293–9.

Eichler, R., Lenz, O., Strecker, T., and Garten, W. (2003b). Signal peptide of Lassa virus glycoprotein GP-C exhibits an unusual length. *FEBS Lett* **538**(1–3), 203–6.

Fearns, R., Peeples, M. E., and Collins, P. L. (1997). Increased expression of the N protein of respiratory syncytial virus stimulates minigenome replication but does not alter the balance between the synthesis of mRNA and antigenome. *Virology* **236**(1), 188–201.

Fischer, S. A., Graham, M. B., Kuehnert, M. J., Kotton, C. N., Srinivasan, A., Marty, F. M., Comer, J. A., Guarner, J., Paddock, C. D., DeMeo, D. L., Shieh, W. J., Erickson, B. R., Bandy, U., DeMaria, A., Jr., Davis, J. P., Delmonico, F. L., Pavlin, B., Likos, A., Vincent, M. J., Sealy, T. K., Goldsmith, C. S., Jernigan, D. B., Rollin, P. E., Packard, M. M., Patel, M., Rowland, C., Helfand, R. F., Nichol, S. T., Fishman, J. A., Ksiazek, T., and Zaki, S. R. (2006). Transmission of lymphocytic choriomeningitis virus by organ transplantation. *N Engl J Med* **354**(21), 2235–49.

Flatz, L., Bergthaler, A., de la Torre, J. C., and Pinschewer, D. D. (2006). Recovery of an arenavirus entirely from RNA polymerase I/II-driven cDNA. *Proc Natl Acad Sci USA* **103**(12), 4663–8.

Freed, E. O. (2002). Viral late domains. *J Virol* **76**(10), 4679–87.

Freedman, D. O., and Woodall, J. (1999). Emerging infectious diseases and risk to the traveler. *Med Clin North Am* **83**(4), 865–83.

Froeschke, M., Basler, M., Groettrup, M., and Dobberstein, B. (2003). Long-lived signal peptide of lymphocytic choriomeningitis virus glycoprotein pGP-C. *J Biol Chem* **278**(43), 41914–20.

Fulhorst, C. F., Bowen, M. D., Salas, R. A., Duno, G., Utrera, A., Ksiazek, T. G., De Manzione, N. M., De Miller, E., Vasquez, C., Peters, C. J., and Tesh, R. B. (1999). Natural rodent host associations of Guanarito and Pirital viruses (Family Arenaviridae) in central Venezuela. *Am J Trop Med Hyg* **61**(2), 325–30.

Fulhorst, C. F., Charrel, R. N., Weaver, S. C., Ksiazek, T. G., Bradley, R. D., Milazzo, M. L., Tesh, R. B., and Bowen, M. D. (2001). Geographic distribution and genetic diversity of Whitewater Arroyo virus in the southwestern United States. *Emerg Infect Dis* **7**(3), 403–7.

Gallaher, W. R., DiSimone, C., and Buchmeier, M. J. (2001). The viral transmembrane superfamily: possible divergence of arenavirus and filovirus glycoproteins from a common RNA virus ancestor. *BMC Microbiol* **1**(1), 1.

Garcia, J. B., Morzunov, S. P., Levis, S., Rowe, J., Calderon, G., Enria, D., Sabattini, M., Buchmeier, M. J., Bowen, M. D., and St Jeor, S. C. (2000). Genetic diversity of the Junin virus in Argentina: geographic and temporal patterns. *Virology* **272**(1), 127–36.

Garcin, D., and Kolakofsky, D. (1990). A novel mechanism for the initiation of Tacaribe arenavirus genome replication. *J Virol* **64**(12), 6196–203.

Garcin, D., and Kolakofsky, D. (1992). Tacaribe arenavirus RNA synthesis in vitro is primer dependent and suggests an unusual model for the initiation of genome replication. *J Virol* **66**(3), 1370–6.

Geisbert, T. W., and Jahrling, P. B. (2004). Exotic emerging viral diseases: progress and challenges. *Nat Med* **10**(12 Suppl), S110–21.

Geisbert, T. W., Jones, S., Fritz, E. A., Shurtleff, A. C., Geisbert, J. B., Liebscher, R., Grolla, A., Stroher, U., Fernando, L., Daddario, K. M., Guttieri, M. C., Mothe, B. R., Larsen, T., Hensley, L. E., Jahrling, P. B., and Feldmann, H. (2005). Development of a new vaccine for the prevention of Lassa fever. *PLoS Med* **2**(6), e183.

Hass, M., Golnitz, U., Muller, S., Becker-Ziaja, B., and Gunther, S. (2004). Replicon system for Lassa virus. *J Virol* **78**(24), 13793–803.

Hass, M., Westerkofsky, M., Muller, S., Becker-Ziaja, B., Busch, C., and Gunther, S. (2006). Mutational analysis of the lassa virus promoter. *J Virol* **80**(24), 12414–9.

Holmes, G. P., McCormick, J. B., Trock, S. C., Chase, R. A., Lewis, S. M., Mason, C. A., Hall, P. A., Brammer, L. S., Perez-Oronoz, G. I., McDonnell, M. K., et al. (1990). Lassa fever in the United States. Investigation of a case and new guidelines for management. *N Engl J Med* **323**(16), 1120–3.

Isaacson, M. (2001). Viral hemorrhagic fever hazards for travelers in Africa. *Clin Infect Dis* **33**(10), 1707–12.

Jacamo, R., Lopez, N., Wilda, M., Franze-Fernandez, M. T. (2003). Tacaribe virus Z protein interacts with the L polymerase protein to inhibit viral RNA synthesis. *J Virol* **77**(19), 10383–93.

Jahrling, P. B., and Peters, C. J. (1992). Lymphocytic choriomeningitis virus. A neglected pathogen of man. *Arch Pathol Lab Med* **116**(5), 486–8.

Kawaoka, Y. (2004). Biology of negative strand RNA viruses. 1st ed. *In* Current Topics in Microbiology and Immunology, Vol. 283. Springer-Verlag, Berlin, Heidelberg.

Kentsis, A., Dwyer, E. C., Perez, J. M., Sharma, M., Chen, A., Pan, Z. Q., and Borden, K. L. (2001). The RING domains of the promyelocytic leukemia protein PML and the arenaviral protein Z repress translation by directly inhibiting translation initiation factor eIF4E. *J Mol Biol* **312**(4), 609–23.

Kunz, S., Borrow, P., and Oldstone, M. B. (2002). Receptor structure, binding, and cell entry of arenaviruses. *Curr Top Microbiol Immunol* **262**, 111–37.

Kunz, S., Edelmann, K. H., de la Torre, J.-C., Gorney, R., and Oldstone, M. B. A. (2003). Mechanisms for lymphocytic choriomeningitis virus glycoprotein cleavage, transport, and incorporation into virions. *Virology* **314**(1), 168–78.

Kunz, S., Sevilla, N., Rojek, J. M., and Oldstone, M. B. (2004). Use of alternative receptors different than alpha-dystroglycan by selected isolates of lymphocytic choriomeningitis virus. *Virology* **325**(2), 432–45.

Lee, K. J., Novella, I. S., Teng, M. N., Oldstone, M. B., and de La Torre, J. C. (2000). NP and L proteins of lymphocytic choriomeningitis virus (LCMV) are sufficient for efficient transcription and replication of LCMV genomic RNA analogs. *J Virol* **74**(8), 3470–7.

Lee, K. J., Perez, M., Pinschewer, D. D., and de la Torre, J. C. (2002). Identification of the lymphocytic choriomeningitis virus (LCMV) proteins required to rescue LCMV RNA analogs into LCMV-like particles. *J Virol* **76**(12), 6393–7.

Licata, J. M., Simpson-Holley, M., Wright, N. T., Han, Z., Paragas, J., and Harty, R. N. (2003). Overlapping motifs (PTAP and PPEY) within the Ebola virus VP40 protein function independently as late budding domains: involvement of host proteins TSG101 and VPS-4. *J Virol* **77**(3), 1812–9.

Lopez, N., and Franze-Fernandez, M. T. (2007). A single stem-loop structure in Tacaribe arenavirus intergenic region is essential for transcription termination but is not required for a correct initiation of transcription and replication. *Virus Res* **124**(1–2), 237–44.

Lopez, N., Jacamo, R., and Franze-Fernandez, M. T. (2001). Transcription and RNA replication of tacaribe virus genome and antigenome analogs require N and L proteins: z protein is an inhibitor of these processes. *J Virol* **75**(24), 12241–51.

Lukashevich, I. S., Patterson, J., Carrion, R., Moshkoff, D., Ticer, A., Zapata, J., Brasky, K., Geiger, R., Hubbard, G. B., Bryant, J., and Salvato, M. S. (2005). A live attenuated vaccine for Lassa fever made by reassortment of Lassa and Mopeia viruses. *J Virol* **79**(22), 13934–42.

Martinez-Sobrido, L., Zuniga, E. I., Rosario, D., Garcia-Sastre, A., and de la Torre, J. C. (2006). Inhibition of the type I interferon response by the nucleoprotein of the prototypic arenavirus lymphocytic choriomeningitis virus. *J Virol* **80**(18), 9192–9.

McCormick, J. B., and Fisher-Hoch, S. P. (2002). Lassa fever. *In* Arenaviruses I (M. B. Oldstone, Ed.), Vol. 262, pp. 75–110. Springer-Verlag, Berlin, Heidelberg, New York.

Merkler, D., Horvath, E., Bruck, W., Zinkernagel, R. M., Del la Torre, J. C., and Pinschewer, D. D. (2006). "Viral deja vu" elicits organ-specific immune disease independent of reactivity to self. *J Clin Invest* **116**(5), 1254–63.

Mets, M. B., Barton, L. L., Khan, A. S., and Ksiazek, T. G. (2000). Lymphocytic choriomeningitis virus: an underdiagnosed cause of congenital chorioretinitis. *Am J Ophthalmol* **130**(2), 209–15.

Meyer, B. J., de La Torre, J. C., and Southern, P. J. (2002). Arenaviruses: genomic RNAs, transcription, and replication. *In* Arenaviruses I (M. B. Oldstone, Ed.), Vol. 262, pp. 139–149. Springer-Verlag, Berlin Heidelberg.

Meyer, B. J., and Southern, P. J. (1994). Sequence heterogeneity in the termini of lymphocytic choriomeningitis virus genomic and antigenomic RNAs. *J Virol* **68**(11), 7659–64.

Muller, G., Bruns, M., Martinez Peralta, L., and Lehmann-Grube, F. (1983). Lymphocytic choriomeningitis virus. IV. Electron microscopic investigation of the virion. *Arch Virol* **75**(4), 229–42.

Neuman, B. W., Adair, B. D., Burns, J. W., Milligan, R. A., Buchmeier, M. J., and Yeager, M. (2005). Complementarity in the supramolecular design of arenaviruses and retroviruses revealed by electron cryomicroscopy and image analysis. *J Virol* **79**(6), 3822–30.

Neumann, G., Whitt, M. A., and Kawaoka, Y. (2002). A decade after the generation of a negative-sense RNA virus from cloned cDNA – what have we learned? *J Gen Virol* **83**(Pt 11), 2635–62.

Oldstone, M. B. (2002). Biology and pathogenesis of lymphocytic choriomeningitis virus infection. *In* Arenaviruses (M. B. Oldstone, Ed.), Vol. 263, pp. 83–118.

Perez, M., Craven, R. C., and de la Torre, J. C. (2003). The small RING finger protein Z drives arenavirus budding: implications for antiviral strategies. *Proc Natl Acad Sci USA* **100**(22), 12978–83.

Perez, M., and de la Torre, J. C. (2002). Characterization of the genomic promoter of the prototypic arenavirus lymphocytic choriomeningitis virus (LCMV). *J Virol* **77**(2), 1184–94.

Perez, M., Greenwald, D. L., and de la Torre, J. C. (2004). Myristoylation of the RING finger Z protein is essential for arenavirus budding. *J Virol* **78**(20), 11443–8.

Peters, C. J. (2002). Human infection with arenaviruses in the Americas. *In* Arenaviruses I (M. B. Oldstone, Ed.), Vol. 262, pp. 65–74. Springer-Verlag, Berlin Heidelberg.

Peters, C. J. (2006). Lymphocytic choriomeningitis virus – an old enemy up to new tricks. *N Engl J Med* **354**(21), 2208–11.

Pinschewer, D. D., Perez, M., and de la Torre, J. C. (2003a). Role of the virus nucleoprotein in the regulation of lymphocytic choriomeningitis virus transcription and RNA replication. *J Virol* **77**(6), 3882–7.

Pinschewer, D. D., Perez, M., and de la Torre, J. C. (2005). Dual role of the lymphocytic choriomeningitis virus intergenic region in transcription termination and virus propagation. *J Virol* **79**(7), 4519–26.

Pinschewer, D. D., Perez, M., Jeetendra, E., Bachi, T., Horvath, E., Hengartner, H., Whitt, M. A., de la Torre, J. C., and Zinkernagel, R. M. (2004). Kinetics of protective antibodies are determined by the viral surface antigen. *J Clin Invest* **114**(7), 988–93.

Pinschewer, D. D., Perez, M., Sanchez, A. B., and de la Torre, J. C. (2003b). Recombinant lymphocytic choriomeningitis virus expressing vesicular stomatitis virus glycoprotein. *Proc Natl Acad Sci USA* **100**(13), 7895–900.

Radoshitzky, S. R., Abraham, J., Spiropoulou, C. F., Kuhn, J. H., Nguyen, D., Li, W., Nagel, J., Schmidt, P. J., Nunberg, J. H., Andrews, N. C., Farzan, M., and Choe, H. (2007). Transferrin receptor 1 is a cellular receptor for New World haemorrhagic fever arenaviruses. *Nature* **446**(7131), 92–6.

Raju, R., Raju, L., Hacker, D., Garcin, D., Compans, R., and Kolakofsky, D. (1990). Nontemplated bases at the 5′ ends of Tacaribe virus mRNAs. *Virology* **174**(1), 53–9.

Salvato, M. S. (1993a). The Arenaviridae. 1st ed. *In* The Viruses (F.-C. H. a. W. R. R., Ed.) Plenum Press, New York.

Salvato, M. S. (1993b). Molecular biology of the prototype arenavirus, lymphocytic choriomeningitis virus. *In* The Arenaviridae (M. S. Salvato, Ed.), Vol. 1, pp. 133–56. Plenum, New York.

Salvato, M. S., Schweighofer, K. J., Burns, J., and Shimomaye, E. M. (1992). Biochemical and immunological evidence that the 11 kDa zinc-binding protein of lymphocytic choriomeningitis virus is a structural component of the virus. *Virus Res* **22**(3), 185–98.

Salvato, M., Shimomaye, E., and Oldstone, M. B. (1989). The primary structure of the lymphocytic choriomeningitis virus L gene encodes a putative RNA polymerase. *Virology* **169**(2), 377–84.

Sanchez, A. B., and de la Torre, J. C. (2005). Genetic and biochemical evidence for an oligomeric structure of the functional L polymerase of the prototypic arenavirus lymphocytic choriomeningitis virus. *J Virol* **79**(11), 7262–8.

Sanchez, A. B., and de la Torre, J. C. (2006). Rescue of the prototypic Arenavirus LCMV entirely from plasmid. *Virology* **350**(2), 370–80.

Sevilla, N., Domingo, E., and de la Torre, J. C. (2002). Contribution of LCMV towards deciphering biology of quasispecies in vivo. *Curr Top Microbiol Immunol* **263**, 197–220.

Smallwood, S., Cevik, B., and Moyer, S. A. (2002). Intragenic complementation and oligomerization of the L subunit of the sendai virus RNA polymerase. *Virology* **304**(2), 235–45.

Smallwood, S., and Moyer, S. A. (2004). The L polymerase protein of parainfluenza virus 3 forms an oligomer and can interact with the heterologous Sendai virus L, P and C proteins. *Virology* **318**(1), 439–50.

Southern, P. J. (1996). Arenaviridae: the viruses and their replication. 3rd ed. Trans. n/a. *In* Fields Virology (D. M. K. Bernard N. Fields, and Peter M. Howley, Eds.), Vol. 2, pp. 1505–51. Lippincott-Raven Publishers, Philadelphia.

Spiropoulou, C. F., Kunz, S., Rollin, P. E., Campbell, K. P., and Oldstone, M. B. (2002). New World arenavirus clade C, but not clade A and B viruses, utilizes alpha-dystroglycan as its major receptor. *J Virol* **76**(10), 5140–6.

Strecker, T., Eichler, R., Meulen, J., Weissenhorn, W., Dieter Klenk, H., Garten, W., and Lenz, O. (2003). Lassa virus Z protein is a matrix protein and sufficient for the release of virus-like particles [corrected]. *J Virol* **77**(19), 10700–5.

Strecker, T., Maisa, A., Daffis, S., Eichler, R., Lenz, O., and Garten, W. (2006). The role of myristoylation in the membrane association of the Lassa virus matrix protein Z. *Virol J* **3**, 93.

Tordo, N., De Haan, P., Goldbach, R., and Poch, O. (1992). Evolution of negative-stranded RNA genomes. *Semin Virol* **3**, 341–57.

Tortorici, M. A., Albarino, C. G., Posik, D. M., Ghiringhelli, P. D., Lozano, M. E., Rivera Pomar, R., and Romanowski, V. (2001). Arenavirus nucleocapsid protein displays a transcriptional antitermination activity in vivo. *Virus Res* **73**(1), 41–55.

Urata, S., Noda, T., Kawaoka, Y., Yokosawa, H., and Yasuda, J. (2006). Cellular factors required for Lassa virus budding. *J Virol* **80**(8), 4191–5.

Vieth, S., Torda, A. E., Asper, M., Schmitz, H., and Gunther, S. (2004). Sequence analysis of L RNA of Lassa virus. *Virology* **318**(1), 153–68.

Weaver, S. C., Salas, R. A., de Manzione, N., Fulhorst, C. F., Duno, G., Utrera, A., Mills, J. N., Ksiazek, T. G., Tovar, D., and Tesh, R. B. (2000). Guanarito virus (Arenaviridae) isolates from endemic and outlying localities in Venezuela: sequence comparisons among and within strains isolated from Venezuelan hemorrhagic fever patients and rodents. *Virology* **266**(1), 189–95.

Weaver, S. C., Salas, R. A., de Manzione, N., Fulhorst, C. F., Travasos da Rosa, A. P., Duno, G., Utrera, A., Mills, J. N., Ksiazek, T. G., Tovar, D., Guzman, H., Kang, W., and Tesh, R. B. (2001). Extreme genetic diversity among Pirital virus (Arenaviridae) isolates from western Venezuela. *Virology* **285**(1), 110–8.

Welsh, R. M., and Pfau, C. J. (1972). Determinants of lymphocytic choriomeningitis interference. *J Gen Virol* **14**(2), 177–87.

Yarnell, W. S., and Roberts, J. W. (1999). Mechanism of intrinsic transcription termination and antitermination. *Science* **284**(5414), 611–5.

York, J., Romanowski, V., Lu, M., and Nunberg, J. H. (2004). The signal peptide of the Junin arenavirus envelope glycoprotein is myristoylated and forms an essential subunit of the mature G1–G2 complex. *J Virol* **78**(19), 10783–92.

Young, P. R., and Howard, C. R. (1983). Fine structure analysis of Pichinde virus nucleocapsids. *J Gen Virol* **64**(Pt 4), 833–42.

Zinkernagel, R. M. (2002). Lymphocytic choriomeningitis virus and immunology. *Curr Top Microbiol Immunol* **263**, 1–5.

Chapter 11
Core-Associated Genome Replication Mechanisms of dsRNA Viruses

Sarah M. McDonald and John T. Patton

Introduction

The double-stranded RNA (dsRNA) viruses are a diverse group, which infect a wide assortment of prokaryotic and eukaryotic hosts. Although not all dsRNA viruses are considered pathogens, many cause devastating disease in their hosts and have widespread medical, veterinary, and agricultural impacts. Currently, the International Committee for the Taxonomy of Viruses (ICTV) recognizes seven distinct families of dsRNA viruses (Hypoviridae, Totiviridae, Birnaviridae, Partitiviridae, Cystoviridae, Chrysoviridae, and Reoviridae) (Table 11.1) (http://phene.cpmc.columbia.edu). Of these families, Reoviridae is composed of the largest number of individual species, which are categorized into 12 separate genera, and includes some of the most severe dsRNA viral pathogens of humans and domestic animals (Mellor and Boorman, 1995; Parashar et al., 2003). In particular, rotaviruses are members of the Reoviridae family and a leading cause of lethal gastroenteritis in young children and infants (Parashar et al., 2003). As such, the Reoviridae family has been studied in detail, providing insights into the general strategies dsRNA viruses use to propagate. Members of the Totiviridae and Cystoviridae families, which infect fungi and bacteria, respectively, have replication strategies similar to Reoviridae and are often viewed as models for understanding dsRNA virus biology (Mindich, 2004; Poranen and Tuma, 2004; Wickner, 1996). Together, studies of these three virus families have elucidated several common themes in dsRNA virus replication: (i) RNA synthesis occurs within a protected core via an anchored RNA-dependent RNA polymerase (RdRp); (ii) genome replication and capsid assembly occur simultaneously; and (iii) *cis*-acting elements in the viral RNA determine template specificity. This chapter will explore these themes regarding the core-associated genome replication of dsRNA viruses by reviewing structural and biochemical studies of individual members of Totiviridae, Cystoviridae, and Reoviridae families.

S.M. McDonald (✉)
Department of Laboratory of Infectious Disease, National Institute of Allergy and Infectious Diseases, National Institutes of Health, Bethesda, MD 20892, USA
e-mail: mcdonaldsa@niaid.nih.gov

C.E. Cameron et al. (eds.), *Viral Genome Replication*,
DOI 10.1007/b135974_11,

Table 11.1 dsRNA virus families, genome segments, and particle types

Family	# Genome segments	Particle type
Hypoviridae	1 (unpackaged)	Enveloped; no protein shell
Totiviridae*	1 (packaged singly)	Single-shelled; pseudo $T = 1$
Partitiviridae	2 (packaged separately)	Single-shelled; unknown structure
Birnaviridae	2 (co-packaged)	Single-shelled; $T = 13$
Cystoviridae*	3 (co-packaged)	Triple-shelled; pseudo $T = 1$ core, Middle $T = 13$ layer, and outer membrane
Chrysoviridae	4 (packaged separately)	Single-shelled; classic $T = -1$ core
Reoviridae*	10, 11, or 12 (co-packaged)	Single, double, or triple-shelled; pseudo $T = 1$ core; Middle and outer $T = 13$ layers

*dsRNA virus families that exhibit a pseudo $T = 1$ core.

RNA Synthesis Occurs Inside a Protected Core

To replicate successfully in the cytoplasm of a eukaryotic host cell, a dsRNA virus must overcome elaborate intracellular defense mechanisms (Garcia-Sastre and Biron, 2006). Specifically, the detection of viral dsRNA by the cell triggers an antiviral response that drastically impedes viral replication (Garcia-Sastre and Biron, 2006; Levy and Garcia-Sastre, 2001). Consequently, dsRNA viruses have evolved to escape this antiviral response by confining their genomes, throughout the entire course of infection, within one to three concentric protein shells (Lawton et al., 2000; Mertens, 2004). The innermost protein shell not only houses the segments of viral genomic dsRNA but also encases the viral RdRp and other enzymes necessary for mediating RNA synthesis (Ahlquist, 2006). Together, the proteins and RNA of the innermost shell make up a proteinaceous structure referred to as the viral core (Fig. 11.1). During the entry of a dsRNA virus into a cell, the outer layers of the virion are sequentially lost, triggering the enzymes within the core to begin viral transcription ((+)RNA synthesis) using the endogenous dsRNA genome as template (Fig. 11.1A) (Mertens and Diprose, 2004; Mindich, 2004; Patton, 2001; Patton and Spencer, 2000; Patton et al., 2004, 2007). Following transcription, the (+)RNA molecules are extruded from the virion core and into the host cell cytoplasm where they are translated into viral proteins (Mertens and Diprose, 2004; Mindich, 2004; Patton, 2001; Patton and Spencer, 2000; Patton et al., 2004, 2007). For Reoviridae, newly synthesized viral proteins accumulate in large cytoplasmic inclusions where the initial stages of virion particle assembly occur simultaneously with genome replication (dsRNA synthesis) (Patton et al., 2006, 2007; Roy and Noad, 2006). Particularly, viral core proteins assemble into intermediate structures, which package (+)RNA molecules at the same time as the core-associated viral enzymes convert them into dsRNA (Fig. 11.1B) (Patton et al., 2007; Roy and Noad, 2006). Because viral cores contain all the components necessary for transcription and genome replication, these processes occur within a protected environment that sequesters the precious viral dsRNA genome away from host cell antiviral sentries (Lawton et al., 2000; Mertens, 2004).

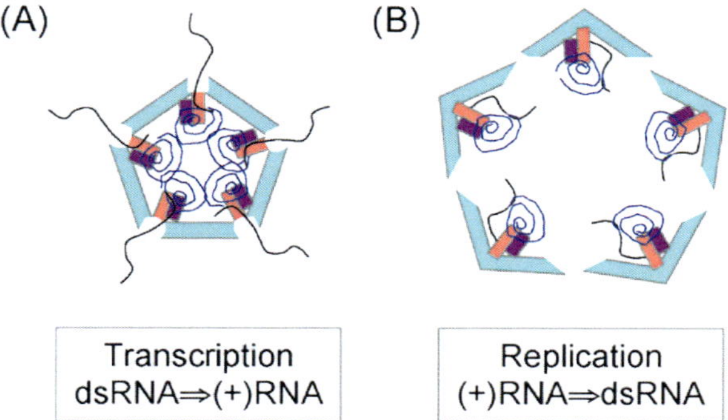

Fig. 11.1 Core-associated transcription and genome replication of dsRNA viruses. The schematics shown above diagram the two stages of RNA synthesis for dsRNA virus. Core components are not drawn to scale. (**A**) Transcription. Entry of a dsRNA virus into a cell triggers the enzymes (*pink* and *purple*) within the core shell (*light blue*) to begin (+)RNA synthesis using the endogenous dsRNA genome (*blue spirals*) as template. Following transcription, the (+)RNA molecules (*black lines*) are extruded from the virion core through channels at the fivefold axes. (**B**) Replication. Viral core proteins assemble into intermediate structures, which package (+)RNA molecules at the same time as the core-associated viral enzymes convert them into dsRNA.

Core Components of Model dsRNA Viruses

Viruses belonging to the Totiviridae, Cystoviridae, and Reoviridae families have been studied in great detail, yielding significant information about the structure and function of dsRNA viral cores. The prototypical member of Totiviridae is L-A virus, a pathogen of the yeast *Saccharomyces cerevisiae*. L-A is one of the simplest dsRNA viruses, having only one genome segment, which is encased by a single shell made up of the viral coat protein (Gag) (Fig. 11.2A) (Wickner, 1996). The viral RdRp (Pol) is expressed from the genome as a Gag–Pol fusion protein due to a –1 ribosomal frameshift and is incorporated into L-A particles (Dinman et al., 1991). Pol is required for mediating the concerted replication and packaging of the viral genome segment while anchored inside the core; however, Gag alone is sufficient for particle formation (Ribas and Wickner, 1992). The observation that Totiviridae members do not have outer layers is likely a reflection of their obligate intracellular life cycle. These viruses do not exit their host fungal cell, but rather spread via cytoplasmic mixing (Wickner, 1996). Despite their simplicity, L-A particles are strikingly similar in structure to cores of the Reoviridae and Cystoviridae families (Naitow et al., 2002). Furthermore, because L-A particles can synthesize dsRNA in vitro using exogenous templates, they remain a straightforward and elegant system for studying dsRNA viral genome replication (Fujimura and Wickner, 1989).

The bacteriophage phi 6 (Φ6) is the best-characterized member of the Cystoviridae family (Mindich, 2004; Poranen and Tuma, 2004). The Φ6 virion is a double-layered nucleocapsid (NC) surrounded by a host cell-derived lipid envelope, which is embedded with several viral proteins (Poranen and Tuma, 2004). The outer protein layer of the NC shows $T = 13$ icosahedral symmetry and is composed entirely

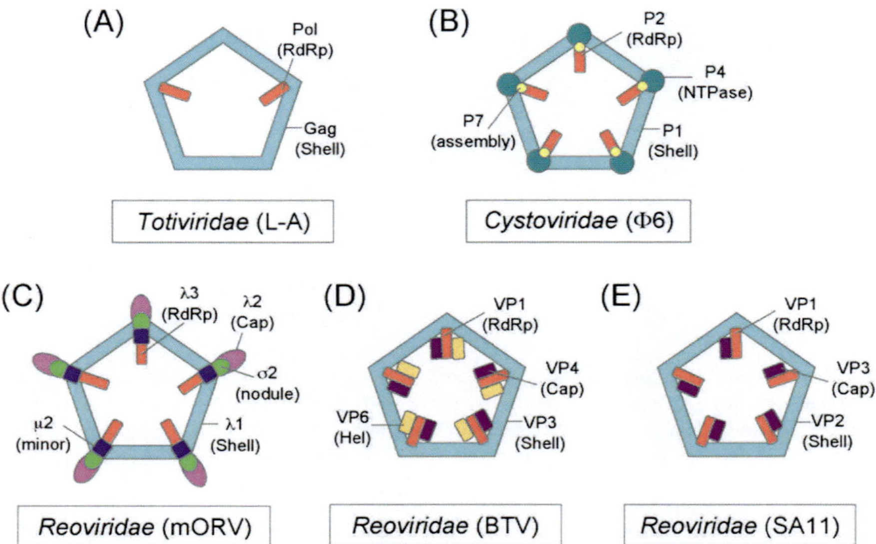

Fig. 11.2 Core components of model dsRNA viruses. The schematics above show the approximate locations of viral core components and are not drawn to scale. (**A**) L-A virus is the prototypical Totiviridae member. (**B**) The bacteriophage phi 6 (Φ6) is the best-characterized member of the Cystoviridae family. (**C**) Mammalian orthoreovirus (mORV) is the type species for turreted members of Reoviridae. (**D**) The well-studied non-turreted Reoviridae members are the orbivirus, blue tongue virus (BTV), and (**E**) the rotavirus simian agent 11 (SA11).

of the viral P8 protein (Butcher et al., 1997; Huiskonen et al., 2006; Jaalinoja et al., 2007; Kainov et al., 2003). Also referred to as the polymerase complex (PC), the Φ6 core consists of four viral proteins: a shell protein (P1), a nodule-like hexameric NTPase (P4), an assembly cofactor (P7), and an internally anchored viral RdRp (P2) (Fig. 11.2B) (Makeyev and Grimes, 2004). Inside the Φ6 core are one copy each of three dsRNA genome segments: small (S), medium (M), and large (L) (Mindich, 2004). Because the Φ6 core fully reproduces RNA packaging in vitro, in addition to template-dependent in vitro RNA synthesis, it has become an important model system (Makeyev and Grimes, 2004; Mindich, 2004). Even more, a high-resolution structure of the catalytically active Φ6 RdRp has suggested mechanisms for semi-conservative transcription and the initiation of viral RNA synthesis (Butcher et al., 2001).

The virion architecture of Reoviridae family members is similar to Totiviridae and Cystoviridae. Yet, this family is more complex due to the increased number of genome segments and capsid proteins. Despite a few rare exceptions, the cores of Reoviridae family members encapsidate 10, 11, or 12 equimolar dsRNA genome segments and are surrounded by two $T = 13$ icosahedral proteins shells (Mertens, 2004). Some Reoviridae genera have turrets composed of the viral capping enzyme(s) (5′-triphosphatase, guanylytransferase, methyltransferase, etc.) that protrude outward from their core shells at each fivefold axis (Mertens, 2004). The type species for turreted members of Reoviridae is the mammalian orthoreovirus (mORV), a ubiquitous non-pathogenic animal virus. The mORV core is composed of five proteins, many with multiple functions (Fig. 11.2C). The core shell protein (λ1) also possesses NTPase, 5′-triphosphatase, and helicase activities. The nodule-forming clamp protein (σ2) helps stabilize the capsid, whereas the turret-forming (λ2) mediates most of the roles related to capping the (+)RNA following transcription. Inside the core resides the anchored RdRp (λ3) and another protein (μ2) that acts as a cofactor during RNA synthesis (Mertens, 2004). In contrast, other genera of Reoviridae are described as being non-turreted because they lack protruding core structures and retain their capping enzymes within their cores. Two well-characterized non-turreted Reoviridae members are the orbivirus, blue tongue virus (BTV), and the rotavirus, simian agent 11 (SA11). BTV is a deadly pathogen of ruminants and has a viral core that consists of a shell protein (VP3) with internal, anchored enzyme complexes composed of an RdRp (VP1), an RNA capping enzyme (VP4), and a helicase (VP6) (Fig. 11.2D) (Mertens and Diprose, 2004; Roy and Noad, 2006). The core of the prototypic rotavirus SA11 is similar to BTV, but lacks the helicase protein. The SA11 core has a shell protein (VP2) that surrounds anchored enzyme complexes made of an RdRp (VP1) and an RNA capping enzyme (VP3) (Fig. 11.2E) (Jayaram et al., 2004). In vitro RdRp activities have been described for mORV λ3 and BTV VP1, and a high-resolution structure of λ3 has been determined (Boyce et al., 2004; Tao et al., 2002). Undoubtedly, these discoveries have greatly enhanced our understanding of Reoviridae genome replication. Nonetheless, SA11 VP1 is the only Reoviridae RdRp with in vitro activities that recapitulate those needed to support virus replication in vivo, making it the most well-studied RdRp of the family (Patton, 2001; Patton and Spencer, 2000; Patton et al., 2004, 2007).

Core Shells Exhibit Pseudo $T = 1$ Icosahedral Symmetry

The structures of viral cores from individual members of the Totiviridae, Cystoviridae, and Reoviridae families have been analyzed to various degrees using cryo-electron microscopy (cryo-EM), X-ray crystallography, and three-dimensional (3D) image reconstruction (Butcher et al., 1997; Caston et al., 2006, 1997; Cheng et al., 1994; Fang et al., 2005; Grimes et al., 1998, 1997; Hewat et al., 1992; Hill et al., 1999; Huiskonen et al., 2006; Jaalinoja et al., 2007; Kainov et al., 2003; Lawton et al., 1997b; Lu et al., 1998; Metcalf et al., 1991; Naitow et al., 2002; Nakagawa et al., 2003; Prasad et al., 1996; Reinisch et al., 2000; Xia et al., 2003; Yeager et al., 1990; Zhang et al., 2005, 2003; Zhou et al., 2003). Although these core proteins show minimal primary amino acid sequence similarities, their architectural organizations are markedly conserved, even among diverse virus families. Specifically, the viral core shells are simple, relatively smooth, icosahedrons composed of 120 subunits of a thin protein (Fig. 11.3) (Jayaram et al., 2004; Kim et al., 2004; Mertens and

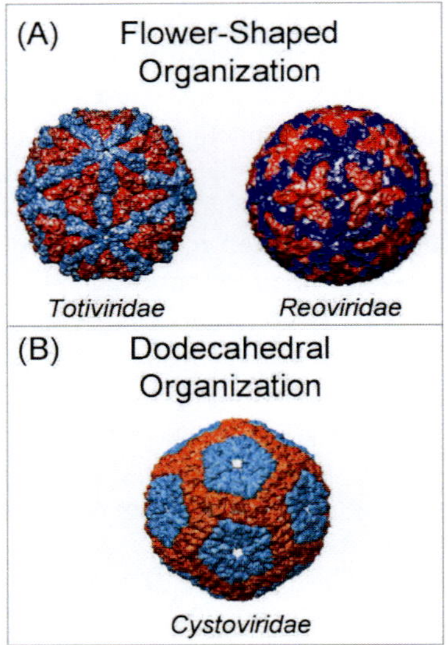

Fig. 11.3 Organization of pseudo $T = 1$ core shell proteins as visualized using cryo-EM. The core shells of Totiviridae, Cystoviridae, and Reoviridae are simple icosahedrons composed of 120 subunits of a thin protein organized as 60 asymmetric dimers. (**A**) Flower-like organization. The core proteins of Totiviridae and Reoviridae family members are arranged as a decamer around each icosahedral fivefold axis like the petals of a flower. Five copies of the A-form of the shell proteins (*blue*) cluster around the fivefold axis, while five copies of the structurally distinct B-form (*red*) are situated further away from the vertex, interdigitated between the A-forms. Cryo-EM images of L-A virus and the mORV core were generated using VIPERdb (Shepherd et al., 2006). (**B**) Dodecahedral organization. For the Cystoviridae family, five copies of the A-form (*blue*) tightly encircle the fivefold axis, whereas the B-form (*red*) makes a dodecahedral skeleton bordering the A-forms. The cryo-EM image of the Φ6 core was adapted with permission from S.J. Butcher (Huiskonen et al., 2006).

Diprose, 2004). The stoichiometry of these cores indicates that they have a forbidden triangulation number of $T = 2$ (Caspar and Klug, 1962). However, the 120 subunits are organized in the core as 60 asymmetric dimers, allowing these structures to be more accurately described as having pseudo $T = 1$ symmetry (Caspar and Klug, 1962). The arrangement of the core shell proteins in such a manner requires that the individual monomers within a dimer unit adopt slightly different conformations (Caspar and Klug, 1962; Steven et al., 1997). Thus, even though the monomers are chemically identical, they are structurally quasi-equivalent molecules. Large aqueous channels traverse the core through each fivefold axis, providing portals for entry of nucleotides and divalent cations and conduits for the exit of viral (+)RNA following transcription (Jayaram et al., 2004; Kim et al., 2004; Mertens and Diprose, 2004). This arrangement of quasi-equivalent core shell proteins to yield pseudo $T = 1$ icosahedrons is a general feature of many dsRNA viruses, but is unique in that it is not seen elsewhere in nature (Steven et al., 1997).

Structural analyses demonstrate that the overall architectures of the pseudo $T = 1$ viral cores of dsRNA viruses are the same; yet, the organization of quasi-equivalent protein dimers is slightly different among some families. For example, the core proteins of Totiviridae and Reoviridae family members, such as L-A Gag, reovirus λ1, BTV VP3, and SA11 VP2, are arranged as a decamer around each icosahedral fivefold axis like the petals of a cupped, inverted flower (Fig. 11.3A) (Grimes et al., 1998; Naitow et al., 2002; Prasad et al., 1996; Reinisch et al., 2000). Precisely, five copies of one type of shell protein (A-form) cluster around the fivefold axis, while five copies of the second, structurally distinct type of the same protein (B-form) are situated further away from the vertex, interdigitated between the A-forms (Fig. 11.3A). In contrast to this flower-like organization seen in Totiviridae and Reoviridae, the core shell proteins of the Cystoviridae Φ6 show a dodecahedral organization (Fig. 11.3B) (Huiskonen et al., 2006; Jaalinoja et al., 2007; Kainov et al., 2003). Specifically, five copies of the A-form of Φ6 P1 tightly encircle the fivefold axis, whereas the B-form makes a dodecahedral skeleton bordering the A-forms (Fig. 11.3B). While the functional significance of these two types of pseudo $T = 1$ arrangements is not known, it has been proposed that the distinctive manner in which Φ6 P1 is organized allows for a generous expansion of the core upon RNA packaging (Huiskonen et al., 2006).

Locations of Viral Enzymes and dsRNA Inside the Core

One of the major functions of the pseudo $T = 1$ core shell of a dsRNA virus is to serve as a platform to which the viral RdRp and associated enzymes are attached. For the Totiviridae member L-A, the viral RdRp Pol is covalently linked to the pseudo $T = 1$ core shell protein Gag as a result of a translational fusion event (Wickner, 1996). The efficiency of this event suggests that two copies of Gag–Pol are incorporated into assembled L-A particles (Wickner, 1996). Although the Pol domains are not visualized in the L-A core structure, the orientation of Gag termini dictates that these RdRps be situated within the viral particle, essentially fixed to the inner wall and proximal to the icosahedral fivefold axis (Naitow et al., 2002). For the more complex Cystoviridae and Reoviridae families, there is an emerging view that the locations of the viral enzymes mirror what is seen for L-A, with the RdRp-containing complexes positioned underneath the vertices of the core shell. In support of this idea, the enzyme complexes of the SA11 have been visualized in cryo-EM image reconstructions as densities beneath the core shell layer at each of the 12 fivefold axes (Fig. 11.4) (Grimes et al., 1998; Prasad et al., 1996). Moreover, the high-resolution crystal structure of the mORV RdRp (λ3) has been modeled into the cryo-EM density map of the core (Zhang et al., 2003). These results are consistent with the other viruses, showing that mORV λ3 is anchored to the inner surface of the λ1 shell, slightly off-center from each fivefold axis (Zhang et al., 2003). Furthermore, the structural organization of dsRNA has been determined for several viruses and, in all cases, is seen as dodecahedral tubules packed in a larger radius around the viral enzymes (Fig. 11.4) (Grimes et al., 1998; Huiskonen et al.,

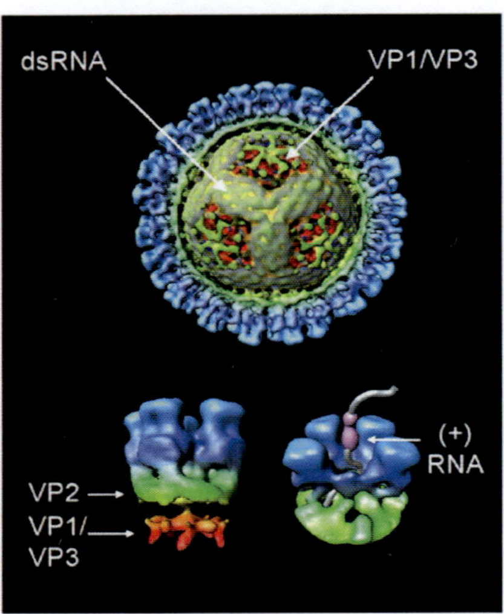

Fig. 11.4 Locations of viral enzymes and dsRNA inside the core. The enzyme complexes (*red*) of the rotavirus SA11 have been visualized in cryo-EM image reconstructions as densities beneath the VP2 core shell layer (*green*) at each of the 12 fivefold axes, and dsRNA (*yellow–gray*) is seen as dodecahedral tubules packed in a larger radius around the viral enzymes. A newly transcribed (+)RNA (*gray*) would acquire a 5′-cap (*purple*) prior to extrusion from the core via a fivefold channel. Images were modified with permission from B.V.V. Prasad (Prasad et al., 1996).

2006; Naitow et al., 2002; Prasad et al., 1996). This observation suggests that, for Cystoviridae and Reoviridae, each of the 12 enzyme complexes is dedicated to transcribing and replicating a single genome segment. Thus, for dsRNA viruses with less than 12 genome segments, some of the vertices will have orphaned enzyme complexes. Alternatively, it is possible that such viruses package only one enzyme complex per dsRNA segment, meaning that they would have some empty vertices. Either way, viral cores may be viewed as a collection of RdRp units, which operate independently and simultaneously during viral replication.

The structural location of the viral enzymes suggests a system for RNA synthesis in which a freely moving template RNA is pulled through a tethered RdRp. During genome replication, the RdRp would be bound to an assembling core shell intermediate, such as a decamer or pro-core (see later section). The (+)RNA would either be pulled into the core during the assembly process or be inserted into an already assembled core-like structure. While anchored near the vertex of the core shell intermediate, the RdRp catalyzes (–)RNA strand synthesis, converting the (+)RNA template into a complete dsRNA genome segment. During transcription, the RdRp uses the (–)RNA strand of the endogenous dsRNA segment, which encircles the enzyme, as template for (+)RNA synthesis. The position of the RdRp near the fivefold axis would allow the nascent (+)RNA to pass directly into the channel along or near the fivefold axis en route to virion exit. Although Totiviridae and Cystoviridae synthesize uncapped (+)RNAs during transcription, the transcripts of Reoviridae family members must acquire a

5′-cap in order to be efficient templates for translation. For mORV, nascent (+)RNAs would be capped as they navigate the hollow chamber of each projecting fivefold λ2 turret (Zhang et al., 2003). Because non-turreted BTV and SA11 have their RNA capping enzymes juxtaposed to their RdRps, a newly synthesized (+)RNA molecule would obtain a 5′-cap prior to being extruded from the core (Fig. 11.4) (Grimes et al., 1998; Prasad et al., 1996). A reconstruction of SA11 and BTV particles in the act of transcription revealed that they have the capacity to synthesize high levels of (+)RNA for several hours, indicating that the RdRp efficiently reengages the dsRNA template numerous times (Diprose et al., 2001; Lawton et al., 1997a). Indeed, the core shell protein might serve as a scaffold on which the dsRNA duplex is melted and repeatedly transcribed by the anchored RdRp. Nonetheless, the precise protein–protein and protein–RNA interactions that govern the placement of enzymes and RNA within the core shell are not fully understood.

RdRp Structures Highlight Mechanisms of RNA Synthesis

The recently solved high-resolution crystal structures of two dsRNA viral RdRps, Cystoviridae Φ6 P2 and Reoviridae mORV λ3, have greatly enhanced our understanding of how these viruses mediate RNA synthesis (Butcher et al., 2001; Tao et al., 2002). The overall fold of these proteins is analogous to that of all known RdRps and can be described as resembling a hollow, cupped right hand with fingers, palm, and thumb sub-domains and an internally located active site (Fig. 11.5) (O'Reilly and Kao, 1998). However, the structures of Φ6 P2 and mORV λ3 have embellishments on the basic RdRp architecture. Specifically, Φ6 P2 has a small amino-terminal extension that straps together the finger and thumb sub-domains, essentially closing the enzyme (Fig. 11.5A) (Butcher et al., 2001). The Φ6 P2 structure also has a carboxy-terminal loop that protrudes into the central cavity of the enzyme (Fig. 11.5A) (Butcher et al., 2001). This loop is referred to as the initiation platform and is important for the de novo initiation of RNA synthesis (see below). The mORV λ3 structure shows large amino- and carboxy-terminal elaborations that form a cage around the catalytic right hand (Fig. 11.5B–E). Like that of Φ6 P2, the amino-terminal domain of mORV λ3 reinforces the bridge between the fingers and thumb, also supporting the closure of the polymerase (Fig. 11.5D) (Tao et al., 2002). Yet, the carboxy terminus of mORV λ3 forms a ring-shaped bracelet that is entirely absent in all other RdRps whose structures are known (Fig. 11.5E) (Tao et al., 2002). This bracelet domain is reminiscent of the clamps of DNA polymerases, which open and close upon templates (Mossi and Hubscher, 1998). Even so, unlike DNA polymerase clamps, the bracelet domain of λ3 likely remains closed and slides along the template during polymerization. The structures of Φ6 P2 and mORV λ3 also show several hollow tunnels that allow the RNA template, nucleotides, and divalent cations to access the catalytic site and to permit the exit of nascent RNA (Fig. 11.6). Both enzymes have a single nucleotide entry tunnel on one side as well as a single template entry tunnel approximately 90° away near the top of the protein (Fig. 11.6). Nonetheless, Φ6 P2 and mORV λ3 have different

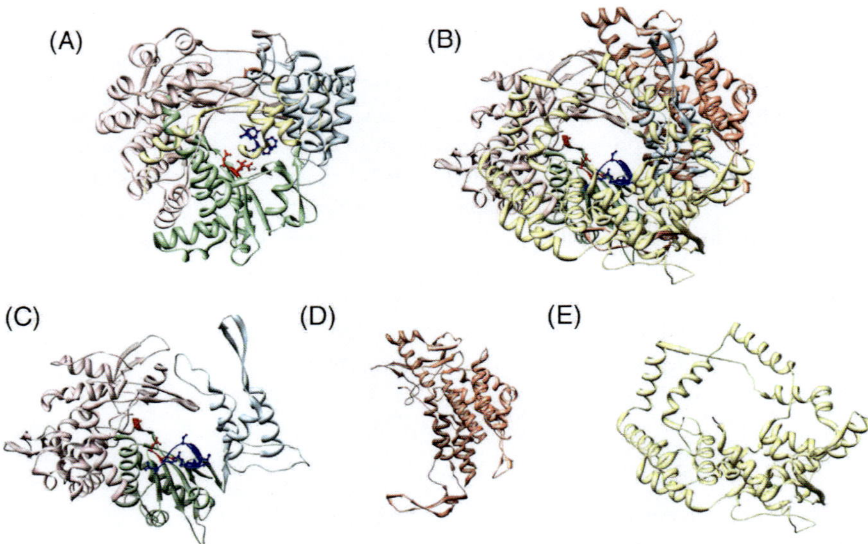

Fig. 11.5 Structures of the Φ6 and mORV RdRps. The overall fold of the enzymes resembles a hollow, cupped right hand with fingers (*light pink*), palm (*green*), and thumb (*light blue*) sub-domains and an internally located active site (catalytic aspartic acids; *red*). Flexible regions of the enzymes involved in initiation complex formation are shown in *blue*. (**A**) Φ6 P2 structure. The small amino-terminal extension that straps together the finger and thumb sub-domains, closing the enzyme is shown at the top of the molecule in *dark pink*. The carboxy-terminal loop that protrudes into the central cavity of the enzyme is shown in *yellow*. (**B**) mORV λ3 structure. The large amino- and carboxy-terminal elaborations that form a cage around the catalytic right hand are shown in *dark pink* and *yellow*, respectively. (**C**) mORV λ3 right-hand pol domain structure with fingers, palm, and thumb sub-domains. (**D**) mORV λ3 amino-terminal elaboration. (**E**) mORV λ3 carboxy-terminal bracelet. The PDB numbers for Φ6 P2 and mORV λ3 are 1MUK and 1HI8, respectively (Butcher et al., 2001; Tao et al., 2002). Images were generated using Chimera computer program (Petterson et al., 2004).

numbers of RNA exit tunnels, reflecting the divergent transcription mechanisms of these viruses. Because Cystoviridae transcription occurs using a semi-conservative mechanism, Φ6 P2 has a single tunnel for the exit of a dsRNA product, making it a three-tunneled RdRp (Fig. 11.6A). In contrast, members of the Reoviridae family use a fully conservative mechanism of transcription, meaning that the RdRp separates the dsRNA product into the nascent (+)RNA and parental (–)RNA strands prior to their exit. This separation requires mORV λ3 to have two RNA exit tunnels, making it a four-tunneled RdRp (Fig. 11.6B).

The semi-conservative and conservative transcription mechanisms of Cystoviridae and Reoviridae, respectively, both use the endogenous dsRNA genome as template for (+)RNA synthesis (Fig. 11.7A and B). For mORV and Φ6, a dsRNA genome segment is separated into a (–)RNA strand, which accesses the RdRp via the template entry tunnel, and a (+)RNA strand that is "peeled-off" away from the enzyme (Fig. 11.7A and B). During Φ6 transcription, the parental (+)RNA strand is shuttled out of the core, while the parental (–)RNA strand is used as a template

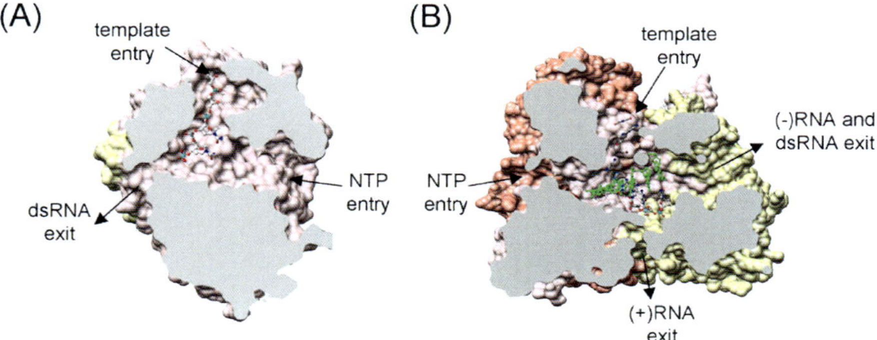

Fig. 11.6 Locations of tunnels within the Φ6 P2 and mORV λ3 structures. The structures of Φ6 P2 and mORV λ3 show several hollow tunnels that allow the RNA template, nucleotides, and divalent cations to access the catalytic site and to permit the exit of nascent RNA. Protein domains are colored as described in Fig. 11.5. An RNA template is shown in the entry tunnels of both structures and is element colored. (**A**) Φ6 P2 has a single tunnel for the exit of a dsRNA product, making it a three-tunneled RdRp. (**B**) mORV λ3 has two RNA exit tunnels, one for (+)RNA and one for (–)RNA and dsRNA, making it a four-tunneled RdRp. A nascent RNA strand is shown in *green*. The PDB numbers for Φ6 P2 and mORV λ3 are 1HHT and 1N35, respectively (Butcher et al., 2001; Tao et al., 2002). Images were generated using the Chimera computer program (Petterson et al., 2004).

for nascent (+)RNA strand synthesis (Fig. 11.7A). The product of semi-conservative Φ6 transcription is a dsRNA duplex composed of nascent the (+)RNA strand paired with the parental (–)RNA strand, which is released from P2 via the single RNA exit tunnel (Fig. 11.7A). This dsRNA molecule is separated again, the (+)RNA transcript is shuttled out of the core, and the (–)RNA strand is used as a template for another round of transcription. On the contrary, during fully conservative mORV transcription, the parental (+)RNA strand that is "peeled-off" the dsRNA segment stays inside the core and waits to reanneal with its complementary (–)RNA strand (Fig. 11.7B). Meanwhile, the parental (–)RNA strand enters λ3 and is used as a template for nascent (+)RNA strand synthesis, made initially as a dsRNA duplex (Fig. 11.7B). Unlike what is described for Φ6 P2, however, mORV λ3 quickly separates the strands of the newly made dsRNA duplex, allowing the parental (–)RNA strand and the nascent (+)RNA strand to exit the enzyme via individual tunnels (Fig. 11.7B). Following release of the two strands from the enzyme, the parental (–)RNA strand base pairs with its initial (+)RNA partner to reform the original dsRNA segment, while the nascent (+)RNA transcript acquires a 5′-cap as it is extruded from the core (Fig. 11.7B). By allowing the parental (+)RNA strand to remain in the core and reanneal with its (–)RNA partner, mORV ensures multiple rounds of transcription from the same dsRNA genome segment. Although dsRNA viral RdRps catalyze several cycles of transcription from a single dsRNA segment, packaged (+)RNA strand is used just once as template during genome replication. This (+)RNA template accesses the active site via the template entry channel and serves to catalyze (–)RNA strand synthesis, forming a dsRNA duplex. For Cystoviridae family members, the dsRNA duplex exits the RdRp in the same manner as

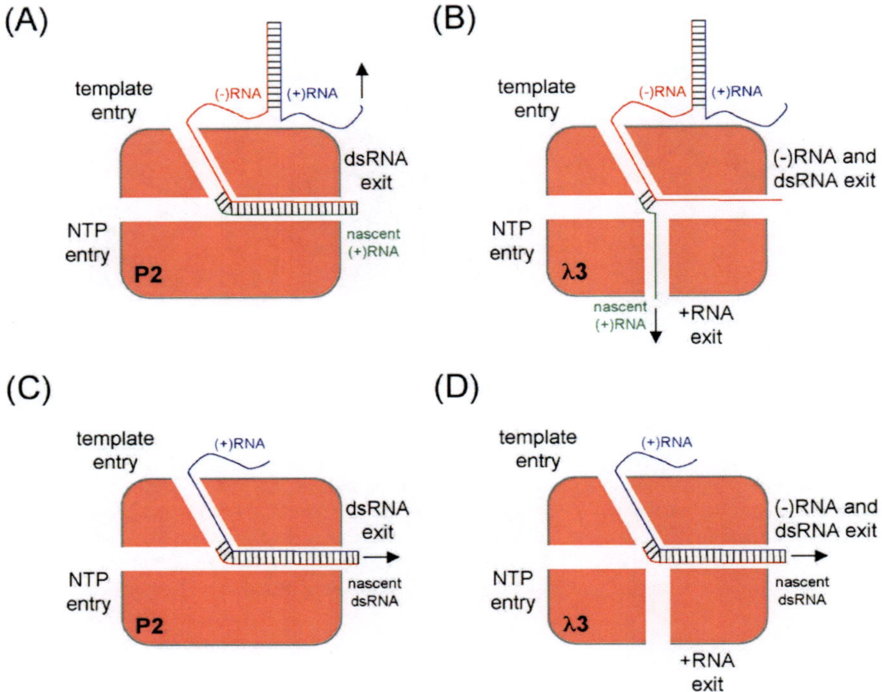

Fig. 11.7 Semi-conservative and conservative mechanisms of RNA synthesis. The schematic shown above illustrates the RNA synthesis mechanisms of Φ6 P2 and mORV λ3. (**A**) Φ6 P2 semi-conservative transcription. A dsRNA genome segment is separated into a (–)RNA strand (*red*), which accesses the RdRp via the template entry tunnel, and a (+)RNA strand (*blue*) that is "peeled-off" away from the enzyme. The parental (+)RNA strand is shuttled out of the core (*arrow*), while the parental (–)RNA strand (*red*) is used as a template for nascent (+)RNA strand synthesis (*green*). The product of semi-conservative Φ6 transcription is a dsRNA duplex, which is released from P2 via the single RNA exit tunnel. (**B**) mORV λ3 conservative transcription. The parental (+)RNA strand (*blue*) that is "peeled-off" the dsRNA segment stays inside the core; meanwhile, the parental (–)RNA strand (*red*) enters λ3 and is used as a template for nascent (+)RNA strand synthesis (*green*), made initially as a dsRNA duplex that is quickly separated. The parental (–)RNA strand (*red*) and the nascent (+)RNA strand (*green*) exit the enzyme via individual tunnels. The nascent (+)RNA transcript (*green*) acquires a 5′-cap as it is extruded from the core (*arrow*). (**C**) Φ6 P2 genome replication. The (+)RNA template (*blue*) accesses the active site via the template entry channel and serves to catalyze (–)RNA strand synthesis (*red*), forming a dsRNA duplex that exits the RdRp in the same manner as during transcription. (**D**) mORV λ3 genome replication. The RdRp uses the (+)RNA template (*blue*) to catalyze (–)RNA strand synthesis (*red*), forming a dsRNA duplex that leaves the enzyme using the (–)RNA exit tunnel.

during transcription (Fig. 11.7C). The Reoviridae members allow the dsRNA duplex to leave the enzyme using the tunnel that, during transcription, is designated for (–)RNA exit (Fig. 11.7D). Cystoviridae and Reoviridae RdRps differentiate among transcription and replication templates by recognizing *cis*-acting RNA elements with different affinities (see later section).

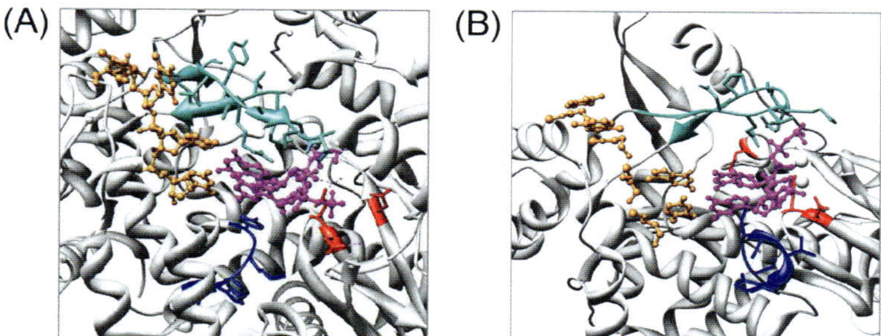

Fig. 11.8 Initiation complex formation for Φ6 P2 and mORV λ3. The Φ6 and mORV RdRps each have a region of the protein that functions as a "stage" on which an initiation complex is constructed. (**A**) The Φ6 P2 initiation platform. During Φ6 RNA synthesis, incoming initiatory nucleotides (*pink*) are stabilized by the P2 carboxy-terminal plug (*blue*) and a motif-F-like structure (*teal*), allowing them to base pair with the RNA template (*gold*) near the catalytic aspartic acids (*red*). (**B**) The mORV λ3 priming loop. For mORV RNA synthesis, the RNA template (*gold*) forms tight stacking interactions against the priming loop (*blue*), which is formed by the residues in the tip of the fingers and the palm sub-domains and stabilized by motif-F (*teal*). The priming loop allows the template (*gold*) to base pair with incoming nucleotides (*pink*). The PDB numbers for Φ6 P2 and mORV λ3 are 1HIO and 1NIH, respectively (Butcher et al., 2001; Tao et al., 2002). Images were generated using the Chimera computer program (Petterson et al., 2004).

The structures of Φ6 P2 and mORV λ3, complexed with nucleotides and RNA, also suggest a mechanism for the de novo initiation of RNA polymerization (Fig. 11.8) (Butcher et al., 2001; Tao et al., 2002). The defining features of de novo RNA synthesis are that no information in the viral genome is lost, and that additional proteins for priming are unnecessary (Makeyev and Grimes, 2004). However, initiation without a primer requires specific molecular interactions to occur between the template and incoming nucleotides in order to keep them correctly positioned at the RdRp active site (Makeyev and Grimes, 2004). To aid in forming these stable interactions, many viral RdRps have a region of the protein that functions as a "stage" on which an initiation complex is constructed. For the Cystoviridae member Φ6, the carboxy-terminal plug provides such a "stage" and is referred to as the initiation platform (Butcher et al., 2001). During Φ6 RNA synthesis, the RNA template enters P2 and is stabilized by the plug, allowing it to base pair with incoming initiatory nucleotides near the active site (Fig. 11.8A). In the course of elongation, the carboxy-terminal plug presumably moves to allow the dsRNA product to egress from the RdRp active site. For mORV, the incoming nucleotides enter λ3 and are stabilized against the priming loop, which is formed by the residues in the tip of the fingers and the palm sub-domains (Fig. 11.8B) (Tao et al., 2002). Thus, the priming loop functions as a "stage" for λ3, allowing the incoming nucleotides to base pair with the RNA template. Like the Φ6 P2 plug, the λ3 priming loop shifts its location following initiation of phosphodiester bond formation so as not to block the elongating dsRNA duplex. Thus, the structures of Φ6 P2 and mORV λ3 demonstrate that these viruses have evolved different strategies for forming a stable initiation complex that allows effective RNA synthesis in the absence of a primer.

Biochemical Studies of dsRNA Viral Genome Replication

In addition to structural studies, elegant biochemical experiments have elucidated many important details regarding Totiviridae, Cystoviridae, and Reoviridae genome replication. The catalytic process of viral dsRNA synthesis has predominantly been studied using core particles that have been disrupted by incubation in hypotonic conditions. For the Totiviridae and Reoviridae members, L-A and SA11, such disrupted cores possess RdRp activity when incubated in the presence of nucleotides, divalent cations, and exogenous (+)RNA (Chen et al., 1994; Fujimura and Wickner, 1989). These assays were instrumental in locating the *cis*-acting replication signals in (+)RNA templates for these viruses (see later section). However, the open core systems of L-A and SA11 fail to package the dsRNA products of replication, suggesting that they do not fully recapitulate in vivo processes. Moreover, these systems do not allow for directed mutagenesis of individual core protein components to study their roles during dsRNA synthesis. The discovery that recombinant SA11 VP1 is capable of catalyzing dsRNA synthesis was a breakthrough for clarifying the functions of proteins during each stage of genome replication (Patton et al., 1997). An interesting feature of this enzyme is that it requires the core shell protein VP2 for biochemical activity (Patton et al., 1997). This phenomenon suggests that in vitro VP1 catalyzes dsRNA synthesis in a manner connected to core assembly. Using these recombinant proteins, the molar ratio of VP1:VP2 required for maximum dsRNA synthesis was determined to be 1:10 (Patton et al., 1997; Tortorici et al., 2003). This ratio mimics that of each decamer of the rotavirus core, indicating that activation of the RdRp might require the formation of an assembly intermediate. Deletion mutagenesis studies have shown that the amino terminus of VP2 contains a domain critical for interactions with VP1, VP3, and RNA. Specifically, an amino-terminally truncated VP2 fails to induce dsRNA synthesis and does not encapsidate the enzyme complex or RNA, but does assemble into core-like particles (Labbe et al., 1994; Lawton et al., 1997b; Patton et al., 1997; Zeng et al., 1998). Although the precise mechanism by which VP2 triggers the function of VP1 is unknown, it is possible that the amino terminus of VP2 forms an internal platform inside the core at the fivefold axis and on which the RdRp operates.

Biochemical studies of SA11 dsRNA synthesis using open cores or recombinant proteins have characterized the requirements for the initiation of genome replication. Particularly, the formation of a (–)RNA strand initiation complex was shown to be a salt-sensitive process that requires the RdRp VP1, the core shell protein VP2, rGTP, Mg^{2+}, and template (+)RNA (Chen and Patton, 2000; Tortorici et al., 2003). The need for rGTP is likely a reflection of the 3′-terminal nucleotides (CC) of all rotavirus template (+)RNAs (see later section). The divalent cation Mg^{2+} is a common cofactor for RdRps, but in these reactions, the addition of Mn^{2+} stimulates the activity of recombinant VP1 (O'Reilly and Kao, 1998; Patton et al., 2004; Tortorici et al., 2003). Importantly, the observation that VP2 must be preincubated with VP1 prior to the elongation step suggests that this protein functions at the initiation step of dsRNA synthesis. VP1 has been shown to interact with (+)RNA and rNTPs in the absence of VP2 and in the presence of salt, demonstrating that (i) VP2 does not merely function to bring together the template, nucleotides, and enzyme and (ii) the salt-sensitivity of complex formation is not due to the lack of template or nucleotide recognition by VP1. Still, it remains unclear exactly how VP2 activates VP1 to initiate dsRNA synthesis (Chen and Patton, 2000; Patton, 1996; Tortorici et al., 2003). Currently, there is no atomic structure for VP1, but sequence comparisons would lead to the prediction that this enzyme is very similar to mORV λ3 (Patton et al., 2007; Tao et al., 2002). The RdRp activity of recombinant mORV λ3, while very minimal in comparison to both mORV open cores and SA11 VP1, does not require the presence of the core shell protein λ1. It will be interesting to determine whether the structures of SA11 VP1 and mORV λ3 show differences in the locations of residues required for initiating RNA synthesis.

The Cystoviridae Φ6 RdRp P2 is capable of catalyzing both RNA replication and transcription activities in vitro as a single viral protein (Makeyev and Grimes, 2004). The minimal requirements for in vitro initiation and elongation using purified P2 are nucleotides, Mg^{2+}, and (+)RNA template (Ojala and Bamford, 1995). The recombinant enzyme shows a high specific activity, but exhibits decreased template specificity in comparison to the Φ6 open core system. Still, purified P2 self-assembles together with P1, P4, and P7 into viral cores that are fully functional for both RNA packaging and genome replication (Mindich, 2004; Poranen and Tuma, 2004). No such packaging system exists for the Totiviridae and Reoviridae families, making studies of Φ6 important to our understanding of how dsRNA synthesis is connected to capsid assembly.

Genome Replication and Capsid Assembly Occur Simultaneously

To protect newly made dsRNA from the host cell antiviral response, (+)RNA packaging into a core-like intermediate is thought to precede genome replication. Core assembly intermediates of the Cystoviridae member Φ6, called pro-cores, can be made using recombinant proteins P1, P2, P4, and P7 (Mindich, 2004; Poranen and Tuma, 2004). These pro-cores are stimulated to package and replicate the (+)RNA templates (S+, M+, and L+) by incubation in polyethylene glycol, ADP, Mg^{2+}, and rNTPs (Mindich, 2004; Poranen and Tuma, 2004). The reaction is consecutive in that S+ is packaged first, followed by M+ and then L+. A hexamer of the NTPase protein (P4) mediates the bulk of (+)RNA packaging at a single fivefold axis, but a cofactor protein (P7) enhances the efficiency of this process (Mindich, 2004; Poranen and Tuma, 2004). Studies indicate that P4 functions like a molecular motor, powering the entry of the (+)RNA molecules into the pro-core. Only after all three templates are packaged inside the pro-core does (–)RNA strand synthesis begin, converting the (+)RNAs into the full-length genome segments (S, M, and L). Forceful expansion of the core shell as a result of RNA packaging is thought to trigger dsRNA synthesis. The prerequisite packaging of (+)RNA into pre-formed core intermediates undoubtedly links this process with genome replication.

Much less is known about how Totiviridae and Reoviridae members package their (+)RNA templates, but limited studies support the hypothesis that this process is linked to genome replication. The fact the L-A viral RdRp (Pol) is fused to the core shell protein (Gag) makes is difficult to argue against the model in which the single (+)RNA template is bound by Gag–Pol during Totiviridae particle formation (Wickner, 1996). For the Reoviridae, however, the presence of 10–12 different (+)RNAs and numerous separate proteins makes understanding the precise pathway of assembly and replication more difficult. Attempts have been made to define this pathway for the rotavirus SA11 by isolating replication intermediates (RIs) from infected cells (Gallegos and Patton, 1989; Patton and Gallegos, 1990). These studies suggest that an initial interaction occurs between the RdRp VP1, the capping enzyme VP3, and a single (+)RNA template, forming a pre-core RI that lacks polymerase activity. Thereafter, a VP2 decamer interacts with a pre-core RI to form a core RI, which is capable of initiating dsRNA synthesis. It is unclear whether genome replication occurs prior to, at the same time as, or after the core RIs close into a complete pseudo $T = 1$ icosahedron. Treatment of core RIs with RNase causes the degradation of (+)RNA templates and abolishes genome replication. In contrast, replicated dsRNAs are protected from RNase degradation, suggesting that the products of replication are protected, possibly in a closed core. In further support of the idea that packaging precedes dsRNA synthesis, free dsRNA has never been detected in rotavirus-infected cells and the *cis*-acting packaging signals are located in the (+)RNAs, but are masked in the dsRNA products (Patton et al., 2007). Importantly, the requirement of SA11 VP2 for binding VP1, VP3, and RNA, and for triggering genome replication ensures that dsRNAs are not produced until cores are available for their protection. Nonetheless, it remains a complete mystery how Reoviridae family members incorporate one of each genome segment in equimolar amounts into viral cores. Also, the functions of SA11 nonstructural proteins during packaging and replication are unclear.

The role of the BTV core shell protein (VP3) during the early stages of particle assembly has been analyzed using deletion mutagenesis (Kar et al., 2004). These studies showed that deletion of residues at either the amino or carboxy terminus of VP3 did not affect its interactions with the enzyme complex proteins (VP1, VP4, and VP6), but that these mutant cores were extremely unstable. Still, deletion of the VP3 carboxy terminus abolished VP3–RNA binding and the formation of complete icosahedrons. These results suggest that the amino terminus of BTV VP3 is dispensable for encapsidating the RdRp-containing enzyme complex and RNA during assembly. The results further suggest that the RNA-binding domain of BTV lies in the carboxy terminus of the protein, overlapping with an oligomerization domain. These results with BTV are in contrast to what has been determined for SA11 VP2, which forms stable cores in the absence of its amino terminus, but does not bind VP1, VP3, or RNA (Labbe et al., 1994; Lawton et al., 1997b; Patton et al., 1997; Zeng et al., 1998). Because the BTV RdRp VP1 has in vitro activity in the absence of its core shell protein, it remains unknown how these assembly mutations alter genome replication. Yet, the reported differences in the core protein–protein interactions between BTV and SA11 suggest that individual Reoviridae family members might have distinct pathways of forming core replication intermediates.

Cis-Acting RNA Signals Determine Template Specificity

In the course of packaging and replication, a dsRNA virus must pick the correct viral (+)RNA molecules from a sea of cellular ones. This specificity is attributed to the presence of *cis*-acting signals that selectively channel the viral RNAs into the assembly and replication complexes. For L-A, a stem-loop structure (internal site) in the (+)RNA forms the packaging signal that is recognized by the Pol domain of the Gag–Pol fusion protein during assembly (Esteban et al., 1989). Conversion of this packaged (+)RNA to dsRNA requires this internal site, as well as sequences in the 3′-end of the template (Fujimura et al., 1992; Wickner et al., 1986). Because the internal site is 400 nucleotides from the 3′-end, it is thought that secondary and tertiary folding of the molecule brings these two sites together during L-A replication. The Cystoviridae member Φ6 specifically recognizes its RNA template based on a conserved 18-nt sequence at the 5′-end, as well as an upstream *pac* sequence that is unique in each segment. The *pac* sequence of each segment folds into a distinctive stem-loop structure required for organized packaging (Gottlieb et al., 1994; Mindich, 2004). Efficient Φ6 (–)RNA strand synthesis follows packaging and requires that the templates have the 3′-sequence 5′-CUCUCUCUCU-3′ (Mindich, 2004; Onodera et al., 1993). Template RNAs lacking this 3′-sequence are packaged, but not replicated, demonstrating that an additional level of specificity occurs during dsRNA synthesis (Onodera et al., 1993).

Little is known regarding the precise *cis*-acting signals underlying gene-specific packaging of (+)RNA for the Reoviridae family. Recent studies using an in vitro rescue system have provided evidence that the packaging signals of mORV reside at the 5′-end of (+)RNAs (Roner and Steele, 2007). Specifically, chimeric (+)RNA molecules that contain the 5′-end of mORV m1 or s2 genes fused to the open reading frame (ORF) of reporter genes are specifically packaged and replicated by helper virus cores (Roner and Steele, 2007). The 5′-ends of the m1 and s2 genes show little sequence similarities, but each is predicted to form a stem-loop structure that might serve as a recognition signal for core proteins. For rotavirus, sequences in the 5′- and 3′-untranslated regions (UTRs) of homologous genes from distantly related strains maintain a high level of conservation (Patton et al., 2007). This sequence conservation is observed for the same gene of different strains, even when the ORF is extremely variable, suggesting that UTRs might contain important *cis*-acting signals for packaging. Moreover, the observation that UTRs from heterologous genes of the same virus are different suggests that these regions form gene-specific signals. Though it is not well understood, the mechanism of (+)RNA packaging for Reoviridae is undoubtedly a strictly regulated process. The result of this meticulous process is the formation of a viral core that contains exactly one copy each of the 10–12 dsRNA genome segments.

The SA11 open core system has allowed considerable progress in defining the *cis*-acting signals in (+)RNAs that support genome replication for the Reoviridae family. These studies have identified two elements within all SA11 template RNAs that promote efficient dsRNA synthesis: (i) a 3′-terminal consensus sequence (3′CS) 5′-UGUGACC-3′ and (ii) a panhandle structure formed by sequences in the 5′- and 3′-UTRs (Chen et al., 2001; Patton et al., 1999, 1996; Tortorici et al., 2003, 2006; Wentz et al., 1996). Of these two elements, the highly conserved 3′CS is the most important, as a deletion of this region in the context of a viral (+)RNA template completely abolishes replication (Patton et al., 1996). The 3′CS is composed of two partially overlapping determinants that mediate specific template recognition by VP1 and productive initiation complex formation. As defined by electromobility shift assays, the UGUG(A) portion of the 3′CS drives high-affinity interactions with VP1 (Chen et al., 2001; Tortorici et al., 2003). In contrast, the terminal ACC nucleotides of the 3′CS are dispensable for VP1 binding, but are important for the formation of a (–)RNA strand initiation complex in vitro (Chen et al., 2001; Tortorici et al., 2003). In addition to the 3′CS, the formation of a panhandle structure, as a result of base pairing between the 5′-UTR and 3′-UTR, is important for genome replication. Specifically, mutations that prevent the panhandle structure from forming a free single-stranded 3′-end completely inhibit replication, even in the presence of an intact 3′CS (Chen and Patton, 1998; Tortorici et al., 2006). This result suggests that the 3′CS must be presented to the RdRp as a single-stranded template, allowing it to be sterically accessible for initiation complex formation. It is also possible that these panhandle structures reveal the putative gene-specific packaging signals in UTRs. Thus, besides determining specificity during dsRNA synthesis, the panhandle structure might also promote the proper assortment of the (+)RNAs during packaging.

Indeed, these *cis*-acting signals in (+)RNA templates are important to efficient (–)RNA strand synthesis during SA11 genome replication. However, it is important to note that SA11 VP1 is capable of using (–)RNA templates, which lack both a 3′CS and a panhandle structure, for multiple rounds of transcription. The 3′-ends of SA11 (–)RNA strands show a less conserved sequence of 5′-$(A/U)_6$AGCC-3′ that is thought to be recognized by VP1 during transcription, but with a lower affinity than the 3′CS of (+)RNAs (Patton et al., 2007). From a biological standpoint, this possibility is realistic because the (–)RNA strand of a genomic dsRNA segment would already be paired with the RdRp inside the core. Thus, there would be no need for VP1 to be selective about which template to use during transcription; essentially, the enzyme has only one choice. Quite the opposite occurs during genome replication when the RdRp must identify the correct (+)RNA among a cadre of cellular RNAs. This requires that the RdRp be very particular, as the wrong choice would cause the generation of a defective particle. The idea that the affinity of template binding determines specificity is supported experimentally for RdRps of the Reoviridae family, yet these principles are likely to apply to the Totiviridae and Cystoviridae RdRps as well.

Summary and Conclusions

Structural and biochemical studies of prototypical members of the Totiviridae, Cystoviridae, and Reoviridae families have identified common themes in the replication strategies of dsRNA viruses. From these studies, it is clear that viral RNA synthesis occurs within a protected core via an anchored RdRp. The evidence to date is also consistent with the ideas that genome replication and capsid assembly occur simultaneously and *cis*-acting elements in the viral RNA determine template specificity. Such precise regulation of core-associated RNA synthesis ensures that the precious dsRNA genome is protected during the entire replication cycle of these viruses. However, several unanswered questions remain about the exact mechanisms Totiviridae, Cystoviridae, and Reoviridae members use to mediate RNA packaging, core assembly, and genome replication. What region(s) within the core shell proteins are important for interactions with the viral enzymes and RNA? Which domain(s) of the RdRp directly engage the core shell and/or other viral enzymes? Do the viral enzymes remain tethered to the inside of the core shell during all stages of viral RNA synthesis? What changes occur inside the core following viral entry/uncoating that trigger transcription? What changes occur inside the core during RNA packaging that trigger genome replication? How do segmented dsRNA viruses package equimolar ratios of genome segments? Is RNA packaging coordinated by protein–RNA interactions only or are RNA–RNA interactions among segments important too? What are the roles of viral nonstructural proteins during packaging and replication? Certainly, future studies addressing these important questions related to dsRNA replication are warranted.

This chapter focuses on the replication strategies of dsRNA viruses that retain their genomes within a pseudo $T = 1$ core, as they are the most numerous in terms of individual species. Yet, it is important to mention that not all dsRNA virus families have this pseudo $T = 1$ core-associated mechanism of genome replication; instead, there are some intriguing exceptions to the described themes (Table 11.1). For example, the core shells of Chrysoviridae family members are similar to pseudo $T = 1$ cores, but are composed of 60 protein subunits instead of dimers, making them classic $T = 1$ structures (Mertens, 2004). Interestingly, viruses within this family package their four dsRNA segments in separate core shells, rather than together, circumventing the need for gene-specific packaging signals (Mertens, 2004). In addition, viruses belonging to the Birnaviridae family have single-shelled particles that show $T = 13$ icosahedral symmetry and are nearly identical to the structure of Reoviridae outer virion layers (Coulibaly et al., 2005). The Birnaviridae members also have a VPg-like protein linked to the 5′-ends of their bisegmented genome, a feature that is seen in several positive-strand RNA viruses. Furthermore, the Hypoviridae family members have a replication strategy that is more similar to that of positive-strand RNA viruses than to other dsRNA viruses. Specifically, members of this family completely lack a core shell and replicate their single dsRNA genome segment in association with cellular membranes (Jacob-Wilk et al., 2006). It is thought that positive-strand RNA viruses mediate RNA synthesis in association with vesicular or invaginated membranes to protect their dsRNA replication intermediates from detection by the host cell antiviral system (Ahlquist, 2006). Therefore, membranous positive-strand viral replication complexes and dsRNA viral cores can be thought of as functionally analogous structures. Although these shared features cannot distinguish divergent from convergent evolution, these parallels suggest that positive-strand RNA and dsRNA viruses might have an ancestral linkage. Ongoing and future studies of viral genome replication are sure to reveal more unifying themes that link seemingly diverse families to each other. Such studies will not only enhance our understanding of how viruses spread but will also help identify important targets for limiting the impact of viral diseases.

References

Ahlquist, P. (2006) Parallels among positive-strand RNA viruses, reverse-transcribing viruses and double-stranded RNA viruses. Nature Reviews 4(5), 371–82.

Boyce, M., Wehrfritz, J., Noad, R. and Roy, P. (2004) Purified recombinant bluetongue virus VP1 exhibits RNA replicase activity. Journal of Virology 78(8), 3994–4002.

Butcher, S.J., Dokland, T., Ojala, P.M., Bamford, D.H. and Fuller, S.D. (1997) Intermediates in the assembly pathway of the double-stranded RNA virus phi6. The EMBO Journal 16(14), 4477–87.

Butcher, S.J., Grimes, J.M., Makeyev, E.V., Bamford, D.H. and Stuart, D.I. (2001) A mechanism for initiating RNA-dependent RNA polymerization. Nature 410(6825), 235–40.

Caspar, D.L. and Klug, A. (1962) Physical principles in the construction of regular viruses. Cold Spring Harbor Symposia on Quantitative Biology 27, 1–24.

Caston, J.R., Luque, D., Trus, B.L., Rivas, G., Alfonso, C., Gonzalez, J.M., Carrascosa, J.L., Annamalai, P. and Ghabrial, S.A. (2006) Three-dimensional structure and stoichiometry of Helmintosporium victoriae190S totivirus. Virology 347(2), 323–32.

Caston, J.R., Trus, B.L., Booy, F.P., Wickner, R.B., Wall, J.S. and Steven, A.C. (1997) Structure of L-A virus: a specialized compartment for the transcription and replication of double-stranded RNA. The Journal of Cell Biology 138(5), 975–85.

Chen, D., Barros, M., Spencer, E. and Patton, J.T. (2001) Features of the 3′-consensus sequence of rotavirus mRNAs critical to minus strand synthesis. Virology 282(2), 221–9.

Chen, D. and Patton, J.T. (1998) Rotavirus RNA replication requires a single-stranded 3′ end for efficient minus-strand synthesis. Journal of Virology 72(9), 7387–96.

Chen, D. and Patton, J.T. (2000) De novo synthesis of minus strand RNA by the rotavirus RNA polymerase in a cell-free system involves a novel mechanism of initiation. RNA (New York, NY) 6(10), 1455–67.

Chen, D., Zeng, C.Q., Wentz, M.J., Gorziglia, M., Estes, M.K. and Ramig, R.F. (1994) Template-dependent, in vitro replication of rotavirus RNA. Journal of Virology 68(11), 7030–9.

Cheng, R.H., Caston, J.R., Wang, G.J., Gu, F., Smith, T.J., Baker, T.S., Bozarth, R.F., Trus, B.L., Cheng, N., Wickner, R.B., et al. (1994) Fungal virus capsids, cytoplasmic compartments for the replication of double-stranded RNA, formed as icosahedral shells of asymmetric Gag dimers. Journal of Molecular Biology 244(3), 255–8.

Coulibaly, F., Chevalier, C., Gutsche, I., Pous, J., Navaza, J., Bressanelli, S., Delmas, B. and Rey, F.A. (2005) The birnavirus crystal structure reveals structural relationships among icosahedral viruses. Cell 120(6), 761–72.

Dinman, J.D., Icho, T. and Wickner, R.B. (1991) A –1 ribosomal frameshift in a double-stranded RNA virus of yeast forms a gag–pol fusion protein. Proceedings of the National Academy of Sciences of the United States of America 88(1), 174–8.

Diprose, J.M., Burroughs, J.N., Sutton, G.C., Goldsmith, A., Gouet, P., Malby, R., Overton, I., Zientara, S., Mertens, P.P., Stuart, D.I. and Grimes, J.M. (2001) Translocation portals for the substrates and products of a viral transcription complex: the bluetongue virus core. The EMBO Journal 20(24), 7229–39.

Esteban, R., Fujimura, T. and Wickner, R.B. (1989) Internal and terminal cis-acting sites are necessary for in vitro replication of the L-A double-stranded RNA virus of yeast. The EMBO Journal 8(3), 947–54.

Fang, Q., Shah, S., Liang, Y. and Zhou, Z.H. (2005) 3D reconstruction and capsid protein characterization of grass carp reovirus. Science in China 48(6), 593–600.

Fujimura, T., Ribas, J.C., Makhov, A.M. and Wickner, R.B. (1992) Pol of gag–pol fusion protein required for encapsidation of viral RNA of yeast L-A virus. Nature 359(6397), 746–9.

Fujimura, T. and Wickner, R.B. (1989) Reconstitution of template-dependent in vitro transcriptase activity of a yeast double-stranded RNA virus. The Journal of Biological Chemistry 264(18), 10872–7.

Gallegos, C.O. and Patton, J.T. (1989) Characterization of rotavirus replication intermediates: a model for the assembly of single-shelled particles. Virology 172(2), 616–27.

Garcia-Sastre, A. and Biron, C.A. (2006) Type 1 interferons and the virus-host relationship: a lesson in detente. Science (New York, NY) 312(5775), 879–82.

Gottlieb, P., Qiao, X., Strassman, J., Frilander, M. and Mindich, L. (1994) Identification of the packaging regions within the genomic RNA segments of bacteriophage phi 6. Virology 200(1), 42–7.

Grimes, J.M., Burroughs, J.N., Gouet, P., Diprose, J.M., Malby, R., Zientara, S., Mertens, P.P. and Stuart, D.I. (1998) The atomic structure of the bluetongue virus core. Nature 395(6701), 470–8.

Grimes, J.M., Jakana, J., Ghosh, M., Basak, A.K., Roy, P., Chiu, W., Stuart, D.I. and Prasad, B.V. (1997) An atomic model of the outer layer of the bluetongue virus core derived from X-ray crystallography and electron cryomicroscopy. Structure (London, England) 5(7), 885–93.

Hewat, E.A., Booth, T.F. and Roy, P. (1992) Structure of bluetongue virus particles by cryoelectron microscopy. Journal of Structural Biology 109(1), 61–9.

Hill, C.L., Booth, T.F., Prasad, B.V., Grimes, J.M., Mertens, P.P., Sutton, G.C. and Stuart, D.I. (1999) The structure of a cypovirus and the functional organization of dsRNA viruses. Nature Structural Biology 6(6), 565–8.

Huiskonen, J.T., de Haas, F., Bubeck, D., Bamford, D.H., Fuller, S.D. and Butcher, S.J. (2006) Structure of the bacteriophage phi6 nucleocapsid suggests a mechanism for sequential RNA packaging. Structure (London, England) 14(6), 1039–48.

Jaalinoja, H.T., Huiskonen, J.T. and Butcher, S.J. (2007) Electron cryomicroscopy comparison of the architectures of the enveloped bacteriophages phi6 and phi8. Structure (London, England) 15(2), 157–67.

Jacob-Wilk, D., Turina, M. and Van Alfen, N.K. (2006) Mycovirus cryphonectria hypovirus 1 elements cofractionate with trans-Golgi network membranes of the fungal host Cryphonectria parasitica. Journal of Virology 80(13), 6588–96.

Jayaram, H., Estes, M.K. and Prasad, B.V. (2004) Emerging themes in rotavirus cell entry, genome organization, transcription and replication. Virus Research 101(1), 67–81.

Kainov, D.E., Butcher, S.J., Bamford, D.H. and Tuma, R. (2003) Conserved intermediates on the assembly pathway of double-stranded RNA bacteriophages. Journal of Molecular Biology 328(4), 791–804.

Kar, A.K., Ghosh, M. and Roy, P. (2004) Mapping the assembly pathway of Bluetongue virus scaffolding protein VP3. Virology 324(2), 387–99.

Kim, J., Tao, Y., Reinisch, K.M., Harrison, S.C. and Nibert, M.L. (2004) Orthoreovirus and Aquareovirus core proteins: conserved enzymatic surfaces, but not protein–protein interfaces. Virus Research 101(1), 15–28.

Labbe, M., Baudoux, P., Charpilienne, A., Poncet, D. and Cohen, J. (1994) Identification of the nucleic acid binding domain of the rotavirus VP2 protein. The Journal of General Virology 75 (Pt 12), 3423–30.

Lawton, J.A., Estes, M.K. and Prasad, B.V. (1997a) Three-dimensional visualization of mRNA release from actively transcribing rotavirus particles. Nature Structural Biology 4(2), 118–21.

Lawton, J.A., Estes, M.K. and Prasad, B.V. (2000) Mechanism of genome transcription in segmented dsRNA viruses. Advances in Virus Research 55, 185–229.

Lawton, J.A., Zeng, C.Q., Mukherjee, S.K., Cohen, J., Estes, M.K. and Prasad, B.V. (1997b) Three-dimensional structural analysis of recombinant rotavirus-like particles with intact and amino-terminal-deleted VP2: implications for the architecture of the VP2 capsid layer. Journal of Virology 71(10), 7353–60.

Levy, D.E. and Garcia-Sastre, A. (2001) The virus battles: IFN induction of the antiviral state and mechanisms of viral evasion. Cytokine & Growth Factor Reviews 12(2–3), 143–56.

Lu, G., Zhou, Z.H., Baker, M.L., Jakana, J., Cai, D., Wei, X., Chen, S., Gu, X. and Chiu, W. (1998) Structure of double-shelled rice dwarf virus. Journal of Virology 72(11), 8541–9.

Makeyev, E.V. and Grimes, J.M. (2004) RNA-dependent RNA polymerases of dsRNA bacteriophages. Virus Research 101(1), 45–55.

Mellor, P.S. and Boorman, J. (1995) The transmission and geographical spread of African horse sickness and bluetongue viruses. Annals of Tropical Medicine and Parasitology 89(1), 1–15.

Mertens, P. (2004) The dsRNA viruses. Virus Research 101(1), 3–13.

Mertens, P.P. and Diprose, J. (2004) The bluetongue virus core: a nano-scale transcription machine. Virus Research 101(1), 29–43.

Metcalf, P., Cyrklaff, M. and Adrian, M. (1991) The three-dimensional structure of reovirus obtained by cryo-electron microscopy. The EMBO Journal 10(11), 3129–36.

Mindich, L. (2004) Packaging, replication and recombination of the segmented genome of bacteriophage Phi6 and its relatives. Virus Research 101(1), 83–92.

Mossi, R. and Hubscher, U. (1998) Clamping down on clamps and clamp loaders – the eukaryotic replication factor C. European Journal of Biochemistry/FEBS 254(2), 209–16.

Naitow, H., Tang, J., Canady, M., Wickner, R.B. and Johnson, J.E. (2002) L-A virus at 3.4 A resolution reveals particle architecture and mRNA decapping mechanism. Nature Structural Biology 9(10), 725–8.

Nakagawa, A., Miyazaki, N., Taka, J., Naitow, H., Ogawa, A., Fujimoto, Z., Mizuno, H., Higashi, T., Watanabe, Y., Omura, T., Cheng, R.H. and Tsukihara, T. (2003) The atomic structure of rice dwarf virus reveals the self-assembly mechanism of component proteins. Structure (Cambridge, Mass.) 11(10), 1227–38.

O'Reilly, E.K. and Kao, C.C. (1998) Analysis of RNA-dependent RNA polymerase structure and function as guided by known polymerase structures and computer predictions of secondary structure. Virology 252(2), 287–303.

Ojala, P.M. and Bamford, D.H. (1995) In vitro transcription of the double-stranded RNA bacteriophage phi 6 is influenced by purine NTPs and calcium. Virology 207(2), 400–8.

Onodera, S., Qiao, X., Gottlieb, P., Strassman, J., Frilander, M. and Mindich, L. (1993) RNA structure and heterologous recombination in the double-stranded RNA bacteriophage phi 6. Journal of Virology 67(8), 4914–22.

Parashar, U.D., Hummelman, E.G., Bresee, J.S., Miller, M.A. and Glass, R.I. (2003) Global illness and deaths caused by rotavirus disease in children. Emerging Infectious Diseases 9(5), 565–72.

Patton, J.T. (1996) Rotavirus VP1 alone specifically binds to the 3′ end of viral mRNA, but the interaction is not sufficient to initiate minus-strand synthesis. Journal of Virology 70(11), 7940–7.

Patton, J.T. (2001) Rotavirus RNA replication and gene expression. Novartis Foundation Symposium 238, 64–77; discussion 77–81.

Patton, J.T., Chnaiderman, J. and Spencer, E. (1999) Open reading frame in rotavirus mRNA specifically promotes synthesis of double-stranded RNA: template size also affects replication efficiency. Virology 264(1), 167–80.

Patton, J.T. and Gallegos, C.O. (1990) Rotavirus RNA replication: single-stranded RNA extends from the replicase particle. The Journal of General Virology 71 (Pt 5), 1087–94.

Patton, J.T., Jones, M.T., Kalbach, A.N., He, Y.W. and Xiaobo, J. (1997) Rotavirus RNA polymerase requires the core shell protein to synthesize the double-stranded RNA genome. Journal of Virology 71(12), 9618–26.

Patton, J.T., Silvestri, L.S., Tortorici, M.A., Vasquez-Del Carpio, R. and Taraporewala, Z.F. (2006) Rotavirus genome replication and morphogenesis: role of the viroplasm. Current Topics in Microbiology and Immunology 309, 169–87.

Patton, J.T. and Spencer, E. (2000) Genome replication and packaging of segmented double-stranded RNA viruses. Virology 277(2), 217–25.

Patton, J.T., Vasquez-Del Carpio, R. and Spencer, E. (2004) Replication and transcription of the rotavirus genome. Current Pharmaceutical Design 10(30), 3769–77.

Patton, J.T., Vasquez-Del Carpio, R., Tortorici, M.A. and Taraporewala, Z.F. (2007) Coupling of rotavirus genome replication and capsid assembly. Advances in Virus Research 69, 167–201.

Patton, J.T., Wentz, M., Xiaobo, J. and Ramig, R.F. (1996) cis-Acting signals that promote genome replication in rotavirus mRNA. Journal of Virology 70(6), 3961–71.

Petterson, E.F., Goddard, T.D., Huang, C.C., Couch, G.S., Greenblatt, D.M., Meng, E.C. and Ferrin, T.E. (2004) UCSF Chimera-A visualization system for exploratory research and analysis. Journal of Computational Chemistry 25(13), 1605–12.

Poranen, M.M. and Tuma, R. (2004) Self-assembly of double-stranded RNA bacteriophages. Virus Research 101(1), 93–100.

Prasad, B.V., Rothnagel, R., Zeng, C.Q., Jakana, J., Lawton, J.A., Chiu, W. and Estes, M.K. (1996) Visualization of ordered genomic RNA and localization of transcriptional complexes in rotavirus. Nature 382(6590), 471–3.

Reinisch, K.M., Nibert, M.L. and Harrison, S.C. (2000) Structure of the reovirus core at 3.6 A resolution. Nature 404(6781), 960–7.

Ribas, J.C. and Wickner, R.B. (1992) RNA-dependent RNA polymerase consensus sequence of the L-A double-stranded RNA virus: definition of essential domains. Proceedings of the National Academy of Sciences of the United States of America 89(6), 2185–9.

Roner, M.R. and Steele, B.G. (2007) Localizing the reovirus packaging signals using an engineered m1 and s2 ssRNA. Virology 358(1), 89–97.

Roy, P. and Noad, R. (2006) Bluetongue virus assembly and morphogenesis. Current Topics in Microbiology and Immunology 309, 87–116.

Shepherd, C.M., Borelli, I.A., Lander, G., Natarajan, P., Siddavanahalli, V., Bajaj, C., Johnson, J.E., Brooks, C.L., 3rd and Reddy, V.S. (2006) VIPERdb: a relational database for structural virology. Nucleic Acids Research 34(Database issue), D386–9.

Steven, A.C., Trus, B.L., Booy, F.P., Cheng, N., Zlotnick, A., Caston, J.R. and Conway, J.F. (1997) The making and breaking of symmetry in virus capsid assembly: glimpses of capsid biology from cryoelectron microscopy. The FASEB Journal 11(10), 733–42.

Tao, Y., Farsetta, D.L., Nibert, M.L. and Harrison, S.C. (2002) RNA synthesis in a cage – structural studies of reovirus polymerase lambda3. Cell 111(5), 733–45.

Tortorici, M.A., Broering, T.J., Nibert, M.L. and Patton, J.T. (2003) Template recognition and formation of initiation complexes by the replicase of a segmented double-stranded RNA virus. The Journal of Biological Chemistry 278(35), 32673–82.

Tortorici, M.A., Shapiro, B.A. and Patton, J.T. (2006) A base-specific recognition signal in the 5′ consensus sequence of rotavirus plus-strand RNAs promotes replication of the double-stranded RNA genome segments. RNA (New York, NY) 12(1), 133–46.

Wentz, M.J., Patton, J.T. and Ramig, R.F. (1996) The 3′-terminal consensus sequence of rotavirus mRNA is the minimal promoter of negative-strand RNA synthesis. Journal of Virology 70(11), 7833–41.

Wickner, R.B. (1996) Double-stranded RNA viruses of *Saccharomyces cerevisiae*. Microbiological Reviews 60(1), 250–65.

Wickner, R.B., Fujimura, T. and Esteban, R. (1986) Overview of double-stranded RNA replication in *Saccharomyces cerevisiae*. Basic Life Sciences 40, 149–63.

Xia, Q., Jakana, J., Zhang, J.Q. and Zhou, Z.H. (2003) Structural comparisons of empty and full cytoplasmic polyhedrosis virus. Protein–RNA interactions and implications for endogenous RNA transcription mechanism. The Journal of Biological Chemistry 278(2), 1094–100.

Yeager, M., Dryden, K.A., Olson, N.H., Greenberg, H.B. and Baker, T.S. (1990) Three-dimensional structure of rhesus rotavirus by cryoelectron microscopy and image reconstruction. The Journal of Cell Biology 110(6), 2133–44.

Zeng, C.Q., Estes, M.K., Charpilienne, A. and Cohen, J. (1998) The N terminus of rotavirus VP2 is necessary for encapsidation of VP1 and VP3. Journal of Virology 72(1), 201–8.

Zhang, X., Tang, J., Walker, S.B., O'Hara, D., Nibert, M.L., Duncan, R. and Baker, T.S. (2005) Structure of avian orthoreovirus virion by electron cryomicroscopy and image reconstruction. Virology 343(1), 25–35.

Zhang, X., Walker, S.B., Chipman, P.R., Nibert, M.L. and Baker, T.S. (2003) Reovirus polymerase lambda 3 localized by cryo-electron microscopy of virions at a resolution of 7.6 A. Nature Structural Biology 10(12), 1011–8.

Zhou, Z.H., Zhang, H., Jakana, J., Lu, X.Y. and Zhang, J.Q. (2003) Cytoplasmic polyhedrosis virus structure at 8 A by electron cryomicroscopy: structural basis of capsid stability and mRNA processing regulation. Structure (London, England) 11(6), 651–63.

Chapter 12
Poxviruses

Kathleen Boyle and Paula Traktman

Introduction

The Poxviridae family comprises large, complex DNA viruses that infect a wide variety of vertebrate and invertebrate hosts (chordopoxviruses and entomopoxviruses, respectively). Within the chordopoxvirus subfamily, the orthopoxvirus genus is best known for containing variola (VARV), the etiological agent of smallpox, and vaccinia (VACV), the virus used as the vaccine in the successful campaign to eradicate smallpox. Vaccinia has also served as the prototype for experimental investigation of poxvirus biology, and the following review will focus on vaccinia replication. Despite possessing DNA genomes, these viruses replicate exclusively in the cytoplasm of the infected host cell. This unusual physical autonomy from the host nucleus is accompanied by genetic autonomy; the ~200 viral gene products encode a repertoire of proteins that mediate three temporally regulated phases of gene expression, genome replication, and virion morphogenesis. Poxviruses also encode a plethora of proteins that intersect with, and modulate, many cellular signaling cascades and components of the innate immune response.

Life Cycle

The poxvirus life cycle is shown schematically in Fig. 12.1. Poxvirus virions ($\sim$360 nm $\times$ 270 nm $\times$ $\sim$250 nm) are quite complex, containing ~75 distinct proteins (Condit et al., 2006). They are surrounded by a membrane (protein-rich lipid bilayer) and contain an internal core which houses the viral genome and a complete transcriptional apparatus. The processes of virion binding and entry are still being elucidated, but it appears that poxviruses can enter cells either by direct fusion of the virion membrane with the plasma membrane or by endocytic uptake. In either case, the internal virion core is then released into the cytoplasm and traffics

P. Traktman (✉)
Department of Microbiology and Molecular Genetics, Medical College of Wisconsin, 8701 Watertown Plank Road, Milwaukee, WI 53226, USA
e-mail: ptrakt@mcw.edu

C.E. Cameron et al. (eds.), *Viral Genome Replication*,
DOI 10.1007/b135974_12, © Springer Science+Business Media, LLC 2009

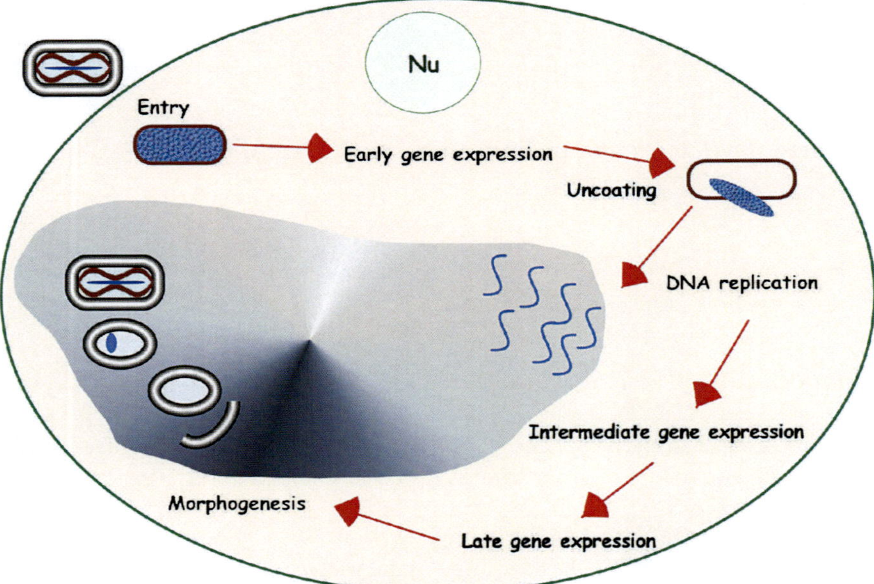

Fig 12.1 *Poxvirus life cycle*. Virion entry results in the deposition of the viral core into the cytoplasm. The remainder of the life cycle is restricted to the cytoplasm and is localized within viral factories. Early gene expression occurs within the core, resulting in the production of proteins involved in DNA replication and intermediate gene transcription. An uncoating event releases the viral genome into the cytoplasm, and viral DNA replication ensues. Intermediate and late gene expression commence, and the complex and highly regulated process of morphogenesis leads to the production of infectious progeny.

on microtubules to a peri-nuclear site (Carter et al., 2003; Mallardo et al., 2001). Within minutes, the encapsidated early transcription machinery is activated and early gene expression initiates within the viral core. Capped and polyadenylated mRNAs, which represent approximately one half of the genome, are extruded into the cytoplasm and translated on host polysomes. Among the early proteins are those needed for genome replication and for the transcription of intermediate mRNAs. Early gene transcription peaks at 1–2 h post-infection, and ceases when "uncoating" occurs, compromising the integrity of the core and releasing the genome into the cytoplasm where it then undergoes replication. Replication continues from ~2 h post-infection until ~12 h post-infection, providing a large pool of progeny genomes (~10,000 per cell, one half of which are encapsidated into nascent virions) (Joklik and Becker, 1964; Salzman, 1960). DNA replication serves as a switch that enables the onset of intermediate, and then late, gene expression. Each phase of gene expression utilizes a unique set of *cis*- and *trans*-acting factors. The synthesis of viral mRNAs, DNAs, and proteins occurs in cytoplasmic domains known as "factories" or "virosomes". Once the late phase of gene expression is underway, the complex process of virion morphogenesis initiates. One of the earliest hallmarks of virion assembly is the appearance of rigid, membranous crescents that enlarge until they become spherical, enclosing proteins destined to form the internal virion core. These spherical particles are known as immature virions; prior to their closure, the genome is encapsidated and an electron-dense nucleoid appears. Maturation of the immature virions into mature, infectious, virions is accompanied by a complex series of proteolytic cleavages and ultrastructural rearrangements. Recent reviews of virion morphogenesis, and the process whereby a subset of virions acquire additional membrane wrappings and undergo release from the cell to mediate intercellular and distal spread, are available (Condit et al., 2006; Smith et al., 2002).

Genome Structure

Poxviral genomes vary significantly in their AT:GC content, although the orthopoxviruses have a very high AT content; the DNA is not methylated. The large genomes (for VACV, 192 kb) possess a number of unique and distinguishing structural characteristics. A schematic representation of the chromosomal structure is depicted in Fig. 12.2. The telomeres of the double-stranded linear DNA genome are composed of highly AT-rich hairpins which contain extrahelical bases (EHB);

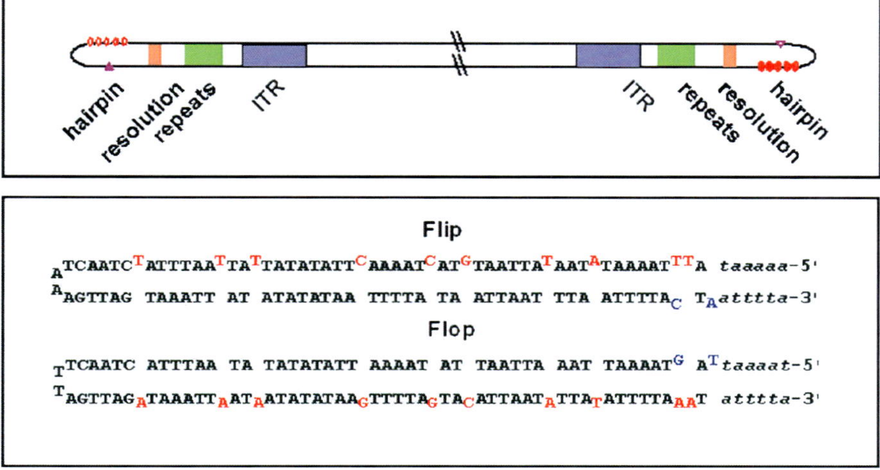

Fig. 12.2 *Poxvirus genome structure*. Top panel: The 192 kb vaccinia genome is a linear DNA duplex flanked by highly AT-rich hairpins which contains 12 extrahelical bases (represented as five *circles* and a *triangle*). Approximately 87 bp regions adjacent to the hairpin contain the motifs required for concatemer resolution. Adjacent to this region are sets of tandem repeats. These motifs, as well as a few genes, are present at both termini of the genome and are referred to as the ITR (inverted terminal repeats). The remainder of the genome encodes ~200 proteins. Bottom panel: The hairpin sequences exist in two isoforms (flip and flop) that are inverted and complementary to each other; the extrahelical bases are shown in color.

in vaccinia virus, the hairpin is 104 nt in length and has 12 extrahelical bases (10 on one strand, 2 on the other). The presence of EHB is conserved in all poxvirus genomes, although the precise number and position of the bases vary. The hairpin sequences are found in two isoforms, known as flip and flop, that are inverted and complementary with respect to one other. Adjacent to the hairpin loop are 87 bp that are essential for both replication (Du and Traktman, 1996) and concatemer resolution (resolution repeats) (DeLange and McFadden, 1990). Beyond these motifs lie several sets of tandem repeats whose function is not known, although a role in mediating intergenomic recombination has been proposed. This entire region is found at both ends of the genome; in some viruses, the repeated region (inverted terminal repeat [ITR]) also contains protein-encoding genes, which are hence diploid in their inheritance.

The remainder of the poxviral genome is tightly packed with genes; in keeping with the cytoplasmic localization of the life cycle, the genes contain no introns and the transcripts do not undergo splicing. The 5′ and 3′ untranslated regions are minimal, as are the intergenic spaces. In the central portion of the genome, the polarity of the genes is somewhat random, but toward the ends of the genome, transcription is almost always oriented toward the telomere. There is no clustering of the genes by temporal class or protein function.

The complete sequences for 27 individual poxviruses (not counting different strain variants) have been determined (see http://poxvirus.org). The first poxvirus genome to be solved was the Copenhagen strain of vaccinia virus (Goebel et al., 1990), upon which the current nomenclature is based. The open reading frames are named according to their location within a given *Hin*dIII restriction fragment (A–P) and their transcriptional orientation (R or L) relative to the standard genomic map. For example, the DNA polymerase gene (E9L) is the ninth open reading frame in the *Hin*dIII E restriction fragment and it is transcribed in a leftward orientation. The protein name uses the same notation without the transcriptional polarity (E9 protein). For complete genomes that have been sequenced more recently, the open reading frames are often numbered sequentially, but most investigators in the field refer to orthologous genes using the Copenhagen nomenclature.

Analysis of Viral Replication Within Infected Cells

Pioneering electron microscopy provided the first evidence that poxviruses establish a distinct region in the cytoplasm of cells, referred to as the "virosome" or "DNA factory", where viral DNA replication occurs (Cairns, 1960). These viral DNA factories have since been visualized by electron microscopy, staining with fluorescent dyes such as DAPI or Hoechst, or indirect immunofluorescent detection using anti-sera directed against BrdU or the viral single-stranded DNA-binding protein I3 (Domi and Beaud, 2000; Rochester and Traktman, 1998; Welsch et al., 2003). Cytoplasmic sites of replication have also been visualized by using DNA templates that bear multiple copies of the *Escherichia coli* lac operator and co-expressing a

GFP-lac repressor (De Silva and Moss, 2005). Recent data suggest that the perinuclear sites of replication are surrounded by membranes derived from the endoplasmic reticulum (Tolonen et al., 2001); clearly, much remains to be learned about how poxviruses establish sub-cellular compartments that enable efficient transcription, replication, and assembly.

Replication of viral DNA in vivo can be quantitated using one of several techniques. Metabolic labeling of nascent DNA with ^{3}H-thymidine can be used to assess the rate of viral DNA synthesis and is helpful in monitoring the immediate impact of adding pharmacological inhibitors or shifting the incubation temperature during infections with temperature-sensitive mutants. However, due to changes in the nucleotide pool that occur as infection progresses, and because of the feedback inhibition of the thymidine kinase (TK, see below), this technique cannot provide an accurate picture of the full extent of viral DNA accumulation. Furthermore, this method of analysis is limited to TK+ viruses, and many recombinants used for experimental analysis are TK–. To assess the steady-state levels of viral DNA synthesis, dot-blot Southern hybridization can be used. Accurate data can be obtained from ~5 h to 24 h post-infection. Most recently, real-time PCR has been used to detect viral DNA sequences; this technique is likely to have a larger dynamic range than those described above.

In addition to monitoring the replication of viral genomes introduced by infection, investigators have also introduced exogenous templates in order to dissect the *cis*- and *trans*-acting components required for optimal replication. One approach that has been taken involves the generation of minichromosomes that mimic the topology of the viral genome and its unique telomeric features (Du and Traktman, 1996). In this assay, a linear plasmid stuffer sequence was flanked by telomeric variants ranging in size from 65 bp to 3000 bp. 150–200 bp were shown to be necessary and sufficient for optimal replication, as assessed by the accumulation of *Dpn*I-resistant DNA. Within this region are the terminal hairpin with its extrahelical bases (~50 bp) and the motifs known to be essential for concatemer resolution (~70 nt). Minichromosomes bearing these telomeres were shown to replicate 10-fold better than supercoiled plasmids. However, other investigators have reported that supercoiled plasmids lacking any viral sequences can be replicated as efficiently as minichromosomes (De Silva and Moss, 2005). Thus, further investigation is required in order to definitively assess the contributions, if any, that *cis*-acting sequences make to the efficient replication of the viral genome. Interestingly, analyses of plasmid replication within infected cells have revealed that both processes rely on the same repertoire of viral replication proteins, which will be described below in greater detail.

Working Model for Poxvirus DNA Replication

Poxviral DNA replication is thought to initiate with the introduction of a nick near one or both genomic termini (Fig. 12.3, step 1). Introduction of this nick is inferred from the observed increase in topological freedom as well as the change in the

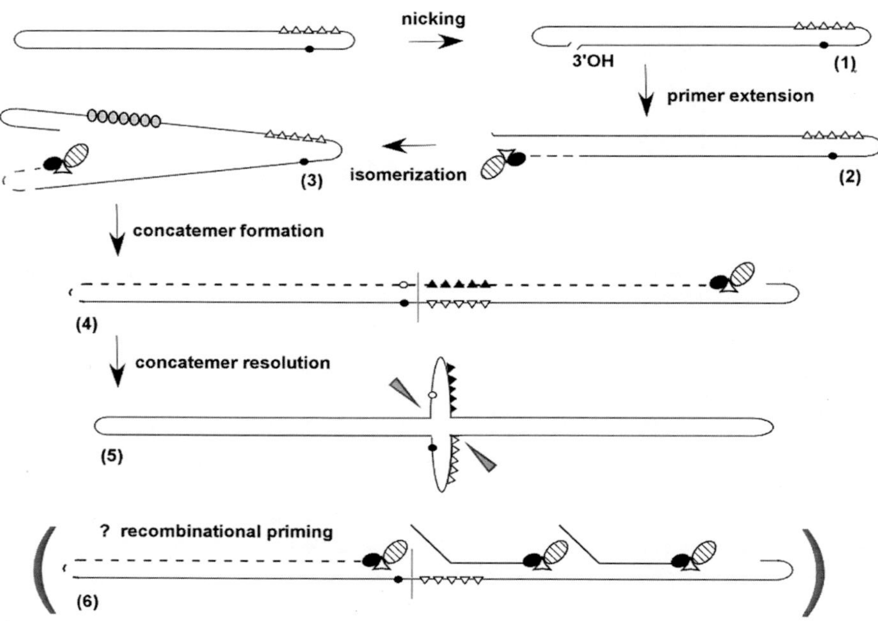

Fig 12.3 *Working model of poxvirus DNA replication.* The linear genome with covalently closed hairpin termini is shown; for convenience, the extrahelical bases (*triangles* and *circle*) are only shown for the right telomere. Replication is thought to begin with the introduction of a nick which exposes a free 3′ OH group that serves as a primer for the trimeric DNA polymerase holoenzyme. The nascent (*dashed lines*) and displaced strands (coated with viral SSB, represented by the *light gray circles*) can each form self-complementary hairpins, which allows leading strand synthesis to replicate the entire molecule. This process generates a tail:tail dimer; a fully duplexed, imperfect palindrome is found at the concatemer junction. The extended duplex form of the palindrome can undergo cruciform extrusion, which creates hairpins that contain extrahelical bases and resemble the viral telomeres. The virally encoded resolvase then cleaves this Holliday junction-like structure, generating monomeric genomes. During later phases of replication, it is proposed that recombination priming events may occur.

sedimentation properties of viral DNA at the onset of replication (Pogo, 1977). Furthermore, the initial site of ^{3}H-thymidine incorporation was mapped to a region within 150 bp from the telomeres (Pogo et al., 1984). The enzyme involved in this nicking event remains elusive, although it is proposed that the nick must leave a free 3′ hydroxyl group to serve as a primer terminus for the viral DNA polymerase. Strand displacement synthesis proceeds toward the hairpin terminus, yielding an intermediate in which the termini of both the nascent and template strands are self-complementary (step 2). This intermediate could assume a conformation that would generate a self-priming hairpin structure (step 3), facilitating the replication of the remainder of both the top and bottom strands of the genome. The initial product would be a tail–tail dimer (step 4). Larger tetrameric molecules may be formed if this process was repeated. Recombinational priming of replicating molecules has also been posited (step 6). Indeed, electrophoretic analysis of replicating DNA has detected the accumulation of large concatemeric and branched intermediates (DeLange, 1989; DeLange and McFadden, 1990; Merchlinsky et al., 1988; Moyer and Graves, 1981).

This current model for poxvirus replication proposes that only leading strand synthesis is employed. This is consistent with the significant levels of single-stranded DNA that have been observed during infection (Esteban and Holowczak, 1977; Pogo et al., 1981). However, early reports suggested that both leading and lagging strand synthesis were involved since it was observed that short nascent DNA strands could be chased into larger forms suggestive of Okazaki fragment formation (Esteban and Holowczak, 1977). Resolution of this issue is an area for further study.

As stated above, the initial products of replication are concatemeric intermediates. The concatemer junction is an imperfect palindrome that can isomerize from a lineform structure to a cruciform structure (step 5). Conversion of the concatemer to monomers that preserve the unique telomeric structure involves a virally encoded Holliday-junction resolvase; this process will be described below in greater detail.

Core Replication Machinery

The repertoire of proteins that mediate vaccinia virus DNA replication has been identified by a blend of genetic, genomic, and biochemical analyses (Table 12.1). Temperature-sensitive (*ts*) viruses, generated by either chemical mutagenesis or targeted clustered charge-to-alanine mutagenesis, reveal that five genes are essential for DNA replication in tissue culture. These encode the DNA polymerase (E9), the two components of the processivity factor (A20 and D4), a nucleic acid-independent NTPase (D5), and a serine/threonine protein kinase (B1).

Table 12.1 Vaccinia-encoded proteins with known or predicted roles in DNA replication are shown

Protein	Function
E9	DNA polymerase
A20	Processivity factor
D4	Uracil DNA glycosylase; processivity factor
D5	DNA independent NTPase; primase; superfamily III helicase homology
B1	Ser/thr protein kinase
I3	Single-strand DNA-binding protein
A22	Holliday-junction resolvase
I6	Telomere-binding protein
A32	Putative ATPase
H6	Topoisomerase I
A50	DNA ligase
F2	dUTPase
F4, I4	Ribonucleotide reductase
J2	Thymidine kinase
A48	Thymidylate kinase

E9, The Catalytic DNA Polymerase

The E9L gene encodes the catalytic DNA polymerase, the core of the trimeric polymerase complex. The 116 kDa E9 protein is expressed early during infection and contains conserved motifs found within the α family of replicative DNA polymerases (Earl et al., 1986; Taddie and Traktman, 1993). E9 has both 5′–3′ polymerization and 3′–5′ exonuclease activities, but lacks intrinsic strand displacement activity (Challberg and Englund, 1979a, b). Structure–function analysis of the protein has been facilitated by phenotypic characterization of two *ts* mutants and drug-resistant mutants that confer resistance to aphidicolin, phosphonoacetic acid, and cytosine arabinoside (DeFilippes, 1984, 1989; Taddie and Traktman, 1991, 1993; Traktman et al., 1989). Aphidicolin-resistant mutants presented a mutator phenotype (Taddie and Traktman, 1991). Lastly, mutations within the E9L gene conferred resistance to the broad spectrum antiviral agent, cidofovir (CDV), a dCMP analog (Kornbluth et al., 2006). It has also been shown that purified E9 can incorporate CDV into the nascent strand adjacent from a G residue, promoting chain termination (Magee et al., 2005). Exploiting E9 as an enzymatic target is an active area of antipoxviral research.

Purified E9 is an inherently distributive enzyme under physiological conditions (McDonald and Traktman, 1994), synthesizing only 10 nt per primer/template-binding event. Since this behavior would not be conductive to efficient and faithful duplication of an ∼200 kb genome in vivo, it was not surprising that cytoplasm extracts of infected cells contained a highly processive form of the polymerase that is able to catalyze the synthesis of >7000 nt per primer/template-binding event (McDonald et al., 1997). As described below, the processive polymerase contains the A20 and D4 proteins as well as E9.

Genetic analyses have shown that there is a clear overlap between the proteins involved in viral replication and homologous recombination. Consistent with these findings, the viral DNA polymerase has been shown to participate in both single-strand annealing and duplex-strand joining reactions (Hamilton and Evans, 2005; Willer et al., 1999; Willer et al., 2000).

A20, A Component of the Polymerase Processivity Factor

Chromatographic purification of the processive form of the viral DNA polymerase revealed that an ∼48 kDa protein was an intrinsic component of this complex (Klemperer et al., 2001). This protein was identified as the product of the A20 gene. A20 and E9 were shown to interact by co-immunoprecipitation experiments, and overexpression of both A20 and E9 during infection led to increased levels of processive polymerase activity. Site-directed, clustered charge-to-alanine mutagenesis of the A20 gene resulted in the generation of *ts* mutants that were defective in DNA replication and defective in the formation of processive polymerase activity (Ishii and Moss, 2001; Punjabi et al., 2001). However, purification of recombinant

A20 protein proved elusive, and it seemed plausible that A20 might interact with, and require, additional proteins to fulfill its role as a processivity factor. Indeed, A20 was shown to interact with the D5 protein (see below), the H5 protein (an abundant phosphoprotein implicated in transcription and morphogenesis), and the D4 protein (see below) in the yeast two-hybrid assay (Ishii and Moss, 2002; McCraith et al., 2000).

D4, Uracil DNA Glycosylase and a Component of the Polymerase Processivity Factor

The product of the D4 gene is an enzymatically active uracil DNA glycosylase (UDG) (Scaramozzino et al., 2003; Stuart et al., 1993; Upton et al., 1993). Conservation of motifs associated with enzymatic activity in other UDG molecules has enabled the generation of viral recombinants that express stable, soluble, but catalytically inactive variants of UDG. These mutants are viable in tissue culture, but attenuated in vivo, suggesting that the repair function of UDG is important during infection of key cell types in vivo (De Silva and Moss, 2003; Stanitsa et al., 2006). In contrast, deletion of the UDG gene is not compatible with viability in tissue culture, underscoring the fact that the UDG protein plays an essential role in viral replication that is independent of its catalytic activity (De Silva and Moss, 2003; Stanitsa et al., 2006).

Phenotypic analysis of two *ts* mutants with lesions in the D4 gene revealed that, like A20, impairment of UDG led to a defect in DNA replication and a defect in the assembly of processive DNA polymerase activity (Stanitsa et al., 2006; Stuart et al., 1993). As mentioned, D4 and A20 interact tightly in the absence of other viral proteins, as assessed by yeast two-hybrid and co-immunoprecipitation assays (Ishii and Moss, 2002; McCraith et al., 2000). Overexpression of 3XFLAG-UDG, A20, and DNA Pol in the context of infected cells has enabled the purification of dimeric UDG/A20 and trimeric UDG/A20/Pol complexes (Stanitsa et al. 2006). These complexes are quite stable, remaining intact in the presence of 750 mM NaCl. The trimeric complex has processive polymerase activity, and the addition of purified 3XFLAG-UDG/A20 to a purified preparation of E9 leads to the reconstitution of processive polymerase activity. In sum, these data support the conclusion that the vaccinia DNA polymerase holoenzyme is comprised of E9, A20, and D4 (Stanitsa et al., 2006). While analysis of the UDG crystal structure has revealed the propensity of UDG to form dimers (Schormann et al., 2007), the higher order structure of the UDG/A20 and UDG/A20/Pol complexes remains to be determined, as does the dissection of the protein:protein interfaces. It is known that the D4:A20 interaction involves the N' terminal 25 amino acids of the A20 protein (Ishii and Moss, 2002). The participation of a UDG as an essential component of a processivity factor is unique to processes and raises the intriguing possibility that replication and repair may be coupled.

I3, Single-Stranded DNA-Binding Protein

A variety of data suggest that the I3 protein is the replicative single-stranded DNA-binding protein (SSB), although this has not yet been proven definitively. I3 is a highly abundant 34 kDa phosphoprotein that is expressed at both early and intermediate times post-infection (Rochester and Traktman, 1998). Purified I3, both endogenous and recombinant, binds to single-stranded DNA with high affinity and specificity. Gel shift assays have derived a binding site size of approximately 10 nt/I3 molecule, and formation of beaded protein/DNA structures can be seen by electron microscopy (Rochester and Traktman, 1998; Tseng et al., 1999). Depletion of I3 during infection using short-interfering RNA (siRNA) technology leads to a reduction in DNA accumulation as assessed by both immunofluorescence and dot-blot hybridization analysis (PT, unpublished). Definitive proof that I3 is the replicative SSB awaits genetic data. Interestingly, the I3 protein has also been shown to bind to the viral ribonucleotide reductase (see below); this interaction could enhance the efficiency of replication by enabling nucleotide precursors to be synthesized at the site of replication (Davis and Mathews, 1993).

D5, Nucleoside Triphosphatase

The 90 kDa D5 protein possesses intrinsic nucleoside triphosphatase activity (NTPase) that is neither dependent on nor stimulated by nucleic acid cofactors (Evans et al., 1995). Genetic analysis of *ts* mutants with lesions in D5 has revealed an essential role for D5 during DNA replication (Boyle et al., 2007; Evans and Traktman, 1992). Temperature shift experiments demonstrate that DNA synthesis arrests immediately (~5 min) after the shift of *ts*D5-infected cultures to the non-permissive temperature, which is indicative of a role of D5 at the replication fork (Evans and Traktman, 1992; McFadden and Dales, 1980). Evidence that D5 can interact with A20 provides additional indirect evidence that D5 might function at the replication fork (McCraith et al., 2000). Although *ts*D5 mutants are defective in DNA replication, they are not defective in the assembly of a processive DNA polymerase complex.

D5 homologs can be identified in every poxvirus sequenced to date, but the protein bears no strong similarity to any cellular protein. Indeed, the D5 protein appears to be a defining member of the D5-like helicase family, which can be found in all poxviruses, certain bacteriophages, and the 1.2 megabase genome of mimivirus (Iyer et al., 2001; Raoult et al., 2004). The sequence of the D5 protein contains conserved motifs that have led to its classification as a peripheral member of the AAA+ family of proteins, within the subfamily of superfamily III DNA helicases (Iyer et al., 2001). This latter group contains several viral proteins that function as replicative helicases. Phenotypic characterization of D5 variants containing amino acid substitutions at key positions within these family-defining motifs has revealed that these motifs are essential, for both NTPase activity and biological function

(Boyle et al., 2007). Structure–function analyses have also revealed that multimerization of the D5 protein is a prerequisite for enzymatic activity and that catalytic activity is necessary but not sufficient for biological activity (Boyle et al., 2007). The catalytic domain comprises the C' terminal half of the protein, while the function of the N' terminal half of the protein, which contains most of the residues affected in the four available *ts*D5 mutants, has not been defined. Given D5's essential role in DNA replication, its NTPase activity, and its similarity to the superfamily III group of helicases, it is tempting to postulate that D5 functions as a helicase in vivo. There is no experimental data to support this hypothesis. However, the sequence of D5 also has motifs associated with DNA primases (Iyer et al., 2005). Indeed, purified D5 has recently been shown to have primase activity in vitro, and the motifs associated with primase activity are important for the biological function of D5 in vivo (De Silva et al., 2007). Finally, it is quite possible that D5 may also participate in homologous recombination, since marker rescue experiments using small (<2 kb) intragenic fragments were impaired during *ts*D5 infections (Evans and Traktman, 1992).

B1, Serine/Threonine Protein Kinase

The importance of dynamic protein phosphorylation in the regulation of the viral life cycle is illustrated by the fact that vaccinia encodes two protein kinases and one protein phosphatase. The B1R gene encodes a 34 kDa serine/threonine protein kinase that is expressed early in infection and encapsidated at low levels in virions. Analysis of two *ts*-B1 mutants revealed a temperature-dependent defect in DNA synthesis whose severity varied with host cell (Boyle and Traktman, 2004; Rempel et al., 1990). The proteins encoded by both mutants had greatly diminished kinase activity and appeared to be labile in vivo at all temperatures. A defect in intermediate gene expression upon impairment of B1 was also reported (Kovacs et al., 2001). Clarification of the role of B1 in the viral life cycle proved elusive for many years; although B1 was shown to phosphorylate the viral H5 protein and two ribosomal proteins, none of the components of the replication machinery appeared to be B1 substrates. Recently, bioinformatic analysis indicated that B1 shows significant similarity to a group of cellular kinases classified as peripheral branch of the casein kinase family. These kinases are now known as vaccinia-related kinases (VRKs) due to the high degree of identity within the catalytic domains of the viral and cellular proteins (Nezu et al., 1997; Nichols and Traktman, 2004; Zelko et al., 1998). This sequence identity implied that the proteins might have overlapping substrate specificities, and indeed, the incorporation of an hVRK1 cDNA into the genome of *ts*B1 fully rescued the DNA replication defect (Boyle and Traktman, 2004). Complementation was not obtained when a catalytically inert variant of hVRK1 was used. The endogenous VRK1 is found only in the nucleus and therefore is not available for complementation.

We now know that the cellular BAF protein is the key substrate whose phosphorylation by B1 regulation is imperative for viral DNA replication to ensue (Wiebe and Traktman, 2007). BAF is an abundant protein, found in both the nucleus and the cytoplasm, that binds to double-stranded DNA avidly and without sequence specificity. Binding of BAF dimers to DNA leads to cross-bridging and condensation. However, when BAF is phosphorylated on key residues within its N' terminus, its ability to bind DNA is abrogated (Nichols et al., 2006). Since vaccinia DNA replication occurs in the cytoplasm, it is accessible to BAF, which can associate with viral DNA within the factories and blocks its replication. BAF appears to be a heretofore unknown host defense against foreign DNA; BAF's repressive action can be reversed by B1-mediated phosphorylation. Recruitment of BAF to the factories does not occur during wild-type infections, but is readily observed during *ts*B1 infections. If BAF is depleted using lentivirus-mediated RNAi, the temperature-sensitive phenotype of the *ts*B1 viruses is largely reversed. Thus the primary, if not only, role of B1 in tissue culture cells is to combat the repressive effect of BAF.

Viral Proteins Involved in Nucleotide Biosynthesis and Precursor Metabolism

Thymidine Kinase and Thymidylate Kinase

Thymidine biosynthesis is directed by two viral enzymes, the thymidine kinase (J2) and thymidylate kinase (A48), neither of which is essential for propagation in tissue culture (Buller et al., 1985; Hughes et al., 1991). The 19 kDa viral thymidine kinase (TK) is a homotetramer whose activity is subject to feedback inhibition with high levels of TTP or TDP (Hruby, 1985; Wilson et al., 1989). This feedback inhibition loop can be bypassed, without compromising enzymatic activity, by altering the key glutamine residue in conserved domain IV (Black and Hruby, 1992). The ability of the viral TK to also utilize BrdU as a substrate has been exploited in the generation of recombinant vaccinia viruses. Insertion of foreign DNA into the TK locus inactivates TK, leading to a BrdU-resistant phenotype which can be selected for by propagation on human 143 TK– cells. TK– viruses are attenuated in vivo (Buller et al., 1985), underscoring the importance of the precursor biosynthetic machinery in key cell types in vivo.

The 23 kDa viral thymidylate kinase functions as a dimer to phosphorylate dTMP. The viral dTMP kinase shares ~40% identity to the human enzyme, retaining several conserved catalytic motifs and the overall conserved core structure (Topalis et al., 2005). The high degree of homology between the viral and cellular enzymes is well illustrated by the finding that A48 can complement the dTMP-deficient *Saccharomyces cerevisiae* cdc8 mutant (Hughes et al., 1991).

Ribonucleotide Reductase

The vaccinia ribonucleotide reductase (RNR) is composed of small (F4) and large (I4) subunits; it catalyzes the reduction of rNDPs to the corresponding dNDPs (Slabaugh et al., 1988; Tengelsen et al., 1988). The allosteric behavior and regulatory mechanism of the viral enzyme are comparable to those of mammalian RNRs (Slabaugh and Mathews, 1984). The large subunit (I4) contains binding sites for nucleotide substrates and allosteric effectors (Slabaugh et al., 1984; Slabaugh and Mathews, 1984), while the small subunit (F4) carries the tyrosyl radical required for catalysis (Howell et al., 1992). It is this tyrosyl radical that is affected by the DNA synthesis inhibitor hydroxyurea (HU) (Ehrenberg and Reichard, 1972). While it would be reasonable to assume that there is an equivalent depletion in the rates of formation of all four dNDPs, experimental evidence has shown that this is not the case. The most significant decrease is seen for dATP levels, and HU-mediated inhibition can be largely overcome by the addition of exogenous deoxyadenosine in the presence of an adenosine deaminase inhibitor (Slabaugh et al., 1991).

RNR-deficient vaccinia mutants replicate normally in tissue culture (Child et al., 1990), suggesting that the host enzyme can provide sufficient precursors for viral replication to proceed; these mutants are somewhat attenuated in vivo. In tissue culture, HU inhibits vaccinia replication; HU-resistant mutants can be isolated, and these display amplification (2–15 copies) of the F4L gene (Slabaugh et al., 1988).

dUTPase

As mentioned above, the D4 protein of vaccinia virus is an active uracil DNA glycosylase, which removes uracil moieties in DNA that can arise either from cytosine deamination or from the incorporation of dUTP by the viral DNA polymerase. The importance of controlling the presence of dUMP within DNA is underscored by the fact that vaccinia also encodes a dUTPase. The F2 protein of vaccinia acts as a trimer to catalyze the hydrolysis of dUTP to dUMP (Broyles, 1993; Roseman et al., 1996). In addition to minimizing the concentration of dUTP within infected cells, dUTPase activity provides increased levels of dUMP, which serves as a precursor in the synthesis of TTP. The F2 sequence shows a high degree of similarity to the human enzyme, and comparisons between the cellular and viral enzymes are facilitated by the recent solution of the crystal structure of the viral dUTPase (Samal et al., 2007).

DNA Replication Accessory Proteins

Topoisomerase

The vaccinia virus H6 protein is type 1B topoisomerase that is expressed late during infection and is encapsidated into the virion core (Bauer et al., 1977; Shaffer and Traktman, 1987; Shuman and Moss, 1987). The poxvirus topoisomerase has a number of unique properties that distinguish it from other enzymes of this class: resistance to the drug camptothecin, sensitivity to the DNA gyrase inhibitors novobiocin and coumermycin, and remarkable specificity for a pentapyrimidine target sequence for DNA cleavage (Hwang et al., 1998; Klemperer et al., 1995; Sekiguchi et al., 1996; Shuman et al., 1988; Shuman and Moss, 1987; Shuman and Prescott, 1990). An extraordinary amount of insight has been gained into the biochemical mechanism used by this enzyme; in contrast, discerning the biological role of the protein proved difficult. Roles in transcription, replication, recombination, and concatemer resolution were posited. Initially, the inability to isolate an H6 deletion mutant led to the conclusion that H6 was essential in tissue culture (Shuman et al., 1989); more recently, however, such a virus has been isolated using a different selection protocol. The topoisomerase-null virus exhibits reduced infectivity due to a diminution in early transcription; neither DNA replication nor genome maturation appears to be affected (Da Fonseca and Moss, 2003). This data suggest that the primary biological role for the vaccinia topoisomerase is to facilitate early gene transcription.

DNA Ligase

The vaccinia virus A50 protein is a 61 kDa ATP-dependent DNA ligase (Kerr and Smith, 1989) with significant similarity to the mammalian type II and III DNA ligases (Chen et al., 1995; Husain et al., 1995; Wang et al., 1994; Wei et al., 1995). The functional redundancy between the viral and cellular ligases was confirmed by the demonstration that the vaccinia DNA ligase can compensate for the loss of the *S. cerevisiae* DNA ligase (Kerr et al., 1991). A significant amount of biochemical and structure/function analyses has been performed on the viral ligase. An A50 deletion mutant has been constructed: the enzyme is not essential for propagation in tissue culture but the deletion mutant is attenuated in vivo (Colinas et al., 1990; Kerr and Smith, 1991). In some contexts, however, deletion of the ligase does compromise viral replication (Parks et al., 1998). Deletion of A50 leads to UV- and bleomycin sensitivity but to etoposide resistance; increasing the copy number of the ligase gene confers etoposide sensitivity (DeLange et al., 1995; Kerr et al., 1991). Clearly, we do not fully appreciate the role(s) that the viral DNA ligase might play during infection.

Genome Maturation

Cis-*Acting Sequences*

As described above, the initial product of genomic replication is the generation of head-to-head or tail-to-tail concatemers (DeLange, 1989; DeLange and McFadden, 1990; Merchlinsky et al., 1988; Moyer and Graves, 1981). These concatemers contain an imperfect palindromic junction that can be extruded into a cruciform structure that mimics a DNA Holliday junction (Dickie et al., 1987, 1988; Merchlinsky et al., 1988). Indeed, within infected cells, plasmids containing these inverted repeats are processed into linear minichromosomes with hairpin termini (DeLange and McFadden, 1987; DeLange et al., 1986; Merchlinsky, 1990a, b;

Merchlinsky et al., 1988; Merchlinsky and Moss, 1986, 1989a). At the concatemer junction, the extended duplex version of the hairpin sequence is flanked on each side by sequences required for resolution. Within these sequences, four *cis*-acting motifs have been defined that are either essential (domains 1A and 1) or stimulatory (domains II and III) in this resolution process. The canonical resolution sequence within domains 1A and 1 has been identified, and site-directed mutagenesis has revealed that resolution is exquisitely dependent on its sequence (DeLange and McFadden, 1987; Merchlinsky, 1990a; Merchlinsky and Moss, 1989a). The resolution sequences are strongly conserved among poxviruses (Merchlinsky, 1990a). Interestingly, the resolution sequence can also function as a late promoter, suggesting that transcription through the region might stimulate resolution (Parsons and Pickup, 1990; Stuart et al., 1991).

Resolvase

The observation that concatemer resolution fails to occur when late protein synthesis is blocked provided the initial hint that a late viral protein might mediate concatemer resolution (DeLange, 1989; Merchlinsky and Moss, 1989b). The A22 protein was subsequently identified as a likely candidate because of its homology to a known Holliday-junction resolvase, the *E. coli* Ruv C protein (Garcia et al., 2000). A22 is a 23 kDa protein that is expressed at late times during infection (Garcia and Moss, 2001), and purified recombinant A22 can bind to and cleave a synthetic Holliday junction, yielding nicked duplex molecules (Garcia et al., 2000). The protein forms a dimer, which most likely mediates the symmetrical cleavage of the concatemeric junction (Garcia et al., 2006). In vitro, the viral enzyme exhibits only weak sequence specificity, which does not explain the strict sequence specificity that characterizes resolution in vivo (Culyba et al., 2006; Garcia et al., 2006). Nevertheless, repression of A22 expression leads to a significant block to concatemer resolution: the majority of newly synthesized viral DNA remains concatemeric and virion morphogenesis is not completed (Garcia and Moss, 2001). Clearly, A22 is necessary for concatemer resolution; whether other viral proteins impart specificity to the reaction remains to be resolved.

Genome Encapsidation

To complete the infectious cycle, progeny DNA genomes must be encapsidated into nascent virions. Genome encapsidation is associated with the appearance of an electron-dense nucleoid within immature virions. Although much remains to be learned about this process, we have gained some insights into some of the proteins that participate in this process (Condit et al., 2006). Repression of the 8 kDa A13 protein, which is a component of the virion membrane, leads to a morphogenesis arrest characterized by the accumulation of immature virions lacking nucleoids and the accumulation of DNA crystalloids in the cytoplasm (Unger and Traktman, 2004). Two components of the virion core are more directly associated with genome encapsidation. The A32L gene encodes a 34 kDa protein with some sequence similarity to proteins involved in adenovirus and bacteriophage DNA encapsidation (Koonin et al., 1993). Based on conserved Walker A and Walker B motifs, the protein has been predicted to have ATPase activity; although no empirical evidence of this activity has been reported. When expression of the A32 protein is repressed, the biochemical events of the viral life cycle progress normally. The late stages of morphogenesis are aberrant; however, resulting in the production of abnormal spherical particles which are devoid of viral DNA (Cassetti et al., 1998). The 44 kDa I6 protein is also directly implicated in genome encapsidation; this protein binds to the telomeric hairpins of the genome with great specificity and stability (DeMasi et al., 2001). During nonpermissive infections with a temperature-sensitive I6 mutant, the viral life cycle again progresses normally until the later stages of morphogenesis. Again, aberrant spherical particles are formed which lack viral DNA, and DNA crystalloids accumulate in the cytoplasm (Grubisha and Traktman, 2003). The I6 protein encoded by this mutant appears to be defective in its telomere-binding capacity (Traktman, unpublished). The A32 protein is encapsidated at wild-type levels in the DNA-deficient particles that assemble during *ts*I6 infections, whereas I6 is not encapsidated in the A32-deficient particles (Traktman, unpublished). It seems plausible to propose that the binding of I6 to the telomeres of the viral genome imparts specificity to the encapsidation process and that entry of the genome maybe facilitated by I6-A32 interactions or by A32-mediated ATP hydrolysis.

Questions for Future Study

Poxvirus DNA replication is unique in that it occurs within the cytoplasm of infected cells and relies almost exclusively on viral proteins. Much has been learned about this fascinating process (Moss and De Silva, 2006; Traktman, 1996), but many unsolved puzzles remain. The development and organization of the cytoplasmic factories in which replication occurs are poorly understood. Replication is thought to initiate with the introduction of a nick, but neither the *cis-* nor *trans*-acting factors that contribute to this initiation process have been identified. Confirmation that replication relies solely on leading strand synthesis is also pending, and disparities between the specificity of concatemer resolution in vivo and the lack of specificity of the resolvase in vitro need to be understood. Processivity is conferred upon the catalytic subunit of the polymerase (E9) by a heterodimeric protein complex (A20 + UDG), one component of which is an active repair enzyme. This is a unique model for a processive holoenzyme, and both the mechanism by which processivity is engendered and the possible coupling of replication and repair remain to be elucidated. The D5 NTPase is likely to be the replicative helicase, and the I3 protein is likely to be the replicative SSB, but further studies are required to support these predictions. The viral B1 kinase plays a unique role in replication, serving to combat the antiviral action of the cellular BAF protein, which binds to the viral genome and prevents replication unless it is disarmed by phosphorylation. Encapsidation appears to rely upon some components of the virion membrane, a telomere-binding protein, and a putative ATPase, and further study of this important process is also needed.

References

Bauer, W. R., Ressner, E. C., Kates, J., and Patzke, J. V. (1977). A DNA nicking-closing enzyme encapsidated in vaccinia virus: partial purification and properties. *Proceedings of the National Academy of Sciences of the USA* **74**, 1841–1845.

Black, M. E. and Hruby, D. E. (1992). A single amino acid substitution abolishes feedback inhibition of vaccinia virus thymidine kinase. *Journal of Biological Chemistry* **267**, 9743–9748.

Boyle, K. A., Arps, L., and Traktman, P. (2007). Biochemical and genetic analysis of the vaccinia virus d5 protein: Multimerization-dependent ATPase activity is required to support viral DNA replication. *Journal of Virology* **81**, 844–859.

Boyle, K. and Traktman, P. (2004). Members of a novel family of mammalian protein kinases complement the DNA$^-$ phenotype of a vaccinia virus *ts* mutant defective in the B1 kinase. *Journal of Virology* **78**.

Broyles, S. S. (1993). Vaccinia virus encodes a functional dUTPase. *Virology* **195**, 863–865.

Buller, R. M., Smith, G. L., Cremer, K., Notkins, A. L., and Moss, B. (1985). Decreased virulence of recombinant vaccinia virus expression vectors is associated with a thymidine kinase-negative phenotype. *Nature* **317**, 813–815.

Cairns, J. (1960). The initiation of vaccinia infection. *Virology* **11**, 603–623.

Carter, G. C., Rodger, G., Murphy, B. J., Law, M., Krauss, O., Hollinshead, M., and Smith, G. L. (2003). Vaccinia virus cores are transported on microtubules. *Journal of General Virology* **84**, 2443–2458.

Cassetti, M. C., Merchlinsky, M., Wolffe, E. J., Weisberg, A. S., and Moss, B. (1998). DNA packaging mutant: repression of the vaccinia virus A32 gene results in noninfectious, DNA-deficient, spherical, enveloped particles. *Journal of Virology* **72**, 5769–5780.

Challberg, M. D. and Englund, P. T. (1979a). Purification and properties of the deoxyribonucleic acid polymerase induced by vaccinia virus. *Journal of Biological Chemistry* **254**, 7812–7819.

Challberg, M. D. and Englund, P. T. (1979b). The effect of template secondary structure on vaccinia DNA polymerase. *Journal of Biological Chemistry* **254**, 7820–7826.

Chen, J., Tomkinson, A. E., Ramos, W., Mackey, Z. B., Danehower, S., Walter, C. A., Schultz, R. A., Besterman, J. M., and Husain, I. (1995). Mammalian DNA ligase III: molecular cloning, chromosomal localization, and expression in spermatocytes undergoing meiotic recombination. *Molecular and Cellular Biology* **15**, 5412–5422.

Child, S. J., Palumbo, G. J., Buller, R. M., and Hruby, D. E. (1990). Insertional inactivation of the large subunit of ribonucleotide reductase encoded by vaccinia virus is associated with reduced virulence in vivo. *Virology* **174**, 625–629.

Colinas, R. J., Goebel, S. J., Davis, S. W., Johnson, G. P., Norton, E. K., and Paoletti, E. (1990). A DNA ligase gene in the Copenhagen strain of vaccinia virus is nonessential for viral replication and recombination. *Virology* **179**, 267–275.

Condit, R. C., Moussatche, N., and Traktman, P. (2006). In a nutshell: structure and assembly of the vaccinia virion. *Advances in Virus Research* **66**, 31–124.

Culyba, M. J., Harrison, J. E., Hwang, Y., and Bushman, F. D. (2006). DNA cleavage by the A22R resolvase of vaccinia virus. *Virology* **352**, 466–476.

Da Fonseca, F. and Moss, B. (2003). Poxvirus DNA topoisomerase knockout mutant exhibits decreased infectivity associated with reduced early transcription. *Proceedings of the National Academy of Sciences of the USA* **100**, 11291–11296.

Davis, R. E. and Mathews, C. K. (1993). Acidic C terminus of vaccinia virus DNA-binding protein interacts with ribonucleotide reductase. *Proceedings of the National Academy of Sciences of the USA* **90**, 745–749.

De Silva, F. S. and Moss, B. (2003). Vaccinia virus uracil DNA glycosylase has an essential role in DNA synthesis that is independent of its glycosylase activity: catalytic site mutations reduce virulence but not virus replication in cultured cells. *Journal of Virology* **77**, 159–166.

De Silva, F. S. and Moss, B. (2005). Origin-independent plasmid replication occurs in vaccinia virus cytoplasmic factories and requires all five known poxvirus replication factors. *Virology Journal* **2**, 23.

De Silva, F. S., Lewis, W., Berglund, P., Koonin, E. V., and Moss, B. (2007). Poxvirus DNA Primase. Proceedings of the National Academy of Sciences of the USA **104**, 18724–18729.

DeFilippes, F. M. (1984). Effect of aphidicolin on vaccinia virus: isolation of an aphidicolin-resistant mutant. *Journal of Virology* **52**, 474–482.

DeFilippes, F. M. (1989). Site of the base change in the vaccinia virus DNA polymerase gene which confers aphidicolin resistance. *Journal of Virology* **63**, 4060–4063.

DeLange, A. M. (1989). Identification of temperature-sensitive mutants of vaccinia virus that are defective in conversion of concatemeric replicative intermediates to the mature linear DNA genome. *Journal of Virology* **63**, 2437–2444.

DeLange, A. M., Carpenter, M. S., Choy, J., and Newsway, V. E. (1995). An etoposide-induced block in vaccinia virus telomere resolution is dependent on the virus-encoded DNA ligase. *Journal of Virology* **69**, 2082–2091.

DeLange, A. M. and McFadden, G. (1987). Efficient resolution of replicated poxvirus telomeres to native hairpin structures requires two inverted symmetrical copies of a core target DNA sequence. *Journal of Virology* **61**, 1957–1963.

DeLange, A. M. and McFadden, G. (1990). The role of telomeres in poxvirus DNA replication. *Current Topics in Microbiology and Immunology* **163**, 71–92.

DeLange, A. M., Reddy, M., Scraba, D., Upton, C., and McFadden, G. (1986). Replication and resolution of cloned poxvirus telomeres in vivo generates linear minichromosomes with intact viral hairpin termini. *Journal of Virology* **59**, 249–259.

DeMasi, J., Du, S., Lennon, D., and Traktman, P. (2001). Vaccinia virus telomeres: interaction with the viral I1, I6, and K4 proteins. *Journal of Virology* **75**, 10090–10105.

Dickie, P., Morgan, A. R., and McFadden, G. (1987). Cruciform extrusion in plasmids bearing the replicative intermediate configuration of a poxvirus telomere. *Journal of Molecular Biology* **196**, 541–558.

Dickie, P., Morgan, A. R., and McFadden, G. (1988). Conformational isomerization of the Holliday junction associated with a cruciform during branch migration in supercoiled plasmid DNA. *Journal of Molecular Biology* **201**, 19–30.

Domi, A. and Beaud, G. (2000). The punctate sites of accumulation of vaccinia virus early proteins are precursors of sites of viral DNA synthesis. *Journal of General Virology* **81 Pt 5**, 1231–1235.

Du, S. and Traktman, P. (1996). Vaccinia virus DNA replication: two hundred base pairs of telomeric sequence confer optimal replication efficiency on minichromosome templates. *Proceedings of the National Academy of Sciences of the USA* **93**, 9693–9698.

Earl, P. L., Jones, E. V., and Moss, B. (1986). Homology between DNA polymerases of poxviruses, herpesviruses, and adenoviruses: nucleotide sequence of the vaccinia virus DNA polymerase gene. *Proceedings of the National Academy of Sciences of the USA* **83**, 3659–3663.

Ehrenberg, A. and Reichard, P. (1972). Electron spin resonance of the iron-containing protein B2 from ribonucleotide reductase. *Journal of Biological Chemistry* **247**, 3485–3488.

Esteban, M. and Holowczak, J. A. (1977). Replication of vaccinia DNA in mouse L cells. III. Intracellular forms of viral DNA. *Virology* **82**, 308–322.

Evans, E., Klemperer, N., Ghosh, R., and Traktman, P. (1995). The vaccinia virus D5 protein, which is required for DNA replication, is a nucleic acid-independent nucleoside triphosphatase. *Journal of Virology* **69**, 5353–5361.

Evans, E. and Traktman, P. (1992). Characterization of vaccinia virus DNA replication mutants with lesions in the D5 gene. *Chromosoma* **102**, S72–S82.

Garcia, A. D., Aravind, L., Koonin, E. V., and Moss, B. (2000). Bacterial-type DNA Holliday junction resolvases in eukaryotic viruses [see comments]. *Proceedings of the National Academy of Sciences of the USA* **97**, 8926–8931.

Garcia, A. D. and Moss, B. (2001). Repression of vaccinia virus Holliday junction resolvase inhibits processing of viral DNA into unit-length genomes. *Journal of Virology* **75**, 6460–6471.

Garcia, A. D., Otero, J., Lebowitz, J., Schuck, P., and Moss, B. (2006). Quaternary structure and cleavage specificity of a poxvirus Holliday junction resolvase. *Journal of Biological Chemistry* **281**, 11618–11626.

Goebel, S. J., Johnson, G. P., Perkus, M. E., Davis, S. W., Winslow, J. P., and Paoletti, E. (1990). The complete DNA sequence of vaccinia virus. *Virology* **179**, 247–263.

Grubisha, O. and Traktman, P. (2003). Genetic analysis of the vaccinia virus I6 telomere-binding protein uncovers a key role in genome encapsidation. *Journal of Virology* **77**, 10929–10942.

Hamilton, M. D. and Evans, D. H. (2005). Enzymatic processing of replication and recombination intermediates by the vaccinia virus DNA polymerase. *Nucleic Acids Research* **33**, 2259–2268.

Howell, M. L., Sanders-Loehr, J., Loehr, T. M., Roseman, N. A., Mathews, C. K., and Slabaugh, M. B. (1992). Cloning of the vaccinia virus ribonucleotide reductase small subunit gene. Characterization of the gene product expressed in Escherichia coli. *Journal of Biological Chemistry* **267**, 1705–1711.

Hruby, D. E. (1985). Inhibition of vaccinia virus thymidine kinase by the distal products of its own metabolic pathway. *Virus Research* **2**, 151–156.

Hughes, S. J., Johnston, L. H., de Carlos, A., and Smith, G. L. (1991). Vaccinia virus encodes an active thymidylate kinase that complements a cdc8 mutant of Saccharomyces cerevisiae. *Journal of Biological Chemistry* **266**, 20103–20109.

Husain, I., Tomkinson, A. E., Burkhart, W. A., Moyer, M. B., Ramos, W., Mackey, Z. B., Besterman, J. M., and Chen, J. (1995). Purification and characterization of DNA ligase III from bovine testes. Homology with DNA ligase II and vaccinia DNA ligase. *Journal of Biological Chemistry* **270**, 9683–9690.

Hwang, Y., Wang, B., and Bushman, F. D. (1998). Molluscum contagiosum virus topoisomerase: purification, activities, and response to inhibitors. *Journal of Virology* **72**, 3401–3406.

Ishii, K. and Moss, B. (2001). Role of Vaccinia Virus A20R Protein in DNA Replication: Construction and Characterization of Temperature-Sensitive Mutants. *Journal of Virology* **75**, 1656–1663.

Ishii, K. and Moss, B. (2002). Mapping interaction sites of the A20R protein component of the vaccinia virus DNA replication complex. *Virology* **303**, 232–239.

Iyer, L. M., Aravind, L., and Koonin, E. V. (2001). Common origin of four diverse families of large eukaryotic DNA viruses. *Journal of Virology* **75**, 11720–11734.

Iyer, L. M., Koonin, E. V., Leipe, D. D., and Aravind, L. (2005). Origin and evolution of the archaeo-eukaryotic primase superfamily and related palm-domain proteins: structural insights and new members. *Nucleic Acids Research* **33**, 3875–3896.

Joklik, W. K. and Becker, Y. (1964). The replication and coating of vaccinia DNA. *Journal of Molecular Biology* **10**, 452–474.

Kerr, S. M., Johnston, L. H., Odell, M., Duncan, S. A., Law, K. M., and Smith, G. L. (1991). Vaccinia DNA ligase complements Saccharomyces cerevisiae cdc9, localizes in cytoplasmic factories and affects virulence and virus sensitivity to DNA damaging agents. *EMBO Journal* **10**, 4343–4350.

Kerr, S. M. and Smith, G. L. (1989). Vaccinia virus encodes a polypeptide with DNA ligase activity. *Nucleic Acids Research* **17**, 9039–9050.

Kerr, S. M. and Smith, G. L. (1991). Vaccinia virus DNA ligase is nonessential for virus replication: recovery of plasmids from virus-infected cells. *Virology* **180**, 625–632.

Klemperer, N., Lyttle, D. J., Tauzin, D., Traktman, P., and Robinson, A. J. (1995). Identification and characterization of the orf virus type I topoisomerase. *Virology* **206**, 203–215.

Klemperer, N., McDonald, W., Boyle, K., Unger, B., and Traktman, P. (2001). The A20R protein is a stoichiometric component of the processive form of vaccinia virus DNA polymerase. *Journal of Virology* **75**, 12298–12307.

Koonin, E. V., Senkevich, T. G., and Chernos, V. I. (1993). Gene A32 product of vaccinia virus may be an ATPase involved in viral DNA packaging as indicated by sequence comparisons with other putative viral ATPases. *Virus Genes* **7**, 89–94.

Kornbluth, R. S., Smee, D. F., Sidwell, R. W., Snarsky, V., Evans, D. H., and Hostetler, K. Y. (2006). Mutations in the E9L polymerase gene of cidofovir-resistant vaccinia virus strain WR are associated with the drug resistance phenotype. *Antimicrobial Agents and Chemotherapy* **50**, 4038–4043.

Kovacs, G. R., Vasilakis, N., and Moss, B. (2001). Regulation of viral intermediate gene expression by the vaccinia virus B1 protein kinase. *Journal of Virology* **75**, 4048–4055.

Magee, W. C., Hostetler, K. Y., and Evans, D. H. (2005). Mechanism of inhibition of vaccinia virus DNA polymerase by cidofovir diphosphate. *Antimicrobial Agents and Chemotherapy* **49**, 3153–3162.

Mallardo, M., Schleich, S., and Krijnse, L. J. (2001). Microtubule-dependent organization of vaccinia virus core-derived early mRNAs into distinct cytoplasmic structures. *Molecular Biology of the Cell* **12**, 3875–3891.

McCraith, S., Holtzman, T., Moss, B., and Fields, S. (2000). Genome-wide analysis of vaccinia virus protein-protein interactions. *Proceedings of the National Academy of Sciences of the USA* **97**, 4879–4884.

McDonald, W. F., Klemperer, N., and Traktman, P. (1997). Characterization of a processive form of the vaccinia virus DNA polymerase. *Virology* **234**, 168–175.

McDonald, W. F. and Traktman, P. (1994). Vaccinia virus DNA polymerase. In vitro analysis of parameters affecting processivity. *Journal of Biological Chemistry* **269**, 31190–31197.

McFadden, G. and Dales, S. (1980). Biogenesis of poxviruses: preliminary characterization of conditional lethal mutants of vaccinia virus defective in DNA synthesis. *Virology* **103**, 68–79.

Merchlinsky, M. (1990a). Mutational analysis of the resolution sequence of vaccinia virus DNA: essential sequence consists of two separate AT-rich regions highly conserved among poxviruses. *Journal of Virology* **64**, 5029–5035.

Merchlinsky, M. (1990b). Resolution of poxvirus telomeres: processing of vaccinia virus concatemer junctions by conservative strand exchange. *Journal of Virology* **64**, 3437–3446.

Merchlinsky, M., Garon, C. F., and Moss, B. (1988). Molecular cloning and sequence of the concatemer junction from vaccinia virus replicative DNA. Viral nuclease cleavage sites in cruciform structures. *Journal of Molecular Biology* **199**, 399–413.

Merchlinsky, M. and Moss, B. (1986). Resolution of linear minichromosomes with hairpin ends from circular plasmids containing vaccinia virus concatemer junctions. *Cell* **45**, 879–884.

Merchlinsky, M. and Moss, B. (1989a). Nucleotide sequence required for resolution of the concatemer junction of vaccinia virus DNA. *Journal of Virology* **63**, 4354–4361.

Merchlinsky, M. and Moss, B. (1989b). Resolution of vaccinia virus DNA concatemer junctions requires late-gene expression. *Journal of Virology* **63**, 1595–1603.

Moss, B. and De Silva, F. S. (2006). Poxvirus. *In* "DNA Replication and Human Disease" (DePamphilis, Ed.), pp. 707–727. Cold Spring Harbor Laboratory Press, Cold Spring Harbor, NY.

Moyer, R. W. and Graves, R. L. (1981). The mechanism of cytoplasmic orthopoxvirus DNA replication. *Cell* **27**, 391–401.

Nezu, J., Oku, A., Jones, M. H., and Shimane, M. (1997). Identification of two novel human putative serine/threonine kinases, VRK1 and VRK2, with structural similarity to vaccinia virus B1R kinase. *Genomics* **45**, 327–331.

Nichols, R. J. and Traktman, P. (2004). Characterization of three paralogous members of the Mammalian vaccinia related kinase family. *Journal of Biological Chemistry* **279**, 7934–7946.

Nichols, R. J., Wiebe, M. S., and Traktman, P. (2006). The vaccinia-related kinases phosphorylate the N′ terminus of BAF, regulating its interaction with DNA and its retention in the nucleus. *Molecular Biology of the Cell* **17**, 2451–2464.

Parks, R. J., Winchcombe-Forhan, C., DeLange, A. M., Xing, X., and Evans, D. H. (1998). DNA ligase gene disruptions can depress viral growth and replication in poxvirus-infected cells. *Virus Research* **56**, 135–147.

Parsons, B. L. and Pickup, D. J. (1990). Transcription of orthopoxvirus telomeres at late times during infection. *Virology* **175**, 69–80.

Pogo, B. G. (1977). Elimination of naturally occurring crosslinks in vaccinia virus DNA after viral penetration into cells. *Proceedings of the National Academy of Sciences of the USA* **74**, 1739–1742.

Pogo, B. G., Berkowitz, E. M., and Dales, S. (1984). Investigation of vaccinia virus DNA replication employing a conditional lethal mutant defective in DNA. *Virology* **132**, 436–444.

Pogo, B. G., O'Shea, M., and Freimuth, P. (1981). Initiation and termination of vaccinia virus DNA replication. *Virology* **108**, 241–248.

Punjabi, A., Boyle, K., DeMasi, J., Grubisha, O., Unger, B., Khanna, M., and Traktman, P. (2001). Clustered charge-to-alanine mutagenesis of the vaccinia virus A20 gene: temperature-sensitive mutants have a DNA-minus phenotype and are defective in the production of processive DNA polymerase activity. *Journal of Virology* **75**, 12308–12318.

Raoult, D., Audic, S., Robert, C., Abergel, C., Renesto, P., Ogata, H., La Scola, B., Suzan, M., and Claverie, J. M. (2004). The 1.2-megabase genome sequence of Mimivirus. *Science* **306**, 1344–1350.

Rempel, R. E., Anderson, M. K., Evans, E., and Traktman, P. (1990). Temperature-sensitive vaccinia virus mutants identify a gene with an essential role in viral replication. *Journal of Virology* **64**, 574–583.

Rochester, S. C. and Traktman, P. (1998). Characterization of the single-stranded DNA binding protein encoded by the vaccinia virus I3 gene. *Journal of Virology* **72**, 2917–2926.

Roseman, N. A., Evans, R. K., Mayer, E. L., Rossi, M. A., and Slabaugh, M. B. (1996). Purification and characterization of the vaccinia virus deoxyuridine triphosphatase expressed in Escherichia coli. *Journal of Biological Chemistry* **271**, 23506–23511.

Salzman, N. P. (1960). The rate of formation of vaccinia deoxyribonucleic acid and vaccinia virus. *Virology* **10**, 150–152.

Samal, A., Schormann, N., Cook, W. J., DeLucas, L. J., and Chattopadhyay, D. (2007). Structures of vaccinia virus dUTPase and its nucleotide complexes. *Acta Crystallogr D Biol Crystallogr* **63**, 571–580.

Scaramozzino, N., Sanz, G., Crance, J. M., Saparbaev, M., Drillien, R., Laval, J., Kavli, B., and Garin, D. (2003). Characterisation of the substrate specificity of homogeneous vaccinia virus uracil-DNA glycosylase. *Nucleic Acids Research* **31**, 4950–4957.

Schormann, N., Grigorian, A., Samal, A., Krishnan, R., DeLucas, L., and Chattopadhyay, D. (2007). Crystal structure of vaccinia virus uracil-DNA glycosylase reveals dimeric assembly. *BMC Structural Biology* **7**, 45.

Sekiguchi, J., Stivers, J. T., Mildvan, A. S., and Shuman, S. (1996). Mechanism of inhibition of vaccinia DNA topoisomerase by novobiocin and coumermycin. *Journal of Biological Chemistry* **271**, 2313–2322.

Shaffer, R. and Traktman, P. (1987). Vaccinia virus encapsidates a novel topoisomerase with the properties of a eucaryotic type I enzyme. *Journal of Biological Chemistry* **262**, 9309–9315.

Shuman, S., Golder, M., and Moss, B. (1988). Characterization of vaccinia virus DNA topoisomerase I expressed in Escherichia coli. *Journal of Biological Chemistry* **263**, 16401–16407.

Shuman, S., Golder, M., and Moss, B. (1989). Insertional mutagenesis of the vaccinia virus gene encoding a type I DNA topoisomerase: evidence that the gene is essential for virus growth. *Virology* **170**, 302–306.

Shuman, S. and Moss, B. (1987). Identification of a vaccinia virus gene encoding a type I DNA topoisomerase. *Proceedings of the National Academy of Sciences of the USA* **84**, 7478–7482.

Shuman, S. and Prescott, J. (1990). Specific DNA cleavage and binding by vaccinia virus DNA topoisomerase I. *Journal of Biological Chemistry* **265**, 17826–17836.

Slabaugh, M. B., Howell, M. L., Wang, Y., and Mathews, C. K. (1991). Deoxyadenosine reverses hydroxyurea inhibition of vaccinia virus growth. *Journal of Virology* **65**, 2290–2298.

Slabaugh, M. B., Johnson, T. L., and Mathews, C. K. (1984). Vaccinia virus induces ribonucleotide reductase in primate cells. *Journal of Virology* **52**, 507–514.

Slabaugh, M. B. and Mathews, C. K. (1984). Vaccinia virus-induced ribonucleotide reductase can be distinguished from host cell activity. *Journal of Virology* **52**, 501–506.

Slabaugh, M. B., Roseman, N., and Mathews, C. (1988). Vaccinia virus encoded ribonucleotide reductase: sequence conservation of the gene for the small subunit and its amplification in hydroxyurea resistant mutants. *Journal of Virology* **62**, 519–527.

Smith, G. L., Vanderplasschen, A., and Law, M. (2002). The formation and function of extracellular enveloped vaccinia virus. *Journal of General Virology* **83**, 2915–2931.

Stanitsa, E. S., Arps, L., and Traktman, P. (2006). Vaccinia virus uracil DNA glycosylase interacts with the A20 protein to form a heterodimeric processivity factor for the viral DNA polymerase. *Journal of Biological Chemistry* **281**, 3439–3451.

Stuart, D., Graham, K., Schreiber, M., Macaulay, C., and McFadden, G. (1991). The target DNA sequence for resolution of poxvirus replicative intermediates is an active late promoter. *Journal of Virology* **65**, 61–70.

Stuart, D. T., Upton, C., Higman, M. A., Niles, E. G., and McFadden, G. (1993). A poxvirus-encoded uracil DNA glycosylase is essential for virus viability. *Journal of Virology* **67**, 2503–2512.

Taddie, J. A. and Traktman, P. (1991). Genetic characterization of the vaccinia virus DNA polymerase: identification of point mutations conferring altered drug sensitivities and reduced fidelity. *Journal of Virology* **65**, 869–879.

Taddie, J. A. and Traktman, P. (1993). Genetic characterization of the vaccinia virus DNA polymerase: cytosine arabinoside resistance requires a variable lesion conferring phosphonoacetate resistance in conjunction with an invariant mutation localized to the 3′-5′ exonuclease domain. *Journal of Virology* **67**, 4323–4336.

Tengelsen, L. A., Slabaugh, M. B., Bibler, J. K., and Hruby, D. E. (1988). Nucleotide sequence and molecular genetic analysis of the large subunit of ribonucleotide reductase encoded by vaccinia virus. *Virology* **164**, 121–131.

Tolonen, N., Doglio, L., Schleich, S., and Krijnse-Locker, J. (2001). Vaccinia virus DNA replication occurs in endoplasmic reticulum-enclosed cytoplasmic mini-nuclei. *Molecular Biology of the Cell* **12**, 2031–2046.

Topalis, D., Collinet, B., Gasse, C., Dugue, L., Balzarini, J., Pochet, S., and Deville-Bonne, D. (2005). Substrate specificity of vaccinia virus thymidylate kinase. *FEBS Journal* **272**, 6254–6265.

Traktman, P. (1996). Poxvirus DNA replication. *In* "DNA Replication in Eukaryotic Cells" pp. 775–798.

Traktman, P., Kelvin, M., and Pacheco, S. (1989). Molecular genetic analysis of vaccinia virus DNA polymerase mutants. *Journal of Virology* **63**, 841–846.

Tseng, M., Palaniyar, N., Zhang, W., and Evans, D. H. (1999). DNA binding and aggregation properties of the vaccinia virus I3L gene product. *Journal of Biological Chemistry* **274**, 21637–21644.

Unger, B. and Traktman, P. (2004). Vaccinia virus morphogenesis: A13 phosphoprotein is required for assembly of mature virions. *Journal of Virology* **78**, 8885–8901.

Upton, C., Stuart, D. T., and McFadden, G. (1993). Identification of a poxvirus gene encoding a uracil DNA glycosylase. *Proceedings of the National Academy of Sciences of the USA* **90**, 4518–4522.

Wang, Y. C., Burkhart, W. A., Mackey, Z. B., Moyer, M. B., Ramos, W., Husain, I., Chen, J., Besterman, J. M., and Tomkinson, A. E. (1994). Mammalian DNA ligase II is highly homologous with vaccinia DNA ligase. Identification of the DNA ligase II active site for enzyme-adenylate formation. *Journal of Biological Chemistry* **269**, 31923–31928.

Wei, Y. F., Robins, P., Carter, K., Caldecott, K., Pappin, D. J., Yu, G. L., Wang, R. P., Shell, B. K., Nash, R. A., and Schar, P. (1995). Molecular cloning and expression of human cDNAs encoding a novel DNA ligase IV and DNA ligase III, an enzyme active in DNA repair and recombination. *Molecular and Cellular Biology* **15**, 3206–3216.

Welsch, S., Doglio, L., Schleich, S., and Krijnse, L. J. (2003). The vaccinia virus I3L gene product is localized to a complex endoplasmic reticulum-associated structure that contains the viral parental DNA. *Journal of Virology* **77**, 6014–6028.

Wiebe, M. S. and Traktman, P. (2007). Poxviral B1 Kinase Overcomes Barrier to Autointegration Factor, a Host Defense against Virus Replication. *Cell Host and Microbe* **1**, 187–197.

Willer, D. O., Mann, M. J., Zhang, W., and Evans, D. H. (1999). Vaccinia virus DNA polymerase promotes DNA pairing and strand-transfer reactions. *Virology* **257**, 511–523.

Willer, D. O., Yao, X. D., Mann, M. J., and Evans, D. H. (2000). In vitro concatemer formation catalyzed by vaccinia virus DNA polymerase [In Process Citation]. *Virology* **278**, 562–569.

Wilson, E. M., Franke, C. A., Black, M. E., and Hruby, D. E. (1989). Expression vector pT7:TKII for the synthesis of authentic biologically active RNA encoding vaccinia virus thymidine kinase. *Gene* **77**, 69–78.

Zelko, I., Kobayashi, R., Honkakoski, P., and Negishi, M. (1998). Molecular cloning and characterization of a novel nuclear protein kinase in mice. *Archives of Biochemistry and Biophysics.* **352**, 31–36.

Chapter 13
Herpesvirus Genome Replication

Sandra K. Weller

The Herpesviruses

The Herpesviridae are a large family of enveloped, double-stranded DNA viruses that are responsible for many human and veterinary diseases. Herpesviruses can infect mammals, birds, and reptiles, and so far, eight distinct family members have been found which infect humans including herpes simplex virus type 1 (HSV-1) and herpes simplex virus type 2 (HSV-2), Epstein–Barr virus (EBV), cytomegalovirus (HCMV), varicella-zoster virus (VZV), human herpesvirus 6 (HHV-6), human herpesvirus 7 (HHV-7), and Kaposi's sarcoma herpesvirus (KSHV). All members of the family are capable of both lytic and latent infections although they differ greatly in tissue tropism and in many aspects of their interactions with their hosts. These viruses share many aspects of virion structure (they all are $T = 16$), genomic organization, mechanisms of DNA replication and life cycle. Herpes simplex viruses 1 and 2 (HSV-1 and HSV-2) are the most extensively studied of all the herpesviruses, in part because they are most amenable to genetic and biochemical approaches. This chapter will focus primarily on herpes simplex virus type 1 (HSV-1); however, other human herpesviruses will be discussed when their replication strategy differs in significant detail from simplex viruses.

Overall Virus Life Cycle

HSV initiates viral infection by specific binding of viral glycoproteins to cell surface glycosaminoglycans and cellular receptors (Spear 2004). Following entry, capsids have been shown to translocate along microtubules (Sodeik et al. 1997) to the nuclear pores where they dock and presumably eject their genomes into the nucleus (Ojala et al. 2000). A tightly regulated cascade of gene expression occurs

S.K. Weller (✉)
Department of Molecular, Microbial and Structural Biology, University of Connecticut Health Center, 263 Farmington Avenue, Farmington, CT 06030-3205, USA
e-mail: Weller@nso2.uchc.edu

C.E. Cameron et al. (eds.), *Viral Genome Replication*,
DOI 10.1007/b135974_13, © Springer Science+Business Media, LLC 2009

consisting of three well-defined kinetic classes of genes: immediate early, early, and late. Viral gene expression, DNA replication, and encapsidation occur within globular domains in the nucleus termed replication compartments (Knipe 1989; Lamberti and Weller 1996). Herpesvirus genomes replicate in the nucleus through the formation of longer-than-unit length concatemers. Monomeric units are cleaved from concatemers during a packaging reaction which occurs in conjunction with the uptake of genomes into preassembled capsids. DNA-containing capsids exit the nucleus by budding through the nuclear membrane, and they acquire their final envelope and mature glycoproteins by a series of envelopment and de-envelopment steps (reviewed in [Baines and Weller 2004]). Unlike other enveloped viruses, herpesviruses are known to package approximately 20–30 proteins between the capsid and the envelope in a region termed the tegument or "skin". Many tegument proteins are known to play active roles in the earliest stages of infection as they are brought in with infecting virions into newly infected cells and play regulatory roles in shut off of host protein synthesis and the stimulation of immediate-early viral gene expression.

Herpesvirus Genomic Structure

Herpesvirus genomes vary in size from 125 to almost 250 kbp and contain both unique and repeated regions. Most herpesviruses contain direct terminal repeats and many also contain repeated sequences in an inverse orientation internally. The HSV genome (152 kbp) contains two unique regions (U_L and U_S) flanked by inverted repeat sequences (Hayward et al. 1975). The *a* sequence is present in three locations in the viral genome: at both termini and an inverted copy is present at the U_L–U_S junction; the *b* sequences flank the U_L segment and the *c* sequences flank U_S (Fig. 13.1A). The HSV genome undergoes genomic inversions in which the unique regions, U_L and U_S, invert with respect to one another during replication. Inversion events may be related to the propensity of HSV to undergo recombination described below.

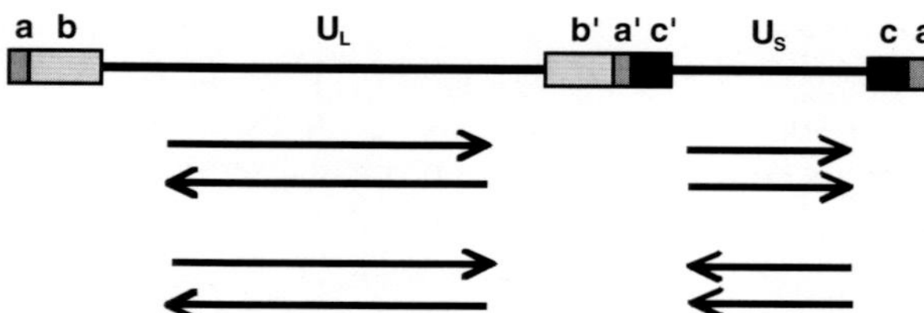

Fig. 13.1 HSV genome. The HSV-1 genome consists of two unique regions UL and US flanked by repeated sequences. UL is flanked by *ab* and *b'a'*, and US is flanked by *a'c'* and *ca*. During infection, the two unique regions invert relative to each other. The arrows reflect possible orientations of the UL and US segments as a result of genomic inversion.

Gene Expression and Regulation

HSV-1 is believed to encode over 80 open reading frames, and gene expression is very tightly temporally controlled. Viral genes are classified as immediate early, early, or late and are transcribed from both strands of the viral genome. Most of the immediate-early genes encode regulatory proteins while the early gene products are primarily involved in viral DNA replication. Many of the structural proteins are encoded as late genes which are not expressed until viral DNA synthesis has occurred.

Cis- and *Trans*-Acting DNA Replication Factors

Origins of Replication

Three origins of replication have been mapped on the HSV-1 genome: two copies of OriS within the repeated *c* region and one copy of OriL within the unique long region (reviewed in [Challberg 1996]). Both origins contain binding sites for the origin-recognition protein, UL9. OriS consists of a 45-bp palindrome containing an A/T-rich region flanked by two recognition sites for UL9 (box I and box II). A third weaker binding site (box III) is located to the left of box 1 outside the palindromic region (Fig. 13.2A). OriL consists of a longer perfect palindrome, 144 bp in length, that contains four recognition sites for UL9 (two copies of box I and two copies if box III) (Fig. 13.2B). Both oriL and oriS are located in the promoter-regulatory regions of divergently transcribed genes: oriS is positioned between two immediate-early genes, ICP4 and ICP22/27, while oriL is located between two early genes, ICP8 (UL29) and the catalytic subunit of the polymerase (UL30). OriL and one copy of oriS can be deleted without affecting the ability of the virus to multiply, suggesting that viral replication can occur in genomes containing only one copy of the origin (Igarashi et al. 1993; Polvino-Bodnar et al. 1987). The origins of replication of other alpha-herpesviruses are similar to those of HSV-1 and HSV-2; however, the origins of replication of the beta- and gamma-herpesviruses are much longer and more complex (Anders and McCue 1996; Yates 1996).

Trans-*Acting HSV-1 Replication Proteins*

The HSV-1 genome encodes seven essential replication proteins and several non-essential replication proteins. The seven essential replication proteins include a single-strand DNA-binding protein (known as ICP8 or UL29), a two subunit DNA polymerase (UL30 and UL42), a three subunit helicase/primase complex (UL5, UL8, and UL52), and an origin-binding protein UL9 (reviewed in [Chattopadhyay et al. 2006; Marintcheva and Weller 2001a; Weller and Coen 2006]). Interestingly, homologs of the first six of these are also encoded by all other human and animal

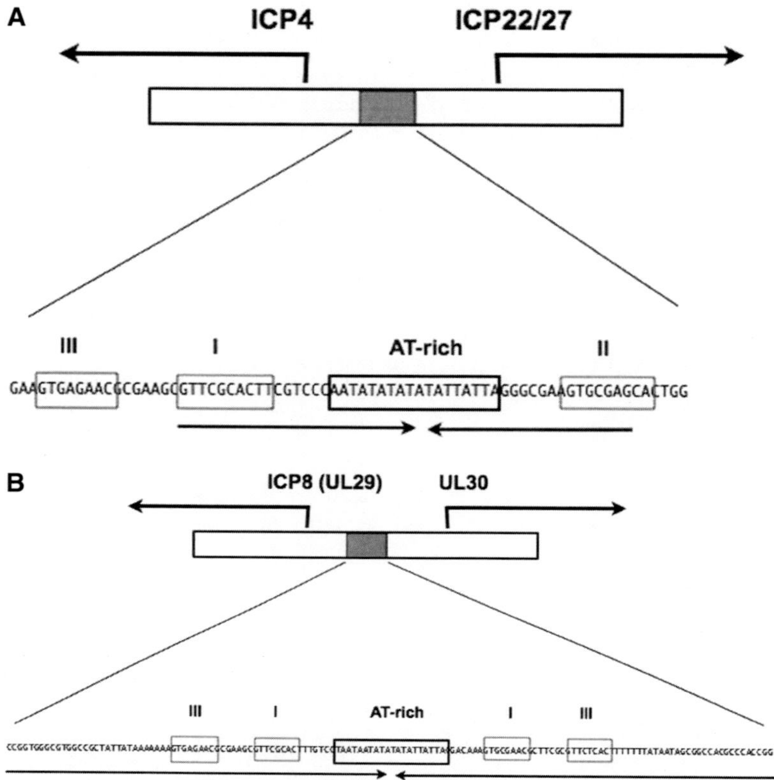

Fig. 13.2 Origins of DNA replication. **A**. OriS can be depicted with three recognition sites for UL9, the origin-binding protein (boxes I, II and III, marked in *gray*) and an AT-rich linker (marked in *black*) positioned between boxes I and II. OriS is positioned between two immediate-early transcripts. **B**. OriL is positioned between two divergently transcribed early mRNAs. Recognition sites for UL9 are boxed in *gray*, and the AT-rich regions are boxed in *black*. Since OriL is a perfect palindrome, the recognition sequences are designated I and III on each side of the palindrome.

herpesviruses and are considered to function as the "core" replication proteins with all the necessary enzymatic activities to stimulate DNA replication on a primed in vitro replication substrate. The conservation of functions suggests that the overall strategy of lytic viral DNA replication is conserved within this family. The mechanism and regulation of initiation of viral DNA synthesis, however, is probably shared only by the alpha-herpesviruses. As mentioned above, the origins of replication for beta- and gamma-herpesviruses are more complex, and in addition, no clear UL9 homologs have been identified.

In addition to the six core replication proteins and the origin-recognition protein, HSV encodes a number of proteins which are not essential for viral DNA synthesis including enzymes involved in nucleotide biosynthesis and DNA metabolism such as thymidine kinase, ribonucleotide reductase, uracil–DNA–glycosylase,

Table 13.1 Auxiliary HSV DNA replication genes

Gene	Alternate abbreviation	Major function	Essential for DNA replication in cultured cells?
UL23	TK	Thymidine kinase	No
UL39	RR1	Large subunit of ribonucleotide reductase	No
UL40	RR2	Small subunit of ribonucleotide reductase	No
UL2	UNG	Uracil-DNA glycosylase	No
UL50	dUTPase	Deoxyuridine triphosphatase	No
UL12	Alkaline exonuclease	Putative viral recombinase subunit	No

deoxyuridine triphosphatase, and the alkaline nuclease (Table 13.1) (reviewed in [Weller and Coen 2006]). Ribonucleotide reductase consists of two subunits, RR1 and RR2, and both are needed for enzymatic activity whose function is to produce dNTPs used in DNA synthesis. Recent reports raise the interesting possibility that the HSV RR1 subunit may also act as a chaperone perhaps to prevent the induction of apoptosis and/or to promote the assembly of the translational machinery (Chabaud et al. 2003; Langelier et al. 2002; Perkins et al. 2002; Walsh and Mohr 2006). The deoxyuridine triphosphatase (dUTPase) and uracil–DNA–glycosylase may be important in preventing misincorporation of uracil residues into the viral genome. The HSV alkaline nuclease (UL12) is a 5′ to 3′ exonuclease; in combination with ICP8, UL12 is capable of a strand exchange activity in vitro, and these two proteins may play a role in single-strand annealing (SSA) during infection (Reuven et al. 2004a, 2003) (see below).

HSV-1 Origin-Binding Protein, UL9 (94 kDa)

Genetic analysis indicates that the origin-binding protein UL9 is essential for viral DNA replication (Carmichael et al. 1988). The analysis of temperature-sensitive (*ts*) mutants indicates, however, that UL9 is required early in HSV-1 infection but not late in infection, once DNA synthesis has initiated (Schildgen et al. 2005). Activities associated with UL9 include nucleoside triphosphatase, DNA helicase on partially double stranded substrates, ability to form dimers in solution and ability to bind cooperatively to viral origins (reviewed in (Chattopadhyay et al. 2006; Weller & Coen 2006)). The seven conserved helicase motif characteristic of the SF2 family of helicases reside in the N-terminal domain (residues 1–534) while the domain responsible for specific origin binding has been mapped to the C-terminal one-third of UL9 (residues 564–832) (reviewed in [Chattopadhyay et al. 2006]). Genetic analysis demonstrated that the conserved helicase motifs are essential for both the in vivo and the in vitro ATPase and helicase activities of UL9 (Malik and Weller 1996; Marintcheva and Weller 2003a, b, 2001b). UL9 has been reported to interact with several other viral proteins including ICP8, UL8 (a component of the trimeric helicase/primase), and UL42 (the DNA polymerase accessory protein) (reviewed in [Weller and Coen 2006]).

UL30/UL42

HSV DNA polymerase comprises a catalytic subunit (UL30) and an accessory subunit which stimulates processivity (UL42) (Anders and McCue 1996). The catalytic subunit, UL30, is a member of the alpha-like DNA polymerase family and possesses a 3′–5′ exonuclease activity. Thus, UL30 contains motifs conserved in other polymerases, and mutations in these motifs have been shown to affect binding of dNTPs and/or their incorporation (Huang et al. 1999). The crystal structure of UL30 has been solved confirming that this protein resembles other alpha-like DNA polymerases (Liu et al. 2006).

The processivity subunit UL42 functions by an unusual mechanism. While other processivity subunits such as PCNA operate as "sliding clamps", which form multimeric rings in solution and are loaded onto DNA with the aid of clamp loaders, UL42 binds to DNA by itself as a monomer with relatively high affinity (Randell and Coen 2004; Weisshart et al. 1999). UL42 can diffuse linearly (slide) on DNA despite its high affinity for DNA (Randell and Coen 2001) and has recently been shown to affect replication fidelity (Jiang et al. 2007). Interestingly, the crystal structure of UL42 suggests that it shares a similar overall structural fold with the processivity factors which function as sliding clamps (Zuccola et al. 2000).

UL5/UL8/UL52

The HSV-1 helicase/primase is a heterotrimer consisting of the products of the UL5, UL8, and UL52 genes, and genetic data indicate that all three are essential for viral DNA replication in cell culture (reviewed in [Marintcheva and Weller 2001a; Weller and Coen 2006]). The HSV-1 UL5/8/52 complex exhibits DNA-dependent ATPase, primase, and helicase activities (reviewed in [Chattopadhyay et al. 2006; Weller and Coen 2006]). UL5 and UL52 together possess all known enzymatic activities while UL8 appears to play a stimulatory role in addition to being able to interact with other members of the replication machinery including UL9, HSV Pol, and ICP8 (reviewed in [Chattopadhyay et al. 2006; Weller and Coen 2006]). The sequence of UL5 reveals seven motifs found in a large helicase superfamily, SF1, and genetic analysis indicates that these motifs are essential for viral DNA replication (Zhu and Weller 1992) and for helicase and ATPase activities in vitro (Graves-Woodward et al. 1997; Graves-Woodward and Weller 1996). UL52 contains a DXD signature motif conserved in many primases, and this motif is essential for viral DNA replication and for primase activity in vitro (Dracheva et al. 1995; Klinedinst and Challberg 1994). The C-terminus of HSV-1 UL52 contains a putative zinc-binding motif, which is also present in prokaryotic and eukaryotic primases (Carrington-Lawrence and Weller 2003; Chen et al. 2005, 2007). The presence of these motifs suggests that the helicase and primase activities of the complex likely reside in the UL5 and UL52 subunits, respectively; however, several lines of evidence suggest a more complex interaction between these two subunits. Mutations in the putative zinc finger at the C-terminus of UL52 have been shown to abrogate not only primase but also ATPase, helicase, and DNA-binding activities of the UL5/UL52 subcomplex (Biswas and Weller 1999, 2001; Chen et al. 2005). These results suggest that UL52 binding to DNA via the zinc finger may be necessary for loading UL5 onto DNA. Alternatively, it is possible that UL5 and UL52 share a DNA-binding site created by interaction between the two subunits.

UL29 or ICP8

ICP8, encoded by the UL29 gene, is the HSV major single-strand DNA-binding protein and is essential for DNA replication. ICP8 (SSB) is a 130-kDa zinc metalloprotein that preferentially binds ssDNA in a non-sequence specific and cooperative manner. ICP8 exhibits helix-destabilizing activities which are thought to play a role in unwinding duplex DNA during DNA synthesis. ICP8 interacts with many viral and cellular proteins (reviewed in [Challberg 1996]). In particular, ICP8 has been reported to enhance biochemical activities of UL9, the helicase/primase, and polymerase ([Arana et al. 2001; Boehmer 1998; Hamatake et al. 1997] and refs therein). ICP8 also interacts with the viral nuclease UL12 and together these proteins exhibit a strand-annealing reaction ([Reuven and Weller 2005] and refs therein). ICP8 has also been reported to regulate viral gene expression by repressing transcription from the parental genome and stimulating late gene expression from progeny genomes ([McNamee et al. 2000] and refs therein).

The structure of ICP8 lacking its C-terminal 60 residues was recently solved, revealing two separate domains, the N- (aa 9–1038) and the C- (aa 1049–1129) terminal domain, connected by a short linker region of loose electron density (Mapelli et al. 2005). The N-terminal domain forms three regions: the head, neck, and shoulder which in turn are formed from non-contiguous secondary structural elements.

Overview of HSV DNA Replication

Although several *cis*- and *trans*-acting elements have been shown to be required for HSV DNA replication as described above, very little is known about the actual mechanisms of HSV DNA replication. The model presented below is for the most part consistent with existing data; however, validation will require additional experimental support.

Fate of Incoming Viral DNA

HSV-1 DNA genomes enter the nucleus following docking of the capsid at a nuclear pore. Once the linear viral DNA is released it is thought that these linear DNA molecules lose their free ends by a process that does not require de novo protein synthesis (Poffenberger and Roizman 1985). The simplest interpretation of these results is that the viral genome loses its free ends through the formation of a covalently closed circular molecule leading to the model that linear virion DNA is rapidly circularized in infected cells. Aspects of this model have recently been challenged experimentally (Jackson and DeLuca 2003), and considerable controversy still surrounds the fate of the incoming viral DNA soon after infection (Strang and Stow 2005). It is possible that the viral genome adopts an endless configuration by an intra- or intermolecular homologous recombination event; however, additional experimentation will be needed to determine the precise fate of the incoming viral genome.

Initiation of Viral DNA Replication

Current models suggest that the origin-recognition protein, UL9, binds one or more of the origins of replication and recruits the rest of the replication machinery to the origin, stimulating DNA synthesis. Although DNA synthesis is probably initiated by a UL9-dependent step, replication at the origins of HSV alone would not be sufficient to generate the observed head-to-tail concatemers; therefore, we have proposed that HSV DNA replication occurs in two stages (Fig. 13.3) (Weller and Coen 2006; Wilkinson and Weller 2003). We have proposed that UL9 is essential during the first stage; however, later in infection DNA replication appears to proceed in an origin-independent manner (Blumel et al. 2000; Blumel and Matz 1995), which may proceed by rolling circle replication and/or a recombination-dependent replication step (Wilkinson and Weller 2003). The involvement of recombination would be analogous to the replication program of the bacteriophages T4, lambda, and many other linear dsDNA bacteriophage.

The precise role of UL9 in the initiation of HSV DNA replication remains uncertain. Although UL9 is known to possess ATPase and helicase activities in addition to its origin-binding properties, it is unable to unwind blunt-ended linear or circular double-stranded DNA containing an origin of replication. The failure of UL9 to unwind duplex origin DNA remains a major impediment to the establishment of an in vitro DNA replication system. Although the details are not yet known, it is believed that UL9 acts in conjunction with ICP8 to distort or perturb the region, followed by the recruitment of the HSV-1 helicase/primase (H/P) complex (Fig. 13.4). We have demonstrated that an active primase is needed to recruit the HSV DNA polymerase to viral foci in infected cells. Once the HSV DNA polymerase is recruited to the fork, it is believed to be responsible for both leading and lagging strand DNA syntheses. The first stage of DNA synthesis may be bidirectional proceeding from one or more of the three origins of replication. The end products of DNA replication are longer-than-unit-length head-to-tail DNA concatemers.

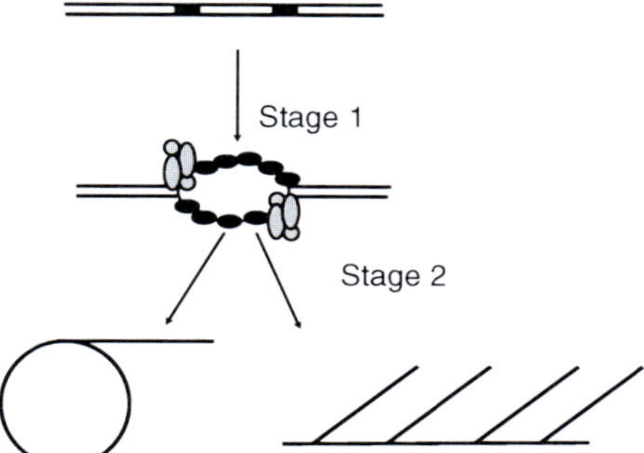

Fig. 13.3 Model for UL9-dependent and UL9-independent HSV DNA replication. According to this model the first stage of HSV DNA replication involves origin unwinding and bidirectional DNA replication (*Stage I*). In this step, UL9 likely acts in conjunction with ICP8. Following the opening, the helicase/primase can be recruited to the complex followed by either recombination-dependent or rolling circle replication (or both) (*Stage II*). Light gray symbols represent UL9 and dark gray symbols represent ICP8, the single-strand DNA-binding protein.

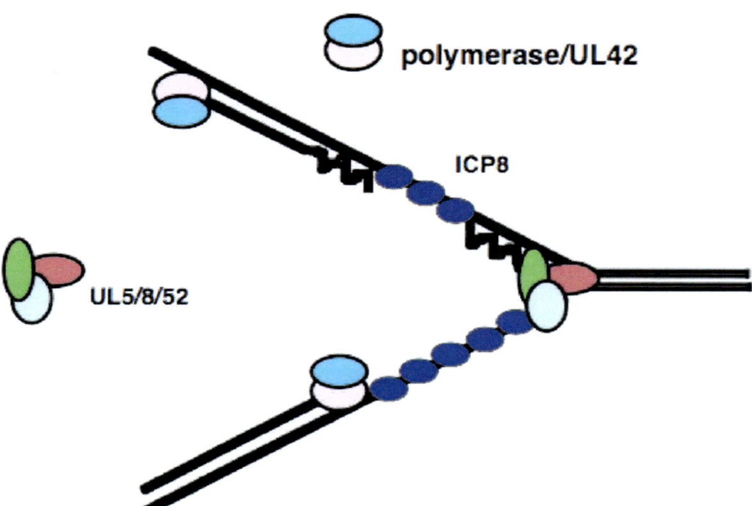

Fig. 13.4 HSV-1 replication fork. An HSV-1 replication fork would be expected to contain the helicase/primase complex (UL5/UL52/UL8) at the fork: UL5 would be expected to unwind duplex DNA ahead of the fork and UL52 would be expected to lay down RNA primers which could then be extended by the two subunit DNA polymerase (UL30 and UL42). The HSV-1 pol would also be expected to carry out leading strand synthesis. ICP8 (UL29, SSB) would be expected to bind to ssDNA generated during HSV DNA synthesis.

Replication Intermediates Are Complex and Branched

Although direct evidence is lacking, several lines of evidence suggest that recombination-dependent replication occurs during the second stage of viral DNA synthesis. (i) We and others have shown that replication intermediates in HSV-1-infected cells are present in a non-linear structure which cannot enter a pulsed field gel, even after digestion with a restriction enzyme which recognizes a single-restriction site within the HSV genome (reviewed in [Wilkinson and Weller 2003]). This complex, perhaps branched, structure is consistent with recombination-dependent replication. (ii) Inversion of the unique regions of the HSV genome has occurred at the earliest times that replicated DNA can be detected (Lamberti and Weller 1996). (iii) Severini et al. (1996) isolated DNA from the well of a pulsed field gel, and following restriction enzyme digestion, fragments were subjected to two-dimensional gel electrophoresis; both Y-shaped arches and X-shaped junctions were observed (Severini et al. 1996). (iv) It has been shown that SV40 DNA replicates by theta replication, resulting in interlocked circles. SV40 DNA replicated by the HSV-1 core replication machinery in infected cells, however, adopts a complex-branched DNA indistinguishable from that of replicating HSV DNA (Blumel et al. 2000). These results taken together suggest that replicating DNA adopts a complex, most likely branched, structure consistent with the involvement of recombination-dependent replication.

The demonstration that HSV encodes a two subunit complex consisting of ICP8 and UL12 which can perform strand exchange is also consistent with a recombination-dependent replication mechanism of DNA replication. This two subunit complex may play a role analogous to that of the red recombinase system encoded by bacteriophage lambda (Reuven et al. 2003; Reuven and Weller 2005; Reuven et al. 2004b). Furthermore, ICP8 has been reported to interact either directly or indirectly with over 50 cellular and viral proteins, some of which play roles, such as repair and recombination (Taylor and Knipe 2004). In addition, we have shown that UL12 interacts with Nbs1, a component of the MRN complex known to play an important role in cellular homologous recombination (Balasubramanian and Weller, manuscript in preparation). Taken together, the ability of ICP8 and UL12 to mediate strand annealing and strand transfer and to interact with viral and cellular repair and recombination proteins is consistent with the suggestion that DNA recombination plays a role in HSV DNA replication (reviewed in [Wilkinson and Weller 2003]). Further experimental evidence will be required to test this model.

HSV DNA Replication at the Cellular Level

HSV-1 DNA replication occurs in large globular replication compartments (RCs) in the nucleus of infected cells, and viral gene expression and DNA replication are thought to occur within these domains (Knipe 1989). ICP8 is believed to play a major role in nuclear events leading to the formation of RCs (Taylor and

Knipe 2003), and recent work by Everett and colleagues has led to a more refined model of the earliest steps of infection (Everett and Murray 2005; Everett et al. 2007, 2004). HSV genomes enter the nucleus and appear to cause the recruitment of cellular proteins of cellular ND10 proteins. ND10 (nuclear domains 10, also known as promyelocytic leukemia nuclear bodies or PODs) are defined by the accumulation of PML and many other cellular proteins involved in growth control, gene expression, and possibly DNA recombination and the DNA damage response (Dellaire and Bazett-Jones 2007, 2004). The genomes of several DNA viruses which replicate in the nucleus including SV40, adenoviruses, and other herpesviruses such as HCMV also appear to recruit ND10 proteins (reviewed in [Everett 2006]). The formation of ND10-like foci at viral genomes may reflect an antiviral cellular mechanism to repress expression of the viral genome, consistent with reports that many ND10 proteins are transcriptional repressors and are found at silenced regions of the chromatin (regions of heterochromatin) (Fernandez-Capetillo et al. 2003; Turner 2007; Turner et al. 2005). In support of this notion, PML was found to play a role in mediating the antiviral effects of IFN treatment in both HSV and HCMV infections (Everett 2006; Tavalai et al. 2006). Alternatively or in addition, it is possible that ND10 components are attracted to viral genomes because the genome is seen as a damaged DNA molecule in need of repair; many ND10 proteins have functions in DNA repair and recombination (Dellaire and Bazett-Jones 2007, 2004). Thus ND10 components may play dual roles in responding to damage and induction of silencing.

HSV Induces Disruption of ND10s

HSV is believed to counter the antiviral action of ND10 recruitment by disrupting ND10 by the action of ICP0, an immediate-early gene product. ICP0 is an E3 ubiquitin ligase (Boutell et al. 2003, 2002), and its ability to disrupt ND10 is thought to be a result of its ability to degrade sumoylated isoforms of PML and SP100, two components of ND10. The disruption of ND10 and the removal of at least some ND10 proteins may relieve the repressive activities of some ND10 proteins and thus provide an environment conducive to viral gene expression and viral DNA replication.

Prereplicative Sites

Replication compartments (RC) form rapidly during infection, and the only way to identify subassemblies of viral proteins important for RC formation is to freeze the progression of infection either by infection with viruses bearing mutations in replication proteins or in the presence of pharmacological agents which inhibit viral DNA synthesis (Burkham et al. 1998; Carrington-Lawrence and Weller 2003; Wilkinson and Weller 2004). We have defined five stages of infection based on the intracellular localization of viral and cellular proteins which are believed to play a role in the formation of RCs. Stage I is defined by the recruitment of ND10 proteins to viral genomes as described above, and during this stage, no ICP8 foci can be detected by immunofluorescence microscopy. In cells that are in Stage II, ND10 have been disrupted and ICP8 can be detected. Although we originally reported that ICP8 is diffusely localized in Stage II, more sensitive microscopy has now revealed that microfoci of ICP8 are present at this stage and that these ICP8 microfoci are positioned adjacent to ICP4 foci (Livingston et al. 2008). Cells in Stage III contain a limited number of ICP8-containing foci whose formation is dependent on the presence of UL5, UL8, UL9, and UL52 (Burkham et al. 1998). Stage III can be divided into two: Stage IIIa foci contain the five viral proteins ICP8, UL5, UL8, UL9, and UL52, whereas Stage IIIb foci contain these five proteins along with HSV Pol, UL42, and PML. As mentioned above, the recruitment of the polymerase holoenzyme to the five protein scaffold requires the presence of an active primase subunit (Carrington-Lawrence and Weller 2003). If replication is allowed to proceed, replication compartments are observed which can be detected with both the ICP8 and the PML antibodies (stage IV). Thus, it appears that viral and cellular proteins assemble to prereplicative sites in an ordered manner to initiate viral DNA replication.

Host–Cell Interactions

Although the viral *cis*- and *trans*-acting factors necessary for viral replication have been identified, we know very little about the role of host proteins in viral DNA replication. Several cellular proteins are rearranged following viral infection: some are recruited into replication compartments and some are sequestered in foci adjacent to RCs called VICE (virus-induced chaperone-enriched) domains (Burch and Weller 2004; Wilkinson and Weller 2006). Several questions remain, however. Do the cellular proteins which are recruited to RCs such as RPA, Rad51, MRN proteins, and hsp90 play a direct role in viral DNA replication (Wilkinson and Weller 2006)? What is the function of the VICE domains, and why are some cellular proteins sequestered there, including the phosphorylated form of RPA, the ATR interaction protein (ATRIP), and the heat-shock protein hsc70? What are the roles of host proteins which have been identified as interaction partners for viral-replication proteins, such as the transcriptional coactivator HCF-1 (T. Kristie, personal communication), polymerase alpha-primase (Lee et al. 1995), and a neural F-box protein NFB42 which may play a role in ubiquitin-dependent degradation (Eom et al. 2004; Eom and Lehman 2003)? The biological significance of these interactions is not clear. It is possible that several cellular proteins play direct roles in either the initiation or the later stages of viral DNA replication. It is intriguing to speculate that the reason it has not been possible to recapitulate origin-dependent HSV DNA synthesis in vitro is that one or more host cellular proteins may be required. Cellular proteins which interact with viral proteins may be co-opted by the virus for various purposes such as subversion of antiviral defenses or the prevention of apoptosis. With the advent of siRNA technology, it should be possible to address some of these unanswered questions about the involvement of host proteins in viral DNA replication.

Encapsidation of Viral Genomes

Viral DNA packaging into virions is a multistep process involving resolution of replication and/or recombination intermediates, specific cleavage events, packaging into preassembled capsids. The steps involved in this process are highly analogous to those of the more extensively studied DNA bacteriophages including (i) the formation of a procapsid intermediate consisting of a capsid shell initially supported by an internal scaffold, (ii) replacement of the internal scaffold with viral DNA, (iii) insertion of DNA through a unique portal vertex, and (iv) generation of unit length molecules by endonucleolytic cleavage of complex DNA concatemers by the activity of a two-component terminase. Several HSV gene products are involved in these steps including a terminase composed of the UL15 and UL28 proteins and a portal protein (UL6), which forms an oligomeric ring through which the viral DNA is taken up during the packaging reaction (reviewed in [Baines and Weller 2004]).

Summary

Although *cis*- and *trans*-acting viral proteins have been identified and their functions determined, many questions about the actual mechanism of HSV DNA replication and the involvement of host proteins remain unanswered. It is important to address these questions in part because viral proteins required for viral DNA replication provide very attractive targets for antiviral chemotherapy, and agents such as acyclovir and its derivatives which target the viral thymidine kinase and the viral DNA polymerase have been very successful. As with many other therapies, however, drug resistance is a very real threat which limits efficacy. Because most of the replication proteins discussed in this chapter are common to all the Herpesviridae, it is anticipated that new information generated here will be applicable to all herpesviruses. The helicase/primase has already been exploited as an antiviral target: two classes of highly potent helicase/primase inhibitors have been reported recently (Kleymann, 2003 #1608).

References

Anders, D. G., and L. A. McCue. 1996. The human cytomegalovirus genes and proteins required for DNA synthesis. Intervirology 39: 378–388.

Arana, M. E., B. Haq, N. Tanguy Le Gac, and P. E. Boehmer. 2001. Modulation of the herpes simplex virus type-1 UL9 DNA helicase by its cognate single-strand DNA-binding protein, icp8. J Biol Chem 276: 6840–6845.

Baines, J., and S. K. Weller. 2004. Cleavage and packaging of herpes simplex virus 1 DNA. In: C. Catalano (ed.) Virus packaging No. in press. Landes Bioscience, Georgetown.

Biswas, N., and S. K. Weller. 1999. A mutation in the c-terminal putative zn2+ finger motif of UL52 severely affects the biochemical activities of the hsv-1 helicase-primase subcomplex. J Biol Chem 274: 8068–8076.

Biswas, N., and S. K. Weller. 2001. The UL5 and UL52 subunits of the herpes simplex virus type 1 helicase-primase subcomplex exhibit a complex interdependence for DNA binding. J Biol Chem 276: 17610–17619.

Blumel, J., S. Graper, and B. Matz. 2000. Structure of simian virus 40 DNA replicated by herpes simplex virus type 1. Virology 276: 445–454.

Blumel, J., and B. Matz. 1995. Thermosensitive UL9 gene function is required for early stages of herpes simplex virus type 1 DNA synthesis. J Gen Virol 76 (Pt 12): 3119–3124.

Boehmer, P. E. 1998. The herpes simplex virus type-1 single-strand DNA-binding protein, ICP8, increases the processivity of the UL9 protein DNA helicase. J Biol Chem 273: 2676–2683.

Boutell, C., A. Orr, and R. D. Everett. 2003. Pml residue lysine 160 is required for the degradation of pml induced by herpes simplex virus type 1 regulatory protein ICP0. J Virol 77: 8686–8694.

Boutell, C., S. Sadis, and R. D. Everett. 2002. Herpes simplex virus type 1 immediate-early protein ICP0 and is isolated ring finger domain act as ubiquitin e3 ligases in vitro. J Virol 76: 841–850.

Burch, A. D., and S. K. Weller. 2004. Nuclear sequestration of cellular chaperone and proteasomal machinery during herpes simplex virus type 1 infection. J Virol 78: 7175–7185.

Burkham, J., D. M. Coen, and S. K. Weller. 1998. Nd10 protein pml is recruited to herpes simplex virus type 1 prereplicative sites and replication compartments in the presence of viral DNA polymerase. J Virol 72: 10100–10107.

Carmichael, E. P., M. J. Kosovsky, and S. K. Weller. 1988. Isolation and characterization of herpes simplex virus type 1 host range mutants defective in viral DNA synthesis. J Virol 62: 91–99.

Carrington-Lawrence, S. D., and S. K. Weller. 2003. Recruitment of polymerase to herpes simplex virus type 1 replication foci in cells expressing mutant primase (UL52) proteins. J Virol 77: 4237–4247.

Chabaud, S. et al. 2003. The R1 subunit of herpes simplex virus ribonucleotide reductase has chaperone-like activity similar to hsp27. FEBS Lett 545: 213–218.

Challberg, M. 1996. Herpesvirus DNA replication. In: M. DePamphilis (ed.) DNA replication in eukaryotic cells. p 721–750. Cold Spring Harbor Press, Cold Spring Harbor.

Chattopadhyay, S., Y. Chen, and S. K. Weller. 2006. The two helicases of herpes simplex virus type 1 (HSV-1). Front Biosci 11: 2213–2223.

Chen, Y., S. D. Carrington-Lawrence, P. Bai, and S. K. Weller. 2005. Mutations in the putative zinc-binding motif of UL52 demonstrate a complex interdependence between the UL5 and UL52 subunits of the human herpes simplex virus type 1 helicase/primase complex. J Virol 79: 9088–9096.

Chen, Y., C. M. Livingston, S. D. Carrington-Lawrence, P. Bai, and S. K. Weller. 2007. A mutation in the human herpes simplex virus type 1 UL52 zinc finger motif results in defective primase activity but can recruit viral polymerase and support viral replication efficiently. J Virol 81: 8742–8751.

Dellaire, G., and D. P. Bazett-Jones. 2004. Pml nuclear bodies: Dynamic sensors of DNA damage and cellular stress. Bioessays 26: 963–977.

Dellaire, G., and D. P. Bazett-Jones. 2007. Beyond repair foci: Subnuclear domains and the cellular response to DNA damage. Cell Cycle 6: 1864–1872.

Dracheva, S., E. V. Koonin, and J. J. Crute. 1995. Identification of the primase active site of the herpes simplex virus type 1 helicase-primase. J Biol Chem 270: 14148–14153.

Eom, C. Y., W. D. Heo, M. L. Craske, T. Meyer, and I. R. Lehman. 2004. The neural f-box protein nfb42 mediates the nuclear export of the herpes simplex virus type 1 replication initiator protein (UL9 protein) after viral infection. Proc Natl Acad Sci USA 101: 4036–4040.

Eom, C. Y., and I. R. Lehman. 2003. Replication-initiator protein (UL9) of the herpes simplex virus 1 binds nfb42 and is degraded via the ubiquitin-proteasome pathway. Proc Natl Acad Sci USA 100: 9803–9807.

Everett, R. D. 2006. Interactions between DNA viruses, ND10 and the DNA damage response. Cell Microbiol 8: 365–374.

Everett, R. D., and J. Murray. 2005. ND10 components relocate to sites associated with herpes simplex virus type 1 nucleoprotein complexes during virus infection. J Virol 79: 5078–5089.

Everett, R. D., J. Murray, A. Orr, and C. M. Preston. 2007. Herpes simplex virus type 1 genomes are associated with ND10 nuclear sub-structures in quiescently infected human fibroblasts. J Virol. 81: 10991–11004.

Everett, R. D., G. Sourvinos, C. Leiper, J. B. Clements, and A. Orr. 2004. Formation of nuclear foci of the herpes simplex virus type 1 regulatory protein ICP4 at early times of infection: Localization, dynamics, recruitment of ICP27, and evidence for the de novo induction of ND10-like complexes. J Virol 78: 1903–1917.

Fernandez-Capetillo, O. et al. 2003. H2ax is required for chromatin remodeling and inactivation of sex chromosomes in male mouse meiosis. Dev Cell 4: 497–508.

Graves-Woodward, K. L., J. Gottlieb, M. D. Challberg, and S. K. Weller. 1997. Biochemical analyses of mutations in the hsv-1 helicase-primase that alter ATP hydrolysis, DNA unwinding, and coupling between hydrolysis and unwinding. J Biol Chem 272: 4623–4630.

Graves-Woodward, K. L., and S. K. Weller. 1996. Replacement of gly815 in helicase motif v alters the single-stranded DNA-dependent ATPase activity of the herpes simplex virus type 1 helicase-primase. J Biol Chem 271: 13629–13635.

Hamatake, R. K., M. Bifano, W. W. Hurlburt, and D. J. Tenney. 1997. A functional interaction of ICP8, the herpes simplex virus single-stranded DNA-binding protein, and the helicase-primase complex that is dependent on the presence of the UL8 subunit. J Gen Virol 78 (Pt 4): 857–865.

Hayward, G. S., R. J. Jacob, S. C. Wadsworth, and B. Roizman. 1975. Anatomy of herpes simplex virus DNA: Evidence for four populations of molecules that differ in the relative orientations of their long and short components. Proc Natl Acad Sci USA 72: 4243–4247.

Huang, L. et al. 1999. The enzymological basis for resistance of herpesvirus DNA polymerase mutants to acyclovir: Relationship to the structure of alpha-like DNA polymerases. Proc Natl Acad Sci USA 96: 447–452.

Igarashi, K., R. Fawl, R. J. Roller, and B. Roizman. 1993. Construction and properties of a recombinant herpes simplex virus 1 lacking both s-component origins of DNA synthesis. J Virol 67: 2123–2132.

Jackson, S. A., and N. A. DeLuca. 2003. Relationship of herpes simplex virus genome configuration to productive and persistent infections. Proc Natl Acad Sci USA 100: 7871–7876.

Jiang, C., Y. T. Hwang, J. C. Randell, D. M. Coen, and C. B. Hwang. 2007. Mutations that decrease DNA binding of the processivity factor of the herpes simplex virus DNA polymerase reduce viral yield, alter the kinetics of viral DNA replication, and decrease the fidelity of DNA replication. J Virol 81: 3495–3502.

Klinedinst, D. K., and M. D. Challberg. 1994. Helicase-primase complex of herpes simplex virus type 1: A mutation in the UL52 subunit abolishes primase activity. J Virol 68: 3693–3701.

Knipe, D. M. 1989. The role of viral and cellular nuclear proteins in herpes simplex virus replication. Adv Virus Res 37: 85–123.

Lamberti, C., and S. K. Weller. 1996. The herpes simplex virus type 1 UL6 protein is essential for cleavage and packaging but not for genomic inversion. Virology 226: 403–407.

Langelier, Y. et al. 2002. The R1 subunit of herpes simplex virus ribonucleotide reductase protects cells against apoptosis at, or upstream of, caspase-8 activation. J Gen Virol 83: 2779–2789.

Lee, S. S., Q. Dong, T. S. Wang, and I. R. Lehman. 1995. Interaction of herpes simplex virus 1 origin-binding protein with DNA polymerase alpha. Proc Natl Acad Sci USA 92: 7882–7886.

Livingston, C. M., N. Deluca, D. E. Wilkinson, and S. K. Weller. 2008. The formation of foci of ICP8, the single strand DNA binding protein of HSV-1, requires the oligomerization of ICP4. J. Virol. 82: 6324–6336.

Liu, S. et al. 2006. Crystal structure of the herpes simplex virus 1 DNA polymerase. J Biol Chem 281: 18193–18200.

Malik, A. K., and S. K. Weller. 1996. Use of transdominant mutants of the origin-binding protein (UL9) of herpes simplex virus type 1 to define functional domains. J Virol 70: 7859–7866.

Mapelli, M., S. Panjikar, and P. A. Tucker. 2005. The crystal structure of the herpes simplex virus 1 ssdna-binding protein suggests the structural basis for flexible, cooperative single-stranded DNA binding. J Biol Chem 280: 2990–2997.

Marintcheva, B., and S. K. Weller. 2001a. A tale of two hsv-1 helicases: Roles of phage and animal virus helicases in DNA replication and recombination. Prog Nucleic Acid Res Mol Biol 70: 77–118.

Marintcheva, B., and S. K. Weller. 2001b. Residues within the conserved helicase motifs of UL9, the origin-binding protein of herpes simplex virus-1, are essential for helicase activity but not for dimerization or origin binding activity. J Biol Chem 276: 6605–6615.

Marintcheva, B., and S. K. Weller. 2003a. Existence of transdominant and potentiating mutants of UL9, the herpes simplex virus type 1 origin-binding protein, suggests that levels of UL9 protein may be regulated during infection. J Virol 77: 9639–9651.

Marintcheva, B., and S. K. Weller. 2003b. Helicase motif Ia is involved in single-strand DNA-binding and helicase activities of the herpes simplex virus type 1 origin-binding protein, UL9. J Virol 77: 2477–2488.

McNamee, E. E., T. J. Taylor, and D. M. Knipe. 2000. A dominant-negative herpesvirus protein inhibits intranuclear targeting of viral proteins: Effects on DNA replication and late gene expression. J Virol 74: 10122–10131.

Ojala, P. M., B. Sodeik, M. W. Ebersold, U. Kutay, and A. Helenius. 2000. Herpes simplex virus type 1 entry into host cells: Reconstitution of capsid binding and uncoating at the nuclear pore complex in vitro. Mol Cell Biol 20: 4922–4931.

Perkins, D., E. F. Pereira, M. Gober, P. J. Yarowsky, and L. Aurelian. 2002. The herpes simplex virus type 2 R1 protein kinase (ICP10 pk) blocks apoptosis in hippocampal neurons, involving activation of the mek/mapk survival pathway. J Virol 76: 1435–1449.

Poffenberger, K. L., and B. Roizman. 1985. A noninverting genome of a viable herpes simplex virus 1: Presence of head-to-tail linkages in packaged genomes and requirements for circularization after infection. J Virol 53: 587–595.

Polvino-Bodnar, M., P. K. Orberg, and P. A. Schaffer. 1987. Herpes simplex virus type 1 oril is not required for virus replication or for the establishment and reactivation of latent infection in mice. J Virol 61: 3528–3535.

Randell, J. C., and D. M. Coen. 2001. Linear diffusion on DNA despite high-affinity binding by a DNA polymerase processivity factor. Mol Cell 8: 911–920.

Randell, J. C., and D. M. Coen. 2004. The herpes simplex virus processivity factor, UL42, binds DNA as a monomer. J Mol Biol 335: 409–413.

Reuven, N. B., S. Antoku, and S. K. Weller. 2004a. The UL12.5 gene product of herpes simplex virus type 1 exhibits nuclease and strand exchange activities but does not localize to the nucleus. J Virol 78: 4599–4608.

Reuven, N. B., A. E. Staire, R. S. Myers, and S. K. Weller. 2003. The herpes simplex virus type 1 alkaline nuclease and single-stranded DNA binding protein mediate strand exchange in vitro. J Virol 77: 7425–7433.

Reuven, N. B., and S. K. Weller. 2005. Herpes simplex virus type 1 single-strand DNA binding protein ICP8 enhances the nuclease activity of the UL12 alkaline nuclease by increasing its processivity. J Virol 79: 9356–9358.

Reuven, N. B., S. Willcox, J. D. Griffith, and S. K. Weller. 2004b. Catalysis of strand exchange by the hsv-1 UL12 and ICP8 proteins: Potent ICP8 recombinase activity is revealed upon resection of dsdna substrate by nuclease. J Mol Biol 342: 57–71.

Schildgen, O., S. Graper, J. Blumel, and B. Matz. 2005. Genome replication and progeny virion production of herpes simplex virus type 1 mutants with temperature-sensitive lesions in the origin-binding protein. J Virol 79: 7273–7278.

Severini, A., D. G. Scraba, and D. L. J. Tyrrel. 1996. Branched structures in the intracellular DNA of herpes simplex virus type 1. J. Virol. 70: 3169–3175.

Sodeik, B., M. W. Ebersold, and A. Helenius. 1997. Microtubule-mediated transport of incoming herpes simplex virus 1 capsids to the nucleus. J Cell Biol 136: 1007–1021.

Spear, P. G. 2004. Herpes simplex virus: Receptors and ligands for cell entry. Cell Microbiol 6: 401–410.

Strang, B. L., and N. D. Stow. 2005. Circularization of the herpes simplex virus type 1 genome upon lytic infection. J Virol 79: 12487–12494.

Tavalai, N., P. Papior, S. Rechter, M. Leis, and T. Stamminger. 2006. Evidence for a role of the cellular ND10 protein pml in mediating intrinsic immunity against human cytomegalovirus infections. J Virol 80: 8006–8018.

Taylor, T. J., and D. M. Knipe. 2003. C-terminal region of herpes simplex virus ICP8 protein needed for intranuclear localization. Virology 309: 219–231.

Taylor, T. J., and D. M. Knipe. 2004. Proteomics of herpes simplex virus replication compartments: Association of cellular DNA replication, repair, recombination, and chromatin remodeling proteins with ICP8. J Virol 78: 5856–5866.

Turner, J. M. 2007. Meiotic sex chromosome inactivation. Development 134: 1823–1831.

Turner, J. M. et al. 2005. Silencing of unsynapsed meiotic chromosomes in the mouse. Nat Genet 37: 41–47.

Walsh, D., and I. Mohr. 2006. Assembly of an active translation initiation factor complex by a viral protein. Genes Dev 20: 461–472.

Weisshart, K., C. S. Chow, and D. M. Coen. 1999. Herpes simplex virus processivity factor UL42 imparts increased DNA-binding specificity to the viral DNA polymerase and decreased dissociation from primer-template without reducing the elongation rate. J Virol 73: 55–66.

Weller, S. K., and D. M. Coen. 2006. Herpes simplex virus. In: M. L. DePamphilis (ed.) DNA replication and human disease. p 663–686. Cold Spring Harbor Laboratory Press, Cold Spring Harbor, NY.

Wilkinson, D. E., and S. K. Weller. 2003. The role of DNA recombination in herpes simplex virus DNA replication. IUBMB Life 55: 451–458.

Wilkinson, D. E., and S. K. Weller. 2004. Recruitment of cellular recombination and repair proteins to sites of herpes simplex virus type 1 DNA replication is dependent on the composition of viral proteins within prereplicative sites and correlates with the induction of the DNA damage response. J Virol 78: 4783–4796.

Wilkinson, D. E., and S. K. Weller. 2006. Herpes simplex virus type I disrupts the atr-dependent DNA-damage response during lytic infection. J Cell Sci 119: 2695–2703.

Yates, J. L. 1996. Epstein-Barr virus DNA replication. In: M. L. DePamphilis (ed.) DNA replication in eukaryotic cells. p 751–773. Cold Spring Harbor Laboratory Press, Cold Spring Harbor, New York.

Zhu, L. A., and S. K. Weller. 1992. The six conserved helicase motifs of the UL5 gene product, a component of the herpes simplex virus type 1 helicase-primase, are essential for its function. J Virol 66: 469–479.

Zuccola, H. J., D. J. Filman, D. M. Coen, and J. M. Hogle. 2000. The crystal structure of an unusual processivity factor, herpes simplex virus UL42, bound to the *c* terminus of its cognate polymerase. Mol Cell 5: 267–278.

Chapter 14
Host Factors Promoting Viral RNA Replication

Peter D. Nagy and Judit Pogany

Abstract Plus-stranded RNA viruses, the largest group among eukaryotic viruses, are capable of reprogramming host cells by subverting host proteins and membranes, by co-opting and modulating protein and ribonucleoprotein complexes, and by altering cellular pathways during infection. To achieve robust replication, plus-stranded RNA viruses interact with numerous cellular molecules via protein–protein, RNA–protein, and protein–lipid interactions using molecular mimicry and other means. These interactions lead to the transformation of the host cells into viral "factories" that can produce 10,000–1,000,000 progeny RNAs per infected cell. This chapter presents the progress that was made largely in the last 15 years in understanding virus–host interactions during RNA virus replication. The most commonly employed approaches to identify host factors that affect plus-stranded RNA virus replication are described. In addition, we discuss many of the identified host factors and their proposed roles in RNA virus replication. Altogether, host factors are key determinants of the host range of a given virus and affect virus pathology, host–virus interactions, as well as virus evolution. Studies on host factors also contribute insights into their normal cellular functions, thus promoting understanding of the basic biology of the host cell. The knowledge obtained in this fast-progressing area will likely stimulate the development of new antiviral methods as well as novel strategies that could make plus-stranded RNA viruses useful in bio- and nanotechnology.

Introduction

Plus-stranded (+)RNA viruses replicate their genomes by manipulating host cells and transforming them into viral "factories." Unraveling the interactions between viruses and their host cells as a function of time can contribute greatly to our

P.D. Nagy (✉)
Department of Plant Pathology, University of Kentucky, 201F Plant Science Bulding, Lexington, KY 40546, USA
e-mail: pdnagy2@uky.edu

C.E. Cameron et al. (eds.), *Viral Genome Replication*,
DOI 10.1007/b135974_14, © Springer Science+Business Media, LLC 2009

understanding of the dynamics of viral infections. (+)RNA viruses replicate their genomes in a two-step process: through the production of minus-strand replication intermediates, followed by the production of (+)RNA progeny via the use of the (–)RNA template. Interestingly, replication is an asymmetric process leading to a 20- to 100-fold excess of the new (+)RNA progeny. All known (+)RNA viruses assemble their own replicase complexes (RCs), likely containing both viral- and host-coded proteins (Ahlquist, 2002; Ahlquist et al., 2003; Buck, 1996; Nagy and Pogany, 2006; Noueiry and Ahlquist, 2003; Shi and Lai, 2005; Strauss and Strauss, 1999). In addition, replication takes place in membraneous compartments derived from intracellular organelles, such as the endoplasmatic reticulum (ER), mitochondrion, vacuole, Golgi, chloroplast, and peroxisome (Salonen et al., 2005). Some viruses actively induce the formation of novel cytoplasmic vesicular compartments, using COPII-coated or possibly autophagosomal membranes (Cherry et al., 2006; Egger and Bienz, 2005; Kirkegaard and Jackson, 2005; Rust et al., 2001). Thus, replication of (+)RNA viruses is a complex process that involves numerous interactions among viral RNA, viral-coded, and host-coded proteins and host membranes (lipids). Dissecting the functions of the various replication-associated or replication-modulating molecules in (+)RNA replication is one of the major frontiers in current virus research. The picture emerging is that the mechanism of genome replication, and the functions of viral and host factors, might be somewhat analogous among various (+)RNA viruses in spite of their diverse genome organizations and gene expression strategies. Also, most of the previously identified host factors are conserved genes, suggesting that (+)RNA viruses might selectively target conserved host functions as opposed to species-specific factors. Such a strategy would help viruses broaden their host range by expanding infections to new host species. Altogether, host factors play crucial roles in all steps of (+)RNA replication. Host factors are also key determinants of the host range of a given virus and affect virus pathology, host–virus interactions, as well as the evolution of the virus. Host factors could also be potent antiviral targets. Studies on host factors also contribute insights into their normal cellular functions, thus promoting understanding of the basic biology of the host cell.

The host factors characterized to date play diverse roles during (+)RNA replication, including mediating intracellular transport of viral proteins and viral RNA, as chaperones facilitating correct folding of viral proteins, as helicases or RNA chaperones assisting the folding of the viral RNA, facilitating the switch from translation to replication by promoting template recognition/selection, and as lipid metabolism enzymes driving membrane proliferation.

This review provides an overview of our current understanding of the role of host factors that facilitate (+)RNA virus replication. Major challenges remain to resolve further what roles the identified host factors play during (+)RNA virus replication. Studies aimed at identifying and dissecting all the replication-associated factors are expected to increase the number and efficiency of our methods to interfere with successful viral replication/infection.

Approaches

Since (+)RNA viruses can potentially co-opt most of the ~20,000–30,000 host proteins (whether animal or plant) for their replication, it is a daunting task to identify those proteins, which are actually subverted by a given (+)RNA virus. Numerous approaches have been developed during recent years to identify host factors, and we will briefly describe only a selected number of ways that yielded the most fruitful hits.

Systems Biology

Possibly, the most powerful means to identify host factors involved in (+)RNA replication are based on genome-wide approaches. Intensive (high-throughput) screens include systematic analysis of most genes available in the genome of a particular host. Yeast (*Saccharomyces cerevisiae*) is particularly useful for this approach since it has a small genome-size (~5,800 genes) and a reduced level of redundancy among host genes. In addition, yeast is the best-known model for eukaryotic cell with the highest percentage of characterized genes in the genome useful to study the aspects of virus–cell interactions. Also, the genome-wide screens can be performed with single-gene deletion (YKO) and the essential gene (yTHC) libraries of yeast. For this strategy to succeed, it is necessary to launch (+)RNA virus replication in yeast, usually based on plasmid-driven expression of viral RNA/proteins. Yeast as a host for Brome mosaic virus (BMV) was pioneered by the Ahlquist laboratory, and subsequently was adapted for studying replication of Flock house virus (FHV) and tombusviruses of plants, such as Tomato bushy stunt virus (TBSV) (Ishikawa et al., 1997b; Janda and Ahlquist, 1993; Panavas and Nagy, 2003; Panaviene et al., 2004; Pantaleo et al., 2003; Price et al., 2000). The genome-wide screens of 80% of yeast genes with BMV and 95% of yeast genes with TBSV led to the identification of over 100 yeast genes for both viruses that affected their replication (Jiang et al., 2006; Kushner et al., 2003; Panavas et al., 2005b). Interestingly, only a small set of yeast genes for BMV and TBSV overlapped, suggesting that these distantly related (+)RNA viruses use and/or are being affected by mostly different set of host genes for their replication (Jiang et al., 2006; Panavas et al., 2005b). Genome-wide screens with the yeast libraries have also been performed to identify host factors affecting RNA recombination in the case of TBSV, further illustrating the usefulness of this approach (Cheng et al., 2006; Serviene et al., 2006, 2005).

A different genome-wide approach was performed with Drosophila C virus (DCV) based on RNA interference (RNAi) and led to down-regulation of the expression of 21,000 (91%) of the *Drosophila* genes (Cherry et al., 2005). This resulted in the identification of 112 host genes affecting DCV replication (Cherry et al., 2005, 2006). More than half of the identified genes were ribosomal genes, suggesting that DCV replication depends greatly on the host translation machinery. Another large-scale RNAi-based approach was performed with hepatitis C virus (HCV) (Ng et al., 2007). Among the 4,000 human genes targeted, Ng et al. found nine cellular genes whose depletion led to 60% or more inhibition of HCV replication.

Genomic Random Mutagenesis

Extensive mutagenesis of the host genome via treatment with a mutagenic chemical, fast-neutron irradiation, UV-treatment, or transposon insertions have also been used to obtain libraries of mutated hosts. Testing for viral (+)RNA replication, followed by positional cloning, genetic complementation, or other approaches can lead to identification of host genes affecting virus replication. Indeed, this approach was used to identify several yeast genes affecting BMV replication and *Arabidopsis* TOM1/2/3 genes affecting Tomato mosaic virus (ToMV) accumulation (Diez et al., 2000; Ishikawa et al., 1997a; Lee et al., 2001; Tsujimoto et al., 2003; Yamanaka et al., 2000).

Proteomics

Recent technological advances with mass-spectrometry (MS)-based identification of proteins present in ribonucleoprotein complexes have made it possible to dissect the composition of purified viral replicase complexes. Two-step affinity purification of a tombusvirus replicase complex from yeast, followed by 2D gel-electrophoresis and MALDI-TOF analysis of proteins cut from 2D gels led to the identification of four host proteins, namely the Ssa1/2p molecular chaperone (a yeast homologue of Hsp70 proteins), Tdh2/3p (glyceraldehyde-3-phosphate dehydrogenase, a metabolic protein with RNA-binding activity), Pdc1p (pyruvate decarboxylase), and Cdc34p E2 ubiquitin-conjugating enzyme (the later is unpublished, Li and Nagy), which were missing in the control samples (Serva and Nagy, 2006).

Similarly, a purified Tobacco mosaic virus (TMV) replicase preparation contained at least four host proteins, based on silver-staining of SDS-PAGE gels (Osman and Buck, 1997). Western blotting led to the identification of an RNA-binding protein, GCD10, which is one of the subunits of the ten-component eIF-3 complex (Osman and Buck, 1997). Another host protein in the TMV replicase might be translation elongation factor 1A (eEF1A), which was found to bind to the TMV 126K replication protein based on co-immunoprecipitation (Yamaji et al., 2006).

A highly purified replicase preparation for BMV contained two viral-coded and ~10 host proteins, based on silver-stained SDS-PAGE analysis (Quadt et al., 1993). One of the host proteins identified was the p41 subunit of the eIF-3 complex. The function of p41 in the BMV replicase is currently unknown.

To identify host proteins interacting with the non-structural protein 3 (nsP3) replication protein of Sindbis virus (SIN), an alphavirus, a green fluorescent protein (GFP)-tagged nsP3 was expressed from the viral genome, followed by immunoaffinity purification with anti-GFP antibody on magnetic beads and mass-spectrometry of the isolated proteins (Cristea et al., 2006). This, and a similar approach by

Frolova et al. (2006), led to the identification of 59 cellular proteins, of which 35 were specific to nsP3. Additional research confirmed a role for the following proteins in SIN replication: G3BP1 and G3BP2a (Ras-GTPase activating protein SH3-domain-binding proteins), hnRNP-1A, -A3, -A2/B1, and -G (heteronuclear ribonucleoproteins), and the 14-3-3 family of proteins (tyrosine-3-monooxygenase/tryptophan 5-monooxygenase activation proteins), which are phosphoserine-binding adaptor proteins (Cristea et al., 2006). The actual functions of the identified nsP3–host protein interactions are not yet known in SIN replication.

Yeast Two-Hybrid (YTH) Screens

Due to the large number of studies using YTH screens, we cannot cover the entire area. Instead, we demonstrate the usefulness of this approach by discussing findings with HCV and poliovirus (PV). For example, NS5A of HCV was screened against an interferon-induced human hepatocyte cDNA library, leading to the discovery of the 33kDa human vesicle-associated membrane protein-associated protein (hVAP-A, also called hVAP-33) (Tu et al., 1999). hVAP-A also interacts with NS5B of HCV (Tu et al., 1999). Another YTH screen with the HCV NS5A, using human brain and liver cDNA libraries, led to the identification of the immunophilin, termed FKBP8, which is a human FK506-binding protein (Okamoto et al., 2006). When the NS5B RdRp protein was used as a bait, p68 helicase was identified as an interactor from a human spleen cDNA library (Goh et al., 2004). A screen with the HCV NS5B, using a human liver cDNA library, led to the identification of eIF4AII (a DEAD box RNA helicase) that partially co-localized with NS5B in infected cells (Kyono et al., 2002). Another YTH screen with the HCV NS5B protein, using a human hepatocyte cDNA library, led to the identification of ubiquitin-like protein hPLIC1, which functionally connects the ubiquitination machinery to the proteosome and could be involved in regulation of NS5B stability and HCV replication (Gao et al., 2003). Using a human thymus cDNA library against HCV NS5B RNA-dependent RNA polymerase (RdRp) as a bait, Kim et al. identified eukaryotic translation initiation factor 4A (isoform 2, named eIF4AII) and septin 6, a GTP-binding protein involved in membrane dynamics, trafficking and cytoskeletal remodeling (Kim et al., 2007). Depletion of septin 6 by RNAi decreased HCV replication, suggesting that septin 6 is a necessary host factor for HCV (Kim et al., 2007). In the case of PV, Kirkegaard et al. found that Sam68 (Src-associated in mitosis, 68kDa) interacted with the 3D RdRp protein, which resulted in re-localization of Sam68 to PV-induced cytoplasmic vesicles/membranes, the site of PV replication (McBride et al., 1996).

Co-purification of RNA–Protein Complexes

One of the most popular approaches for identifying RNA-binding host proteins is based on purification of the viral RNA (with or without UV-crosslinking), followed by MS-based identification of the host proteins. Using the untranslated regions (UTRs) of a coronavirus RNA for UV-crosslinking, Lai, Leibowitz, Hogue, and their colleagues identified numerous RNA-binding host proteins, including heterogeneous nuclear ribonucleoprotein (hnRNP) A1, polypyrimidine-tract-binding (PTB) protein, poly(A)-binding protein (PABP), and mitochondrial aconitase (m-aconitase) (Li et al., 1997; Nanda and Leibowitz, 2001; Shi and Lai, 2005; Spagnolo and Hogue, 2000). Also, the hnRNP-C that interacts with the (–)RNA of PV has been identified (Brunner et al., 2005; Roehl and Semler, 1995). Using West Nile virus (WNV) (+) and (–)RNAs, Brinton et al. identified several host proteins that were specifically UV-crosslinked to the viral RNA (Li et al., 2002). These cellular proteins included T-cell intracellular antigen-1 (TIA-1), TIA-1-related protein (TIAR), and eEF1A. In the case of bovine viral diarrhea virus (BVDV), a pestivirus related to HCV, various portions of the BVDV genomic (g)RNA were used in UV-crosslinking experiments with bovine and human cellular extracts (Isken et al., 2003). This work led to the identification of three RNA-binding proteins, named NF90 (nuclear factor 90, also called NFAR-1), NF45, and RNA helicase A (RHA), which are RNA-binding proteins containing double-stranded RNA-binding motifs.

Chemical Virology

A cell culture-based screen with a library of chemicals can lead to potent antiviral inhibitors, which, in turn, can be used to identify the potential cellular or viral targets of these inhibitors (Watashi and Shimotohno, 2007). For example, studies with the immunosuppressant cyclosporin A (CsA) that suppresses HCV RNA replication, led to the discovery of cellular peptidyl–prolyl *cis–trans* isomerase (PPIase), termed CyPB, which is a cellular cofactor for HCV replication (see below) (Watashi et al., 2005).

Additional approaches, such as MS-based profiling for up- and down-regulated proteins in FHV-infected cells (Go et al., 2006), protoarrays with thousands of purified recombinant host proteins (Zhu et al., 2007), and the numerous DNA microarrays used to identify up- and down-regulated cellular genes by viral infections will not be discussed here.

Molecular Interaction Between the Host and (+)RNA Virus During Replication

Most of the known host factors inhibit replication when absent, or present in reduced amount, suggesting that these genes facilitate (+)RNA virus replication by providing useful functions, directly or indirectly (see below). Identified host factors are known to be involved in various cellular processes, such as metabolism/modifications of RNAs, lipids, and proteins; in protein intracellular transport/targeting; or, in general metabolism (Kushner et al., 2003; Panavas et al., 2005b). It is intriguing to note that a large set of host genes affecting (+)RNA virus replication is unique for any given virus, suggesting that (+)RNA viruses have developed different ways to utilize the immense resources of cells. In spite of the differences in the host genes involved, we predict that many of the different genes might provide mechanistically similar functions during replication of various (+)RNA viruses. For example, molecular chaperones, albeit different members of the chaperone family, have been found to affect BMV, TBSV, FHV, HCV, and coronavirus replication (Tomita et al., 2003; Castorena et al., 2007; Kampmueller and Miller, 2005; Nanda et al., 2004; Okamoto et al., 2006; Serva and Nagy, 2006).

The genome-wide screens with BMV, TBSV, and DCV confirmed that (+)RNA viruses depend greatly on the intracellular components of infected hosts for robust viral replication. It appears that the interactions between host cells and (+)RNA viruses are complex, likely including numerous replication-associated host factors with direct or indirect roles. Overall, the identified host factors likely belong to one of the following three groups. (i) Those that may directly interact with the viral RNA(s) or replication proteins and perform essential functions for the virus. This group of host factors also include host membranes/lipids, various components of the intracellular transport and trafficking system, the translation apparatus, and possibly intracellular compartments, such as the ER, peroxisome, and vesicles, which (+)RNA viruses require and/or utilize to complete their replication. For example, the Hsp70 molecular chaperone, which is present in the tombusviral RC, might be directly involved in replicase assembly (Serva and Nagy, 2006). (ii) Those host factors that indirectly affect (+)RNA virus replication via influencing the amount and/or activity of those host factors, which are directly involved [group (i) above]. These indirect host factors may affect the competition between the virus and the host for limited cellular resources, host proteins, and intracellular compartments. For example, transcription factors could affect the amount of host factors available in the infected cells, thus indirectly affecting (+)RNA virus replication. (iii) The third group of host factors includes direct inhibitory factors, such as components of the host innate and general antiviral defense mechanisms, which affect virus replication by destroying/modifying viral RNAs or viral replication proteins in targeted or in general manners. For example, Ngl2p endoribonuclease, the yeast Xrn1p, and the *Arabidopsis* Xrn4p 5'–3' exoribonucleases were found to affect degradation, and thus stability, of TBSV RNA (Cheng et al., 2007, 2006; Serviene et al., 2005). The second and third groups of host factors will not be discussed further in this chapter.

The Replication Cycle of (+)RNA Viruses Consists of Six Distinct Steps

The viral (+)RNA has to participate in three-to-five competing processes required for successful infection. Activities include translation to produce viral proteins, replication, transcription [to produce subgenomic (sg)RNA for some viruses], encapsidation, and cell-to-cell movement (in the case of plant RNA viruses). These processes are highly regulated and compartmentalized to avoid collision between

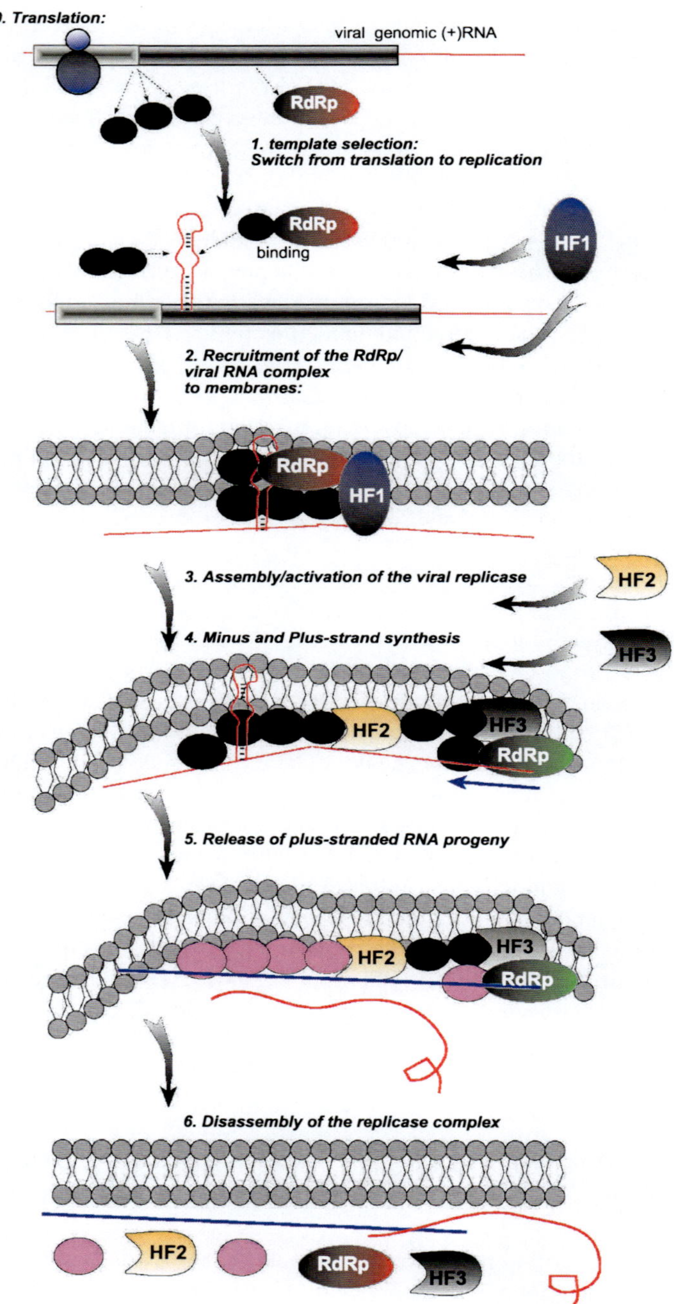

Fig. 14.1 (continued)

the ribosome and the viral RC. Recent advances in knowledge about (+)RNA virus replication have resulted in the division of the replication cycle into six separate steps (Fig. 14.1). These include (1) recruitment/selection of the viral (+)RNA template for replication (the switch of viral RNA from translation to replication); (2) targeting of viral replication proteins to the site of replication; (3) assembly of the functional viral replicase complex on intracellular membraneous surfaces; (4) synthesis of viral RNA progeny; (5) release of viral (+)RNA progeny from the site of replication; and (6) disassembly of the viral RC. In the following subchapter, we will discuss our current knowledge on the role of host factors in each of those steps.

Roles of Host Factors in Various Steps of Viral RNA Replication

Step 1. Roles of Host Proteins in the Selection of Viral (+)RNA Template for Replication and the Switch from Translation to Replication

During translation, the viral (+)RNA is used as a mRNA to produce replication proteins and other viral-coded proteins. It has been estimated that a single viral RNA is translated 1,000–10,000 times (Quinkert et al., 2005). However, unlike host mRNAs, the viral (+)RNA has to be recruited for replication and saved from RNA degradation. Both host- and viral-coded proteins have been documented to participate in the regulation of the switch from translation to replication. Translation of the replication proteins and the selection of the viral gRNA for replication likely takes place in the cytoplasm, whereas replication of (+)RNA viruses occurs on the cytoplasmic faces of various organelle-derived membrane surfaces (Ahlquist et al., 2003; Buck, 1996; Salonen et al., 2005). Therefore, the viral gRNA, together with viral and host factors, must be transported/recruited to the site of replication. Current models involve specific template selection (specific binding of viral and/or host proteins to the template RNA) as a key step in regulation of (+)RNA replication.

The recently emerging picture is that viral replication proteins can bind selectively to viral (+)RNA, which likely leads to selection/recruitment of the viral RNA

Fig. 14.1 A general model showing six separate steps during (+)RNA virus replication. The RC is shown schematically, but the RC likely contains more protein components. The *red* coloring for the RdRp suggests that the RdRp is inactive (replication incompetent) at the beginning, while it gets activated (shown in *green* color) during the replicase assembly process, possibly with the help of host factors and the viral (+)RNA. Viral-coded auxiliary replication proteins are shown as *black* circles, whereas they are shown as *pink* circles after their putative inactivation, possibly via phosphorylation. The entire RC may become inactivated by phosphorylation and/or ubiquitinaton prior to disassembly. *Red line* indicates plus-stranded, while *blue line* shows the minus-stranded viral RNAs. *Small circles* show the subcellular membraneous compartment used for replication. RdRp is the viral-coded RNA-dependent RNA polymerase, whereas HF stands for a host factor.

from translation into replication. This has been shown for the 1a protein of BMV, the p33 replication protein in tombusviruses, the 126K protein of TMV, and the 3CD protein of PV (Gamarnik and Andino, 2000; Osman and Buck, 2003; Pogany et al., 2005; Wang et al., 2005b). In the case of TBSV, highly specific binding of the p33 replication protein to a conserved sequence forming a long hairpin structure with an internal C•C mismatch in the TBSV gRNA (termed the p33 recognition element, p33RE) is essential for TBSV RNA replication in yeast or virus replication in plants (Monkewich et al., 2005; Pogany et al., 2005). Additional in vitro and in vivo data firmly support the role of the p33:p33RE interaction as the major factor in selection of TBSV RNA for replication from the diverse RNA pool present in the host cell. Interestingly, the selection of viral RNA templates for replication is mechanistically similar to viral RNA encapsidation, which is also based on the requirement of a specialized viral protein, the coat protein, leading to packaging of the viral RNA into virions that are mostly inaccessible for other processes.

The involvement of the viral replication proteins in selection of the viral RNA for replication does not exclude host proteins also from contributing to this early step in replication. For example, recruitment of BMV RNAs for replication is affected by Lsm1p, which belongs to the seven member Sm-like family of proteins (Diez et al., 2000; Noueiry et al., 2003). Lsm1p, together with Lsm2-7p, form the Lsm1p-7p heptameric ring, which is involved in mRNA turnover. Diez et al. found that Lsm1p-7p, which together with Pat1p and Dhh1p, is called the decapping activator complex, that moves cellular mRNAs from translation to degradation in the cytoplasmic processing bodies (termed P-bodies), and that might function as a key regulator to switch BMV RNAs from translation to replication (Mas et al., 2006). According to this model, the Lsm1p–7p/Pat1p/Dhh1p complex could refold the BMV RNAs allowing separation from ribosomes and translation factors and the binding of the BMV 1a replication protein to viral RNA. Then, the viral (+)RNA, likely in association with the replication proteins and host factors, ends up in the P-body that may serve as a place to pre-assemble the BMV RC prior to transport to the ER, where maturation likely occurs on the membranes (the site of BMV replication) (Beckham et al., 2007).

In the case of PV, a central role for the host poly(rC)-binding protein-2 (PCBP-2) in RNA template recruitment has been postulated. Based on this model, the full-length PCBP-2 can bind to stem-loop 4, which is part of the internal ribosome entry site (IRES), in PV RNA as well as to the host protein SRp20 (involved in RNA trafficking) to promote the cap-independent translation of the PV RNA. Due to protease activity of the PV-coded 3CD or 3C, PCBP-2 is cleaved proteolytically, removing the KH3 RNA-binding site. The cleaved form of PCBP-2, however, still contains KH1-2 RNA-binding sites and it retains its ability to bind to stem-loop 1 in the PV RNA, which is critical for replication by bringing together the 3′ and 5′ ends of the PV RNA via interaction with the poly(A)-binding protein (PABP) (Perera et al., 2007; Walter et al., 2002). PCBP also binds to the spacer sequence located between the cloverleaf and the IRES, which is required for the replication (Toyoda et al., 2007). In addition to the cellular proteins, the PV-coded 3CD RdRp precursor regulates RNA template selection and switching of the viral RNA from translation to replication by binding to the coverleaf-like structure at the 5′ UTR and affecting genome circularization (Gamarnik and Andino, 2000; Herold and Andino, 2001). The resulting genome circularization is predicted to be critical for PV minus-strand synthesis (Gamarnik and Andino, 1998, 2000; Herold and Andino, 2001; Perera et al., 2007).

Polypyrimidine-tract-binding protein (PTB) may play a central role in translation and replication of HCV. Since PTB was found to bind to the HCV RNA and since it is re-localized to detergent-resistant membrane fractions, the sites of HCV replication, Lai et al. suggested that PTB or its truncated (proteolytically cleaved) versions might facilitate the switch from translation to replication (Aizaki et al., 2006). In addition, PTB is also critical for HCV RNA synthesis, because antibody-based depletion of PTB inhibited HCV RNA synthesis in a cell-free system and small interfering (si)RNA-based depletion of PTB inhibited HCV accumulation in infected cells (Aizaki et al., 2006; Chang and Luo, 2006).

The switch of coronavirus RNA from translation to replication may depend on two host factors, the PTB protein and hnRNP A1, which, via their interaction with each other, were proposed to mediate 5′–3′ crosstalk of the UTRs in the coronavirus (+)RNA (Choi et al., 2002; Li et al., 1997; Shi and Lai, 2005). The resulting genome circularization might facilitate RNA replication, as suggested above for PV. An additional hnRNP protein, termed synaptotagmin-binding cytoplasmic RNA-interacting protein (SYNCRIP), may also be involved in the switch to replication, via inducing conformational changes in the highly structured coronavirus RNA that could facilitate RNA replication (Choi et al., 2004). Interestingly, PTB and hnRNP-A1 have also been found to play roles in HCV replication (Aizaki et al., 2006; Kim et al., 2007). Thus, it is possible that these host factors could play similar roles in gRNA circularization or the structural rearrangement of the viral RNA during HCV replication, as proposed above for coronaviruses.

Recruitment of BVDV RNA from translation into replication was proposed to depend on three interacting cellular factors: NF90, NF40, and RHA (Isken et al., 2003). Based on RNA competition studies that included various regions of the (+)RNA genome of BVDV, Behrens et al. proposed that NF90/NF40/RHA bind to both the 3′ and the 5′ UTRs, leading to circularization of the BVDV (+)RNA (Isken et al., 2003). This could be the signal to switch from translation to replication. Indeed, RHA has helicase activity that may facilitate the refolding of the viral RNA to make it suitable for replication. Also, NF90/NF40/RHA proteins are part of the antiviral response, thus their recruitment for BVDV replication in the cytoplasm can weaken the antiviral response that takes place in the nucleus. The NF90/NF40/RHA proteins together with NF110, might also promote the circularization of HCV (+)RNA via facilitating the interaction between the 5′ UTR and the 3′ UTR. Moreover, these proteins are re-localized to the site of HCV replication, suggesting that their roles could be similar in HCV as in BVDV RNA replication (Isken et al., 2007).

Overall, the selection of the viral (+)RNA for replication in concert with switching from translation to replication (i) guarantees that the viral (+)RNA avoids degradation; (ii) prevents collision between the ribosome and the viral RC, which must use the same RNA, though progressing in opposite directions; and (iii) facilitates the specific selection of the authentic viral (+)RNA for replication.

Step 2. Targeting of Essential Viral Replication Proteins and the Viral RNA to the Site of Replication

Translation and then selection of the viral (+)RNA for replication likely take place in the cytoplasm, whereas replication of (+)RNA viruses occurs on the cytoplasmic surfaces of various intracellular membranes (Ahlquist, 2002; Ahlquist et al., 2003; Burgyan et al., 1996; Egger and Bienz, 2005; Navarro et al., 2004; Panavas et al., 2005a; Rubino and Russo, 1998; Salonen et al., 2005). Therefore, the viral (+)RNA and the viral-coded replication proteins must be targeted to those subcellular compartments (Fig. 14.1, step 2). The viral gRNA is likely recruited in *cis* by the newly produced replication proteins, which could bind to the gRNA present in the same location (*cis*-binding) (Neeleman and Bol, 1999; Novak and Kirkegaard, 1994; Oster et al., 1998; Panaviene et al., 2003; Vlot et al., 2003; Weiland and Dreher, 1993) during natural infections. Altogether, recruitment in *cis* may be a more efficient process than recruitment in *trans* that could secure the effective transportation of the limited amount of viral gRNA to the site of replication. The picture emerging from recent studies with several (+)RNA viruses is that both viral- and host-coded proteins participate in the intracellular targeting/transport step.

In the case of some plant (+)RNA viruses, existing evidence supports a master role for the replication proteins, such as the BMV 1a protein and the TBSV p33 replication protein, in intracellular targeting of the viral RdRp as well as the viral (+)RNA, likely in the form of multimolecular complexes, to the site of replication (Fig. 14.1). Formation of the multimolecular complexes including the replication proteins, viral RNA, and some host factors, in the cytoplasm would likely facilitate efficient transport and co-localization of all these essential components to the same replication sites maximizing the assembly of fully functional RCs. Such organization can greatly increase the probability of successful replicase assembly. The pre-organization of replication factors into multicomponent complexes could be especially important at the beginning of infections when limited amounts of viral factors are available.

Recruitment of the viral replication proteins to the intracellular membranes, the sites of (+)RNA virus replication, is likely guided by host proteins involved in intracellular transport. For example, replication of ToMV depends on two *Arabidopsis* proteins, termed TOM1 and TOM3, seven-pass membrane proteins, which interact with the 130K helicase-like replication protein of TMV (Yamanaka et al., 2002, 2000). TOM1/3 are also part of the ToMV RC and they likely act as anchors of the RC to the membrane via their transmembrane domains. Mutations or RNAi-based depletion of TOM1/3 severely inhibited ToMV replication in *Arabidopsis* plants (Yamanaka et al., 2002, 2000). Interestingly, the RNAi-driven depletion of the homologous proteins in tobacco greatly inhibited the accumulation of several tobamoviruses related to ToMV, but not the more distantly related Cucumber mosaic virus, suggesting that TOM1/3 genes have conserved functions in diverse hosts (Asano et al., 2005).

Transport of BMV replication proteins and the (+)RNAs to the sites of replication is affected by Lsm1p-7p together with Pat1p and Dhh1p, which might facilitate localization of protein and the viral RNAs to the P-bodies, where pre-assembly of the BMV RC could take place prior to final transport to the ER, which is the site of BMV replication (Beckham et al., 2007).

hVAP-A, a SNARE-like protein, was shown to interact with the HCV NS5A and NS5B replication proteins (Tu et al., 1999). The interactions may be important for the association of these viral replication proteins with intracellular membranes, the site of HCV replication, due to hVAP-A serving as a membrane receptor. Also, a geranylgeranylated cellular protein, named FBL2, might be involved in the recruitment of the HCV NS5A replication protein to the intracellular membranes. The interaction between the geranylgeranylated FBL2 and the HCV NS5A is essential for HCV replication, based on depletion of FBL2 by RNAi and the expression of the dominant negative mutant of FBL2 that both led to inhibition of HCV accumulation (Wang et al., 2005a).

Recruitment of coronavirus RNA might be affected by interaction with molecular chaperones. Leibowitz et al. have shown that mitochondrial aconitase (m-aconitase), mitochondrial HSP70 (mtHSP70), HSP60, and HSP40 form a complex, probably in the cytosol, which may be involved in stabilizing the viral RNA and/or its recruitment to replication via binding to the coronavirus 3′ UTR (Nanda et al., 2004; Nanda and Leibowitz, 2001). Though m-aconitase and mtHSP70 are mostly localized to the mitochondria, these proteins are translated in the cytoplasm and they could be hijacked by coronavirus RNA prior to their translocation to mitochondria.

At this time, our knowledge on the transport/trafficking of the viral (+)RNA and viral replication proteins is incomplete. Future works employing cellular and biochemical approaches are expected to advance this area of research for many (+)RNA viruses.

Step 3. Role of Host Factors in the Assembly of the Viral Replication Complex

(+)RNA virus replication takes place on intracellular membraneous compartments (Salonen et al., 2005) or via the active induction of the formation of novel cytoplasmic vesicular compartments (Cherry et al., 2006; Egger and Bienz, 2005; Kirkegaard and Jackson, 2005). These compartments (virosomes) contain the membrane-bound viral RC, which is the key enzyme in (+)RNA replication. The RC has to perform many functions during replication, including recognition of minus- and plus-strand initiation promoters located at the 3’ terminus of (+) or (–)RNA, de novo (primer-independent) or primer-dependent initiation, as well as the synthesis of complementary RNA strands, strand separation, and the repair of viral

RNAs with damaged termini (Dreher, 1999; Kao et al., 2001; Nagy et al., 1997; Rao et al., 1989). In addition, the viral RC has to recognize additional regulatory RNA elements, such as any replication silencer or replication enhancer, which either down- or up-regulate RNA synthesis (Nagy and Pogany, 2006). Also, the viral RC is involved in the production of sgRNAs in some viruses (Buck, 1996, 1999; White, 2002). Moreover, the activity of the RC leads to genetic mutations and RNA recombination via template-switching during RNA synthesis, which affects RNA virus evolution (Cheng and Nagy, 2003; Kim and Kao, 2001; Nagy et al., 1995; Nagy and Simon, 1997; Nagyet al., 1998; Roossinck, 2003; Wierzchoslawski and Bujarski, 2006). Although it is currently unknown how the viral RC can perform so many activities, it is possible that the multi-functionality is due to the elaborate composition of the viral RC.

Molecular Composition of the Viral Replicase Complex

The viral RC shows surprising complexity as it contains both viral- and host-derived components (Serva and Nagy, 2006). Quantitative analysis of the HCV RC revealed that hundreds/thousands of viral nonstructural (NS) proteins are present in a protease/nuclease-resistant spherule containing ~1 minus-strand and ~5 plus-stranded RNAs (Quinkert et al., 2005). Electron microscopy (EM) and immuno-EM images of the sites of BMV replication, containing the active viral RC, revealed that the sites of BMV replication consist of 50–60nm spherule-like structures with cellular membranes surrounding the replication proteins and the viral RNA (Schwartz et al., 2002). Interestingly, individual spherules contain small openings (membraneous necks), which likely serve as gates for transportation of molecules between the spherules and the cytoplasm (Schwartz et al., 2002). In a broad sense, we can regard one separate spherule as one active/matured RC. Studies on the molecular composition of a single spherule based on immuno-labeled images revealed that one spherule could contain 25-fold more 1a replication protein than $2a^{pol}$, whereas the actual number and nature of host molecules within single spherules are currently unknown. Current models predict that highly organized protein:protein and protein:RNA complexes, with the help of cellular membranes, facilitate the formation of an active (+)RNA virus RC.

Regulation of the Activities of the Viral Replicase Complex

Some RdRp proteins, such as the tombusvirus $p92^{pol}$, the BMV $2a^{pol}$, Alfalfa mosaic virus (AlMV) P2, and the HCV NS5B, are nonfunctional before the assembly of the RC (Panaviene et al., 2005; Panaviene et al., 2004; Quadt et al., 1995; Vlot et al., 2001), suggesting that these RdRps have to become "activated" in cells, likely during the assembly process (Fig. 14.1, step 3). Therefore, assembly of the viral replicase could be an important regulatory step in (+)RNA virus replication. Moreover, co-expression of (+)RNA has been shown to enhance replicase assembly/activity by ~40- to 100-fold for TBSV and was required for the BMV replicase in yeast (Panaviene et al., 2005, 2004; Quadt et al., 1995) and the AlMV replicase in plants

(Vlot et al., 2001). Thus, the viral (+)RNA may serve as a platform to facilitate replicase assembly. It is also possible that initiation of minus-strand synthesis might lead to the stabilization of the viral RC.

Roles of Host Factors Within the Viral Replicase Complex

In addition to the *cis*-acting RNA factors, *trans*-acting factors are needed for the assembly of the functional viral RC. These include both viral and host factors that likely contribute to replicase assembly. For example, cellular chaperones may play essential roles in the assembly of viral replicases. The documented cases include Ydj1p, J-domain protein of the Hsp40 chaperone family that was found to affect minus-strand synthesis of BMV in yeast, but not the 1a-mediated recruitment of viral RNA or $2a^{pol}$ to the site of replication (Tomita et al., 2003). Also, Ydj1p affected the amount of $2a^{pol}$ present as a soluble protein and a reduced amount of Ydj1p in yeast led to aggregation of a fraction of $2a^{pol}$. Based on these data and the known property of Ydj1p, Ahlquist and co-authors proposed that Ydj1p is involved in the assembly and/or activation of the BMV RC prior to minus-strand synthesis (Tomita et al., 2003). Serva and Nagy proposed a similar function for the Hsp70 proteins (termed Ssa1p and Ssa2p in yeast) in folding/activation of tombusvirus RC (Serva and Nagy, 2006). Indeed, deletion of *SSA1/2* led to fourfold decrease in TBSV replication in yeast and reduced the amount of the replication proteins, suggesting that Hsp70 might stabilize the replication proteins. Also, Ssa1/2p were found to bind to the p33 replication protein and be present within the highly purified tombusvirus RC (Serva and Nagy, 2006).

Another example for the role of a cellular chaperone has been found with FHV, which requires Hsp90 for production of protein A RdRp (Castorena et al., 2007; Kampmueller and Miller, 2005). When Hsp90 was inhibited with geldanamycin, protein A was present at $\sim$20% of its normal level, leading to inefficient assembly of the FHV RC and a reduced rate of RNA synthesis (Castorena et al., 2007). The actual mechanism of Hsp90-assisted translation/stabilization of protein A is currently unknown.

Moreover, the human FKBP8 immunophilin has been shown to interact with the HCV NS5A, which is a multifunctional phosphoprotein, and with the Hsp90 chaperone (Okamoto et al., 2006). These interactions were postulated to play a role in stabilization of the HCV RC. RNAi-based knock-down experiments with FKBP8 and the use of an inhibitor of Hsp90 demonstrated significant reduction in HCV replication in human hepatoma cell lines (Okamoto et al., 2006).

Picornaviruses, such as PV or DCV, induce a cytoplasmic vesicular compartment where RNA replication takes place (Cherry et al., 2006; Egger and Bienz, 2005). The formation of the vesicular compartment by DCV depends on the COPI host protein (Cherry et al., 2006). Depletion of COPI via RNAi reduced cytoplasmic vesicles by 2.5-fold, suggesting that the COPI coatomer is critical for the formation of the DCV-induced vesicular compartment. The induction of COPI-based vesicle formation could be regulated by small GTPase ADP-ribosylation factor 1 (Arf1). Indeed, the PV RC recruits Arfs, which, in GTP-bound form (the membrane-associated active form), could recruit other cellular proteins to change membrane curvature, induce transport vesicles from intracellular membrane organelles, and stimulate phospholipase D activity that modifies the lipid composition of membranes (Belov and Ehrenfeld, 2007; Belov et al., 2007). PV 3A has been shown to recruit GBF1, a GEF (guanine exchange factor) to facilitate Arf1-activation/localization to the membrane, whereas the PV-coded 3CD protein binds to BIG1/2 GEFs (Belov et al., 2007). Thus, it is likely that PV 3A and 3CD proteins act synergistically to modify the cellular membrane traffic pathway to facilitate the formation of the PV RC (Belov and Ehrenfeld, 2007). Additional data from Wessels et al. revealed that modulation of the Arf-activating pathway by PV might also lead to inhibition of cellular secretion to ensure survival in the infected animal (Wessels et al., 2006a,b). Moreover, reticulon 3 cellular protein, which interacts with the 2A protein of several picornaviruses, might also be involved in modulating the assembly of the viral RC (Tang et al., 2007).

The assembly of the functional tobamovirus RC likely depends on several *Arabidopsis* proteins, including the membrane proteins TOM1 and TOM3, the four-pass membrane protein TOM2A, and the basic protein TOM2B (Tsujimoto et al., 2003; Yamanaka et al., 2002, 2000). Whereas TOM1/3 have been proposed to participate in the recruitment of the TMV 130K replication protein via direct interaction; TOM2A/2B do not interact with 130K, but they interact with TOM1/3. All these host proteins are likely part of the TMV RC on the vacuolar membrane (i.e., tonoplast in plants) based on co-localization studies (Hagiwara et al., 2003). While TOM1/3 might act as anchors for the 130K and 180K RdRp to the membrane via their transmembrane domains, the roles of TOM2A/2B remain uncharacterized (Yamanaka et al., 2002, 2000). TOM2A might be involved in organization of the TMV RC via its interaction with TOM1/3 and the membrane or, alternatively, it could facilitate recruitment of additional, as yet unknown, host factors (Hagiwara et al., 2003; Tsujimoto et al., 2003).

Assembly of the HCV RC might take place on the surfaces of detergent-insoluble lipid rafts, consisting of cholesterol- and sphingolipid-rich microdomains within the subcellular membranes. The cellular hVAP-A, a vesicle membrane transport protein associated with HCV NS4B, NS5A, and NS5B replication proteins, was shown to play a critical role in the assembly of the HCV RC on lipid rafts (Gao et al., 2004). hVAP-A is partially associated with lipid rafts and dominant negative versions of hVAP-A or RNAi-driven depletion of hVAP-A resulted in relocation of NS5B from detergent-resistant to detergent-sensitive membranes. In addition, the NS5A and hVAP-A interaction is also critical for HCV RC assembly, based on the observation that hypophosphorylated NS5A can interact with hVAP-A, leading to efficient replication, whereas the hyperphosphorylated NS5A cannot interact with hVAP-A resulting in reduced replication (Evans et al., 2004). More recently, hVAP-B, a ubiquitous VAP-A-like protein, has also been implicated in HCV replication (Hamamoto et al., 2005). hVAP-B interacts with both NS5A and NS5B of HCV and it is likely part of the HCV RC because specific antibody against VAP-B inhibited HCV replication in vitro. Furthermore, depletion of hVAP-B by RNAi inhibited HCV replication. Unlike overexpression of hVAP-A, the overexpression of hVAP-B enhanced HCV replication, suggesting that hVAP-B is the limiting factor in the formation of the heterodimer between hVAP-A and hVAP-B, which could be the most effective form in facilitating HCV replication (Hamamoto et al., 2005). Overall, hVAP-A and likely hVAP-B play roles in targeting NS5B to the membrane and the assembly of the RC on lipid rafts (Gao et al., 2004).

The assembly of the HCV RC on membranes might also be affected by Rab5, an early endosome protein, which was found to associate with an "ER-derived membraneous web" (re-arranged membrane structure), the site of HCV replication. Rab5, which is a Ras-type small GTPase involved in membrane fusion, was found to associate with the HCV NS4B, the web-inducing protein. Reduction of Rab5 protein expression reduced HCV replication and web formation (Stone et al., 2007).

Another major group of host factors that affect the assembly of the viral RC is those proteins that affect the lipid composition of intracellular membranes used by particular (+)RNA viruses. For example, Ole1p of yeast, a $\Delta 9$ fatty acid desaturase, is known to affect the amount of unsaturated fatty acids, which are important for membrane flexibility/fluidity (Lee and Ahlquist, 2003; Lee et al., 2001). Consistent with this observation, in the *ole1*-mutant yeast, the membranes surrounding the BMV 1a-induced spherules, the site of BMV replication, were stained poorly by osmium tetroxide, which specifically binds to unsaturated fatty acids. Additional work defined that the BMV 1a was still able to induce membrane proliferation by 25% in the *ole1*-mutant yeast, but the activity of the assembled RC was reduced, likely due to the altered binding of the BMV 1a replication protein to the membrane with reduced ratio of unsaturated fatty acids (Lee and Ahlquist, 2003).

The generation of new membrane surfaces via the fatty acid biosynthesis pathway also has a major effect on DCV replication (Cherry et al., 2006). For example, depletion of the regulator of fatty acid metabolism, HLH106, and the fatty acid synthase CG3523 (performing a rate limiting step in fatty acid biosynthesis) blocked the formation of the DCV-induced vesicular compartment and altered Golgi structure. Also, carulenin, a fatty acid synthase inhibitor, blocked DCV and PV replication in tissue culture (Cherry et al., 2006). The cellular Arfs proteins also regulate the enzymatic activities of proteins involved in lipid metabolism and cytoskeleton function. Down-regulation of COPI and several genes involved in fatty acid metabolism also affected TBSV replication in yeast (Jiang et al., 2006; Panavas et al., 2005b). Altogether, these observations support the role of de novo fatty acid/membrane synthesis in picornavirus replication, where a twofold increase in total membrane surface area within infected cells has been noted.

Role of Viral Replicase Complex Assembly in Template Specificity

The elaborate assembly of the viral replicase could be an important specificity factor (a secondary, safe-guarding step behind the template selection step, see above) to prevent efficient replication of some defective viral RNAs, cellular, and/or heterologous viral RNAs. Regulation of RC assembly might ensure that efficient replication could occur only for those RNAs that contain all the required *cis*-acting elements for the assembly of the viral replicase. Thus, RC assembly might be a "safeguard" mechanism against wasting limited viral/host components on amplification of defective viral RNA templates, which, when present in large amounts, could also trigger antiviral responses (Szittya et al., 2002).

Altogether, the assembly of functional replicase seems to be a highly regulated event during (+)RNA virus infections (Fig. 14.1, step 3). Both *cis*-acting RNA elements and *trans*-acting viral and host factors contribute to the fidelity of the assembly process, which is highly controlled in order to guarantee the presence of appropriate factors and that the assembly takes place only at the correct intracellular space and at the right time.

Step 4. The Synthesis of Viral RNA Progeny

After the assembly of the viral RC, the replicase must efficiently recognize *cis*-acting elements in the viral RNA to be able to synthesize the progeny RNA in a regulated fashion. First, the full-length minus-stranded complementary RNA is produced, then, the plus-stranded RNA progeny are synthesized in a 20- to 100-fold excess amount. Production of both (–) and (+)RNAs requires initiation of RNA synthesis at specific sites, called promoter (initiation) elements, by the viral RC either de novo (i.e., independent of primers) or via using protein primers (Dreher, 1999; Kao et al., 2001; Nagy and Pogany, 2000; Paul et al., 2000, 1998; White and Nagy, 2004). Although all (+)RNA viruses code for their own RdRp and usually one-to-six auxiliary proteins, such as helicase and methyltransferase, host-coded factors are predicted to participate in each step of RNA synthesis as described for selected (+)RNA viruses below.

Host Factors Affecting RNA Binding by the Viral RdRp

Host factors may potentially affect the conformation of the viral RdRp, which in turn could alter the template activity of the RdRp. Indeed, cyclophilin B (CyPB), a cellular peptidyl–prolyl *cis–trans* isomerase (PPIase), was found to regulate the RNA-binding ability of the HCV NS5B RdRp protein (Watashi et al., 2005). RNAi-based knock-down of the CyPB level inhibited HCV RNA accumulation by fivefold. Moreover, both a cyclosporin A and a point mutation in NS5B inhibited the interaction of NS5B with CyPB, causing a significant decrease in HCV replication. These findings argue that CyPB interaction with NS5B is critical during HCV replication.

Host Factors Affecting Minus-Strand RNA Synthesis

eEF1A, a eukaryotic translation elongation factor, was found to bind to the 3′ UTR of (+)RNA of WNV, a flavivirus, based on RNase footprinting and nitrocellulose filter-binding assays (Blackwell and Brinton, 1997; Brinton, 2001). Deleterious RNA mutations introduced at the eEF1A binding sites decreased minus-strand synthesis, strongly suggesting that eEF1A plays a critical role in WNV replication.

Moreover, eEF1A is co-localized with the WNV RC in the infected cells, indicating that eEF1A facilitates the interaction between the viral replicase and the 3′ UTR of the viral gRNA (Davis et al., 2007). Similarly, eEF1A binds to the 3′ end of Turnip yellow mosaic virus (+)RNA, which enhances translation, but represses minus-strand synthesis (Matsuda et al., 2004). The role of eEF1A in virus replication might be rather broad since eEF1A was reported to bind to TMV (+)RNA, the TMV replicase (Yamaji et al., 2006), to the NS5A replication protein of BVDV (Johnson et al., 2001), NS4A of HCV (Kou et al., 2006), Gag polyprotein of HIV-1 (Cimarelli and Luban, 1999), the TBSV RNA, and p33 replication protein (Li and Nagy, unpublished). It is also part of the RC for vesicular stomatitis virus, a negative-stranded RNA virus (Qanungo et al., 2004). Therefore, the highly conserved eEF1A could play a major role in the replication of several RNA viruses via its interactions with viral RNAs and viral replication proteins. The high abundance of eEF1A in cells might facilitate its recruitment into virus replication.

A human RNA helicase, p68, is redistributed from the nucleus to the cytosol during HCV infections due to its interaction with NS5B RdRp of HCV (Goh et al., 2004). RNAi-based depletion of p68 was found to lead to a reduction in minus-strand HCV RNA accumulation, suggesting that p68 might act as a transcription factor for HCV replication (Goh et al., 2004).

Host Factors Regulating Plus-Strand RNA Synthesis

hnRNP-C binds specifically to the 3′ end of the PV (–)RNA and to poly(U) and oligo(U) stretches in the PV RNA, via its RRM domain, and it also interacts with the replication protein 3CD (Brunner et al., 2005). Immunoprecipitation of hnRNP-C led to co-purification of both (+) and (–) PV RNAs. Although hnRNP-C is a nuclear protein, it is redistributed to the cytoplasm, probably due to PV-infection driven alterations in nucleocytoplasmic trafficking (Gustin and Sarnow, 2001). Semler et al. suggested that hnRNP-C and PV protein 2C NTPase/helicase and/or 2BC precursor bind to the PV stem-loop structure at the 3′ terminus of (–)RNA. They also proposed that hnRNP-C might recruit the 3CD to the (–)RNA template. The RNA chaperone activity of hnRNP-C could facilitate the folding of (–)RNA or the replicative dsRNA intermediate into a replication-compatible conformation that leaves the two terminal adenines exposed via binding to a 3′ proximal uridine stretch (Brunner et al., 2005).

The stress granule proteins named T-cell intracellular antigen-1 (TIA-1) as well as the TIA-1-related protein (TIAR), containing three RRM RNA-binding motifs, have been found to bind specifically to the 3′ terminal stem-loop in WNV and to the NS3 replication protein (Emara and Brinton, 2007; Li et al., 2002). Brinton et al. proposed that specific binding of TIAR and, to lesser extent, TIA-1 could promote the binding of the WNV to the minus-strand template, which could lead to efficient plus-strand synthesis. In addition to the proposed transcription factor roles of TIAR and TIA-1, the recruitment/sequestration of these proteins for WNV replication might also inhibit the formation of stress granules and P-bodies, thus potentially inhibiting the shut-off of host protein translation and mRNA degradation (Emara and Brinton, 2007).

Regulation of Asymmetrical RNA Synthesis by Host Factors

One of the hallmark features of (+)RNA viruses is the asymmetric nature of viral RNA synthesis, which leads to 10- to 100-fold more (+)RNA progeny than the (–)RNA replication intermediate (Buck, 1996). Wang and Nagy (unpublished) found that the key co-opted host protein for TBSV replication that regulates asymmetrical RNA synthesis is glyceraldehyde-3-phosphate dehydrogenase (GAPDH, or Tdh2/3p in yeast). GAPDH, which is present in the tombusvirus RC (Serva and Nagy, 2006), is a highly conserved, very abundant protein that is ubiquitous in nature (Sirover, 1999). GAPDH is a key component of cytosolic energy production, but it also displays many additional cellular activities, including roles in apoptosis, endocytosis, nuclear tRNA transport, vesicular secretory transport, nuclear membrane fusion, modulation of the cytoskeleton, DNA replication and repair, maintenance of telomere structure, and transcriptional control of histone gene expression (Sirover, 1999, 2005). GAPDH also binds to various RNAs, such as AU-rich sequences at the 3′ terminus of mRNAs, which can lead to stabilization of the RNA in the cell (Bonafe et al., 2005). Interestingly, down-regulation of GAPDH inhibited TBSV replication in yeast and in plants and resulted in the production of (+) and (–)RNAs in a 1:1 ratio, instead of the hallmark asymmetric RNA synthesis. Thus, the replication of TBSV became double-stranded RNA virus like when only a limiting amount of GAPDH was present in the viral RC. Moreover, GAPDH was re-localized from the cytosol to the peroxisomal membrane surface, the site of TBSV RNA synthesis. Based on in vitro and in vivo data, it has been proposed that GAPDH promotes asymmetric RNA synthesis by selectively retaining the (–)RNA template in the viral RC, thus allowing efficient access of the (–)RNA replication intermediate to the viral RC. On the contrary, (+)RNA progeny, which are not bound by GAPDH, get released from the RC into the cytosol (Wang and Nagy, unpublished). Thus, a cellular metabolic enzyme can regulate asymmetric viral RNA synthesis, explaining this hallmark feature of (+)RNA viruses.

Regulation of Subgenomic RNA Synthesis by Host Factors

A number of (+)RNA viruses express a set of their genes via producing sgRNAs from the gRNA (via a minus-stranded intermediate) (Lin and White, 2004; Pasternak et al., 2006, 2001; Sawicki et al., 2007; Snijder, 2001; White, 2002). In addition to the viral replication proteins, host proteins are also predicted to participate in regulation of sgRNA synthesis. Accordingly, Hardy and his colleagues found that hnRNP-K, a predominantly nuclear poly(C)-binding protein with 3 KH RNA-binding domains, co-precipitated with the SIN sgRNA, but not with the gRNA (Burnham et al., 2007). hnRNP-K also interacted with SIN nonstructural proteins and it was redistributed to the membrane fraction and co-localized with the SIN RC in infected cells. Interestingly, the amount of mitogen-activated protein kinase (MAPK) is up-regulated in SIN-infected cells and MAPK-based phosphorylation of hnRNP-K can lead to cytosolic accumulation of hnRNP-K, suggesting that the MAPK pathway could be involved in the re-localization of hnRNP-K for SIN replication. Overall, the interaction of hnRNP-K with the sgRNA and the SIN non-structural proteins might lead to up-regulation of the sgRNA synthesis that produces ~fourfold more sgRNA than gRNA (Burnham et al., 2007).

Step 5. Release of Viral RNA from Replication

After completion of the new (+)RNAs, they must be released from replication in order to participate in additional functions, such as new rounds of translation, replication, or packaging (Fig. 14.1, step 5). Time point studies with TBSV RNA revealed that a significant portion of the (+)RNA was associated with the sites of replication at an early time point (12 hours), while most (+)RNA was cytosolic, thus released from replication, at a later time (48 hours) (Panavas et al., 2005a). On the other hand, the (–)RNA was associated with the p33 replication protein on the peroxisomal membrane surface at both early and late time points (Panavas et al., 2005a). These observations suggest that the release of RNA progeny from replication is a highly regulated event.

Unfortunately, the escape mechanism of viral (+)RNA from replication is currently not known. It has been proposed that post-translational modification, such as phosphorylation, of host or viral proteins within the RC might play a role. For example, phosphorylation of serine/threonine residues in the vicinity of the RNA-binding domain of the TBSV p33 replication protein was shown to reduce RNA-binding capacity of p33 (Shapka et al., 2005; Stork et al., 2005). If phosphorylation takes place reversibly, then the same RC could release the viral (+)RNA progeny (Fig. 14.1, step 5), followed by new rounds of RNA synthesis and release. The cellular kinase involved in phosphorylation of TBSV p33 might be protein kinase-C (PKC) like, based on in vitro experiments. Overall, the above model does not exclude that other processes, such as replicase disassembly might also play a role in viral (+)RNA release.

Step 6. Disassembly of the Viral Replicase

The viral RC likely becomes inactivated and goes through disassembly at the end of replication (Fig. 14.1, step 6). Also, disassembly of the RC might promote the release of the (+)RNA progeny from replication. Phosphorylation and/or additional post-translational modifications might be involved in RC disassembly. A genome-wide screen in yeast identified host genes that could affect the ubiquitination pathway, such as *BRE1* and *RAD6* (Panavas et al., 2005b). Thus, ubiquitination of replication proteins may alter the stability of the virus RC. Also, it has been postulated that an unidentified cellular kinase that phosphorylates the HCV NS5A protein could inhibit its interaction with hVAP-33, thus leading to disassembly of the HCV RC (Evans et al., 2004). Subsequent work identified the human p70S6K and related kinases as the likely candidates for hyperphosphorylation of NS5A (Coito et al., 2004). Overall, our current knowledge is poor about RC disassembly.

Future Directions

In spite of recent major efforts in studies of host–virus interactions and viral replication, our knowledge in many areas is still incomplete. However, the application of systems biology approaches and the availability of complete sequences for many host genomes, in combination with development of in vitro approaches and yeast as a model host, will likely lead to rapid advances in identification and characterization of host factors involved in (+)RNA virus replication. The combined use of genetics, biochemistry, and cell biology will help dissect the detailed functions of subverted host proteins. Also, proteomics-based analysis of the viral RC is expected to lead to identification of host proteins recruited into (+)RNA virus replication. Determination of 3D structures of viral RdRp and the auxiliary replication proteins with bound RNAs, as well as high resolution imaging of viral RCs, will likely unravel the mechanism and regulation of (+)RNA replication. Proteomics approaches should also accelerate identification of various post-translational modifications of viral and host proteins that could affect and/or regulate their functions during the replication process. These advances will lead to a better understanding of virus replication and host:(+)RNA virus interactions, which are key aspects of viral pathogenesis.

Acknowledgments The authors thank Dr. David Smith and members of the Nagy lab for discussion. The authors apologize to those colleagues whose research on replication of plus-strand RNA viruses was not mentioned in this review due to page restrictions. This work was supported by NIH-NIAID.

References

Ahlquist, P. (2002). RNA-dependent RNA polymerases, viruses, and RNA silencing. *Science* **296**(5571)**,** 1270–3.

Ahlquist, P., Noueiry, A. O., Lee, W. M., Kushner, D. B., and Dye, B. T. (2003). Host factors in positive-strand RNA virus genome replication. *J Virol* **77**(15)**,** 8181–6.

Aizaki, H., Choi, K. S., Liu, M., Li, Y. J., and Lai, M. M. (2006). Polypyrimidine-tract-binding protein is a component of the HCV RNA replication complex and necessary for RNA synthesis. *J Biomed Sci* **13**(4)**,** 469–80.

Asano, M., Satoh, R., Mochizuki, A., Tsuda, S., Yamanaka, T., Nishiguchi, M., Hirai, K., Meshi, T., Naito, S., and Ishikawa, M. (2005). Tobamovirus-resistant tobacco generated by RNA interference directed against host genes. *FEBS Lett* **579**(20)**,** 4479–84.

Beckham, C. J., Light, H. R., Nissan, T. A., Ahlquist, P., Parker, R., and Noueiry, A. (2007). Interactions between brome mosaic virus RNAs and cytoplasmic processing bodies. *J Virol* **81**(18)**,** 9759–68.

Belov, G. A., and Ehrenfeld, E. (2007). Involvement of cellular membrane traffic proteins in poliovirus replication. *Cell Cycle* **6**(1)**,** 36–8.

Belov, G. A., Habbersett, C., Franco, D., and Ehrenfeld, E. (2007). Activation of cellular Arf GTPases by poliovirus protein 3CD correlates with virus replication. *J Virol* **81**(17)**,** 9259–67.

Blackwell, J. L., and Brinton, M. A. (1997). Translation elongation factor-1 alpha interacts with the 3′ stem-loop region of West Nile virus genomic RNA. *J Virol* **71**(9), 6433–44.

Bonafe, N., Gilmore-Hebert, M., Folk, N. L., Azodi, M., Zhou, Y., and Chambers, S. K. (2005). Glyceraldehyde-3-phosphate dehydrogenase binds to the AU-Rich 3′ untranslated region of colony-stimulating factor-1 (CSF-1) messenger RNA in human ovarian cancer cells: possible role in CSF-1 posttranscriptional regulation and tumor phenotype. *Cancer Res* **65**(9), 3762–71.

Brinton, M. A. (2001). Host factors involved in West Nile virus replication. *Ann NY Acad Sci* **951**, 207–19.

Brunner, J. E., Nguyen, J. H., Roehl, H. H., Ho, T. V., Swiderek, K. M., and Semler, B. L. (2005). Functional interaction of heterogeneous nuclear ribonucleoprotein C with poliovirus RNA synthesis initiation complexes. *J Virol* **79**(6), 3254–66.

Buck, K. W. (1996). Comparison of the replication of positive-stranded RNA viruses of plants and animals. *Adv Virus Res* **47**, 159–251.

Buck, K. W. (1999). Replication of tobacco mosaic virus RNA. *Philos Trans R Soc Lond B Biol Sci* **354**(1383), 613–27.

Burgyan, J., Rubino, L., and Russo, M. (1996). The 5′-terminal region of a tombusvirus genome determines the origin of multivesicular bodies. *J Gen Virol* **77**(Pt 8), 1967–74.

Burnham, A. J., Gong, L., and Hardy, R. W. (2007). Heterogeneous nuclear ribonuclear protein K interacts with Sindbis virus nonstructural proteins and viral subgenomic mRNA. *Virology* **367**(1), 212–21.

Castorena, K. M., Weeks, S. A., Stapleford, K. A., Cadwallader, A. M., and Miller, D. J. (2007). A functional heat shock protein 90 chaperone is essential for efficient flock house virus RNA polymerase synthesis in Drosophila cells. *J Virol* **81**(16), 8412–20.

Chang, K. S., and Luo, G. (2006). The polypyrimidine tract-binding protein (PTB) is required for efficient replication of hepatitis C virus (HCV) RNA. *Virus Res* **115**(1), 1–8.

Cheng, C. P., Jaag, H. M., Jonczyk, M., Serviene, E., and Nagy, P. D. (2007). Expression of the Arabidopsis Xrn4p 5′-3′ exoribonuclease facilitates degradation of tombusvirus RNA and promotes rapid emergence of viral variants in plants. *Virology* **368**(2), 238–48.

Cheng, C. P., and Nagy, P. D. (2003). Mechanism of RNA recombination in carmo- and tombusviruses: evidence for template switching by the RNA-dependent RNA polymerase in vitro. *J Virol* **77**(22), 12033–47.

Cheng, C. P., Serviene, E., and Nagy, P. D. (2006). Suppression of viral RNA recombination by a host exoribonuclease. *J Virol* **80**(6), 2631–40.

Cherry, S., Doukas, T., Armknecht, S., Whelan, S., Wang, H., Sarnow, P., and Perrimon, N. (2005). Genome-wide RNAi screen reveals a specific sensitivity of IRES-containing RNA viruses to host translation inhibition. *Genes Dev* **19**(4), 445–52.

Cherry, S., Kunte, A., Wang, H., Coyne, C., Rawson, R. B., and Perrimon, N. (2006). COPI activity coupled with fatty acid biosynthesis is required for viral replication. *PLoS Pathog* **2**(10), e102.

Choi, K. S., Huang, P., and Lai, M. M. (2002). Polypyrimidine-tract-binding protein affects transcription but not translation of mouse hepatitis virus RNA. *Virology* **303**(1), 58–68.

Choi, K. S., Mizutani, A., and Lai, M. M. (2004). SYNCRIP, a member of the heterogeneous nuclear ribonucleoprotein family, is involved in mouse hepatitis virus RNA synthesis. *J Virol* **78**(23), 13153–62.

Cimarelli, A., and Luban, J. (1999). Translation elongation factor 1-alpha interacts specifically with the human immunodeficiency virus type 1 Gag polyprotein. *J Virol* **73**(7), 5388–401.

Coito, C., Diamond, D. L., Neddermann, P., Korth, M. J., and Katze, M. G. (2004). High-throughput screening of the yeast kinome: identification of human serine/threonine protein kinases that phosphorylate the hepatitis C virus NS5A protein. *J Virol* **78**(7), 3502–13.

Cristea, I. M., Carroll, J. W., Rout, M. P., Rice, C. M., Chait, B. T., and MacDonald, M. R. (2006). Tracking and elucidating alphavirus-host protein interactions. *J Biol Chem* **281**(40), 30269–78.

Davis, W. G., Blackwell, J. L., Shi, P. Y., and Brinton, M. A. (2007). Interaction between the Cellular Protein eEF1A and the 3′-Terminal Stem-Loop of West Nile Virus Genomic RNA Facilitates Viral Minus-Strand RNA Synthesis. *J Virol* **81**(18), 10172–87.

Diez, J., Ishikawa, M., Kaido, M., and Ahlquist, P. (2000). Identification and characterization of a host protein required for efficient template selection in viral RNA replication. *Proc Natl Acad Sci USA* **97**(8)**,** 3913–8.

Dreher, T. W. (1999). Functions of the 3′-Untranslated Regions of Positive Strand Rna Viral Genomes. *Annu Rev Phytopathol* **37,** 151–174.

Egger, D., and Bienz, K. (2005). Intracellular location and translocation of silent and active poliovirus replication complexes. *J Gen Virol* **86**(Pt 3)**,** 707–18.

Emara, M. M., and Brinton, M. A. (2007). Interaction of TIA-1/TIAR with West Nile and dengue virus products in infected cells interferes with stress granule formation and processing body assembly. *Proc Natl Acad Sci USA* **104**(21)**,** 9041–6.

Evans, M. J., Rice, C. M., and Goff, S. P. (2004). Phosphorylation of hepatitis C virus nonstructural protein 5A modulates its protein interactions and viral RNA replication. *Proc Natl Acad Sci USA* **101**(35)**,** 13038–43.

Frolova, E., Gorchakov, R., Garmashova, N., Atasheva, S., Vergara, L. A., and Frolov, I. (2006). Formation of nsP3-specific protein complexes during Sindbis virus replication. *J Virol* **80**(8)**,** 4122–34.

Gamarnik, A. V., and Andino, R. (1998). Switch from translation to RNA replication in a positive-stranded RNA virus. *Genes Dev* **12**(15)**,** 2293–304.

Gamarnik, A. V., and Andino, R. (2000). Interactions of viral protein 3CD and poly(rC) binding protein with the 5′ untranslated region of the poliovirus genome. *J Virol* **74**(5)**,** 2219–26.

Gao, L., Aizaki, H., He, J. W., and Lai, M. M. (2004). Interactions between viral nonstructural proteins and host protein hVAP-33 mediate the formation of hepatitis C virus RNA replication complex on lipid raft. *J Virol* **78**(7)**,** 3480–8.

Gao, L., Tu, H., Shi, S. T., Lee, K. J., Asanaka, M., Hwang, S. B., and Lai, M. M. (2003). Interaction with a ubiquitin-like protein enhances the ubiquitination and degradation of hepatitis C virus RNA-dependent RNA polymerase. *J Virol* **77**(7)**,** 4149–59.

Go, E. P., Wikoff, W. R., Shen, Z., O'Maille, G., Morita, H., Conrads, T. P., Nordstrom, A., Trauger, S. A., Uritboonthai, W., Lucas, D. A., Chan, K. C., Veenstra, T. D., Lewicki, H., Oldstone, M. B., Schneemann, A., and Siuzdak, G. (2006). Mass spectrometry reveals specific and global molecular transformations during viral infection. *J Proteome Res* **5**(9)**,** 2405–16.

Goh, P. Y., Tan, Y. J., Lim, S. P., Tan, Y. H., Lim, S. G., Fuller-Pace, F., and Hong, W. (2004). Cellular RNA helicase p68 relocalization and interaction with the hepatitis C virus (HCV) NS5B protein and the potential role of p68 in HCV RNA replication. *J Virol* **78**(10)**,** 5288–98.

Gustin, K. E., and Sarnow, P. (2001). Effects of poliovirus infection on nucleo-cytoplasmic trafficking and nuclear pore complex composition. *Embo J* **20**(1–2)**,** 240–9.

Hagiwara, Y., Komoda, K., Yamanaka, T., Tamai, A., Meshi, T., Funada, R., Tsuchiya, T., Naito, S., and Ishikawa, M. (2003). Subcellular localization of host and viral proteins associated with tobamovirus RNA replication. *Embo J* **22**(2)**,** 344–53.

Hamamoto, I., Nishimura, Y., Okamoto, T., Aizaki, H., Liu, M., Mori, Y., Abe, T., Suzuki, T., Lai, M. M., Miyamura, T., Moriishi, K., and Matsuura, Y. (2005). Human VAP-B is involved in hepatitis C virus replication through interaction with NS5A and NS5B. *J Virol* **79**(21)**,** 13473–82.

Herold, J., and Andino, R. (2001). Poliovirus RNA replication requires genome circularization through a protein–protein bridge. *Mol Cell* **7**(3)**,** 581–91.

Ishikawa, M., Diez, J., Restrepo-Hartwig, M., and Ahlquist, P. (1997a). Yeast mutations in multiple complementation groups inhibit brome mosaic virus RNA replication and transcription and perturb regulated expression of the viral polymerase-like gene. *Proc Natl Acad Sci USA* **94**(25)**,** 13810–5.

Ishikawa, M., Janda, M., Krol, M. A., and Ahlquist, P. (1997b). In vivo DNA expression of functional brome mosaic virus RNA replicons in Saccharomyces cerevisiae. *J Virol* **71**(10)**,** 7781–90.

Isken, O., Baroth, M., Grassmann, C. W., Weinlich, S., Ostareck, D. H., Ostareck-Lederer, A., and Behrens, S. E. (2007). Nuclear factors are involved in hepatitis C virus RNA replication. *Rna* **13**(10)**,** 1675–92.

Isken, O., Grassmann, C. W., Sarisky, R. T., Kann, M., Zhang, S., Grosse, F., Kao, P. N., and Behrens, S. E. (2003). Members of the NF90/NFAR protein group are involved in the life cycle of a positive-strand RNA virus. *Embo J* **22**(21), 5655–65.

Janda, M., and Ahlquist, P. (1993). RNA-dependent replication, transcription, and persistence of brome mosaic virus RNA replicons in S. cerevisiae. *Cell* **72**(6), 961–70.

Jiang, Y., Serviene, E., Gal, J., Panavas, T., and Nagy, P. D. (2006). Identification of essential host factors affecting tombusvirus RNA replication based on the yeast Tet promoters Hughes Collection. *J Virol* **80**(15), 7394–404.

Johnson, C. M., Perez, D. R., French, R., Merrick, W. C., and Donis, R. O. (2001). The NS5A protein of bovine viral diarrhoea virus interacts with the alpha subunit of translation elongation factor-1. *J Gen Virol* **82**(Pt 12), 2935–43.

Kampmueller, K. M., and Miller, D. J. (2005). The cellular chaperone heat shock protein 90 facilitates Flock House virus RNA replication in Drosophila cells. *J Virol* **79**(11), 6827–37.

Kao, C. C., Singh, P., and Ecker, D. J. (2001). De novo initiation of viral RNA-dependent RNA synthesis. *Virology* **287**(2), 251–60.

Kim, M. J., and Kao, C. (2001). Factors regulating template switch in vitro by viral RNA-dependent RNA polymerases: implications for RNA-RNA recombination. *Proc Natl Acad Sci USA* **98**(9), 4972–7.

Kim, C. S., Seol, S. K., Song, O. K., Park, J. H., and Jang, S. K. (2007). An RNA-binding protein, hnRNP A1, and a scaffold protein, septin 6, facilitate hepatitis C virus replication. *J Virol* **81**(8), 3852–65.

Kirkegaard, K., and Jackson, W. T. (2005). Topology of double-membraned vesicles and the opportunity for non-lytic release of cytoplasm. *Autophagy* **1**(3), 182–4.

Kou, Y. H., Chou, S. M., Wang, Y. M., Chang, Y. T., Huang, S. Y., Jung, M. Y., Huang, Y. H., Chen, M. R., Chang, M. F., and Chang, S. C. (2006). Hepatitis C virus NS4A inhibits cap-dependent and the viral IRES-mediated translation through interacting with eukaryotic elongation factor 1A. *J Biomed Sci* **13**(6), 861–74.

Kushner, D. B., Lindenbach, B. D., Grdzelishvili, V. Z., Noueiry, A. O., Paul, S. M., and Ahlquist, P. (2003). Systematic, genome-wide identification of host genes affecting replication of a positive-strand RNA virus. *Proc Natl Acad Sci USA* **100**(26), 15764–9.

Kyono, K., Miyashiro, M., and Taguchi, I. (2002). Human eukaryotic initiation factor 4AII associates with hepatitis C virus NS5B protein in vitro. *Biochem Biophys Res Commun* **292**(3), 659–66.

Lee, W. M., and Ahlquist, P. (2003). Membrane synthesis, specific lipid requirements, and localized lipid composition changes associated with a positive-strand RNA virus RNA replication protein. *J Virol* **77**(23), 12819–28.

Lee, W. M., Ishikawa, M., and Ahlquist, P. (2001). Mutation of host delta9 fatty acid desaturase inhibits brome mosaic virus RNA replication between template recognition and RNA synthesis. *J Virol* **75**(5), 2097–106.

Li, W., Li, Y., Kedersha, N., Anderson, P., Emara, M., Swiderek, K. M., Moreno, G. T., and Brinton, M. A. (2002). Cell proteins TIA-1 and TIAR interact with the 3′ stem-loop of the West Nile virus complementary minus-strand RNA and facilitate virus replication. *J Virol* **76**(23), 11989–2000.

Li, H. P., Zhang, X., Duncan, R., Comai, L., and Lai, M. M. (1997). Heterogeneous nuclear ribonucleoprotein A1 binds to the transcription-regulatory region of mouse hepatitis virus RNA. *Proc Natl Acad Sci USA* **94**(18), 9544–9.

Lin, H. X., and White, K. A. (2004). A complex network of RNA-RNA interactions controls subgenomic mRNA transcription in a tombusvirus. *Embo J* **23**(16), 3365–74.

Mas, A., Alves-Rodrigues, I., Noueiry, A., Ahlquist, P., and Diez, J. (2006). Host deadenylation-dependent mRNA decapping factors are required for a key step in brome mosaic virus RNA replication. *J Virol* **80**(1), 246–51.

Matsuda, D., Yoshinari, S., and Dreher, T. W. (2004). eEF1A binding to aminoacylated viral RNA represses minus strand synthesis by TYMV RNA-dependent RNA polymerase. *Virology* **321**(1), 47–56.

McBride, A. E., Schlegel, A., and Kirkegaard, K. (1996). Human protein Sam68 relocalization and interaction with poliovirus RNA polymerase in infected cells. *Proc Natl Acad Sci USA* **93**(6), 2296–301.

Monkewich, S., Lin, H. X., Fabian, M. R., Xu, W., Na, H., Ray, D., Chernysheva, O. A., Nagy, P. D., and White, K. A. (2005). The p92 Polymerase Coding Region Contains an Internal RNA Element Required at an Early Step in Tombusvirus Genome Replication. *J Virol* **79**(8), 4848–58.

Nagy, P. D., Carpenter, C. D., and Simon, A. E. (1997). A novel 3′-end repair mechanism in an RNA virus. *Proc Natl Acad Sci USA* **94**(4), 1113–8.

Nagy, P. D., Dzianott, A., Ahlquist, P., and Bujarski, J. J. (1995). Mutations in the helicase-like domain of protein 1a alter the sites of RNA-RNA recombination in brome mosaic virus. *J Virol* **69**(4), 2547–56.

Nagy, P. D., and Pogany, J. (2000). Partial purification and characterization of Cucumber necrosis virus and Tomato bushy stunt virus RNA-dependent RNA polymerases: similarities and differences in template usage between tombusvirus and carmovirus RNA-dependent RNA polymerases. *Virology* **276**(2), 279–88.

Nagy, P. D., and Pogany, J. (2006). Yeast as a model host to dissect functions of viral and host factors in tombusvirus replication. *Virology* **344**(1), 211–20.

Nagy, P. D., and Simon, A. E. (1997). New insights into the mechanisms of RNA recombination. *Virology* **235**(1), 1–9.

Nagy, P. D., Zhang, C., and Simon, A. E. (1998). Dissecting RNA recombination in vitro: role of RNA sequences and the viral replicase. *Embo J* **17**(8), 2392–403.

Nanda, S. K., Johnson, R. F., Liu, Q., and Leibowitz, J. L. (2004). Mitochondrial HSP70, HSP40, and HSP60 bind to the 3′ untranslated region of the Murine hepatitis virus genome. *Arch Virol* **149**(1), 93–111.

Nanda, S. K., and Leibowitz, J. L. (2001). Mitochondrial aconitase binds to the 3′ untranslated region of the mouse hepatitis virus genome. *J Virol* **75**(7), 3352–62.

Navarro, B., Rubino, L., and Russo, M. (2004). Expression of the Cymbidium ringspot virus 33-kilodalton protein in Saccharomyces cerevisiae and molecular dissection of the peroxisomal targeting signal. *J Virol* **78**(9), 4744–52.

Neeleman, L., and Bol, J. F. (1999). Cis-acting functions of alfalfa mosaic virus proteins involved in replication and encapsidation of viral RNA. *Virology* **254**(2), 324–33.

Ng, T. I., Mo, H., Pilot-Matias, T., He, Y., Koev, G., Krishnan, P., Mondal, R., Pithawalla, R., He, W., Dekhtyar, T., Packer, J., Schurdak, M., and Molla, A. (2007). Identification of host genes involved in hepatitis C virus replication by small interfering RNA technology. *Hepatology* **45**(6), 1413–21.

Noueiry, A. O., and Ahlquist, P. (2003). Brome Mosaic Virus RNA Replication: Revealing the Role of the Host in RNA Virus Replication. *Annu Rev Phytopathol.*

Noueiry, A. O., Diez, J., Falk, S. P., Chen, J., and Ahlquist, P. (2003). Yeast Lsm1p-7p/Pat1p deadenylation-dependent mRNA-decapping factors are required for brome mosaic virus genomic RNA translation. *Mol Cell Biol* **23**(12), 4094–106.

Novak, J. E., and Kirkegaard, K. (1994). Coupling between genome translation and replication in an RNA virus. *Genes Dev* **8**(14), 1726–37.

Okamoto, T., Nishimura, Y., Ichimura, T., Suzuki, K., Miyamura, T., Suzuki, T., Moriishi, K., and Matsuura, Y. (2006). Hepatitis C virus RNA replication is regulated by FKBP8 and Hsp90. *Embo J* **25**(20), 5015–25.

Osman, T. A., and Buck, K. W. (1997). The tobacco mosaic virus RNA polymerase complex contains a plant protein related to the RNA-binding subunit of yeast eIF-3. *J Virol* **71**(8), 6075–82.

Osman, T. A., and Buck, K. W. (2003). Identification of a region of the tobacco mosaic virus 126- and 183-kilodalton replication proteins which binds specifically to the viral 3′-terminal tRNA-like structure. *J Virol* **77**(16), 8669–75.

Oster, S. K., Wu, B., and White, K. A. (1998). Uncoupled expression of p33 and p92 permits amplification of tomato bushy stunt virus RNAs. *J Virol* **72**(7), 5845–51.

Panavas, T., Hawkins, C. M., Panaviene, Z., and Nagy, P. D. (2005a). The role of the p33:p33/p92 interaction domain in RNA replication and intracellular localization of p33 and p92 proteins of Cucumber necrosis tombusvirus. *Virology*.

Panavas, T., and Nagy, P. D. (2003). Yeast as a model host to study replication and recombination of defective interfering RNA of Tomato bushy stunt virus. *Virology* **314**(1), 315–25.

Panavas, T., Serviene, E., Brasher, J., and Nagy, P. D. (2005b). Yeast genome-wide screen reveals dissimilar sets of host genes affecting replication of RNA viruses. *Proc Natl Acad Sci USA* **102**(20), 7326–31.

Panaviene, Z., Baker, J. M., and Nagy, P. D. (2003). The overlapping RNA-binding domains of p33 and p92 replicase proteins are essential for tombusvirus replication. *Virology* **308**(1), 191–205.

Panaviene, Z., Panavas, T., and Nagy, P. D. (2005). Role of an internal and two 3′-terminal RNA elements in assembly of tombusvirus replicase. *J Virol* **79**(16), 10608–18.

Panaviene, Z., Panavas, T., Serva, S., and Nagy, P. D. (2004). Purification of the cucumber necrosis virus replicase from yeast cells: role of coexpressed viral RNA in stimulation of replicase activity. *J Virol* **78**(15), 8254–63.

Pantaleo, V., Rubino, L., and Russo, M. (2003). Replication of Carnation Italian ringspot virus defective interfering RNA in Saccharomyces cerevisiae. *J Virol* **77**(3), 2116–23.

Pasternak, A. O., Spaan, W. J., and Snijder, E. J. (2006). Nidovirus transcription: how to make sense...? *J Gen Virol* **87**(Pt 6), 1403–21.

Pasternak, A. O., van den Born, E., Spaan, W. J., and Snijder, E. J. (2001). Sequence requirements for RNA strand transfer during nidovirus discontinuous subgenomic RNA synthesis. *Embo J* **20**(24), 7220–8.

Paul, A. V., Rieder, E., Kim, D. W., van Boom, J. H., and Wimmer, E. (2000). Identification of an RNA hairpin in poliovirus RNA that serves as the primary template in the in vitro uridylylation of VPg. *J Virol* **74**(22), 10359–70.

Paul, A. V., van Boom, J. H., Filippov, D., and Wimmer, E. (1998). Protein-primed RNA synthesis by purified poliovirus RNA polymerase. *Nature* **393**(6682), 280–4.

Perera, R., Daijogo, S., Walter, B. L., Nguyen, J. H., and Semler, B. L. (2007). Cellular protein modification by poliovirus: the two faces of poly(rC)-binding protein. *J Virol* **81**(17), 8919–32.

Pogany, J., White, K. A., and Nagy, P. D. (2005). Specific Binding of Tombusvirus Replication Protein p33 to an Internal Replication Element in the Viral RNA Is Essential for Replication. *J Virol* **79**(8), 4859–69.

Price, B. D., Roeder, M., and Ahlquist, P. (2000). DNA-Directed expression of functional flock house virus RNA1 derivatives in Saccharomyces cerevisiae, heterologous gene expression, and selective effects on subgenomic mRNA synthesis. *J Virol* **74**(24), 11724–33.

Qanungo, K. R., Shaji, D., Mathur, M., and Banerjee, A. K. (2004). Two RNA polymerase complexes from vesicular stomatitis virus-infected cells that carry out transcription and replication of genome RNA. *Proc Natl Acad Sci USA* **101**(16), 5952–7.

Quadt, R., Ishikawa, M., Janda, M., and Ahlquist, P. (1995). Formation of brome mosaic virus RNA-dependent RNA polymerase in yeast requires coexpression of viral proteins and viral RNA. *Proc Natl Acad Sci USA* **92**(11), 4892–6.

Quadt, R., Kao, C. C., Browning, K. S., Hershberger, R. P., and Ahlquist, P. (1993). Characterization of a host protein associated with brome mosaic virus RNA-dependent RNA polymerase. *Proc Natl Acad Sci USA* **90**(4), 1498–502.

Quinkert, D., Bartenschlager, R., and Lohmann, V. (2005). Quantitative analysis of the hepatitis C virus replication complex. *J Virol* **79**(21), 13594–605.

Rao, A. L., Dreher, T. W., Marsh, L. E., and Hall, T. C. (1989). Telomeric function of the tRNA-like structure of brome mosaic virus RNA. *Proc Natl Acad Sci USA* **86**(14), 5335–9.

Roehl, H. H., and Semler, B. L. (1995). Poliovirus infection enhances the formation of two ribonucleoprotein complexes at the 3′ end of viral negative-strand RNA. *J Virol* **69**(5), 2954–61.

Roossinck, M. J. (2003). Plant RNA virus evolution. *Curr Opin Microbiol* **6**(4), 406–9.

Rubino, L., and Russo, M. (1998). Membrane targeting sequences in tombusvirus infections. *Virology* **252**(2), 431–7.

Rust, R. C., Landmann, L., Gosert, R., Tang, B. L., Hong, W., Hauri, H. P., Egger, D., and Bienz, K. (2001). Cellular COPII proteins are involved in production of the vesicles that form the poliovirus replication complex. *J Virol* **75**(20)**,** 9808–18.

Salonen, A., Ahola, T., and Kaariainen, L. (2005). Viral RNA replication in association with cellular membranes. *Curr Top Microbiol Immunol* **285,** 139–73.

Sawicki, S. G., Sawicki, D. L., and Siddell, S. G. (2007). A contemporary view of coronavirus transcription. *J Virol* **81**(1)**,** 20–9.

Schwartz, M., Chen, J., Janda, M., Sullivan, M., den Boon, J., and Ahlquist, P. (2002). A positive-strand RNA virus replication complex parallels form and function of retrovirus capsids. *Mol Cell* **9**(3)**,** 505–14.

Serva, S., and Nagy, P. D. (2006). Proteomics analysis of the tombusvirus replicase: Hsp70 molecular chaperone is associated with the replicase and enhances viral RNA replication. *J Virol* **80**(5)**,** 2162–9.

Serviene, E., Jiang, Y., Cheng, C. P., Baker, J., and Nagy, P. D. (2006). Screening of the yeast yTHC collection identifies essential host factors affecting tombusvirus RNA recombination. *J Virol* **80**(3)**,** 1231–41.

Serviene, E., Shapka, N., Cheng, C. P., Panavas, T., Phuangrat, B., Baker, J., and Nagy, P. D. (2005). Genome-wide screen identifies host genes affecting viral RNA recombination. *Proc Natl Acad Sci USA* **102**(30)**,** 10545–50.

Shapka, N., Stork, J., and Nagy, P. D. (2005). Phosphorylation of the p33 replication protein of Cucumber necrosis tombusvirus adjacent to the RNA binding site affects viral RNA replication. *Virology* **343**(1)**,** 65–78.

Shi, S. T., and Lai, M. M. (2005). Viral and cellular proteins involved in coronavirus replication. *Curr Top Microbiol Immunol* **287,** 95–131.

Sirover, M. A. (1999). New insights into an old protein: the functional diversity of mammalian glyceraldehyde-3-phosphate dehydrogenase. *Biochim Biophys Acta* **1432**(2)**,** 159–84.

Sirover, M. A. (2005). New nuclear functions of the glycolytic protein, glyceraldehyde-3-phosphate dehydrogenase, in mammalian cells. *J Cell Biochem* **95**(1)**,** 45–52.

Snijder, E. J. (2001). Arterivirus RNA synthesis dissected. Nucleotides, membranes, amino acids, and a bit of zinc. *Adv Exp Med Biol* **494,** 241–53.

Spagnolo, J. F., and Hogue, B. G. (2000). Host protein interactions with the 3′ end of bovine coronavirus RNA and the requirement of the poly(A) tail for coronavirus defective genome replication. *J Virol* **74**(11)**,** 5053–65.

Stone, M., Jia, S., Heo, W. D., Meyer, T., and Konan, K. V. (2007). Participation of rab5, an early endosome protein, in hepatitis C virus RNA replication machinery. *J Virol* **81**(9)**,** 4551–63.

Stork, J., Panaviene, Z., and Nagy, P. D. (2005). Inhibition of in vitro RNA binding and replicase activity by phosphorylation of the p33 replication protein of Cucumber necrosis tombusvirus. *Virology* **343**(1)**,** 79–92.

Strauss, J. H., and Strauss, E. G. (1999). Viral RNA replication. With a little help from the host. *Science* **283**(5403)**,** 802–4.

Szittya, G., Molnar, A., Silhavy, D., Hornyik, C., and Burgyan, J. (2002). Short defective interfering RNAs of tombusviruses are not targeted but trigger post-transcriptional gene silencing against their helper virus. *Plant Cell* **14**(2)**,** 359–72.

Tang, W. F., Yang, S. Y., Wu, B. W., Jheng, J. R., Chen, Y. L., Shih, C. H., Lin, K. H., Lai, H. C., Tang, P., and Horng, J. T. (2007). Reticulon 3 binds the 2C protein of enterovirus 71 and is required for viral replication. *J Biol Chem* **282**(8)**,** 5888–98.

Tomita, Y., Mizuno, T., Diez, J., Naito, S., Ahlquist, P., and Ishikawa, M. (2003). Mutation of host DnaJ homolog inhibits brome mosaic virus negative-strand RNA synthesis. *J Virol* **77**(5)**,** 2990–7.

Toyoda, H., Franco, D., Fujita, K., Paul, A. V., and Wimmer, E. (2007). Replication of poliovirus requires binding of the poly(rC) binding protein to the cloverleaf as well as to the adjacent C-rich spacer sequence between the cloverleaf and the internal ribosomal entry site. *J Virol* **81**(18)**,** 10017–28.

Tsujimoto, Y., Numaga, T., Ohshima, K., Yano, M. A., Ohsawa, R., Goto, D. B., Naito, S., and Ishikawa, M. (2003). Arabidopsis TOBAMOVIRUS MULTIPLICATION (TOM) 2 locus encodes a transmembrane protein that interacts with TOM1. *Embo J* **22**(2), 335–43.

Tu, H., Gao, L., Shi, S. T., Taylor, D. R., Yang, T., Mircheff, A. K., Wen, Y., Gorbalenya, A. E., Hwang, S. B., and Lai, M. M. (1999). Hepatitis C virus RNA polymerase and NS5A complex with a SNARE-like protein. *Virology* **263**(1), 30–41.

Vlot, A. C., Laros, S. M., and Bol, J. F. (2003). Coordinate replication of alfalfa mosaic virus RNAs 1 and 2 involves cis- and trans-acting functions of the encoded helicase-like and polymerase-like domains. *J Virol* **77**(20), 10790–8.

Vlot, A. C., Neeleman, L., Linthorst, H. J., and Bol, J. F. (2001). Role of the 3′-untranslated regions of alfalfa mosaic virus RNAs in the formation of a transiently expressed replicase in plants and in the assembly of virions. *J Virol* **75**(14), 6440–9.

Walter, B. L., Parsley, T. B., Ehrenfeld, E., and Semler, B. L. (2002). Distinct poly(rC) binding protein KH domain determinants for poliovirus translation initiation and viral RNA replication. *J Virol* **76**(23), 12008–22.

Wang, C., Gale, M., Jr., Keller, B. C., Huang, H., Brown, M. S., Goldstein, J. L., and Ye, J. (2005a). Identification of FBL2 as a geranylgeranylated cellular protein required for hepatitis C virus RNA replication. *Mol Cell* **18**(4), 425–34.

Wang, X., Lee, W. M., Watanabe, T., Schwartz, M., Janda, M., and Ahlquist, P. (2005b). Brome mosaic virus 1a nucleoside triphosphatase/helicase domain plays crucial roles in recruiting RNA replication templates. *J Virol* **79**(21), 13747–58.

Watashi, K., Ishii, N., Hijikata, M., Inoue, D., Murata, T., Miyanari, Y., and Shimotohno, K. (2005). Cyclophilin B is a functional regulator of hepatitis C virus RNA polymerase. *Mol Cell* **19**(1), 111–22.

Watashi, K., and Shimotohno, K. (2007). Chemical genetics approach to hepatitis C virus replication: cyclophilin as a target for anti-hepatitis C virus strategy. *Rev Med Virol* **17**(4), 245–52.

Weiland, J. J., and Dreher, T. W. (1993). Cis-preferential replication of the turnip yellow mosaic virus RNA genome. *Proc Natl Acad Sci USA* **90**(13), 6095–9.

Wessels, E., Duijsings, D., Lanke, K. H., van Dooren, S. H., Jackson, C. L., Melchers, W. J., and van Kuppeveld, F. J. (2006a). Effects of picornavirus 3A Proteins on Protein Transport and GBF1-dependent COP-I recruitment. *J Virol* **80**(23), 11852–60.

Wessels, E., Duijsings, D., Niu, T. K., Neumann, S., Oorschot, V. M., de Lange, F., Lanke, K. H., Klumperman, J., Henke, A., Jackson, C. L., Melchers, W. J., and van Kuppeveld, F. J. (2006b). A viral protein that blocks Arf1-mediated COP-I assembly by inhibiting the guanine nucleotide exchange factor GBF1. *Dev Cell* **11**(2), 191–201.

White, K. A. (2002). The premature termination model: a possible third mechanism for subgenomic mRNA transcription in (+)-strand RNA viruses. *Virology* **304**(2), 147–54.

White, K. A., and Nagy, P. D. (2004). Advances in the molecular biology of tombusviruses: gene expression, genome replication, and recombination. *Prog Nucleic Acid Res Mol Biol* **78,** 187–226.

Wierzchoslawski, R., and Bujarski, J. J. (2006). Efficient in vitro system of homologous recombination in brome mosaic bromovirus. *J Virol* **80**(12), 6182–7.

Yamaji, Y., Kobayashi, T., Hamada, K., Sakurai, K., Yoshii, A., Suzuki, M., Namba, S., and Hibi, T. (2006). In vivo interaction between Tobacco mosaic virus RNA-dependent RNA polymerase and host translation elongation factor 1A. *Virology* **347**(1), 100–8.

Yamanaka, T., Imai, T., Satoh, R., Kawashima, A., Takahashi, M., Tomita, K., Kubota, K., Meshi, T., Naito, S., and Ishikawa, M. (2002). Complete inhibition of tobamovirus multiplication by simultaneous mutations in two homologous host genes. *J Virol* **76**(5), 2491–7.

Yamanaka, T., Ohta, T., Takahashi, M., Meshi, T., Schmidt, R., Dean, C., Naito, S., and Ishikawa, M. (2000). TOM1, an Arabidopsis gene required for efficient multiplication of a tobamovirus, encodes a putative transmembrane protein. *Proc Natl Acad Sci USA* **97**(18), 10107–12.

Zhu, J., Gopinath, K., Murali, A., Yi, G., Hayward, S. D., Zhu, H., and Kao, C. (2007). RNA-binding proteins that inhibit RNA virus infection. *Proc Natl Acad Sci USA* **104**(9), 3129–34.

Chapter 15
Host Factors that Restrict Retrovirus Replication

Mark D. Stenglein, April J. Schumacher, Rebecca S. LaRue, and Reuben S. Harris

Abstract/Primer Over the past several decades, it has become clear that a variety of cellular proteins actively restrict retrovirus replication. Two families of proteins in particular, the TRIMs and the APOBEC3s, coordinate a robust innate defense to retrovirus infection. The TRIM proteins, led by TRIM5alpha, impose a replication block after entry, such that the invading retrovirus is degraded prior to integration. The APOBEC3 proteins, notably APOBEC3G, inhibit the replication of retroviruses by a mutagenic mechanism that is associated with degradation of viral DNA. Retroviruses have evolved means of avoiding their host's TRIM and APOBEC3 defenses. Often, however, this leaves the virus susceptible to TRIMs and APOBECs from other species. Thus, these restriction systems limit the cross-species mobility of retroviruses. The prospects of developing new antiviral therapies that exploit these innate host defenses are promising.

Abbreviations AIDS, acquired immunodeficiency syndrome; AGM, African green monkey; APOBEC1, apolipoprotein B mRNA editing catalytic subunit 1; APOBEC3G, apolipoprotein B mRNA editing enzyme, catalytic polypeptide-like 3G; CA, retroviral capsid protein; CypA, cyclophilin A; Fv, Friend virus; HIV-1, human immunodeficiency virus type 1; Lv1, lentivirus susceptibility factor 1; MLV, murine leukemia virus; Ref1, restriction factor 1; SIV, simian immunodeficiency virus; SIVmac, SIV derived from rhesus macaque; TRIM, tripartite motif protein; Vif, virion/viral infectivity factor; Z-motif, APOBEC zinc-binding motif.

R.S. Harris (✉)
Department of Biochemistry, Molecular Biology and Biophysics,
University of Minnesota, 321 Church Street South East,
6-155 Jackson Hall, Minneapolis, MN 55455, USA
e-mail: rsh@umn.edu

C.E. Cameron et al. (eds.), *Viral Genome Replication*,
DOI 10.1007/b135974_15,

Introduction

Overview

Most (if not all) organisms are vulnerable to viral infections and, consequently, many systems have been developed for protection. Many bacteria, for instance, utilize site-specific restriction endonucleases to cleave foreign DNA that is not marked as 'self' (reviewed by Tock and Dryden, 2005). The B and T lymphocytes of vertebrates are capable of specialized adaptive immune responses, recognizing foreign invaders (antigens), and neutralizing them through specific antibody- and cell-mediated responses. In addition to adaptive immunity, vertebrates have innate defense systems that recognize and eliminate invading pathogens. Key molecules include Toll-like receptors, antimicrobial peptides, interferons, and many others, which are beyond the scope of this chapter (reviewed in Chapter 9 of Fields et al., 2007). Here we focus on mammalian proteins, termed retrovirus restriction factors, which limit the infectivity of a broad and growing number of viruses (reviewed by Bieniasz, 2004; Goff, 2004; Mangeat and Trono, 2005; Chiu and Greene, 2006; Haché et al., 2006; Holmes et al., 2007b; Towers, 2007).

Brief History

A longstanding rule is that any given retrovirus can infect only certain cell types (e.g., Friend, 1957; Lilly, 1967). Cells in which the virus can replicate are termed 'permissive' and other cells are termed 'non-permissive'. One obvious reason for this is that many non-permissive cells lack machinery that the virus requires for replication. For instance, viruses often employ receptors located on the cell surface to facilitate their entry into the cell. An example of this is the CD4 and the CXCR4 or CXCR5 polypeptides that the human immunodeficiency virus-1 (HIV-1) envelope protein (gp120) recognizes and uses for particle entry. This in part explains why CD4+ T cells are a favored HIV-1 reservoir, and why HIV-1 cannot efficiently infect CD4–. Many other cellular factors have integral roles in nearly every stage of the retrovirus life cycle [see Chapter 6 in this text and an excellent review by Goff (2007)].

Another decades-old observation is that some cell types that seem to contain the necessary complement of positively acting factors are nevertheless resistant to retroviral infection. One example of such a scenario is the 'resting' CD4+ T cell, which is resistant to HIV-1 infection (Stevenson et al., 1990; Zack et al., 1990). A second type of resistance can be observed when cross-species viral infections are attempted, even between similar species. For instance, although HIV-1 efficiently infects most humans, it is unable to infect several closely related primate species, for example, the rhesus macaque (Shibata et al., 1995; Himathongkham and Luciw, 1996; Hofmann et al., 1999). These observations can be explained by hypothesizing that certain cells express dominant factors that interfere with retroviral replication. Several such proteins have now been described and they are called restriction factors.

To date, the discovery and characterization of all retroviral restriction factors has followed a similar general storyline. First, an investigator notices that a particular virus is able to infect some type of cell or organism but is unable to infect a closely related organism or cell type (in the best-case scenario these cell lines are clonally related). Second, the nature of this difference in infectivity is determined. Is the lack of infection due to the lack of a necessary co-factor or the presence of an inhibitor? At what stage of the viral life cycle does replication halt? Is the block to infection genetically dominant? Finally, cloning of the gene in question precipitates a major advance in the understanding of the molecular mechanism of the restriction and elucidation of additional steps of the viral life cycle.

The First Described Restriction Factors: Fv1 and Fv4

The origin of the hunt for retroviral restriction factors can be traced back to 1956, when Charlotte Friend isolated a virus (now called the Friend murine leukemia virus or Friend MLV) that was able to infect and cause leukemia in some mouse strains but not others (Friend, 1957). Over the next several decades, many scientists investigated the basis of this difference (for additional reviews on this topic, see Jolicoeur, 1979; Goff, 1996; Bieniasz, 2003, 2004; Goff, 2004). These investigations have focused most extensively on two genes, Fv1 and Fv4 (Lilly, 1967; Suzuki, 1975).

The Fv4 gene is found in a Japanese mouse strain and genetic crosses using this strain demonstrated that Fv4 confers dominant resistance to Friend MLV infection (Suzuki, 1975). Cells isolated from Fv4+ mice and cultured in vitro are also resistant to infection, suggesting that resistance due to Fv4 does not depend on a complex immune response (Kai et al., 1976). Ikeda and coworkers discovered the molecular identity of Fv4, which is an envelope gene of a truncated, integrated, MLV-like provirus (Ikeda et al., 1985). When expressed on the cell surface, this envelope-like protein competes with the envelope proteins of incoming virus particles for binding with their cognate receptor (Kai et al., 1986; Ikeda and Sugimura, 1989). This prevents the retrovirus from entering the cell, and therefore from replicating.

Fv1 is another locus that dominantly confers resistance to MLV infection (Lilly, 1967). Like Fv4, Fv1 confers resistance in cells grown in tissue culture (Hartley et al., 1970). Fv1 and Fv4 segregate independently in genetic crosses, and thus are not identical (Suzuki, 1975). Crossing experiments revealed that there are two major Fv1 alleles: $Fv1^N$ and $Fv1^B$ (Hartley et al., 1970). NIH Swiss mice are homozygous for the $Fv1^N$ allele, and Balb/c mice are homozygous carriers of the $Fv1^B$ allele. A mouse's Fv1 genotype controls its susceptibility to infection by different strains of MLV. So, N-tropic virus is able to replicate in mice carrying the $Fv1^N$ allele (NIH mice), but not on mice with a B allele (Balb/c). Conversely, B-tropic virus can replicate on mice with the $Fv1^B$ allele. The N and B alleles confer resistance dominantly; a heterozygous mouse, with genotype $Fv1^{B/N}$ is resistant to both N-tropic and B-tropic virus (Pincus et al., 1971). This dominance suggests that the Fv1 gene product is not a necessary co-factor for viral replication but rather an inhibitory factor.

Another informative characteristic of Fv1-mediated resistance is that it is saturable (Decleve et al., 1975; Pincus et al., 1975). This means that the resistance can be overcome by increasing the titer of virus used for inoculation. This is true even if the inoculum consists of replication-defective viral particles (Bassin et al., 1978; Boone et al., 1990). Models to explain this phenomenon propose that saturating viral particles titrate out the machinery on which Fv1-mediated restriction relies.

Fv-1 blocks infection relatively early after a virus enters a cell and appears to involve an interaction with the viral capsid. In cells exhibiting Fv1 restriction, normal levels of reverse transcription products are detectable but nuclear forms of viral DNA, including circular viral DNA and integrated proviruses, are not (Jolicoeur and Baltimore, 1976; Jolicoeur and Rassart, 1980) (Fig. 15.1). This suggests that

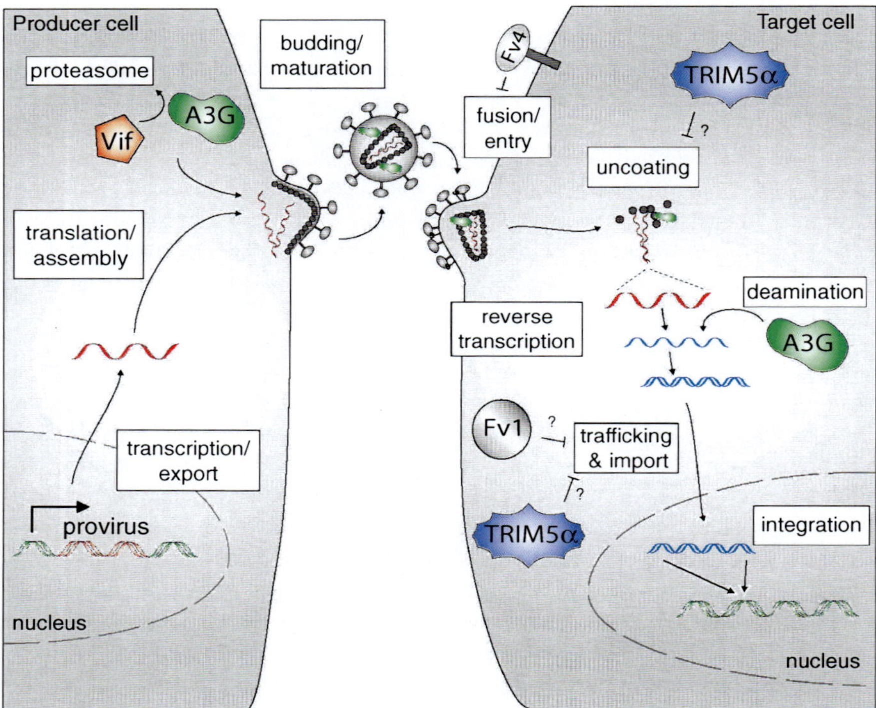

Fig. 15.1 The retroviral life cycle and the stages at which restriction factors act. This cartoon depicts key steps in the retroviral life cycle, and the stages at which restriction factors are thought to act. These are transcription and nuclear export; translation and particle assembly (APOBEC3G incorporation and Vif-mediated APOBEC3G degradation); budding; binding and fusion with the target cell (Fv4 block); uncoating (TRIM5alpha block?); reverse transcription (APOBEC3G deamination, TRIM5alpha block?); trafficking and nuclear import (Fv1 and TRIM5alpha block?); and integration. Note that the processes following entry of the virus particle into the target cell are thought to occur within a protected environment bounded by the viral capsid. TRIM5α designates TRIM5alpha.

Fv1-restricted viruses enter the cell and reverse transcribe their genome, but fail to transit to the nucleus and integrate their genome. A single amino acid (110) in the viral capsid protein (CA) defines the difference in Fv1 susceptibility between N- and B-tropic MLV strains, implying that Fv1 targets the viral capsid (DesGroseillers and Jolicoeur, 1983; Kozak and Chakraborti, 1996).

The molecular identity of Fv1 was determined in 1996 by the Stoye lab, who showed that Fv1 was a gene with sequence similarity to the *gag* gene of endogenous retroviruses of the HERV-L/MERV-L family (Best et al., 1996). This suggests that the Fv1 protein may be directly engaging the retroviral particle, but the details of this restriction remain to be determined.

TRIM5alpha and Related Proteins

Retroviral Restriction Factors in Mammalian Cells: Ref1, Lv1, and TRIM5alpha

The first of the two major families of mammalian restriction factors that this chapter will focus on is the TRIM5s. As was the case with Fv1 and Fv4, many of the characteristics of TRIM5-mediated restriction were described several years before its molecular identity was discovered.

As discussed above, the mouse Fv1 and Fv4 genes confer resistance to infection by certain MLV strains. In 2000, Towers et al. reasoned that other mammalian cells might express similar factors that confer resistance to MLV infection (Towers et al., 2000). Indeed, many mammalian cell lines proved resistant to infection by N-tropic MLV (although most remained relatively susceptible to infection by B-MLV). The authors attributed this resistance to cellular inhibitors of viral replication similar to Fv1, and named the putative factor Ref1 (*re*striction *f*actor 1).

Around the same time, evidence was mounting that retroviruses other than MLV, including primate lentiviruses were limited in their host range (Shibata et al., 1995; Himathongkham and Luciw, 1996; Hofmann et al., 1999). For example, many non-human primates could not be productively infected with HIV-1. It was uncertain whether this limitation was due to the absence of factors in these cells that the viruses required to replicate or to dominant antiviral factors such as Fv1 and Ref1. This question was resolved by several key studies in 2002, which showed that many primate cells express a dominant factor that, like Ref1, blocks lentiviral infection (Besnier et al., 2002; Cowan et al., 2002; Munk et al., 2002). This factor was dubbed Lv1 (*l*enti*v*irus susceptibility factor 1) (Cowan et al., 2002).

The original studies describing Ref1 and Lv1 and those that followed closely on their heels began to paint a clearer picture of these restriction factors and their mechanism of inhibition (Towers et al., 2000; Besnier et al., 2002; Cowan et al., 2002; Munk et al., 2002; Towers et al., 2002; Besnier et al., 2003; Hatziioannou et al., 2003; Kootstra et al., 2003; Berthoux et al., 2004; Hatziioannou et al., 2004b). Most tested cell lines restricted infection of some retroviruses, especially retroviruses from other species. For example, human cells restricted infection by N-MLV and equine infectious anemia virus (EIAV) but did not restrict HIV-1, HIV-2, or the rhesus macaque simian immunodeficiency virus (SIVmac) (Hatziioannou et al., 2003).

These studies also showed that Ref1 and Lv1 blocked viral infection in a way that was reminiscent of the Fv1 block in murine cells. Furthermore, several key characteristics of Fv1-mediated and Ref1-mediated restriction were identical. First, restriction by Ref1 and Lv1 was dominant in heterokaryon assays, in which permissive and non-permissive cells are fused (Cowan et al., 2002; Munk et al., 2002). Secondly, for Lv1 and Ref1, as was for Fv1, the viral capsid protein determines whether a virus is sensitive to restriction (Kozak and Chakraborti, 1996; Towers et al., 2000; Cowan et al., 2002; Hatziioannou et al., 2004b; Ikeda et al., 2004; Dodding et al., 2005; Lassaux et al., 2005). Typically, the capsid of a particular retrovirus confers resistance to restriction by the Ref1 or Lv1 of the virus's host species. For example, replacing the CA and adjoining P2 region of the SIVmac *gag* gene with the corresponding HIV-1 sequence results in a virus that is restricted like HIV-1 (Cowan et al., 2002). Thirdly, Ref1 and Lv1 block replication at a similar stage of the viral life cycle. Judging by the failure to detect reverse-transcribed linear viral DNA, this block occurs after the virus has entered the target cell but before it has reverse transcribed its genome (Fig. 15.1) (Towers et al., 2000; Besnier et al., 2002; Cowan et al., 2002; Munk et al., 2002). An important distinction is that this is slightly earlier than the Fv1-mediated block in mouse cells, which occurs after reverse transcription but before nuclear import of the viral pre-integration complex (Jolicoeur and Rassart, 1980; Yang et al., 1980; Towers et al., 2000). A final similarity between Ref1 and Lv1 was that, as was the case with Fv1, restriction could be saturated, i.e., overwhelmed by increasing the viral titer or by pre-treating the cells with virus particles (Hartley et al., 1970; Boone et al., 1990; Besnier et al., 2002; Cowan et al., 2002; Munk et al., 2002; Towers et al., 2002; Kootstra et al., 2003; Dodding et al., 2005). Some viruses were even found capable of cross-saturation. For example, African green monkey cells normally restrict HIV-1 and SIVmac, but pretreatment of these cells with HIV-1 particles increases the efficiency of a subsequent SIVmac infection and vice versa (Cowan et al., 2002). These data suggested that a single mechanism inhibits both HIV-1 and SIVmac in these cells.

Because of the striking similarities between Lv1 and Ref1, it was speculated that they were species-specific alleles of the same gene (Cowan et al., 2002; Serhan et al., 2002). The cloning of the gene responsible for both activities validated this prediction.

The Cloning of TRIM5alpha/Ref1/Lv1

As explained above, HIV-1 is capable of infecting human cells but unable to infect many other primate cells. This inability was attributed to a restriction activity named Lv1, but the molecular identity of Lv1 was unknown. In a breakthrough study in 2004, Stremlau and colleagues identified a cDNA from a rhesus macaque library that conferred to human cells resistance to HIV-1 infection (Stremlau et al., 2004).

The cDNA contained a rhesus gene named TRIM5alpha. Several characteristics of TRIM5alpha-mediated restriction indicated that it might be the Lv1/Ref1 gene. The block occurred prior to reverse transcription, as evidenced by a severe decrease in the accumulation of reverse transcription products (Fig. 15.1). Chimeric SIV/HIV viruses showed that the viral determinants of the block mapped to the capsid (CA) gene. Knockdown of endogenous TRIM5alpha in primary rhesus monkey lung fibroblasts resulted in an increase in the efficiency of HIV-1 infection.

It was not long before it was confirmed that species-specific alleles of TRIM5alpha were identical to (or at least essential for) Lv1 in monkey cells and Ref1 in human cells (Hatziioannou et al., 2004a; Keckesova et al., 2004; Perron et al., 2004; Yap et al., 2004). In accordance with this idea, human TRIM5alpha expression limited infection by N-MLV but not B-MLV. Similarly, expression of non-human primate TRIM5alpha variants limited infectivity by a variety of retroviruses. In all cases, heterologous expression of a TRIM5alpha allele conferred a restriction activity normally associated with cells of the species from which the allele had been derived (Hatziioannou et al., 2004a; Keckesova et al., 2004; Perron et al., 2004; Yap et al., 2004; Nakayama et al., 2005; Saenz et al., 2005; Song et al., 2005a; Ylinen et al., 2005; Kaumanns et al., 2006). Conversely, knockdown of endogenous TRIM5alpha increased cells' susceptibility to infection by normally restricted viruses.

The overarching conclusion from these studies was that many primate cells express TRIM5alpha proteins that block infection by a variety of retroviruses. Typically, however, TRIM5alpha of a given species is ineffective against that species' retroviruses. For instance human TRIM5alpha exhibits only a modest inhibitory effect against HIV-1 (Stremlau et al., 2004). This suggests that retroviruses have evolved ways around the TRIM5alpha-imposed restriction in their host's cells.

The Mechanism of TRIM5alpha-Mediated Retroviral Restriction

In humans, TRIM5alpha is just one of a large family of TRIM genes (there are nearly 70 in humans) (Reymond et al., 2001; Nisole et al., 2005). The TRIM proteins are named after their characteristic *tri*partite *m*otif (Fig. 15.2) (Reddy et al., 1992; Borden, 1998; Reymond et al., 2001). This motif, also known as the RBCC motif, consists of a *R*ING domain, one or two *B* boxes, and a *c*oiled-*c*oil domain.

The RING domain (*r*eally *i*nteresting *n*ew *g*ene) contains a number of conserved cysteines and histidines that coordinate two zinc atoms (Saurin et al., 1996). Some RING domains have been reported to have E3 ubiquitin ligase activity, and some TRIM proteins (including TRIM5alpha) have been shown to ubiquitinate themselves in a RING domain-dependent manner (Xu et al., 2003; Diaz-Griffero et al., 2006a).

The B box is another zinc-binding domain with a three-dimensional structure similar to a RING domain, but with a poorly characterized function (Reddy and Etkin, 1991; Reddy et al., 1992; Massiah et al., 2007). TRIM5alpha has a single B box, but other TRIMs have one or two of these domains.

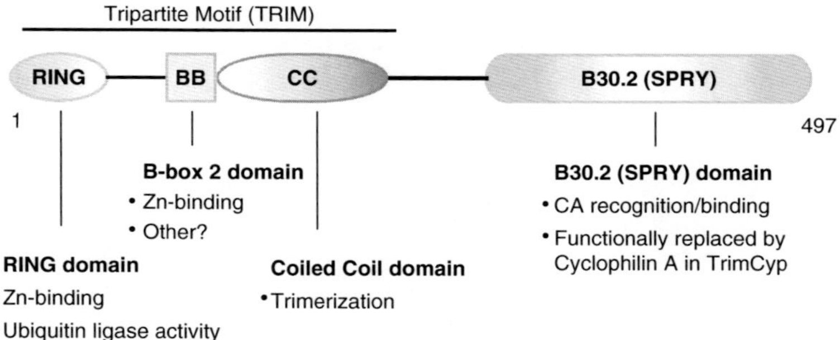

Fig. 15.2 A cartoon depicting the domain organization of rhesus macaque TRIM5alpha.

The coiled-coil domain has been implicated in multimerization of the TRIM proteins, and TRIM5alpha forms a trimer (Reymond et al., 2001; Mische et al., 2005; Perez-Caballero et al., 2005; Javanbakht et al., 2006a). Truncated TRIM5alpha proteins lacking the coiled-coil domain fail to multimerize, demonstrating that this domain is required for multimerization (Mische et al., 2005). However, the TRIM5gamma isoform, which lacks the carboxy-terminal B30.2(SPRY) domain, forms dimers, not trimers, suggesting that multiple domains influence the oligomeric state of the protein.

In TRIM5alpha, the tripartite (RBCC) motif is followed by a B30.2(SPRY) domain (Fig. 15.2) (Henry et al., 1997; Rhodes et al., 2005; Woo et al., 2006; James et al., 2007). SPRY domains have been implicated in protein–protein interactions.

All of the domains of TRIM5alpha are necessary to block retroviral infection. Truncated proteins lacking the amino-terminal RING, B box, or coiled-coil domains have severely attenuated antiviral activity (Javanbakht et al., 2005; Perez-Caballero et al., 2005). Mutant proteins with a disrupted coiled-coil domain fail to trimerize and also fail to block infection (Javanbakht et al., 2006a). A similar inability to block infection is observed with some point mutants in these domains (Stremlau et al., 2004; Javanbakht et al., 2005, 2006a). Similarly, truncated proteins lacking the SPRY domain show that the tripartite motif (RBCC domains) by itself is inactive against retroviruses (Stremlau et al., 2004; Perez-Caballero et al., 2005). A model to explain the contribution of TRIM5alpha's domains proposes that the B30.2(SPRY) domain provides the capsid interaction surface, the coiled-coil domain promotes trimerization, and the RING and B box domains provide some unknown effector function, perhaps involving the RING domain-associated ubiquitin ligase activity (Fig. 15.2).

Different species' TRIM5alphas restrict different retroviruses. Four variable loops in the B30.2(SPRY) domain determine this difference in efficacy (Nakayama et al., 2005; Stremlau et al., 2005; Yap et al., 2005; Ohkura et al., 2006; Perron et al., 2006). Primate species' TRIM5alpha alleles differ most in these variable regions (Song et al., 2005b). These regions, called V1–V4, are predicted to be surface-exposed loops (Ohkura et al., 2006; Perron et al., 2006; Woo et al., 2006; James et al., 2007). It has been proposed that these loops form the capsid interaction surface, and that different loop configurations enable interactions with different viral capsids. A striking finding is that the specificity can be determined by as little as a single amino acid (Stremlau et al., 2005; Yap et al., 2005; Li et al., 2006b). Amino acid 332 of TRIM5alpha is arginine in humans and proline in rhesus macaques. Mutating the arginine to proline in human TRIM5alpha confers strong anti-HIV-1 activity to the human protein (Stremlau et al., 2005; Yap et al., 2005; Li et al., 2006b). An intriguing hypothesis is that TRIM5alpha alleles, differing in their B30.2(SPRY) variable loops, become fixed in populations in response to pandemic retroviral infections (Kaiser et al., 2007). In this case, the ability of the fixed allele to ward off the exigent pathogen is counterbalanced by its inactivity against other retroviruses. Such a scenario has been proposed to explain why the human TRIM5alpha allele is inactive against HIV-1 (Kaiser et al., 2007).

The TRIM5alpha gene is found in a cluster of TRIM genes on chromosome 11p15 (Reymond et al., 2001). From this locus six alternatively spliced TRIM5 isoforms are expressed (Reymond et al., 2001). TRIM5alpha is the longest of these, and it encodes a protein of 497 amino acids (rhesus macaque TRIM5alpha). Other TRIM5 isoforms lack at least one of the above-mentioned domains and those tested lack antiviral activity (Stremlau et al., 2004; Perez-Caballero et al., 2005).

TRIM5s are expressed ubiquitously in adult tissues (Reymond et al., 2001). TRIM5alpha localizes to the cytoplasm of cells, in a diffuse cytosolic manner and in bodies that appear as bright dots in fluorescent microscopy (Reymond et al., 2001; Xu et al., 2003; Campbell et al., 2007). This localization pattern does not appear to be functionally required for blocking retroviral infection, although it is consistent with the requirement that the protein be able to engage incoming viral particles in the cytoplasm (Perez-Caballero et al., 2005;, Song et al., 2005c).

The precise mechanism by which TRIM5alpha acts remains unclear. Several candidate mechanisms have been put forward, the general theme of which is that TRIM5alpha acts by modulating the stability or activity of the viral core (Fig. 15.3). All of the mechanisms involve a TRIM5alpha trimer binding the hexameric lattice of the capsid of an incoming virus. Indeed, TRIM5alpha–capsid binding has been demonstrated in vitro, and mutations to either protein that disrupt this interaction correlate with a loss of restriction (Sebastian and Luban, 2005; Chatterji et al., 2006; Li et al., 2006b; Stremlau et al., 2006b). The difference between the proposed mechanisms is the fate of the TRIM5-bound viral capsid (Fig. 15.3). In the simplest case, TRIM5alpha binds to the capsid and renders it incompetent to perform reverse transcription. This could be because TRIM5alpha blocks necessary cellular factors from accessing the viral core. A corollary of this model is that TRIM5alpha covalently modifies the capsid, perhaps by ubiquitinating or sumoylating it (Stremlau et al., 2004). In another proposed model, TRIM5alpha binding triggers the degradation of the capsid (Chatterji et al., 2006). This destruction of the capsid would preclude the completion of the virus replication cycle. A final model proposes that TRIM5alpha mediates the faster than normal uncoating of the capsid from the viral core (Stremlau et al., 2006b; Perron et al., 2007). In this scenario, the accelerated uncoating disrupts the normal progression of events required for a successful infection (Forshey et al., 2002).

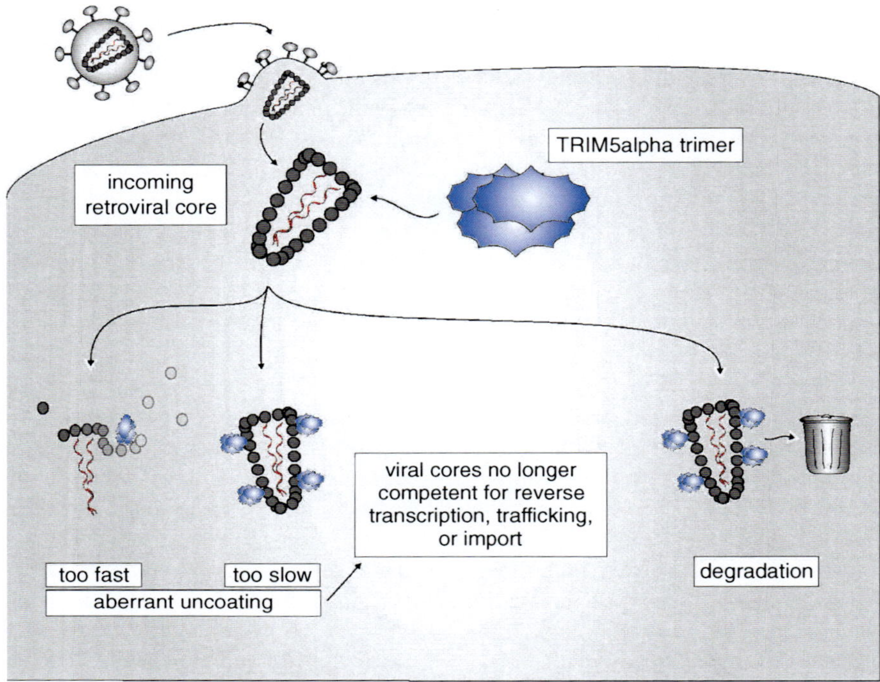

Fig. 15.3 Proposed mechanisms of TRIM5alpha-mediated retroviral restriction. All of the proposed mechanisms involve a TRIM5alpha trimer binding to the viral capsid. This binding causes either the degradation of the viral capsid or disrupts the normal uncoating process in such a way that the virus core particle is incompetent for subsequent steps of replication.

Several papers have argued that TRIM5alpha blocks infection at two distinct stages (Berthoux et al., 2004; Anderson et al., 2006; Wu et al., 2006). These arguments are based on the observation that when cells are treated with proteasome inhibitors, TRIM5alpha still restricts infection, but it no longer blocks reverse transcription. In these cells, reverse transcription products are detectable, but nuclear forms of viral DNA are not. This suggests that TRIM5alpha may have a dual effect in cells (Fig. 15.3). First, in a manner that is proteasome-dependent, TRIM5alpha blocks reverse transcription. Second it prevents the nuclear import of viral pre-integration complexes.

In all of the models discussed above, target cell TRIM5alpha engages the capsid of an incoming virus. However, a recent intriguing paper from the Ikeda lab offered evidence that TRIM5alpha may also target outgoing virus in producer cells (Sakuma et al., 2007b). Expression of rhesus TRIM5alpha in cells producing virus was found to reduce the viral titer. TRIM5alpha expression correlated with a reduction in the half-life of viral Gag protein in the producer cells, suggesting that the mechanism by which TRIM5alpha reduces the viral titer is by destabilizing Gag.

In summary, several mechanisms have been put forth to model TRIM5alpha-mediated retrovirus restriction. Each model has supporting evidence. Moreover, it is important to note that these mechanisms are not necessarily mutually exclusive. Indeed, it would behoove the cell to target the virus at as many points as possible. As data continue to accumulate, the mechanism or mechanisms of TRIMalpha-mediated restriction will become apparent.

Cylclophilin A and TRIM

It is impossible to discuss TRIM5alpha-mediated retroviral restriction without discussing its intimate relationship with cyclophilin A (reviewed by Nisole et al., 2005; Luban, 2007; Towers, 2007). Cyclophilin A (CypA) is a peptidyl–prolyl isomerase encoded by the *PPIA* gene (Fischer et al., 1998). It was identified as a HIV-1 capsid-binding protein in a yeast two-hybrid screen (Luban et al., 1993). CypA interacts with a proline-rich surface of the HIV-1 capsid protein, and it catalyzes the isomerization of the peptidyl–prolyl bond between residues glycine 89 and proline 90 (Gamble et al., 1996; Bosco et al., 2002). The CypA–CA interaction has been shown to be important for HIV-1 infectivity, as disruption of this interaction results in a substantial decrease in infectivity (Thali et al., 1994; Braaten et al., 1996c; Braaten et al., 1996a,b; Braaten and Luban, 2001). The target cell (i.e., the cell that is being newly infected) provides the functionally important CypA (Franke et al., 1994; Sokolskaja et al., 2004; Hatziioannou et al., 2005).

CypA modulates TRIM5alpha's antiviral activity in certain cell types. In rhesus macaques and African green monkey cells, TRIM5alpha restriction requires CypA (Berthoux et al., 2005a; Chatterji et al., 2005; Keckesova et al., 2006; Sokolskaja et al., 2006). It has been proposed that in these simian cells, CypA isomerization of the capsid renders it sensitive to TRIM5alpha. In contrast to this, in human cells CypA is required for HIV-1 infection. There, CypA binds to the HIV-1 capsid and this binding is thought to protect the virus from the activity of a restriction factor. This restriction factor was originally thought to be human TRIM5alpha, but now it is believed that CypA shields the capsid from an as yet to be discovered factor (Sayah and Luban, 2004; Keckesova et al., 2006; Sokolskaja et al., 2006; Stremlau et al., 2006a).

The relevance of CypA to TRIM5-mediated restriction is further highlighted (in astounding fashion) by the existence of a TRIMCyp fusion gene in several primate species. The TRIMCyp gene appears to have arisen when a CypA mRNA was transposed into the TRIM5 locus by LINE-1 retrotransposon machinery. The Luban lab originally identified TRIMCyp while investigating an apparent CypA-mediated antiviral activity in owl monkey cells (Sayah et al., 2004). The TRIM-Cyp gene encodes a protein similar to TRIM5alpha but with CypA replacing the B30.2(SPRY) domain (Fig. 15.2). In this fusion protein, CypA supplies the capsid-binding activity previously provided by the B30.2(SPRY) domain (Nisole et al., 2004; Diaz-Griffero et al., 2006b). This allele is obviously an effective functional replacement for TRIM5alpha, as it has become fixed in all owl monkey species

(Ribeiro et al., 2005). Since the original discovery of the (new world) owl monkey TRIMCyp, a nearly equivalent TRIMCyp gene has been described in (old world) macaques (Liao et al., 2007; Brennan et al., 2008; Newman et al., 2008; Virgen et al., 2008; Wilson et al., 2008). These TRIMCyp genes appear to have evolved independently in the two primate lineages. This unlikely occurrence represents a beautiful example of convergent evolution and suggests that the TRIMCyp fusion protein may provide a strong selective advantage.

The Evolution of TRIMs as Antiviral Defenses

In primates, TRIM5alpha is just one member of a large protein family, and several studies have assessed the antiviral activity of some of the other primate TRIMs, namely TRIMs 1, 4, 6, 18, 19, 21, 22, 27, and 34 (Yap et al., 2004, 2005; Li et al., 2006a; Zhang et al., 2006; Li et al., 2007). These TRIMs were chosen for study because they have exhibited activity against other viruses, because they have the most similar sequence or overall domain structure to TRIM5alpha, or because their genes are at the same locus on chromosome 11 as TRIM5alpha. Apart from two monkey species' TRIM1s, which restricts N-MLV (Yap et al., 2004, 2005), these other TRIMs failed to exhibit significant activity against a variety of retroviruses. However, using chimeric proteins, two studies demonstrated that it is possible to functionally replace the RBCC domain of TRIM5alpha (or TRIMCyp) with those of several other TRIMs (Li et al., 2006a; Yap et al., 2006). This suggests that the amino-terminal RBCC domains of these TRIMs are competent for restriction but they are not effectively targeted to the viral capsid by their carboxy-terminal domains.

Truncated TRIM6 and TRIM34 proteins lacking their carboxy-terminal B30.2(SPRY) domain have a dominant negative effect on human TRIM5alpha's ability to restrict N-MLV infection (Zhang et al., 2006). These proteins may hetero-multimerize with TRIM5alpha, thereby preventing it from functioning normally. Various TRIM5 proteins exhibit a similar dominant negative effect on each other, especially in cross-species contexts or when truncated proteins are used (Stremlau et al., 2004; Berthoux et al., 2005b; Perez-Caballero et al., 2005).

It has been proposed that other TRIM proteins play roles in the defense against viral infection. The *pro*m*y*elocytic *l*eukemia protein, PML, also known as TRIM19, and TRIM22 are such family members (for further discussion, please see Nisole et al., 2005; Everett and Chelbi-Alix, 2007; Towers, 2007).

In addition to the primate TRIM5alphas, there are orthologs in other mammalian species. So far, TRIM5alpha or TRIM5alpha-like TRIM proteins with antiretroviral activity have been described in cows and rabbits (Si et al., 2006; Ylinen et al., 2006; Schaller et al., 2007). There is also a murine TRIM5 ortholog, but its antiviral activity remains to be confirmed (Hoffman et al., 2006; Noser et al., 2006). It is likely that primate TRIM5s and these other mammalian homologs are derived from an ancestral TRIM with antiretroviral activity.

There is convincing evidence that selective pressures due to pathogenic retroviruses have driven the evolution of the TRIM5s. As described above, several variable capsid-interacting loops in the B30.2(SPRY) domain determine the antiviral specificity of different species' TRIM5alpha proteins. Evolutionary analyses have shown that the regions of the TRIM5 genes encoding these loops bear strong signatures of episodic positive selection (Liu et al., 2005; Sawyer et al., 2005; Ortiz et al., 2006). Such signatures are attributable to pathogen-induced selection. A compelling study showed that two Old World Monkey species (sooty mangabeys and rhesus macaques) are maintaining multiple TRIM5alpha alleles (Newman et al., 2006). Phylogenetic analysis argues that these alleles were present in the common ancestor of these species and that therefore balancing selection has maintained these alleles for millions of years. The complex evolutionary history of TRIM5s suggests that these proteins have been entangled in long-term struggles against pathogens including retroviruses.

TRIM Frontiers

Although significant progress has been made toward understanding TRIM5alpha-mediated retroviral restriction, many key questions remain unanswered. One significant point is that there has been no demonstration yet of TRIM5alpha's effectiveness in vivo. All of the studies to date have used tissue culture systems (although it should be mentioned that the in vitro results mirror the reality of many viruses' limited host ranges). Several studies have examined the relationship between naturally occurring TRIM5 polymorphisms, the protein's antiviral activity, and clinical measures of HIV disease and epidemiology (Goldschmidt et al., 2006; Javanbakht et al., 2006b; Sawyer et al., 2006; Speelmon et al., 2006; Nakayama et al., 2007; Vigano et al., 2007). These studies have not produced convincing associations. Another important question is whether TRIM5alpha-related antiviral therapies can be developed. Two pilot studies have explored the possible use of TRIM5alpha in gene therapy (Anderson and Akkina, 2005; Sakuma et al., 2007a). The therapeutic expression of rhesus TRIM5alpha in human cells would be predicted to make them resistant to HIV-1 infection. Drugs that modulate the TRIM5alpha–capsid interaction represent another therapeutic possibility. For example, small molecules that increased human TRIM5alpha's affinity for the HIV-1 capsid might empower the protein to better restrict the virus. All in all, our understanding of this important antiviral defense system has rocketed forward in recent years, but many important questions still await an answer.

The APOBEC3 Proteins of Mammals

Discovery

The second of the two major families of mammalian restriction factors that this chapter will focus on is the APOBEC3s. In particular, the focus will be on the antiretroviral activity of family member APOBEC3G (apolipoprotein B mRNA editing enzyme, catalytic polypeptide-like 3G, **A3G**). As was the case with TRIM5alpha, the existence of APOBEC3G was inferred before its identity was unveiled. In this case, APOBEC3G's discovery was rooted in the observation that HIV-1 molecular clones that lacked the virion infectivity factor (Vif) accessory gene were able to replicate only in a subset of human T-cell lines. The Vif-proficient parental virus replicated normally in these cells (e.g., Fisher et al., 1987; Strebel et al., 1987; Nara and Fischinger, 1988; Gabuzda et al., 1992; von Schwedler et al., 1993; Simon and Malim, 1996). Hybrid T-cell lines, derived from fusing permissive and non-permissive lines, also failed to replicate Vif-defective viruses (Madani and Kabat, 1998; Simon et al., 1998). These experiments argued against models that suggested that Vif was required to compensate for a cellular factor lacking in non-permissive cells. Instead, the results were consistent with the existence of a dominant cellular factor in the non-permissive cells that prevented replication of Vif-defective HIV-1. The fact that Vif is required for HIV-1 replication on primary human cells and the fact that Vif-deficient SIV fails to replicate or cause disease in rhesus macaques highlight the importance of the non-permissive condition and the likelihood that it approximates the cellular environment in vivo (Fisher et al., 1987; Strebel et al., 1987; Gabuzda et al., 1992; von Schwedler et al., 1993; Gabuzda et al., 1994; Desrosiers et al., 1998; Victoria and Robinson, 2005).

This non-permissive versus permissive dichotomy was particularly striking for the permissive CEM T-cell line and its non-permissive derivative, CEM-SS (Simon and Malim, 1996). Malim and coworkers reasoned that the dominant cellular factor would be more highly expressed in non-permissive cells, and performed subtractive hybridization experiments to isolate messages present in CEM but absent in CEM-SS (Sheehy et al., 2002). Among the many differentially expressed mRNAs so identified, one termed APOBEC3G (initially called CEM15) was expressed in CEM but not in CEM-SS. Expression of the APOBEC3G gene alone rendered CEM-SS cells non-permissive for Vif-defective virus replication (Sheehy et al., 2002). APOBEC3G showed sequence similarity to the mRNA cytosine-to-uracil (C-to-U) editing protein APOBEC1, which leads to suggestions that APOBEC3G functioned by editing the message of a cellular protein and thereby endowing it with anti-HIV activity. This was just one hypothesis, however, and the discovery by Malim and coworkers inspired a number of investigations that answered two key questions: (1) what is the molecular mechanism underlying the antiviral effect of APOBEC3G and (2) how does Vif permit HIV-1 to replicate in the presence of this potent antiviral protein?

A Deamination-Dependent Mechanism

As described above, initial speculation on APOBEC3G's antiviral mechanism focused on the protein's potential as an mRNA editor. This speculation was soon discarded in favor of a model wherein APOBEC3G acts as a DNA mutator. Neuberger and colleagues provided evidence in favor of this by demonstrating that APOBEC3G possesses a DNA (rather than an RNA) cytosine deaminase enzymatic activity (Harris et al., 2002). Prior work from the Neuberger laboratory had shown that a related protein, activation-induced deaminase (AID), was a DNA cytosine-to-uracil (C-to-U) deaminase. The evidence for this was that AID expression in *E. coli* caused an increase in mutation frequencies and a corresponding C/G-to-T/A transition mutation bias (Petersen-Mahrt et al., 2002). Moreover, these effects were more pronounced in cells lacking uracil excision repair, strengthening the conclusion that DNA cytosines were being deaminated (Petersen-Mahrt et al., 2002). The AID-catalyzed uracils would become fixed as C/G-to-T/A mutations when they templated the incorporation of adenines during replication. Similar experiments showed that APOBEC3G triggered a mutator phenotype in *E. coli*, indicating that it is a DNA cytosine deaminase (Harris et al., 2002).

Several facts combined to suggest a model in which APOBEC3G blocked HIV replication by deaminating (mutating) viral cDNA during reverse transcription: the anti-HIV-1 activity of APOBEC3G, the demonstration that APOBEC3G could mutate DNA cytosines, and numerous reports of hypermutated viral sequences from patients (i.e., high levels of (minus strand) C-to-T transition mutations). This spurred several groups to test this model (Fig. 15.4) (Harris et al., 2003a; Lecossier

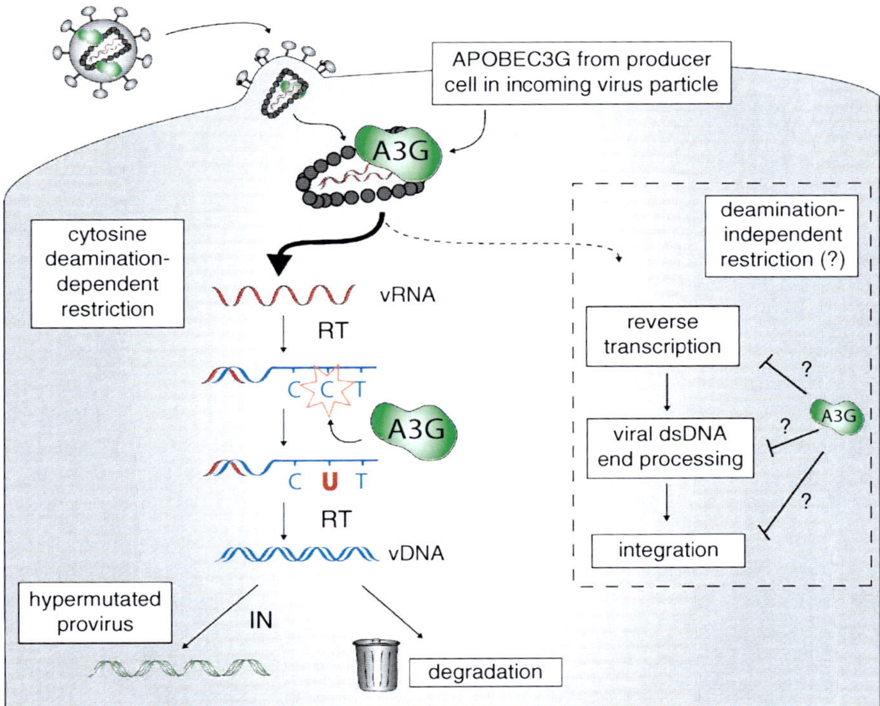

Fig. 15.4 The mechanism of APOBEC3G-mediated retroviral restriction. In the producer cell (not shown), APOBEC3G is incorporated into budding virus particles. During reverse transcription in the target cell (shown), APOBEC3G deaminates the viral cDNA while it is transiently single stranded, converting cytosines to uracils. This results in either degradation of the viral DNA or integration of a hypermutated provirus. Possible modes of deamination-independent restriction are depicted. RT, viral reverse transcriptase; IN, viral integrase; vRNA, viral RNA; vDNA, viral DNA.

et al., 2003; Mangeat et al., 2003; Zhang et al., 2003). When APOBEC3G was expressed in cells producing virus, the viral titer was not reduced, but the infectivity of the resulting particles was severely attenuated. Also, in strong support of the model, proviral DNA that accumulated in the presence of APOBEC3G exhibited massive increases in plus-strand G-to-A hypermutation. This hypermutation could be attributed to the deamination of minus-strand (cDNA) cytosines during reverse transcription (Fig. 15.4) (Harris et al., 2003a; Mangeat et al., 2003; Zhang et al., 2003). Thus, it seemed that a major part of the mechanism of APOBEC3G-mediated restriction of Vif-deficient HIV-1 was due to DNA cytosine deamination. This was further supported by experiments indicating that the putative zinc-coordinating residues of the active site were required for antiviral activity (Mangeat et al., 2003).

In addition to triggering hypermutation of the viral genome, APOBEC3G also causes degradation of viral cDNA (Fig. 15.4) (Goncalves et al., 1996; Simon and Malim, 1996; Bishop et al., 2006; Holmes et al., 2007a; Mbisa et al., 2007). APOBEC3G's enzymatic activity is required for this (Mbisa et al., 2007). The mechanism of degradation has not been explained because, despite initial predictions, the most obvious cellular DNA repair pathway – UNG2-dependent base excision repair – does not appear to be involved (Harris et al., 2003a,b; Kaiser and Emerman, 2006; Mbisa et al., 2007; Schumacher et al., 2008). However, a potentially important clue was published independently by the Pathak and Yu groups, who showed that APOBEC3G interferes with integration by inhibiting integrase and/or by creating aberrant cDNA ends that cannot be properly engaged by integrase (Luo et al., 2007; Mbisa et al., 2007). Future studies may clarify whether the degradation is an active reaction or simply the result of a failure to produce integration-competent viral DNA structures (which would eventually be degraded by cellular nucleases).

The human APOBEC3G protein is 384 amino acids long. It is composed of two similar domains arranged in tandem, each of which contains a characteristic zinc-binding motif (Fig. 15.5). It is impossible to discern by simply examining the amino acid sequences of these motifs whether one or both harbor the protein's enzymatic activity. A number of studies have shown that the two domains provide complementary activities. Studies with chimeric and mutant proteins showed that the carboxy-terminal zinc-binding domain harbors the protein's enzymatic activity (Haché et al., 2005; Navarro et al., 2005; Newman et al., 2005; Iwatani et al., 2006; Chen et al., 2007, 2008). In contrast, mutations to the conserved residues in the amino-terminal 'pseudo-active' site demonstrated that it is essential for nucleic acid binding and efficient incorporation into the viral particle (Fig. 15.5). It is thought that this amino-terminal domain allows APOBEC3G to gain access to the viral particle by binding a viral ribonucleoprotein complex that includes the viral genome and the viral nucleocapsid protein (e.g., Alce and Popik, 2004; Cen et al., 2004; Luo et al., 2004; Schafer et al., 2004; Svarovskaia et al., 2004; Zennou et al., 2004; Khan et al., 2005; Iwatani et al., 2006; Burnett and Spearman, 2007). Once so incorporated, APOBEC3G travels along in the virus particle to another cell that will be infected. There, APOBEC3G's carboxy-terminal-mediated enzymatic activity is brought to bear during reverse transcription. In this manner, both domains contribute to antiviral activity.

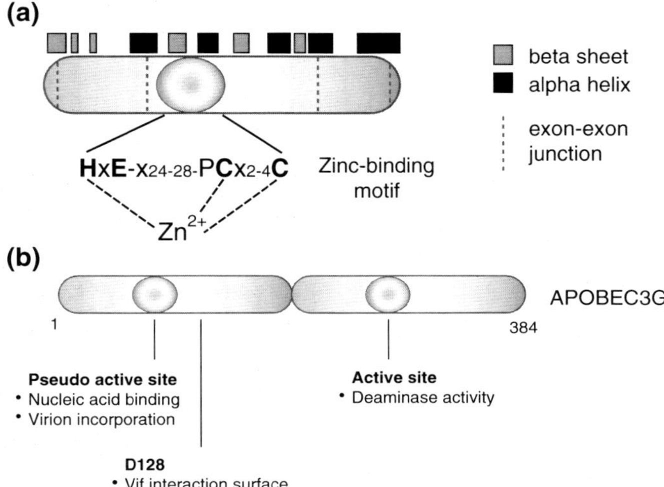

Fig. 15.5 Domain organization of the APOBEC3 proteins. (**a**) A schematic of the carboxy-terminal half of APOBEC3G, representing a typical APOBEC3 domain. *Dashed lines* delineate the exon/exon boundaries. The NMR structure-based secondary structure elements are depicted as *black* (alpha helix) and *gray* (beta sheet) boxes (Chen et al., 2008). The characteristic zinc-binding motif is shown. (**b**) The 'double-domain' structure of human APOBEC3G, consisting of two APOBEC3 domains as depicted in (a).

A Deamination-Independent Mechanism?

A number of reports have suggested that APOBEC3G may also harbor an antiviral activity that does not depend on the protein's enzymatic activity (Shindo et al., 2003; Dutko et al., 2005; Navarro et al., 2005; Newman et al., 2005; Iwatani et al., 2006). One series of studies tested the HIV-1 restriction activity of deaminase-defective APOBEC3G mutants. These mutants exhibited nearly wild-type antiviral activity in single-cycle infectivity assays (Shindo et al., 2003; Navarro et al., 2005; Newman et al., 2005; Iwatani et al., 2006). APOBEC3G derivatives with analogous substitutions in the amino-terminal, pseudo-catalytic zinc-coordinating motif also appeared capable of HIV-1 restriction. In contrast, the restriction activity of APOBEC3G was fully compromised when both the amino- and carboxy-terminal domains were mutated (Shindo et al., 2003; Newman et al., 2005; Iwatani et al., 2006). These studies concluded therefore that APOBE3G could exert an antiviral effect via carboxy-terminal-mediated deamination or by an unspecified activity provided by the amino-terminal zinc-binding motif.

These studies have spawned a debate about whether APOBEC3G's cytosine deaminase activity is strictly required for HIV restriction. Several lines of evidence suggest that it is indeed required. One line is based on the fact that many of the studies purporting to show that APOBEC3G's enzymatic activity is dispensable for restriction were carried out under high expression conditions. In contrast to these, several recent studies have included careful titrations of protein expression levels (Holmes et al., 2007a; Mbisa et al., 2007; Miyagi et al., 2007; Schumacher et al., 2008). Under high expression conditions, wild-type and mutant constructs exhibited similar anti-HIV-1 activity. In contrast, under low expression conditions, only wild-type APOBEC3G exerted significant antiviral activity. A high expression level is potentially non-physiological, and under such conditions a nucleic acid-binding protein like APOBEC3G could derail HIV-1 infectivity even when catalytically inert. Corroborating the results of these titration experiments is an experiment involving the stable expression of near-physiological levels of wild-type and mutant APOBEC3G in CEM-SS T-cell clones (Miyagi et al., 2007; Haché et al., 2008; Schumacher et al., 2008). In this experiment, growth of a Vif-defective virus was inhibited completely by APOBEC3G (as observed originally by Malim and co-workers; Sheehy et al., 2002), but the same virus preparation replicated normally in cells expressing the deaminase-deficient protein (Miyagi et al., 2007; Schumacher et al., 2008). Together, these series of experiments bolster the argument that the catalytic activity of APOBEC3G is indeed required for HIV-1 restriction.

It should be noted that it is not necessarily possible to generalize data related to APOBEC3G's ability to restrict HIV-1 to other viruses (and vice versa). APOBEC3G has shown clear deaminase-independent activity against hepatitis B virus (HBV), against porcine endogenous retrovirus (PERV), and against the retrotransposon Alu (Seppen, 2004; Turelli et al., 2004; Rosler et al., 2005; Suspene et al., 2005; Hulme et al., 2007; Jonsson et al., 2007; Nguyen et al., 2007). Similarly, data from other APOBEC3 proteins cannot be generalized to APOBEC3G (and vice versa); other human APOBEC3 proteins have exhibited DNA deaminase-independent activity. The next decade of research will undoubtedly demonstrate which APOBEC3 proteins target which retroelements (exogenous viruses and/or endogenous retrotransposons) physiologically. Furthermore, it may be revealed that different APOBEC3 proteins employ different mechanisms to inhibit these various retroelements. All mechanisms, however, must account for the fact that the zinc-coordinating DNA cytosine deaminase domain is the major defining and evolutionarily conserved activity of this protein family (a point highlighted by structural comparisons; e.g., Chen et al., 2008).

How Vif Counteracts APOBEC3G

Before APOBEC3G was identified, several groups had demonstrated the necessity of Vif for growth in non-permissive cell lines including primary human T cells and macrophages (e.g., Fisher et al., 1987; Strebel et al., 1987; Gabuzda et al., 1992; von Schwedler et al., 1993; Simon and Malim, 1996). It was also clear from trans-expression experiments that Vif was required in virus-producing cells and not in target cells. Additional advances were hampered by the fact that Vif is a highly basic, 23 kDa protein that has repeatedly resisted biochemical and structural studies. Despite these technical hurdles, several groups were able to use genetic approaches to independently converge on a common explanation: that Vif functions by triggering the proteasome-dependent destruction of human APOBEC3G (Fig. 15.1) (Conticello et al., 2003; Marin et al., 2003; Sheehy et al., 2003; Yu et al., 2003; Liu et al., 2004; Mehle et al., 2004). Moreover, the responsible molecules were quickly identified by the Yu group, who used an epitope-tagged HIV-1 Vif protein to affinity purify an E3 ubiquitin ligase complex consisting of CUL5, ELONGIN B, ELONGINC and RBX1 (Yu et al., 2003). Cells depleted for these proteins or expressing dominant-negative variants preserved APOBEC3G and resisted HIV-1 (Vif+) infection (Yu et al., 2003; Mehle et al., 2004).

The importance of the Vif-mediated APOBEC3G degradation is highlighted by the fact that it is conserved. For instance, the Vif from a virus that infects African green monkeys (SIVagm) degrades that species' APOBEC3G (Bogerd et al., 2004). Although this mechanism of APOBEC3G neutralization is conserved, the APOBEC3G–Vif interaction is remarkably species specific (Mariani et al., 2003). For instance, Vif derived from viruses that infect monkeys (SIV infecting rhesus macaques or African green monkeys) is unable to degrade human APOBEC3G, which is therefore able to inhibit the growth of these viruses (Bogerd et al., 2004; Mangeat et al., 2004; Schrofelbauer et al., 2004). Correspondingly, Vif from HIV-1 does not degrade rhesus macaque or African green monkey APOBEC3G, which inhibit the human virus. This species specificity can be mapped to a single amino acid. Mutating residue 128 from aspartic acid to lysine (D128K) enables human APOBEC3G to resist degradation by HIV-1 Vif, and sensitizes it to degradation by SIVmac or SIVagm Vif (Bogerd et al., 2004; Mangeat et al., 2004; Schrofelbauer et al., 2004; Xu et al., 2004). This specificity suggests that each retroviral *vif* gene has evolved to optimally counteract its host species' APOBEC3 proteins.

It should be noted that Vif might employ more than one mechanism to neutralize APOBEC3G (Stopak et al., 2003; Kao et al., 2004; Santa-Marta et al., 2005). One study indicated that Vif could directly inhibit the DNA deaminase activity of APOBEC3G (Santa-Marta et al., 2005). A second study suggested that Vif could impair APOBEC3G translation (Stopak et al., 2003). A third identified a Vif variant that restored infectivity of HIV-1 but did not cause obvious APOBEC3G degradation (Kao et al., 2007). In any event, despite the obvious appeal of a direct inhibition mechanism (in addition to proteosome-dependent degradation), a consensus opinion has yet to emerge and more research is clearly needed in this area.

Broad Functionality of the Mammalian APOBEC3 Protein Family

As alluded to above, APOBEC3G is one member of a larger family of related proteins, a fact that can be appreciated through a phylogenetic overview (Fig. 15.6). Humans encode a total of 11 family members: seven APOBEC3 proteins, APOBEC3A, -3B, -3C, -3DE, -3F, -3G, and -3H, from a single locus on chromosome 22p13, APOBEC4 from 1q25.3, APOBEC2 from 6p21, APOBEC1 from 12p13.1, and AID from 12p13. APOBEC2 and AID are evolutionarily the oldest, as they are the only ones found in all vertebrates. The physiological function of APOBEC2 remains a mystery, but AID has a pivotal role in B lymphocytes as the

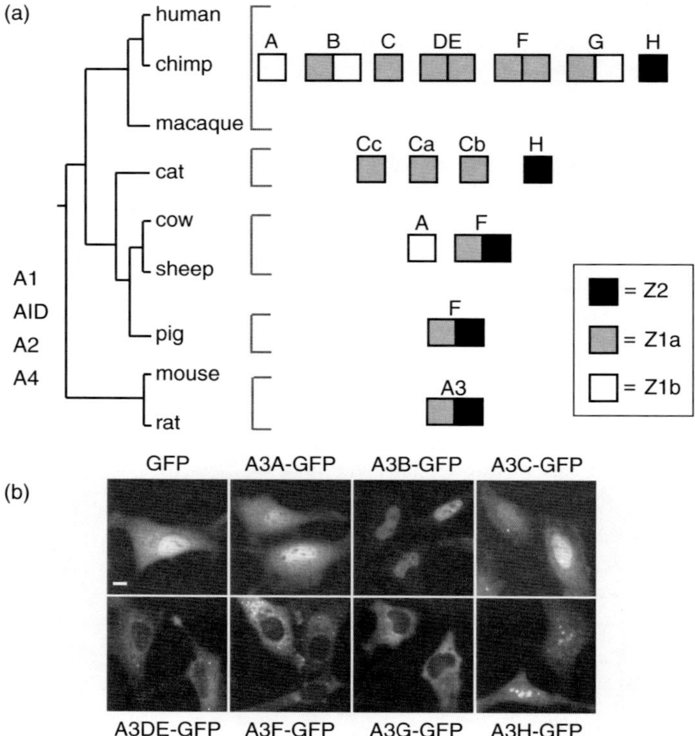

Fig. 15.6 Phylogeny and localization of the APOBEC3s. (**a**) A phylogenetic tree showing the APOBEC3 repertoire of several mammalian species. Z1a, Z1b, and Z2 designate three classes of evolutionarily related zinc-binding APOBEC3 domains, as described in Conticello et al. (2005). The cat APOBEC3 locus is described in Munk et al. (2008); the artiodactyl locus in LaRue and Harris (2008). Note that the lineage leading to primates has undergone a dramatic expansion in the number of APOBEC3 genes. Note also that some APOBEC3s, such as APOBEC3G, are double-domain proteins, whereas others consist of a single domain (e.g., APOBEC3A). (**b**) The sub-cellular localization of the human APOBEC3 proteins. These images are of live HeLa cells expressing GFP-tagged proteins. The scale bar indicates 10 μm.

DNA cytosine deaminase that triggers several distinct antibody gene diversification reactions (Mikl et al., 2005; Longerich et al., 2006; Di Noia and Neuberger, 2007). APOBEC4 is specific to higher vertebrates and its function is also unknown (Rogozin et al., 2005). In contrast, APOBEC1 and the APOBEC3 proteins are found only in mammals (Fujino et al., 1998; Jarmuz et al., 2002; Wedekind et al., 2003; Harris and Liddament, 2004; Conticello et al., 2005, 2007).

The similarity of APOBEC1 and APOBEC3 proteins to AID (in both sequence and enzymatic activity) has led to models in which duplications of an ancestral AID gene gave rise to the precursors of the present day APOBEC1 and APOBEC3 genes. Several subsequent duplication events must have occurred to generate the present day primate APOBEC3 locus (Fig. 15.6). Although the broader family of polynucleotide cytosine deaminases appears to have expanded gradually during vertebrate evolution, the APOBEC3s have experienced a much more dramatic and recent expansion in mammals. At the heart of this expansion is the conserved, zinc-coordinating motif (Fig. 15.5). This motif is encoded by a single exon and it consists of at least four key residues, one histidine, one glutamate, and two cysteines, $H–X_1–E–X_{23–28}–C–X_{2–4}–C$ (X indicates a non-conserved residue) (Jarmuz et al., 2002; Wedekind et al., 2003; Harris and Liddament, 2004; Conticello et al., 2005, 2007). The histidine and the cysteines directly coordinate zinc. The glutamate participates indirectly by binding a water molecule, which in turn binds zinc and serves as the nucleophile for cytosine deamination. Some APOBEC3 proteins have two conserved zinc-coordinating (Z) motifs, whereas other family members have only one.

A recent burst of near-complete mammalian genome sequences has revealed that primates encode the greatest number of APOBEC3 zinc-coordinating motifs (11 of these so-called Z-motifs encoded by seven genes) and the rodents have the smallest (two Z-motifs encoded by one gene) (Fig. 15.6). These zinc(Z)-coordinating motifs segregate phylogenetically into three sub-groups, Z1a, Z1b, and Z2 (Conticello et al., 2005). The Z1s and the Z2 differ at many positions, including an obvious serine or threonine that precedes the conserved cysteines, *S*WSPC$X_{2–4}$C (Z1s) or *T*WSPC$X_{2–4}$C (Z2). The Z1a and Z1b motifs also differ at more than 20 positions, but a key identifier (WF in Z1a domains and X(V/I) in Z1b proteins) can be found in six residues carboxy-terminal to the conserved HXE motif. Interestingly, mammals that branch phylogenetically between rodents and humans have intermediate APOBEC3 gene numbers (Jonsson et al., 2006; Munk et al., 2008 and Fig. 15.6). For instance, artiodactyls such as sheep and cattle have one A3A-like gene and one A3F-like gene (three Z-motifs in two genes) (Jonsson et al., 2006; LaRue and Harris, 2008). And felines have three A3C-like genes and one A3H-like gene (Munk et al., 2008). These expansions are entirely attributable to the Z1-motif, as the Z2-motif is single-copy in all mammals. Species within a larger order/family or members of a particular species may have fewer Z-motifs. For instance, at least one deletion must have occurred to cause domesticated pigs (another artiodactyl) to have only two Z-motifs (one APOBEC3 gene due to an APOBEC3A-3F deletion) and some humans to have nine Z-motifs (six APOBEC3 genes due to an APOBEC3A-3B deletion) (Kidd et al., 2007). The functional significance of these deletions is not yet clear, but it is certainly worth investigating.

In addition to the dramatic Z-motif expansions, the primate APOBEC3 genes contain signatures of strong positive selection (Sawyer et al., 2004; Zhang and Webb, 2004). These two features combine to indicate that this locus is under a powerful and ongoing selective pressure, which appears to intensify in specific mammalian lineages (e.g., primates). Although the precise selective pressures have not been (and may never be) identified, accumulating evidence suggests that retroviruses and endogenous retrotransposons may provide the major driving forces. This is supported by many reports of APOBEC3 proteins inhibiting endogenous retroelements, observations that endogenized mice retroelements bear scars of APOBEC3-like deamination events, and the fact that the overall number of active retroelements appears to be considerably less in humans than mice (Lander et al., 2001; Esnault et al., 2005; Schumacher et al., 2005; Bogerd et al., 2006a, b; Chiu et al., 2006; Esnault et al., 2006; Stenglein and Harris, 2006; Hulme et al., 2007; Jern et al., 2007; Jonsson et al., 2007; Kinomoto et al., 2007; Schumacher et al., 2008). In other words, there is a compelling inverse correlation between the number of APOBEC3 genes and retroelement mobility.

APOBEC3 Frontiers

The number of APOBEC3G PubMed occurrences has risen from 4 in 2002 to 327 currently, and this rate shows no signs of diminishing. Several frontiers are wide open. First, which APOBEC3 proteins are physiologically relevant to the restriction of HIV-1 and other medically relevant viruses? The answer to this important question could come from a variety of sources, including analyses of variations in human APOBEC3 genes, simian experiments, and cell-based experiments. Currently, APOBEC3G and APOBEC3F are the leading candidates because their mutational signatures (determined by the nucleotide preceding the deaminated cytosine) are observed in patient-derived HIV-1 sequences (e.g., Liddament et al., 2004). Moreover, these proteins appear co-expressed and are the only two human APOBEC3 proteins that HIV-1 Vif can inhibit significantly (aforementioned references for APOBEC3G and Bishop et al., 2004; Liddament et al., 2004; Wiegand et al., 2004; Zheng et al., 2004; Simon et al., 2005; Holmes et al., 2007a).

Second, a structural understanding of the APOBEC3G–Vif interaction would be of great benefit. Advances in this area have been hindered by the fact that both APOBEC3G and Vif are poorly soluble. Indeed, Vif is the only HIV-1 protein for which there is no high-resolution structural information. However, a recent solution structure of the APOBEC3G catalytic domain has indicated that single-strand DNA is recognized by a positively charged, arginine-rich brim, which facilitates deamination by flipping out the target cytosine base such that it can be accommodated by an active site pocket (Chen et al., 2008). Moreover, this structure has facilitated a model of the full-length APOBEC3G protein, which offers several testable predictions that relate to the Vif-interacting region (defined by D128, Fig. 15.5). Overall, this first structure will provide the foundation for many experiments and will help answer important questions, for example, what is the mechanism of Vif binding? What determines nucleic acid substrate specificity? And, what is the stoichiometry of the APOBEC3G–Vif interaction?

Third, not much is known about APOBEC3 gene regulation. APOBEC3G and -3F appear broadly and constitutively expressed, but other APOBEC3 genes like APOBEC3B and -3DE appear to be less abundant and tissue restricted (Harris and Liddament, 2004, Liddament et al., 2004; Wiegand et al., 2004). Moreover, several of the APOBEC3 genes appear to be interferon inducible, suggesting that APOBEC3 proteins may play a role in the innate immune response (Rose et al., 2004b; Taylor et al., 2004; Bonvin et al., 2006; Chen et al., 2006; Peng et al., 2006; Sarkis et al., 2006; Tanaka et al., 2006; Komohara et al., 2007; Stopak et al., 2007; Ying et al., 2007). It is therefore highly likely that several of these genes will be induced by virus infection, which can be a very potent inducer of the interferon response.

Finally, moving back to the APOBEC3 proteins themselves. Common sense alone would dictate that such potent DNA mutating enzymes would be subject to tight post-translational control within the cell. Failure to do so could be catastrophic and/or contribute to carcinogenesis. APOBEC3G, for instance, appears predominantly cytoplasmic, safely away from the genomic DNA (Fig. 15.6) (Mangeat et al., 2003; Rose et al., 2004a; Wichroski et al., 2005; Jonsson et al., 2006; Kozak et al., 2006; Stenglein and Harris, 2006; Wichroski et al., 2006). Although several groups have identified a plethora of candidate APOBEC3G-interacting proteins (and RNAs) that may be important for post-translational regulation and retroelement restriction, the best evidence to date was provided by Greene's laboratory. Chiu et al. reported that APOBEC3G resides in enzymatically inactive, high-molecular-mass (HMM) ribonucleoprotein complexes in activated CD4+ T cells and in enzymatically active, low-molecular-mass (LMM) ribonucleoprotein complex in resting CD4+ T cells (Chiu et al., 2005, 2006; Kozak et al., 2006; Gallois-Montbrun et al., 2007). This finding demonstrates that the state of the cell can directly impact the nature of the APOBEC3G-associated factors, and that this in turn can determine the permissiveness of the cell for virus infection. Moreover, taken together with the fact that human APOBEC3 proteins can occupy nearly every sub-cellular compartment, it is likely that these proteins are subject to multiple layers of regulation (Fig. 15.6).

Other Retrovirus Restriction Factors

Although APOBEC3G and TRIM5alpha are the most studied and best understood restriction systems, several others have been described, and the existence of undiscovered factors has been inferred. The Goff lab described the zinc finger antiviral protein (ZAP) (Gao et al., 2002). ZAP is a rat protein that limits retroviral replication by targeting cytoplasmic viral RNA for degradation (Guo et al., 2004, 2007). Another restriction factor that blocks infection by HIV-2, termed Lv2, has been described (Schmitz et al., 2004). Lv2 remains an inferred activity and an Lv2 gene has yet to be cloned. Additionally, as described above, CypA seems to protect the HIV-1 capsid from a restriction factor whose molecular identity remains undetermined. Thus, the mammalian repertoire of restriction factors continues to expand.

Despite the increasing number of cellular restriction factors, the central importance of TRIM5alpha and APOBEC3G is highlighted by two recent reports. These demonstrated that simian barriers to HIV-1 infection could be overcome by replacing the capsid and vif-coding regions of HIV-1 with the corresponding SIV genes (Hatziioannou et al., 2006; Kamada et al., 2006). Recall that Vif and Capsid are the primary viral determinants of susceptibility to APOBEC3G and TRIM5alpha restriction, respectively. The resulting chimeric virus (over 90% HIV-1) was able to replicate in normally non-permissive monkey cells. The prospect of using such viruses as HIV-1/AIDS disease models is very good. Moreover, these studies imply that the TRIM- and APOBEC3-mediated barriers are critical cellular barriers that function to limit the zoonotic transmission of lentiviruses and, perhaps, of retroviruses in general. Such barriers must be overcome before a TRIM and/or APOBEC3 susceptible virus can colonize a host and potentially cause disease.

Conclusions and Future Directions

In the past several decades it has become increasingly clear that the natural host range of many retroviruses is extremely limited, and that this is due in large part to retrovirus restriction factors. For a retrovirus to be able to infect a particular host, it must circumvent that host's restriction factors. Usually, this ability comes at a cost, as it typically means that the virus will be susceptible to restriction by other species' restriction factors. This conflict between virus and host defenses has left traces of positive selection on the genes involved. In terms of human health, the ultimate goal of studying restriction factors is to enable the development of novel antiviral therapies. These therapies could take several forms, for instance, restriction factors could be used in gene therapy. Alternatively, small molecule drugs could tip the balance in favor of the restriction factor, for example, by inhibiting the APOBEC3G–Vif interaction. As our understanding of restriction factors continues to increase, the prospects of developing such therapies will improve proportionately.

Acknowledgments We thank N. Somia, L. Mansky, M. Huseby, and several laboratory members for thoughtful comments. Studies in the authors' laboratory are supported by grants from the National Institutes of Health (AI064046 and GM080437), the Medica Foundation (Minnesota Partnership for Biotechnology and Medical Genomics) and the University of Minnesota (Leukemia Research Fund and Cancer Center Brainstorm Program), and the Cancer Biology Training Grant (CA009138).

References

Alce, T. M. and Popik, W. 2004. APOBEC3G is incorporated into virus-like particles by a direct interaction with HIV-1 Gag nucleocapsid protein. J Biol Chem. 279(33): 34083–6.

Anderson, J. and Akkina, R. 2005. TRIM5alpharh expression restricts HIV-1 infection in lentiviral vector-transduced CD34+-cell-derived macrophages. Mol Ther. 12(4): 687–96.

Anderson, J. L., Campbell, E. M., Wu, X., Vandegraaff, N., Engelman, A. and Hope, T. J. 2006. Proteasome inhibition reveals that a functional preintegration complex intermediate can be generated during restriction by diverse TRIM5 proteins. J Virol. 80(19): 9754–60.

Bassin, R. H., Duran-Troise, G., Gerwin, B. I. and Rein, A. 1978. Abrogation of Fv-1b restriction with murine leukemia viruses inactivated by heat or by gamma irradiation. J Virol. 26(2): 306–15.

Berthoux, L., Sebastian, S., Sokolskaja, E. and Luban, J. 2004. Lv1 inhibition of human immunodeficiency virus type 1 is counteracted by factors that stimulate synthesis or nuclear translocation of viral cDNA. J Virol. 78(21): 11739–50.

Berthoux, L., Sebastian, S., Sokolskaja, E. and Luban, J. 2005a. Cyclophilin A is required for TRIM5{alpha}-mediated resistance to HIV-1 in Old World monkey cells. Proc Natl Acad Sci USA. 102(41): 14849–53.

Berthoux, L., Sebastian, S., Sayah, D. M. and Luban, J. 2005b. Disruption of human TRIM5alpha antiviral activity by nonhuman primate orthologues. J Virol. 79(12): 7883–8.

Besnier, C., Takeuchi, Y. and Towers, G. 2002. Restriction of lentivirus in monkeys. Proc Natl Acad Sci USA. 99(18): 11920–5.

Besnier, C., Ylinen, L., Strange, B., Lister, A., Takeuchi, Y., Goff, S. P. and Towers, G. J. 2003. Characterization of murine leukemia virus restriction in mammals. J Virol. 77(24): 13403–6.

Best, S., Le Tissier, P., Towers, G. and Stoye, J. P. 1996. Positional cloning of the mouse retrovirus restriction gene Fv1. Nature. 382(6594): 826–9.

Bieniasz, P. D. 2003. Restriction factors: a defense against retroviral infection. Trends Microbiol. 11(6): 286–91.

Bieniasz, P. D. 2004. Intrinsic immunity: a front-line defense against viral attack. Nat Immunol. 5(11): 1109–15.

Bishop, K. N., Holmes, R. K., Sheehy, A. M., Davidson, N. O., Cho, S. J. and Malim, M. H. 2004. Cytidine deamination of retroviral DNA by diverse APOBEC proteins. Curr Biol. 14(15): 1392–6.

Bishop, K. N., Holmes, R. K. and Malim, M. H. 2006. Antiviral potency of APOBEC proteins does not correlate with cytidine deamination. J Virol. 80(17): 8450–8.

Bogerd, H. P., Doehle, B. P., Wiegand, H. L. and Cullen, B. R. 2004. A single amino acid difference in the host APOBEC3G protein controls the primate species specificity of HIV type 1 virion infectivity factor. Proc Natl Acad Sci USA. 101(11): 3770–4.

Bogerd, H. P., Wiegand, H. L., Doehle, B. P., Lueders, K. K. and Cullen, B. R. 2006a. APOBEC3A and APOBEC3B are potent inhibitors of LTR-retrotransposon function in human cells. Nucleic Acids Res. 34(1): 89–95.

Bogerd, H. P., Wiegand, H. L., Hulme, A. E., Garcia-Perez, J. L., O'Shea, K. S., Moran, J. V. and Cullen, B. R. 2006b. Cellular inhibitors of long interspersed element 1 and Alu retrotransposition. Proc Natl Acad Sci USA. 103(23): 8780–5.

Bonvin, M., Achermann, F., Greeve, I., Stroka, D., Keogh, A., Inderbitzin, D., Candinas, D., Sommer, P., Wain-Hobson, S., Vartanian, J. P. and Greeve, J. 2006. Interferon-inducible expression of APOBEC3 editing enzymes in human hepatocytes and inhibition of hepatitis B virus replication. Hepatology. 43(6): 1364–74.

Boone, L. R., Innes, C. L. and Heitman, C. K. 1990. Abrogation of Fv-1 restriction by genome-deficient virions produced by a retrovirus packaging cell line. J Virol. 64(7): 3376–81.

Borden, K. L. 1998. RING fingers and B-boxes: zinc-binding protein–protein interaction domains. Biochem Cell Biol. 76(2–3): 351–8.

Bosco, D. A., Eisenmesser, E. Z., Pochapsky, S., Sundquist, W. I. and Kern, D. 2002. Catalysis of cis/trans isomerization in native HIV-1 capsid by human cyclophilin A. Proc Natl Acad Sci USA. 99(8): 5247–52.

Braaten, D., Franke, E. K. and Luban, J. 1996a. Cyclophilin A is required for an early step in the life cycle of human immunodeficiency virus type 1 before the initiation of reverse transcription. J Virol. 70(6): 3551–60.

Braaten, D., Franke, E. K. and Luban, J. 1996b. Cyclophilin A is required for the replication of group M human immunodeficiency virus type 1 (HIV-1) and simian immunodeficiency virus SIV(CPZ)GAB but not group O HIV-1 or other primate immunodeficiency viruses. J Virol. 70(7): 4220–7.

Braaten, D., Aberham, C., Franke, E. K., Yin, L., Phares, W. and Luban, J. 1996c. Cyclosporine A-resistant human immunodeficiency virus type 1 mutants demonstrate that Gag encodes the functional target of cyclophilin A. J Virol. 70(8): 5170–6.

Braaten, D. and Luban, J. 2001. Cyclophilin A regulates HIV-1 infectivity, as demonstrated by gene targeting in human T cells. Embo J. 20(6): 1300–9.

Brennan, G., Kozyrev, Y. and Hu, S. L. 2008. TRIMCyp expression in Old World primates Macaca nemestrina and Macaca fascicularis. Proc Natl Acad Sci USA. 105(9): 3569–74.

Burnett, A. and Spearman, P. 2007. APOBEC3G Multimers Are Recruited to the Plasma Membrane for Packaging into Human Immunodeficiency Virus Type 1 Virus-Like Particles in an RNA-Dependent Process Requiring the NC Basic Linker. J Virol. 81(10): 5000–13.

Campbell, E. M., Dodding, M. P., Yap, M. W., Wu, X., Gallois-Montbrun, S., Malim, M. H., Stoye, J. P. and Hope, T. J. 2007. TRIM5 alpha cytoplasmic bodies are highly dynamic structures. Mol Biol Cell. 18(6): 2102–11.

Cen, S., Guo, F., Niu, M., Saadatmand, J., Deflassieux, J. and Kleiman, L. 2004. The interaction between HIV-1 Gag and APOBEC3G. J Biol Chem. 279(32): 33177–84.

Chatterji, U., Bobardt, M. D., Stanfield, R., Ptak, R. G., Pallansch, L. A., Ward, P. A., Jones, M. J., Stoddart, C. A., Scalfaro, P., Dumont, J. M., Besseghir, K., Rosenwirth, B. and Gallay, P. A. 2005. Naturally occurring capsid substitutions render HIV-1 cyclophilin A independent in human cells and TRIM-cyclophilin-resistant in Owl monkey cells. J Biol Chem. 280(48): 40293–300.

Chatterji, U., Bobardt, M. D., Gaskill, P., Sheeter, D., Fox, H. and Gallay, P. A. 2006. Trim5alpha accelerates degradation of cytosolic capsid associated with productive HIV-1 entry. J Biol Chem. 281(48): 37025–33.

Chen, K., Huang, J., Zhang, C., Huang, S., Nunnari, G., Wang, F. X., Tong, X., Gao, L., Nikisher, K. and Zhang, H. 2006. Alpha interferon potently enhances the anti-human immunodeficiency virus type 1 activity of APOBEC3G in resting primary CD4 T cells. J Virol. 80(15): 7645–57.

Chen, K. M., Martemyanova, N., Lu, Y., Shindo, K., Matsuo, H. and Harris, R. S. 2007. Extensive mutagenesis experiments corroborate a structural model for the DNA deaminase domain of APOBEC3G. FEBS Lett. 581(24): 4761–6.

Chen, K. M., Harjes, E., Gross, P. J., Fahmy, A., Lu, Y., Shindo, K., Harris, R. S. and Matsuo, H. 2008. Structure of the DNA deaminase domain of the HIV-1 restriction factor APOBEC3G. Nature. 452(7183): 116–9.

Chiu, Y. L., Soros, V. B., Kreisberg, J. F., Stopak, K., Yonemoto, W. and Greene, W. C. 2005. Cellular APOBEC3G restricts HIV-1 infection in resting CD4+ T cells. Nature. 435(7038): 108–14.

Chiu, Y. L. and Greene, W. C. 2006. Multifaceted antiviral actions of APOBEC3 cytidine deaminases. Trends Immunol. 27(6): 291–7.

Chiu, Y. L., Witkowska, H. E., Hall, S. C., Santiago, M., Soros, V. B., Esnault, C., Heidmann, T. and Greene, W. C. 2006. High-molecular-mass APOBEC3G complexes restrict Alu retrotransposition. Proc Natl Acad Sci USA. 103(42): 15588–93.

Conticello, S. G., Harris, R. S. and Neuberger, M. S. 2003. The Vif protein of HIV triggers degradation of the human antiretroviral DNA deaminase APOBEC3G. Curr Biol. 13(22): 2009–13.

Conticello, S. G., Thomas, C. J., Petersen-Mahrt, S. and Neuberger, M. S. 2005. Evolution of the AID/APOBEC Family of Polynucleotide (Deoxy)Cytidine Deaminases. Mol Biol Evol. 22(2): 367–77.

Conticello, S. G., Langlois, M. A., Yang, Z. and Neuberger, M. S. 2007. DNA deamination in immunity: AID in the context of its APOBEC relatives. Adv Immunol. 94: 37–73.

Cowan, S., Hatziioannou, T., Cunningham, T., Muesing, M. A., Gottlinger, H. G. and Bieniasz, P. D. 2002. Cellular inhibitors with Fv1-like activity restrict human and simian immunodeficiency virus tropism. Proc Natl Acad Sci USA. 99(18): 11914–9.

Decleve, A., Niwa, O., Gelmann, E. and Kaplan, H. S. 1975. Replication kinetics of N- and B-tropic murine leukemia viruses on permissive and nonpermissive cells in vitro. Virology. 65(2): 320–32.

DesGroseillers, L. and Jolicoeur, P. 1983. Physical mapping of the Fv-1 tropism host range determinant of BALB/c murine leukemia viruses. J Virol. 48(3): 685–96.

Desrosiers, R. C., Lifson, J. D., Gibbs, J. S., Czajak, S. C., Howe, A. Y., Arthur, L. O. and Johnson, R. P. 1998. Identification of highly attenuated mutants of simian immunodeficiency virus. J Virol. 72(2): 1431–7.

Di Noia, J. M. and Neuberger, M. S. 2007. Molecular mechanisms of antibody somatic hypermutation. Annu Rev Biochem. 76: 1–22.

Diaz-Griffero, F., Li, X., Javanbakht, H., Song, B., Welikala, S., Stremlau, M. and Sodroski, J. 2006a. Rapid turnover and polyubiquitylation of the retroviral restriction factor TRIM5. Virology. 349(2): 300–15.

Diaz-Griffero, F., Vandegraaff, N., Li, Y., McGee-Estrada, K., Stremlau, M., Welikala, S., Si, Z., Engelman, A. and Sodroski, J. 2006b. Requirements for capsid-binding and an effector function in TRIMCyp-mediated restriction of HIV-1. Virology. 351(2): 404–19.

Dodding, M. P., Bock, M., Yap, M. W. and Stoye, J. P. 2005. Capsid processing requirements for abrogation of Fv1 and Ref1 restriction. J Virol. 79(16): 10571–7.

Dutko, J. A., Schafer, A., Kenny, A. E., Cullen, B. R. and Curcio, M. J. 2005. Inhibition of a yeast LTR retrotransposon by human APOBEC3 cytidine deaminases. Curr Biol. 15(7): 661–6.

Esnault, C., Heidmann, O., Delebecque, F., Dewannieux, M., Ribet, D., Hance, A. J., Heidmann, T. and Schwartz, O. 2005. APOBEC3G cytidine deaminase inhibits retrotransposition of endogenous retroviruses. Nature. 433(7024): 430–3.

Esnault, C., Millet, J., Schwartz, O. and Heidmann, T. 2006. Dual inhibitory effects of APOBEC family proteins on retrotransposition of mammalian endogenous retroviruses. Nucleic Acids Res. 34(5): 1522–31.

Everett, R. D. and Chelbi-Alix, M. K. 2007. PML and PML nuclear bodies: implications in antiviral defence. Biochimie. 89(6–7): 819–30.

Fields, B. N., Knipe, D. M. and Howley, P. M. 2007. Fields' virology (5th). Philadelphia: Wolters Kluwer Health/Lippincott Williams & Wilkins.

Fischer, G., Tradler, T. and Zarnt, T. 1998. The mode of action of peptidyl prolyl cis/trans isomerases in vivo: binding vs. catalysis. FEBS Lett. 426(1): 17–20.

Fisher, A. G., Ensoli, B., Ivanoff, L., Chamberlain, M., Petteway, S., Ratner, L., Gallo, R. C. and Wong-Staal, F. 1987. The sor gene of HIV-1 is required for efficient virus transmission in vitro. Science. 237(4817): 888–93.

Forshey, B. M., von Schwedler, U., Sundquist, W. I. and Aiken, C. 2002. Formation of a human immunodeficiency virus type 1 core of optimal stability is crucial for viral replication. J Virol. 76(11): 5667–77.

Franke, E. K., Yuan, H. E. and Luban, J. 1994. Specific incorporation of cyclophilin A into HIV-1 virions. Nature. 372(6504): 359–62.

Friend, C. 1957. Cell-free transmission in adult Swiss mice of a disease having the character of a leukemia. J Exp Med. 105(4): 307–18.

Fujino, T., Navaratnam, N. and Scott, J. 1998. Human apolipoprotein B RNA editing deaminase gene (APOBEC1). Genomics. 47(2): 266–75.

Gabuzda, D. H., Lawrence, K., Langhoff, E., Terwilliger, E., Dorfman, T., Haseltine, W. A. and Sodroski, J. 1992. Role of Vif in replication of human immunodeficiency virus type 1 in CD4+ T lymphocytes. J Virol. 66(11): 6489–95.

Gabuzda, D. H., Li, H., Lawrence, K., Vasir, B. S., Crawford, K. and Langhoff, E. 1994. Essential role of vif in establishing productive HIV-1 infection in peripheral blood T lymphocytes and monocyte/macrophages. J Acquir Immune Defic Syndr. 7(9): 908–15.

Gallois-Montbrun, S., Kramer, B., Swanson, C. M., Byers, H., Lynham, S., Ward, M. and Malim, M. H. 2007. Antiviral protein APOBEC3G localizes to ribonucleoprotein complexes found in P bodies and stress granules. J Virol. 81(5): 2165–78.

Gamble, T. R., Vajdos, F. F., Yoo, S., Worthylake, D. K., Houseweart, M., Sundquist, W. I. and Hill, C. P. 1996. Crystal structure of human cyclophilin A bound to the amino-terminal domain of HIV-1 capsid. Cell. 87(7): 1285–94.

Gao, G., Guo, X. and Goff, S. P. 2002. Inhibition of retroviral RNA production by ZAP, a CCCH-type zinc finger protein. Science. 297(5587): 1703–6.

Goff, S. P. 1996. Operating under a Gag order: a block against incoming virus by the Fv1 gene. Cell. 86(5): 691–3.

Goff, S. P. 2004. Retrovirus restriction factors. Mol Cell. 16(6): 849–59.

Goff, S. P. 2007. Host factors exploited by retroviruses. Nat Rev Microbiol. 5(4): 253–63.

Goldschmidt, V., Bleiber, G., May, M., Martinez, R., Ortiz, M. and Telenti, A. 2006. Role of common human TRIM5alpha variants in HIV-1 disease progression. Retrovirology. 3: 54.

Goncalves, J., Korin, Y., Zack, J. and Gabuzda, D. 1996. Role of Vif in human immunodeficiency virus type 1 reverse transcription. J Virol. 70(12): 8701–9.

Guo, X., Carroll, J. W., Macdonald, M. R., Goff, S. P. and Gao, G. 2004. The zinc finger antiviral protein directly binds to specific viral mRNAs through the CCCH zinc finger motifs. J Virol. 78(23): 12781–7.

Guo, X., Ma, J., Sun, J. and Gao, G. 2007. The zinc-finger antiviral protein recruits the RNA processing exosome to degrade the target mRNA. Proc Natl Acad Sci USA. 104(1): 151–6.

Haché, G., Liddament, M. T. and Harris, R. S. 2005. The retroviral hypermutation specificity of APOBEC3F and APOBEC3G is governed by the C-terminal DNA cytosine deaminase domain. J Biol Chem. 280(12): 10920–4.

Haché, G., Mansky, L. M. and Harris, R. S. 2006. Human APOBEC3 proteins, retrovirus restriction, and HIV drug resistance. AIDS Rev. 8(3): 148–57.

Haché, G., Shindo, K., Albin, J. S. and Harris, R. S. 2008. Evolution of HIV-1 Isolates that Use a Novel Vif-Independent Mechanism to Resist Restriction by Human APOBEC3G. Curr Biol. 18(11): 819–24.

Harris, R. S., Petersen-Mahrt, S. K. and Neuberger, M. S. 2002. RNA editing enzyme APOBEC1 and some of its homologs can act as DNA mutators. Mol Cell. 10(5): 1247–53.

Harris, R. S., Bishop, K. N., Sheehy, A. M., Craig, H. M., Petersen-Mahrt, S. K., Watt, I. N., Neuberger, M. S. and Malim, M. H. 2003a. DNA deamination mediates innate immunity to retroviral infection. Cell. 113(6): 803–9.

Harris, R. S., Sheehy, A. M., Craig, H. M., Malim, M. H. and Neuberger, M. S. 2003b. DNA deamination: not just a trigger for antibody diversification but also a mechanism for defense against retroviruses. Nat Immunol. 4(7): 641–3.

Harris, R. S. and Liddament, M. T. 2004. Retroviral restriction by APOBEC proteins. Nat Rev Immunol. 4(11): 868–77.

Hartley, J. W., Rowe, W. P. and Huebner, R. J. 1970. Host-range restrictions of murine leukemia viruses in mouse embryo cell cultures. J Virol. 5(2): 221–5.

Hatziioannou, T., Cowan, S., Goff, S. P., Bieniasz, P. D. and Towers, G. J. 2003. Restriction of multiple divergent retroviruses by Lv1 and Ref1. Embo J. 22(3): 385–94.

Hatziioannou, T., Perez-Caballero, D., Yang, A., Cowan, S. and Bieniasz, P. D. 2004a. Retrovirus resistance factors Ref1 and Lv1 are species-specific variants of TRIM5alpha. Proc Natl Acad Sci USA. 101(29): 10774–9.

Hatziioannou, T., Cowan, S., Von Schwedler, U. K., Sundquist, W. I. and Bieniasz, P. D. 2004b. Species-specific tropism determinants in the human immunodeficiency virus type 1 capsid. J Virol. 78(11): 6005–12.

Hatziioannou, T., Perez-Caballero, D., Cowan, S. and Bieniasz, P. D. 2005. Cyclophilin interactions with incoming human immunodeficiency virus type 1 capsids with opposing effects on infectivity in human cells. J Virol. 79(1): 176–83.

Hatziioannou, T., Princiotta, M., Piatak, M., Jr., Yuan, F., Zhang, F., Lifson, J. D. and Bieniasz, P. D. 2006. Generation of simian-tropic HIV-1 by restriction factor evasion. Science. 314(5796): 95.

Henry, J., Ribouchon, M. T., Offer, C. and Pontarotti, P. 1997. B30.2-like domain proteins: a growing family. Biochem Biophys Res Commun. 235(1): 162–5.

Himathongkham, S. and Luciw, P. A. 1996. Restriction of HIV-1 (subtype B) replication at the entry step in rhesus macaque cells. Virology. 219(2): 485–8.

Hofmann, W., Schubert, D., LaBonte, J., Munson, L., Gibson, S., Scammell, J., Ferrigno, P. and Sodroski, J. 1999. Species-specific, postentry barriers to primate immunodeficiency virus infection. J Virol. 73(12): 10020–8.

Hoffman, B. G., Williams, K. L., Tien, A. H., Lu, V., de Algara, T. R., Ting, J. P. and Helgason, C. D. 2006. Identification of novel genes and transcription factors involved in spleen, thymus and immunological development and function. Genes Immun. 7(2): 101–12.

Holmes, R. K., Koning, F. A., Bishop, K. N. and Malim, M. H. 2007a. APOBEC3F can inhibit the accumulation of HIV-1 reverse transcription products in the absence of hypermutation. Comparisons with APOBEC3G. J Biol Chem. 282(4): 2587–95.

Holmes, R. K., Malim, M. H. and Bishop, K. N. 2007b. APOBEC-mediated viral restriction: not simply editing? Trends Biochem Sci. 32(3): 118–28.

Hulme, A. E., Bogerd, H. P., Cullen, B. R. and Moran, J. V. 2007. Selective inhibition of Alu retrotransposition by APOBEC3G. Gene. 390(1–2): 199–205.

Ikeda, H., Laigret, F., Martin, M. A. and Repaske, R. 1985. Characterization of a molecularly cloned retroviral sequence associated with Fv-4 resistance. J Virol. 55(3): 768–77.

Ikeda, H. and Sugimura, H. 1989. Fv-4 resistance gene: a truncated endogenous murine leukemia virus with ecotropic interference properties. J Virol. 63(12): 5405–12.

Ikeda, Y., Ylinen, L. M., Kahar-Bador, M. and Towers, G. J. 2004. Influence of gag on human immunodeficiency virus type 1 species-specific tropism. J Virol. 78(21): 11816–22.

Iwatani, Y., Takeuchi, H., Strebel, K. and Levin, J. G. 2006. Biochemical activities of highly purified, catalytically active human APOBEC3G: correlation with antiviral effect. J Virol. 80(12): 5992–6002.

James, L. C., Keeble, A. H., Khan, Z., Rhodes, D. A. and Trowsdale, J. 2007. Structural basis for PRYSPRY-mediated tripartite motif (TRIM) protein function. Proc Natl Acad Sci USA. 104(15): 6200–5.

Jarmuz, A., Chester, A., Bayliss, J., Gisbourne, J., Dunham, I., Scott, J. and Navaratnam, N. 2002. An anthropoid-specific locus of orphan C to U RNA-editing enzymes on chromosome 22. Genomics. 79(3): 285–96.

Javanbakht, H., Diaz-Griffero, F., Stremlau, M., Si, Z. and Sodroski, J. 2005. The contribution of RING and B-box 2 domains to retroviral restriction mediated by monkey TRIM5alpha. J Biol Chem. 280(29): 26933–40.

Javanbakht, H., Yuan, W., Yeung, D. F., Song, B., Diaz-Griffero, F., Li, Y., Li, X., Stremlau, M. and Sodroski, J. 2006a. Characterization of TRIM5alpha trimerization and its contribution to human immunodeficiency virus capsid binding. Virology. 353(1): 234–46.

Javanbakht, H., An, P., Gold, B., Petersen, D. C., O'Huigin, C., Nelson, G. W., O'Brien, S. J., Kirk, G. D., Detels, R., Buchbinder, S., Donfield, S., Shulenin, S., Song, B., Perron, M. J., Stremlau, M., Sodroski, J., Dean, M. and Winkler, C. 2006b. Effects of human TRIM5alpha polymorphisms on antiretroviral function and susceptibility to human immunodeficiency virus infection. Virology. 354(1): 15–27.

Jern, P., Stoye, J. P. and Coffin, J. 2007. Role of APOBEC3 in Genetic Diversity among Endogenous Murine Leukemia Viruses. PLoS Genetics. preprint(2007): e183.eor.

Jolicoeur, P. and Baltimore, D. 1976. Effect of Fv-1 gene product on proviral DNA formation and integration in cells infected with murine leukemia viruses. Proc Natl Acad Sci USA. 73(7): 2236–40.

Jolicoeur, P. 1979. The Fv-1 gene of the mouse and its control of murine leukemia virus replication. Curr Top Microbiol Immunol. 86: 67–122.

Jolicoeur, P. and Rassart, E. 1980. Effect of Fv-1 gene product on synthesis of linear and supercoiled viral DNA in cells infected with murine leukemia virus. J Virol. 33(1): 183–95.

Jonsson, S. R., Hache, G., Stenglein, M. D., Fahrenkrug, S. C., Andresdottir, V. and Harris, R. S. 2006. Evolutionarily conserved and non-conserved retrovirus restriction activities of artiodactyl APOBEC3F proteins. Nucleic Acids Res. 34(19): 5683–94.

Jonsson, S. R., Larue, R. S., Stenglein, M. D., Fahrenkrug, S. C., Andresdottir, V. and Harris, R. S. 2007. The Restriction of Zoonotic PERV Transmission by Human APOBEC3G. PLoS ONE. 2(9): e893.

Kai, K., Ikeda, H., Yuasa, Y., Suzuki, S. and Odaka, T. 1976. Mouse strain resistant to N-, B-, and NB-tropic murine leukemia viruses. J Virol. 20(2): 436–40.

Kai, K., Sato, H. and Odaka, T. 1986. Relationship between the cellular resistance to Friend murine leukemia virus infection and the expression of murine leukemia virus-gp70-related glycoprotein on cell surface of BALB/c-Fv-4wr mice. Virology. 150(2): 509–12.

Kaiser, S. M. and Emerman, M. 2006. Uracil DNA glycosylase is dispensable for human immunodeficiency virus type 1 replication and does not contribute to the antiviral effects of the cytidine deaminase Apobec3G. J Virol. 80(2): 875–82.

Kaiser, S. M., Malik, H. S. and Emerman, M. 2007. Restriction of an extinct retrovirus by the human TRIM5alpha antiviral protein. Science. 316(5832): 1756–8.

Kamada, K., Igarashi, T., Martin, M. A., Khamsri, B., Hatcho, K., Yamashita, T., Fujita, M., Uchiyama, T. and Adachi, A. 2006. Generation of HIV-1 derivatives that productively infect macaque monkey lymphoid cells. Proc Natl Acad Sci USA. 103(45): 16959–64.

Kao, S., Miyagi, E., Khan, M. A., Takeuchi, H., Opi, S., Goila-Gaur, R. and Strebel, K. 2004. Production of infectious human immunodeficiency virus type 1 does not require depletion of APOBEC3G from virus-producing cells. Retrovirology. 1(1): 27.

Kao, S., Goila-Gaur, R., Miyagi, E., Khan, M. A., Opi, S., Takeuchi, H. and Strebel, K. 2007. Production of infectious virus and degradation of APOBEC3G are separable functional properties of human immunodeficiency virus type 1 Vif. Virology. 369(2): 329–39.

Kaumanns, P., Hagmann, I. and Dittmar, M. T. 2006. Human TRIM5alpha mediated restriction of different HIV-1 subtypes and Lv2 sensitive and insensitive HIV-2 variants. Retrovirology. 3: 79.

Keckesova, Z., Ylinen, L. M. and Towers, G. J. 2004. The human and African green monkey TRIM5alpha genes encode Ref1 and Lv1 retroviral restriction factor activities. Proc Natl Acad Sci USA. 101(29): 10780–5.

Keckesova, Z., Ylinen, L. M. and Towers, G. J. 2006. Cyclophilin A renders human immunodeficiency virus type 1 sensitive to Old World monkey but not human TRIM5 alpha antiviral activity. J Virol. 80(10): 4683–90.

Khan, M. A., Kao, S., Miyagi, E., Takeuchi, H., Goila-Gaur, R., Opi, S., Gipson, C. L., Parslow, T. G., Ly, H. and Strebel, K. 2005. Viral RNA is required for the association of APOBEC3G with human immunodeficiency virus type 1 nucleoprotein complexes. J Virol. 79(9): 5870–4.

Kidd, J. M., Newman, T. L., Tuzun, E., Kaul, R. and Eichler, E. E. 2007. Population Stratification of a Common APOBEC Gene Deletion Polymorphism. PLoS Genet. 3(4): e63.

Kinomoto, M., Kanno, T., Shimura, M., Ishizaka, Y., Kojima, A., Kurata, T., Sata, T. and Tokunaga, K. 2007. All APOBEC3 family proteins differentially inhibit LINE-1 retrotransposition. Nucleic Acids Res. 35(9): 2955–64.

Komohara, Y., Suekane, S., Noguchi, M., Matsuoka, K., Yamada, A. and Itoh, K. 2007. Expression of APOBEC3G in kidney cells. Tissue Antigens. 69(1): 95–8.

Kootstra, N. A., Munk, C., Tonnu, N., Landau, N. R. and Verma, I. M. 2003. Abrogation of postentry restriction of HIV-1-based lentiviral vector transduction in simian cells. Proc Natl Acad Sci USA. 100(3): 1298–303.

Kozak, C. A. and Chakraborti, A. 1996. Single amino acid changes in the murine leukemia virus capsid protein gene define the target of Fv1 resistance. Virology. 225(2): 300–5.

Kozak, S. L., Marin, M., Rose, K. M., Bystrom, C. and Kabat, D. 2006. The anti-HIV-1 editing enzyme APOBEC3G binds HIV-1 RNA and messenger RNAs that shuttle between polysomes and stress granules. J Biol Chem. 281(39): 29105–19.

Lander, E. S., Linton, L. M., Birren, B., Nusbaum, C., Zody, M. C., Baldwin, J., Devon, K., Dewar, K., Doyle, M., FitzHugh, W., Funke, R., Gage, D., Harris, K., Heaford, A., Howland, J., Kann, L., Lehoczky, J., LeVine, R., McEwan, P., McKernan, K., Meldrim, J., Mesirov, J. P., Miranda, C., Morris, W., Naylor, J., Raymond, C., Rosetti, M., Santos, R., Sheridan, A., Sougnez, C., Stange-Thomann, N., Stojanovic, N., Subramanian, A., Wyman, D., Rogers, J., Sulston, J., Ainscough, R., Beck, S., Bentley, D., Burton, J., Clee, C., Carter, N., Coulson, A., Deadman, R., Deloukas, P., Dunham, A., Dunham, I., Durbin, R., French, L., Grafham, D., Gregory, S., Hubbard, T., Humphray, S., Hunt, A., Jones, M., Lloyd, C., McMurray, A., Matthews, L., Mercer, S., Milne, S., Mullikin, J. C., Mungall, A., Plumb, R., Ross, M., Shownkeen, R., Sims, S., Waterston, R. H., Wilson, R. K., Hillier, L. W., McPherson, J. D., Marra, M. A., Mardis, E. R., Fulton, L. A., Chinwalla, A. T., Pepin, K. H., Gish, W. R., Chissoe, S. L., Wendl, M. C., Delehaunty, K. D., Miner, T. L., Delehaunty, A., Kramer, J. B., Cook, L. L., Fulton, R. S., Johnson, D. L., Minx, P. J., Clifton, S. W., Hawkins, T., Branscomb, E., Predki, P., Richardson, P., Wenning, S., Slezak, T., Doggett, N., Cheng, J. F., Olsen, A., Lucas, S., Elkin, C., Uberbacher, E., Frazier, M., Gibbs, R. A., Muzny, D. M., Scherer, S. E., Bouck, J. B., Sodergren, E. J., Worley, K. C., Rives, C. M., Gorrell, J. H., Metzker, M. L.,

Naylor, S. L., Kucherlapati, R. S., Nelson, D. L., Weinstock, G. M., Sakaki, Y., Fujiyama, A., Hattori, M., Yada, T., Toyoda, A., Itoh, T., Kawagoe, C., Watanabe, H., Totoki, Y., Taylor, T., Weissenbach, J., Heilig, R., Saurin, W., Artiguenave, F., Brottier, P., Bruls, T., Pelletier, E., Robert, C., Wincker, P., Smith, D. R., Doucette-Stamm, L., Rubenfield, M., Weinstock, K., Lee, H. M., Dubois, J., Rosenthal, A., Platzer, M., Nyakatura, G., Taudien, S., Rump, A., Yang, H., Yu, J., Wang, J., Huang, G., Gu, J., Hood, L., Rowen, L., Madan, A., Qin, S., Davis, R. W., Federspiel, N. A., Abola, A. P., Proctor, M. J., Myers, R. M., Schmutz, J., Dickson, M., Grimwood, J., Cox, D. R., Olson, M. V., Kaul, R., Shimizu, N., Kawasaki, K., Minoshima, S., Evans, G. A., Athanasiou, M., Schultz, R., Roe, B. A., Chen, F., Pan, H., Ramser, J., Lehrach, H., Reinhardt, R., McCombie, W. R., de la Bastide, M., Dedhia, N., Blocker, H., Hornischer, K., Nordsiek, G., Agarwala, R., Aravind, L., Bailey, J. A., Bateman, A., Batzoglou, S., Birney, E., Bork, P., Brown, D. G., Burge, C. B., Cerutti, L., Chen, H. C., Church, D., Clamp, M., Copley, R. R., Doerks, T., Eddy, S. R., Eichler, E. E., Furey, T. S., Galagan, J., Gilbert, J. G., Harmon, C., Hayashizaki, Y., Haussler, D., Hermjakob, H., Hokamp, K., Jang, W., Johnson, L. S., Jones, T. A., Kasif, S., Kaspryzk, A., Kennedy, S., Kent, W. J., Kitts, P., Koonin, E. V., Korf, I., Kulp, D., Lancet, D., Lowe, T. M., McLysaght, A., Mikkelsen, T., Moran, J. V., Mulder, N., Pollara, V. J., Ponting, C. P., Schuler, G., Schultz, J., Slater, G., Smit, A. F., Stupka, E., Szustakowski, J., Thierry-Mieg, D., Thierry-Mieg, J., Wagner, L., Wallis, J., Wheeler, R., Williams, A., Wolf, Y. I., Wolfe, K. H., Yang, S. P., Yeh, R. F., Collins, F., Guyer, M. S., Peterson, J., Felsenfeld, A., Wetterstrand, K. A., Patrinos, A., Morgan, M. J., de Jong, P., Catanese, J. J., Osoegawa, K., Shizuya, H., Choi, S. and Chen, Y. J. 2001. Initial sequencing and analysis of the human genome. Nature. 409(6822): 860–921.

LaRue, R. S., Jónsson, S. R., Silverstein, K. A., Lajoie, M., Bertrand, D., El-Mabrouk, N., Hötzel, I., Andrésdóttir, V., Smith, T. P., Harris, R. S. 2008. The artiodactyl APOBEC3 innate immune repertoire shows evidence for a multi-functional domain organization that existed in the ancestor of placental mammals. BMC Mol. Biol. 9: 104.

Lassaux, A., Sitbon, M. and Battini, J. L. 2005. Residues in the murine leukemia virus capsid that differentially govern resistance to mouse Fv1 and human Ref1 restrictions. J Virol. 79(10): 6560–4.

Lecossier, D., Bouchonnet, F., Clavel, F. and Hance, A. J. 2003. Hypermutation of HIV-1 DNA in the absence of the Vif protein. Science. 300(5622): 1112.

Li, X., Li, Y., Stremlau, M., Yuan, W., Song, B., Perron, M. and Sodroski, J. 2006a. Functional replacement of the RING, B-box 2, and coiled-coil domains of tripartite motif 5alpha (TRIM5alpha) by heterologous TRIM domains. J Virol. 80(13): 6198–206.

Li, Y., Li, X., Stremlau, M., Lee, M. and Sodroski, J. 2006b. Removal of arginine 332 allows human TRIM5alpha to bind human immunodeficiency virus capsids and to restrict infection. J Virol. 80(14): 6738–44.

Li, X., Gold, B., O'HUigin, C., Diaz-Griffero, F., Song, B., Si, Z., Li, Y., Yuan, W., Stremlau, M., Mische, C., Javanbakht, H., Scally, M., Winkler, C., Dean, M. and Sodroski, J. 2007. Unique features of TRIM5alpha among closely related human TRIM family members. Virology. 360(2): 419–33.

Liao, C. H., Kuang, Y. Q., Liu, H. L., Zheng, Y. T. and Su, B. 2007. A novel fusion gene, TRIM5-Cyclophilin A in the pig-tailed macaque determines its susceptibility to HIV-1 infection. Aids. 21 Suppl 8: S19–26.

Liddament, M. T., Brown, W. L., Schumacher, A. J. and Harris, R. S. 2004. APOBEC3F properties and hypermutation preferences indicate activity against HIV-1 in vivo. Curr Biol. 14(15): 1385–91.

Lilly, F. 1967. Susceptibility to two strains of Friend leukemia virus in mice. Science. 155(761): 461–2.

Liu, B., Yu, X., Luo, K., Yu, Y. and Yu, X. F. 2004. Influence of primate lentiviral Vif and proteasome inhibitors on human immunodeficiency virus type 1 virion packaging of APOBEC3G. J Virol. 78(4): 2072–81.

Liu, H. L., Wang, Y. Q., Liao, C. H., Kuang, Y. Q., Zheng, Y. T. and Su, B. 2005. Adaptive evolution of primate TRIM5alpha, a gene restricting HIV-1 infection. Gene. 362: 109–16.

Longerich, S., Basu, U., Alt, F. and Storb, U. 2006. AID in somatic hypermutation and class switch recombination. Curr Opin Immunol. 18(2): 164–74.

Luban, J., Bossolt, K. L., Franke, E. K., Kalpana, G. V. and Goff, S. P. 1993. Human immunodeficiency virus type 1 Gag protein binds to cyclophilins A and B. Cell. 73(6): 1067–78.

Luban, J. 2007. Cyclophilin A, TRIM5, and resistance to human immunodeficiency virus type 1 infection. J Virol. 81(3): 1054–61.

Luo, K., Liu, B., Xiao, Z., Yu, Y., Yu, X., Gorelick, R. and Yu, X. F. 2004. Amino-terminal region of the human immunodeficiency virus type 1 nucleocapsid is required for human APOBEC3G packaging. J Virol. 78(21): 11841–52.

Luo, K., Wang, T., Liu, B., Tian, C., Xiao, Z., Kappes, J. and Yu, X. F. 2007. Cytidine deaminases APOBEC3G and APOBEC3F interact with human immunodeficiency virus type 1 integrase and inhibit proviral DNA formation. J Virol. 81(13): 7238–48.

Madani, N. and Kabat, D. 1998. An endogenous inhibitor of human immunodeficiency virus in human lymphocytes is overcome by the viral Vif protein. J Virol. 72(12): 10251–5.

Mangeat, B., Turelli, P., Caron, G., Friedli, M., Perrin, L. and Trono, D. 2003. Broad antiretroviral defence by human APOBEC3G through lethal editing of nascent reverse transcripts. Nature. 424(6944): 99–103.

Mangeat, B., Turelli, P., Liao, S. and Trono, D. 2004. A single amino acid determinant governs the species-specific sensitivity of APOBEC3G to Vif action. J Biol Chem. 279(15): 14481–3.

Mangeat, B. and Trono, D. 2005. Lentiviral vectors and antiretroviral intrinsic immunity. Hum Gene Ther. 16(8): 913–20.

Mariani, R., Chen, D., Schrofelbauer, B., Navarro, F., Konig, R., Bollman, B., Munk, C., Nymark-McMahon, H. and Landau, N. R. 2003. Species-specific exclusion of APOBEC3G from HIV-1 virions by Vif. Cell. 114(1): 21–31.

Marin, M., Rose, K. M., Kozak, S. L. and Kabat, D. 2003. HIV-1 Vif protein binds the editing enzyme APOBEC3G and induces its degradation. Nat Med. 9(11): 1398–403.

Massiah, M. A., Matts, J. A., Short, K. M., Simmons, B. N., Singireddy, S., Yi, Z. and Cox, T. C. 2007. Solution structure of the MID1 B-box2 CHC(D/C)C(2)H(2) zinc-binding domain: insights into an evolutionarily conserved RING fold. J Mol Biol. 369(1): 1–10.

Mbisa, J. L., Barr, R., Thomas, J. A., Vandegraaff, N., Dorweiler, I. J., Svarovskaia, E. S., Brown, W. L., Mansky, L. M., Gorelick, R. J., Harris, R. S., Engelman, A. and Pathak, V. K. 2007. Human immunodeficiency virus type 1 cDNAs produced in the presence of APOBEC3G exhibit defects in plus-strand DNA transfer and integration. J Virol. 81(13): 7099–110.

Mehle, A., Strack, B., Ancuta, P., Zhang, C., McPike, M. and Gabuzda, D. 2004. Vif overcomes the innate antiviral activity of APOBEC3G by promoting its degradation in the ubiquitin-proteasome pathway. J Biol Chem. 279(9): 7792–8.

Mikl, M. C., Watt, I. N., Lu, M., Reik, W., Davies, S. L., Neuberger, M. S. and Rada, C. 2005. Mice deficient in APOBEC2 and APOBEC3. Mol Cell Biol. 25(16): 7270–7.

Mische, C. C., Javanbakht, H., Song, B., Diaz-Griffero, F., Stremlau, M., Strack, B., Si, Z. and Sodroski, J. 2005. Retroviral restriction factor TRIM5alpha is a trimer. J Virol. 79(22): 14446–50.

Miyagi, E., Opi, S., Takeuchi, H., Khan, M., Goila-Gaur, R., Kao, S. and Strebel, K. 2007. Enzymatically active APOBEC3G is required for efficient inhibition of human immunodeficiency virus type 1. J Virol. 81(24): 13346–53.

Munk, C., Brandt, S. M., Lucero, G. and Landau, N. R. 2002. A dominant block to HIV-1 replication at reverse transcription in simian cells. Proc Natl Acad Sci USA. 99(21): 13843–8.

Munk, C., Beck, T., Zielonka, J., Hotz-Wagenblatt, A., Chareza, S., Battenberg, M., Thielebein, J., Cichutek, K., Bravo, I. G., O'Brien, S. J., Lochelt, M. and Yuhki, N. 2008. Functions, structure, and read-through alternative splicing of feline APOBEC3 genes. Genome Biol. 9(3): R48.

Nakayama, E. E., Miyoshi, H., Nagai, Y. and Shioda, T. 2005. A specific region of 37 amino acid residues in the SPRY (B30.2) domain of African green monkey TRIM5alpha determines species-specific restriction of simian immunodeficiency virus SIVmac infection. J Virol. 79(14): 8870–7.

Nakayama, E. E., Carpentier, W., Costagliola, D., Shioda, T., Iwamoto, A., Debre, P., Yoshimura, K., Autran, B., Matsushita, S. and Theodorou, I. 2007. Wild type and H43Y variant of human TRIM5alpha show similar anti-human immunodeficiency virus type 1 activity both in vivo and in vitro. Immunogenetics. 59(6): 511–5.

Nara, P. L. and Fischinger, P. J. 1988. Quantitative infectivity assay for HIV-1 and-2. Nature. 332(6163): 469–70.

Navarro, F., Bollman, B., Chen, H., Konig, R., Yu, Q., Chiles, K. and Landau, N. R. 2005. Complementary function of the two catalytic domains of APOBEC3G. Virology. 333(2): 374–86.

Newman, E. N., Holmes, R. K., Craig, H. M., Klein, K. C., Lingappa, J. R., Malim, M. H. and Sheehy, A. M. 2005. Antiviral function of APOBEC3G can be dissociated from cytidine deaminase activity. Curr Biol. 15(2): 166–70.

Newman, R. M., Hall, L., Connole, M., Chen, G. L., Sato, S., Yuste, E., Diehl, W., Hunter, E., Kaur, A., Miller, G. M. and Johnson, W. E. 2006. Balancing selection and the evolution of functional polymorphism in Old World monkey TRIM5alpha. Proc Natl Acad Sci USA. 103(50): 19134–9.

Newman, R. M., Hall, L., Kirmaier, A., Pozzi, L. A., Pery, E., Farzan, M., O'Neil, S. P. and Johnson, W. 2008. Evolution of a TRIM5-CypA splice isoform in old world monkeys. PLoS Pathog. 4(2): e1000003.

Nguyen, D. H., Gummuluru, S. and Hu, J. 2007. Deamination-independent inhibition of hepatitis B virus reverse transcription by APOBEC3G. J Virol. 81(9): 4465–72.

Nisole, S., Lynch, C., Stoye, J. P. and Yap, M. W. 2004. A Trim5-cyclophilin A fusion protein found in owl monkey kidney cells can restrict HIV-1. Proc Natl Acad Sci USA. 101(36): 13324–8.

Nisole, S., Stoye, J. P. and Saib, A. 2005. TRIM family proteins: retroviral restriction and antiviral defence. Nat Rev Microbiol. 3(10): 799–808.

Noser, J. A., Towers, G. J., Sakuma, R., Dumont, J. M., Collins, M. K. and Ikeda, Y. 2006. Cyclosporine increases human immunodeficiency virus type 1 vector transduction of primary mouse cells. J Virol. 80(15): 7769–74.

Ohkura, S., Yap, M. W., Sheldon, T. and Stoye, J. P. 2006. All three variable regions of the TRIM5alpha B30.2 domain can contribute to the specificity of retrovirus restriction. J Virol. 80(17): 8554–65.

Ortiz, M., Bleiber, G., Martinez, R., Kaessmann, H. and Telenti, A. 2006. Patterns of evolution of host proteins involved in retroviral pathogenesis. Retrovirology. 3: 11.

Peng, G., Lei, K. J., Jin, W., Greenwell-Wild, T. and Wahl, S. M. 2006. Induction of APOBEC3 family proteins, a defensive maneuver underlying interferon-induced anti-HIV-1 activity. J Exp Med. 203(1): 41–6.

Perez-Caballero, D., Hatziioannou, T., Yang, A., Cowan, S. and Bieniasz, P. D. 2005. Human tripartite motif 5alpha domains responsible for retrovirus restriction activity and specificity. J Virol. 79(14): 8969–78.

Perron, M. J., Stremlau, M., Song, B., Ulm, W., Mulligan, R. C. and Sodroski, J. 2004. TRIM5alpha mediates the postentry block to N-tropic murine leukemia viruses in human cells. Proc Natl Acad Sci USA. 101(32): 11827–32.

Perron, M. J., Stremlau, M. and Sodroski, J. 2006. Two surface-exposed elements of the B30.2/SPRY domain as potency determinants of N-tropic murine leukemia virus restriction by human TRIM5alpha. J Virol. 80(11): 5631–6.

Perron, M. J., Stremlau, M., Lee, M., Javanbakht, H., Song, B. and Sodroski, J. 2007. The human TRIM5alpha restriction factor mediates accelerated uncoating of the N-tropic murine leukemia virus capsid. J Virol. 81(5): 2138–48.

Petersen-Mahrt, S. K., Harris, R. S. and Neuberger, M. S. 2002. AID mutates *E. coli* suggesting a DNA deamination mechanism for antibody diversification. Nature. 418(6893): 99–103.

Pincus, T., Hartley, J. W. and Rowe, W. P. 1971. A major genetic locus affecting resistance to infection with murine leukemia viruses. I. Tissue culture studies of naturally occurring viruses. J Exp Med. 133(6): 1219–33.

Pincus, T., Hartley, J. W. and Rowe, W. P. 1975. A major genetic locus affecting resistance to infection with murine leukemia viruses. IV. Dose-response relationships in Fv-1-sensitive and resistant cell cultures. Virology. 65(2): 333–42.

Reddy, B. A. and Etkin, L. D. 1991. A unique bipartite cysteine-histidine motif defines a subfamily of potential zinc-finger proteins. Nucleic Acids Res. 19(22): 6330.

Reddy, B. A., Etkin, L. D. and Freemont, P. S. 1992. A novel zinc finger coiled-coil domain in a family of nuclear proteins. Trends Biochem Sci. 17(9): 344–5.

Reymond, A., Meroni, G., Fantozzi, A., Merla, G., Cairo, S., Luzi, L., Riganelli, D., Zanaria, E., Messali, S., Cainarca, S., Guffanti, A., Minucci, S., Pelicci, P. G. and Ballabio, A. 2001. The tripartite motif family identifies cell compartments. Embo J. 20(9): 2140–51.

Rhodes, D. A., de Bono, B. and Trowsdale, J. 2005. Relationship between SPRY and B30.2 protein domains. Evolution of a component of immune defence? Immunology. 116(4): 411–7.

Ribeiro, I. P., Menezes, A. N., Moreira, M. A., Bonvicino, C. R., Seuanez, H. N. and Soares, M. A. 2005. Evolution of cyclophilin A and TRIMCyp retrotransposition in New World primates. J Virol. 79(23): 14998–5003.

Rogozin, I. B., Basu, M. K., Jordan, I. K., Pavlov, Y. I. and Koonin, E. V. 2005. APOBEC4, a new member of the AID/APOBEC family of polynucleotide (deoxy)cytidine deaminases predicted by computational analysis. Cell Cycle. 4(9): 1281–5.

Rose, K. M., Marin, M., Kozak, S. L. and Kabat, D. 2004a. The viral infectivity factor (Vif) of HIV-1 unveiled. Trends Mol Med. 10(6): 291–7.

Rose, K. M., Marin, M., Kozak, S. L. and Kabat, D. 2004b. Transcriptional regulation of APOBEC3G, a cytidine deaminase that hypermutates human immunodeficiency virus. J Biol Chem. 279(40): 41744–9.

Rosler, C., Kock, J., Kann, M., Malim, M. H., Blum, H. E., Baumert, T. F. and von Weizsacker, F. 2005. APOBEC-mediated interference with hepadnavirus production. Hepatology. 42(2): 301–9.

Saenz, D. T., Teo, W., Olsen, J. C. and Poeschla, E. M. 2005. Restriction of feline immunodeficiency virus by Ref1, Lv1, and primate TRIM5alpha proteins. J Virol. 79(24): 15175–88.

Sakuma, R., Noser, J. A., Ohmine, S. and Ikeda, Y. 2007a. Inhibition of HIV-1 replication by simian restriction factors, TRIM5alpha and APOBEC3G. Gene Ther. 14(2): 185–9.

Sakuma, R., Noser, J. A., Ohmine, S. and Ikeda, Y. 2007b. Rhesus monkey TRIM5alpha restricts HIV-1 production through rapid degradation of viral Gag polyproteins. Nat Med. 13(5): 631–5.

Santa-Marta, M., da Silva, F. A., Fonseca, A. M. and Goncalves, J. 2005. HIV-1 Vif can directly inhibit apolipoprotein B mRNA-editing enzyme catalytic polypeptide-like 3G-mediated cytidine deamination by using a single amino acid interaction and without protein degradation. J Biol Chem. 280(10): 8765–75.

Sarkis, P. T., Ying, S., Xu, R. and Yu, X. F. 2006. STAT1-independent cell type-specific regulation of antiviral APOBEC3G by IFN-alpha. J Immunol. 177(7): 4530–40.

Saurin, A. J., Borden, K. L., Boddy, M. N. and Freemont, P. S. 1996. Does this have a familiar RING? Trends Biochem Sci. 21(6): 208–14.

Sawyer, S. L., Emerman, M. and Malik, H. S. 2004. Ancient adaptive evolution of the primate antiviral DNA-editing enzyme APOBEC3G. PLoS Biol. 2(9): E275.

Sawyer, S. L., Wu, L. I., Emerman, M. and Malik, H. S. 2005. Positive selection of primate TRIM5alpha identifies a critical species-specific retroviral restriction domain. Proc Natl Acad Sci USA. 102(8): 2832–7.

Sawyer, S. L., Wu, L. I., Akey, J. M., Emerman, M. and Malik, H. S. 2006. High-frequency persistence of an impaired allele of the retroviral defense gene TRIM5alpha in humans. Curr Biol. 16(1): 95–100.

Sayah, D. M. and Luban, J. 2004. Selection for loss of Ref1 activity in human cells releases human immunodeficiency virus type 1 from cyclophilin A dependence during infection. J Virol. 78(21): 12066–70.

Sayah, D. M., Sokolskaja, E., Berthoux, L. and Luban, J. 2004. Cyclophilin A retrotransposition into TRIM5 explains owl monkey resistance to HIV-1. Nature. 430(6999): 569–73.

Schafer, A., Bogerd, H. P. and Cullen, B. R. 2004. Specific packaging of APOBEC3G into HIV-1 virions is mediated by the nucleocapsid domain of the gag polyprotein precursor. Virology. 328(2): 163–8.

Schaller, T., Hue, S. and Towers, G. J. 2007. An active TRIM5 protein in rabbits indicates a common antiviral ancestor for mammalian TRIM5 proteins. J Virol. 81(21): 11713–21.

Schmitz, C., Marchant, D., Neil, S. J., Aubin, K., Reuter, S., Dittmar, M. T. and McKnight, A. 2004. Lv2, a novel postentry restriction, is mediated by both capsid and envelope. J Virol. 78(4): 2006–16.

Schrofelbauer, B., Chen, D. and Landau, N. R. 2004. A single amino acid of APOBEC3G controls its species-specific interaction with virion infectivity factor (Vif). Proc Natl Acad Sci USA. 101(11): 3927–32.

Schumacher, A. J., Nissley, D. V. and Harris, R. S. 2005. APOBEC3G hypermutates genomic DNA and inhibits Ty1 retrotransposition in yeast. Proc Natl Acad Sci USA. 102(28): 9854–9.

Schumacher, A. J., Hache, G., Macduff, D. A., Brown, W. L. and Harris, R. S. 2008. The DNA deaminase activity of human APOBEC3G is required for Ty1, MusD, and human immunodeficiency virus type 1 restriction. J Virol. 82(6): 2652–60.

Sebastian, S. and Luban, J. 2005. TRIM5alpha selectively binds a restriction-sensitive retroviral capsid. Retrovirology. 2: 40.

Seppen, J. 2004. Unedited inhibition of HBV replication by APOBEC3G. J Hepatol. 41(6): 1068–9.

Serhan, F., Jourdan, N., Saleun, S., Moullier, P. and Duisit, G. 2002. Characterization of producer cell-dependent restriction of murine leukemia virus replication. J Virol. 76(13): 6609–17.

Sheehy, A. M., Gaddis, N. C., Choi, J. D. and Malim, M. H. 2002. Isolation of a human gene that inhibits HIV-1 infection and is suppressed by the viral Vif protein. Nature. 418(6898): 646–50.

Sheehy, A. M., Gaddis, N. C. and Malim, M. H. 2003. The antiretroviral enzyme APOBEC3G is degraded by the proteasome in response to HIV-1 Vif. Nat Med. 9(11): 1404–7.

Shibata, R., Sakai, H., Kawamura, M., Tokunaga, K. and Adachi, A. 1995. Early replication block of human immunodeficiency virus type 1 in monkey cells. J Gen Virol. 76(11): 2723–30.

Shindo, K., Takaori-Kondo, A., Kobayashi, M., Abudu, A., Fukunaga, K. and Uchiyama, T. 2003. The enzymatic activity of CEM15/Apobec-3G is essential for the regulation of the infectivity of HIV-1 virion but not a sole determinant of its antiviral activity. J Biol Chem. 278(45): 44412–6.

Si, Z., Vandegraaff, N., O'Huigin, C., Song, B., Yuan, W., Xu, C., Perron, M., Li, X., Marasco, W. A., Engelman, A., Dean, M. and Sodroski, J. 2006. Evolution of a cytoplasmic tripartite motif (TRIM) protein in cows that restricts retroviral infection. Proc Natl Acad Sci USA. 103(19): 7454–9.

Simon, J. H. and Malim, M. H. 1996. The human immunodeficiency virus type 1 Vif protein modulates the postpenetration stability of viral nucleoprotein complexes. J Virol. 70(8): 5297–305.

Simon, J. H., Gaddis, N. C., Fouchier, R. A. and Malim, M. H. 1998. Evidence for a newly discovered cellular anti-HIV-1 phenotype. Nat Med. 4(12): 1397–400.

Simon, V., Zennou, V., Murray, D., Huang, Y., Ho, D. D. and Bieniasz, P. D. 2005. Natural Variation in Vif: Differential Impact on APOBEC3G/3F and a Potential Role in HIV-1 Diversification. PLoS Pathog. 1(1): e6.

Sokolskaja, E., Sayah, D. M. and Luban, J. 2004. Target cell cyclophilin A modulates human immunodeficiency virus type 1 infectivity. J Virol. 78(23): 12800–8.

Sokolskaja, E., Berthoux, L. and Luban, J. 2006. Cyclophilin A and TRIM5alpha independently regulate human immunodeficiency virus type 1 infectivity in human cells. J Virol. 80(6): 2855–62.

Song, B., Javanbakht, H., Perron, M., Park, D. H., Stremlau, M. and Sodroski, J. 2005a. Retrovirus restriction by TRIM5alpha variants from Old World and New World primates. J Virol. 79(7): 3930–7.

Song, B., Gold, B., O'Huigin, C., Javanbakht, H., Li, X., Stremlau, M., Winkler, C., Dean, M. and Sodroski, J. 2005b. The B30.2(SPRY) domain of the retroviral restriction factor TRIM5alpha exhibits lineage-specific length and sequence variation in primates. J Virol. 79(10): 6111–21.

Song, B., Diaz-Griffero, F., Park, D. H., Rogers, T., Stremlau, M. and Sodroski, J. 2005c. TRIM5alpha association with cytoplasmic bodies is not required for antiretroviral activity. Virology. 343(2): 201–11.

Speelmon, E. C., Livingston-Rosanoff, D., Li, S. S., Vu, Q., Bui, J., Geraghty, D. E., Zhao, L. P. and McElrath, M. J. 2006. Genetic association of the antiviral restriction factor TRIM5alpha with human immunodeficiency virus type 1 infection. J Virol. 80(5): 2463–71.

Stenglein, M. D. and Harris, R. S. 2006. APOBEC3B and APOBEC3F inhibit L1 retrotransposition by a DNA deamination-independent mechanism. J Biol Chem. 281(25): 16837–41.

Stevenson, M., Stanwick, T. L., Dempsey, M. P. and Lamonica, C. A. 1990. HIV-1 replication is controlled at the level of T cell activation and proviral integration. Embo J. 9(5): 1551–60.

Stopak, K., de Noronha, C., Yonemoto, W. and Greene, W. C. 2003. HIV-1 Vif blocks the antiviral activity of APOBEC3G by impairing both its translation and intracellular stability. Mol Cell. 12(3): 591–601.

Stopak, K. S., Chiu, Y. L., Kropp, J., Grant, R. M. and Greene, W. C. 2007. Distinct patterns of cytokine regulation of APOBEC3G expression and activity in primary lymphocytes, macrophages, and dendritic cells. J Biol Chem. 282(6): 3539–46.

Strebel, K., Daugherty, D., Clouse, K., Cohen, D., Folks, T. and Martin, M. A. 1987. The HIV 'A' (sor) gene product is essential for virus infectivity. Nature. 328(6132): 728–30.

Stremlau, M., Owens, C. M., Perron, M. J., Kiessling, M., Autissier, P. and Sodroski, J. 2004. The cytoplasmic body component TRIM5alpha restricts HIV-1 infection in Old World monkeys. Nature. 427(6977): 848–53.

Stremlau, M., Perron, M., Welikala, S. and Sodroski, J. 2005. Species-specific variation in the B30.2(SPRY) domain of TRIM5alpha determines the potency of human immunodeficiency virus restriction. J Virol. 79(5): 3139–45.

Stremlau, M., Song, B., Javanbakht, H., Perron, M. and Sodroski, J. 2006a. Cyclophilin A: an auxiliary but not necessary cofactor for TRIM5alpha restriction of HIV-1. Virology. 351(1): 112–20.

Stremlau, M., Perron, M., Lee, M., Li, Y., Song, B., Javanbakht, H., Diaz-Griffero, F., Anderson, D. J., Sundquist, W. I. and Sodroski, J. 2006b. Specific recognition and accelerated uncoating of retroviral capsids by the TRIM5alpha restriction factor. Proc Natl Acad Sci USA. 103(14): 5514–9.

Suspene, R., Guetard, D., Henry, M., Sommer, P., Wain-Hobson, S. and Vartanian, J. P. 2005. Extensive editing of both hepatitis B virus DNA strands by APOBEC3 cytidine deaminases in vitro and in vivo. Proc Natl Acad Sci USA. 102(23): 8321–6.

Suzuki, S. 1975. FV-4: a new gene affecting the splenomegaly induction by Friend leukemia virus. Jpn J Exp Med. 45(6): 473–8.

Svarovskaia, E. S., Xu, H., Mbisa, J. L., Barr, R., Gorelick, R. J., Ono, A., Freed, E. O., Hu, W. S. and Pathak, V. K. 2004. Human apolipoprotein B mRNA-editing enzyme-catalytic polypeptide-like 3G (APOBEC3G) is incorporated into HIV-1 virions through interactions with viral and nonviral RNAs. J Biol Chem. 279(34): 35822–8.

Tanaka, Y., Marusawa, H., Seno, H., Matsumoto, Y., Ueda, Y., Kodama, Y., Endo, Y., Yamauchi, J., Matsumoto, T., Takaori-Kondo, A., Ikai, I. and Chiba, T. 2006. Anti-viral protein APOBEC3G is induced by interferon-alpha stimulation in human hepatocytes. Biochem Biophys Res Commun. 341(2): 314–9.

Taylor, M. W., Grosse, W. M., Schaley, J. E., Sanda, C., Wu, X., Chien, S. C., Smith, F., Wu, T. G., Stephens, M., Ferris, M. W., McClintick, J. N., Jerome, R. E. and Edenberg, H. J. 2004. Global effect of PEG-IFN-alpha and ribavirin on gene expression in PBMC in vitro. J Interferon Cytokine Res. 24(2): 107–18.

Thali, M., Bukovsky, A., Kondo, E., Rosenwirth, B., Walsh, C. T., Sodroski, J. and Gottlinger, H. G. 1994. Functional association of cyclophilin A with HIV-1 virions. Nature. 372(6504): 363–5.

Tock, M. R. and Dryden, D. T. 2005. The biology of restriction and anti-restriction. Curr Opin Microbiol. 8(4): 466–72.

Towers, G., Bock, M., Martin, S., Takeuchi, Y., Stoye, J. P. and Danos, O. 2000. A conserved mechanism of retrovirus restriction in mammals. Proc Natl Acad Sci USA. 97(22): 12295–9.

Towers, G., Collins, M. and Takeuchi, Y. 2002. Abrogation of Ref1 retrovirus restriction in human cells. J Virol. 76(5): 2548–50.

Towers, G. J. 2007. The control of viral infection by tripartite motif proteins and cyclophilin A. Retrovirology. 4: 40.

Turelli, P., Mangeat, B., Jost, S., Vianin, S. and Trono, D. 2004. Inhibition of hepatitis B virus replication by APOBEC3G. Science. 303(5665): 1829.

Victoria, J. G. and Robinson, W. E., Jr. 2005. Disruption of the putative splice acceptor site for SIV(mac239)Vif reveals tight control of SIV splicing and impaired replication in Vif non-permissive cells. Virology. 338(2): 281–91.

Vigano, A., Saresella, M., Schenal, M., Erba, P., Piacentini, L., Tornaghi, R., Naddeo, V., Giacomet, V., Borelli, M., Trabattoni, D. and Clerici, M. 2007. Immune activation and normal levels of endogenous antivirals are seen in healthy adolescents born of HIV-infected mothers. Aids. 21(2): 245–8.

Virgen, C. A., Kratovac, Z., Bieniasz, P. D. and Hatziioannou, T. 2008. Independent genesis of chimeric TRIM5-cyclophilin proteins in two primate species. Proc Natl Acad Sci USA. 105(9): 3563–8.

von Schwedler, U., Song, J., Aiken, C. and Trono, D. 1993. Vif is crucial for human immunodeficiency virus type 1 proviral DNA synthesis in infected cells. J Virol. 67(8): 4945–55.

Wedekind, J. E., Dance, G. S., Sowden, M. P. and Smith, H. C. 2003. Messenger RNA editing in mammals: new members of the APOBEC family seeking roles in the family business. Trends Genet. 19(4): 207–16.

Wichroski, M. J., Ichiyama, K. and Rana, T. M. 2005. Analysis of HIV-1 viral infectivity factor-mediated proteasome-dependent depletion of APOBEC3G: correlating function and subcellular localization. J Biol Chem. 280(9): 8387–96.

Wichroski, M. J., Robb, G. B. and Rana, T. M. 2006. Human retroviral host restriction factors APOBEC3G and APOBEC3F localize to mRNA processing bodies. PLoS Pathog. 2(5): e41.

Wiegand, H. L., Doehle, B. P., Bogerd, H. P. and Cullen, B. R. 2004. A second human antiretroviral factor, APOBEC3F, is suppressed by the HIV-1 and HIV-2 Vif proteins. Embo J. 23(12): 2451–8.

Wilson, S. J., Webb, B. L., Ylinen, L. M., Verschoor, E., Heeney, J. L. and Towers, G. J. 2008. Independent evolution of an antiviral TRIMCyp in rhesus macaques. Proc Natl Acad Sci USA. 105(9): 3557–62.

Woo, J. S., Imm, J. H., Min, C. K., Kim, K. J., Cha, S. S. and Oh, B. H. 2006. Structural and functional insights into the B30.2/SPRY domain. Embo J. 25(6): 1353–63.

Wu, X., Anderson, J. L., Campbell, E. M., Joseph, A. M. and Hope, T. J. 2006. Proteasome inhibitors uncouple rhesus TRIM5alpha restriction of HIV-1 reverse transcription and infection. Proc Natl Acad Sci USA. 103(19): 7465–70.

Xu, L., Yang, L., Moitra, P. K., Hashimoto, K., Rallabhandi, P., Kaul, S., Meroni, G., Jensen, J. P., Weissman, A. M. and D'Arpa, P. 2003. BTBD1 and BTBD2 colocalize to cytoplasmic bodies with the RBCC/tripartite motif protein, TRIM5delta. Exp Cell Res. 288(1): 84–93.

Xu, H., Svarovskaia, E. S., Barr, R., Zhang, Y., Khan, M. A., Strebel, K. and Pathak, V. K. 2004. A single amino acid substitution in human APOBEC3G antiretroviral enzyme confers resistance to HIV-1 virion infectivity factor-induced depletion. Proc Natl Acad Sci USA. 101(15): 5652–7.

Yang, W. K., Kiggans, J. O., Yang, D. M., Ou, C. Y., Tennant, R. W., Brown, A. and Bassin, R. H. 1980. Synthesis and circularization of N- and B-tropic retroviral DNA Fv-1 permissive and restrictive mouse cells. Proc Natl Acad Sci USA. 77(5): 2994–8.

Yap, M. W., Nisole, S., Lynch, C. and Stoye, J. P. 2004. Trim5alpha protein restricts both HIV-1 and murine leukemia virus. Proc Natl Acad Sci USA. 101(29): 10786–91.

Yap, M. W., Nisole, S. and Stoye, J. P. 2005. A single amino acid change in the SPRY domain of human Trim5alpha leads to HIV-1 restriction. Curr Biol. 15(1): 73–8.

Yap, M. W., Dodding, M. P. and Stoye, J. P. 2006. Trim-cyclophilin A fusion proteins can restrict human immunodeficiency virus type 1 infection at two distinct phases in the viral life cycle. J Virol. 80(8): 4061–7.

Ying, S., Zhang, X., Sarkis, P. T., Xu, R. and Yu, X. 2007. Cell-specific Regulation of APOBEC3F by Interferons. Acta Biochim Biophys Sin (Shanghai). 39(4): 297–304.

Ylinen, L. M., Keckesova, Z., Wilson, S. J., Ranasinghe, S. and Towers, G. J. 2005. Differential restriction of human immunodeficiency virus type 2 and simian immunodeficiency virus SIVmac by TRIM5alpha alleles. J Virol. 79(18): 11580–7.

Ylinen, L. M., Keckesova, Z., Webb, B. L., Gifford, R. J., Smith, T. P. and Towers, G. J. 2006. Isolation of an active Lv1 gene from cattle indicates that tripartite motif protein-mediated innate immunity to retroviral infection is widespread among mammals. J Virol. 80(15): 7332–8.

Yu, X., Yu, Y., Liu, B., Luo, K., Kong, W., Mao, P. and Yu, X. F. 2003. Induction of APOBEC3G ubiquitination and degradation by an HIV-1 Vif-Cul5-SCF complex. Science. 302(5647): 1056–60.

Zack, J. A., Arrigo, S. J., Weitsman, S. R., Go, A. S., Haislip, A. and Chen, I. S. 1990. HIV-1 entry into quiescent primary lymphocytes: molecular analysis reveals a labile, latent viral structure. Cell. 61(2): 213–22.

Zennou, V., Perez-Caballero, D., Gottlinger, H. and Bieniasz, P. D. 2004. APOBEC3G incorporation into human immunodeficiency virus type 1 particles. J Virol. 78(21): 12058–61.

Zhang, H., Yang, B., Pomerantz, R. J., Zhang, C., Arunachalam, S. C. and Gao, L. 2003. The cytidine deaminase CEM15 induces hypermutation in newly synthesized HIV-1 DNA. Nature. 424(6944): 94–8.

Zhang, J. and Webb, D. M. 2004. Rapid evolution of primate antiviral enzyme APOBEC3G. Hum Mol Genet. 13(16): 1785–91.

Zhang, F., Hatziioannou, T., Perez-Caballero, D., Derse, D. and Bieniasz, P. D. 2006. Antiretroviral potential of human tripartite motif-5 and related proteins. Virology. 353(2): 396–409.

Zheng, Y. H., Irwin, D., Kurosu, T., Tokunaga, K., Sata, T. and Peterlin, B. M. 2004. Human APOBEC3F is another host factor that blocks human immunodeficiency virus type 1 replication. J Virol. 78(11): 6073–6.

Part II
Elements, Factors and Enzymes: Structure-Function and Mechanism

Chapter 16
T4 Phage Replisome

Scott W. Nelson, Zhihao Zhuang, Michelle M. Spiering, and Stephen J. Benkovic

The bacteriophage T4 DNA replisome has been a useful model system for studying cellular DNA replication. Several decades of studies revealed that despite the variation in number and nature of individual proteins in the T4 replisome as compared to other model systems (*Escherichia coli* and yeast *Saccharomyces cerevisiae*), the fundamental components that constitute a functional replication fork in bacteriophage T4 faithfully represent other more complex replication systems (Fig. 16.1; Table 16.1). Therefore what has been learned from the T4 replisome can be extended to the replisomes of other organisms, including higher eukaryotes. One advantage of the T4 replisome as a model system is its manipulable complexity. The eight proteins that constitute the T4 replisome are the DNA polymerase (gp43), the clamp loader and clamp proteins (gp44/62 and gp45), the single-stranded DNA-binding protein (gp32), the primase and the helicase (gp61 and gp41), and the helicase-loading protein (gp59). These proteins form sub-complexes including DNA polymerase holoenzyme (gp43, gp45, and gp44/62), primosome (gp61, gp41, and gp59), and single-stranded DNA-binding protein (gp32). In this chapter we will discuss the properties of individual sub-complexes as well as the structural and functional aspects of their components. We will address how these complexes are assembled from individual proteins and how their functions are coordinated to ensure the efficient duplication of the T4 phage genome.

DNA Polymerase Holoenzyme

In bacteriophage T4 the DNA replisome is responsible for the rapid and accurate synthesis of genomic DNA. The core of the T4 replisome is the DNA polymerase (gp43, 898 a.a.), which catalyzes the incorporation of deoxynucleotides in the 5′ to 3′ direction. Like many other replicative DNA polymerases T4 gp43 also possesses a 3′ to 5′ exonuclease activity. The polymerase alone is not processive, i.e.,

S.J. Benkovic (✉)
Department of Chemistry, Pennsylvania State University, 414 Wartik Laboratory, University Park, PA 16802, USA
e-mail: sjb1@psu.edu

C.E. Cameron et al. (eds.), *Viral Genome Replication*,
DOI 10.1007/b135974_16,

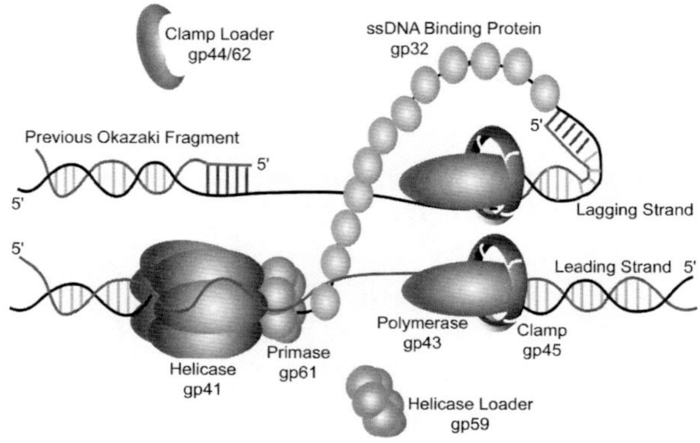

Fig. 16.1 Architecture of the bacteriophage T4 DNA replication complex. The T4 replication complex is composed of eight proteins that interact to synthesize DNA. In the current model, a helicase (gp41) and primase (gp61) form stacked hexameric rings that encircle the lagging DNA strand. This primosome complex is assembled with the aid of a helicase assembly protein, gp59. The helicase unwinds duplex DNA ahead of the polymerase while the primase synthesizes pentaribonucleotide primers for use by the lagging strand polymerase (gp43). Single-strand regions of DNA created from helicase activity are bound by gp32, a single-stranded DNA binding protein. Two trimeric gp45 proteins (which are loaded by the gp44/62 clamp loader complex) bind to the leading and lagging strand polymerases and increase their processivity.

Table 16.1 Replication and recombination proteins from T4 phage, *E. coli*, and *S. cerevisiae*

	T4 phage	*E. coli*	*S. cerevisiae*
Polymerase	gp43	α, ε, θ	Pol3, 31, 32, Pol2, Dpb(2–4)
Clamp	gp45	β	PCNA
Clamp loader	gp44, 62	γ, τ, δ, δ', χ, Ψ	RFC(1–5)
Helicase	gp41	DnaB	Possibly Mcm(2–7)
Helicase loader	gp59	DnaC	Unknown
Primase	gp61	DnaG	pol1, 12, and Pri2, 1
ssDNA-binding protein	gp32	SSB	RPA, RPA70, 30, 14

it produces short DNA product strands during a single DNA-binding event (Mace and Alberts 1984). Efficient replication of the 168 kb T4 genome necessitates the formation of the DNA holoenzyme complex that has greatly increased stability on DNA. The formation of polymerase holoenzyme requires accessory proteins gp45 (228 a.a.) and gp44/62 (319/187 a.a.). gp45 is a toroid-shaped processivity factor with an inner diameter large enough to encircle duplex DNA. Once loaded onto duplex DNA the clamp acts as a platform that tethers gp43 to DNA through a topological linkage between the C-terminus of gp43 and one specific face of the gp45 toroid (Latham et al. 1997a,b; Goodrich et al. 1997). The loading of gp45 onto duplex DNA requires the clamp loader gp44/62. In T4 the clamp loader is a complex of four gp44 and one gp62 subunits (Janzen et al. 1999). Besides the role of clamp loading gp44/62 also functions as a molecular chaperon for recruiting gp43 to assemble the functional holoenzyme (Trakselis et al. 2003a).

gp43 DNA Polymerase

To date the high-resolution structure of the T4 gp43 has not been reported. Nonetheless the X-ray crystal structure of the RB69 polymerase gp43 that shares 61% sequence identity to the T4 gp43 was solved by Steitz and coworkers. Such a high degree of amino acid sequence identity strongly argues for the conservation of three-dimensional structures between the two polymerases. Indeed a T4 gp43 structural model was generated with high confidence through a threading program (Xi et al. 2005a) and we will use the RB69 gp43 structure for an account of the structure and function of the T4 gp43.

Three RB69 gp43 structures were reported, including an apo form (Wang et al. 1997), a binary complex of gp43 and a primer–template DNA (Shamoo and Steitz 1999), and a ternary complex of gp43 with primer–template DNA and dTTP bound (Franklin et al. 2001). The gp43 polymerase belongs to the Pol α family. The structure of gp43 can be divided into five domains within a polypeptide chain of 903 residues. The N-terminal half of gp43 (residues 1–380) can be separated from the C-terminal half (residues 381–903) by truncation of the full-length protein (Lin et al. 1994). The N-terminal half of gp43 consists of an N-terminal domain and an exonuclease domain. The C-terminal half of gp43 adopts the common right-handed shape observed for DNA polymerases with three domains named finger, palm, and thumb. Overall the five domains form a disk with a noticeable hole in the center. The palm domain is the most conserved domain compared to other polymerases from the Pol I, Pol α, and reverse transcriptase families. The palm domain of gp43 is formed by a β-sheet that is flanked by two α-helices on one side. The three conserved aspartate residues (D411, D621, and D623) that are important for polymerase activity are located on three β-strands of the palm domain. Two of the conserved residues (D411 and D623) contribute to the binding of two metal ions. The phosphoryl transfer reaction catalyzed by gp43 is most likely through an associative transition state following the formation of a closed ternary complex of polymerase, DNA, and incoming dNTP. The ternary X-ray structure suggests that residue K560 in the finger domain together with the two metal ions bound by the conserved acidic residues in the palm domain help to stabilize the pentacovalent geometry and neutralize the additional negative charge on the equatorial oxygens developing in the transition state.

The finger domain of gp43 is formed by two long antiparallel helices (residues 471–572). In the apo-gp43 structure the finger domain adopts an open conformation protruding away from the backside of the disk formed by palm, thumb, and exonuclease domains. In the ternary complex of gp43 bound with primer–template DNA and dTTP, a drastic conformational change of the finger domain is evident where the finger domain rotates toward the palm domain by 60°. As a result the conserved basic residues in the finger domain move closer to the active site and interact with the incoming dNTP. The conformation of the thumb domain also changes upon the binding of primer–template by rotating closer to the palm domain. The thumb domain thus wraps around the primer–template DNA and makes close contact with the DNA minor groove.

A comparison of gp43 in the polymerizing mode (in the gp43–DNA–dTTP ternary structure) to the editing mode (in the gp43–DNA binary structure) provides us with a view of polymerase structural changes accompanying the switching between the two modes. The thumb domain plays an important role in the transition by holding DNA firmly with its tip. A rotation within the thumb domain between the tip and the base is thought to guide the DNA transition on a confined path between the polymerase active site and the exonuclease active site. With a 40° rotation of the double-stranded DNA, the primer terminus of the DNA travels 40 Å from the polymerase to the exonuclease active site. The structures reveal a cleft between the thumb domain and the exonuclease domain. Upon switching to the polymerizing mode, the two domains move toward each other and close the cleft.

gp45 Clamp Protein

The important function of the clamp protein in DNA replication is evident by its widespread presence in prokaryote (T4 phage and *E. coli*), eukaryote (human and yeast), and archaeon as well. The T4 gp45 crystal structure reveals a homotrimer with an inner diameter of ca. 35 Å and a thickness of ca. 25 Å (Moarefi et al. 2000). Each monomer comprises two domains that adopt a similar fold. The neighboring subunits are held together by four pairs of hydrogen bonds formed between two β-strands donated by each subunit. Although the T4 clamp has an overall negative charge, its inner surface shows positive electrostatic potentials. This unusual charge distribution favors a mechanism in which duplex DNA threads through the ring and interacts with the interior of the ring through electrostatic forces.

The high-resolution X-ray crystal structure of gp45 provided the initial view of the bacteriophage sliding clamp. To probe the clamp structure in solution, both chemical cross-linking and FRET approaches were utilized. Based on the T4 gp45 crystal structure, two unique cysteines (R86C and T167C) were introduced across the subunit interfaces by site-directed mutagenesis allowing efficient cross-linking of the neighboring subunits through disulfide bond formation. It was found that only two of the three possible disulfide bonds were formed per gp45 trimer, suggesting that one subunit interface remained open (Alley et al. 1999b). This notion was further tested by introducing two unique cysteines in the interdomain loop to form an intrasubunit cross-link. The flexibility of the interdomain loop is thought to be required for clamp opening. The separation distance across the subunit interface can be measured by introducing a FRET pair comprised of W91 and V162C-CPM (7-diethylamino-3-(4′-maleimidyl phenyl)-4-methylcoumarin). The subunit interface distance of 19 Å measured in the intrasubunit cross-linked clamp was close to 14 Å derived from the X-ray structure of gp45, suggesting that the cross-linking of the intradomain loop facilitates the closing of the clamp. However in the absence of cross-linking the same clamp mutant demonstrated a reduced energy transfer efficiency between the FRET pair with a calculated donor–acceptor distance of ca. 38 Å. Therefore in solution the T4 clamp exists in an open conformation. Further hydrodynamic analyses of T4 gp45 corroborated this conclusion (Alley et al. 1996).

The steady-state FRET experiment may represent an average of two equilibrating states of gp45, i.e., a closed clamp and a clamp with one open interface, or alternatively a single open state. To distinguish between these two possibilities the time-resolved FRET was measured between W92 and V163C-CPM. The results indicated that gp45 exists in only one state with one open and two closed subunit interfaces (Millar et al. 2004).

Despite the low sequence identities the crystal structures of the three clamps from bacteriophage T4 (gp45), *E. coli* (β-subunit), and *S. cerevisiae* (PCNA) all form closed rings in X-ray crystal structures. However the solution structure of gp45 is unique in that it exists as an opened ring. In comparison PCNA exists as a closed ring in solution (Zhuang et al. 2006b), which is likely to be the case for the β-subunit as well. These observations agree with the differences in the oligomer stability of the three clamps with the T4 gp45 being the least stable dissociating into subunits with a K_d of 250 nM compared to the dissociation of PCNA and β-subunit (K_d ~21 nM and <60 pM, respectively) (Yao et al. 1996). An X-ray crystal structure determination of the β-subunit suggests the existence of a so-called "spring tension" in the closed ring (Jeruzalmi et al. 2001). Thus an open T4 clamp in solution is attributed to looser interactions between the trimer subunit interfaces that cannot maintain an intact ring under "tension".

An earlier study showed that a deletion of the C-terminal six amino acids of gp43 abolished the interaction between gp43 and its cognate clamp gp45 (Berdis et al. 1996). This observation suggests that gp43 interacts with gp45 through the polymerase C-terminal tail. A structural view of this notion is provided by the cocrystal structure of the RB69 gp45, which shares 78% sequence identity to the T4 gp45, bound with a gp43 C-terminal peptide (Shamoo and Steitz 1999). The binding site on gp45 is located midway between the two homologous sub-domains of the individual clamp monomer. The binding interactions are mostly hydrophobic with residues L897, M900, and F901 in the peptide bound to a hydrophobic pocket on gp45. A holoenzyme model (Fig. 16.2) that contains polymerase and clamp was built by docking the C-terminal tail of the RB69 gp43 into the known binding pocket on clamp with the guidance of a duplex DNA, which binds to the polymerase active site and at the same time threads through the interior ring of gp45. The gp43 C-terminal peptide was found to adopt different conformations in different crystals. Such flexibility in the link between gp43 and gp45 may be beneficial for processive DNA synthesis by the polymerase holoenzyme since the movement of the bound DNA between the polymerase and exonuclease sites would be well tolerated (comparing the polymerizing to the editing modes in Fig. 16.2).

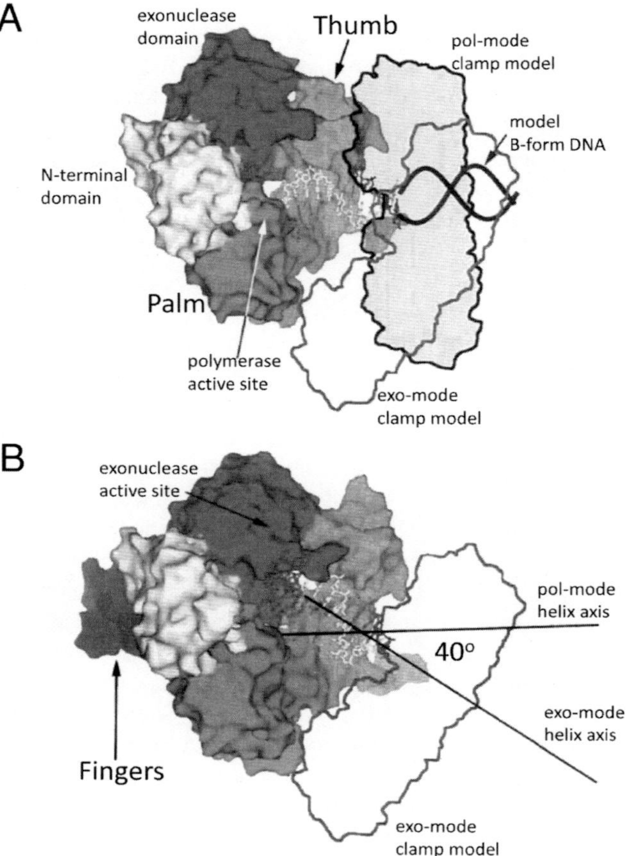

Fig. 16.2 gp43 holoenzyme model depicting the switching between polymerizing (A) and editing (B) modes. The polymerase is shown as a molecular surface model. The individual domains are labeled. The primer–template DNA in the crystal structure is shown in stick form. The modeled B-form extension of this DNA is shown as a backbone worm. The docked clamp is shown as a *black outline* to compare its position in polymerizing and editing modes. Figure adapted from (Franklin et al. 2001).

gp44/62 Clamp Loader

Loading of the clamp requires its cognate clamp loader. The T4 clamp loader is a binary complex of four gp44 subunits and one gp62 subunit. The 36 kDa gp44 subunit contains the Walker A motif and the SRC motif that are responsible for ATP binding and hydrolysis. In contrast the 21 kDa gp62 subunit contains neither motif. At present the subunit arrangement of gp44/62 complex is not clear. Nonetheless a circular structure was proposed based on analogy drawn between gp44/62 and the *E. coli* γ complex. The four gp44 subunits are thought to be equivalent to γ

and δ' subunits in the γ complex, with the gp62 subunit equivalent to the δ subunit. Mutagenesis analysis of the *E. coli* γ complex showed that ATP hydrolysis by the clamp loader involves the SRC motif of the neighboring subunit. Since the SRC motif is absent in the gp62 subunit, one can infer that only three of the four ATP sites of gp44/62 are competent in catalyzing ATP hydrolysis if gp44/62 adopts a similar ring structure as the γ complex. Alternatively the four gp44 subunits could form a symmetric tetramer with gp62 subunit attached at an unknown site in the complex. This molecular architecture, albeit different from what is observed in the γ complex, would allow four competent ATPase sites.

The Holoenzyme Assembly Pathways

The assembly of the T4 polymerase holoenzyme requires the coordinated actions of all three holoenzyme components (gp43, gp44/62, and gp45). The order of events leading to holoenzyme formation has been investigated extensively using a combination of rapid chemical quench and stopped-flow fluorescence approaches (Alley et al. 2000; Trakselis et al. 2003a, 2001). Early efforts have focused on an assembly pathway starting with the formation of a gp44/62–gp45 binary complex in the presence of ATP. Subsequent binding of this complex to the primer–template DNA triggers the loading of gp45 onto DNA. In the last step of holoenzyme formation a physical linkage between gp43 and gp45 is established with the departure of gp44/62 from the assembled complex. Therefore gp44/62 acts as a molecular chaperon in this assembly process.

To relate ATP hydrolysis to various steps in clamp loading and subsequent holoenzyme formation, a rapid-quench technique was used to probe the fast ATP hydrolysis and to determine the stoichiometry of ATP consumption at various steps. It was found that two equivalents of ATP are hydrolyzed upon interaction with gp45 to form the gp44/62–gp45 complex. In the next step of clamp loading two more equivalents of ATP are hydrolyzed once the gp44/62–gp45 complex interacts with primer–template DNA. It should be noted that another rapid-quench experiment found one equivalent of ATP was hydrolyzed upon mixing DNA with gp44/62 and gp45 (Pietroni et al. 2001), although it was later found that the reduced ATP equivalents arise from a defect in the quenching protocol (Trakselis et al. 2003a).

Another important feature of the clamp-loading pathway is the conformational changes of gp45 catalyzed by gp44/62. To directly probe the dynamic opening and closing of gp45, the FRET signal derived from a pair of fluorophores (W92 as the donor and CPM attached to V163C as the acceptor) introduced across the clamp interface was followed using stopped-flow fluorescence (Alley et al. 2000). Upon binding to gp44/62 one gp45 subunit interface is opened further from 40 Å to greater than 45 Å as measured between W92 and V163C-CPM (note that the distance measured for W92–V163C-CPM FRET pair is larger than the distance between the closest amino acids across the open subunit) with accompanying ATP hydrolysis. Substituting ATP with ATPγS resulted in no further increase in the gp45 subunit interface distance indicating ATP hydrolysis is required. Following the introduction of DNA to the gp44/62–gp45 binary complex, the clamp subunit interface is closed around DNA to a W92 to V163C-CPM distance of 35 Å. The DNA polymerase holoenzyme can then be formed after the addition of gp43. It has been shown that both gp44/62 and gp43 interact with the same face of the gp45 toroid (Latham et al. 1997b) and that gp44/62 departs from a gp44/62·gp43·gp45·DNA complex to form the holoenzyme (Trakselis et al. 2003a).

The opening of gp45 could be either in-plane or out-of-plane. To obtain precise information on clamp conformational changes in the various steps of clamp loading and holoenzyme formation, a total of three FRET pairs were introduced across the clamp interface to triangulate the distance changes across the subunit interfaces (Trakselis et al. 2001). Residue W92 was chosen as the fixed FRET donor, while site-specific mutations were introduced at V163C, S158C, and T168C individually to conjugate CPM as the acceptor. The stopped-flow FRET measurements revealed 10 steps in the clamp-loading process (Fig. 16.3). The distance information obtained for the individual steps was then used to derive the directionality of the gp45 opening and closing. Combined with computer modeling, the open gp45 trimer structure was constructed by adjusting the torsion angles within the interdomain loop. In state $gp45^{D}$, the clamp exists in an in-plane open conformation. Binding of DNA to the $gp45^{D}$–gp44/62 complex resulted in an initial in-plane closing of gp45 through step G followed by an out-of-plane reorientation to a spiral configuration ($gp45^{H}$). Introduction of gp43 resulted in a final in-plane closing of gp45 to an interface distance smaller than the initial distance observed for gp45 alone in solution. At this final state the interface distance remains 11 Å, consistent with accommodation of the C-terminus of gp43 within the gp45 subunit interface (Alley et al. 1999a). One can speculate that the insertion of the C-terminal peptide of gp43 to the gp45 interface can effectively relieve the "tension" incurred with a fully closed gp45 clamp.

The observation of an opened gp45 clamp in solution raised an interesting question regarding the role of gp44/62 in clamp loading and the need for ATP hydrolysis. Since gp45 is open with an interface separation large enough to allow the passage of double-strand DNA (B-form DNA has a diameter of ca. 20 Å), a question exists as to why gp44/62 is needed to further open the clamp with the consumption of two ATPs in the rapid initial steps of clamp loading. Inquiry into this question led to the discovery of an alternative pathway of gp45 loading catalyzed by gp44/62, where the clamp loader was found to bind specifically to the primer–template DNA and mark the locus of clamp loading (Zhuang et al. 2006a). This gp44/62–DNA interaction requires the hydrolysis of one equivalent of ATP to form the binary complex. Following the formation of the gp44/62–DNA complex, an open gp45 clamp is recruited to the 3′ end of the DNA primer for loading. However closing of gp45 onto DNA does not require rapid ATP hydrolysis suggesting that the electrostatic interactions between the positively charged inner rim of gp45 and the negatively charged DNA backbone likely drives the closing of gp45.

It is unique that the T4 clamp loading may occur through more than one pathway. Previous studies demonstrated that in *E. coli* and yeast the clamp could only be loaded following one pathway, i.e., the clamp loader interacts with clamp to

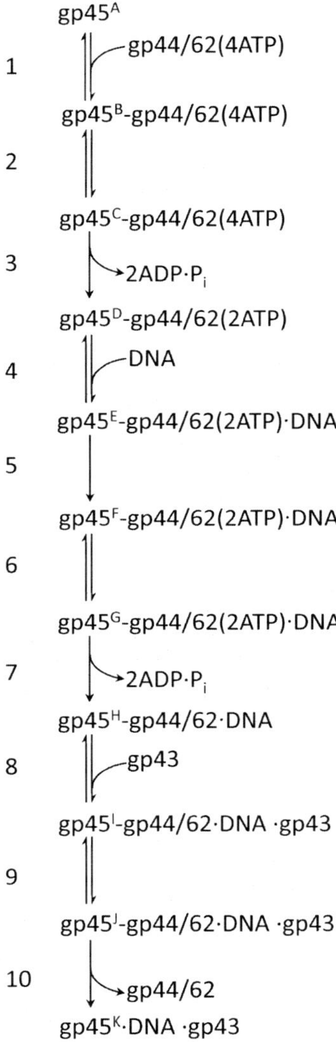

Fig. 16.3 A 10-step model depicting the holoenzyme assembly process. The different states of gp45 are designated by the superscript A through K. The equivalents of ATP bound to gp44/62 are indicated in parenthesis. See text for details.

form a binary complex, which interacts with DNA to effect the clamp loading. The difference can be understood in light of the unique solution structure of the T4 clamp and inferred from the structure of the clamp loader–clamp complex as represented by the yeast RFC–PCNA complex (Bowman et al. 2004). Five ATPase subunits of the RFC complex adopt a spiral configuration and form an inner chamber that presumably accommodates the duplex DNA. Three of the five ATPase subunits make contact with the PCNA ring. It is evident from this structure that the initial formation of the RFC–DNA binary complex will make it topologically impossible to establish a productive interaction with PCNA, thus hindering the clamp opening. However, this may not present a problem for T4 since an opened gp45 in solution can readily encircle duplex DNA and establish interactions with the gp44/62 ATPase domains simultaneously. Therefore the open clamp conformation of T4 lends flexibility to the order of clamp loading and holoenzyme formation.

A single-molecule approach was applied to investigate in greater detail the T4 holoenzyme assembly process (Smiley et al. 2006). Individual holoenzyme components, gp43 and gp45, were labeled with fluorescent dyes Alexa Fluor 488 and 555, which are amenable for single-molecule fluorescence detection because of their high fluorescent intensity and photochemical stability. A forked DNA substrate for clamp loading was immobilized on a glass slide surface through a biotin–streptavidin interaction. The fluorescence signal from single molecules was detected with total internal reflection optics and three microscope filter sets were selected for the observation of emission from donor Alexa Fluor 488, acceptor Alexa Fluor 555 (excited at 488 and 514 nm, respectively) and emission due to FRET between the two fluorophores (excited at 488 nm). The observation of a FRET signal between labeled gp43 and gp45 suggested the close juxtaposition of holoenzyme components, accompanying the formation of the polymerase holoenzyme. The activity of the assembled holoenzyme was ensured by demonstrating strand displacement synthesis by the complex upon the addition of required deoxynucleotides. The results from these single-molecule experiments provided further support for the previously demonstrated assembly pathways and led to the discovery of additional assembly pathways including (1) gp45 binding to DNA followed by gp44/62 and then gp43 and (2) gp43 binding to DNA followed by gp44/62–gp45 complex. In all cases MgATP was required for holoenzyme assembly. The existence of the multiple holoenzyme assembly pathways revealed by in vitro experiments underscores the remarkable flexibility of the T4 holoenzyme complex, which is likely required to cope with the diverse DNA structures (D-loop, R-loop, and lagging strand DNA primed with short RNA pentamer) encountered in T4 DNA replication.

Polymerase Exchange

One salient property of the T4 DNA polymerase holoenzyme is its high processivity during DNA replication with a dissociation half-life of ca. 9 min (Yang et al. 2004). However, counterintuitively, it was demonstrated that during normal DNA replication an active exchange process takes place frequently between the gp43 in solution and the gp43 within the polymerase holoenzyme (Yang et al. 2004). Given an estimated in vivo gp43 concentration of ca. 600 nM the polymerase exchange on any given replisome would occur on average once every 10 s, or approximately 90 events per replication fork during the 15 min time span for copying the T4 genome.

A model for the T4 polymerase exchange process was proposed based on molecular modeling using the available X-ray crystal structures of gp45 and RB69 gp43

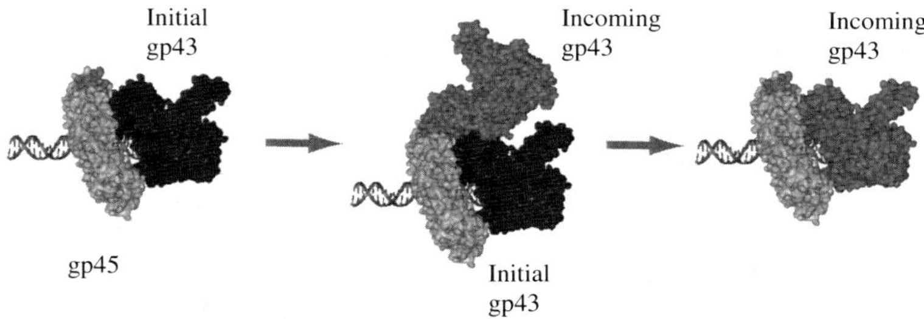

Fig. 16.4 Solution structure models of polymerase exchange. The clamp protein gp45 is shown as a toroid encircling DNA. The initial and incoming polymerases gp43 are both depicted as molecular surface models.

(Fig. 16.4). In this model, gp45 acts as a platform for the exchange process. The incoming polymerase likely binds to a second polymerase-binding site on gp45 to avoid a steric clash with the resident gp43. Given the frequency of stalling of the DNA replisome, the polymerase exchange process uncovered for the T4 polymerase holoenzyme may serve to overcome the replication barriers. Given that the clamp-polymerase structure and interaction appears to be widely conserved across all three branches of life, the polymerase exchange property observed for T4 may be applicable to other replication systems.

gp32 ssDNA-Binding Protein

The T4 phage ssDNA-binding protein (gp32, 301 a.a.) is considered to be a prototype for ssDNA-binding proteins. gp32 is absolutely required for DNA replication in vivo (Curtis and Alberts 1976) and reconstitution of coupled leading and lagging strand DNA synthesis in vitro (Huberman et al. 1971; Yang et al. 2003). Additionally, gp32 is required for homologous recombination, DNA repair, transcription, and DNA packaging (Miller et al. 2003). The binding of gp32 to ssDNA is very tight ($K_d = 0.1\ \mu M$ on long ssDNA strands), highly cooperative, and sequence non-specific (Kelly et al. 1976). The binding site size has been determined to be 7–9 bases of ssDNA and the cooperativity parameter (ω) is greater than 1000 (Kowalczykowski et al. 1980). gp32 contains three distinct domains, which were initially isolated and characterized using limited proteolysis. The N-terminal domain (referred to as domain B for "basic") is involved in cooperative ssDNA binding, such that removal of residues 1–21 completely eliminates binding cooperativity (Giedroc et al. 1990). The major function of the highly acidic C-terminal domain (residues 254–301, referred to as domain A for "acidic") is to interact with other T4 proteins. Affinity chromatography using gp32-agarose has detected interactions between gp32 and itself, gp43, gp45, and gp59 (Formosa et al. 1983; Morrical et al. 1996). gp32 also has been shown to co-purify with gp61 (Burke et al. 1985). The structure of the gp32 core domain (residues 22–253) in complex with a six-base ssDNA oligonucleotide has been solved using X-ray crystallography (Shamoo et al. 1995). The core domain contains the ssDNA-binding site and is made up of three smaller sub-domains. The sub-domains are (1) a zinc-finger motif that is thought to be important for protein stability; (2) a five-stranded β-sheet containing several aromatic side chains involved in DNA base recognition; and (3) a connecting region bridging sub-domains 1 and 2. The affinity of the core domain for short ssDNA oligos (which can only accommodate a single gp32 monomer) is essentially identical to that of the full-length protein. However, due to the loss of cooperative interactions, the affinity of the core domain for long ssDNA strands is greatly reduced compared to the full-length protein (Giedroc et al. 1990).

gp59 Helicase-Loading Protein

gp59 helicase-loading protein (gp59, 217 a.a) is a basic protein that plays a central role in the assembly and possibly the function of the T4 phage replisome. In vivo, gp59 is necessary for recombination-dependent initiation of replication and gp59 mutants display the classic DNA arrest phenotype that is indicative of that defect (Wu and Yeh 1975; Dudas and Kreuzer 2005). gp59 was first isolated and described as a helicase assembly protein by the Alberts laboratory (Barry and Alberts 1994), making it the final member to the eight protein set that makes up the T4 replisome (Barry and Alberts 1994).

Equilibrium fluorescence and gel mobility shift experiments have demonstrated that gp59 binds to a variety of ss and dsDNA substrates in a sequence-independent fashion (Lefebvre and Morrical 1997; Mueser et al. 2000; Jones et al. 2000). However, gp59 shows a distinct preference for forked DNA structures over substrates that contain only ssDNA or dsDNA (Jones et al. 2000). These preferred structures are D-loops, R-loops, three- and four-stranded Holliday junctions, and model replication forks (Nelson et al. 2006; Nossal et al. 2001). When binding to a model replication fork, gp59 requires at least six bases of ssDNA before the ss/ds junction (Jones et al. 2000). This is consistent with the number of single-strand DNA bases bound by the gp59 monomer as determined by fluorescence enhancement of etheno-modified ssDNA (Lefebvre and Morrical 1997).

The oligomeric state of the functional form of gp59 is unclear. Kinetic evidence indicates that a 1:1 ratio of gp59 to gp41 subunits (which is hexameric) provides the maximal enhancement of gp41 loading (Raney et al. 1996). Additionally, cross-linking studies have shown that gp59 can induce the oligomerization of gp41 and that gp59 forms up to pentamers in the presence of DNA or gp32 (Ishmael et al. 2002, 2001). Together, these data suggest that gp59 may be hexameric at the replication fork. On the other hand, sedimentation velocity and glycerol gradient centrifugation experiments have shown that gp59 in solution is monomeric in the absence of DNA substrate (Yonesaki 1994; Xu et al. 2001).

The structure of gp59 was determined in the Nossal and Mueser laboratories (gp59 subunit of Figs. 16.5 and 16.6; Mueser et al. 2000). gp59 is essentially an alpha-helical protein with two domains, the N-domain (residues 1–109) and the

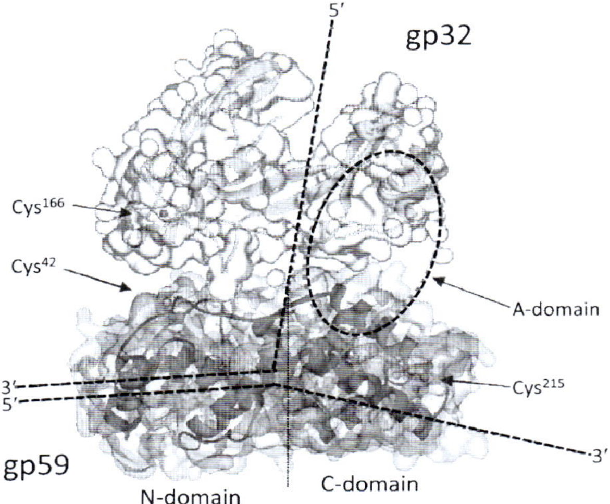

Fig. 16.5 Model of the gp32–gp59 complex. gp59 is predicted to associate with the replication fork such that the N-terminal domain binds to the duplex and the lagging strand traverses a groove between the N- and C-terminal domains (Mueser et al. 2000). This orientation of gp59 on DNA places Cys-42 in close contact with Cys-166 of gp32, which is bound to ssDNA. The A-domain of gp32 most likely interacts with the C-terminal domain of gp59. Figure adapted from (Ishmael et al. 2001).

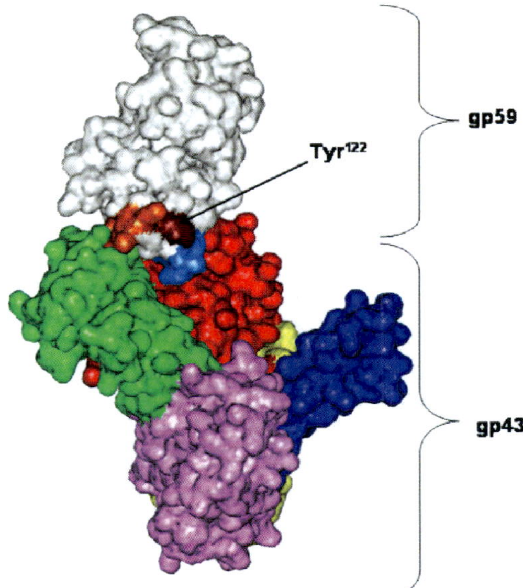

Fig. 16.6 Interaction model of the gp43–gp59 complex. The proposed interaction model (Xi et al. 2005a) showing the location of Y122A near the interface between gp43 and gp59. The N-terminal, exonuclease, palm, fingers, and thumb domains of gp43 are colored as *yellow*, *red*, *magenta*, *blue*, and *green*, respectively. gp59 is colored *gray* except for residue Y122 (*maroon*), N202-K217 (*cyan*), and K126 to E134 (*gold*).

C-domain (residues 110–217). A portion (residues 9–65) of the N-domain shows structural similarity to members of the high-mobility-group (HMG) family of DNA minor groove-binding proteins. Based on this structural similarity a speculative model for the interaction of monomeric gp59 with the ss/ds junction of a replication fork has been proposed (Mueser et al. 2000). The HMG region of gp59 is postulated to bind at the duplex region ahead of the fork in much the same manner as HMG1 binds to the fork region of a four-stranded Holliday junction (Hardman et al. 1995). The ssDNA that would make up the lagging strand arm behind the fork binds in a shallow groove located in the junction between the N- and C-domains of the protein and a hydrophobic region on the surface of the C-domain interacts with the other ssDNA arm (leading strand) behind the replication fork. Site-specific mutants designed to test this model have revealed a single residue, I87, that when mutated reduces the affinity of gp59 for both forked and single-stranded DNA (Jones et al. 2004a). In the proposed model for the interaction of gp59 with fDNA, I87 is positioned at the site where the ssDNA arms begin to separate from the duplex (fork region). The other mutations that were located in the duplex region, leading strand arm, and lagging strand arm had little effect on the affinity for forked DNA structures. It is possible that the loss of a single interaction point, except for those located precisely in the fork region, does not cause a loss of binding affinity great enough to be detected in gel mobility shift experiments or alternatively, the proposed model does not accurately reflect the gp59–fDNA structure given the possibility that gp59 is in a higher order oligomeric state when bound to fDNA.

gp41 Helicase

Helicases, like the T4 phage helicase (gp41, 475 a.a.), are enzymes that play an essential role in nearly all DNA metabolic processes, catalyzing the transient opening of DNA duplexes. Mutations in gene 41 strongly reduce the amount of DNA replication in phage-infected cells (Epstein et al. 1963) and eliminate synthesis on the lagging strand (Kreuzer and Morrical 1994). The purified gp41 exhibits a DNA-dependent ATPase or GTPase activity, which is strongly stimulated by long pieces of ssDNA rather than short pieces of ssDNA or dsDNA (Liu and Alberts 1981a). Nucleotide hydrolysis powers DNA unwinding by gp41 (Venkatesan et al. 1982) or translocation of gp41 on ssDNA (Young et al. 1994) in a unidirectional 5′-to-3′ manner. The preferred DNA substrates of gp41 are preformed forked DNA hybrids with an absolute requirement for a 5′-ssDNA tail of 32 nt or more (Venkatesan et al. 1982). Optimal DNA unwinding also requires a 3′-ssDNA extension longer than 29 nt suggesting that the gp41 protein interacts with both the leading strand and lagging strand templates at the replication fork as it unwinds the duplex region (Richardson and Nossal 1989). The length of the 5′-ssDNA tail and the necessity for the 3′-ssDNA tail are reduced in the presence of gp59 (Jones et al. 2000; Yonesaki 1994). Nucleotide triphosphate binding, but not hydrolysis, is necessary for stable gp41–ssDNA complex formation (Liu and Alberts 1981a; Richardson and Nossal 1989). A binding site size of 12–20 nt per gp41 monomer has been determined by electrophoretic mobility shift assays with gp41 and dT_{12-20} in the presence of nucleotide (Young et al. 1994).

The gp41 protein alone will translocate on ssDNA at a rate of 400 nt/s with an association half-life of 1 min (Liu and Alberts 1981a). The DNA unwinding rate of gp41, measured at 30 bp/s in the presence of gp59 (Raney et al. 1996), and the processivity of unwinding (ca. 650 nt) in the presence of gp61 on long duplexes (Richardson and Nossal 1989) are much lower. However, the gp41 protein alone is necessary and sufficient to establish a high processive rate (~250 bp/s) of unwinding during DNA synthesis on the leading strand by the polymerase holoenzyme (Alberts et al. 1980; Cha and Alberts 1989). In a complete replisome, a dissociation half-life for gp41 of 11 min was measured by dilution experiments, revealing that the 41 protein is sufficiently processive to finish replicating the entire T4 genome (168 kb) at the observed replication rate of ~400 nt/s (Schrock and Alberts 1996).

The oligomerization of gp41 has been observed by numerous groups using a variety of techniques. Oligomerization of gp41 was first reported by Liu and Alberts (1981a) who detected a sigmoidal dependence of nucleotide hydrolysis on gp41 concentration and an increased sedimentation rate with GTPγS in sucrose gradients. Further studies have shown that gp41 exists as a monomer/dimer equilibrium when free in solution (primarily as a dimer at physiological protein concentrations) and forms a hexamer upon activation by ATP or ATPγS binding with the assembly of dimers to tetramers to hexamers (Dong et al. 1995). Cryoelectron microscopy (Dong et al. 1995) and protein cross-linking (Morris and Raney 1999) suggest these hexamers are toroidal rings with ssDNA probably passing through the center like other hexameric helicases (Stasiak et al. 1994; Egelman et al. 1995). The most recent electron microscopy of gp41 reveals two distinct forms of gp41 hexamers, termed "open" and "closed", which may be important for the assembly of gp41 onto ssDNA (Norcum et al. 2005).

gp61 Primase

The T4 phage primase protein (gp61, 342 a.a.) is essential for normal DNA synthesis in vivo. T4 phage defective in gene 61 has a reduced rate of DNA synthesis (Yegian et al. 1971) and accumulates abnormal amounts of ssDNA (Gold et al. 1976). gp61 catalyzes the synthesis of short RNA molecules used as primers for DNA polymerase on the lagging strand of a replication fork. The in vivo recognition site for priming is 5′-GTT-3′, additionally 5′-GCT-3′ can be used in vitro. Although the 3′-T is required for recognition, it is not copied into the resulting primer; the primers therefore have the sequence 5′-pppACNNN-3′ and 5′-pppGCNNN-3′, respectively (Cha and Alberts 1986). High concentrations of gp61 alone synthesize primers, mostly dimers and a small number of primers 5–45 nt long, mainly from the 5′-GCT-3′ priming site (Cha and Alberts 1986; Hinton and Nossal 1987). The combination of gp41 and gp61 greatly increases the overall primer synthesis rate favoring a pentaribonucleotide product over the dimer and the use of the 5′-GTT-3′ rather than the 5′-GCT-3′ recognition site (Hinton and Nossal 1987; Cha and Alberts

1990). Including optimal levels of gp32, gp59, and gp41 have produced a priming rate of almost one primer per second per primosome, a rate sufficient for the in vivo operation of the replisome (Valentine et al. 2001).

Similar to other prokaryotic primase proteins, gp61 is made up of three domains. The N-terminal domain contains a zinc-binding domain known as a zinc ribbon. The single zinc ion is bound by residues C37, C40, C65, and C68 (Valentine et al. 2001). When the zinc ribbon sequence in the T7 primase was replaced with those from *E. coli* and T4, the resulting chimeric proteins were active but primed at sequences different than either of the parent proteins (Kusakabe and Richardson 1996), strongly suggesting that some, but not all, of the priming site recognition is afforded by the zinc ribbon motif. The middle domain of the primase is the catalytic domain, responsible for ribonucleotide polymerization and the C-terminal domain is involved in protein–protein interactions with the helicase (Jing et al. 1999).

Analytical ultracentrifugation or isothermal titration calorimetry (ITC) displayed no evidence for primase self-association suggesting that gp61 alone in solution is monomeric (Valentine et al. 2001). In the presence of DNA and/or gp41, a study using gel mobility shift analysis suggested a monomeric primase as the active component of the primosome (Dong and von Hippel 1996). However, in more recent investigations, gp61 has been observed to bind to short ssDNA oligos primarily as a trimer by ITC and chemical cross-linking techniques (Valentine et al. 2001) and as a hexamer by fluorescence anisotropy with dye-labeled ssDNA with a K_d of 50–100 nM (Yang et al. 2005). Furthermore, a three-dimensional reconstruction from electron microscopic images indicates a ring-like structure and a hexameric stoichiometry for gp61 when complexed with ssDNA (Norcum et al. 2005). Defective primases either missing the N-terminal zinc-binding domain (deletion mutant) or full-length catalytically inactive (active site mutant) proteins without priming activity could be mixed to create oligomeric primases with restored catalytic activity suggesting that oligomers of gp61 may also be functional.

Interaction Between gp32 and gp59

The initial characterization of gp59 established its role as an accessory protein required for the loading of the gp41 helicase onto gp32-coated ssDNA (Barry and Alberts 1994; Morrical et al. 1994). gp59-agarose affinity chromatography also indicated that gp32 and gp59 interact with each other (Yonesaki 1994). Based on these and other data it was proposed that a direct interaction between gp59 and gp32 facilitated the loading of gp41 helicase onto DNA (Barry and Alberts 1994). Since those initial observations, the interaction between gp32 and gp59 has been extensively characterized using a wide variety of techniques such as protein cross-linking, ensemble and single-molecule FRET, analytical ultracentrifugation, and in vitro replication kinetic assays.

Thiol–thiol cross-linking has been used to identify the specific points of interaction between gp32 and gp59 (Ishmael et al. 2001). It was found that C166 on gp32 and C42 on gp59 must be within 6 Å of each other. These experiments also revealed higher order oligomeric states of gp59. In the presence of either ssDNA or gp32, species of gp59 that ranged up to a pentameric subunit composition were observed. Further experiments showed that the A-domain of gp32 is required for this gp32-induced gp59 oligomerization. However, while the A-domain greatly increases the interaction between gp32 and gp59, it is not absolutely required. Based on these cross-linking results, the model of the gp59–fDNA complex (Mueser et al. 2000) was expanded to include gp32 (Fig. 16.5). Here, gp32 is bound to the lagging strand ssDNA and interacts with gp59 at both its N- and C-domains. The gp32 core domain, which contains C166, makes contact with the N-domain of gp59, whereas the gp32 A-domain interacts with the C-domain.

While protein cross-linking experiments are highly informative with regard to the specific sites of interactions between two proteins, fluorescent experiments are better suited for the determination of stoichiometry and binding affinities. Labeling of gp59 at C42 with the fluorescent dye rhodamine has allowed the anisotropy of gp59 and gp59 in complex with gp32 truncation mutants to be monitored (full-length gp32 was not tested due to insolubility problems) (Xu et al. 2001). Both gp32-B (gp32 missing the N-terminal B domain) and gp32A (only the A-domain) bind gp59 with a 1:1 stoichiometry (Xu et al. 2001). The K_d for the interaction of gp59-rhodamine with gp32-B and gp32A are 10 and 3 nM, respectively. In addition to fluorescent anisotropy, ensemble FRET experiments have found a 1:1 stoichiometry between gp59 and gp32 on both single-stranded and forked DNA substrates (Zhang et al. 2005) and the Morrical lab has used etheno-modified ssDNA to demonstrate that gp59 and gp32 simultaneously co-occupy ssDNA (Lefebvre et al. 1999).

Reconstitution of replisome-mediated DNA synthesis has revealed the necessity of the gp32–gp59 complex for productive loading of gp41 helicase (Jones et al. 2004b; Ma et al. 2004). Consequently, when gp59 loads gp41 onto a replication fork, an interaction between gp32 and gp59 must occur. If gp41 loads without the aid of gp59, the presence of gp32 is not required (Jones et al. 2004b). Mutational analysis of gp32 indicates that the ability of gp32 to bind both gp59 and ssDNA is necessary for this effect (Ma et al. 2004). Based on these results and others, the Morrical lab has proposed that a gp59–gp32 cluster (presumably in a 6:6 ratio) forms a condensed-coil structure termed the helicase-loading complex (HLC). The HLC recruits gp41 dimers that form hexamers at the fork and is proposed to slide along ssDNA, thus allowing it to remain with a moving replication fork (Ma et al. 2004).

Interaction Between gp43 and gp59

Almost immediately after gp59 was discovered, it was found that high concentrations of gp59 inhibited both helicase-dependent and helicase-independent DNA synthesis in vitro (Barry and Alberts 1994). Strong inhibition of replication was also observed in vivo when gp59 was over-expressed from an IPTG-inducible plasmid (Spacciapoli and Nossal 1994). More recently, gp59 was shown to inhibit in vitro replication when initiating from an R-loop in the absence of gp41 helicase (Nossal et al. 2001). Several possible explanations for the inhibition of DNA synthesis by gp59 have been proposed, such as acting as a steric block, sequestering protein off DNA, and interacting directly with a component of the replisome. The latter has been proven to be correct.

A site-specific cross-link has been observed between gp59 and gp43 polymerase in the presence of a forked DNA substrate and was confirmed using FRET experiments (Ishmael et al. 2003). gp43 polymerase was labeled N-terminally with Oregon Green dye and gp59 was labeled at C42 with CPM dye. When the two labeled proteins were mixed in the presence of forked DNA, a small but significant amount of energy transfer was observed (Ishmael et al. 2003; Xi et al. 2005a). The data from these two complementary experiments clearly demonstrate an interaction between gp43 and gp59. Further experiments were done to map the site of interaction between gp43 and gp59 (Xi et al. 2005a). The cross-linked protein was subjected to in-gel digestion followed by MALDI-TOF mass spectrometry. A unique fragment was identified and the site of interaction was determined to be C215 of gp59 and C169 of gp43. This information, coupled with FRET distance measurements, was used to evaluate the top 30 models from a series of computer-generated (ClusPro) interaction models for the gp43–gp59 complex (Xi et al. 2005a). The first constraint applied was that gp59-C215 and gp43-C169 must be within 10.2 Å of each other (the length of the cross-linker). The second constraint was that gp59-C42 and the N-terminus of gp43 must be about 50 Å apart (based on FRET efficiency). These two constraints reduced the number of possible models from 30 to 1 (Fig. 16.6). In this model gp59-C215 and gp43-C169 are 8 Å apart and gp59-C42 and the N-terminus of gp43 are located 55 Å apart. The interaction model rationalizes the inhibition of gp43 by gp59. The C-terminal helix of gp59 is inserted into a cleft between the thumb and exonuclease domains of gp43. Presumably, this insertion prevents the closing of the cleft and therefore interferes with the switching of the polymerase from its exonuclease to polymerase modes. As predicted from this model, gp59 inhibits both the polymerase activity and the exonuclease activity of gp43 (Xi et al. 2005a; Nelson et al. 2006). This result also demonstrates that gp59 is not merely acting as a block impeding the forward movement of the polymerase, but it forms a discrete complex with gp43 and inhibits both forward and backward movements by a direct interaction.

In a mutational screen for putative protein–protein interaction hotspots located on the surface of gp59, a single mutation Y122A was found to have drastically altered properties compared to the wild-type enzyme (Nelson et al. 2006). The defect most relevant to the interaction model is that Y122A is unable to inhibit the exonuclease activity of gp43. Additional FRET experiments similar to those described above indicated that the interaction between gp59 and gp43 was disrupted by the mutation. The interaction model of the gp59–gp43 complex places Y122 directly between two regions (helix H7 and loop H6–H7) of gp59 that are predicted to interact with gp43 (Fig. 16.6). Based on this central location, the role of Y122 may be to stabilize helix H13 of gp43 and the loop H6–H7 of gp59 in their interactions with gp43. This explanation is strengthened by the fact that helix H7 of gp59, which contains Y122, makes extensive contacts with both helix H6 and H13 of gp43.

A thorough study of the in vivo effects of gp59 deficiency has revealed a physiological function of gp43 inhibition by gp59 (Dudas and Kreuzer 2005). In both the presence and absence of gp59, DNA replication that initiates from R-loop-containing origins occurs with coupled leading and lagging strand DNA synthesis. After a short delay, a second replication fork with opposite directionality to the first (retrograde synthesis) is initiated. In wild-type cells, retrograde replication occurs in a similar fashion as the first replication fork with coupled leading and lagging strand DNA synthesis. However, in the absence of gp59, lagging strand synthesis does not occur in the retrograde replication fork. Based on these results, a model in which gp59 inhibits the leading strand polymerase until the helicase and primase are loaded onto the leading strand template was proposed (Dudas and Kreuzer 2005). This is in complete agreement with the model based on in vitro data (Xi et al. 2005a).

Interaction Between gp59 and gp41

The primary function of gp59 is to load gp41 helicase onto the lagging strand DNA template, therefore a large number of studies have documented the functional interaction between gp41 and gp59. This section will focus on the direct physical interaction between the two proteins.

The architecture of the gp41–gp59 complex both on and off DNA has been studied using protein cross-linking (Ishmael et al. 2002). A cysteine residue was added to the C-terminus of gp41 by cleavage of a gp41–intein fusion with cysteine. This cysteine residue was then modified with a photoactivable cross-linker and incubated with gp59 under UV light. It was found that the C-terminus of gp41 and gp59 is in close proximity in the presence and absence of ATP and with or without DNA substrate. A similar experiment was performed with gp41 labeled at its N-terminus with a photoactivable cross-linker. With the cross-linker in this position, the N-terminus of gp41 cross-linked to gp59 both on and off DNA in the presence and absence of ATP. Next, thiol–thiol cross-linking revealed that the C-terminus of gp41 is in close proximity to C215 of gp59 in the presence or absence of ATP. However, when DNA was included in the cross-linking reaction, no cross-link between gp41 and gp59 was observed. When coupled with the previous result demonstrating that gp41 and gp59 do indeed interact on DNA, this result indicates that a conformational change occurs when the gp41–gp59 complex binds to the replication fork. Defining the exact nature of this conformational change will require a high-resolution structure of the gp59–gp41 complex.

As discussed above, gp59 inhibits the polymerase and exonuclease activities of gp43 through a direct protein–protein interaction (Xi et al. 2005a). This specific interaction has been observed using single-molecule FRET (Xi et al. 2005a,b). It was found that addition of gp41 and ATP to the gp43–gp59–fDNA complex results in the release of gp59 from the replication fork. Ensemble studies indicate that the polymerase is "unlocked" from its gp43–gp59 interaction and is capable of DNA synthesis following the displacement of gp59 by gp41. The gp43–gp59 complex was not disrupted when gp41 was loaded in the presence of the non-hydrolyzable analog ATPγS, suggesting that translocation by gp41 is necessary for the effect. The fate of gp59 following displacement by gp41 is still unclear. Several studies using small model substrates indicate that gp59 is released into solution following displacement by gp41; however, electron microscopy studies using larger DNA substrates indicate that gp59 remains behind the replication fork in complex with gp32 on the ssDNA of the lagging strand template (Chastain et al. 2003). The effect of this complex on the replication fork appears to be minimal (if any) since once the replisome is assembled, leading and lagging strand synthesis is unaffected by the absence of gp59 (Xi et al. 2005b).

Interaction Between gp41 and gp61

When first identified and isolated, it was unclear which protein, gp41 or gp61, was the T4 helicase or primase and whether they only had activity as a complex (Nossal 1980; Liu and Alberts 1980, 1981b). It was not until higher protein amounts were obtained that we learned each protein had a different activity and that they could function independent of one another. However, their activities remain closely associated since the unwinding rate and processivity of gp41 are increased in the presence of gp61. Also, gp41 greatly increases the overall priming rate and influences the sequence of primers made by gp61. A gp41–gp61 complex on ssDNA, which requires nucleotide binding to form, has been detected by electrophoretic mobility shift assays (Richardson and Nossal 1989; Jing, Beechem and Patton 2004) and by single-molecule FRET studies (Zhang et al. 2005). The interaction between gp41 and gp61 as measured by isothermal titration calorimetry is 10-fold stronger on ssDNA than in solution (Valentine et al. 2001). Protein kinase protection experiments with gp61 tagged with a phosphorylation sequence at either the N- or C-terminus have demonstrated that the N-terminal domain of primase is protected from phosphorylation by binding ssDNA, both the N- and C-terminus are equally protected when complexed with gp41, and almost no phosphorylation occurs in the ternary complex of gp61, gp41, and ssDNA (Jing et al. 1999). Despite all this evidence, a gp41–gp61 complex has not been isolated to date.

Summary of Replisome Assembly

Based on all the available evidence, a likely (although speculative) pathway for T4 phage replisome assembly is described. We have chosen to describe assembly at a D-loop since the majority of T4 DNA replication originates from this type of structure (Kreuzer 2000). Immediately after D-loop formation, gp32 coats the displaced ssDNA and recruits gp59 monomers to the fork region of the D-loop. The exact number of gp59 monomers is dependent on the length of the displaced strand and the number of bound gp32 monomers. Several (presumably six) of the gp32–gp59 complexes nearest the fork condense to form the coiled helicase-loading complex (HLC). Next, the clamp loader in complex with a clamp protein binds to the ss/dsDNA junction and the clamp loader chaperones the loading of the clamp onto the dsDNA. The holoenzyme is formed when the polymerase binds to the clamp protein and displaces the clamp loader from the complex. As noted above, there are several possible pathways for holoenzyme assembly and the one given here is not necessarily dominant. At this stage, the progression of the holoenzyme is prevented through a direct interaction between a gp59 monomer and the polymerase. Following holoenzyme formation, gp41 dimers are recruited to the replication fork by the HLC and each dimer forms an interaction with two gp59 monomers. Before hexameric helicase can form, the single gp59 monomer that is in complex with the polymerase must break its interaction with the polymerase and form a new interaction with gp41. Once all gp59 monomers are in complex with gp41, gp41 is fully loaded and translocation begins with ATP hydrolysis. ATP hydrolysis by gp41 either displaces gp59 from the replication fork or gp59 slides behind the fork onto the ssDNA. Leading strand replication begins immediately after the gp59–gp43 interaction is broken and gp41 begins to unwind the DNA duplex. Lagging strand replication is initiated when gp61 binds to gp41 helicase and begins to synthesize RNA pentamers on the lagging strand template. Finally, the lagging strand holoenzyme is formed at the RNA primer and lagging strand DNA synthesis proceeds. At this point, replisome assembly is complete and coordinated leading and lagging strand replication begins.

Coordination of Leading and Lagging Strand Synthesis

Once the replisome is assembled, the individual proteins act in concert to simultaneously replicate both strands of the DNA duplex. During replication the replisome undergoes a repeated remodeling process where specific proteins enter and exit the replisome as the replication fork travels down the DNA duplex. Dilution and protein trapping experiments indicate that the clamp, clamp loader, primase, and gp32 proteins dissociate from the actively replicating replisome and exchange with proteins in solution before returning to the replication fork (Kadyrov and Drake 2001; Trakselis et al. 2003b). Similar experiments have demonstrated the processivity of the helicase and both of the leading and lagging strand polymerases (Alberts et al. 1983; Yang et al. 2004). The half-life of the helicase and polymerases are on the order of several minutes, which enables complete replication of the 168 kb phage genome (Kaboord and Benkovic 1993; Yang et al. 2004).

Because of the high processivity of the lagging strand polymerase, coupled with its 5′ to 3′ direction of DNA synthesis, one problem faced by all DNA replication machineries is how to simultaneously and coordinately replicate two antiparallel DNA strands. The major hypothesis put forth to reconcile this problem is that the replisome and replication fork form a specific structure so that the two holoenzyme complexes are in close proximity to each other and the lagging strand DNA template is folded back to form a loop (see Fig. 16.1). This enables both polymerases to synthesize DNA in the 5′ to 3′ direction while allowing the replication fork to move in the same direction (Alberts et al. 1983). The lagging strand loop that is predicted by the trombone model has been directly visualized using electron microscopy (Chastain et al. 2000). Although the lagging strand polymerase is processive, lagging strand DNA synthesis is discontinuous and results in short Okazaki fragments with an average length of 1–2 kb (Chastain et al. 2000). This discontinuous yet processive synthesis requires a specific mechanism for the release of the lagging strand polymerase from its template and its recycling to the newly synthesized pentameric RNA primer. Several mechanisms have been proposed to serve as the trigger for the release and recycling of the lagging strand polymerase. Among them, two mechanisms have gained the most experimental support. In the first model (the collision model), RNA primer synthesis and lagging strand holoenzyme release and recycling is triggered by the collision of the lagging strand polymerase into the end of the previous Okazaki fragment (Alberts et al. 1983). In support of this model, the off rate of the holoenzyme is greatly increased by a hairpin structure (Hacker and Alberts 1994) or an annealed DNA or RNA designed to mimic the 5′-end of the previous Okazaki fragment (Carver et al. 1997). In the second model (signaling model), lagging strand polymerase releases and recycles as the result of distinct macromolecular interaction events involved in repetitive lagging strand cycles, such as the association of the primase with the replisome, the RNA primer synthesis, or the loading of the clamp onto the newly synthesized primer (Wu et al. 1992). Recently, evidence for this model has come from a study demonstrating that under conditions that slow the lagging strand polymerase relative to the leading strand polymerase ssDNA gaps form between Okazaki fragments (Yang et al. 2006). The existence of gaps between successive Okazaki fragments cannot be accommodated within the confines of the collision model. In order for ssDNA gaps to form, the lagging strand polymerase must release and recycle prior to reaching the 5′-end of the previous Okazaki fragment. The nature of the signal for this release is still unclear. However, the modulation of RNA primer utilization and Okazaki fragment size by clamp and clamp loader concentrations may indicate that the loading of the clamp onto the newly synthesized RNA primer signals lagging strand polymerase recycling (Yang et al. 2006). Due to the stochastic nature of the recycling signal, in some cases the polymerase will reach the end of the lagging strand template before the signal is sent. In these situations, the polymerase may release and recycle via the collision mechanism.

The signaling model as presented above requires that the transfer of the RNA primer to the lagging strand polymerase from the primase occurs via the clamp and clamp loader. This is reminiscent of the indirect primer transfer mechanism that is well documented in the *E. coli* replication system (Yuzhakov et al. 1999). However, in the T4 phage system, solid evidence supporting the indirect transfer of the RNA primer to the lagging strand polymerase is lacking, as it is only inferred from the signaling model. Additionally, since the collision mechanism of lagging strand polymerase release and recycling may also occur, a direct transfer of the primer to the polymerase (before clamp loading) may be possible in these situations (Kato et al. 2001).

Future Directions

The T4 replisome has been proven to be an invaluable system for studying the molecular mechanism of DNA replication. We expect it will continue to provide useful new information. Several prominent questions regarding DNA replication in general can be addressed using the reconstituted T4 replisome as a model system. Central to efficient DNA replication is the coordinated DNA synthesis in both leading and lagging strands. More challenges lie in the understanding of the discontinuous lagging strand DNA synthesis. One prominent question is how lagging strand DNA polymerase is disengaged from the ongoing Okazaki fragment synthesis and initiates the next round of lagging strand synthesis. Future studies should shed light on what drives the gymnastic movement of the DNA polymerase in the lagging strand milieu. At present, most attention has been focused on the protein portion of the T4 replisome. It will be equally important to obtain a more precise picture of the DNA configuration at the replication fork, for example, how DNA threads through the protein complexes, especially the primosome. The DNA within the replisome is likely to be highly dynamic and the understanding of its movement in a confined space is crucial to our understanding of the lagging strand recycling process. Another challenge is to study the replication process in vivo with the recent advance of fluorescent approaches that make it possible to image single proteins or single complexes with high temporal and spatial resolution in live cells.

References

Alberts, B. M., Barry, J., Bedinger, P., Burke, R. L., Hibner, U., Liu, C.-C., and Sheridian, R. 1980. In Mechanistic Studies of DNA replication and Genetic Recombination, ed. B. M. Alberts, pp. 449–471. New York: Academic Press

Alberts, B. M., Barry, J., Bedinger, P., Formosa, T., Jongeneel, C. V., and Kreuzer, K. N. 1983. Studies on DNA replication in the bacteriophage T4 in vitro system. Cold Spring Harb. Symp. Quant. Biol. 47:655–668.

Alley, S. C., Jones, A. D., Soumillion, P., and Benkovic, S. J. 1999a. The carboxyl terminus of the bacteriophage T4 DNA polymerase contacts its sliding clamp at the subunit interface. J. Biol. Chem. 274:24485–24489.

Alley, S. C., Shier, V. K., Abel-Santos, E., Sexton, D. J., Soumillion, P., and Benkovic, S. J. 1999b. Sliding clamp of the bacteriophage T4 polymerase has open and closed subunit interfaces in solution. Biochemistry 38:7696–7709.

Alley, S. C., Abel-Santos, E., and Benkovic, S. J. 2000. Tracking sliding clamp opening and closing during bacteriophage T4 DNA polymerase holoenzyme assembly. Biochemistry 39:3076–3090.

Barry, J., and Alberts, B. 1994. Purification and characterization of bacteriophage T4 gene 59 protein. A DNA helicase assembly protein involved in DNA replication. J. Biol. Chem. 269:33049–33062.

Berdis, A. J., Soumillion, P., and Benkovic, S. J. 1996. The carboxyl terminus of the bacteriophage T4 DNA polymerase is required for holoenzyme complex formation. Proc. Natl. Acad. Sci. USA 93:12822–12827.

Bowman, G. D., O'Donnell, M., and Kuriyan, J. 2004. Structural analysis of a eukaryotic sliding DNA clamp-clamp loader complex. Nature 429:724–730.

Burke, R. L., Munn, M., Barry, J., and Alberts, B. M. 1985. Purification and properties of the bacteriophage T4 gene 61 RNA priming protein. J. Biol. Chem. 260:1711–1722.

Carver, T. E. Jr, Sexton, D. J., and Benkovic, S. J. 1997. Dissociation of bacteriophage T4 DNA polymerase and its processivity clamp after completion of Okazaki fragment synthesis. Biochemistry 36:14409–14417.

Cha, T.-A., and Alberts, B. M. 1986. Studies of the DNA helicase-RNA primase unit from bacteriophage T4. A trinucleotide sequence on the DNA template starts RNA primer synthesis. J. Biol. Chem. 261:7001–7010.

Cha, T.-A., and Alberts, B. M. 1989. The bacteriophage T4 DNA replication fork: only DNA helicase is required for leading strand DNA synthesis by the DNA polymerase holoenzyme. J. Biol. Chem. 264:12220–12225.

Cha, T.-A., and Alberts, B. M. 1990. Effects of the bacteriophage T4 gene 41 and gene 32 proteins on RNA primer synthesis: coupling of leading- and lagging-strand DNA synthesis at a replication fork. Biochemistry 29:1791–1798.

Chastain, P. D. 2nd, Makhov, A. M., Nossal, N. G., and Griffith, J. D. 2000. Analysis of the Okazaki fragment distributions along single long DNAs replicated by the bacteriophage T4 proteins. Mol. Cell. 6:803–814.

Chastain, P. D. 2nd, Makhov, A. M., Nossal, N. G., and Griffith, J. 2003. Architecture of the replication complex and DNA loops at the fork generated by the bacteriophage T4 proteins. J. Biol. Chem. 278:21276–21285.

Curtis, M. J., and Alberts, B. M. 1976. Studies on the structure of intracellular bacteriophage T4 DNA. J. Mol. Biol. 102:793–816.

Dong, F., Gogol, E. P., and von Hippel, P. H. 1995. The phage T4-coded DNA replication helicase (gp41) forms a hexamer upon activation by nucleoside triphosphate. J. Biol. Chem. 270:7462–7473.

Dong, F., and von Hippel, P.H. 1996. The ATP-activated hexameric helicase of bacteriophage T4 (gp41) forms a stable primosome with a single subunit of T4-coded primase (gp61). J. Biol. Chem. 271:19625–19631.

Dudas, K. C., and Kreuzer, K. N. 2005. Bacteriophage T4 helicase loader protein gp59 functions as gatekeeper in origin-dependent replication in vivo. J. Biol. Chem. 280: 21561–21569.

Egelman, E. H., Yu, X., Wild, R., Hingorani, M. M., and Patel, S. S. 1995. Bacteriophage T7 helicase/primase proteins form rings around single-stranded DNA that suggest a general structure for hexameric helicases. Proc. Natl. Acad. Sci. USA 92:3869–3873.

Epstein, R. H., Bolle, A., Steinberg, C. M., Kellenberger, E., Boy de la Tour, E., Chevalley, R., Edgar, R. S., Susman, M., Denhardt, G. H., and Lielausis, A. 1963. Physiological studies of conditional lethal mutants of bacteriophage T4D. Cold Spring Harbor Symp. Quant. Biol. 28:375–392.

Formosa, T., Burke, R. L., and Alberts, B. M. 1983. Affinity purification of bacteriophage T4 proteins essential for DNA replication and genetic recombination. Proc. Nat. Acad. Sci. USA 80:2442–2446.

Franklin, M. C., Wang, J., and Steitz, T. A. 2001. Structure of the replicating complex of a pol alpha family DNA polymerase. Cell 105:657–667.

Giedroc, D. P., Khan, R., and Barnhart, K. 1990. Overexpression, purification, and characterization of recombinant T4 gene 32 protein 22–301 (g32P-B). J. Biol. Chem. 265:11444–11455.

Gold, L., O'Farrell, P. Z., and Russel, M. 1976. Regulation of gene 32 expression during bacteriophage T4 infection of *Escherichia coli*. J. Biol. Chem. 251:7251–7262.

Goodrich, L. D., Lin, T. C., Spicer, E. K., Jones, C., and Konigsberg, W. H. 1997. Residues at the carboxy terminus of T4 DNA polymerase are important determinants for interaction with the polymerase accessory proteins. Biochemistry 36:10474–10481.

Hacker, K. J., and Alberts, B. M. 1994. The rapid dissociation of the T4 DNA polymerase holoenzyme when stopped by a DNA hairpin helix. A model for polymerase release following the termination of each Okazaki fragment. J. Biol. Chem. 269:24221–24228.

Hardman, C. H., Broadhurst, R. W., Raine, A. R., Grasser, K. D., Thomas, J. O., and Laue, E. D. 1995. Structure of the A-domain of HMG1 and its interaction with DNA as studied by heteronuclear three- and four-dimensional NMR spectroscopy. Biochemistry 34:16596–16607.

Hinton, D. M., and Nossal, N. G. 1987. Bacteriophage T4 DNA primase-helicase. Characterization of oligomer synthesis by T4 61 protein alone and in conjunction with T4 41 protein. J. Biol. Chem. 262:10873–10878.

Huberman, J. A., Kornberg, A., and Alberts, B. M. 1971. Stimulation of T4 bacteriophage DNA polymerase by the protein product of T4 gene 32. J. Mol. Biol. 62:39–52.

Ishmael, F. T., Alley, S. C., and Benkovic, S. J. 2001. Identification and mapping of protein-protein interactions between gp32 and gp59 by cross-linking. J. Biol. Chem. 276:25236–25242.

Ishmael, F. T., Alley, S. C., and Benkovic, S. J. 2002. Assembly of the bacteriophage T4 helicase: architecture and stoichiometry of the gp41–gp59 complex. J. Biol. Chem. 277:20555–20562.

Ishmael, F. T., Trakselis, M. A., and Benkovic, S. J. 2003. Protein–protein interactions in the bacteriophage T4 replisome. The leading strand holoenzyme is physically linked to the lagging strand holoenzyme and the primosome. J. Biol. Chem. 278:3145–3152.

Janzen, D. M., Torgov, M. Y., and Reddy, M. K. 1999. In vitro reconstitution of the bacteriophage T4 clamp loader complex (gp44/62). J. Biol. Chem. 274:35938–35943.

Jeruzalmi, D., Yurieva, O., Zhao, Y., Young, M., Stewart, J., Hingorani, M., O'Donnell M., and Kuriyan, J. 2001. Mechanism of processivity clamp opening by the delta subunit wrench of the clamp loader complex of *E. coli* DNA polymerase III. Cell 106:417–428.

Jing, D. H., Dong, F., Latham, G. J., and von Hippel, P. H. 1999. Interactions of bacteriophage T4-coded primase (gp61) with the T4 replication helicase (gp41) and DNA in primosome formation. J. Biol. Chem. 274:27287–27298.

Jing, D., Beechem, J. M., and Patton, W. F. 2004. The utility of a two-color fluorescence electrophoretic mobility shift assay procedure for the analysis of DNA replication complexes. Electrophoresis 25:2439–2446.

Jones, C. E., Mueser, T. C., and Nossal, N. G. 2000. Interaction of the bacteriophage T4 gene 59 helicase loading protein and gene 41 helicase with each other and with fork, flap and cruciform DNA. J. Biol. Chem. 275:27145–27154.

Jones, C. E., Green, E. M., Stephens, J. A., Mueser, T. C., and Nossal, N. G. 2004a. Mutations of bacteriophage T4 59 helicase loader defective in binding fork DNA and in interactions with T4 32 single-stranded DNA-binding protein. J. Biol. Chem. 279:25721–25728.

Jones, C. E., Mueser, T. C., and Nossal, N. G. 2004b. Bacteriophage T4 32 protein is required for helicase-dependent leading strand synthesis when the helicase is loaded by the T4 59 helicase-loading protein. J. Biol. Chem. 279:12067–12075.

Kaboord, B. F., and Benkovic, S. J. 1993. Rapid assembly of the bacteriophage T4 core replication complex on a linear primer/template construct. Proc. Natl. Acad. Sci. USA 90:10881–10885.

Kadyrov, F. A., and Drake, J. W. 2001. Conditional coupling of leading-strand and lagging-strand DNA synthesis at bacteriophage T4 replication forks. J. Biol. Chem. 276:29559–29566.

Kato, M., Frick, D. N., Lee, J., Tabor, S., Richardson, C. C., and Ellenberger, T. 2001. A complex of the bacteriophage T7 primase-helicase and DNA polymerase directs primer utilization. J. Biol. Chem. 276:21809–21820.

Kelly, R. C., Jensen, D. E., and von Hippel, P. H. 1976. DNA "melting" proteins. IV. Fluorescence measurements of binding parameters for bacteriophage T4 gene 32-protein to mono-, oligo-, and polynucleotides. J. Biol. Chem. 251:7240–7250.

Kowalczykowski, S. C., Lonberg, N., Newport, J. W., Paul, L. S., and von Hippel, P. H. 1980. On the thermodynamics and kinetics of the cooperative binding of bacteriophage T4-coded gene 32 (helix destabilizing) protein to nucleic acid lattices. Biophys. J. 32:403–418.

Kreuzer, K. N., and Morrical, S. W. 1994. In Molecular Biology of Bacteriophage T4, ed. J. Karan, pp. 28–42. Washington, D. C.: American Society for Microbiology.

Kreuzer, K. N. 2000. Recombination-dependent DNA replication in phage T4. Trends. Biochem. Sci. 25:165–173.

Kusakabe, T., and Richardson, C. C. 1996. The role of the zinc motif in sequence recognition by DNA primases. J. Biol. Chem. 271:19563–19570.

Latham, G. J., Bacheller, D. J., Pietroni, P., and von Hippel, P. H. 1997a. Structural analyses of gp45 sliding clamp interactions during assembly of the bacteriophage T4 DNA polymerase holoenzyme. II. The Gp44/62 clamp loader interacts with a single defined face of the sliding clamp ring. J. Biol. Chem. 272:31677–31684.

Latham, G. J., Bacheller, D. J., Pietroni, P., and von Hippel, P. H. 1997b. Structural analyses of gp45 sliding clamp interactions during assembly of the bacteriophage T4 DNA polymerase holoenzyme. III. The Gp43 DNA polymerase binds to the same face of the sliding clamp as the clamp loader. J. Biol. Chem. 272:31685–31692.

Lefebvre, S. D., and Morrical, S. W. 1997. Interactions of the bacteriophage T4 gene 59 protein with single-stranded polynucleotides: binding parameters and ion effects. J. Mol. Biol. 272:312–326.

Lefebvre, S. D., Wong, M. L., and Morrical, S. W. 1999. Simultaneous interactions of bacteriophage T4 DNA replication proteins gp59 and gp32 with single-stranded (ss) DNA. Co-modulation of ssDNA binding activities in a DNA helicase assembly intermediate. J. Biol. Chem. 274:22830–22838.

Lin, T. C., Karam, G., and Konigsberg, W. H. 1994. Isolation, characterization, and kinetic properties of truncated forms of T4 DNA polymerase that exhibit 3′-5′ exonuclease activity. J. Biol. Chem. 269:19286–19294.

Liu, C.-C., and Alberts, B. M. 1980. Pentaribonucleotides of mixed sequence are synthesized and efficiently prime de novo DNA chain starts in the T4 bacteriophage DNA replication system. Proc. Natl. Acad. Sci. USA 77:5698–5702.

Liu, C.-C., and Alberts, B. M. 1981a. Characterization of the DNA-dependent GTPase activity of T4 gene 41 protein, and essential component of the T4 bacteriophage DNA replication apparatus. J. Biol. Chem. 256:2813–2820.

Liu, C.-C., and Alberts, B. M. 1981b. Characterization of RNA primer synthesis in the T4 bacteriophage in vitro DNA replication system. J. Biol. Chem. 256:2821–2829.

Ma, Y., Wang, T., Villemain, J. L., Giedroc, D. P., and Morrical, S. W. 2004. Dual functions of single-stranded DNA-binding protein in helicase loading at the bacteriophage T4 DNA replication fork. J. Biol. Chem. 279:19035–19045.

Mace, D. C., and Alberts, B. M. 1984. T4 DNA polymerase rates and processivity on single-stranded DNA templates. J. Mol. Biol. 177:295–311.

Millar, D., Trakselis, M. A., and Benkovic, S. J. 2004. On the solution structure of the T4 sliding clamp (gp45). Biochemistry 43:12723–12727.

Miller, E. S., Kutter, E., Mosig, G., Arisaka, F., Kunisawa, T., and Ruger, W. 2003. Bacteriophage T4 genome. Microbiol. Mol. Biol. Rev. 67:86–156.

Moarefi, I., Jeruzalmi, D., Turner, J., O′Donnell, M., and Kuriyan, J. 2000. Crystal structure of the DNA polymerase processivity factor of T4 bacteriophage. J. Mol. Biol. 296:1215–1223.

Morrical, S. W., Hempstead, K., and Morrical, M. D. 1994. The gene 59 protein of bacteriophage T4 modulates the intrinsic and single–stranded DNA-stimulated ATPase activities of gene 41 protein, the T4 replicative DNA helicase. J. Biol. Chem. 269:33069–33081.

Morrical, S. W., Beernink, H. T., Dash, A., and Hempstead, K. 1996. The gene 59 protein of bacteriophage T4. Characterization of protein-protein interactions with gene 32 protein, the T4 single-stranded DNA binding protein. J. Biol. Chem. 271:20198–20207.

Morris, P. D., and Raney, K. D. 1999. DNA helicases displace streptaviding from biotin-labeled oligonucleotides. Biochemistry 38:5164–5171.

Mueser, T. C., Jones, C. E., Nossal, N. G., and Hyde, C. C. 2000. Bacteriophage T4 gene 59 helicase assembly protein binds replication fork DNA. The 1.45 Å resolution crystal structure reveals a novel alpha-helical two-domain fold. J. Mol. Biol. 296:597–612.

Nelson, S. W., Yang, J., and Benkovic, S. J. 2006. Site-directed mutations of T4 helicase loading protein (gp59) reveal multiple modes of DNA polymerase inhibition and the mechanism of unlocking by gp41 helicase. J. Biol. Chem. 281:8697–8706.

Norcum, M. T., Warrington, J. A., Spiering, M. M., Ishmael, F. T., Trakselis, M. A., and Benkovic, S. J. 2005. Architecture of the bacteriophage T4 primosome: electron microscopy studies of helicase (gp41) and primase (gp61). Proc. Natl. Acad. Sci. USA 102:3623–3626.

Nossal, N. G. 1980. RNA priming of DNA replication by bacteriophage T4 proteins. J. Biol. Chem. 255:2176–2182.

Nossal, N. G., Dudas, K. C., and Kreuzer, K. N. 2001. Bacteriophage T4 proteins replicate plasmids with a preformed R loop at the T4 ori(uvsY) replication origin in vitro. Mol. Cell 7:31–41.

Pietroni, P., Young, M. C., Latham, G. J., and von Hippel, P. H. 2001. Dissection of the ATP-driven reaction cycle of the bacteriophage T4 DNA replication processivity clamp loading system. J. Mol. Biol. 309:869–891.

Raney, K. D., Carver, T. E., and Benkovic, S. J. 1996. Stoichiometry and DNA unwinding by the bacteriophage T4 41:59 helicase. J. Biol. Chem. 271:14074–14081.

Richardson, C. C., and Nossal, N. G. 1989. Characterization of the bacteriophage T4 gene 41 DNA helicase. J. Biol. Chem. 264:4725–4731.

Stasiak, A., Tsaneva, I. R., West, S. C., Benson, C. J., Yu, S., and Egelman, E. H. 1994. The *Escherichia coli* RuvB branch migration protein forms double hexameric rings around DNA. Proc. Natl. Acad. Sci. USA 91:7618–7622.

Schrock, R. D., and Alberts, B. M. 1996. Processivity of the gene 41 DNA helicase at the bacteriophage T4 DNA replication fork. J. Biol. Chem. 271:16678–16682.

Shamoo, Y., Friedman, A. M., Parsons, M. R., Konigsberg, W. H., and Steitz, T.A. 1995. Crystal structure of a replication fork single-stranded DNA binding protein (T4 gp32) complexed to DNA. Nature 376:362–366.

Shamoo, Y., and Steitz, T. A. 1999. Building a replisome from interacting pieces: sliding clamp complexed to a peptide from DNA polymerase and a polymerase editing complex. Cell 99:155–166.

Smiley, R. D., Zhuang, Z., Benkovic, S. J., and Hammes, G. G. 2006. Single-molecule investigation of the T4 bacteriophage DNA polymerase holoenzyme: multiple pathways of holoenzyme formation. Biochemistry 45:7990–7997.

Spacciapoli, P., and Nossal, N. G. 1994. Interaction of DNA polymerase and DNA helicase within the bacteriophage T4 DNA replication complex. Leading strand synthesis by the T4 DNA polymerase mutant A737V (tsL141) requires the T4 gene 59 helicase assembly protein. J. Biol. Chem. 269:447–455.

Trakselis, M. A., Alley, S. C., Abel-Santos, E., and Benkovic, S. J. 2001. Creating a dynamic picture of the sliding clamp during T4 DNA polymerase holoenzyme assembly by using fluorescence resonance energy transfer. Proc. Natl. Acad. Sci. USA 98:8368–8375.

Trakselis, M. A., Berdis, A. J., and Benkovic, S. J. 2003a. Examination of the role of the clamp-loader and ATP hydrolysis in the formation of the bacteriophage T4 polymerase holoenzyme. J. Mol. Biol. 326:435–451.

Trakselis, M. A., Roccasecca, R. M., Yang, J., Valentine, A. M., and Benkovic, S. J. 2003b. Dissociative properties of the proteins within the bacteriophage T4 replisome. J. Biol. Chem. 278:49839–49849.

Valentine, A. M., Ishmael, F. T., Shier, V. K., and Benkovic, S. J. 2001. A zinc ribbon protein in DNA replication: primer synthesis and macromolecular interactions by the bacteriophage T4 primase. Biochemistry 40:15074–15085.

Venkatesan, M., Silver, L. L., and Nossal, N. G. 1982. Bacteriophage T4 gene 41 protein, required for the synthesis of RNA primers, is also a DNA helicase. J. Biol. Chem. 257:12426–12434.

Wang, J., Sattar, A. K., Wang, C. C., Karam, J. D., Konigsberg, W. H., and Steitz, T. A. 1997. Crystal structure of a pol alpha family replication DNA polymerase from bacteriophage RB69. Cell 89:1087–1099.

Wu, J. R., and Yeh, Y. C. 1975. New Late Gene, *dar*, Involved in DNA Replication of Bacteriophage T4 I. Isolation, Characterization, and Genetic Location. J. Virol. 15:1096–1106.

Wu, C. A., Zechner, E. L., Reems, J. A., McHenry, C. S., and Marians, K. J. 1992. Coordinated leading- and lagging-strand synthesis at the *Escherichia coli* DNA replication fork. V. Primase action regulates the cycle of Okazaki fragment synthesis. J. Biol. Chem. 267: 4074–4083.

Xi, J., Zhuang, Z., Zhang, Z., Selzer, T., Spiering, M. M., Hammes, G. G., and Benkovic, S. J. 2005a. Interaction between the T4 helicase-loading protein (gp59) and the DNA polymerase (gp43): a locking mechanism to delay replication during replisome assembly. Biochemistry 44:2305–2318.

Xi, J., Zhang, Z., Zhuang, Z., Yang, J., Spiering, M. M., Hammes, G. G., and Benkovic, S. J. 2005b. Interaction between the T4 helicase loading protein (gp59) and the DNA polymerase (gp43): unlocking of the gp59-gp43-DNA complex to initiate assembly of a fully functional replisome. Biochemistry 44:7747–7756.

Xu, H., Wang, Y., Bleuit, J. S., and Morrical, S. W. 2001. Helicase assembly protein Gp59 of bacteriophage T4: fluorescence anisotropy and sedimentation studies of complexes formed with derivatives of Gp32, the phage ssDNA binding protein. Biochemistry 40:7651–7661.

Yang, J., Trakselis, M. A., Roccasecca, R. M., and Benkovic, S. J. 2003. The application of a minicircle substrate in the study of the coordinated T4 DNA replication. J. Biol. Chem. 278:49828–49838.

Yang, J., Zhuang, Z., Roccasecca, R. M., Trakselis, M. A., and Benkovic, S. J. 2004. The dynamic processivity of the T4 DNA polymerase during replication. Proc. Natl. Acad. Sci. USA 101:8289–8294.

Yang, J., Xi, J., Zhuang, Z., and Benkovic, S. J. 2005. The oligomeric T4 primase is the functional form during replication. J. Biol. Chem. 280:25416–25423.

Yang, J., Nelson, S. W., and Benkovic, S. J. 2006. The control mechanism for lagging strand polymerase recycling during bacteriophage T4 DNA replication. Mol. Cell 21:153–164.

Yao, N., Turner, J., Kelman, Z., Stukenberg, P. T., Dean, F., Shechter, D., Pan, Z. Q., Hurwitz, J., and O'Donnell, M. 1996. Clamp loading, unloading and intrinsic stability of the PCNA, beta and gp45 sliding clamps of human, *E. coli* and T4 replicases. Genes Cells 1:101–113.

Yegian, C. D., Mueller, M., Selzer, G., Russo, V., and Stahl, F. W. 1971. Properties of the DNA-delay mutants of bacteriophage T4. Virology 46:900–919.

Yonesaki, T. 1994. The purification and characterization of gene 59 protein from bacteriophage T4. J. Biol. Chem. 269:1284–1289.

Young, M. C., Schultz, D. E., Ring, D., and von Hippel, P. H. 1994. Kinetic parameters of the translocation of bacteriophage gene 41 protein helicase on single-stranded DNA. J. Mol. Biol. 235:1447–1458.

Yuzhakov, A., Kelman, Z., and O'Donnell, M. 1999. Trading places on DNA – a three-point switch underlies primer handoff from primase to the replicative DNA polymerase. Cell. 96:153–163.

Zhang, Z., Spiering, M. M., Trakselis, M. A., Ishmael, F. T., Xi, J., Benkovic, S. J., and Hammes, G. G. 2005. Assembly of the bacteriophage T4 primosome: single-molecule and ensemble studies. Proc. Natl. Acad. Sci. USA 102:3254–3259.

Zhuang, Z., Berdis, A. J., and Benkovic, S. J. 2006a. An alternative clamp loading pathway via the T4 clamp loader gp44/62-DNA complex. Biochemistry 45:7976–7989.

Zhuang, Z., Yoder, B. L., Burgers, P. M., and Benkovic, S. J. 2006b. The structure of a ring-opened proliferating cell nuclear antigen-replication factor C complex revealed by fluorescence energy transfer. Proc. Natl. Acad. Sci. USA 103:2546–2551.

Chapter 17
Atomic Structure of the Herpes Simplex Virus 1 DNA Polymerase

Shenping Liu and Fred L. Homa

Viral polynucleotide replication is central to the reproduction of all viruses. For DNA viruses, DNA polymerase is the core component of this process. There are at least six major DNA polymerase classes: class A, B, C, D, X, and Y (Koonin 2006, Burgers et al. 2001, Friedberg 2006). DNA replication in both bacterial and animal viruses is carried out by polymerases that belong to archaeal-class B DNA polymerases. These replicative polymerases also include the major eukaryotic DNA polymerases α, δ, ε, and also DNA polymerase ζ (Koonin 2006, Burgers et al. 2001, Friedberg 2006). The basic enzymatic function of a viral DNA polymerase is to catalyze the consecutive incorporation of new nucleotides to the 3′-end of the primer strand DNA using an existing single-strand DNA as the template (Lehman and Boehmer 1999, Knopf 1979, Boehmer and Villani2003, Crute and Lehman 1989). For viral DNA replication to process faithfully, the structure of the polymerase must favor the incorporation of the correct nucleotides (Hwang et al. 1999, Chaudhuri et al. 2003). To further increase the fidelity of the process, many DNA polymerases possess a 3′–5′ exonuclease activity that removes the mismatched nucleotide from the primer DNA strand (Koonin 2006, Burgers et al. 2001, Friedberg 2006, Hwang et al. 1999, Gibbs et al. 1991, Song et al. 2004). There are also other functional proteins that assist replicative polymerases in their DNA replication process, these include helicase/primases that unwind double-strand DNA duplexes (Lehman And Boehmer 1999, Knopf 1979, Boehmer and Villani 2003, Crute and Lehman 1989, Marsden et al. 1997) and accessory factors that increase DNA polymerase processivity so that a single polymerase can elongate the primer strand by thousands of nucleotides before falling off the DNA duplex (Digard et al. 1993, Gottlieb et al. 1990, Parris et al. 1988, Weisshart et al. 1999, Chow and Coen 1995).

In this chapter we will use the crystal structure of the DNA polymerase from herpes simplex virus type 1 (HSV-1) as an example to illustrate the structure–function relationships of viral DNA polymerases. HSV-1 is a member of the Herpesviridae virus family that infects all vertebrates including human (Lehman

S. Liu (✉)
Exploratory Medicinal Sciences, Pfizer Inc., Eastern Point Rd., Groton, CT 06340, USA
e-mail: shenping.liu@pfizer.com

C. E. Cameron et al. (eds.), *Viral Genome Replication*,
DOI 10.1007/b135974_17,

and Boehmer 1999, Knopf 1979, Boehmer and Villani 2003, Crute and Lehman 1989). Humans are the hosts for infection by eight herpesviruses that include several important human pathogens. HSV-1, herpes simplex virus type 2 (HSV-2), and varicella zoster virus (VZV) produce neurotropic infections, such as cutaneous and genital herpes, chickenpox, and shingles. Infections of a lymphotropic nature are caused by cytomegalovirus (CMV), HHV-6, HHV-7, and Epstein–Barr virus. The viral DNA polymerase is required for replication of all herpesviruses and is involved in both protein-primed DNA replication and rolling circle replication (Lehman and Boehmer 1999, Knopf 1979, Boehmer and Villani 2003, Crute and Lehman 1989). Anti-herpes drugs were among the first anti-virus drugs developed and marketed (Elion 1989). Most anti-herpes drugs used today are nucleoside analogs, such as acyclovir, that target the viral DNA polymerase (De Clercq 2004, Coen and Schaffer 2003, Gilbert et al. 2002, Wathen 2002, Thomsen et al. 2003, Huang et al. 1999, Reardon and Spector 1989). While clinically useful, this class of drug exhibits a narrow antiviral spectrum, and resistance to these agents is an emerging problem for disease management (Gilbert et al. 2002, Wathen 2002, Thomsen et al. 2003). New classes of anti-herpes drugs with the potential of inhibiting the replication of multiple herpesvirus are under development. One such class is non-nucleotide inhibitors that target the polymerases of herpesviruses. These novel compounds are active against drug (acyclovir)-resistant HSV mutants (Wathen 2002, Thomsen et al. 2003, Huang et al. 1999, Liu et al. 2006). A better understanding of herpes virus replication will help in the development of new, safe, and effective anti-herpetic drugs.

Based on primary sequence the herpes simplex virus type 1 (HSV POL) is classified as pol α-like polymerase (Lehman and Boehmer 1999, Knopf 1979, Boehmer and Villani 2003, Crute and Lehman 1989). The pol α family includes human polymerase α and polymerases from animals and other viruses (Koonin 2006). The HSV POL has served as a model for POL of other herpes viruses as well as other eukaryotic DNA polymerases, including human pol α polymerase. In HSV-infected cells the polymerase consists of two subunits: UL30 and UL42 (Lehman and Boehmer 1999, Knopf 1979, Boehmer and Villani 2003, Crute and Lehman 1989). UL30 is the catalytic subunit of HSV DNA polymerase that carries out all the enzymatic functions and UL42 is the accessory subunit that enhances UL30's processivity (Digard et al. 1993, Gottlieb et al. 1990, Parris et al. 1988, Weisshart et al. 1999, Chow and Coen 1995).

UL30: DNA Elongation and Proofreading

The catalytic subunit of herpes simplex type 1 DNA polymerase (HSV POL), UL30, is the prototype of the herpesvirus polymerase family. It is composed of 1,235 amino acids and exhibits all the enzymatic functions of a polymerase (Lehman and Boehmer 1999, Knopf 1979, Boehmer and Villani 2003, Crute and Lehman 1989). In addition to a polymerase activity required to extend the DNA primer chain, HSV

POL possesses an intrinsic 3′–5′-exonuclease activity that serves as a proofreading activity to ensure the high fidelity of DNA replication (Hwang et al. 1999, Chaudhuri et al. 2003, Gibbs et al. 1991, Song et al. 2004). HSV POL has also been reported to have an intrinsic RNase H activity that presumably functions in the removal of RNA primers during the processing of Okazaki fragments (Crute and Lehman 1989). The RNase H activity was initially attributed to a 5′–3′-exonuclease function associated with HSV POL; however, this activity was subsequently attributed to the potent 3′–5′-exonuclease of HSV POL (Lehman and Boehmer 1999). In vivo, the extreme C-terminus of the cognate polymerase has been shown at the molecular level to interact with UL42 (Digard et al. 1993), and is critical for UL30's processivity (Digard et al. 1993, Gottlieb et al. 1990, Parris et al. 1988, Weisshart et al. 1999, Chow and Coen 1995). During the predominant rolling circle mode of DNA replication, HSV POL forms a large replication complex with the viral primase/helicase complex (Lehman and Boehmer 1999, Knopf 1979, Boehmer and Villani 2003, Crute and Lehman 1989). It was reported that this replication complex consisting of the HSV POL/HSV primase/helicase complex is formed by interactions between UL30 and the C-terminus of the UL8 component of the viral helicase–primase complex (Marsden et al. 1997). The details of these interactions at a molecular level have yet to be elucidated. Recently, the crystal structure of the apo form of UL30 has been reported (Liu et al. 2006).

UL42: The Processivity Subunit of HSV POL

As a replicative polymerase, HSV POL is capable of polymerizing tens of thousands of nucleotides without dissociating from their DNA templates (Lehman and Boehmer 1999, Knopf 1979, Boehmer and Villani 2003, Crute and Lehman 1989). However, the processivity of HSV POL is critically dependent on the accessory subunit of the HSV POL dimer, UL42 (Digard et al. 1993, Gottlieb et al. 1990, Parris et al. 1988, Weisshart et al. 1999, Chow and Coen 1995). UL42 is a 488-amino acid protein that forms a heterodimer with UL30, the catalytic subunit of HSV POL (Gottlieb et al. 1990, Parris et al. 1988), with an association constant of 1×10^8 M^{-1} (Digard et al. 1993, Gottlieb et al. 1990, Parris et al. 1988, Weisshart et al. 1999, Chow and Coen 1995). It was reported that the extreme C-terminus of UL30 is necessary and sufficient for the functional interaction with UL42 and for viral replication. UL42 also binds the DNA duplex, and is a functional homolog of other DNA polymerase processivity-enhancing factors (Weisshart et al. 1999, Chow and Coen 1995). However, unlike other processivity factors, UL42 binds directly to DNA with high affinity for double-stranded and single-stranded DNA without apparent sequence specificity (Weisshart et al. 1999, Chow and Coen 1995). This intrinsic DNA-binding activity is crucial for the function of UL42 as a processivity factor, as UL42 mutations that specifically abrogate DNA binding drastically impair long-chain DNA synthesis (Chow and Coen 1995). Binding with the DNA duplex and interacting with UL30, UL42 prevents premature dissociation of the polymerase from the template and primer by tethering UL30 to the DNA (Parris et al. 1988, Weisshart et al. 1999, Chow and Coen 1995). By increasing UL30's affinity with the DNA duplex, UL42 somehow increases the fidelity of HSV POL replication (Chaudhuri et al. 2003). The crystal structure of the UL42/UL30 C-terminal peptide complex as well as their analogs in cytomegalovirus DNA polymerase (UL44 and UL44/UL54 peptide complex) has been reported recently (Zuccola et al. 2000, Appleton et al. 2006, 2004).

Overall Structure of HSV POL

All DNA polymerases share some common architectural features (Steitz 1999, Brautigam and Steitz 1998). In addition to domains that perform the DNA polymerase activities, they usually have other functional domains, i.e., the 3′–5′ exonuclease domain that functions to remove mismatched bases from incorrectly incorporated DNA primer strand. The overall arrangements of these domains relative to each other are quite diversified (Steitz 1999, Brautigam and Steitz 1998). However, the common feature of the polymerase domains can be described as a right hand with specific domains referred to as the thumb, palm, and finger domains (Steitz 1999, Brautigam and Steitz 1998).

The crystal structures of several members of the pol α family have been reported, these include the POL α from thermophilic and archaebacteria (Rodriguez et al. 2000, Zhao et al. 1999, Wang et al. 1997, Shamoo and Steitz 1999, Franklin et al. 2001). Most significantly, structures of the apo form, the editing complex and the replication complex of a bacteriophage DNA polymerase, RB69 (RB69 POL), have been obtained (Wang et al. 1997, Shamoo and Steitz 1999, Franklin et al. 2001). These structures not only revealed the general architecture of the pol α polymerases but also provided important information on interactions of the enzyme with DNA and nucleotides. Based on the similar three-dimensional structure and primary sequence of HSV POL compared with other members of pol α family, models of the HSV POL structure at different stages of the polymerase reaction can be constructed, using a three-dimensional structure superimposing program (Liu et al. 2006). We also attempted to build a crude UL30/UL42 heterodimer complex model using the reported structures to provide a general idea how the UL30/UL42 processively replicate HSV DNA strands. The modeled structures of HSV POL with different ligands bound at different stages are helpful to show how HSV POL achieves both high processivity and high fidelity in viral DNA replication. They also provide important information with regard to how anti-herpes agents specifically inhibit HSV POL.

The resemblance of the overall architecture of the UL30 subunit of HSV POL to other POL α structures is apparent, despite being at least 300 amino acid residues longer and exhibiting low sequence homology (range 16–50%) with other members of this family (Gilbert et al. 2002). When compared with other members of the POL α family, six conserved structural domains of HSV POL can be easily identified: a pre-NH_2 domain, an NH_2 domain, a 3′–5′-exonuclease domain, and the polymerase palm, fingers, and thumb domains (Fig. 17.1). The exonuclease domain contains conserved regions exo I, exo II (region IV), and exo III (δ–C region). Regions III and VI belong to the fingers, regions I, II, and VII are located in the palm sub-domain, and the thumb sub-domain contains the conserved region V (Gilbert et al. 2002, Wathen 2002, Thomsen et al. 2003, Huang et al. 1999, Liu et al. 2006). These domains assemble to form a disk-like shape around a central hole, with the NH_2- and C-termini at opposite sides of the protein (Fig. 17.1). Three grooves can be seen emanating from the central hole (Fig. 17.1). The first groove, located at the interface of NH2-terminal and exonuclease domains, is lined with positively charged side chains (Fig. 17.1). The structure of the RB69 POL replicating and editing complex structures shows that this groove binds the 5′-extension of the single-stranded DNA template (Shamoo and Steitz 1999, Franklin et al. 2001). In HSV POL these positively charged side chains are predicted to interact with the phosphate backbone of the single-strand DNA template. The second groove located between the exonuclease domain and the tip of the thumb domain leads to the putative exonuclease active site (Fig. 17.1). The RB69 POL editing complex shows that the unwound

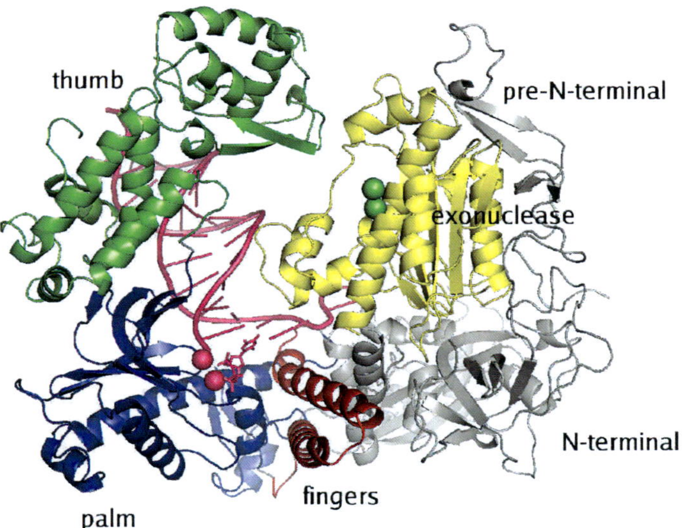

Fig. 17.1 The ribbon diagram of the overall structure of herpes simplex 1 DNA polymerase. Different domains are colored differently and labeled. Fingers, palm, and thumb domains form the polymerase catalytic site, with the incoming nucleotide (stick model) and the two active site metals shown as *magenta balls* indicating the active site. The 3′–5′ exonuclease domain is responsible for correcting mismatched DNA duplexes. The metal ions at the active site of the exonuclease are shown as *green balls*. The N-terminal domain forms part of the single-strand template-binding groove, and the pre-N-terminal domain is proposed to interact with HSV helicase/primase complex during rolling circle DNA replication. The modeled DNA duplex in the replication mode is shown as *magenta ribbons* for backbone strands and ladders for base pairs. Pictures are prepared using the program PyMOL (DeLano 2002).

primer strand binds in this groove (Wang et al. 1997, Shamoo and Steitz 1999). The DNA duplex, in both the editing and replicating complex structures, binds to a third groove formed by the palm and thumb sub-domains (Shamoo and Steitz 1999, Franklin et al. 2001).

The NH_2-terminal 250 residues of HSV POL, which is unique to the herpes DNA polymerases, forms the pre-N-terminal domain and part of the N-terminal domain (Figs. 17.1 and 17.2). Based on its location, the N-terminal domain was proposed to be involved in interactions with the UL8 component of the HSV helicase/primase complex during replication (Liu et al. 2006). A probable ribonuclease H function was assigned to the N-terminal domain of HSV POL (Liu et al. 2006).

Modeling of Different Ligand-Bound HSV POL Structures

As stated above, with the exception of the N-terminal 250 amino acids of HSV POL, the remainder of the sequence can be matched with conserved regions of other pol α family members. This high degree of conservation can be used to build detailed models of the structure of HSV POL bound to different ligands. For instance, to build a model of the DNA editing complex of HSV POL, the 3′–5′ exonuclease domain of HSV POL was compared to that of RB69 POL editing complex. Using homology modeling tools the HSV POL exodomain was first superimposed on the exodomain of RB69 POL and then the palm and thumb domains of HSV POL (which are relevant to DNA editing) were added on top of those of the RB69 POL complex. For the replicating model, we chose the palm domain as a reference, and then overlapped the fingers and the thumb domains. We observed that only moderate conformational changes at peptides that link different domains are needed to assemble those different complexes. Inevitably, these modeling exercises create some discontinuity at peptides that join different domains, but these peptides are all flexible loops and away from the areas of interest. These types of models help us understand the structural features of the relevant HSV POL functions. To ensure the correctness of the models, the modeled structures are checked against the primary sequence of a number of polymerases and against biochemical data, such as mutagenesis and drug-resistant data. As an example (as described in the replication structure section), a few key residues located on the palm and finger domains of the polymerases serve critical roles in the polymerase catalysis reaction. These residues superimpose very well with those of RB69 POL and can explain their roles in the reaction mechanism, confirming the usefulness of these homology models.

HSV POL Replicating Mechanism

The active sites of all polymerases with known structures contain two metal ions. A universal reaction mechanism was proposed that involves metal–phosphate interactions that stabilize the penta-covalent reaction transition state of the α phosphate of the incoming nucleotide (Steitz 1999, Brautigam and Steitz 1998). As described above, a model of the replication complex of HSV POL was constructed by superimposing individual domains of HSV POL onto those of the replication complex of RB69 POL, with the palm domain of the replicating RB69 POL complex as the reference. In the known structures of the apo POL α, as well as the editing complex RB69 POL, there is a kink at the highly conserved residues N815–S816 on the second helix (FB) of the finger domain. In the replicating RB69 POL complex structure the FB helix is continuous and our model was adjusted accordingly. The continuous nature of the FB helix is required to bring the highly conserved Y818–G819 (Y567–G568 into RB69 POL) into close proximity with the catalytic site. No adjustments were made to smooth out the joints between domains after superimposition and only a few side chains of key residues located at the polymerase active site needed adjustments to adopt the conformations present in the RB69 POL replicating complex (Fig. 17.2A). This simple model can easily explain the sequence–activity relationship of HSV POL. The side chains of highly conserved D717, D888, and carbonyl oxygen of F718 residing in the palm domain coordinate two catalytic metal ions (Fig. 17.2A). From the palm domain side, these two metals interact with the triphosphates of the incoming nucleotides and the 3′-OH end of the primer strand. Specifically, metal A interacts with the 3′-OH group of the last nucleotide on the primer strand and the α phosphate of the incoming nucleotide, and metal B interacts with all three phosphates of the incoming nucleotide. From the fingers domain, the positively charged side chains of R785, R789, and K811 interact with the α, β, and γ phosphate groups of the incoming nucleotide, respectively (Fig. 17.2). These charge–charge interactions are important for the positioning of the 3′-OH of the primer end to the phosphates moiety, and for neutralizing the accumulated negative charge of the reaction intermediate (Steitz 1999, Brautigam and Steitz 1998). The single-stranded template would make a sharp turn at the POL active site and extend into the groove formed between NH_2-terminal domain and the exonuclease domain. The DNA duplex would exit through the groove between thumb and palm domains that is close to the C-terminal region that interacts with UL42. After incorporation of the new nucleotide and the dissociation of the pyrophosphate from HSV POL active site, the protein may loose its grip on the DNA duplex allowing it to slide down one nucleotide so that it is in position to incorporate the next nucleotide (Steitz 1999, Brautigam and Steitz 1998).

The polymerase active site of HSV POL is the primary target for nearly all anti-herpes drugs. For example, acyclovir, the gold standard anti-herpes drug, was only active after being first converted to acyclovir triphosphate by viral and cellular enzymes (Elion 1989, Reardon and Spector 1989). Acyclovir triphosphate inhibits HSV POL by competing with the incoming nucleotides for binding at the polymerase active site, but more importantly for its potency, by forming a tight dead-end HSV POL/DNA complex after being incorporated into the primer strand (Reardon and Spector 1989) (Fig. 17.2B). Foscarnet, a pyrophosphate analog and another commonly used anti-herpes drug (De Clercq 2004, Coen and Schaffer 2003), inhibits DNA replication possibly by interacting with the positive charged side chains of R785, R789, and K811 and two catalytic metal ions and slowing

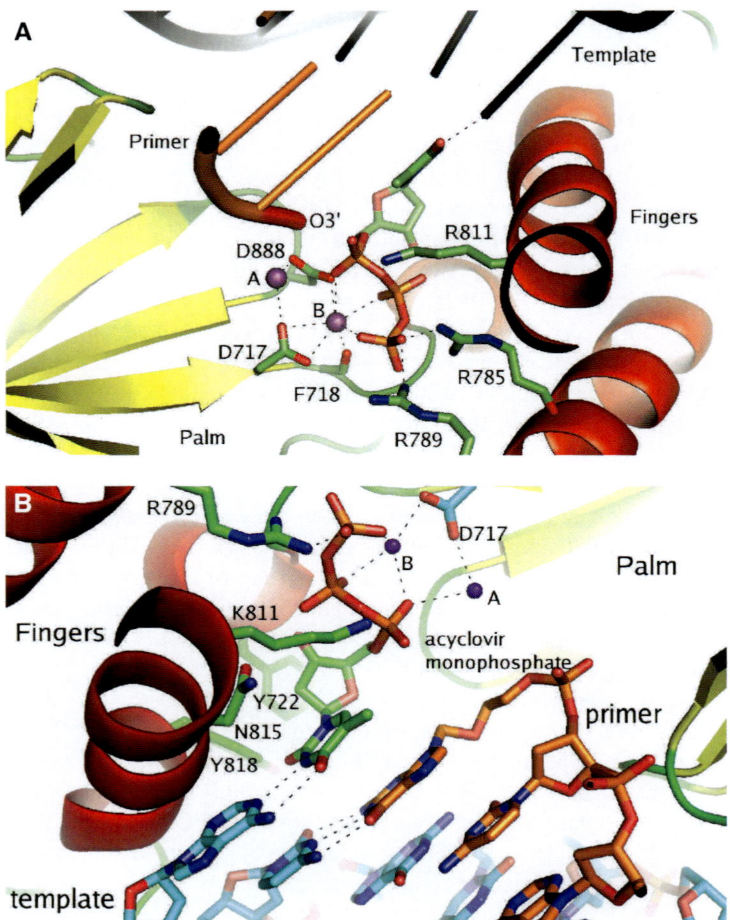

Fig. 17.2 The replicating center structure of HSV POL by homology modeling. **A** The catalytic site of HSV POL is formed by the fingers domain and the palm domain of the protein. The active site metals, metal A and metal B, are coordinated by conserved Asp888 and Asp717, plus a main chain carbonyl oxygen. The base of the incoming nucleotide (stick) binds in a hydrophobic pocket formed by protein residues and DNA bases from the bound DNA duplex, and is specifically recognized by Watson–Crick hydrogen bonds with the first unpaired base on the template strand. The triphosphate of the nucleotide interacts with the metal ions from the palm side of the protein, and positive side chains from the fingers. The DNA duplex binds to the active site in such a way that the 3′-OH is positioned to make an in-line nucleophilic attack on the α-phosphate of the nucleotide. **B** After being converted into triphosphate, acyclovir can be incorporated onto the 3′-end of the DNA primer strand and forms a dead-end HSV POL/DNA complex. The modeled acyclovir/HSV POL/DNA dead-end complex shows that acyclovir can be accommodated at HSV POL's replication center.

down the position shift of the DNA duplex from the active site of HSV POL. A novel class of non-nucleotide POL inhibitors has recently been shown to selectively target herpesvirus polymerases (Wathen 2002, Thomsen et al. 2003). These compounds are competitive inhibitors of the HSV POL that bind the polymerase in the presence of the DNA duplex, forming a POL/DNA dead-end complex (Thomsen et al. 2003, Liu et al. 2006).

Based on models presented here it is interesting to speculate why the K_m for the incoming nucleotide is much higher as a substrate for the incorporation into a normal DNA duplex than the K_d for its binding to an acyclovir monophosphate (Reardon and Spector 1989, Liu et al. 2006). Apparently the 3′-OH of a normal DNA primer strand would be in close contact with the α-phosphate and part of the binding energy would be utilized to overcome the repulse for the incorporation reaction (Fig. 17.2A). On the other hand, a smaller moiety and the lack of a 3′-OH group make acyclovir monophosphate a better group to interact with the incoming nucleotide, thus the DNA strand with the incorporated acyclovir monophosphate forms a dead-end complex with HSV POL when the next incoming nucleotide binds (Fig. 17.2B). This dead-end ternary complex binds much more tightly than the primer strand dissociates and gets excised much slower (Reardon and Spector 1989, Liu et al. 2006).

Structural Elements for HSV POL's High Replicating Fidelity

The high fidelity of DNA replication is crucial for the long-term survival of viruses. HSV POL is less faithful than most replicative DNA polymerases in incorporating the correct nucleotides: it has been reported that for HSV POL the incorporation of the correct dNTP over the incorrect dNTP was 300 (Hwang et al. 1999, Chaudhuri et al. 2003). However, the fidelity of HSV POL was probably increased in part by a kinetic barrier that decreases the rate of extending a mismatch DNA primer (Chaudhuri et al. 2003, Song et al. 2004). The slow extension rate for mismatched primer also increases its probability to be transferred to the exosite where it can be excised by the associated 3′–5′ nuclease activity (Hwang et al. 1999, Chaudhuri et al. 2003, Gibbs et al. 1991, Song et al. 2004, Marsden et al. 1997). There are two aspects of the polymerase fidelity: the faithful synthesis of the DNA primer strand with nucleotide sequence complimentary to the template DNA strand and the preferential incorporation of deoxynucleotides versus their 2′-OH ribose counterpart. The proposed HSV POL replicating complex provides a model to explain how the protein achieves both of these fidelities. In this model, the nucleotide-binding pocket is formed by the template DNA strand, the 3′-end of the primer DNA strand, and protein atoms of the HSV POL (Figs. 17.2 and 17.3). The base of the incoming nucleotide makes a π–π stacking interaction with the base of the last nucleotide of the primer strand and also forms hydrogen bonds that are complimentary to the first unpaired nucleotide on the template strand. On the other face of the base, the side chain of N815 from the fingers domain stacks against the base. One edge of the

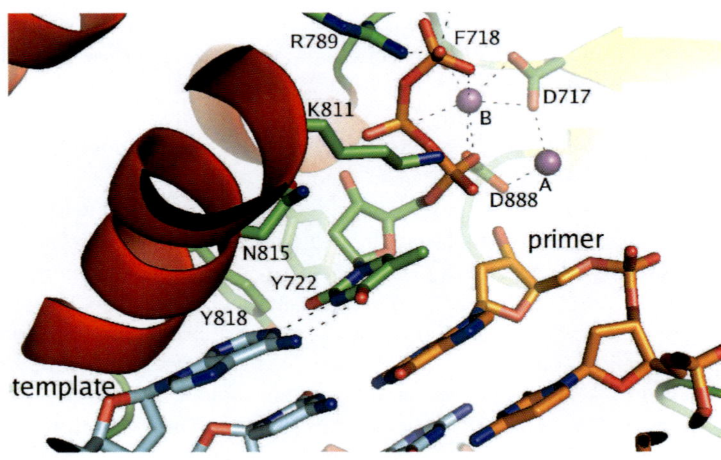

Fig. 17.3 The fidelity of HSV POL is realized at the replication center formed by the polymerase and bases from both the template and primer DNA strands. The correct nucleotides are selected over incorrect ones by hydrogen bond patterns that are complimentary to the sequences of the DNA template strand. The correct nucleotides have a much higher affinity of binding to the active site formed by DNA–POL complex, thus ensuring their higher chances of incorporation into the 3′-end of primer strand.

base also contacts the side chain of the conserved Y818 from the fingers. Because the first unpaired nucleotide on the template strand forms part of the binding pocket for the incoming nucleotides, the sequence of the DNA template strand can preferentially select the right nucleotides with complimentary hydrogen bond patterns for binding and incorporation (Figs. 17.3 and 17.4). The side chain of Y818 prevents the protrusion of mismatched nascent base pairs into the DNA minor groove, reducing the opportunities of mismatched hydrogen bonds, and further ensuring the incorporation of the correct nucleotides (Figs. 17.3 and 17.4). The deoxyribose of the incoming nucleotide stacks on the aromatic ring of conserved residues Y722 and makes hydrophobic interactions with the side chain of T887 and L721. The snuggly fitting of the deoxyribose–ribose into this space is apparently unfavorable for the binding of the 2′-ribose moiety of ribonucleotides (Fig. 17.3). Specifically, the side chain of strictly the conserved Y722 makes a close Van del Waal contact with the C2′ atom of the deoxyribose, and would clash with the hydrophilic 2′-OH group of ribose, providing a strong "steric gating" effect against the binding and incorporation of ribonucleotides. The phosphate tail of the incoming nucleotide extends to the catalytic site that is described in the catalytic mechanism section.

There are structural elements near the HSV POL polymerase active site that contribute to the kinetic barrier of extending the mismatched primer strands. These elements ensure the fidelity of the polymerase reaction when the incorporation of the wrong base does occur (Hwang et al. 1999, Chaudhuri et al. 2003, Gibbs et al. 1991, Song et al. 2004). In the HSV POL replicating model, the conserved KKKY (amino acids 938–941) motif and the side chains of Y818, Y884, and D886 interact

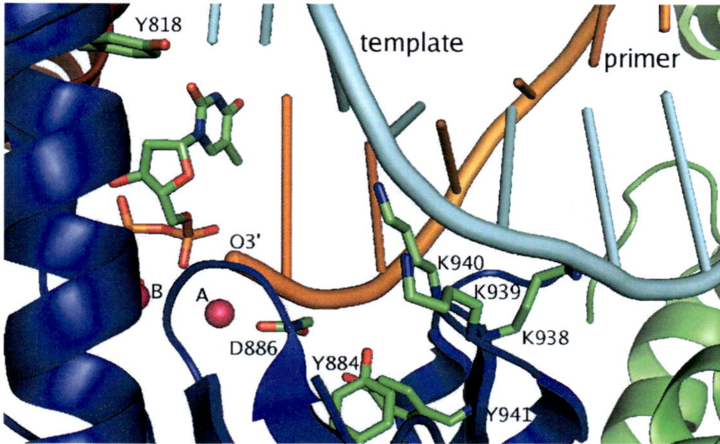

Fig. 17.4 Replication fidelity of HSV POL is further reinforced by interactions of HSV POL amino acids found in the catalytic site with the DNA minor groove. The side chain of Tyr818 directly contacts the base of the incoming nucleotide and prevents the protruding of the nucleotide bases into the minor groove, minimizing the chances of binding and incorporation of the wrong nucleotides by forming the non-conventional hydrogen bonds with template bases. Side chains of strictly conserved Tyr884 and the KKKY (938–941) motif serve the function of sensing mismatches after replication errors happened and extended a few nucleotides. These mismatched residues would weaken the binding of newly formed product DNA with a minor groove that is distorted by the mismatched nucleotides and facilitate the dissociation and unwinding of the primer strands for excision.

with the minor groove of the newly formed DNA duplex (Fig. 17.4). These conserved residues, interacting with the penultimate base pair, would normally serve the function of sensing mismatches in the newly formed DNA duplex (Franklin et al. 2001). Incorporation of mismatched nucleotides would distort the geometry of minor groove of the newly formed product DNA and interfere with interactions with these residues (Fig. 17.4). These would in turn result in an unstable DNA–POL complex and halt the further extension of the DNA duplex. The weakened interactions between the mismatched DNA duplex and HSV POL favor the dissociation of HSV POL/DNA complex and increase the chance of duplex unwinding for the repair at the 3′–5′ exonuclease active site of HSV POL.

HSV POL Mismatch Repairing Model

The error correction mechanism of the 3′–5′ nuclease activity improves the fidelity rate of HSV POL by ˜39-fold (Hwang et al. 1999, Chaudhuri et al. 2003, Gibbs et al. 1991, Song et al. 2004, Marsden et al. 1997). As discussed above, a mismatched DNA duplex has weakened interactions with HSV POL and the POL/DNA complex dissociates easier. Once dissociated, the DNA duplex could unwind and the

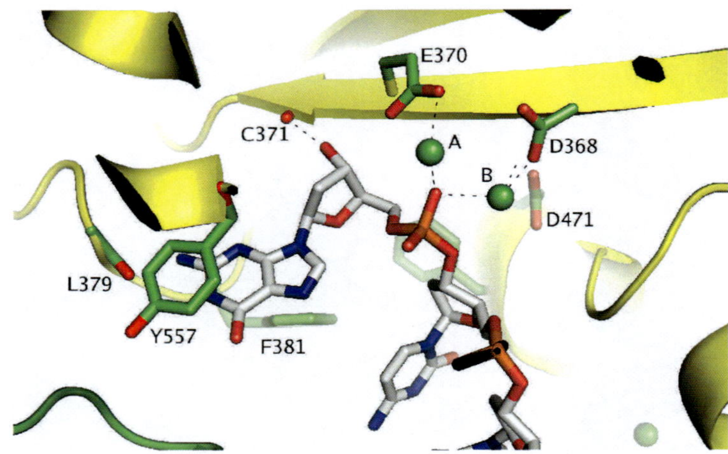

Fig. 17.5 Modeling of the 3′–5′ exonuclease center of HSV POL. The exonuclease catalytic site contains a di-metal cluster that is coordinated by conserved residues, Asp 368, Asp471, and Glu 370. A non-discriminating hydrophobic pocket formed by Tyr557 and Phe381 can accommodate different types of bases. Mismatched primer strands result in unstable replicating DNA–POL complexes that have a much higher chance of being sent to the exonuclease active site for excision.

single-strand primer shifts position to the 3′–5′ nuclease domain for cleavage. It has been demonstrated that an exonuclease-deficient mutant of HSV POL has higher rate for incorporation of the mismatched nucleotides (Hwang et al. 1999, Chaudhuri et al. 2003, Gibbs et al. 1991, Song et al. 2004). The modeled structure of HSV POL editing complex reveals a bi-metal exonuclease active site (Fig. 17.5). The two active metals are coordinated by side chains of conserved D368, E370, and D471 of HSV POL (D114, E116, and E222 in RB69 POL). The two metals interact with the phosphate group, consistent with the bi-metal–phosphate bond cleavage mechanism of nucleases. The ribose of the nucleotide being cleaved would interact with the carbonyl oxygen of C371, and the base located in a hydrophobic pocket formed by Y557, F381 and may form hydrogen bond with the carbonyl of L379. Apparently such a nucleotide pocket will accommodate any base, fulfilling the requirements for universal nucleotide cleavage (Fig. 17.5).

DNA Duplex Binding

The newly synthesized DNA duplex binds to thumb and palm domains. In the structures of RB69 POL the thumb domain interacts with the backbone of the DNA duplex and the tip and base of this domain adopt different conformations in the editing and replication modes (Fig. 17.6). Interestingly, the conformation of the thumb domain of HSV POL is similar to that of the replicating conformation of RB69 POL and thus does not require a large conformational change upon DNA binding

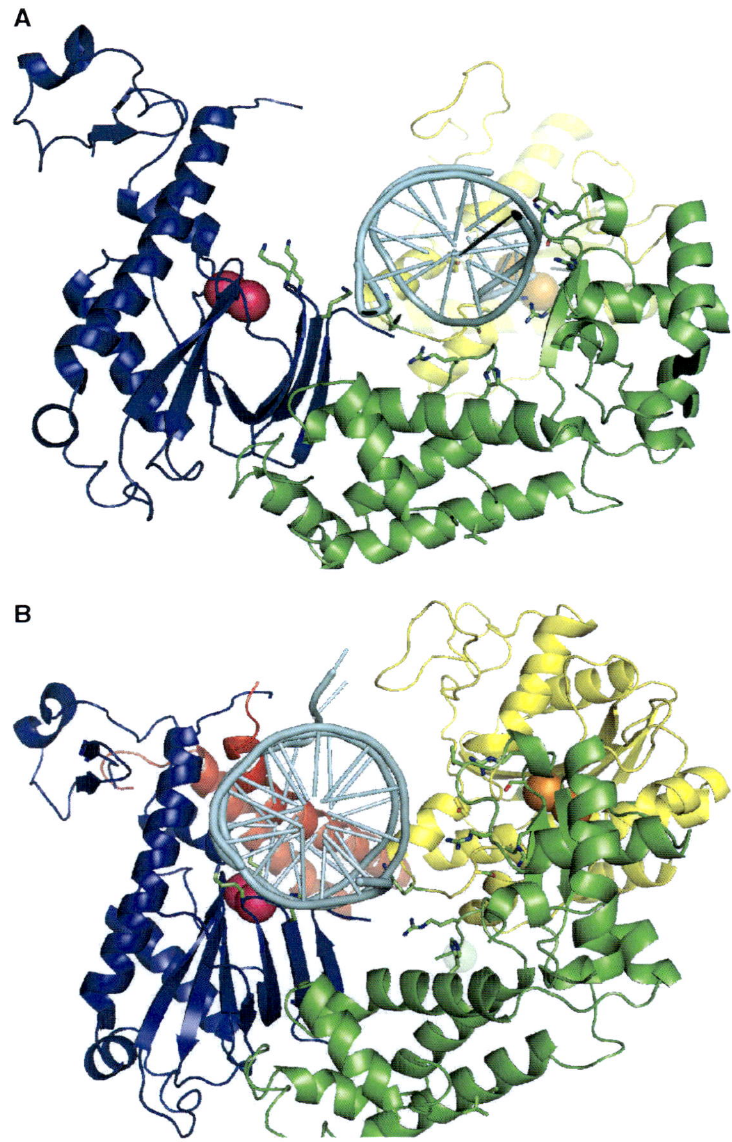

Fig. 17.6 Homology models of binding of DNA duplex with UL30 of HSV POL based on RB69 editing and replicating complex. **A** DNA duplex binding of HSV POL editing complex. **B** DNA duplex-binding mode of HSV POL replicating complex. Visible movements occur at the thumb domain of HSV POL.

(Fig. 17.6B). Sequence conservation in the thumb domain does not need to be high as DNA duplex binding involves collective interactions; however, the sequence conservation of this domain in the herpes POL family as well as the conserved region V across the POL α family does make sense in the context of interacting with DNA duplex backbone (e.g., R959, R1039, H1051, and R1071). Large deletions or insertions in these domains of the POL α family usually occur at loops, or form some additional secondary structures distant from the catalytic centers.

Single-Strand DNA Template Binding

To processively incorporate many nucleotides into the growing primer strand, the catalytic unit of HSV POL, UL30, should be able to bind to single-strand DNA template. In the RB69 POL replicating complex structure, the short 5′-extension of the single-stranded DNA template was shown to bind to a groove at the surface of the polymerase (Shamoo and Steitz 1999, Franklin et al. 2001). The analogous groove exists in HSV POL. This groove, located at the interface of NH_2-terminal and exonuclease domains, is lined with positively charged side chains (Fig. 17.7). In HSV POL these positively charged side chains are predicted to interact with the phosphate backbone of the single-strand DNA template.

Processivity Model

It has been shown that the extreme C-terminal peptide of UL30 is necessary and sufficient for interaction with UL42 (Digard et al. 1993, Gottlieb et al. 1990). This extreme C-terminal peptide of UL30 is unseen in the published HSV POL structure,

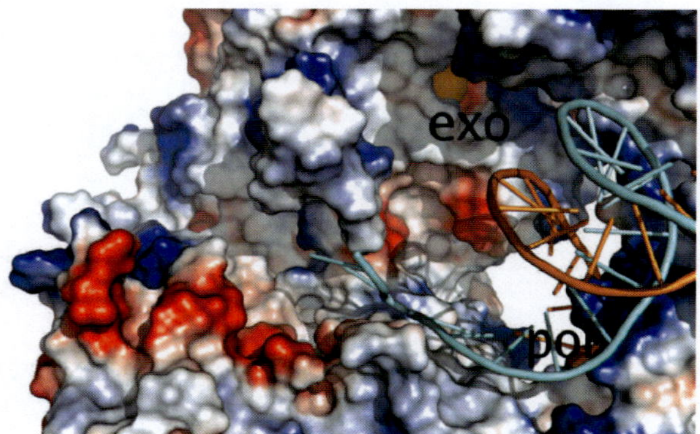

Fig. 17.7 Surface representation of the putative single-strand binding groove of HSV POL, colored by electrostatic potentials, with *red* representing negative and *blue* positive. The lower half of the binding grove is from the N-terminal domain and the upper part is from the 3′–5′ exonuclease domain. The DNA template strand is shown in *cyan ribbon* and the primer strand in *brown*. The active sites of the exonuclease and polymerase domains are marked.

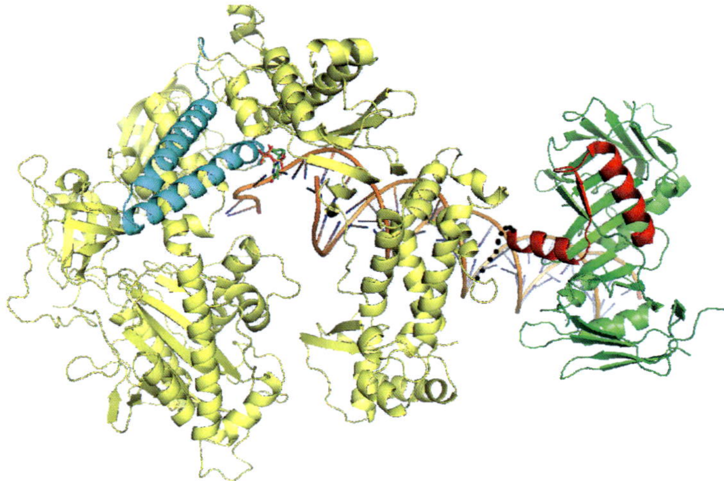

Fig. 17.8 Ribbon diagram of HSV POL heterodimer model. The polymerase subunit (UL30) is on the *left* and the accessory subunit (UL42) is on the *right*. The finger domain of UL30 is colored *cyan*, the C-terminal peptide of UL30 that interacts with UL42 is colored *red*, and the rest of UL30 and UL42 are colored *yellow* and *green*, respectively. The missing peptide linking the C-terminal peptide to UL30 is shown in *dash*. DNA duplex is shown as an *orange ribbon*, with base pairs as ladders. The relative locations of UL42, UL30, and DNA duplex strands are estimated and their orientations are arbitrary.

probably due to disorder (Liu et al. 2006). However, it is clear that the C-terminus is located downstream from the putative DNA duplex-binding site and is found at the protein surface, which is perfect for interacting with UL42. The peptide is ordered in the UL42/UL30 C-peptide complex structure, forming two α-helixes (Zuccola et al. 2000). In the HSV UL42–UL30 peptide complex, the UL30 C-peptide adopts an α–β–α fold and has extensive interactions with UL42. The UL30/UL42/DNA duplex complex was modeled in Fig. 17.8. Interestingly, in the recently reported crystal structures of the cytomegalovirus DNA polymerase processivity subunit UL44, and its complex with the C-terminus from the catalytic subunit UL54, UL44 has a very similar fold as that of UL42, even though these proteins have no obvious sequence homology (Appleton et al. 2006, 2004). However, the interactions between UL42/UL30 of HSV POL and UL44/UL54 of cytomegalovirus DNA polymerase are significantly different, suggesting specific recognitions of the polymerase and its partners (Zuccola et al. 2000, Appleton et al. 2006, 2004). During rolling circle DNA replication (Lehman and Boehmer 1999, Knopf 1979, Boehmer and Villani 2003, Crute and Lehman 1989), HSV POL could potentially interact with the UL8 component of the viral helicase/primase complex through UL30's pre-NH_2 domain and to process the newly unwound single-strand DNA template. The HSV POL accessory protein UL42 increases the processivity of the polymerase by anchoring HSV POL to the newly synthesized DNA duplex through the interactions with the C-terminal region.

Conclusion

Crystal structure of HSV POL and homology modeling of different HSV POL complexes provide structural basis for the polymerase's reaction mechanism. They are valuable in understanding HSV POL's fidelity and helpful in designing new antiherpes drugs.

References

Appleton, B.A., Loregian, A., Filman, D.J., Coen, D.M., and Hogle, J.M. (2004). The cytomegalovirus DNA polymerase subunit UL44 forms a C clamp-shaped dimer. *Mol Cell* **15**, 233–244.

Appleton, B.A., Brooks, J., Loregian, A., Filman, D.J., Coen, D.M., and Hogle, J.M. (2006). Crystal structure of the cytomegalovirus DNA polymerase subunit UL44 in complex with the C terminus from the catalytic subunit. Differences in structure and function relative to unliganded UL44. *J Biol Chem* **281**, 5224–5232.

Boehmer, P.E. and Villani, G. (2003). Herpes simplex virus type-1: a model for genome transactions. *Prog Nucleic Acid Res Mol Biol* **75**, 139–171.

Brautigam, C. and Steitz, T.A. (1998). Structural and functional insights provided by crystal structures of DNA polymerases and their substrate complexes. *Curr Opinion Struct Biol* **8**, 54–63.

Burgers, P., Koonin, E., Bruford, E. et al. (2001). Eukaryotic DNA polymerases: proposal for a revised nomenclature. *J Biol Chem* **276** (47), 43487–43490.

Chaudhuri, M., Song, L., and Parris, D. S. (2003) The Herpes Simplex Virus Type 1 DNA Polymerase Processivity Factor Increases Fidelity without Altering Pre-steady-state Rate Constants for Polymerization or Excision. *J Biol Chem* **278**, 8996–9004.

Chow, C.S. and Coen, D.M. (1995). Mutations that specifically impair the DNA binding activity of the herpes simplex virus protein UL42. *J Virol* **69**, 6965–6971.

Coen, D.M. and Schaffer, P.A. (2003). Antiherpesvirus drugs: a promising spectrum of new drugs and drug targets. *Nat Rev Drug Discov* **2**, 278–288.

Crute, J.J. and Lehman, I.R. (1989). Herpes simplex-1 DNA polymerase. Identification of an intrinsic 5′-3′ exonuclease with ribonuclease H activity. *J Biol Chem* **264**, 19266–19270.

De Clercq, E. (2004). Antiviral drugs in current clinical use. *J Clin Virol* **30**, 115–133.

DeLano, W.L. (2002). The PyMOL molecular Graphics System. DeLano Scientific, San Carlos, CA.

Digard, P., Bebrin, W.R., Weisshart, K., and Coen, D.M. (1993). The extreme C terminus of herpes simplex virus DNA polymerase is crucial for functional interaction with processivity factor UL42 and for viral replication. *J Virol* ***67***, 398–406.

Elion, G.B. (1989). The purine path to chemotherapy. *Science* ***244***, 241–247.

Franklin, M., Wang, J., and Steitz, T.A. (2001). Structure of the replicating complex of a pol alpha family DNA polymerase. *Cell* ***105***, 657–667.

Friedberg, E.C. (2006). The eureka enzyme: the discovery of DNA polymerase. *Nat Rev Mol Cell Biol* **7**, 143–147.

Gibbs, J.S., Weisshart, K., Digard, P., deBruynKops, A., Knipe, D.M., and Coen, D.M.. (1991). Polymerization activity of an alpha-like DNA polymerase requires a conserved 3′-5′ exonuclease active site. *Mol Cell Biol* **11**, 4786–4795.

Gilbert, C., Bestman-Smith, J., and Boivin, G. (2002). Resistance of herpesviruses to antiviral drugs: clinical impacts and molecular mechanisms. *Drug Res Update* **5**, 88–114.

Gottlieb, J., Marcy, A.I., Coen, D.M., and Challberg, M.D. (1990). The herpes simplex virus type 1 UL42 gene product: a subunit of DNA polymerase that functions to increase processivity. *J Virol* **64**, 5976–5987.

Huang, L., Ishi, K.K., Zuccola, H., Gehring, A.M., Hwang C.B.C., Hogle, J., and Coen, D.M. (1999). The enzymological basis for resistance of herpesvirus DNA polymerase mutants to acyclovir: relationship to the structure of alpha-like DNA polymerases. *Proc Natl Acad Sci USA* **96**, 447–452.

Hwang, Y.T., Liu, B.Y., Hong, C.Y., Shillitoe, E.J., and Hwang, C.B.C. (1999). Effects of exonuclease activity and nucleotide selectivity of the herpes simplex virus DNA polymerase on the fidelity of DNA replication in vivo. *J Virol* **73**, 5326–5332.

Knopf, K.W. (1979). Properties of herpes simplex virus DNA polymerase and characterization of its associated exonuclease activity. *Eur J Biochem* **98**, 231–244.

Koonin, E.V. (2006). Temporal order of evolution of DNA replication systems inferred by comparison of cellular and viral DNA polymerases. *Biology Direct* **1**, 39–57.

Lehman, I.R. and Boehmer, P.E. (1999). Replication of herpes simplex virus DNA. *J Bio Chem* **274**, 28059–28062.

Liu, S., Knafels, J.D., Chang, J.S., Waszak, G.A., Baldwin, E.T., Deibel, M.R., Jr., Thomsen, D.R., Homa, F.L., Wells, P.A., Tory, M.C., Poorman, R.A., Gao, H., Qiu, X., and Seddon, A.P. (2006). Crystal structure of the herpes simplex virus 1 DNA polymerase. *J Biol Chem* **281**, 18193–18200.

Marsden, H.S., McLean, G.W., Barnard, E.C., Francis, G.J., MacEachran, K., Murphy, M., McVey, G., Cross, A., Abbotts, A.P., and Stow, N.D. (1997). The catalytic subunit of the DNA polymerase of herpes simplex virus type 1 interacts specifically with the C terminus of the UL8 component of the viral helicase–primase complex. *J Virol* **71**, 6390–6397.

Parris, D.S., Cross, A., Haarr, L., Orr, A., Frame, M.C., Murphy, M., McGeoch, D.J., and Marsden, H.S. (1988). Identification of the gene encoding the 65-kilodalton DNA binding protein of herpes simplex virus type 1. *J Virol* **62**, 818–825.

Reardon, J.E. and Spector, T. (1989). Herpes simplex virus type 1 DNA polymerase. Mechanism of inhibition by acyclovir triphosphate. *J Biol Chem* **264**, 7405–7411.

Rodriguez, A.C., Park, H.W., Mao, C., and Beese, L.S. (2000). Crystal structure of a pol alpha family DNA polymerase from the hyperthermophilic archaeon *Thermococcus* sp. 9 degrees N-7. *J Mol Biol* **299**, 447–462.

Shamoo, Y. and Steitz, T.A. (1999). Building a replisome from interacting pieces: sliding clamp complexed to a peptide from DNA polymerase and a polymerase editing complex. *Cell* **99**, 155–166.

Song, L., Chaudhuri, M., Knopf, C.W., and Parris, D.S. (2004). Contribution of the 3′- to 5′-exonuclease activity of herpes simplex virus type 1 DNA polymerase to the fidelity of DNA synthesis. *J Biol Chem* **279**, 18535–18543.

Steitz, T.A. (1999). DNA polymerases: structural diversity and common mechanisms. *J Biol Chem* **274**, 17395–17398.

Thomsen, D.R., Oien, N.L., Hopkins, T.A., Knechtel, M.L., Brideau, R.J., Wathen, M.W., and Homa, F.L. (2003). Amino acid changes within conserved region III of the herpes simplex virus and human cytomegalovirus DNA polymerases confer resistance to 4-oxo-dihydroquinolines, a novel class of herpesvirus antiviral agents. *J Virol* **77**, 1868–1876.

Wang, J., Sattar, A., Wang, C.C., Karam, J.D., Konigsberg, W.H., and Steitz, T. A. (1997). Crystal structure of a pol alpha family replication DNA polymerase from bacteriophage RB69. *Cell* **89**, 1087–1099.

Wathen, M.W. (2002). Non-nucleoside inhibitors of herpesviruses. *Rev Med Virol* **12**, 167–178.

Weisshart, K., Chow, C.S., and Coen, D.M. (1999). Herpes simplex virus processivity factor UL42 imparts increased DNA-binding specificity to the viral DNA polymerase and decreased dissociation from primer-template without reducing the elongation rate. *J Virol* **73**, 55–66.

Zhao, Y., Jeruzalmi, D., Moarefi, I., Leighton, L., Lasken, R., and Kuriyan, J. (1999). Crystal structure of an archaebacterial DNA polymerase. *Structure* **7**, 1189–1199.

Zuccola, H.J., Filman, D.J., Coen, D.M., and Hogle, J.M. (2000). The crystal structure of an unusual processivity factor, herpes simplex virus UL42, bound to the C terminus of its cognate polymerase. *Mol Cell* **5**, 267–278.

Chapter 18
RNA Virus Polymerases

Cristina Ferrer-Orta and Nuria Verdaguer

Introduction

Genome replication and transcription of RNA viruses are catalyzed by an RNA-dependent RNA polymerase (RdRP). RdRPs synthesize RNA using an RNA template and are normally associated with other virus- or/and host-encoded proteins that modulate RNA polymerization activity and template specificity.

An important feature of RNA-directed RNA replication is the high error frequency compared to DNA-directed replication. This is due, at least in part, to the low fidelity of RdRPs and the absence of error-repair mechanisms in RNA viruses.

The vast majority of RdRPs have been identified on the basis of sequence similarity. Computational analyses identified several motifs common among the putative RdRPs of a broad range of viruses (Koonin, 1991). Catalytic activity has also been demonstrated biochemically in a number of these viral proteins (O'Reilly and Kao, 1998) and actually there is a growing body of information available on polymerase structure and function. To date, the crystal structure of 14 different RdRPs from five families of positive-stranded and double-stranded RNA (dsRNA) viruses has been determined, either isolated or bound to nucleic acid substrates. These include the polymerase NS5B from three members of the Flaviviridae family (Ago et al., 1999; Bressanelli et al., 1999; Choi et al., 2006, 2004; Lesburg et al., 1999; Yap et al., 2007), the RdRPs of three members within the Caliciviridae family (Fullerton et al., 2007; Ng et al., 2002, 2004), the polymerase 3D from three different members of the Picornaviridae family (Appleby et al., 2005; Ferrer-Orta et al., 2006a, 2004; Love et al., 2004; Thompson and Peersen, 2004), the RdRP from the double-stranded RNA bacteriophage ϕ6 (Butcher et al., 2001), and the reovirus λ3 polymerase (Tao et al., 2002). The RdRP structures resemble a cupped right hand composed of "fingers", "palm," and "thumb" sub-domains as in other classes of polymerases such as DNA-dependent RNA polymerases, DNA-dependent DNA polymerases (Klenow fragment), and reverse transcriptases. The right-hand

N. Verdaguer (✉)
Institut de Biologia Molecular de Barcelona (CSIC), Parc Científic de Barcelona, Baldiri i Reixac 15, E-08028 Barcelona, Spain
e-mail: nvmcri@ibmb.csic.es

C.E. Cameron et al. (eds.), *Viral Genome Replication*,
DOI 10.1007/b135974_18,

architecture provides the correct geometrical arrangement of substrate molecules and metal ions at the polymerase active site for catalysis (Brautigam and Steitz, 1998). All nucleic acid polymerases require two divalent cations as cofactors to catalyze phosphoryl transfer. The first metal ion is brought into the active site complexed to the triphosphate moiety of the nucleotide substrate. This metal ion may facilitate formation of the conformation required for nucleophilic attack of the α-phosphorus atom and facilitates the exit of the pyrophosphate product (PPi). The second metal ion is required to lower the pK_a of the primer 3′-hydroxyl group to facilitate formation of the nucleophile required for catalysis. Mg^{2+} is thought to be the divalent cation employed for most polymerases known (Steitz, 1998).

This chapter summarizes structural and biochemical studies of the different RdRPs reported in the past few years

Overall Structure of Viral RDRPs

A unique feature of the RdRP structure is the "closed-hand" conformation, opposed to the "open-hand" found in other polynucleotide polymerases, which is accomplished by interconnecting the fingers and thumb domains through the N-terminal portion of the protein and several loops protruding from fingers, named the fingertips that completely encircle the active site of the enzyme and largely restricts the interdomain mobility (Ferrer-Orta et al., 2006b) (Fig. 18.1). In picornaviruses and flaviviruses the interaction is dominated by the insertion of two aromatic side chains protruding from the fingertips into the hydrophobic pocket at the top of the thumb domain (Thompson et al., 2007). Structural and biochemical analyses of these enzymes suggest that the interdomain interaction between fingers and thumb plays a significant role in the activity of these polymerases (Hobson et al., 2001; Labonte et al., 2002).

Fig. 18.1 Gallery of viral RdRPs, representing one member of the five virus families for which at least one structure is known: (A) RHDV Caliciviridae (PDB id: 1KHV; *top left*), HCV Flaviviridae (PDB id:1NB4; *top right*), bacteriophage ϕ6 Cystoviridae (PDB id: 1HI8; *bottom left*), λ3 Reoviridae (PDB id: 1N35; *bottom right*) and (B) FMDV Picornaviridae (PDB id: 1U09). All molecules are shown in the conventional orientation as if looking into a right hand. In (A), fingers, palm, and thumb domains are colored *blue*, *green*, and *red*, respectively. The fingertips are in *cyan* and the C-terminal protrusions, in HCV and ϕ6 RDRPs, are in *yellow*. Reovirus λ3 polymerase represents the largest RDRP structure determined. In this structure, the fingers, palm, and thumb domains are surrounded by elaborate additional elements derived from long N- and C-terminal extensions in which the figures are also colored in *cyan* and *yellow*, respectively. In (B), the FMDV polymerase is shown in *gray* with the six different conserved structural motifs of palm and fingers domains colored as follows: A, *orange*; B, *green*; C, *red*; D, *orange*; E, *brown*; F, *magenta*; and G, *cyan*. The six conserved residues of the palm that form the nucleotide-binding pocket are shown as *sticks* and *labeled*. The positively charged amino acids of motif F, forming the NTP tunnel, and the conserved Pro and Gly residues of motif G, at the entry of the template channel, are also shown and explicitly labeled.

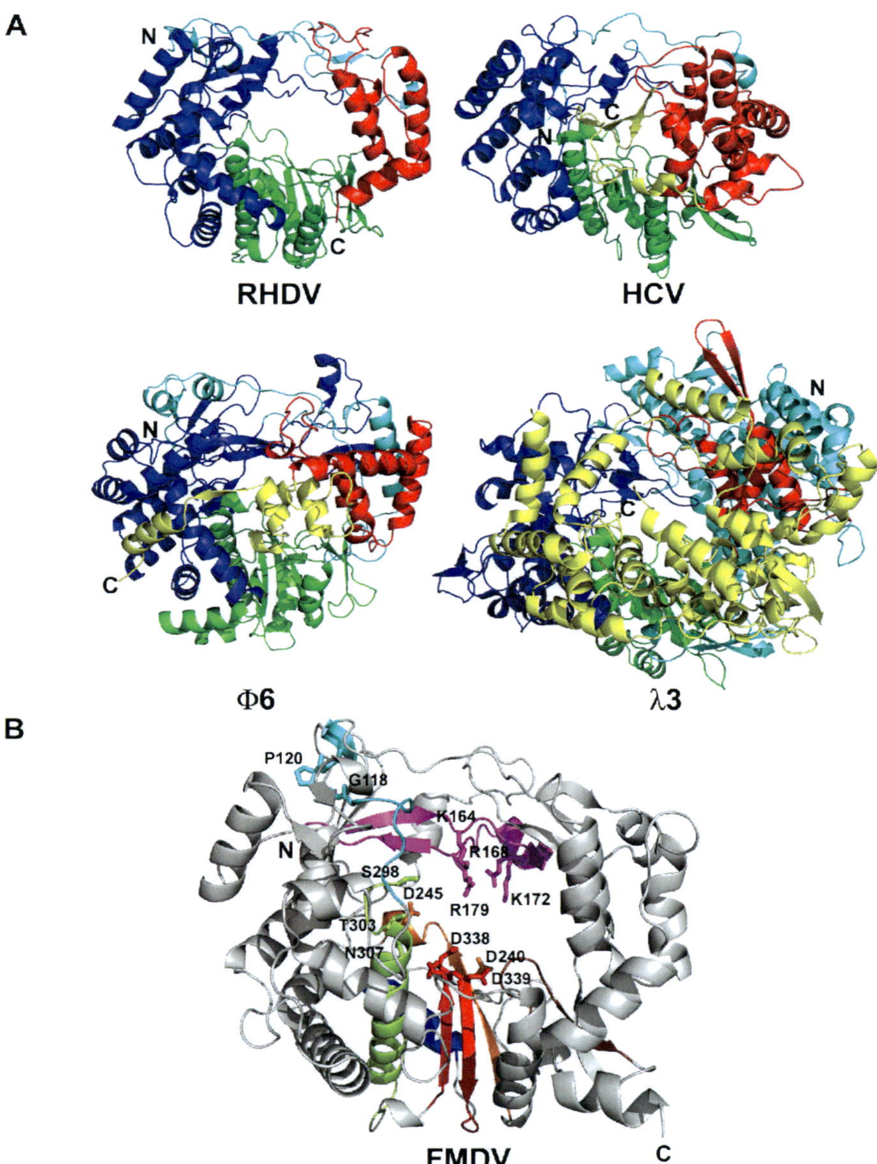

Fig. 18.1 (continued)

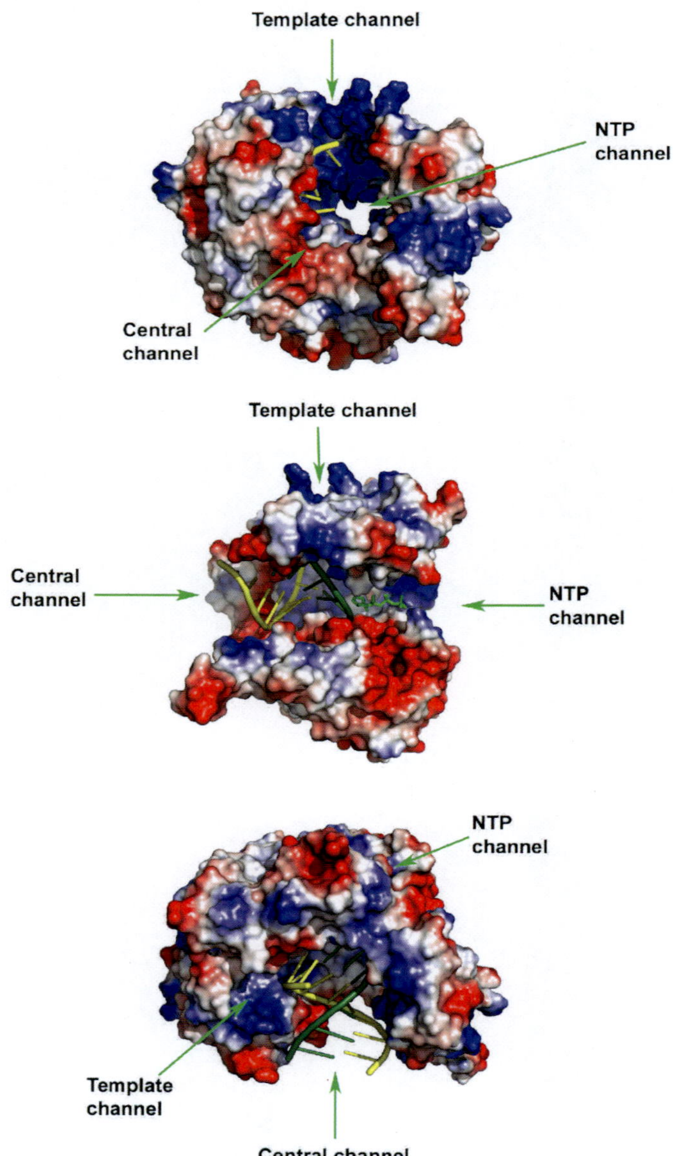

Fig. 18.2 Molecular surfaces of the FMDV 3D polymerase in complex with an RNA template–primer (PDB id: 1WNE) shown in different views: (A) the conventional orientation as if looking into a right hand; (B) side view and (90º clockwise rotation with respect to the orientation in (A)); and (C) top- down view (90º upward rotation with respect to (A)). Surfaces are represented with the electrostatic potential in *blue* for positive charges and *red* for negative charges. The three channels that serve as the entry paths for template (template channel) and for nucleoside triphosphates (NTP channel) and as the exit path for dsRNA product (central channel) are labeled. In (B) and (C), part of the N-terminal and thumb regions are removed to show the substrate cavities.

Three well-defined channels have been identified in the RdRP structures, serving as the entry path for template (template channel) and for nucleoside triphosphates (NTP channel) and the exit path for dsRNA product (central channel) (Fig. 18.2).

Specialized Features Facilitating dsRNA Strand Separation and Template Entry

The structures of the RdRPs from two dsRNA viruses revealed the presence of specialized features that have been proposed to facilitate dsRNA strand separation, providing the basis for the mechanism that ensures feeding of the correct strand to the catalytic site for initiation of RNA synthesis. These include the reovirus 5′-cap-binding site, which was identified on the surface of the λ3 RdRP between the template entrance and exit channels (Tao et al., 2002), and the bacteriophage ϕ6 plough, next to the entry of the template channel (Butcher et al., 2001). The structure of BVDB RdRP also revealed an additional N-terminal region, folded into a separate domain. This domain is located over the thumb and interacts with the fingertip region through a β-hairpin motif. This motif, rich in positive charges, points toward the template channel. It has been proposed that in BVDV polymerase, this positively charged surface may be used to open up RNA secondary structural hairpins before the ssRNA template enters the active center (Choi et al., 2004). Structural comparisons and bioinformatics analysis of the HCV, RHDV, and picornavirus RdRPs also identified a positively charged region at the entrance of the template channel as possibly being involved in unwinding of RNA secondary structures for transcription (Fig. 18.2) (Bruenn, 2003).

Fingers, Palm, and Thumb Sub-Domains

The fingers sub-domain is organized into two regions: an inner region that consists primarily of a bundle of α-helices, surrounding and packed against the palm sub-domain, and an outer region projecting away from the palm. The outer fingers comprise a β-strand-rich region and contain a long insert which, together with the N-terminus of the protein, effectively extends toward the thumb domain forming the fingertip region. The central β-sheet forming the core of the outer fingers is a common feature among RDRPs, although the conformation of the polypeptide chain in fingertips differs considerably among RdRPs (Fig. 18.1). The fingers sub-domain contains two sequence motifs (named F and G), conserved among RdRPs of positive-strand RNA viruses (Bruenn, 2003; Koonin, 1991). Motif G has the consensus sequence $T/SX_{1-2}GP$ (X is a non-conserved residue) and is contained in a loop that forms the template channel entry, allowing the template nucleotides to accede the active site (Fig. 18.2). Mutational experiments in the conserved Gly and Pro of two picornaviruses suggest the critical role of these residues in template binding and translocation of nucleic acid during synthesis (Arias et al., 2005;

Thompson and Peersen, 2004). Motif F is defined as R–X_{1-2}–I/L. The conserved arginine, together with other partially conserved basic residues in the vicinity, forms the rNTP channel (Fig. 18.2). This positively charged tunnel is opened on the back-side of the molecule and serves for nucleotide diffusion, as seen in the structure of different RdRPs in complex with a number of rNTP substrates (Bressanelli et al., 2002, 1999; Butcher et al., 2001; Tao et al., 2002).

The palm sub-domain contains the catalytic site and shows the greatest structural conservation not only in RdRPs but also among all known template-dependent polynucleotide polymerases (TdPPs) (Gorbalenya et al., 2002; O'Reilly and Kao, 1998). The architecture of this region is made of a four-stranded antiparallel core β-sheet flanked by two α-helices in one side and by an additional α-helix on the other side of the β-sheet (Fig. 18.1). The palm sub-domain contains most of the conserved sequence motifs identified for oligonucleotide polymerases by comparative analysis, with A, B, and C being the most prominent. Motif A is located at the end of a β-strand in the central core (Fig. 18.1) and has the consensus sequence $DX_{4-5}D$, while motif C is at the top of a β-hairpin (Fig. 18.1) and contains the GDD tripeptide. In TdPPs other than RdRPs, only the amino terminal Asp residues in motifs A and C are conserved. These aspartic acid residues are spatially juxtaposed, bind divalent cations, and are crucial for catalysis. In RdRPs, the second aspartate of motif A seems to play an important role in the selection of NTPs over 2′-desoxyribonucleotide triphosphates (2′d-NTPs) by hydrogen bonding to the 2′- and 3′-hydroxyl groups of the incoming nucleotide (Gohara et al., 2004) (Fig. 18.3 and see below). Motif B forms an α-helix that packs against one strand of the β-sheet core and contains a conserved Asn residue also involved in the selection of NTPs (Figs. 18.1 and 18.3). RdRPs also share the palm motifs D and E that do not have any consensus sequence. Motif D is comprised by an α-helix and a short loop that bends back around to form the fourth strand of the β-sheet core and provides structural support for motif A (Fig. 18.1). Motif E forms a tight loop which lies at the junction between the palm and thumb sub-domains. The turn of this loop projects into the active site cavity where it has been implicated in helping to position the 3′-end of the primer strand for attack on the α-phosphate of the NTP during phosphoryl transfer (Figs. 18.3 and 18.5B). In the structure of HIV RT, the residues immediately following motif E act as a pivot point for the thumb sub-domain movement upon template–primer binding (Huang et al., 1998). However, in RdRPs, the enclosed active sites make unlikely that such large conformational changes occur. In fact, the structure of the catalytic complexes of reovirus λ3 and FMDV RdRPs revealed that no major domain movements occur in these polymerases when they bind NTPs (Tao et al., 2002) (Ferrer-Orta et al., 2007) and instead, the nucleic acid binding to the fingers domain may play a role in translocation of nucleic acid during synthesis. Despite the closed-hand conformation of RdRPs stabilizes a relatively rigid unit, structural data of RHDV polymerase showed that the thumb domain can rotate, few degrees, relative to the fingers and palm, giving a more open conformation that is believed to be an inactive form of the enzyme (Ng et al., 2002). The recent structure of HCV genotype 2a RdRP also revealed an extreme case of this "open" conformation, where the contact between

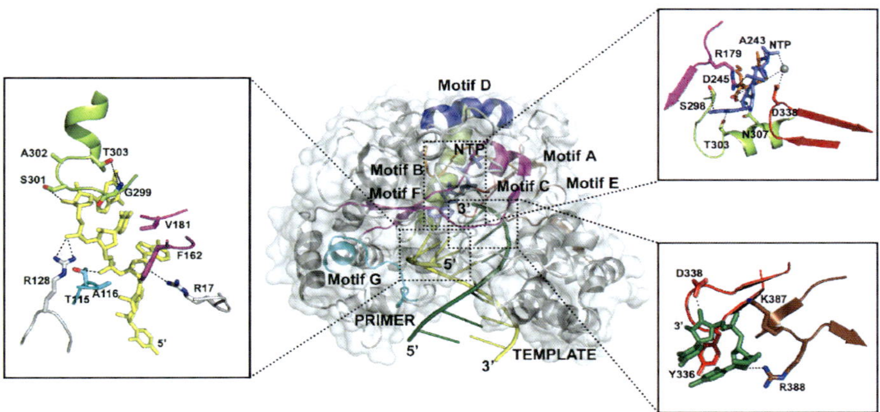

Fig. 18.3 Conserved interactions between the FMDV 3D and the RNA template–primer substrate. The polymerase is shown in *gray* in the central panel with the conserved motifs, involved in contacts with the RNA molecule, highlighted in different colors and explicitly labeled. The template and primer strands of the RNA molecule are shown in *yellow* and *green*, respectively; contacting residues are shown in *sticks* and hydrogen bonds as *dashed lines* in *black*. The 5′-overhang region of the template binds the template channel (*left side panel*), where the different residues of the N-terminal region (*gray*), motif G (*cyan*), and motif F (*magenta*) drive the ssRNA to the active site cavity. In the active site, the position of the template acceptor base is stabilized by different interaction mediated by residues of motif B (*green*). The primer strand interacts with motifs C (*red*) and E (*brown*) of the palm sub-domain, and with different residues in the thumb sub-domain. The incoming NTP is located at the active site, adjacent to the 3′-terminus of the primer, and base paired to the template acceptor base. The position of the NTP base (*top right panel*) is further stabilized by interactions with residues of motif B, while the triphosphate moiety is hydrogen bonded to different residues of motifs A and F and interacts with one metal ion. The 2′-hydroxyl group of the sugar moiety forms a double hydrogen bond with the residues Asp245 of motif A and Asn307 of motif B. The position of the priming nucleotide (*right side panel*) is stabilized by polar and hydrophobic interactions mediated by the catalytic aspartate and a conserved tyrosine of motif C, and by the positively charged residues of motif E which are hydrogen bonded to sugar–phosphate backbone of the primer.

the fingertips and the thumb is partially disrupted (Biswal et al., 2005). Crystallographic studies of complexes between HVC RdRP and various non-nucleoside inhibitors (NNIs) revealed that the different NNIs seem to work by freezing the enzyme in an "open" inactive conformation (De Francesco and Migliaccio, 2005; Di Marco et al., 2005).

The thumb domain, consisting of the C-terminal region of the polypeptide chain, is the most diverse feature among the known viral RdRPs (Fig. 18.1). Picornavirus and calicivirus RdRPs have small thumb domains, built mainly by a four-helix bundle. The small size of the domain contributes to the formation of a large central cleft, of approximately 14 Å across and 22 Å deep, located in the front of the molecule leading to the active site (Figs. 18.1 and 18.2). In contrast, the flavivirus and bacteriophage ϕ6 polymerase thumb domains are significantly larger (Fig. 18.1). The large thumb sub-domain of Flaviviridae RdRPs contain more than twice the number of residues as Picornaviridae 3D polymerases, including three additional α-helices and a β-thumb region which protrudes into the active site (Choi et al., 2004). In addition, ϕ6 and HCV RdRPs contain C-terminal extensions that fold back into the molecule filling most of the active site cavity (Butcher et al., 2001; Leveque et al., 2003) (Fig. 18.1).

Proteolytic Activation of a Picornavirus Polymerase Activity

Picornaviruses are known to use 3CD, an uncleaved precursor of the protease and polymerase generated during polyprotein processing, as a functional intermediate in viral replication (Paul et al., 2000). In poliovirus, the precursor 3CD contains an active 3C protease component while the 3D polymerase remains inactive until protein processing is complete. The crystallographic structure of PV 3D polymerase (Thompson and Peersen, 2004) showed that the N-terminal glycine was buried in a pocket at the base of the fingers region, participating in a network of hydrogen bonds, that was proposed to help in positioning the conserved aspartate of motif A, involved in rNTP selection, into the catalytic site in a correct orientation for interactions with the 2′-hydroxyl group of the incoming NTP. The lack of polymerase activity in poliovirus 3CD was thought to be associated with the removal of the N-terminus of 3D from its binding pocket, as modifications made to the N-terminus of 3D disrupt its polymerase activity (Rothstein et al., 1988; Thompson and Peersen, 2004).

However, the X-ray structure of the PV 3CD protein recently determined (Marcotte et al., 2007) revealed a very similar arrangement of the residues in the active site in both, 3CD and 3Dpol, despite the disruption of a network of interactions proposed to position key residues in the active site. The comparison of the structures suggested that the canonical active site of PV 3D polymerase was mostly preformed and did not require the buried N-terminus. However, in a number of viral polymerases, including all picornaviral RdRPs, the N-terminus and the active site are stabilized via hydrogen bonding networks involving the fingers domain, suggesting that this binding scheme could be relevant for polymerase activity. Putting together all data, Marcotte et al. (2007) proposed that changes in molecular flexibility, rather than large structural rearrangements, would be the determinants of the PV RdRP activity, as observed in other polymerases (Harris et al., 1998; Kim et al., 2006; Kool, 2002). This molecular flexibility would also be regulated by the insertion of the 3Dpol N-terminus in its binding pocket.

Template and Primer Recognition by Picornavirus Polymerases

The structure of the FMDV 3D polymerase in complex with a template–primer RNA showed the first structural evidence of how the physiological substrates bind the large exposed active site of the picornavirus RdRPs (Ferrer-Orta et al., 2004). In the structure, the single-stranded RNA template binds to the template channel, which extends across the face of the fingers domain toward the active site cleft (Figs. 18.2 and 18.3). The basic residues in this channel contact the phosphodiester backbone driving the single-stranded RNA toward the active site cavity. Then, the double-stranded stem that mimics the duplex product stretches from the active site to the C-terminal end of the protein exiting through the large central cavity of the polymerase molecule (Figs. 18.2 and 18.3). The template strand of the duplex product is adjacent to the fingers domain, whereas one α-helix of the thumb domain runs along the phosphodiester backbone of the primer. The structure also revealed the amino acids of FMDV 3D involved in the correct positioning of the template and primer nucleotides (Fig 18.3). The acceptor base of the template strand is located adjacent to the nucleotide-binding site completely accessible to the incoming substrate. In the active site, the 3′-hydroxyl of the primer is hydrogen bonded to the catalytic aspartic acid of motif C. The position of the priming nucleotide is further stabilized by interactions involving a conserved tyrosine residue, also in motif C, and two positively charged residues of motif E (Fig. 18.3). The modeling of an incoming nucleotide substrate in the structure 3D–RNA structure (Ferrer-Orta et al., 2004), as well as the recently determined structures of various 3D–RNA–NTP ternary complexes (Ferrer-Orta et al., 2007 under revision), allowed the identification of different residues in motifs A, B, and F, involved in the recognition and positioning of the incoming substrate for catalysis (Fig. 18.3).

Kinetic and Thermodynamic Mechanism of Nucleotide Incorporation for PV 3Dpol

Cameron and colleagues developed an elegant method to characterize the complete kinetic mechanism for single nucleotide incorporation catalyzed by the PV 3D polymerase (Arnold and Cameron, 2000). Briefly, the analysis revealed that the nucleotide incorporation mechanism would be described by five steps: in step 1, the enzyme–RNA complex (ER_n) binds the incoming NTP to form a ternary complex (ER_nNTP); in step 2, ER_nNTP undergoes a conformational change to reach the competent form for phosphoryl transfer ($^*ER_nNTP$); in step 3, chemistry occurs, forming a ternary product complex ($ER_{n+1}PP_i$); in step 4, the $ER_{n+1}PP_i$ complex isomerizes to form the ternary product complex where the PPi can dissociate; finally, in step 5, PPi dissociates and the ER_{n+1} product complex remains in its competent form for the next cycle of nucleotide incorporation. Further analyses showed that two steps of this mechanism appeared to be rate limiting for the discrimination between correct and incorrect nucleotide incorporation by PV 3D: the conformational change, prior to the phosphoryl transfer (step 2), and the phosphoryl transfer step (step 3) (Arnold and Cameron, 2004; Castro et al., 2005).

Sequence alignments indicated the presence of the six amino acid residues, conserved across all polymerases of positive-strand RNA viruses, mapping to the nucleotide-binding pocket: the two aspartic acids of motif A, the catalytic aspartates of motif C, and the conserved Asn, Ser, and Thr of motif B. These six core residues seemed to be good candidates to interact with the incoming nucleotide substrate (Hansen et al., 1997; Koonin, 1991). Cameron and colleagues determined the importance of these residues for nucleotide selection by site-directed mutagenesis and evaluation of the kinetics of correct and incorrect nucleotide incorporation (Gohara et al., 2004, 2000). All these data led authors to suggest a two-step model for nucleotide binding. In the first step, the incoming nucleotide, in complex with one metal ion, was bound to the polymerase–template–primer complex in a ground-state configuration, where the binding was driven by the metal-complexed triphosphate moiety of the nucleotide. In this state, the ribose moiety of the incoming nucleotide cannot bind in a productive orientation because the interaction between Asp in motif A (Asp_A) and Asn of motif B (Asn_B) occluded the ribose-binding pocket. The mentioned interaction was observed in the crystallographic structures of all picornavirus RdRPs determined in the absence of incoming NTPs (Appleby et al., 2005; Ferrer-Orta et al., 2004; Hansen et al., 1997; Thompson and Peersen, 2004). In the second step, a conformational change occurs to bring the metal-complexed triphosphate moiety into the appropriate position to interact with the conserved catalytic residues of the enzyme, and at the same time, the polymerase organizes the active site for the acceptance of the second metal ion required for catalysis (Arnold and Cameron, 2004; Arnold et al., 2004). Moreover, the stability of the complex in the competent conformation will dictate the efficiency of phosphoryl transfer. This stability would be maintained by an extensive network of hydrogen bonds, involving the polymerase residues lining the ribose-binding pocket and the triphosphate moiety of the substrate (Gohara et al., 2004). Formation of this network required reorientation of Asp_A and Asn_B as well as the interaction of the β-phosphate oxygen with the 3′-hydroxyl group of the NTP. The reoriented residues would stabilize the position of the ribose by direct interactions formed with the 3′- and 2′-hydroxyl groups of the incoming nucleotide (Arnold and Cameron, 2004; Arnold et al., 2004). Then, catalysis occurs. This model is fully supported by the recent structural data of different RNA elongation complexes of the closely related RdRP of FMDV (Ferrer-Orta et al., 2007).

Role of the Divalent Cations

The crucial role of Mg^{2+} ions in the catalysis of phosphodiester bond formation has been discussed at the beginning of this chapter. Furthermore, polymerase activity is also supported by other divalent cations (Arnold et al., 1999; Tabor and Richardson, 1989). Several specific regulatory effects of different metal ions on viral RNA synthesis have been described. In particular, Mn^{2+} is known be an effective divalent cation cofactor for a number of RdRPs including those of HCV (Alaoui-Lsmaili et al., 2000; Zhong et al., 2000), PV (Arnold et al., 1999), and different members of the Cystoviridae family (van Dijk et al., 2004). However, this metal usually alters the biochemical properties of the polymerase, decreasing the stringency of substrate selection and incorporation fidelity (Arnold et al., 1999, 2004; Beckman et al., 1985; Goodman et al., 1983; Huang et al., 1997; Tabor and Richardson, 1989) provided insights into the molecular bases for the destructive effects of Mn^{2+}

on PV polymerase fidelity. Briefly, they found that, by using Mg^{2+} as the divalent cation cofactor, PV 3Dpol can use both the first conformational change step and the second phosphoryl transfer step to distinguish between correct and incorrect nucleotides. However, by using Mn^{2+} as the cofactor, the ability to diminish the rate of phosphoryl transfer for incorrect nucleotides relative to correct nucleotides is lost completely, leaving only the conformational change step for selection of correct nucleotide.

Ca^{2+} is known to inhibit in vitro transcription of reovirus (Sargent and Borsa, 1984) and bacteriophage $\phi 6$ (van Dijk et al., 1995). The crystal structures of $\phi 6$ RdRP initiation complexes with either Mg^{2+} or Ca^{2+} revealed key differences that may explain the inhibitory effect of Ca^{2+} (Butcher et al., 2001; Salgado et al., 2004). In the inhibition complex, the two Mg^{2+} ions that are present in the initiation complex are substituted by Ca^{2+} ions. One of the Ca^{2+} occupies a position equivalent to the corresponding Mg^{2+} in the initiation complex. The other Ca^{2+} has a different coordination sphere from the equivalent Mg^{2+}, altering the geometry of interactions in the catalytic position.

Structural Features Facilitating Initiation of RNA Synthesis

Correct initiation of RNA synthesis is essential for the integrity of the viral genome. There are two main mechanisms by which viral replication can be initiated: primer-independent or de novo, and primer-dependent initiation (Kao et al., 2001; van Dijk et al., 2004). Briefly, in the de novo synthesis, one initiation nucleotide provides the 3′-hydroxyl for the addition of the next nucleotide whereas the primer-dependent initiation requires the use of either an oligonucleotide or a protein primer as provider of the hydroxyl nucleophile. RNA viruses can use either one or sometimes both of these mechanisms for initiation of RNA synthesis. De novo is used by viruses with positive, negative, dsRNA, and ambisense RNA genomes (Kao et al., 2001). Specific examples include the dsRNA viruses such as the Cystoviridae (Makeyev and Bamford, 2000a,b; Yang et al., 2003) and rotavirus (Chen and Patton, 2000), and negative-strand RNA viruses such as vesicular stomatitis virus (VSV) (Testa and Banerjee, 1979). Positive-strand RNA viruses that use de novo initiation include plant alphavirus-like virus (Strauss and Strauss, 1994) and members of the Flaviviridae family, as hepatitis C virus and dengue 2 (Ackermann and Padmanabhan, 2001; Kao et al., 2000). Influenza virus employs a combination of the two mechanisms with the choice being determined by the type of RNA to be synthesized (Honda et al., 1986) while the members of the Picornaviridae family use exclusively the protein-primed mechanism of initiation. In this process, a tyrosine residue provides the hydroxyl group for the formation of a phosphodiester bond with the first nucleotide (Paul et al., 1998).

Comparisons among different RNA-dependent RNA polymerases whose structures have been solved show that those viruses which follow a primer-dependent mechanism of initiation of replication, as picornaviruses and caliciviruses, have a more accessible active site than viruses with a de novo initiation mechanism,

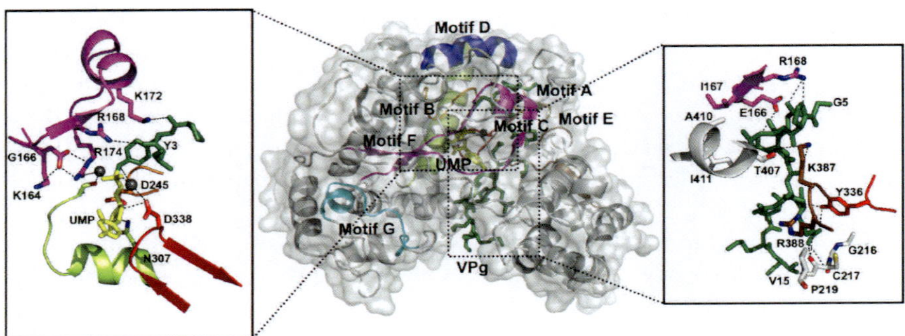

Fig. 18.4 Structure and interactions of the FMDV 3D–VPg–UMP complex (PDB id: 2F8E). The FMDV polymerase is shown in *gray*, the primer protein VPg in *green* and UMP in *yellow*. VPg lines the RNA-binding cleft of the 3D polymerase, positioning its Tyr3 hydroxyl group as a molecular mimic of the free 3′-hydroxyl group of a nucleic acid primer at the active site for nucleotidylylation. (*Left panel*) In the active site, the hydroxyl group of Tyr3 side chain was found covalently attached to a UMP molecule by a phosphodiester linkage (*yellow*). Two metal ions (*gray spheres*) participate in the uridylylation reaction. Metal 1 bridges the catalytic aspartate of motif C (*red*) and the O^- of tyrosine side chain, now covalently bonded to phosphate α of UMP. Metal 2 coordinates the carboxylic group of the catalytic aspartate of motif A (*orange*), the O1 oxygen of phosphate α and the hydroxyl group of the conserved serine of motif B (*green*). The conserved tyrosine of motif C and the positively charged residues of motif F (*magenta*) also participate in the uridylylation process. (*Right panel*) In addition to the interactions in the polymerase active site, different residues of motifs F (*magenta*) and E (*brown*), together with residues within the first helix of the thumb sub-domain interact with the central part of VPg. Finally, the FMDV 3D residues Gly216, Cys217, and Pro219 (*gray*), in the fingers sub-domain, establish hydrophobic contacts with VPg at the exit of the polymerase cavity.

as flaviviruses, reoviruses, and bacteriophage ϕ6 (Ferrer-Orta et al., 2006b). The structure of the FMDV 3D polymerase in complex with its protein primer VPg evidenced how the wide central channel of picornavirus RDRPs is able to accommodate the primer during the initiation stage of replication (Ferrer-Orta et al., 2006a) (Fig. 18.4). The same cavity can also accommodate a template–primer duplex during the phase of RNA elongation (Ferrer-Orta et al., 2004) (Fig. 18.3). In contrast, protruding extensions of the thumb domain of flaviviruses and ϕ6 polymerases partially occlude the active site resulting in a more compact molecule where two narrow positively charged tunnels allow the access of RNA template and NTP substrate to the active site (Fig. 18.5A).

De novo Initiation

Structural and biochemical studies indicated that the C-terminal protrusions of ϕ6 and HCV RdRPs have three distinct functions: (i) stabilize the initiation complexes by interacting with initiating nucleotides; (ii) prevent undesirable back-priming reaction by physically separating the template-binding site from the room reserved

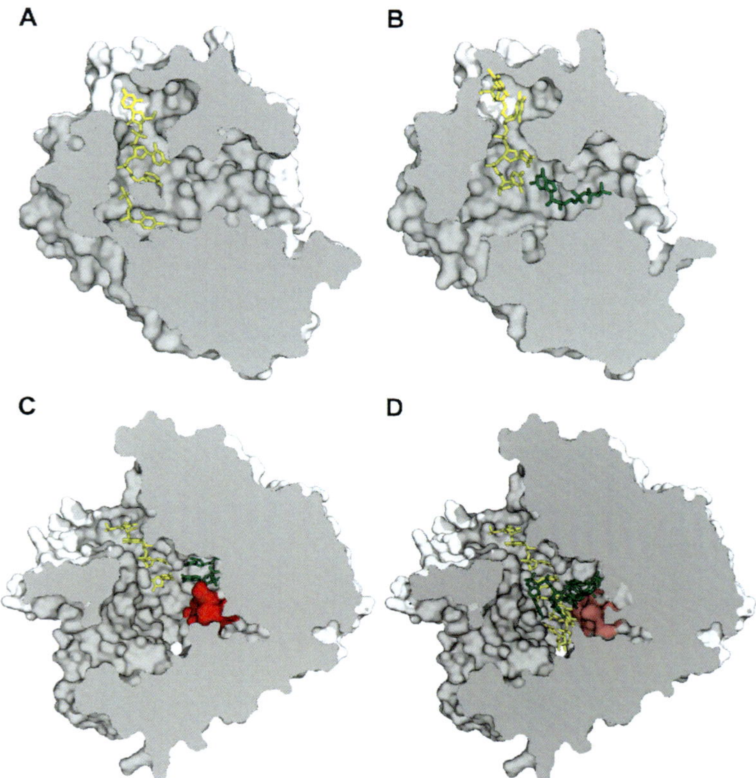

Fig. 18.5 Structural features facilitating de novo initiation of RNA synthesis in φ6 and λ3 RdRPs. The polymerase molecules are represented as sliced molecular surfaces (*gray*) to better show the substrate cavities. RNA template (*yellow*) and priming nucleotides (*green*) are shown in *sticks*. (A) The φ6 RdRP has two positively charged tunnels that, respectively, allow the access of the RNA template and NTP substrates to the active site. The central channel in φ6 is blocked by the C-terminal domain which acts as an initiation platform. From the structure of φ6–DNA complex, Butcher et al. (2001) proposed a sequence of events that could result in the formation of the initiation complex. Template enters the channel and interacts with a specific pocket (PDB id: 1HI0; *left panel*); the incoming NTP binds to the initiation platform and the template ratchets back freeing the template nucleotide 3′-end from the specific pocket (PDB id: 1UVK; *right panel*). Then a second NTP enters to lock the initiation complex into its active form. (B) The reovirus RdRP has a special priming loop (*red*) that supports the stacking of the priming NTP in the initiation complex (PDB id: 1N1H; *left panel*). In the fully active polymerase elongation complex (PDB id: 1N35; *right panel*), this loop (*salmon*) retracts toward the palm with respect to its position in the apo-enzyme, an initiation complex to fit into the minor groove of the product duplex.

for the daughter RNA chain; and (iii) serve as a physical barrier to block the exit of the template tunnel during initiation (Butcher et al., 2001; Leveque et al., 2003; van Dijk et al., 2004) (Fig. 18.5A). Furthermore, these initiation platforms block the path of the elongating RNA product at the level of two or three nucleotides and large conformational rearrangements are required to accommodate longer product chains. These conformational changes would mark the transition from the initiation to the elongation. The reovirus λ3 RdRP also has a special priming loop that supports the stacking of the priming NTP in the formation of the initiation complex (Tao et al., 2002). This loop appears as an insertion within the palm domain and has not been observed in any other polymerase. The structure of a λ3 RdRP elongation complex showed how upon the formation of the first phosphodiester bond, the priming loop retracts toward the palm with respect to its position in the apo-enzyme and in the initiation complex to fit into the minor groove of the product duplex (Fig. 18.5B). This allows the newly synthesized RNA to exit the polymerase and facilitates the transition between the initiation and elongation stages of RNA synthesis (Tao et al., 2002).

Primer-Dependent Initiation in Picornaviruses

Picornavirus RNA replication is initiated by the successive attachment of two UMP molecules to the hydroxyl group of a tyrosine in the terminal protein VPg (3B) (Paul et al., 1998). This reaction is catalyzed by 3Dpol, using an AA-containing RNA template either from the 3′-poly(A) tail or the *cis*-acting replication element (Cre). The precursor 3CD also contribute to the process. In PV, VPg uridylylation templated by Cre is stimulated 20-fold by the addition of 3CD and 10-fold by adding 3C (Pathak et al., 2002; Paul et al., 2000). Furthermore, the uncleaved precursor 3BC is a ninefold better substrate for uridylylation than 3B alone. The uncleaved precursor 3BDC is also a substrate for uridylylation, albeit a poorer one than 3B or 3BC (Marcotte et al., 2007).

Three models for uridylylation have recently been published. Two of the models, based on mutational and computational studies in PV, assume that VPg and the VPg precursor 3AB bind to the same surface of 3Dpol and that VPg enters the polymerases active site from the backside of the polymerase molecule through the NTP channel (Schein et al., 2006; Tellez et al., 2006). In contrast, structural and functional evidences from the FMDV 3D–VPg complexes show an alternative arrangement, where VPg accesses the polymerase catalytic site from the front face of the enzyme through the central RNA-binding cleft (Ferrer-Orta et al., 2006a) (Fig. 18.4). The structure of two complexes between FMDV 3D and VPg is actually available showing both the uridylylated and non-uridylylated forms of VPg. In the two structures, VPg adopts almost the same conformation with little secondary structure (Fig. 18.4). The N-terminal portion of the protein is located close to the NTP channel and projects the side chain of residue Tyr3 into the active site, then the peptide chain snakes through the large RNA-binding cleft toward the thumb domain of 3Dpol, following a similar trajectory to that taken by the RNA primer and duplex product in the 3D–RNA complex (Ferrer-Orta et al., 2006a, 2004) (Fig. 18.4). Conserved residues in the fingers and thumb domains of the polymerase were identified as being responsible for stabilizing VPg in its binding cavity. In the 3D–VPg–UMP complex, the hydroxyl group of Tyr3 side chain was found covalently attached to the UMP molecule by a phosphodiester linkage. Two divalent cations participate in the uridylylation reaction that appears to follow a similar mechanism to that described for the nucleotidyl transfer reaction in other polymerases (Steitz, 1998). The positively charged residues of motif F also participate in the uridylylation process, stabilizing Tyr3 and UMP in a proper conformation for the catalytic reaction (Fig. 18.4).

Mutational analysis of the conserved FMDV 3D residues that strongly interact with VPg; in particular, charged amino acids of motifs E and F and the catalytic Asp of motif C show a drastic defect in VPg uridylylation (Ferrer-Orta et al., 2006a).

The crystal structure of the PV precursor 3CD fully supports the FMDV model (Marcotte et al., 2007). VPg binding to FMDV 3D is likely to be highly analogous to the binding in PV 3D; the interacting residues are strictly conserved and the FMDV enzyme can use the VPg of PV as its substrate (Nayak et al., 2005). Furthermore, mutations made to different PV 3D residues of motifs C and E, which correspond to FMDV residues actively involved in VPg binding, result in an almost complete loss of uridylylation activity (Lyle et al., 2002). In the 3CD crystal structure, the 3D domain makes extensive contacts with the 3C and 3D domains of neighboring molecules and the N-terminus of 3C lies close to the VPg-binding site, as VPg

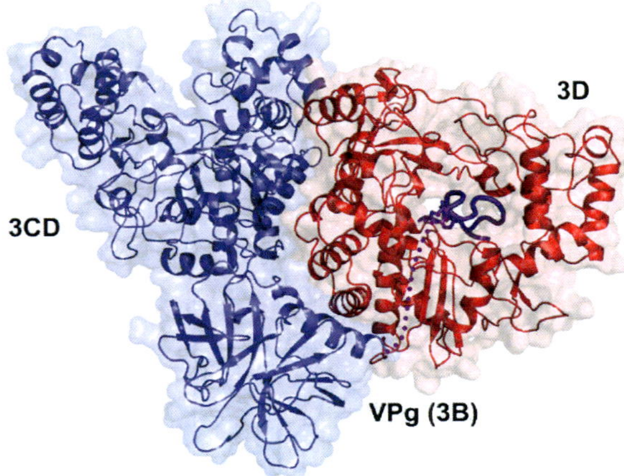

Fig. 18.6 Proposed model of the uridylylation complex from the crystal structures of the PV protein 3CD (PDB id: 2IJD) and the FMDV 3D–VPg complex (PDB id: 2F8E). In the 3CD crystal structure, each 3D domain (*red*) forms extensive interfaces of contact with the 3C and 3D domains of a symmetry-related 3CD molecule (*blue*). Several conserved residues appeared to stabilize these interfaces. These packing interactions showed that the N-terminus of one 3C domain was located close to the VPg-binding site of an adjacent 3D domain of another 3CD molecule. Assuming that the observed interfaces were relevant to 3CD-stimulated uridylylation reaction, authors found only one possible way to connect the VPg-binding site to the N-terminus of 3C by linking across the front face of 3Dpol (*discontinuous lines* in *purple*). VPg (*purple*) has been modeled into the 3D domain using the FMDV 3D–VPg structure as a template.

was seen in the FMDV polymerase complex (Marcotte et al., 2007) (Fig. 18.6). The observed arrangement suggests a possible biological role of the contacting interfaces in forming and regulating the VPg uridylylation complex.

References

Ackermann, M., and Padmanabhan, R. (2001). De novo synthesis of RNA by the dengue virus RNA-dependent RNA polymerase exhibits temperature dependence at the initiation but not elongation phase. J Biol Chem *276*, 39926–39937.

Ago, H., Adachi, T., Yoshida, A., Yamamoto, M., Habuka, N., Yatsunami, K., and Miyano, M. (1999). Crystal structure of the RNA-dependent RNA polymerase of hepatitis C virus. Structure *7*, 1417–1426.

Alaoui-Lsmaili, M. H., Hamel, M., L'Heureux, L., Nicolas, O., Bilimoria, D., Labonte, P., Mounir, S., and Rando, R. F. (2000). The hepatitis C virus NS5B RNA-dependent RNA polymerase activity and susceptibility to inhibitors is modulated by metal cations. J Hum Virol *3*, 306–316.

Appleby, T. C., Luecke, H., Shim, J. H., Wu, J. Z., Cheney, I. W., Zhong, W., Vogeley, L., Hong, Z., and Yao, N. (2005). Crystal structure of complete rhinovirus RNA polymerase suggests front loading of protein primer. J Virol *79*, 277–288.

Arias, A., Agudo, R., Ferrer-Orta, C., Perez-Luque, R., Airaksinen, A., Brocchi, E., Domingo, E., Verdaguer, N., and Escarmis, C. (2005). Mutant viral polymerase in the transition of virus to error catastrophe identifies a critical site for RNA binding. J Mol Biol *353*, 1021–1032.

Arnold, J. J., and Cameron, C. E. (2000). Poliovirus RNA-dependent RNA polymerase (3D(pol)). Assembly of stable, elongation-competent complexes by using a symmetrical primer-template substrate (sym/sub). J Biol Chem *275*, 5329–5336.

Arnold, J. J., and Cameron, C. E. (2004). Poliovirus RNA-dependent RNA polymerase (3Dpol): pre-steady-state kinetic analysis of ribonucleotide incorporation in the presence of Mg2+. Biochemistry *43*, 5126–5137.

Arnold, J. J., Ghosh, S. K., and Cameron, C. E. (1999). Poliovirus RNA-dependent RNA polymerase (3D(pol)). Divalent cation modulation of primer, template, and nucleotide selection. J Biol Chem *274*, 37060–37069.

Arnold, J. J., Gohara, D. W., and Cameron, C. E. (2004). Poliovirus RNA-dependent RNA polymerase (3Dpol): pre-steady-state kinetic analysis of ribonucleotide incorporation in the presence of Mn2+. Biochemistry *43*, 5138–5148.

Beckman, R. A., Mildvan, A. S., and Loeb, L. A. (1985). On the fidelity of DNA replication: manganese mutagenesis in vitro. Biochemistry *24*, 5810–5817.

Biswal, B. K., Cherney, M. M., Wang, M., Chan, L., Yannopoulos, C. G., Bilimoria, D., Nicolas, O., Bedard, J., and James, M. N. (2005). Crystal structures of the RNA-dependent RNA polymerase genotype 2a of hepatitis C virus reveal two conformations and suggest mechanisms of inhibition by non-nucleoside inhibitors. J Biol Chem *280*, 18202–18210.

Brautigam, C. A., and Steitz, T. A. (1998). Structural and functional insights provided by crystal structures of DNA polymerases and their substrate complexes. Curr Opin Struct Biol *8*, 54–63.

Bressanelli, S., Tomei, L., Rey, F. A., and De Francesco, R. (2002). Structural analysis of the hepatitis C virus RNA polymerase in complex with ribonucleotides. J Virol *76*, 3482–3492.

Bressanelli, S., Tomei, L., Roussel, A., Incitti, I., Vitale, R. L., Mathieu, M., De Francesco, R., and Rey, F. A. (1999). Crystal structure of the RNA-dependent RNA polymerase of hepatitis C virus. Proc Natl Acad Sci USA *96*, 13034–13039.

Bruenn, J. A. (2003). A structural and primary sequence comparison of the viral RNA-dependent RNA polymerases. Nucleic Acids Res *31*, 1821–1829.

Butcher, S. J., Grimes, J. M., Makeyev, E. V., Bamford, D. H., and Stuart, D. I. (2001). A mechanism for initiating RNA-dependent RNA polymerization. Nature *410*, 235–240.

Castro, C., Arnold, J. J., and Cameron, C. E. (2005). Incorporation fidelity of the viral RNA-dependent RNA polymerase: a kinetic, thermodynamic and structural perspective. Virus Res *107*, 141–149.

Chen, D., and Patton, J. T. (2000). De novo synthesis of minus strand RNA by the rotavirus RNA polymerase in a cell-free system involves a novel mechanism of initiation. Rna *6*, 1455–1467.

Choi, K. H., Gallei, A., Becher, P., and Rossmann, M. G. (2006). The structure of bovine viral diarrhea virus RNA-dependent RNA polymerase and its amino-terminal domain. Structure *14*, 1107–1113.

Choi, K. H., Groarke, J. M., Young, D. C., Kuhn, R. J., Smith, J. L., Pevear, D. C., and Rossmann, M. G. (2004). The structure of the RNA-dependent RNA polymerase from bovine viral diarrhea virus establishes the role of GTP in de novo initiation. Proc Natl Acad Sci USA *101*, 4425–4430.

De Francesco, R., and Migliaccio, G. (2005). Challenges and successes in developing new therapies for hepatitis C. Nature *436*, 953–960.

Di Marco, S., Volpari, C., Tomei, L., Altamura, S., Harper, S., Narjes, F., Koch, U., Rowley, M., De Francesco, R., Migliaccio, G., and Carfi, A. (2005). Interdomain communication in hepatitis C virus polymerase abolished by small molecule inhibitors bound to a novel allosteric site. J Biol Chem *280*, 29765–29770.

Ferrer-Orta, C., Arias, A., Pérez-Luque, R., Escarmís, C., Domingo, E., and Verdaguer N. (2007). Sequential structures provide insights into the fidelity of RNA replication. Proc Natl Acad Sci U S A. *104*, 9463–9468.

Ferrer-Orta, C., Arias, A., Agudo, R., Perez-Luque, R., Escarmis, C., Domingo, E., and Verdaguer, N. (2006a). The structure of a protein primer-polymerase complex in the initiation of genome replication. Embo J *25*, 880–888.

Ferrer-Orta, C., Arias, A., Escarmis, C., and Verdaguer, N. (2006b). A comparison of viral RNA-dependent RNA polymerases. Curr Opin Struct Biol *16*, 27–34.

Ferrer-Orta, C., Arias, A., Perez-Luque, R., Escarmis, C., Domingo, E., and Verdaguer, N. (2004). Structure of foot-and-mouth disease virus RNA-dependent RNA polymerase and its complex with a template–primer RNA. J Biol Chem *279*, 47212–47221.

Fullerton, S. W., Blaschke, M., Coutard, B., Gebhardt, J., Gorbalenya, A., Canard, B., Tucker, P. A., and Rohayem, J. (2007). Structural and functional characterization of sapovirus RNA-dependent RNA polymerase. J Virol *81*, 1858–1871.

Gohara, D. W., Arnold, J. J., and Cameron, C. E. (2004). Poliovirus RNA-dependent RNA polymerase (3Dpol): kinetic, thermodynamic, and structural analysis of ribonucleotide selection. Biochemistry *43*, 5149–5158.

Gohara, D. W., Crotty, S., Arnold, J. J., Yoder, J. D., Andino, R., and Cameron, C. E. (2000). Poliovirus RNA-dependent RNA polymerase (3Dpol): structural, biochemical, and biological analysis of conserved structural motifs A and B. J Biol Chem *275*, 25523–25532.

Goodman, M. F., Keener, S., Guidotti, S., and Branscomb, E. W. (1983). On the enzymatic basis for mutagenesis by manganese. J Biol Chem *258*, 3469–3475.

Gorbalenya, A. E., Pringle, F. M., Zeddam, J. L., Luke, B. T., Cameron, C. E., Kalmakoff, J., Hanzlik, T. N., Gordon, K. H., and Ward, V. K. (2002). The palm subdomain-based active site is internally permuted in viral RNA-dependent RNA polymerases of an ancient lineage. J Mol Biol *324*, 47–62.

Hansen, J. L., Long, A. M., and Schultz, S. C. (1997). Structure of the RNA-dependent RNA polymerase of poliovirus. Structure *5*, 1109–1122.

Harris, D., Kaushik, N., Pandey, P. K., Yadav, P. N., and Pandey, V. N. (1998). Functional analysis of amino acid residues constituting the dNTP binding pocket of HIV-1 reverse transcriptase. J Biol Chem *273*, 33624–33634.

Hobson, S. D., Rosenblum, E. S., Richards, O. C., Richmond, K., Kirkegaard, K., and Schultz, S. C. (2001). Oligomeric structures of poliovirus polymerase are important for function. Embo J *20*, 1153–1163.

Honda, A., Mizumoto, K., and Ishihama, A. (1986). RNA polymerase of influenza virus. Dinucleotide-primed initiation of transcription at specific positions on viral RNA. J Biol Chem *261*, 5987–5991.

Huang, H., Chopra, R., Verdine, G. L., and Harrison, S. C. (1998). Structure of a covalently trapped catalytic complex of HIV-1 reverse transcriptase: implications for drug resistance. Science *282*, 1669–1675.

Huang, Y., Beaudry, A., McSwiggen, J., and Sousa, R. (1997). Determinants of ribose specificity in RNA polymerization: effects of Mn2+ and deoxynucleoside monophosphate incorporation into transcripts. Biochemistry *36*, 13718–13728.

Kao, C. C., Singh, P., and Ecker, D. J. (2001). De novo initiation of viral RNA-dependent RNA synthesis. Virology *287*, 251–260.

Kao, C. C., Yang, X., Kline, A., Wang, Q. M., Barket, D., and Heinz, B. A. (2000). Template requirements for RNA synthesis by a recombinant hepatitis C virus RNA-dependent RNA polymerase. J Virol *74*, 11121–11128.

Kim, T. W., Brieba, L. G., Ellenberger, T., and Kool, E. T. (2006). Functional evidence for a small and rigid active site in a high fidelity DNA polymerase: probing T7 DNA polymerase with variably sized base pairs. J Biol Chem *281*, 2289–2295.

Kool, E. T. (2002). Active site tightness and substrate fit in DNA replication. Annu Rev Biochem *71*, 191–219.

Koonin, E. V. (1991). The phylogeny of RNA-dependent RNA polymerases of positive-strand RNA viruses. J Gen Virol *72 (Pt 9)*, 2197–2206.

Labonte, P., Axelrod, V., Agarwal, A., Aulabaugh, A., Amin, A., and Mak, P. (2002). Modulation of hepatitis C virus RNA-dependent RNA polymerase activity by structure-based site-directed mutagenesis. J Biol Chem *277*, 38838–38846.

Lesburg, C. A., Cable, M. B., Ferrari, E., Hong, Z., Mannarino, A. F., and Weber, P. C. (1999). Crystal structure of the RNA-dependent RNA polymerase from hepatitis C virus reveals a fully encircled active site. Nat Struct Biol *6*, 937–943.

Leveque, V. J., Johnson, R. B., Parsons, S., Ren, J., Xie, C., Zhang, F., and Wang, Q. M. (2003). Identification of a C-terminal regulatory motif in hepatitis C virus RNA-dependent RNA polymerase: structural and biochemical analysis. J Virol *77*, 9020–9028.

Love, R. A., Maegley, K. A., Yu, X., Ferre, R. A., Lingardo, L. K., Diehl, W., Parge, H. E., Dragovich, P. S., and Fuhrman, S. A. (2004). The crystal structure of the RNA-dependent RNA polymerase from human rhinovirus: a dual function target for common cold antiviral therapy. Structure *12*, 1533–1544.

Lyle, J. M., Clewell, A., Richmond, K., Richards, O. C., Hope, D. A., Schultz, S. C., and Kirkegaard, K. (2002). Similar structural basis for membrane localization and protein priming by an RNA-dependent RNA polymerase. J Biol Chem *277*, 16324–16331.

Makeyev, E. V., and Bamford, D. H. (2000a). The polymerase subunit of a dsRNA virus plays a central role in the regulation of viral RNA metabolism. Embo J *19*, 6275–6284.

Makeyev, E. V., and Bamford, D. H. (2000b). Replicase activity of purified recombinant protein P2 of double-stranded RNA bacteriophage phi6. Embo J *19*, 124–133.

Marcotte, L. L., Wass, A. B., Gohara, D. W., Pathak, H. B., Arnold, J. J., Filman, D. J., Cameron, C. E., and Hogle, J. M. (2007). Crystal structure of poliovirus 3CD protein: virally encoded protease and precursor to the RNA-dependent RNA polymerase. J Virol *81*, 3583–3596.

Nayak, A., Goodfellow, I. G., and Belsham, G. J. (2005). Factors required for the Uridylylation of the foot-and-mouth disease virus 3B1, 3B2, and 3B3 peptides by the RNA-dependent RNA polymerase (3Dpol) in vitro. J Virol *79*, 7698–7706.

Ng, K. K., Cherney, M. M., Vazquez, A. L., Machin, A., Alonso, J. M., Parra, F., and James, M. N. (2002). Crystal structures of active and inactive conformations of a caliciviral RNA-dependent RNA polymerase. J Biol Chem *277*, 1381–1387.

Ng, K. K., Pendas-Franco, N., Rojo, J., Boga, J. A., Machin, A., Alonso, J. M., and Parra, F. (2004). Crystal structure of norwalk virus polymerase reveals the carboxyl terminus in the active site cleft. J Biol Chem *279*, 16638–16645.

O'Reilly, E. K., and Kao, C. C. (1998). Analysis of RNA-dependent RNA polymerase structure and function as guided by known polymerase structures and computer predictions of secondary structure. Virology *252*, 287–303.

Pathak, H. B., Ghosh, S. K., Roberts, A. W., Sharma, S. D., Yoder, J. D., Arnold, J. J., Gohara, D. W., Barton, D. J., Paul, A. V., and Cameron, C. E. (2002). Structure-function relationships of the RNA-dependent RNA polymerase from poliovirus (3Dpol). A surface of the primary oligomerization domain functions in capsid precursor processing and VPg uridylylation. J Biol Chem *277*, 31551–31562.

Paul, A. V., Rieder, E., Kim, D. W., van Boom, J. H., and Wimmer, E. (2000). Identification of an RNA hairpin in poliovirus RNA that serves as the primary template in the in vitro uridylylation of VPg. J Virol *74*, 10359–10370.

Paul, A. V., van Boom, J. H., Filippov, D., and Wimmer, E. (1998). Protein-primed RNA synthesis by purified poliovirus RNA polymerase. Nature *393*, 280–284.

Rothstein, M. A., Richards, O. C., Amin, C., and Ehrenfeld, E. (1988). Enzymatic activity of poliovirus RNA polymerase synthesized in Escherichia coli from viral cDNA. Virology *164*, 301–308.

Salgado, P. S., Makeyev, E. V., Butcher, S. J., Bamford, D. H., Stuart, D. I., and Grimes, J. M. (2004). The structural basis for RNA specificity and Ca2+ inhibition of an RNA-dependent RNA polymerase. Structure *12*, 307–316.

Sargent, M. D., and Borsa, J. (1984). Effects of Ca2+ and Mg2+ on the switch-on of transcriptase function in reovirus in vitro. Can J Biochem Cell Biol *62*, 162–169.

Schein, C. H., Oezguen, N., Volk, D. E., Garimella, R., Paul, A., and Braun, W. (2006). NMR structure of the viral peptide linked to the genome (VPg) of poliovirus. Peptides *27*, 1676–1684.

Steitz, T. A. (1998). A mechanism for all polymerases. Nature *391*, 231–232.

Strauss, J. H., and Strauss, E. G. (1994). The alphaviruses: gene expression, replication, and evolution. Microbiol Rev *58*, 491–562.

Tabor, S., and Richardson, C. C. (1989). Effect of manganese ions on the incorporation of dideoxynucleotides by bacteriophage T7 DNA polymerase and Escherichia coli DNA polymerase I. Proc Natl Acad Sci USA *86*, 4076–4080.

Tao, Y., Farsetta, D. L., Nibert, M. L., and Harrison, S. C. (2002). RNA synthesis in a cage – structural studies of reovirus polymerase lambda3. Cell *111*, 733–745.

Tellez, A. B., Crowder, S., Spagnolo, J. F., Thompson, A. A., Peersen, O. B., Brutlag, D. L., and Kirkegaard, K. (2006). Nucleotide channel of RNA-dependent RNA polymerase used for intermolecular uridylylation of protein primer. J Mol Biol *357*, 665–675.

Testa, D., and Banerjee, A. K. (1979). Initiation of RNA synthesis in vitro by vesicular stomatitis virus. Role of ATP. J Biol Chem *254*, 2053–2058.

Thompson, A. A., Albertini, R. A., and Peersen, O. B. (2007). Stabilization of poliovirus polymerase by NTP binding and fingers-thumb interactions. J Mol Biol *366*, 1459–1474.

Thompson, A. A., and Peersen, O. B. (2004). Structural basis for proteolysis-dependent activation of the poliovirus RNA-dependent RNA polymerase. Embo J *23*, 3462–3471.

van Dijk, A. A., Frilander, M., and Bamford, D. H. (1995). Differentiation between minus- and plus-strand synthesis: polymerase activity of dsRNA bacteriophage phi 6 in an in vitro packaging and replication system. Virology *211*, 320–323.

van Dijk, A. A., Makeyev, E. V., and Bamford, D. H. (2004). Initiation of viral RNA-dependent RNA polymerization. J Gen Virol *85*, 1077–1093.

Yang, H., Makeyev, E. V., Butcher, S. J., Gaidelyte, A., and Bamford, D. H. (2003). Two distinct mechanisms ensure transcriptional polarity in double-stranded RNA bacteriophages. J Virol *77*, 1195–1203.

Yap, T. L., Xu, T., Chen, Y. L., Malet, H., Egloff, M. P., Canard, B., Vasudevan, S. G., and Lescar, J. (2007). The crystal Structure of the Dengue virus RNA-dependent RNA polymerase catalytic domain at 1.85 Å resolution. J Virol *81*, 4753–4765.

Zhong, W., Ferrari, E., Lesburg, C. A., Maag, D., Ghosh, S. K., Cameron, C. E., Lau, J. Y., and Hong, Z. (2000). Template/primer requirements and single nucleotide incorporation by hepatitis C virus nonstructural protein 5B polymerase. J Virol *74*, 9134–9143.

Chapter 19
Human Immunodeficiency Virus Reverse Transcriptase

Michaela Wendeler, Jennifer T. Miller, and Stuart F.J. Le Grice

Introduction

Human immunodeficiency virus type 1 (HIV-1) reverse transcriptase (RT) is an essential enzyme for HIV replication which converts the single-stranded viral RNA into a double-stranded DNA, suitable for integration into the host cell genome (Telesnitsky and Goff 1997). Heterodimeric p66/p51 HIV-1 RT (Fig. 19.1) has both synthetic (DNA polymerase) and degradative activities (ribonuclease H or RNase H), located at the N- and C-terminus of its p66 subunit, respectively. During HIV-1 replication, a complicated series of protein–protein and protein–nucleic acid interactions occur, each with the potential to be targeted by a small-molecule antagonist which might be developed into a potent therapeutic agent. Given the problem of increasing resistance against anti-HIV drugs currently in clinical use and the continued need to identify new drug targets, enzymatic activities and interactions, protein folding, and disruption of nucleic-acid structure are areas where detailed studies promise to unveil novel approaches with the potential for therapeutic intervention.

The Reverse Transcription Cycle of HIV

Reverse transcription is initiated from the 3′-end of a cellular tRNA hybridized to a sequence near the 5′-end of the viral genome designated the primer binding site (PBS) (Fig. 19.2[A]). Each retroviral genus has evolved to use a specific tRNA species, i.e., HIV-1 exploits the $tRNA^{Lys3}$ isoform (Marquet et al. 1995; Le Grice 2003; Kleiman et al. 2004), while Rous sarcoma and Moloney murine leukemia virus use $tRNA^{Trp}$ and $tRNA^{Pro}$, respectively. When RT reaches the 5′-end of the viral genome, the resulting cDNA is termed minus-strand strong stop DNA

S.F.J. Le Grice (✉)
HIV Drug Resistance Program, National Cancer Institute – Frederick, Building 535, Room 312, P.O. Box B, Frederick, MD 21702-1201, USA
e-mail: slegrice@ncifcrf.gov

C.E. Cameron et al. (eds.), *Viral Genome Replication*,
DOI 10.1007/b135974_19,

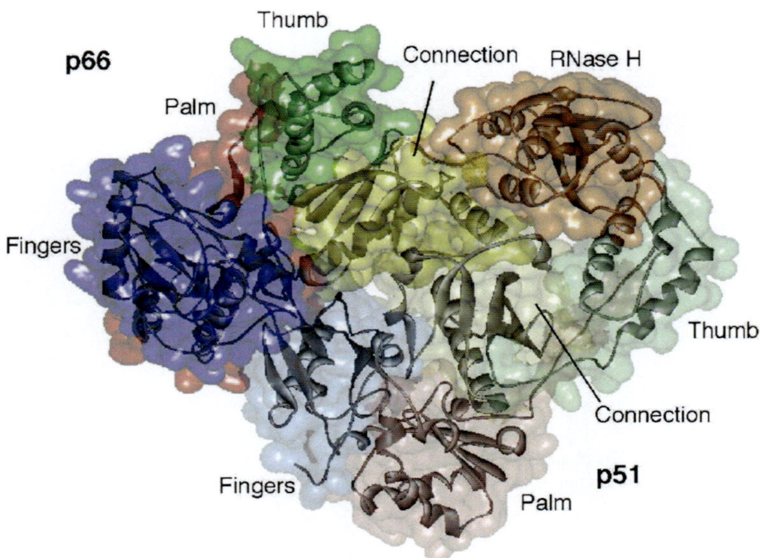

Fig. 19.1 Structure of p66/p51 HIV-1 reverse transcriptase. The p66 and p51 subunits are derived from the same gene, differing in that p51 lacks the C-terminal RNase H domain. Subdomains are denoted fingers (*blue*), palm (*red*), thumb (*green*), and connection (*yellow*), while the RNase H domain is *gold*. p51 subdomains are indicated by lighter shading.

(–ssDNA). RNase H activity concomitantly degrades viral RNA of the resulting DNA/RNA hybrid. Terminal redundancy at the 5′- and 3′-ends of the viral genome facilitates annealing of –ssDNA to the 3′-end of the viral genome, allowing strand transfer (Fig. 19.2[B]) and continued minus-strand DNA synthesis (Telesnitsky and Goff 1997). Again, RT-mediated RNase H activity degrades the RNA genome, with the exception of two purine-rich sequences designated the 3′ polypurine tract (3′PPT), present near the 3′-end of the viral genome and the central PPT (cPPT),

→

Fig. 19.2 The HIV-1 reverse transcription cycle. [**A**] (–) strand DNA is synthesized from the 3′ terminus of PBS-bound $tRNA^{Lys3}$, while RNase H activity concomitantly degrades RNA of the resulting RNA/DNA hybrid. [**B**] Once the replication machinery reaches the 5′ terminus of the (+) RNA genome, it relocates, via a strand-transfer event, to the 3′ terminus, allowing continued (–) strand DNA synthesis. [**C**] While the (+) RNA genome is essentially non-specifically degraded, polypurine tracts at the center (cPPT) and 3′ end of the genome (3′ PPT) are refractory to hydrolysis and serve as primers for (+) strand DNA synthesis. Shortly after initiation of (+) DNA synthesis, both the cPPT and the 3′ PPT are removed. 3′ PPT-mediated DNA synthesis pauses after copying over 18 nt of the tRNA primer, allowing RNase H hydrolysis in the vicinity of the tRNA/(–) DNA junction. [**D**] tRNA release makes (+) strand PBS sequences available for hybridization to the (–) strand complement, supporting plus-strand transfer and continued synthesis from both PPTs. [**E**] Bidirectional DNA synthesis generates the double-stranded DNA provirus containing a discontinuity near the center of the plus strand.

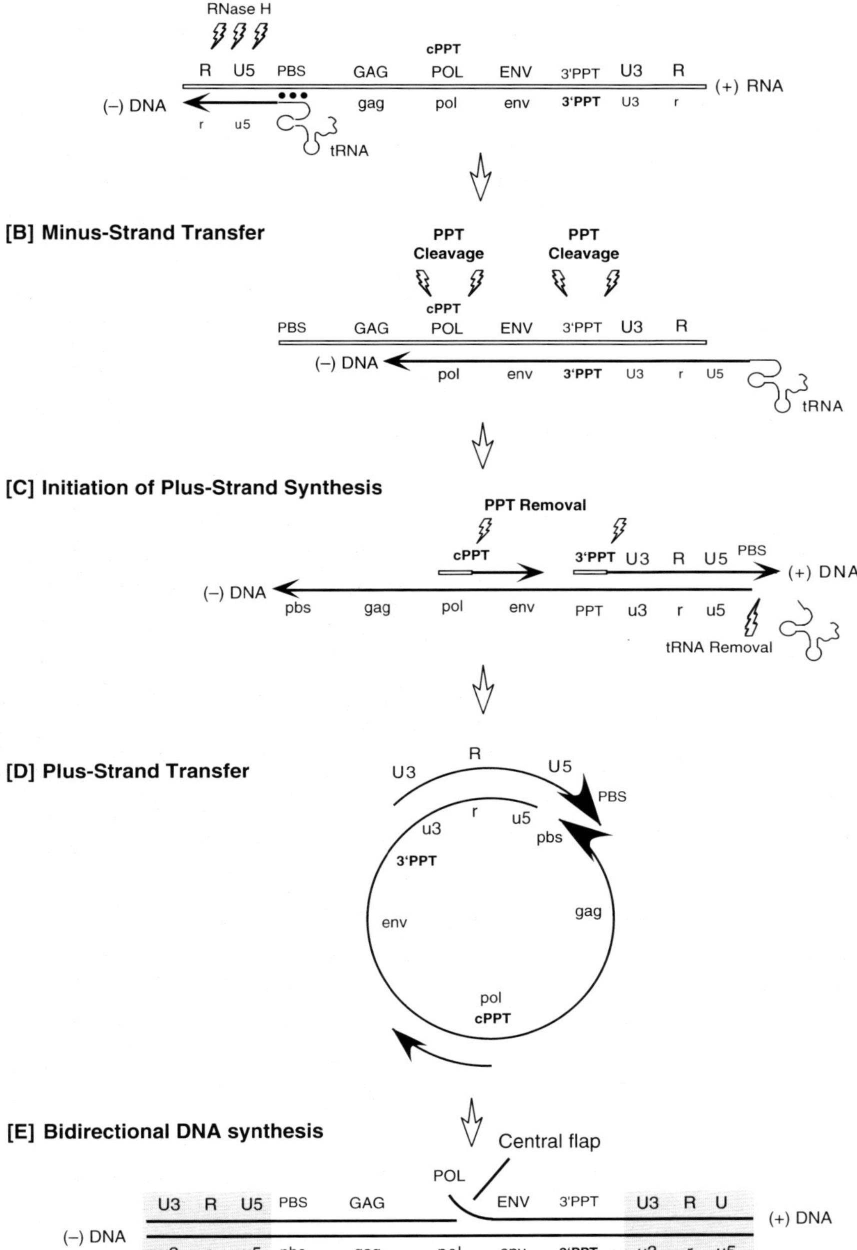

Fig. 19.2 (continued)

an identical sequence within the integrase (IN)-coding region. Both PPTs serve as primers for initiation of the positive (+) strand copy of the cDNA, and synthesis from the 3′ PPT proceeds over the newly synthesized (–) strand copy of the viral cDNA and partially into the tRNA primer (Fig. 19.2[C]). This paused species is designated plus-strand strong-stop DNA (+ssDNA). RNase H activity degrades 3′ terminal tRNA nucleotides to create a single-stranded region complementary to the minus-strand PBS, thereby facilitating the second-strand-transfer event (Fig. 19.2[D]). A circular intermediate is formed, and viral DNA synthesis can now be completed via bidirectional DNA synthesis (Fig. 19.2[E]). For reasons, which are not fully understood, the final product of reverse transcription in HIV contains a plus-strand discontinuity. Each of these steps will be described in more detail below.

Minus-Strand Initiation

Initiation of minus-strand DNA synthesis in HIV-1 utilizes the cellular $tRNA^{Lys3}$ isoform as the replication primer (Mak and Kleiman 1997; Le Grice 2003). This tRNA is recruited into the infected cell prior to particle formation by interactions involving the viral polyprotein precursors Pr55 *gag* and Pr160 *gag-pol*. Portions of the Pr160 *gag-pol* precursor polyprotein, more specifically, the thumb subdomain of RT, have been suggested to interact with $tRNA^{Lys3}$ (Khorchid et al. 2000), while Pr55 *gag* has been proposed to recruit the mitochondrial form of lysine tRNA synthetase (LysRS) bound to $tRNA^{Lys3}$ (Javanbakht et al. 2003; Kovaleski et al. 2006; Kaminska et al. 2007) to assist in its encapsidation. Subsequent to cleavage of the polyprotein, the viral nucleocapsid (NC) protein facilitates annealing of the first 18 nt of $tRNA^{Lys3}$ to the genomic PBS sequence (Barat et al. 1993; Barraud et al. 2007). Preliminary data suggest that the cellular-editing enzyme APOBEC3G can inhibit $tRNA^{Lys3}$-directed priming in *vif* negative viruses, but whether this is due to a direct action with the nucleic acid remains in question (Guo et al. 2006). There is considerable controversy surrounding the structure adopted by the tRNA:viral RNA binary complex. In contrast to early models proposed by Leis and co-workers (Cobrinik et al. 1988; Aiyar et al. 1992; Aiyar et al. 1994), in vitro studies demonstrated a complex series of interactions between HIV-1 viral RNA and $tRNA^{Lys3}$, the most prominent of which involved the tRNA anticodon loop and a viral A-rich loop sequence upstream of the PBS. These interactions were subsequently shown to exist in the HIV-1 Mal isolate and in the A and G HIV-1 subgroups (Isel et al. 1993; Isel et al. 1995), whereas only weak interactions were noted for the Lai or NL4.3 strains, as well as in 86% of HIV-1 isolates (Bajji et al. 2002; Goldschmidt et al. 2004; Miller et al. 2004; Tisne et al. 2004). In vivo, however, when the viral A-rich loop is mutated or deleted, the sequences quickly revert (Liang et al. 1997) and have been shown to be important for maintaining efficient DNA synthesis from heterologous primers (Kang et al. 1996; Ni et al. 2007). In summary, although binary structures may differ, the predominant interaction in the majority of strains occurs primarily between the viral PBS and the 3′-terminal 18 nt of $tRNA^{Lys3}$ (Fig. 19.3).

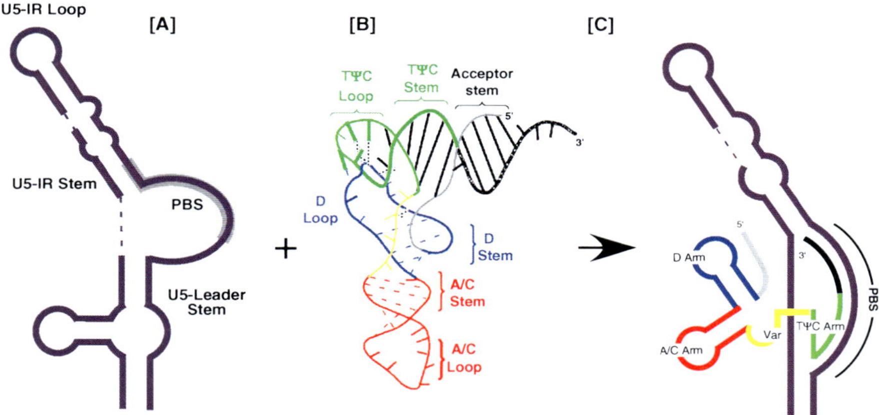

Fig. 19.3 tRNA–viral RNA interactions mediating initiation of reverse transcription in HIV. **[A]** Secondary structure elements of the viral genome in the immediate vicinity of the PBS. In **[B]**, elements of the tRNA primer are color coded. **[C]** Annealing of $tRNA^{Lys3}$ to the PBS involves both its 3' 18 nt (*black*) and the TYC arm (*green*), leaving the D- and anticodon arms (*red and blue*, respectively) intact.

In the presence of NC, an additional interaction pairing the anticodon stem/variable loop of the tRNA with a stretch of viral RNA upstream of the A-rich loop has been demonstrated (Iwatani et al. 2003). Several studies have suggested a role for tRNA–viral RNA interactions that, while transient, influence the kinetics of reverse transcription. Disrupting the A-rich loop–anticodon loop interaction in vitro has been shown to influence the transition from an initiation phase of DNA synthesis that lasts through the addition of the first 5 nt to an elongation phase which imparts significantly higher processivity (Lanchy et al. 1996). Additionally, a sequence denoted the primer activation signal, also upstream of the PBS, has been shown to regulate initiation of reverse transcription via its availability to pair with the TΨC stem/loop of the tRNA primer (Beerens et al. 2001; Beerens and Berkhout 2002; Ooms et al. 2007). In vivo, viral or cellular proteins may facilitate these and other contacts, resulting in highly dynamic binary and tertiary complexes specific to the initiation phase of the reverse transcription cycle.

Minus-Strand DNA Transfer

The HIV-1 minus-strand transfer reaction is mediated by a 97 nt homologous repeat or R region present at each end of the genome. However, prior to cDNA annealing to the acceptor strand, viral RNA must be removed to expose the immediate 3′ terminus of nascent –ssDNA. Once –ssDNA has been synthesized to the 5′ end of the RNA genome, "polymerase-dependent" RNase H cleavage (Furfine and Reardon 1991; Gopalakrishnan et al. 1992; Jacobo-Molina et al. 1993; Gotte et al. 1998) leaves an 18-nt RNA fragment stably attached, which impairs strand transfer, after which "polymerase-independent" RNase H activity (DeStefano 1995) cleaves the 18-nt RNA into smaller fragments that are more likely to dissociate, leaving –ssDNA free to participate in strand transfer (DeStefano et al. 2001; Gao et al. 2001; Wisniewski et al. 2002; Levin et al. 2005). After degradation of the viral RNA, the –ssDNA (to which the tRNA primer is still attached) anneals via its complementary R region to the RNA at the 3′-terminus of the viral genome. This reaction can occur intra- or intermolecularly (van Wamel and Berkhout 1998). The 5′-end of the HIV-1 R region comprises the highly structured transactivation response element (TAR), a stem-loop that must be destabilized to allow annealing. The viral NC protein facilitates the disruption and subsequent annealing of this hairpin (You and McHenry 1994; Guo et al. 2000; Golinelli and Hughes 2003).

Several studies have shown that a higher efficiency of strand transfer correlates with increased homology between donor and acceptor molecules (Luo and Taylor 1990; Andersen et al. 2003). However, mutant acceptor molecules with only 30 nt of R region homology efficiently promoted strand transfer in vivo (Berkhout et al. 1995). Minus-strand transfer has been studied in vitro utilizing single-molecule FRET on a wide variety of substrates. Barbara and colleagues demonstrated that TAR annealing occurs more efficiently through a mechanism involving NC-mediated unwinding at the base of the hairpin (called the Y intermediate), but also can involve a dual-loop interaction called the "kissing" mechanism (Liu et al. 2005; Liu et al. 2007). Likewise, Godet et al. studied annealing kinetics of TAR with fluorescently labeled cTAR derivatives and found that NC-promoted stimulation is correlated with its ability to destabilize the lower half of the TAR stem (Godet et al. 2006). In addition to pause-induced RNase H cleavages, Kim et al. showed that low-dNTP concentrations could influence the rate of strand transfer in mutants with dNTP-binding deficiencies (Operario et al. 2006). Minus-strand transfer is also stimulated in *cis* by sequences 3′ of the PBS, thought to create RNA secondary structures that enhance transfer efficiency, although the mechanism is not known at present (Song et al. 2006).

Polypurine Tract-Primed Initiation of Plus-Strand DNA Synthesis

Following minus-strand DNA transfer, continued polymerization creates a cDNA copy of the plus-strand viral RNA genome. Concomitantly, RNase H activity degrades RNA of the RNA/DNA replication intermediate, with the exception of two purine-rich RNAs: the 3′PPT and cPPT. Both PPTs share the sequence 5′-AAAAGAAAAGGGGGG-3′ and are preceded at their 5′ termini by a run of consecutive U residues. For the 3′PPT, cleavage to create the primer for second strand DNA synthesis occurs 3′ to the $d(C)_6$:$r(G)_6$ tract, defining the 5′ boundary of the U3 sequence (Huber and Richardson 1990). It is vital that this cut is precise, as the first deoxynucleotides of +ssDNA ultimately define the recognition site for the viral integrase protein. Extensive study has focused on structural features that render the PPT resistant to cleavage by the otherwise promiscuous RNase H activity. The resistance is specific to retroviral RNases H, as *Escherichia coli* RNase H readily degrades the PPT primer (Lener et al. 2002). The structure of HIV-1 RT complexed with a PPT-containing RNA/DNA hybrid (Sarafianos et al. 2001) indicates that amino acids in the connection and RNase H domains, termed the RNase H primer grip (RHPG), make specific contacts with the DNA strand (Fig. 19.4). In addition, the structure of the nucleic-acid hybrid shows unpaired and mispaired bases and minor groove compression in the 5′ r(A)$_4$:d(T)$_4$ tract. Probing the unbound RNA/DNA hybrid with $KMNO_4$ showed structural perturbations in the same area, indicating that these features may be inherent to the nucleic acid (Kvaratskhelia et al. 2002). Although it is not clear whether such features are enhanced or modified by RT, they may contribute to resistance to hydrolysis. The sequence of the PPT with respect to cleavage specificity has been extensively studied. In vivo, mutations that disrupt or ablate the PPT severely affect replication (Robson and Telesnitsky 2000; McWilliams et al. 2003; Miles et al. 2005). Model systems confirmed the significance of the d(C)$_6$:(rG)$_6$ tract in primer selection (Pullen et al. 1993) and also showed that neither the sequence context surrounding the primer nor its location in the genome affected its cleavage or utilization (Powell and Levin 1996). A more detailed mutational analysis demonstrated the specific importance of the rG residue at position –2 (Julias et al. 2004; Schultz et al. 2006; Jones and Hughes 2007; Rausch and Le Grice 2007). Recently, nucleoside analogs have been used to dissect the role of base and sugar constituents in recognition, cleavage specificity, and extension of the 3′PPT. Such analogs included (a) non-hydrogen bonding pyrimidine isosteres, to interrogate hydrogen bonding and flexibility (Rausch et al. 2003), (b) abasic tetrahydrofuran linkages, which determine the role of the nucleobase (Yi-Brunozzi and Le Grice 2005; Dash et al. 2006a), the fluorescent analog pyrrolo-dC (Dash et al. 2004a), and (c) locked nucleic-acid (Dash et al. 2004b; Dash et al. 2006b) analogs which provide information on local base pairing and nucleic-acid flexibility. In summary, these studies demonstrate the significance of hydrogen

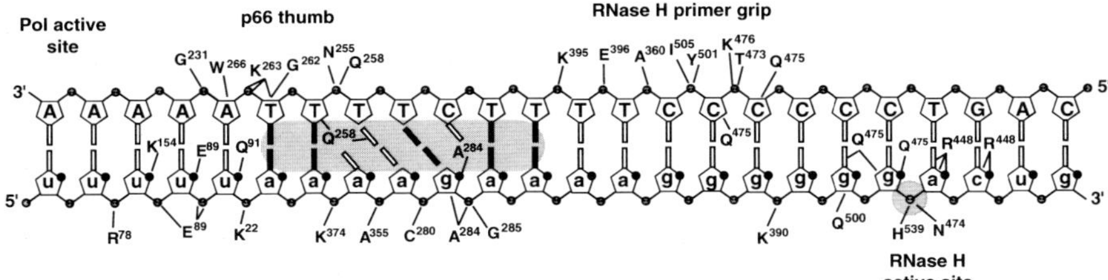

Fig. 19.4 Proposed contacts between the HIV-1 RT and the PPT-containing RNA/DNA hybrid. The model is based on the structure of Sarafianos et al. (2001), and positions His^{539} of the active site over the scissile bond. The *shaded portion* represents anomalous base pairing, comprising weakly paired (*filled bars*), mispaired and unpaired bases, collectively referred to as "unzipped". Structural elements involved in these interactions have been indicated.

bonding at the junction of the $d(T)_4{:}(rA)_4$ and $d(C)_6{:}(rG)_6$ tracts and the necessity for nucleobases in the RNA strand at the PPT/U3 junction for accurate cleavage. A detailed examination utilizing nucleoside analogs to study the importance of exocyclic groups of the purine ring (Rausch and Le Grice 2007) has also established a link between 5′ primer cleavage specificity and the upstream $d(A)_5{:}(rU)_5$ and $d(T)_4{:}(rA)_4$ tracts, whereas the downstream $d(C)_6{:}(rG)_6$ delineates the specificity of initiation from the PPT. Furthermore, introducing substituents at position 2 of −2G or +1A or inserting a $-NH_2$ group at position 6 of −4G greatly influenced cleavage specificity. Clearly, the determinants underlying recognition, cleavage, and extension of the PPT primer are multifaceted, allowing the virus to tightly regulate the sequences that specify the termini of the double-stranded viral preintegrative genome.

For reasons stated above, the 3′PPT primer must be removed from nascent +ssDNA with precision. In a manner that can be likened to abortive initiation during transcription (Steitz 2004), Gotte et al. (2001) have shown that, following addition of 12 nt to the primer, the PPT/U3 junction is recognized and cleaved by the RNase H domain. This is followed by re-positioning of RT and renewed plus-strand synthesis from the 3′-terminus of the short DNA fragment. Although speculative, pausing after synthesis of 12 nt of plus-strand DNA would position the RNA–DNA junction of the chimeric primer within the RNase H primer grip, which might induce transient stalling, allowing dissociation of RT to make the PPT/U3 junction amenable to hydrolysis.

Plus-Strand Transfer

Both the newly synthesized minus-strand DNA and its covalently attached tRNA replication primer serve as templates for 3′PPT-primed plus-strand DNA synthesis. Once U3, R, and U5 sequences are duplicated, HIV-1 RT copies 18 nt into the 3′-end of $tRNA^{Lys3}$, pausing at A57, the first modified riboncleotide encountered (Ben-Artzi et al. 1996) In this paused complex, the RNase H domain is now positioned in the vicinity of the minus-strand DNA/tRNA junction. Surprisingly, model systems duplicating tRNA primer-removal event showed that the DNA–RNA junction is not the preferred site of hydrolysis. Instead, HIV-1 RT cleaves between rC75 and rA76 of $tRNA^{Lys3}$, leaving a single ribonucleotide attached to the minus-strand DNA (Smith et al. 1999). Following cleavage, RNase H activity further degrades the short tRNA/DNA hybrid, allowing dissociation of the fragmented tRNA primer. As a consequence, plus-strand PBS sequences become available for a second, intra-strand, transfer event, mediated through complementarity between minus- and plus-strand PBS sequences.

Termination at the Central PPT

Although bidirectional DNA synthesis should be sufficient to complete the reverse transcription cycle and create a double-stranded, integration-competent DNA provirus, additional use of the cPPT primer in HIV produces a plus-strand discontinuity. These events are outlined schematically in Figs 19.2 [D] and [E]. Following strand transfer, 3′PPT-mediated DNA synthesis continues to the cPPT, thereafter displacing ~100 nt of cPPT-primed DNA. At this point, the replication complex abruptly halts at the "central termination sequence" (CTS), generating a structure designated the "central flap" (Charneau and Clavel 1991; Hungnes et al. 1991; Charneau et al. 1992; Hungnes et al. 1992), a prominent feature of which is phased A-tracts which induce minor groove narrowing (Lavigne et al. 1997; Lavigne and Buc 1999; Lavigne et al. 2001). Since the same A-tracts do not interfere with cPPT-primed DNA synthesis, central termination apparently reflects the replication complex encountering an aberrant duplex DNA structure while catalyzing strand displacement synthesis. The final product of DNA synthesis is a double-stranded DNA containing a "flap" of ~100 single-stranded nucleotides (Kankia and Musier-Forsyth 2007), Fig. 19.2 [E]. In vitro studies of Bambara and colleagues (Rumbaugh et al. 1998) have demonstrated that Fen1, a nuclease involved in processing Okazaki fragments (Kao et al. 2004) will remove the single-stranded overhang, while the discontinuity can be repaired by eukaryotic DNA ligase. Thus, while not demonstrated directly, such studies lay a precedent for host enzyme involvement in creating the intact double-stranded provirus.

The function of the central flap has been the subject of considerable controversy. Several studies show that its mutation or deletion impairs virus replication (Charneau et al. 1992; Charneau et al. 1994; De Rijck and Debyser 2006) and that it confers a positive transduction efficiency to lentiviral vectors (Sirven et al. 2000; Dardalhon et al. 2001; Zennou et al. 2001; Van Maele et al. 2003; Ao et al. 2004). Imaging studies show that "flap-negative" pre-integration complexes accumulate around the nuclear membrane but do not enter the nucleus (Zennou et al. 2000). Charneau and co-workers showed by scanning electron microscopy that reverse transcription occurs within an intact capsid shell and that in the absence of central flap formation, uncoating is impaired and linear DNA remains trapped within the integral shell precluding nuclear import (Arhel et al. 2007). However, other studies found that the flap is dispensable for nuclear import of the pre-integration complex and the effect seen previously may have been cell-type dependent and can be masked by a high multiplicity of infection (Dvorin et al. 2002; Limon et al. 2002; Arhel et al. 2006). Despite controversy on the necessity for a central termination sequence, it is worthwhile noting that a similar element has been found in equine infectious anemia virus (Borroto-Esoda and Boone 1991; Stetor et al. 1999) and feline immunodeficiency virus (Whitwam et al. 2001).

HIV-1 Reverse Transcriptase

The asymmetric RT heterodimer is composed of 66 kDa (p66) and 51 kDa (p51) subunits derived from the gag/pol polyprotein, the smaller of which arises through HIV-1 protease-mediated cleavage of p66 between Phe440 and Tyr441(di Marzo Veronese et al. 1986; Lowe et al. 1988). Both subunits contain four subdomains, designated fingers (residues 1–85 and 118–154), palm (86–117 and 155–242), thumb (255–318), and connection (319–426). In addition, residues 427–560 of p66 comprise the RNase domain (Kohlstaedt et al. 1992; Jacobo-Molina et al. 1993) (Fig. 19.1). Although their amino acid sequences are identical, the subdomains of p66 and p51 adopt a very different fold. Whereas the polymerase domain of p66 folds into an open, extended structure with a large active-site cleft, the equivalent region of p51 forms a closed, compact domain which cannot participate in catalysis (Kohlstaedt et al. 1992; Wang et al. 1994). Both the DNA polymerase and the RNase H activities are accommodated by the p66 subunit, while p51 is proposed to provide structural support to p66 (Le Grice et al. 1991; Kohlstaedt et al. 1992; Jacobo-Molina et al. 1993; Amacker and Hubscher 1998). Additional functional roles of p51 involve facilitating the loading of p66 on the template-primer (Harris et al. 1998) and stabilizing the appropriate conformation of p66 during initiation of reverse transcription from $tRNA^{Lys3}$ (Jacques et al. 1994a; Arts et al. 1996a). In contrast to the p66 subunit, which can undergo large-scale motions (especially its fingers and thumb subdomains), the p51 subunit is essentially rigid (Bahar et al. 1999).

RT Dimerization

The monomeric forms of the RT subunits are catalytically inert, i.e., biological activity strictly requires dimer formation (Muller et al. 1989; Restle et al. 1990, 1992). The p66/p51 dimer interface has recently attracted interest as a potential target for the development of novel antiviral drugs, in particular substituted TSAO-T derivatives (Srivastava et al. 2006). The most stable form is the heterodimer with a dissociation constant in the nanomolar range (Restle et al. 1990; Divita et al. 1993). p66 and p51 can form homodimers in vitro, but are considerably less stable with dissociation constants of 10^{-6} and 10^{-5} M, respectively (Divita et al. 1993; Wang et al. 1994). While the p66 homodimer retains both DNA polymerase and RNase H function, it inefficiently catalyzes tRNA-primed minus-strand DNA synthesis from the PBS (Arts et al. 1996b). In contrast, the p51 homodimer is only poorly active and distributive in nature (Wohrl et al. 1994).

A two-step model for heterodimer formation was proposed from in vitro studies (Divita et al. 1993, 1995; Cabodevilla et al. 2001). The first involves concentration-dependent association of the two subunits to a non-functional enzyme (Tachedjian et al. 2003), and a subsequent slow conformational change gives rise to the mature, active form (Divita et al. 1993, 1995). The maturation process involves interactions of the fingers and thumb subdomains of p51 with the palm subdomain and RNase H domain of p66. Three major contact regions at the dimer interface are involved in subunit interactions, namely (i) the p51 fingers interact with the p66 palm, (ii) the connection of p51 interacts with its p66 counterpart, and (iii) the p51 thumb interacts with the RNase H domain of p66 (Wang et al. 1994). These interactions are mostly hydrophobic (Becerra et al. 1991; Wang et al. 1994) and contributions from residues of p51 may play a more important role in dimer stability (Menendez-Arias et al. 2001).

Among residues of the connection subdomain, the tryptophan-rich motif, spanning residues 398–414, was found to be of central importance for dimerization (Menendez-Arias et al. 2001; Rodriguez-Barrios et al. 2001; Tachedjian et al. 2003).

This cluster of six tryptophan residues (Trp398, Trp401, Trp402, Trp406, Trp410, and Trp414) is highly conserved in primate lentiviral RTs. Interactions between the two connection subdomains are asymmetric in that different residues from p51 and p66 are important for the contacts. A study using the yeast-two hybrid system to analyze the effect of mutations in the tryptophan-rich motif on heterodimer formation demonstrated that mutating Trp401 and Trp414 of p66 impairs dimerization by altering positioning of structural elements between these residues that make important contacts with p51 (Tachedjian et al. 2003). A *trans*-complementation approach for analyzing the RT heterodimer in the context of infectious virions (Mulky et al. 2005) identified residues Trp398, Trp402, Trp406, and Trp414 of p51 as crucial for subunit interaction and dimer stabilization. Important contributions to inter-subunit interactions were further found for amino acids Trp401, Tyr405, and Asn363 of p51 and Trp410 of p66. In addition to residues located directly at the dimer interface, mutagenesis studies identified additional amino acids in both subunits at a distance from the interface that also contribute to dimerization and stability (Goel et al. 1993; Ghosh et al. 1996; Wohrl et al. 1997; Tachedjian et al. 2003).

Nucleic-Acid Binding

The nucleic-acid-binding cleft is formed by fingers, palm, and thumb of p66, and crystal structures of HIV-1 RT bound to duplex DNA and an RNA/DNA hybrid have identified numerous contacts between the enzyme and both strands of primer template (Jacobo-Molina et al. 1993; Ding et al. 1998; Huang et al. 1998; Sarafianos et al. 2001). Interactions occur mainly between the sugar-phosphate backbone of the nucleic acid and highly conserved motifs of the fingers, palm, thumb, and RNase H domain of p66. Superposition of X-ray structures of unliganded and ligand-bound RT underscore the high flexibility of the enzyme, especially in the p66 thumb (Carvalho et al. 2006). In crystal structures of wild-type unliganded RT, this subdomain is folded down into the nucleic-acid-binding cleft (Rodgers et al. 1995; Hsiou et al. 1996). This conformation was also found for unliganded RT in solution by a spin-labeling study (Kensch et al. 2000). Binding of duplex nucleic acid produces large changes in the orientation of the thumb relative to the p66 palm, resulting in a more open conformation. Helix α-H of the p66 thumb is involved in extensive contacts with the primer strand in the minor groove of the DNA (Beard et al. 1994; Bebenek et al. 1995, 1997). Residues Pro227-His235 form the β12-β13 hairpin and are designated the "primer grip". This motif, which is highly conserved among retroviral RTs (Xiong and Eickbush 1990), plays a central role for maintaining the orientation of the primer 3′-OH for nucleophilic attack on the incoming dNTP (Jacobo-Molina et al. 1993). Important primer grip contacts involve the main-chain atoms of Met230 and Gly231 with the primer terminal phosphate (Ding et al. 1998). Mutational studies on the primer grip have been shown to have pleiotropic effects, altering DNA polymerase and RNase H activity as well as reducing dimer stability (Jacques et al. 1994b; Ghosh et al. 1996, 1997; Palaniappan et al. 1997; Powell et al. 1997; Wohrl et al. 1997; Wisniewski et al. 1999). Contacts to the template strand are mediated by the "template grip", comprising elements of the p66 palm (αB-β6 loop, β-strand 9, α-helix E, and the β8-aE connecting loop) and fingers (β-strand 4) (Jacobo-Molina et al. 1993; Ding et al. 1998).

In the crystal structure of Huang et al. (Huang et al. 1998) containing a disulfide-tethered DNA duplex, the single-stranded template overhang ahead of the polymerase active site was not co-linear with the duplex, but bent away and contacted by the p66 fingers. This structure revealed contacts of nucleobases +2 and +3 with Trp24, Pro25, Phe61, and Ile63 and of base +1 with Leu74 (defining +1 as the first unpaired template nucleotide). In addition, Arg78, Lys154, and Glu89 interact with phosphate groups. A study based on nucleoside analog interference (Dash et al. 2006a) demonstrated that alterations to template geometry 1–2 nt ahead of the catalytic center influenced protein–DNA contacts and could interrupt catalysis and that this effect was sensitive to Phe61 substitutions.

The DNA Polymerase Active Site

The DNA polymerase active site resides within the palm subdomain of p66, at the base of the nucleic-acid-binding cleft, and is characterized by the Asp110, Asp185, and Asp186 catalytic triad, a feature conserved in many polymerases (Kohlstaedt et al. 1992; Jacobo-Molina et al. 1993) (Fig. 19.5). Among polymerase families, the structure of the palm domain is also highly conserved, comprising a four- to six-stranded β-sheet flanked on one side by two α-helices (Brautigam and Steitz 1998). Detailed studies have delineated the effect of mutating the catalytic triad on RT polymerase activity (Boyer et al. 1992, 1994; Tantillo et al. 1994).

In crystal structures containing duplex nucleic acid, the catalytic aspartates of HIV-1 RT are close to the 3′-terminus of the primer. Asp185 and Asp186 are part of the conserved –Tyr–Met–Asp–Asp– motif, which adopts an unusual β-turn conformation (Esnouf et al. 1995; Hsiou et al. 1996; Ding et al. 1998), possibly to facilitate correct positioning of the aspartate residues for catalysis. The phenoxy side chain of Tyr183 is involved in hydrogen bonding with nucleobases at position –2 (Ding et al. 1998). Site-specific replacement of Tyr183 with the unnatural analog *nor*-tyrosine was found to significantly decrease RNA-dependent DNA polymerase activity, while DNA-dependent DNA polymerase activity was unaffected (Klarmann et al. 2004). A comparison of crystal structures of DNA-bound RT with unliganded or non-nucleoside-bound enzymes reveals significant conformational differences for the –Tyr–Met–Asp–Asp– quartet, implicating a high degree of structural flexibility for this element (Ding et al. 1998).

The incoming dNTP is tightly coordinated by p66 fingers' residues Lys65 and Arg72, the main-chain NH groups of residues Asp113 and Ala114 and two metal ions. The ribose moiety is accommodated by a pocket lined by Asp113, Tyr115, and Phe116 on one side, and Glu151 and Arg72 on the other. Additional contacts of the dNTP occur through base-pairing and base-stacking with the template-overhang (Huang et al. 1998). Mutagenesis studies have designated Tyr115 the "steric gate",

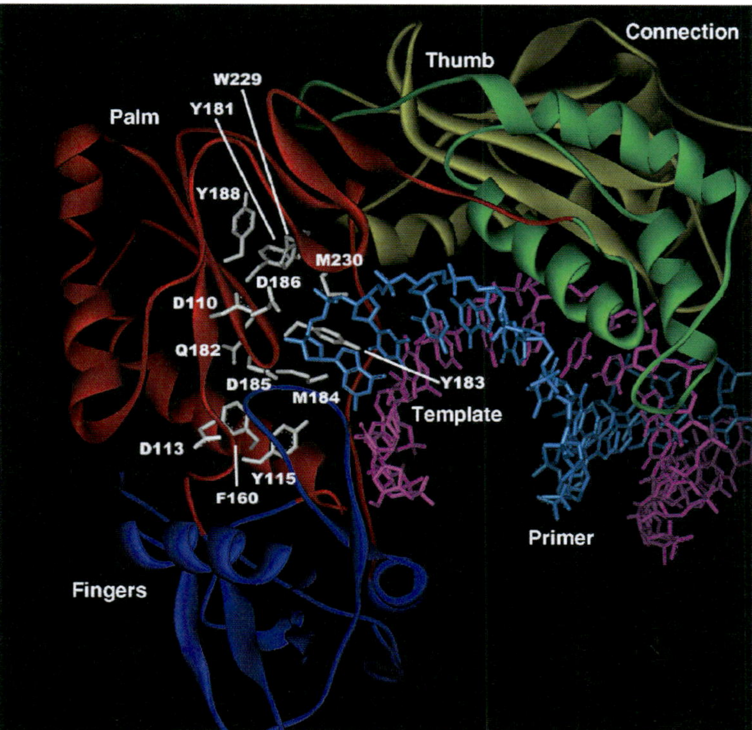

Fig. 19.5 Model of the DNA polymerase active site of p66 HIV-1 RT. For clarity, the p51 subunit has been removed. p66 fingers, palm, thumb, and connection subdomains are indicated in *blue*, *red*, *green*, and *yellow*, respectively. Template nucleobases are represented in magenta and primer nucleobases in *cyan*.

implicating a critical role for this residue in discriminating between deoxy- and ribonucleoside triphosphates (Martin-Hernandez et al. 1996; Cases-Gonzalez et al. 2000). A later study replacing Tyr115 with a variety of unnatural amino acid analogs demonstrated that a Tyr115 -> aminomethyl-Phe115 substitution conferred resistance to the nucleoside analog 3TC by allowing more efficient incorporation of dCTP (Klarmann et al. 2007). In addition, the fidelity of dNTP insertion is critically influenced by interactions of the γ-phosphate moiety with Lys65 (Garforth et al. 2007).

NNRTI-Binding Pocket

The hydrophobic-binding pocket occupied by non-nucleoside RT inhibitors (NNRTIs) is formed by residues of the p66 palm and thumb, at a distance of approximately 10 Å from the DNA polymerase active site (Kohlstaedt et al. 1992; Ding et al. 1995). This binding pocket is not evident in RT in the absence of the inhibitor. NNRTI binding induces conformational changes, and in particular a rearrangement of side chains Tyr181 and Tyr188, and a substantial displacement of the DNA polymerase primer grip that allosterically inhibits catalysis (Esnouf et al. 1995; Rittinger et al. 1995). A recent crystal structure of double mutant Lys103Asn/Tyr181Cys in the absence and presence of the NNRTI HBY 097 suggests that non-nucleoside binding restricts flexibility of the –Tyr–Met–Asp–Asp– loop and prevents correct orientation of catalytic aspartate residues required for metal binding (Das et al. 2007).

DNA Synthesis Mechanism

Kinetic analyses identified an ordered mechanism for DNA polymerase activity, where RT binds first primer-template and subsequently the incoming dNTP (Majumdar et al. 1988). The second step induces a conformational change resulting in closure of the fingers' subdomain (Sarafianos et al. 1999), trapping the substrate and correctly aligning the α-phosphate of the bound dNTP and the 3′-OH group of the primer terminus (Huang et al. 1998; Doublie et al. 1999). Kinetic studies identified this step to be rate-limiting for catalysis (Dahlberg and Benkovic 1991; Spence et al. 1995).

As for all DNA polymerases, the catalytic activity of RT requires divalent metal cations, such as Mg^{2+} or Mn^{2+} (Derbyshire et al. 1988). In addition to serving a catalytic function, the metal ion stabilizes binding to nucleic acids to guarantee precise positioning of the substrates (DeStefano et al. 1993). For HIV-1 RT, Mg^{2+} is the preferred cofactor (Tan et al. 1991), and a two-metal ion mechanism was proposed for DNA polymerases (Beese and Steitz 1991; Steitz 1993, 1998, 1999). A recent crystal structure of a Lys103Asn/Tyr181Cys double mutant (Das et al. 2007) shows octahedral coordination geometry of two Mn^{2+} ions and apparently reflects a trapped reaction intermediate state on the way to nucleotide incorporation. Metal ion A is thought to activate the 3′-OH group of the primer by lowering its pK_a, thus facilitating the nucleophilic attack at the α-phosphate of the bound dNTP. In addition, metal ion A might act to facilitate precise positioning of the scissile phosphodiester bond of the incoming dNTP with respect to the oxygen at the 3′ end of the primer. Together with the three catalytic aspartate residues, the two metal ions in the DNA polymerase active site define the platform for binding the triphosphate and stabilize the transition state. After the new phosphodiester bond is formed, metal ion B is thought to be involved in the release of pyrophosphate (Steitz 1998, 1999). Finally, the metal ions and the pyrophosphate group dissociate, and RT translocates a single position further downstream to clear the nucleotide-binding site for a new cycle of nucleotide incorporation (Steitz et al. 1994; Gotte 2006). This step converts RT from the "pre-translocational state", in which the 3′-end of the primer occupies the nucleotide-binding site, to the "post-translocational state", where the dNTP-binding site is cleared (Gotte 2006). For this movement of RT relative to its nucleic-acid substrate two different mechanisms have been proposed: An "active" model suggests that dNTP hydrolysis and pyrophosphate release together supply the energy for translocation (Steitz and Yin 2004; Temiakov et al. 2004; Yin and Steitz 2004), while a "passive" model postulates that thermal energy is sufficient to enable free oscillation of RT between pre- and post-translational stages and that the incoming dNTP acts like the pawl of a ratchet and traps RT in the post-translocational state (Guajardo and Sousa 1997; Landick 2004; Bar-Nahum et al. 2005; Gotte 2006; Marchand et al. 2007).

Pre-steady-state kinetic analyses (Wohrl et al. 1999) and single-molecule spectroscopy (Rothwell et al. 2003) suggest a third binding mode or "dead end complex" that is not capable of dNTP incorporation. In this complex, the nucleic-acid substrate is postulated to be bound at a site remote from the crystallographically observed nucleic-acid-binding cleft, and nucleotide incorporation can occur only after dissociation followed by reassociation and formation of productive complexes.

For a comprehensive discussion of polymerase inhibitors, we refer the reader to Chapter 26 (*Reverse Transcriptase Inhibitors*) of this volume.

The RNase H Domain

The C-terminal residues 427–560 of p66 RT comprise the RNase H domain. X-ray crystallography of the isolated RNase H domain of HIV-1 RT (Davies et al. 1991) indicated a structural similarity with the *Escherichia coli* enzyme (Katayanagi et al. 1990), differing in that the bacterial enzyme has an additional α-helix (α-C) designated the "basic handle" proposed to mediate binding of the RNA/DNA hybrid. RNases H belong together with retroviral integrases, transposases, and *E. coli* RuvC to the polynucleotidyl transferase superfamily (Rice and Baker 2001). Furthermore, RNase H is structurally related to the PIWI domain of argonaute, the key catalytic component of the RNA-induced silencing complex (RISC) (Song et al. 2004; Ma et al. 2005; Yuan et al. 2005). While RT-associated RNase H cleaves the RNA strand of RNA/DNA hybrids (Molling et al. 1971), the HIV-1 enzyme has also been demonstrated to cleave duplex RNA (Ben-Artzi et al. 1992a, 1992b; Blain and Goff 1993), the significance of which remains to be established. HIV-1 RT-RNase H hydrolyzes RNA/DNA hybrids by a variety of mechanisms. When the 3′ terminus of the DNA strand is recessed, cleavage is initially dictated by the spatial separation between the DNA polymerase and the RNase H active sites (18–19 bp). Subsequent cleavage from this position extends to within 8 bp of the recessed DNA 3′ terminus, after which the remaining RNA/DNA hybrid is presumably unstable and dissociates. The two mechanisms have been designated "polymerization-dependent" and "polymerization-independent" (Furfine and Reardon 1991; Peliska and Benkovic 1992), the necessity for which may be related to the minus- and plus-strand transfer steps of the reverse transcription cycle (see Fig. 19.2). Termination of DNA synthesis at the 5′ end of the plus RNA genome (prior to minus-strand transfer) or pausing of the replication complex when a modified base of the tRNA primer is encountered (prior to plus-strand transfer) would initially result in cleavage of the RNA strand ~18 bp behind the 3′ terminus of nascent DNA. For each scenario, in vitro studies have shown that the 17–18 nt hydrolysis product fails to dissociate, impairing strand transfer (Peliska and Benkovic 1992; Smith et al. 1999). Strand-transfer activity is, however, restored when this RNA "remnant" is reduced to ~8 nt through polymerization-independent RNase H activity. In contrast, when the RNA strand is recessed on a DNA template, as might be expected when genomic RNA is cleaved during DNA synthesis, the RNA 5′ terminus dictates binding of the DNA polymerase domain, presumably mediated by contacts to the "template grip" (Jacobo-Molina et al. 1993). Although the mechanism underlying RT binding stably to a recessed RNA 5′ terminus remains to be established, the ability to make a number of contacts with the ribose 2′ OH may be critical.

The involvement of divalent metals in RNase H-mediated hydrolysis has been a subject of considerable controversy, with plausible models for one- (Cowan et al. 2000) and two-metal-assisted catalysis (Steitz and Steitz 1993) being suggested. Although an equivalent high-resolution structure is not available for the HIV-1 enzyme, crystallization of *Bacillus halodurans* RNase H in presence of substrate by Nowotny and colleagues (Nowotny et al. 2005; Nowotny and Yang 2006; Yang et al. 2006) supports the original 2-metal hypothesis of Steitz and Steitz (Steitz and Steitz 1993). Based on the *B. halodurans* RNase H structure in the presence of substrate, metal A (Mg^{2+}), which is coordinated by Asp443 and Asp549, serves to coordinate a water molecule, aligning this for inline nucleophilic attack on the scissile phosphate; while metal B (Mg^{2+}), coordinated by Asp443, Asp478, and Asp498, is appropriately positioned to stabilize the transition state and leaving group. In addition to the conserved carboxylates, another controversial issue is whether His539 directly participates in hydrolysis. Its counterpart in *B. halodurans* RNase H, Glu188, exhibits a considerable degree of flexibility and may either contribute to metal A coordination or perturb this in such a manner as to assist in product release. The notion that the His539-containing loop of HIV-1 RT also exhibits considerable flexibility in conjunction with mutagenesis studies (Tisdale et al. 1991) would be consistent with this hypothesis. Finally, although RNase H is absolutely required for HIV replication (Tisdale et al. 1991), progress in developing small-molecule antagonists has been surprisingly slow. However, recent reports that N-hydroxyimides (Hang et al. 2004), hydroxylated tropolones (Budihas et al. 2005), and thiophene diketo acids (Shaw-Reid et al. 2003) potently and selectively inhibit HIV RNase H provide encouragement for continuing to exploit this as a therapeutic target.

References

Aiyar, A., D. Cobrinik, et al. (1992). "Interaction between retroviral U5 RNA and the T psi C loop of the tRNA(Trp) primer is required for efficient initiation of reverse transcription." *J Virol* **66**(4): 2464–72.

Aiyar, A., Z. Ge, et al. (1994). "A specific orientation of RNA secondary structures is required for initiation of reverse transcription." *J Virol* **68**(2): 611–8.

Amacker, M. and U. Hubscher (1998). "Chimeric HIV-1 and feline immunodeficiency virus reverse transcriptases: critical role of the p51 subunit in the structural integrity of heterodimeric lentiviral DNA polymerases." *J Mol Biol* **278**(4): 757–65.

Andersen, E. S., R. E. Jeeninga, et al. (2003). "Dimerization and template switching in the 5′ untranslated region between various subtypes of human immunodeficiency virus type 1." *J Virol* **77**(5): 3020–30.

Ao, Z., X. Yao, et al. (2004). "Assessment of the role of the central DNA flap in human immunodeficiency virus type 1 replication by using a single-cycle replication system." *J Virol* **78**(6): 3170–7.

Arhel, N., S. Munier, et al. (2006). "Nuclear import defect of human immunodeficiency virus type 1 DNA flap mutants is not dependent on the viral strain or target cell type." *J Virol* **80**(20): 10262–9.

Arhel, N. J., S. Souquere-Besse, et al. (2007). "HIV-1 DNA Flap formation promotes uncoating of the pre-integration complex at the nuclear pore." *Embo J* **26**(12): 3025–37.

Arts, E. J., M. Ghosh, et al. (1996a). "Restoration of tRNA3Lys-primed(-)-strand DNA synthesis to an HIV-1 reverse transcriptase mutant with extended tRNAs. Implications for retroviral replication." *J Biol Chem* **271**(15): 9054–61.

Arts, E. J., S. R. Stetor, et al. (1996b). "Initiation of (-) strand DNA synthesis from tRNA(3Lys) on lentiviral RNAs: implications of specific HIV-1 RNA-tRNA(3Lys) interactions inhibiting primer utilization by retroviral reverse transcriptases." *Proc Natl Acad Sci USA* **93**(19): 10063–8.

Bahar, I., B. Erman, et al. (1999). "Collective motions in HIV-1 reverse transcriptase: examination of flexibility and enzyme function." *J Mol Biol* **285**(3): 1023–37.

Bajji, A. C., M. Sundaram, et al. (2002). "An RNA complex of the HIV-1 A-loop and tRNA(Lys,3) is stabilized by nucleoside modifications." *J Am Chem Soc* **124**(48): 14302–3.

Bar-Nahum, G., V. Epshtein, et al. (2005). "A ratchet mechanism of transcription elongation and its control." *Cell* **120**(2): 183–93.

Barat, C., O. Schatz, et al. (1993). "Analysis of the interactions of HIV1 replication primer tRNA(Lys,3) with nucleocapsid protein and reverse transcriptase." *J Mol Biol* **231**(2): 185–90.

Barraud, P., C. Gaudin, et al. (2007). "New insights into the formation of HIV-1 reverse transcription initiation complex." *Biochimie* **89**(10): 1204–10.

Beard, W. A., S. J. Stahl, et al. (1994). "Structure/function studies of human immunodeficiency virus type 1 reverse transcriptase. Alanine scanning mutagenesis of an alpha-helix in the thumb subdomain." *J Biol Chem* **269**(45): 28091–7.

Bebenek, K., W. A. Beard, et al. (1995). "Reduced frameshift fidelity and processivity of HIV-1 reverse transcriptase mutants containing alanine substitutions in helix H of the thumb subdomain." *J Biol Chem* **270**(33): 19516–23.

Bebenek, K., W. A. Beard, et al. (1997). "A minor groove binding track in reverse transcriptase." *Nat Struct Biol* **4**(3): 194–7.

Becerra, S. P., A. Kumar, et al. (1991). "Protein-protein interactions of HIV-1 reverse transcriptase: implication of central and C-terminal regions in subunit binding." *Biochemistry* **30**(50): 11707–19.

Beerens, N. and B. Berkhout (2002). "The tRNA primer activation signal in the human immunodeficiency virus type 1 genome is important for initiation and processive elongation of reverse transcription." *J Virol* **76**(5): 2329–39.

Beerens, N., F. Groot, et al. (2001). "Initiation of HIV-1 reverse transcription is regulated by a primer activation signal." *J Biol Chem* **276**(33): 31247–56.

Beese, L. S. and T. A. Steitz (1991). "Structural basis for the 3′-5′ exonuclease activity of Escherichia coli DNA polymerase I: a two metal ion mechanism." *Embo J* **10**(1): 25–33.

Ben-Artzi, H., E. Zeelon, et al. (1992a). "Double-stranded RNA-dependent RNase activity associated with human immunodeficiency virus type 1 reverse transcriptase." *Proc Natl Acad Sci USA* **89**(3): 927–31.

Ben-Artzi, H., E. Zeelon, et al. (1992b). "Characterization of the double stranded RNA dependent RNase activity associated with recombinant reverse transcriptases." *Nucleic Acids Res* **20**(19): 5115–8.

Ben-Artzi, H., J. Shemesh, et al. (1996). "Molecular analysis of the second template switch during reverse transcription of the HIV RNA template." *Biochemistry* **35**(32): 10549–57.

Berkhout, B., J. van Wamel, et al. (1995). "Requirements for DNA strand transfer during reverse transcription in mutant HIV-1 virions." *J Mol Biol* **252**(1): 59–69.

Blain, S. W. and S. P. Goff (1993). "Nuclease activities of Moloney murine leukemia virus reverse transcriptase. Mutants with altered substrate specificities." *J Biol Chem* **268**(31): 23585–92.

Borroto-Esoda, K. and L. R. Boone (1991). "Equine infectious anemia virus and human immunodeficiency virus DNA synthesis in vitro: characterization of the endogenous reverse transcriptase reaction." *J Virol* **65**(4): 1952–9.

Boyer, P. L., A. L. Ferris, et al. (1994). "Mutational analysis of the fingers and palm subdomains of human immunodeficiency virus type-1 (HIV-1) reverse transcriptase." *J Mol Biol* **243**(3): 472–83.

Boyer, P. L., A. L. Ferris, et al. (1992). "Cassette mutagenesis of the reverse transcriptase of human immunodeficiency virus type 1." *J Virol* **66**(2): 1031–9.

Brautigam, C. A. and T. A. Steitz (1998). "Structural and functional insights provided by crystal structures of DNA polymerases and their substrate complexes." *Curr Opin Struct Biol* **8**(1): 54–63.

Budihas, S. R., I. Gorshkova, et al. (2005). "Selective inhibition of HIV-1 reverse transcriptase-associated ribonuclease H activity by hydroxylated tropolones." *Nucleic Acids Res* **33**(4): 1249–56.

Cabodevilla, J. F., L. Odriozola, et al. (2001). "Factors affecting the dimerization of the p66 form of HIV-1 reverse transcriptase." *Eur J Biochem* **268**(5): 1163–72.

Carvalho, A. T., P. A. Fernandes, et al. (2006). "Molecular dynamics model of unliganded HIV-1 reverse transcriptase." *Med Chem* **2**(5): 491–8.

Cases-Gonzalez, C. E., M. Gutierrez-Rivas, et al. (2000). "Coupling ribose selection to fidelity of DNA synthesis. The role of Tyr-115 of human immunodeficiency virus type 1 reverse transcriptase." *J Biol Chem* **275**(26): 19759–67.

Charneau, P., M. Alizon, et al. (1992). "A second origin of DNA plus-strand synthesis is required for optimal human immunodeficiency virus replication." *J Virol* **66**(5): 2814–20.

Charneau, P. and F. Clavel (1991). "A single-stranded gap in human immunodeficiency virus unintegrated linear DNA defined by a central copy of the polypurine tract." *J Virol* **65**(5): 2415–21.

Charneau, P., G. Mirambeau, et al. (1994). "HIV-1 reverse transcription. A termination step at the center of the genome." *J Mol Biol* **241**(5): 651–62.

Cobrinik, D., L. Soskey, et al. (1988). "A retroviral RNA secondary structure required for efficient initiation of reverse transcription." *J Virol* **62**(10): 3622–30.

Cowan, J. A., T. Ohyama, et al. (2000). "Metal-ion stoichiometry of the HIV-1 RT ribonuclease H domain: evidence for two mutually exclusive sites leads to new mechanistic insights on metal-mediated hydrolysis in nucleic acid biochemistry." *J Biol Inorg Chem* **5**(1): 67–74.

Dahlberg, M. E. and S. J. Benkovic (1991). "Kinetic mechanism of DNA polymerase I (Klenow fragment): identification of a second conformational change and evaluation of the internal equilibrium constant." *Biochemistry* **30**(20): 4835–43.

Dardalhon, V., B. Herpers, et al. (2001). "Lentivirus-mediated gene transfer in primary T cells is enhanced by a central DNA flap." *Gene Ther* **8**(3): 190–8.

Das, K., S. G. Sarafianos, et al. (2007). "Crystal structures of clinically relevant Lys103Asn/Tyr181Cys double mutant HIV-1 reverse transcriptase in complexes with ATP and non-nucleoside inhibitor HBY 097." *J Mol Biol* **365**(1): 77–89.

Dash, C., T. S. Fisher, et al. (2006a). "Examining interactions of HIV-1 reverse transcriptase with single-stranded template nucleotides by nucleoside analog interference." *J Biol Chem* **281**(38): 27873–81.

Dash, C., J. P. Marino, et al. (2006b). "Examining Ty3 polypurine tract structure and function by nucleoside analog interference." *J Biol Chem* **281**(5): 2773–83.

Dash, C., J. W. Rausch, et al. (2004a). "Using pyrrolo-deoxycytosine to probe RNA/DNA hybrids containing the human immunodeficiency virus type-1 3′ polypurine tract." *Nucleic Acids Res* **32**(4): 1539–47.

Dash, C., H. Y. Yi-Brunozzi, et al. (2004b). "Two modes of HIV-1 polypurine tract cleavage are affected by introducing locked nucleic acid analogs into the (-) DNA template." *J Biol Chem* **279**(35): 37095–102.

Davies, J. F., 2nd, Z. Hostomska, et al. (1991). "Crystal structure of the ribonuclease H domain of HIV-1 reverse transcriptase." *Science* **252**(5002): 88–95.

De Rijck, J. and Z. Debyser (2006). "The central DNA flap of the human immunodeficiency virus type 1 is important for viral replication." *Biochem Biophys Res Commun* **349**(3): 1100–10.

Derbyshire, V., P. S. Freemont, et al. (1988). "Genetic and crystallographic studies of the 3′,5′-exonucleolytic site of DNA polymerase I." *Science* **240**(4849): 199–201.

DeStefano, J. J. (1995). "The orientation of binding of human immunodeficiency virus reverse transcriptase on nucleic acid hybrids." *Nucleic Acids Res* **23**(19): 3901–8.

DeStefano, J. J., R. A. Bambara, et al. (1993). "Parameters that influence the binding of human immunodeficiency virus reverse transcriptase to nucleic acid structures." *Biochemistry* **32**(27): 6908–15.

DeStefano, J. J., J. V. Cristofaro, et al. (2001). "Physical mapping of HIV reverse transcriptase to the 5′ end of RNA primers." *J Biol Chem* **276**(35): 32515–21.

di Marzo Veronese, F., T. D. Copeland, et al. (1986). "Characterization of highly immunogenic p66/p51 as the reverse transcriptase of HTLV-III/LAV." *Science* **231**(4743): 1289–91.

Ding, J., K. Das, et al. (1998). "Structure and functional implications of the polymerase active site region in a complex of HIV-1 RT with a double-stranded DNA template-primer and an antibody Fab fragment at 2.8 A resolution." *J Mol Biol* **284**(4): 1095–111.

Ding, J., K. Das, et al. (1995). "Structure of HIV-1 reverse transcriptase in a complex with the non-nucleoside inhibitor alpha-APA R 95845 at 2.8 A resolution." *Structure* **3**(4): 365–79.

Divita, G., T. Restle, et al. (1993). "Characterization of the dimerization process of HIV-1 reverse transcriptase heterodimer using intrinsic protein fluorescence." *FEBS Lett* **324**(2): 153–8.

Divita, G., K. Rittinger, et al. (1995). "Dimerization kinetics of HIV-1 and HIV-2 reverse transcriptase: a two step process." *J Mol Biol* **245**(5): 508–21.

Doublie, S., M. R. Sawaya, et al. (1999). "An open and closed case for all polymerases." *Structure* **7**(2): R31–5.

Dvorin, J. D., P. Bell, et al. (2002). "Reassessment of the roles of integrase and the central DNA flap in human immunodeficiency virus type 1 nuclear import." *J Virol* **76**(23): 12087–96.

Esnouf, R., J. Ren, et al. (1995). "Mechanism of inhibition of HIV-1 reverse transcriptase by non-nucleoside inhibitors." *Nat Struct Biol* **2**(4): 303–8.

Furfine, E. S. and J. E. Reardon (1991). "Reverse transcriptase. RNase H from the human immunodeficiency virus. Relationship of the DNA polymerase and RNA hydrolysis activities." *J Biol Chem* **266**(1): 406–12.

Gao, H. Q., S. G. Sarafianos, et al. (2001). "RNase H cleavage of the 5′ end of the human immunodeficiency virus type 1 genome." *J Virol* **75**(23): 11874–80.

Garforth, S. J., T. W. Kim, et al. (2007). "Site-directed mutagenesis in the fingers subdomain of HIV-1 reverse transcriptase reveals a specific role for the beta3-beta4 hairpin loop in dNTP selection." *J Mol Biol* **365**(1): 38–49.

Ghosh, M., P. S. Jacques, et al. (1996). "Alterations to the primer grip of p66 HIV-1 reverse transcriptase and their consequences for template-primer utilization." *Biochemistry* **35**(26): 8553–62.

Ghosh, M., J. Williams, et al. (1997). "Mutating a conserved motif of the HIV-1 reverse transcriptase palm subdomain alters primer utilization." *Biochemistry* **36**(19): 5758–68.

Godet, J., H. de Rocquigny, et al. (2006). "During the early phase of HIV-1 DNA synthesis, nucleocapsid protein directs hybridization of the TAR complementary sequences via the ends of their double-stranded stem." *J Mol Biol* **356**(5): 1180–92.

Goel, R., W. A. Beard, et al. (1993). "Structure/function studies of HIV-1(1) reverse transcriptase: dimerization-defective mutant L289K." *Biochemistry* **32**(48): 13012–8.

Goldschmidt, V., J. C. Paillart, et al. (2004). "Structural variability of the initiation complex of HIV-1 reverse transcription." *J Biol Chem* **279**(34): 35923–31.

Golinelli, M. P. and S. H. Hughes (2003). "Secondary structure in the nucleic acid affects the rate of HIV-1 nucleocapsid-mediated strand annealing." *Biochemistry* **42**(27): 8153–62.

Gopalakrishnan, V., J. A. Peliska, et al. (1992). "Human immunodeficiency virus type 1 reverse transcriptase: spatial and temporal relationship between the polymerase and RNase H activities." *Proc Natl Acad Sci USA* **89**(22): 10763–7.

Gotte, M. (2006). "Effects of nucleotides and nucleotide analogue inhibitors of HIV-1 reverse transcriptase in a ratchet model of polymerase translocation." *Curr Pharm Des* **12**(15): 1867–77.

Gotte, M., M. Kameoka, et al. (2001). "Analysis of efficiency and fidelity of HIV-1 (+)-strand DNA synthesis reveals a novel rate-limiting step during retroviral reverse transcription." *J Biol Chem* **276**(9): 6711–9.

Gotte, M., G. Maier, et al. (1998). "Localization of the active site of HIV-1 reverse transcriptase-associated RNase H domain on a DNA template using site-specific generated hydroxyl radicals." *J Biol Chem* **273**(17): 10139–46.

Guajardo, R. and R. Sousa (1997). "A model for the mechanism of polymerase translocation." *J Mol Biol* **265**(1): 8–19.

Guo, F., S. Cen, et al. (2006). "Inhibition of formula-primed reverse transcription by human APOBEC3G during human immunodeficiency virus type 1 replication." *J Virol* **80**(23): 11710–22.

Guo, J., T. Wu, et al. (2000). "Zinc finger structures in the human immunodeficiency virus type 1 nucleocapsid protein facilitate efficient minus- and plus-strand transfer." *J Virol* **74**(19): 8980–8.

Hang, J. Q., S. Rajendran, et al. (2004). "Activity of the isolated HIV RNase H domain and specific inhibition by N-hydroxyimides." *Biochem Biophys Res Commun* **317**(2): 321–9.

Harris, D., R. Lee, et al. (1998). "The p51 subunit of human immunodeficiency virus type 1 reverse transcriptase is essential in loading the p66 subunit on the template primer." *Biochemistry* **37**(17): 5903–8.

Hsiou, Y., J. Ding, et al. (1996). "Structure of unliganded HIV-1 reverse transcriptase at 2.7 A resolution: implications of conformational changes for polymerization and inhibition mechanisms." *Structure* **4**(7): 853–60.

Huang, H., R. Chopra, et al. (1998). "Structure of a covalently trapped catalytic complex of HIV-1 reverse transcriptase: implications for drug resistance." *Science* **282**(5394): 1669–75.

Huber, H. E. and C. C. Richardson (1990). "Processing of the primer for plus strand DNA synthesis by human immunodeficiency virus 1 reverse transcriptase." *J Biol Chem* **265**(18): 10565–73.

Hungnes, O., E. Tjotta, et al. (1991). "The plus strand is discontinuous in a subpopulation of unintegrated HIV-1 DNA." *Arch Virol* **116**(1–4): 133–41.

Hungnes, O., E. Tjotta, et al. (1992). "Mutations in the central polypurine tract of HIV-1 result in delayed replication." *Virology* **190**(1): 440–2.

Isel, C., C. Ehresmann, et al. (1995). "Initiation of reverse transcription of HIV-1: secondary structure of the HIV-1 RNA/tRNA(3Lys) (template/primer)." *J Mol Biol* **247**(2): 236–50.

Isel, C., R. Marquet, et al. (1993). "Modified nucleotides of tRNA(3Lys) modulate primer/template loop-loop interaction in the initiation complex of HIV-1 reverse transcription." *J Biol Chem* **268**(34): 25269–72.

Iwatani, Y., A. E. Rosen, et al. (2003). "Efficient initiation of HIV-1 reverse transcription in vitro. Requirement for RNA sequences downstream of the primer binding site abrogated by nucleocapsid protein-dependent primer-template interactions." *J Biol Chem* **278**(16): 14185–95.

Jacobo-Molina, A., J. Ding, et al. (1993). "Crystal structure of human immunodeficiency virus type 1 reverse transcriptase complexed with double-stranded DNA at 3.0 A resolution shows bent DNA." *Proc Natl Acad Sci USA* **90**(13): 6320–4.

Jacques, P. S., B. M. Wohrl, et al. (1994a). "Modulation of HIV-1 reverse transcriptase function in "selectively deleted" p66/p51 heterodimers." *J Biol Chem* **269**(2): 1388–93.

Jacques, P. S., B. M. Wohrl, et al. (1994b). "Mutating the "primer grip" of p66 HIV-1 reverse transcriptase implicates tryptophan-229 in template-primer utilization." *J Biol Chem* **269**(42): 26472–8.

Javanbakht, H., R. Halwani, et al. (2003). "The interaction between HIV-1 Gag and human lysyl-tRNA synthetase during viral assembly." *J Biol Chem* **278**(30): 27644–51.

Jones, F. D. and S. H. Hughes (2007). "In vitro analysis of the effects of mutations in the G-tract of the human immunodeficiency virus type 1 polypurine tract on RNase H cleavage specificity." *Virology* **360**(2): 341–9.

Julias, J. G., M. J. McWilliams, et al. (2004). "Effects of mutations in the G tract of the human immunodeficiency virus type 1 polypurine tract on virus replication and RNase H cleavage." *J Virol* **78**(23): 13315–24.

Kaminska, M., V. Shalak, et al. (2007). "Viral hijacking of mitochondrial lysyl-tRNA synthetase." *J Virol* **81**(1): 68–73.

Kang, S. M., J. K. Wakefield, et al. (1996). "Mutations in both the U5 region and the primer-binding site influence the selection of the tRNA used for the initiation of HIV-1 reverse transcription." *Virology* **222**(2): 401–14.

Kankia, B. I. and K. Musier-Forsyth (2007). "The HIV-1 central DNA flap region contains a "flapping" third strand." *Biophys Chem* **127**(1–2): 64–8.

Kao, H. I., J. Veeraraghavan, et al. (2004). "On the roles of Saccharomyces cerevisiae Dna2p and Flap endonuclease 1 in Okazaki fragment processing." *J Biol Chem* **279**(15): 15014–24.

Katayanagi, K., M. Miyagawa, et al. (1990). "Three-dimensional structure of ribonuclease H from E. coli." *Nature* **347**(6290): 306–9.

Kensch, O., T. Restle, et al. (2000). "Temperature-dependent equilibrium between the open and closed conformation of the p66 subunit of HIV-1 reverse transcriptase revealed by site-directed spin labelling." *J Mol Biol* **301**(4): 1029–39.

Khorchid, A., H. Javanbakht, et al. (2000). "Sequences within Pr160gag-pol affecting the selective packaging of primer tRNA(Lys3) into HIV-1." *J Mol Biol* **299**(1): 17–26.

Klarmann, G. J., B. M. Eisenhauer, et al. (2007). "Investigating the "steric gate" of human immunodeficiency virus type 1 (HIV-1) reverse transcriptase by targeted insertion of unnatural amino acids." *Biochemistry* **46**(8): 2118–26.

Klarmann, G. J., B. M. Eisenhauer, et al. (2004). "Site- and subunit-specific incorporation of unnatural amino acids into HIV-1 reverse transcriptase." *Protein Expr Purif* **38**(1): 37–44.

Kleiman, L., R. Halwani, et al. (2004). "The selective packaging and annealing of primer tRNALys3 in HIV-1." *Curr HIV Res* **2**(2): 163–75.

Kohlstaedt, L. A., J. Wang, et al. (1992). "Crystal structure at 3.5 A resolution of HIV-1 reverse transcriptase complexed with an inhibitor." *Science* **256**(5065): 1783–90.

Kovaleski, B. J., R. Kennedy, et al. (2006). "In vitro characterization of the interaction between HIV-1 Gag and human lysyl-tRNA synthetase." *J Biol Chem* **281**(28): 19449–56.

Kvaratskhelia, M., S. R. Budihas, et al. (2002). "Pre-existing distortions in nucleic acid structure aid polypurine tract selection by HIV-1 reverse transcriptase." *J Biol Chem* **277**(19): 16689–96.

Lanchy, J. M., C. Ehresmann, et al. (1996). "Binding and kinetic properties of HIV-1 reverse transcriptase markedly differ during initiation and elongation of reverse transcription." *Embo J* **15**(24): 7178–87.

Landick, R. (2004). "Active-site dynamics in RNA polymerases." *Cell* **116**(3): 351–3.

Lavigne, M. and H. Buc (1999). "Compression of the DNA minor groove is responsible for termination of DNA synthesis by HIV-1 reverse transcriptase." *J Mol Biol* **285**(3): 977–95.

Lavigne, M., L. Polomack, et al. (2001). "DNA synthesis by HIV-1 reverse transcriptase at the central termination site: a kinetic study." *J Biol Chem* **276**(33): 31429–38.

Lavigne, M., P. Roux, et al. (1997). "DNA curvature controls termination of plus strand DNA synthesis at the centre of HIV-1 genome." *J Mol Biol* **266**(3): 507–24.

Le Grice, S. F. (2003). ""In the beginning": initiation of minus strand DNA synthesis in retroviruses and LTR-containing retrotransposons." *Biochemistry* **42**(49): 14349–55.

Le Grice, S. F., T. Naas, et al. (1991). "Subunit-selective mutagenesis indicates minimal polymerase activity in heterodimer-associated p51 HIV-1 reverse transcriptase." *Embo J* **10**(12): 3905–11.

Lener, D., S. R. Budihas, et al. (2002). "Mutating conserved residues in the ribonuclease H domain of Ty3 reverse transcriptase affects specialized cleavage events." *J Biol Chem* **277**(29): 26486–95.

Levin, J. G., J. Guo, et al. (2005). "Nucleic acid chaperone activity of HIV-1 nucleocapsid protein: critical role in reverse transcription and molecular mechanism." *Prog Nucleic Acid Res Mol Biol* **80**: 217–86.

Liang, C., X. Li, et al. (1997). "The importance of the A-rich loop in human immunodeficiency virus type 1 reverse transcription and infectivity." *J Virol* **71**(8): 5750–7.

Limon, A., N. Nakajima, et al. (2002). "Wild-type levels of nuclear localization and human immunodeficiency virus type 1 replication in the absence of the central DNA flap." *J Virol* **76**(23): 12078–86.

Liu, H. W., G. Cosa, et al. (2005). "Single-molecule FRET studies of important intermediates in the nucleocapsid-protein-chaperoned minus-strand transfer step in HIV-1 reverse transcription." *Biophys J* **89**(5): 3470–9.

Liu, H. W., Y. Zeng, et al. (2007). "Insights on the role of nucleic acid/protein interactions in chaperoned nucleic acid rearrangements of HIV-1 reverse transcription." *Proc Natl Acad Sci USA* **104**(13): 5261–7.

Lowe, D. M., A. Aitken, et al. (1988). "HIV-1 reverse transcriptase: crystallization and analysis of domain structure by limited proteolysis." *Biochemistry* **27**(25): 8884–9.

Luo, G. X. and J. Taylor (1990). "Template switching by reverse transcriptase during DNA synthesis." *J Virol* **64**(9): 4321–8.

Ma, J. B., Y. R. Yuan, et al. (2005). "Structural basis for 5′-end-specific recognition of guide RNA by the A. fulgidus Piwi protein." *Nature* **434**(7033): 666–70.

Majumdar, C., J. Abbotts, et al. (1988). "Studies on the mechanism of human immunodeficiency virus reverse transcriptase. Steady-state kinetics, processivity, and polynucleotide inhibition." *J Biol Chem* **263**(30): 15657–65.

Mak, J. and L. Kleiman (1997). "Primer tRNAs for reverse transcription." *J Virol* **71**(11): 8087–95.

Marchand, B., E. P. Tchesnokov, et al. (2007). "The pyrophosphate analogue foscarnet traps the pre-translocational state of HIV-1 reverse transcriptase in a Brownian ratchet model of polymerase translocation." *J Biol Chem* **282**(5): 3337–46.

Marquet, R., C. Isel, et al. (1995). "tRNAs as primer of reverse transcriptases." *Biochimie* **77**(1–2): 113–24.

Martin-Hernandez, A. M., E. Domingo, et al. (1996). "Human immunodeficiency virus type 1 reverse transcriptase: role of Tyr115 in deoxynucleotide binding and misinsertion fidelity of DNA synthesis." *Embo J* **15**(16): 4434–42.

McWilliams, M. J., J. G. Julias, et al. (2003). "Mutations in the 5′ end of the human immunodeficiency virus type 1 polypurine tract affect RNase H cleavage specificity and virus titer." *J Virol* **77**(20): 11150–7.

Menendez-Arias, L., A. Abraha, et al. (2001). "Functional characterization of chimeric reverse transcriptases with polypeptide subunits of highly divergent HIV-1 group M and O strains." *J Biol Chem* **276**(29): 27470–9.

Miles, L. R., B. E. Agresta, et al. (2005). "Effect of polypurine tract (PPT) mutations on human immunodeficiency virus type 1 replication: a virus with a completely randomized PPT retains low infectivity." *J Virol* **79**(11): 6859–67.

Miller, J. T., A. Khvorova, et al. (2004). "Synthetic tRNALys,3 as the replication primer for the HIV-1HXB2 and HIV-1Mal genomes." *Nucleic Acids Res* **32**(15): 4687–95.

Molling, K., D. P. Bolognesi, et al. (1971). "Association of viral reverse transcriptase with an enzyme degrading the RNA moiety of RNA-DNA hybrids." *Nat New Biol* **234**(51): 240–3.

Mulky, A., S. G. Sarafianos, et al. (2005). "Identification of amino acid residues in the human immunodeficiency virus type-1 reverse transcriptase tryptophan-repeat motif that are required for subunit interaction using infectious virions." *J Mol Biol* **349**(4): 673–84.

Muller, B., T. Restle, et al. (1989). "Co-expression of the subunits of the heterodimer of HIV-1 reverse transcriptase in Escherichia coli." *J Biol Chem* **264**(24): 13975–8.

Ni, N., W. Xu, et al. (2007). "Importance of A-loop complementarity with tRNAHis anticodon for continued selection of tRNAHis as the HIV reverse transcription primer." *Virol J* **4**: 4.

Nowotny, M., S. A. Gaidamakov, et al. (2005). "Crystal structures of RNase H bound to an RNA/DNA hybrid: substrate specificity and metal-dependent catalysis." *Cell* **121**(7): 1005–16.

Nowotny, M. and W. Yang (2006). "Stepwise analyses of metal ions in RNase H catalysis from substrate destabilization to product release." *Embo J* **25**(9): 1924–33.

Ooms, M., D. Cupac, et al. (2007). "The availability of the primer activation signal (PAS) affects the efficiency of HIV-1 reverse transcription initiation." *Nucleic Acids Res* **35**(5): 1649–59.

Operario, D. J., M. Balakrishnan, et al. (2006). "Reduced dNTP interaction of human immunodeficiency virus type 1 reverse transcriptase promotes strand transfer." *J Biol Chem* **281**(43): 32113–21.

Palaniappan, C., M. Wisniewski, et al. (1997). "Mutations within the primer grip region of HIV-1 reverse transcriptase result in loss of RNase H function." *J Biol Chem* **272**(17): 11157–64.

Peliska, J. A. and S. J. Benkovic (1992). "Mechanism of DNA strand transfer reactions catalyzed by HIV-1 reverse transcriptase." *Science* **258**(5085): 1112–8.

Powell, M. D., M. Ghosh, et al. (1997). "Alanine-scanning mutations in the "primer grip" of p66 HIV-1 reverse transcriptase result in selective loss of RNA priming activity." *J Biol Chem* **272**(20): 13262–9.

Powell, M. D. and J. G. Levin (1996). "Sequence and structural determinants required for priming of plus-strand DNA synthesis by the human immunodeficiency virus type 1 polypurine tract." *J Virol* **70**(8): 5288–96.

Pullen, K. A., A. J. Rattray, et al. (1993). "The sequence features important for plus strand priming by human immunodeficiency virus type 1 reverse transcriptase." *J Biol Chem* **268**(9): 6221–7.

Rausch, J. W. and S. F. Le Grice (2007). "Purine analog substitution of the HIV-1 polypurine tract primer defines regions controlling initiation of plus-strand DNA synthesis." *Nucleic Acids Res* **35**(1): 256–68.

Rausch, J. W., J. Qu, et al. (2003). "Hydrolysis of RNA/DNA hybrids containing nonpolar pyrimidine isosteres defines regions essential for HIV type 1 polypurine tract selection." *Proc Natl Acad Sci USA* **100**(20): 11279–84.

Restle, T., B. Muller, et al. (1990). "Dimerization of human immunodeficiency virus type 1 reverse transcriptase. A target for chemotherapeutic intervention." *J Biol Chem* **265**(16): 8986–8.

Restle, T., B. Muller, et al. (1992). "RNase H activity of HIV reverse transcriptases is confined exclusively to the dimeric forms." *FEBS Lett* **300**(1): 97–100.

Rice, P. A. and T. A. Baker (2001). "Comparative architecture of transposase and integrase complexes." *Nat Struct Biol* **8**(5): 302–7.

Rittinger, K., G. Divita, et al. (1995). "Human immunodeficiency virus reverse transcriptase substrate-induced conformational changes and the mechanism of inhibition by nonnucleoside inhibitors." *Proc Natl Acad Sci USA* **92**(17): 8046–9.

Robson, N. D. and A. Telesnitsky (2000). "Selection of optimal polypurine tract region sequences during Moloney murine leukemia virus replication." *J Virol* **74**(22): 10293–303.

Rodgers, D. W., S. J. Gamblin, et al. (1995). "The structure of unliganded reverse transcriptase from the human immunodeficiency virus type 1." *Proc Natl Acad Sci USA* **92**(4): 1222–6.

Rodriguez-Barrios, F., C. Perez, et al. (2001). "Identification of a putative binding site for [2′,5′-bis-O-(tert-butyldimethylsilyl)-beta-D-ribofuranosyl]-3′-spiro-5″- (4″-amino-1″,2″-oxathiole-2″,2″-dioxide)thymine (TSAO) derivatives at the p51-p66 interface of HIV-1 reverse transcriptase." *J Med Chem* **44**(12): 1853–65.

Rothwell, P. J., S. Berger, et al. (2003). "Multiparameter single-molecule fluorescence spectroscopy reveals heterogeneity of HIV-1 reverse transcriptase:primer/template complexes." *Proc Natl Acad Sci USA* **100**(4): 1655–60.

Rumbaugh, J. A., G. M. Fuentes, et al. (1998). "Processing of an HIV replication intermediate by the human DNA replication enzyme FEN1." *J Biol Chem* **273**(44): 28740–5.

Sarafianos, S. G., K. Das, et al. (1999). "Touching the heart of HIV-1 drug resistance: the fingers close down on the dNTP at the polymerase active site." *Chem Biol* **6**(5): R137–46.

Sarafianos, S. G., K. Das, et al. (2001). "Crystal structure of HIV-1 reverse transcriptase in complex with a polypurine tract RNA:DNA." *Embo J* **20**(6): 1449–61.

Schultz, S. J., M. Zhang, et al. (2006). "Sequence, distance, and accessibility are determinants of 5′-end-directed cleavages by retroviral RNases H." *J Biol Chem* **281**(4): 1943–55.

Shaw-Reid, C. A., V. Munshi, et al. (2003). "Inhibition of HIV-1 ribonuclease H by a novel diketo acid, 4-[5-(benzoylamino)thien-2-yl]-2,4-dioxobutanoic acid." *J Biol Chem* **278**(5): 2777–80.

Sirven, A., F. Pflumio, et al. (2000). "The human immunodeficiency virus type-1 central DNA flap is a crucial determinant for lentiviral vector nuclear import and gene transduction of human hematopoietic stem cells." *Blood* **96**(13): 4103–10.

Smith, C. M., J. S. Smith, et al. (1999). "RNase H requirements for the second strand transfer reaction of human immunodeficiency virus type 1 reverse transcription." *J Virol* **73**(8): 6573–81.

Song, J. J., S. K. Smith, et al. (2004). "Crystal structure of Argonaute and its implications for RISC slicer activity." *Science* **305**(5689): 1434–7.

Song, M., M. Balakrishnan, et al. (2006). "Stimulation of HIV-1 minus strand strong stop DNA transfer by genomic sequences 3′ of the primer binding site." *J Biol Chem* **281**(34): 24227–35.

Spence, R. A., W. M. Kati, et al. (1995). "Mechanism of inhibition of HIV-1 reverse transcriptase by nonnucleoside inhibitors." *Science* **267**(5200): 988–93.

Srivastava, S., N. Sluis-Cremer, et al. (2006). "Dimerization of human immunodeficiency virus type 1 reverse transcriptase as an antiviral target." *Curr Pharm Des* **12**(15): 1879–94.

Steitz, T. A. (1993). "DNA- and RNA-dependent DNA polymerases." *Curr Opin Struct Biol* **3**: 31–38.

Steitz, T. A. (1998). "A mechanism for all polymerases." *Nature* **391**(6664): 231–2.

Steitz, T. A. (1999). "DNA polymerases: structural diversity and common mechanisms." *J Biol Chem* **274**(25): 17395–8.

Steitz, T. A. (2004). "The structural basis of the transition from initiation to elongation phases of transcription, as well as translocation and strand separation, by T7 RNA polymerase." *Curr Opin Struct Biol* **14**(1): 4–9.

Steitz, T. A., S. J. Smerdon, et al. (1994). "A unified polymerase mechanism for nonhomologous DNA and RNA polymerases." *Science* **266**(5193): 2022–5.

Steitz, T. A. and J. A. Steitz (1993). "A general two-metal-ion mechanism for catalytic RNA." *Proc Natl Acad Sci USA* **90**(14): 6498–502.

Steitz, T. A. and Y. W. Yin (2004). "Accuracy, lesion bypass, strand displacement and translocation by DNA polymerases." *Philos Trans R Soc Lond B Biol Sci* **359**(1441): 17–23.

Stetor, S. R., J. W. Rausch, et al. (1999). "Characterization of (+) strand initiation and termination sequences located at the center of the equine infectious anemia virus genome." *Biochemistry* **38**(12): 3656–67.

Tachedjian, G., H. E. Aronson, et al. (2003). "Role of residues in the tryptophan repeat motif for HIV-1 reverse transcriptase dimerization." *J Mol Biol* **326**(2): 381–96.

Tan, C. K., J. Zhang, et al. (1991). "Functional characterization of RNA-dependent DNA polymerase and RNase H activities of a recombinant HIV reverse transcriptase." *Biochemistry* **30**(10): 2651–5.

Tantillo, C., J. Ding, et al. (1994). "Locations of anti-AIDS drug binding sites and resistance mutations in the three-dimensional structure of HIV-1 reverse transcriptase. Implications for mechanisms of drug inhibition and resistance." *J Mol Biol* **243**(3): 369–87.

Telesnitsky, A. and S. P. Goff (1997). Reverse Transcriptase and the Generation of Retroviral DNA. *Retroviruses*. J. M. Coffin, S. H. Hughes and H. E. Varmus. Plainview, Cold Spring Harbor Laboratory Press: 121–160.

Temiakov, D., V. Patlan, et al. (2004). "Structural basis for substrate selection by t7 RNA polymerase." *Cell* **116**(3): 381–91.

Tisdale, M., T. Schulze, et al. (1991). "Mutations within the RNase H domain of human immunodeficiency virus type 1 reverse transcriptase abolish virus infectivity." *J Gen Virol* **72(Pt 1)**: 59–66.

Tisne, C., B. P. Roques, et al. (2004). "The annealing mechanism of HIV-1 reverse transcription primer onto the viral genome." *J Biol Chem* **279**(5): 3588–95.

Van Maele, B., J. De Rijck, et al. (2003). "Impact of the central polypurine tract on the kinetics of human immunodeficiency virus type 1 vector transduction." *J Virol* **77**(8): 4685–94.

van Wamel, J. L. and B. Berkhout (1998). "The first strand transfer during HIV-1 reverse transcription can occur either intramolecularly or intermolecularly." *Virology* **244**(2): 245–51.

Wang, J., S. J. Smerdon, et al. (1994). "Structural basis of asymmetry in the human immunodeficiency virus type 1 reverse transcriptase heterodimer." *Proc Natl Acad Sci USA* **91**(15): 7242–6.

Whitwam, T., M. Peretz, et al. (2001). "Identification of a central DNA flap in feline immunodeficiency virus." *J Virol* **75**(19): 9407–14.

Wisniewski, M., Y. Chen, et al. (2002). "Substrate requirements for secondary cleavage by HIV-1 reverse transcriptase RNase H." *J Biol Chem* **277**(32): 28400–10.

Wisniewski, M., C. Palaniappan, et al. (1999). "Mutations in the primer grip region of HIV reverse transcriptase can increase replication fidelity." *J Biol Chem* **274**(40): 28175–84.

Wohrl, B. M., K. J. Howard, et al. (1994). "Alternative modes of polymerization distinguish the subunits of equine infectious anemia virus reverse transcriptase." *J Biol Chem* **269**(11): 8541–8.

Wohrl, B. M., R. Krebs, et al. (1999). "Refined model for primer/template binding by HIV-1 reverse transcriptase: pre-steady-state kinetic analyses of primer/template binding and nucleotide incorporation events distinguish between different binding modes depending on the nature of the nucleic acid substrate." *J Mol Biol* **292**(2): 333–44.

Wohrl, B. M., R. Krebs, et al. (1997). "Kinetic analysis of four HIV-1 reverse transcriptase enzymes mutated in the primer grip region of p66. Implications for DNA synthesis and dimerization." *J Biol Chem* **272**(28): 17581–7.

Xiong, Y. and T. H. Eickbush (1990). "Origin and evolution of retroelements based upon their reverse transcriptase sequences." *Embo J* **9**(10): 3353–62.

Yang, W., J. Y. Lee, et al. (2006). "Making and breaking nucleic acids: two-Mg2+-ion catalysis and substrate specificity." *Mol Cell* **22**(1): 5–13.

Yi-Brunozzi, H. Y. and S. F. Le Grice (2005). "Investigating HIV-1 polypurine tract geometry via targeted insertion of abasic lesions in the (-)-DNA template and (+)-RNA primer." *J Biol Chem* **280**(20): 20154–62.

Yin, Y. W. and T. A. Steitz (2004). "The structural mechanism of translocation and helicase activity in T7 RNA polymerase." *Cell* **116**(3): 393–404.

You, J. C. and C. S. McHenry (1994). "Human immunodeficiency virus nucleocapsid protein accelerates strand transfer of the terminally redundant sequences involved in reverse transcription." *J Biol Chem* **269**(50): 31491–5.

Yuan, Y. R., Y. Pei, et al. (2005). "Crystal structure of A. aeolicus argonaute, a site-specific DNA-guided endoribonuclease, provides insights into RISC-mediated mRNA cleavage." *Mol Cell* **19**(3): 405–19.

Zennou, V., C. Petit, et al. (2000). "HIV-1 genome nuclear import is mediated by a central DNA flap." *Cell* **101**(2): 173–85.

Zennou, V., C. Serguera, et al. (2001). "The HIV-1 DNA flap stimulates HIV vector-mediated cell transduction in the brain." *Nat Biotechnol* **19**(5): 446–50.

Chapter 20
Viral Helicases

Vaishnavi Rajagopal and Smita S. Patel

Introduction

Helicases are motor proteins that use the free energy of NTP hydrolysis to catalyze the unwinding of duplex nucleic acids. Helicases participate in almost all processes involving nucleic acids. Their action is critical for replication, recombination, repair, transcription, translation, splicing, mRNA editing, chromatin remodeling, transport, and degradation (Matson and Kaiser-Rogers 1990; Matson et al. 1994; Mendonca et al. 1995; Luking et al. 1998).

A significant number of genes of all organisms encode for helicases. A study in *Salmonella cerevisiae* revealed 134 ORFs encoding for helicases (Shiratori et al. 1999). This would account for about 2% of the yeast genome. Similarly, more than 12 DNA helicases and about 17 RNA helicases have been identified in *Escherchia coli*(Matson 1991; Schmid and Linder 1992; Bird et al. 1998; Egelman 1998; Dreyfus 2006). However, this is not unexpected considering that these enzymes are ubiquitous and are involved in such diverse metabolic roles. As is the case with the bacteria and other higher eukaryotes, most viruses too encode for proteins with conserved helicase motifs.

For viruses whose genomes are comprised of double-stranded DNA or RNA, the presence of a helicase in the virus-encoded genome is conceivable. This is indeed the case for most double-stranded DNA and RNA viruses for which the genome sequence has been reported (Gorbalenya et al. 1988a,b; Gorbalenya and Koonin 1989). Many positive-strand RNA viruses too encode their own helicases presumably to remove any partial duplexes that might exist within the genome and to facilitate viral replication either directly or indirectly (Jeang and Yedavalli 2006). Some viruses have been identified to encode for more than one helicase indicating the role of helicases in other viral processes like packaging (Kadare and Haenni 1997; Luking et al. 1998). Table 20.1 gives a comprehensive list of viral genuses with their

S.S. Patel (✉)
Department of Biochemistry, Robert Wood Johnson Medical School, 675 Hoes Lane, Piscataway, NJ 08854, USA
e-mail: patelss@umdnj.edu

C.E. Cameron et al. (eds.), *Viral Genome Replication*,
DOI 10.1007/b135974_20, © Springer Science+Business Media, LLC 2009

Table 20.1 Viral genuses and their associated helicases

Virus family	Representative common name	Known hosts[a]	Helicase superfamily[b]
Double-strand DNA virus			
Myoviridae	Phage T4	Arc, Eub	SF1
Siphoviridae	Phage1	Arc, Eub	NC
Podoviridae	Phage T7	Eub	DnaB-like
Tectiviridae	Phage PRD1	Eub	NC
Corticoviridae	Phage PM2	Eub	NC
Plasmaviridae	Phage L2	Eub	NC
Lipothrixviridae	Thermoproteus virus 1	Arc	NC
Rudiviridae	Sulfolobus virus SIRV1	Arc	NC
Fuselloviridae	Sulfolobus virus SSV1	Arc	NC
Poxviridae	Vaccinia virus	Inv, Ver	SF2
Asfarviridae	African swine fever virus	Ver	NC
Iridoviridae	Lymphocystis disease virus1	Inv, Ver	NC
Phycodnaviridae	*Paramecium bursaria* chlorella virus1	Inv	NC
Baculoviridae	*Cydia pomonella* granulovirus	Alg	NC
Herpesviridae	Herpes virus	Ver	SF1, SF2
Adenoviridae	Adenovirus	Ver	SF3
Polyomaviridae	Simian Virus 40	Ver	SF3
Papillomaviridae	Human papilloma virus	Ver	DnaB-like
Polydnaviridae	*Campoletis aprilis* ichnovirus	Inv	NC
Ascoviridae	*Diadromus pulchellus* ascovirus	Inv	NC
Single-strand DNA virus			
Inoviridae	Phage M13	Eub	NC
Microviridae	Phage ϕX174	Eub	NC
Geminiviridae	Maize streak virus	Pla	Yet to be classified
Circoviridae	Porcine circovirus	Ver	NC
Parvoviridae	Adeno-Associated virus	Inv, Ver	SF3

Table 20.1 (continued)

Virus family	Representative common name	Known hosts[a]	Helicase superfamily[b]
DNA-RNA reverse-transcribing virus			
Hepadnaviridae	Hepatitis B virus	Ver	NA
Caulimoviridae	Cauliflower mosaic virus	Pla	NA
Pseudoviridae	*Saccharomyces cerevisiae* Ty-1 virus	Fun, Inv, Pla	NA
Metaviridae	*Drosophila melanogaster* gypsy virus	Fun, Inv, Pla	NA
Retroviridae	HIV-1	Ver	
Double-strand RNA virus			
Cystoviridae	Phage f6	Eub	NC
Reoviridae	Rice dwarf virus	Inv, Ver, Pla	Yet to be classified
Birnaviridae	infectious pancreatic necrosis virus	Inv, Ver	NC
Totiviridae	*Giardia lamblia* virus	Fun, Pro	NC
Partiviridae	*Penicillium chrysogenum* virus	Fun, Pla	NC
Hypoviridae	*Chryphonectria* hypovirus	Fun	NC
Single-strand (–) RNA virus			
Bornaviridae	Borna disease virus	Ver	NC
Filoviridae	Zaire ebola virus	Ver	NC
Paramyxoviridae	Mumps virus, measles virus	Pla, Ver	NC
Rhabdoviridae	Rabies virus, Potato yello dwarf virus	Pla, Ver	Yet to be classified
Orthomyxoviridae	Influenza A virus	Ver	NC
Bunyaviridae	Tomato spotted wilt virus	Pla, Ver	SF2
Arenaviridae	Hepatitis δ-virus	Ver	NC

Table 20.1 (continued)

Virus family	Representative common name	Known hosts[a]	Helicase superfamily[b]
Single-strand (+) RNA virus			
Leviviridae	Phage MS2	Eub	NC
Narnaviridae	*Saccharomyces cerevisiae* narnavirus 20 s	Fun	NC
Picornaviridae	Polio virus, Hepatitis A virus	Ver	SF3
Sequiviridae	Parsnip yellow fleck virus	Pla	NC
Comoviridae	Tobacco ringspot virus	pla	NC
Potyviridae	Ryegrass mosaic virus	pla	SF2
Calciviridae	Rabbit haemorrhagic disease virus	Ver	SF3
Astroviridae	Human astrovirus1	Ver	NC
Nodaviridae	Striped jack nervous necrosis virus	Inv, Ver	NC
Tetraviridae	Bean mosaic virus	Inv	NC
Luteoviridae	Barley yellow drarf virus PAV	Pla	NC
Tombusviridae	Oat chlorotic stunt virus	Pla	Yet to be classified
Coronaviridae	Equine torovirus	Ver	SF1
Arteviridae	Equine arteritis virus	Ver	NC
Flaviviridae	Hepatitis C Virus, Dengue Virus	Ver	SF2
Togaviridae	Rubella virus, Tobacco mosaic virus	Pla, Ver	SF1
Bromoviridae	Brome mosaic virus	Pla	SF1
Closteroviridae	Grapevine virus A	Pla	Yet to be classified
Barnaviridae	Mushroom bacilliform virus	Fun	NC

The six classes and their respective families of viruses and their hosts are listed. The table also indicates if any helicases have been identified and characterized in each of these families (Fields et al. 1996; Kadare and Haenni 1997; Mindell 2004; Ackermann 2006)

a – Arc – Archea; Eub – Eubacteria; Pro – Protist ; Alg – Algae; Fun – Fungi; Ver – vertebrate; Inv – Invertebrate; Pla – Plant;

b – NA – No helicase identified; NC – Helicase not characterized corresponding hosts. The table also indicates if the virus encodes a helicase, and if so, the superfamily it has been classified to.

Classification of Helicases

Helicases can be broadly classified into two groups based on their substrate requirements: DNA helicases and RNA helicases. This, however, is not a very stringent classification as many of the DNA helicases can unwind RNA and vice versa (Matson and Kaiser-Rogers 1990; Kadare and Haenni 1997; Luking et al. 1998). Helicases are also classified based on the polarity of their translocation as $3' \rightarrow 5'$ helicases or $5' \rightarrow 3'$ helicases. A $3' \rightarrow 5'$ helicase requires a $3'$single-stranded tail to load onto the nucleic-acid substrate and move unidirectionally toward the $5'$ end of the substrate. Some examples of $3' \rightarrow 5'$ helicase are HCV NS3 (Gwack et al. 1996), *E. coli* UvrD (Matson 1986), the RNA helicase from *Vaccinia* virus NPH-II (Shuman 1993), etc. A $5' \rightarrow 3'$ helicase requires a $5'$ single-stranded tail for it to load and move along the nucleic acid. Some examples of $5' \rightarrow 3'$ helicases are T7 gp4 helicase–primase (Matson et al. 1983), *E. coli* DnaB (LeBowitz and McMacken 1986), phage T4 Dda helicase (Jongeneel et al. 1984), etc. Though most helicases require a single-stranded tail to initiate unwinding, some enzymes like *E. coli* RecBCD can initiate the strand-separation reaction from a blunt-ended duplex (Braedt and Smith 1989).

Based on their oligomeric structure, helicases can be either ring-shaped or non-ring shaped. Although most of the non-ring shaped helicases are $3' \rightarrow 5'$ helicases and the ring-shaped helicases are predominantly $5' \rightarrow 3'$ helicases (Lohman 1993; Hall and Matson 1999; Patel and Picha 2000), there are a few exceptions to the rule; the hexameric E1 helicase from the Papilloma virus – a $3' \rightarrow 5'$ helicase (Hughes and Romanos 1993), and the monomeric Dda helicase from bacteriophage T4 – a $5' \rightarrow 3'$ helicase (Jongeneel et al. 1984) are a couple of examples.

The largest classification of helicases is based on their primary structure. The earliest classification of helicases by Gorbalenya and Koonin, based on amino acid sequence similarities revealed several conserved sequence motifs. Based on the extent of sequence similarity, they classified helicases into three large superfamilies: SF1, SF2, and SF3 (SF standing for superfamily) and smaller families (Fig. 20.1) (Gorbalenya and Koonin 1993). SF1 and SF2 constitute the largest of these superfamilies. The proteins of these superfamilies shared seven conserved motifs (Fig. 20.2). Two of these motifs, designated as the Walker A and the Walker B motifs are conserved among all the helicases and other nucleotide hydrolases as they are implicated in NTP binding and hydrolysis (Gorbalenya et al. 1988a,b). The SF3 superfamily of helicases contained only three conserved motifs including the Walker A and Walker B sequences (Fig. 20.2). This superfamily includes the majority of viral helicases (Gorbalenya et al. 1990). Of the two smaller families, one contained helicases related to the *E. coli* DnaB helicase. These proteins shared three distinct conserved motifs in addition to Walker A and Walker B sequences (Fig. 20.2). Only bacterial and bacteriophage members of this family have been identified so far and all the (putative) helicases have been shown to have a functional and/or physical

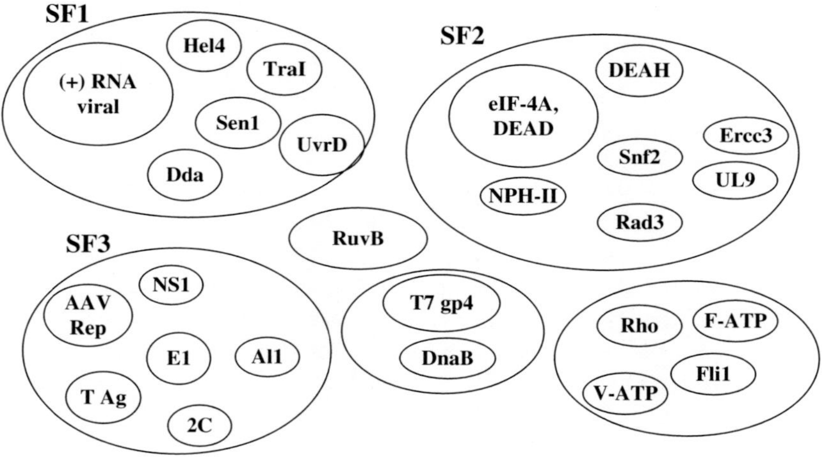

Fig. 20.1 Superfamilial classification of helicases. Helicases can be broadly classified into families and superfamilies based on sequence similarities. The diagram includes both viral and non-viral helicases; the list is not exhaustive and does not include many of the plant helicases (Gorbalenya and Koonin 1993; Levin 2002).

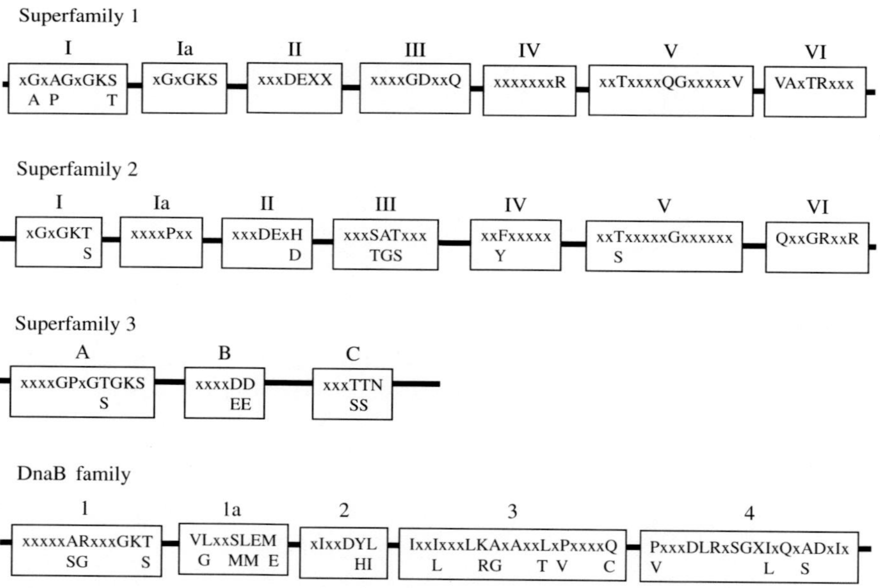

Fig. 20.2 Conserved motifs in superfamilies. Primary sequence analysis has led to the identification of certain sequence motifs that are conserved between many helicases. These sequence similarities have resulted in the classification of the helicases into different superfamilies. Represented here are the consensus sequences for the different superfamilies in the N-terminus to C-terminus orientation. The spaces between the motifs are arbitrary (Gorbalenya and Koonin 1989; Gorbalenya et al. 1989; Ilyina et al. 1992; Hall and Matson 1999).

association with a primase function as well (Ilyina et al. 1992). The last group of proteins showed extensive sequence similarity to the transcription terminator Rho. This group also included the AAA^+ family of ATPases, demonstrating an apparent evolutionary relationship between helicases and non-helicase NTPases (Iyer et al. 2004).

In the past decade, new helicases have been identified and characterized from a variety of different viruses, bacteria, archaebacteria, and plants. Novel protein motifs have been discovered which are conserved among helicases across different species. This called for a revision in the classification of the helicase superfamilies. In the current system of classification suggested by Singleton et al., the helicases are classified into six superfamilies SF1 through SF6. SF1 and SF2 still remain the largest of the superfamilies; SF3 continues to constitute the viral helicases. The DnaB-like family has been renamed as superfamily 4 (SF4), Rho family as SF5. The AAA^+ ATPases are now classified into a stand-alone superfamily of their own in SF6 (Singleton et al. 2007). The conserved sequence motifs among SF1 and SF2 family members have been extended to include the TxGx motif (Pause and Sonenberg 1992), Q-motif (Tanner et al. 2003), motif-4a (Korolev et al. 1998), and TRG motif (Mahdi et al. 2003), some of which are specific to each superfamily or subfamilies therein(Singleton et al. 2007). Singleton et al. also propose to include the directionality of translocation as criterion for classifying the members of the superfamily into subfamilies. In their classification, subfamily A represents $3' \rightarrow 5'$ helicases within the superfamily, while subfamily B represents $5' \rightarrow 3'$ helicases (Singleton et al. 2007).

Recent discoveries have identified helicases that either do not translocate (e.g., Swi/Snf) or translocate along duplex DNA (e.g., EcoR124I). The latter class of helicases has been referred to as translocases in order to differentiate them from bona fide helicases, which translocate along single-stranded substrates and bring about duplex unwinding. In the new system of classification, classic helicases are referred to as sub-type α, while the translocases are referred to as sub-type β. As per the new system of classification, Dda helicase from bacteriophage T4, a $5' \rightarrow 3'$ SF1 helicase, will be classified into SF1Bα, while NS3 helicase of HCV will be classified into SF2Aα.

It is clear that classification by sequence homology does not correlate with other helicase taxonomies and that helicases with different substrate specificities or directionalities could still be classified under the same superfamily (Fig. 20.1). What this means is that minor changes in the amino acid sequence could result in changes in substrate specificity or polarity of the enzyme, but the overall mechanism of enzyme action is more conserved. Thus, sequence similarities between helicases could directly reflect on their conserved enzyme mechanisms.

Structure and Function of Helicases

The classification of helicases into superfamilies laid out the groundwork for most of the structural studies on helicases. With the emergence of crystal structure information, the signature helicase motifs have been extensively characterized not only

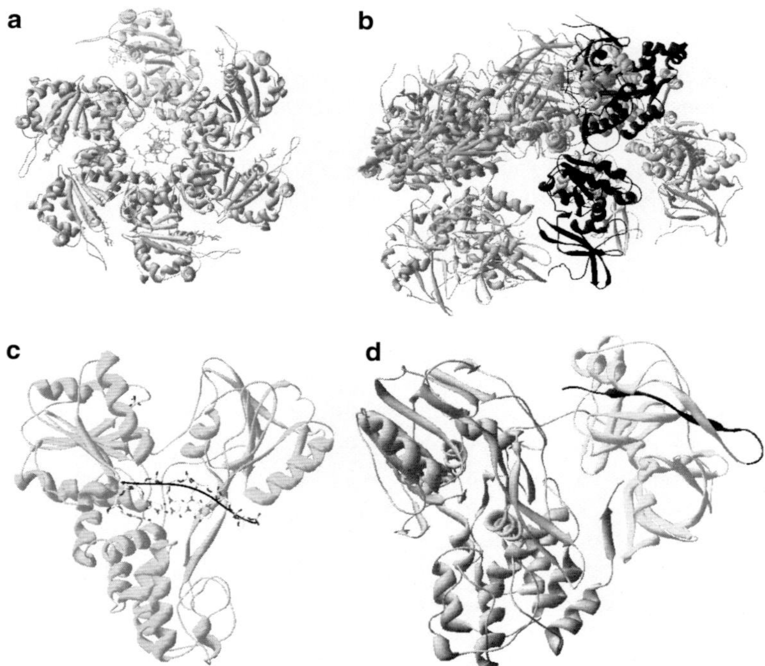

Fig. 20.3 Crystal structures of viral helicases. Many viral helicases have been crystallized either in their apo-forms or bound with their substrates (NTP/nucleic acid). X-ray structures often reveal important information about substrate binding and/or catalysis. In panel (a) is the E1 helicase of Papilloma virus in complex with ssDNA and ADP (PDB ID: 2GXA). In E1 helicase, the ssDNA is bound in the central channel of hexameric helicase and the ADP is co-ordinated at the interface between the subunits. Panel (b) is the structure of the bi-functional primase–helicase, also a hexameric helicase, from bacteriophage T7 (PDB ID:1Q57). In this protein, the primase domain trails behind the helicase domain, giving a distinct two domain organization of the protein. It should also be noted that the primase domain of one subunit interacts with the helicase domain of the neighbouring subunit. Panel (c) is the X-ray structure of helicase domain of the Hepatitis C virus NS3 helicase co-crystallized with ssDNA (PDB ID:1A1V). Unlike E1 and T7 gene4 helicases, NS3 is a monomeric helicase. It binds to the ssDNA in the cleft between all its subdomains. Panel (d) is the crystal structure of the full length NS3 protease-helicase (PDB ID: 1C1U). The protein is a recombinant construct where the 4A peptide, the co-factor for the protease domain (Howe et al. 1999), has been covalently attached to the N-terminus of the protease domain of the full length protease-helicase. All the crystal structures in the figure were rendered in 3D using Expasy's Swiss PDB Viewer (Guex and Peitsch 1997).

at the amino acid level but also at the three-dimensional structure level. Figure 20.3 contains a few of the crystal structures of some representative viral helicases. A list of all helicase structures along with the PDB IDs are given in Table 20.2.

Given that the conserved sequence motifs are short stretches of 4–10 amino acids interspersed with non-conserved segments, it has been hypothesized that the divergent regions are responsible for the individual protein functions while the highly conserved regions are involved in nucleotide binding and/or hydrolysis. Though

Table 20.2 Viral helicases: Crystal and solution structures

Name	Complex	Substrate/Superfamily	PDBID	References
BPV E1	ssDNA Mg-ADP	DNA	2GXA	(Enemark and Joshua-Tor 2006)
Dengue Virus NS3		RNA	2BMF	(Xu et al. 2005)
Dengue Virus NS3		RNA	2BMR	(Xu et al. 2005)
Kunjin virus NS3		RNA	2QEQ	(Mastrangelo et al. 2006)
HCV NS3	ssDNA	RNA/DNA SF2	1A1V, 2F55	(Kim et al. 1998)
HCV NS3 helicase/protease		RNA/DNA SF2	1CU1	(Yao et al. 1999)
HCV NS3		RNA/DNA SF2	1HEI	(Yao et al. 1997)
HCV NS3 Domain 2		RNA/DNA SF2	1JR6	(Liu et al. 2001)
HCV NS3 Domain 2		RNA/DNA SF2	1ONB	(Liu et al. 2001)
HCV NS3		RNA/DNA SF2	8OHM	(Lun et al. 1998)
HPV E1 DNA-binding domain		DNA SF3	1F08	(Enemark et al. 2002)
HPV E1		DNA SF3	1KSX	(Enemark et al. 2002)
HPV E1		DNA SF3	1KSY	(Enemark et al. 2002)
HPV E1	E2	DNA SF3	1TUE	(Abbate et al. 2004)
Phage Phi12P4	$ADPMg^{2+}$	RNA	1W46	(Mancini et al. 2004)
Phage Phi12P4	ADP	RNA	1W44	(Mancini et al. 2004)
Phage Phi12P4	$ADPMn^{2+}$	RNA	1W47	(Mancini et al. 2004)
Phage Phi12P4	AMPCPP	RNA	1W48	(Mancini et al. 2004)
Rep40		DNA SF3	1S9H	(James et al. 2003)
Murray Valley Encephalitis virus NS3		RNA SF2	2V8O	(Mancini et al. 2007)
Rep40	ADP	DNA SF3	1U0J	(James et al. 2004)
Rep40	ADP	DNA SF3	1U0J	(James et al. 2004)
RepA		DNA SF3	1G8Y	(Niedenzu et al. 2001)

Table 20.2 (continued)

Name	Complex	Substrate/Superfamily	PDBID	References
RepA	Sulfate	DNA SF3	1NLF	(Xu et al. 2003)
RepA		DNASF3	1OLO	(Ziegelin et al. 2003)
SV40		DNA SF3	1N25	(Li et al. 2003)
SV40		DNA SF3	1SVO	(Gai et al. 2004)
SV40	ADP	DNA SF3	1SVL	(Gai et al. 2004)
SV40	ATP	DNA SF3	1SVM	(Gai et al. 2004)
T7 Gene 4 Ring helicase		DNA	1CR0	(Sawaya et al. 1999)
T7 Gene 4 Ring helicase	dTTP	DNA	1CR1	(Sawaya et al. 1999)
T7 Gene 4 Ring helicase	dATP	DNA	1CR2	(Sawaya et al. 1999)
T7 Gene 4 Ring helicase	dTDP	DNA	1CR4	(Sawaya et al. 1999)
T7 Gene 4 Ring helicase	ANPNP	DNA	1E0J	(Singleton et al. 2000)
T7 Gene 4 Ring helicase		DNA	1E0K	(Singleton et al. 2000)
T7 Gene 4 Ring helicase		DNA	1NUI	(Kato et al. 2003)
T7 Gene 4 Ring helicase		DNA	1Q57	(Toth et al. 2003)
UvsW			1RIF	(Sickmier et al. 2004)
Yellow Fever Virus NS3	ADP	RNA SF2	1YMF	(Wu et al. 2005)
Japanese encephalitis virus NS3		RNA SF2	2Z83	To be published

Structures of different helicases either in their *apo-* form or as complexes are given. The table also gives the corresponding PDB ID of each of the structures. The superfamily to which the helicase has been classified, and its unwinding/translocation substrate is also indicated.

separated in sequence, the structural data indicate that these conserved regions are close together in space and form one large functional domain (Hall and Matson 1999). Some of these motifs have been biochemically characterized. Given below is a brief description of the structural description of each of the major superfamilies with special reference to these conserved motifs.

Superfamily 1 and Superfamily 2 Helicases

SF1 and SF2 helicases are among the most extensively studied and structurally well-characterized helicases. Though there is currently no high-resolution structural information available for the SF1 viral helicases, their counterpart in bacteria like PcrA (Velankar et al. 1999), Rep (Korolev et al. 1997), etc. have been well studied. Among the SF2 viral helicases, crystal structures are available for the helicases of many members of the flaviviridae family – HCV, Yellow fever virus, Kunjin virus, and Dengue virus (See Table 20.1). Sequence alignment among the members of the SF1 and SF2 superfamilies reveal up to 40% identity within the family members, with ˜90% lying in the conserved domains (Gorbalenya and Koonin 1993). This close relationship between the sequences allows for x-ray data from one enzyme to be extrapolated to the other members of the superfamily.

Structural studies on the SF1 and SF2 helicases indicate that these enzymes share extensive similarities in their stricture. The representative structures of SF1 helicases indicate a four domain structure, with all the conserved sequence motifs concentrated in two of these domains: domain 1A and domain 2A (Subramanya et al. 1996; Velankar et al. 1999). In the case of the SF2 helicases, these sequences reside within two large domains, domain 1 and domain 2, which are homologous to the domain 1A and 2A of SF1 helicases (Yao et al. 1997; Kim et al. 1998).

Motif I, also referred to as the Walker A motif, has a consensus of XGXAGXGKT in SF1 helicases and a consensus of XGXGKT/S in SF2 helicases (Blinov et al. 1989; Tuteja and Tuteja 2004). The conserved lysine residue in both the families is responsible for binding to the β- and γ-phosphates of NTP–Mg complex. Mutation of this lysine residue results in deficiency of the ATPase activity (Hall and Matson 1999). However, it has no effect on nucleic-acid binding (Levin and Patel 2002). Motif Ia of both superfamilies is involved in ssDNA binding (Kim et al. 1998; Lee and Yang 2006). In a recent study on HSV-1 UL9 helicase, mutational analysis of the residues in the Ia motif implicated in DNA binding resulted in moderate to severe defects in single-stranded nucleic-acids binding and ssNA stimulated ATPase activity, while retaining the intrinsic ATPase activity similar to that of wildtype enzyme (Marintcheva and Weller 2003). Motif II has a conserved sequence of XXDEXD/H and is referred to as the Walker B motif (Linder et al. 1989). Proteins carrying a conserved D-E-A-D sequence, also referred to as the DEAD-box proteins, are predominantly RNA helicases (Koonin 1991; Koonin 1992; Linder and Daugeron 2000; Cordin et al. 2006), while proteins carrying variant of the DEAD sequence like DEAH/DEXH are usually DNA helicases (Subramanya et al. 1996;

Linder 2000; Linder and Daugeron 2000). The conserved D of this motif has been shown to interact with the catalytic Mg^{2+} and is important for the NTPase activity. A mutation in this residue affects both the NTPase and the helicase function (Pause and Sonenberg 1992).

Motifs III and VI of the SF1 and SF2 helicases, though not equivalent in sequence or structure, are implicated in coupling ATPase activity to the helicase function. Mutations in the SAT domain of motif III of the SF2 helicases resulted in loss of helicase activity with no effect on the ATPase activity (Pause and Sonenberg 1992; Graves-Woodward et al. 1997), while mutations in motif VI resulted in loss of both ATPase and RNA helicase activity. An invariant arginine in this domain has been shown to be extremely important for the RNA helicase activity of HCV NS3 helicase (Kim et al. 1997). In SF1 helicases, residues in motif III are also involved in nucleic-acid binding through hydrogen-bonding and stacking interactions with the nucleic-acid bases (Hall and Matson 1999). Motifs I, IV, and V of the SF1 helicases have been shown to have direct contact with either the nucleotide in the enzyme-NTP complexes or interact with the nucleic acid through the sugar–phosphate backbone (Korolev et al. 1997; Hall and Matson 1999; Velankar et al. 1999). In the SF2 family, a newly discovered motif, called the Q-motif, owing to its conserved Gln residue has been implicated in adenine recognition of these enzymes (Tanner 2003; Tanner et al. 2003; Tuteja and Tuteja 2004; Killoran and Keck 2006).

Superfamily 3

All SF3 helicases contain the Walker A and Walker B sequences which are important for nucleotide binding and/or hydrolysis. In addition to these they contain the conserved motif C (Bork and Koonin 1993; Gorbalenya and Koonin 1993; Iyer et al. 2004) and a newly discovered motif B′ (Yoon-Robarts et al. 2004). The Walker B motif is atypical and carries a consensus of XXXXEE, while the motif C carries the consensus XXX(S/T)(S/T)N (Hall and Matson 1999; James et al. 2003). The motif C of SF3 helicases is implicated in distinguishing ATP from ADP. The conserved Asn hydrogen bonds to the γ-phosphate of ATP to facilitate this function (James et al. 2003). B′ motif is characterized by a 14-residue long stretch, with a central highly conserved glycine, and positively charged residues on either end of the motif. This motif has been established to be involved in nucleotide binding and unwinding. Mutation of a Lys at one end of the motif abolishes both helicase and ATPase activity, while the mutation of the other Lys eliminates helicase but not ATPase activity (Walker et al. 1997).

DnaB-like Family

This family of hexameric helicases possesses five conserved sequence motifs. Two of the five are the Walker A and Walker B motifs common to all helicases and

NTPases, while the other three are specific to this helicase family. Bacteriophage T7 helicase, the viral representative of this family, has been extensively studied, not only in terms of its structure but also biochemically and mechanistically (Matson et al. 1983; Rosenberg et al. 1992; Hingorani and Patel 1993; Patel and Hingorani 1993; Patel et al. 1994; Egelman et al. 1995; Patel and Hingorani 1995; Hingorani and Patel 1996; Washington et al. 1996; Yu et al. 1996; Ahnert and Patel 1997; Hingorani et al. 1997; Picha and Patel 1998; Sawaya et al. 1999; Ahnert et al. 2000; Patel and Picha 2000; Singleton et al. 2000; Kim et al. 2002; Toth et al. 2003; Jeong et al. 2004). From the crystal structure studies, four of the five conserved domains, 1,1a, 2, and 3 lie in the conserved C-terminal domain of the helicase and are involved in nucleotide binding and hydrolysis. Domain 4 is a part of the DNA-binding surface of the helicase and lines the region which forms the central channel when the hexamer gets assembled (Sawaya et al. 1999; Singleton et al. 2000; Toth et al. 2003).

Helicases and the RecA Fold

The crystal structures of all the helicases from the different superfamilies discussed above reveals an interesting fact: all these proteins share a common fold – the RecA fold (Bird et al. 1998). The basic structural unit of the helicases from the SF1 and SF2 superfamilies is the RecA-like subdomains (Yao et al. 1997; Kim et al. 1998; Velankar et al. 1999; Lee and Yang 2006). Structures of the T7 gene 4 helicase and the SF3 helicase from Adeno-associated virus type-2 also possesses a RecA-like fold in their helicase domains (Sawaya et al. 1999; James et al. 2003), as is the case with SV40 T-antigen (Seif 1982). It has also been shown that the ATP-binding domain of RecA and the F1-ATPase superimpose with a root mean squared deviation of less than 2 Å (Story et al. 1992; Abrahams et al. 1994) (Fig. 20.3).

The conservation of this structural motif in all helicases could mean that this fold is the minimal requirement for all helicases (including the generic NTPases) for NTP binding and hydrolysis. For all other diverse functions that the helicases carry out, this minimal domain needs to be supplemented with additional domains. This observation is exemplified by the eukaryotic transcription initiation factor eIF4a, an SF2 DEAD-box helicase (Rogers et al. 2002). eIF4a protein, which is essentially just the RecA-like motor with no additional domains, is a very poor helicase (Du et al. 2002). However, its helicase activity gets considerable enhanced in the presence of other factors like 4B, 4H, etc. (Rogers et al. 2001).

The commonality in the motor domain of all these proteins could mean that all these proteins couple binding and hydrolysis of nucleotides to conformational changes that in turn affect the affinity of these enzymes for different forms of nucleic acids. However, the disparities between the helicases in terms of their polarity, substrate specificity, oligomeric nature, etc., could be derived from the associated domains and/or proteins. Thus, it is often necessary to study these proteins as a part of the macromolecular complex in which they form the central functional component, rather than in isolation. This is especially true of viral helicases, where the helicase not only plays a central role in genome replication but also in other functions like mRNA capping, recombination, packaging, etc.

In viruses, the minimal replisome consists of a helicase, polymerase, and a single-strand binding protein (SSB). The virus hijacks the host machinery for all other accessory proteins. However, the virus-specific activities lie within the minimal replisome. Thus, many of the virus-encoded proteins are known to be multi-domain with multiple functions. Figure 20.4 gives a few examples of the accessory activities associated with some viral helicases along with their role in viral replication. In addition to the multiple functions that many viral helicase possess, they act in concert with many other proteins to carry out functions that aid viral replications. In some viruses, this kind of multi-protein interactions is required for facilitating basic functions that might reside within a single polypeptide in other viruses. For example, in herpes-simplex virus type-1, the association of three proteins, UL8, UL5, and UL52, constitutes the helicase–primase activity (Crute et al. 1991). Similarly, in papillomaviruses, the E1 protein has to associate with the E2 protein for origin binding and initiation of replication (Seo et al. 1993; Masterson et al. 1998; Gillitzer et al. 2000). Helicases also interact functionally and physically

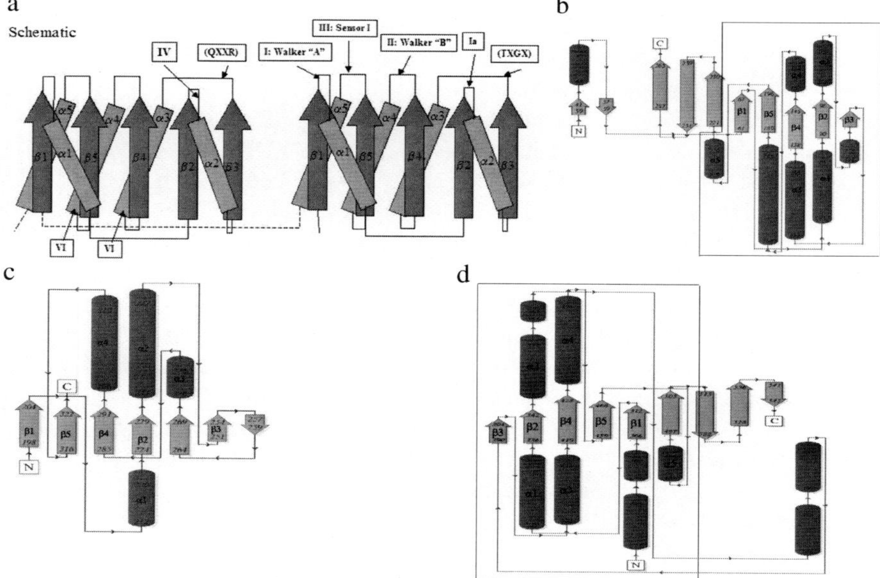

Fig. 20.4 Topology diagrams of viral helicase structures. Viral helicases are structurally homologous to the *E.coli* RecA protein. Panel (**a**) is a schematic representation of the RecA fold with the corresponding conserved motifs (**b**) is a topology diagram of the *E.coli* RecA protein (**c**) is the topological representation of domain 1 of the HCV helicase NS3, an SF2 helicase. (**d**) is the representation of T7 gene4 helicase, a member of the hexameric DnaB family. No structures of any SF1 viral helicase are available so far. The topology diagrams were adapted from EMBL's PDBSum database.

with the polymerases within the replication complex. These polymerases could be either host-derived or virus derived (Smale and Tjian 1986; Gannon and Lane 1990; Park et al. 1994; Notarnicola et al. 1997; Delagoutte and von Hippel 2001; Kato et al. 2001; Piccininni et al. 2002). These interactions could be either direct protein–protein interactions or mediated through an intermediary scaffolding protein. Single-strand binding proteins have also shown to interact with the helicases both in vitro and in vivo. The SSB could once again be host-derived or virus derived (For examples see: (Nakai and Richardson 1988; Hamatake et al. 1997; Kong and Richardson 1998; Lefebvre et al. 1999).

Biochemical studies on hepatitis C virus showed that the eukaryotic RNA helicase p68 and the poly-pyrimidine tract binding protein (PTB) are essential for viral replication (Goh et al. 2004; Zhang et al. 2004; Aizaki et al. 2006; Chang and Luo 2006; Lim et al. 2006). Similar observations have been made for many of the tumor-inducing viruses, which show interactions with the cellular factors which are important in apoptosis and other related metabolic pathways (Barber 2001; Schattner 2002; Lavia et al. 2003; Brechot 2004; Ledwaba et al. 2004; Zhang et al. 2004; Levrero 2006; Strath and Blair 2006). Table 20.3 gives a list of some of the proteins that the viral helicases interact with along with their corresponding function in viral replication and/or infection.

Why are these interactions with the cellular factors important? Viruses are facultative parasites. They have evolved to have some of the smallest genomes, encoding for only those functions that are most essential and are specific to its replication. Thus from the point of view of viral evolution and host-virus specificity, hijacking host proteins for viral replication maximizes the viral perpetuation by re-routing all or most of the cellular metabolism towards virus-directed processes. Protein–protein interactions between helicases and other accessory proteins also have kinetic and thermodynamic implications. The function of most helicases within such assemblies is not merely to catalyze the opening of a dsNA segment, but also to drive rearrangements in which one or both of the ssNA products end up bound to another macromolecular component. Often the inclusion of loading or trapping factors can improve helicase activity. A loading factor facilitates *initiation* of the helicase reaction, while a trapping component (e.g., ssNA binding protein) facilitates *elongation* by stabilizing ssNA intermediates in the reaction as they are formed. In the context of replication, the ssNA thus stabilized can be used by the polymerase for genome replication. Thus, a simplified replisome can be thought to be a combination of at least two motor proteins: the helicase and the polymerase.

In a mathematical treatment by Stukalin et al., it is apparent that when there is coupling between two motor proteins, it results in a much more efficient motor when compared to the individual motors (Stukalin et al. 2005). The increased efficiency could be reflected as an increase in the overall rate of the reaction and/or the processivity of the enzyme(s) (Jarvis et al. 1991; von Hippel and Delagoutte 2001; Delagoutte and von Hippel 2002; Stano et al. 2005). Such a behaviour has also been reported for isolated helicases of the SF1 and SF2 superfamilies, where functional oligomerization of the enzymes resulted in an increase in the processivity of the

Table 20.3 Helicase associated proteins and functions

Helicase	Other associated activities	Protein-protein interactions	Physiological function	References
SARS virus nsp13	5′-RNA triphosphatase	RNA-dependent RNA polymerase	mRNA capping, replication	(Ivanov and Ziebuhr 2004)
HSV UL5/UL52	Not determined	UL8, SSB-ICP8, DNA polymerase	Replication, recombination, chromatin remodelling	(Carrington-Lawrence and Weller 2003; Trego and Parris 2003; Taylor and Knipe 2004)
AAV Rep68	Endonuclease	Rep 78	Viral replication	(Im and Muzyczka 1990)
Papillomavirus E1	Primase	E2	Binding to duplex origin, origin melting, replication	(Seo, Muller et al. 1993)
Bacteriophage T4 gene41		Gene 43 (polymerase), gene 61 (primase), gene 32 (SSB)	Replication	(Jarvis et al. 1990; Jarvis et al. 1991; Jing et al. 1999; Delagoutte and von Hippel 2001)
Bacteriophage T7	Primase	DNA polymerase, host lysozyme, host SSB, gene 2.5 (Viral SSB)	Replication	(Nakai and Richardson 1986; Nakai and Richardson 1988; He and Richardson 2004)
Baculovirus helicase p143	Not Determined	ssDNA-binding protein LEF-3	Replication	(Ito et al. 2004)
Bromovirus helicase	Methyltransferase	RNA-dependent RNA polymerase	mRNA capping, Replication	(Chen and Ahlquist 2000)

Table 20.3 (continued)

Helicase	Other associated activities	Protein-protein interactions	Physiological function	References
Beet yellows Closterovirus helicase	Methyltransferase	RNA-dependent RNA polymerase	mRNA capping, Replication	(Dinant et al. 1993)
Alfalfa mosaic virus helicase	Guanylyl transferase/methyl transferase	RNA-dependent RNA polymerase	mRNA capping, Replication	(Van Der Heijden et al. 2001)
Tobacco mosaic virus helicase	Methyltransferase	RNA-dependent RNA polymerase	mRNA capping, replication	(Goregaoker, Lewandowski et al. 2001)
AAV type2 Rep 40	Endonuclease	Rep 52	Replication, packaging	(King et al. 2001)
EBV helicase	Primase	SSB (BALF2), ZTA-lytic transactivator	Replication, induction of lytic cycle	(Gao et al. 1998)
Human coronavirus helicase	5′-RNA triphosphatase	RNA-dependent RNA polymerase	mRNA capping, replication	(Ivanov et al. 2004)
Dengue virus type-2 NS3	5′-RNA triphosphataseSer-protease	RNA-dependent RNA polymerase	Polyprotein processing, replication, mRNA capping	(Yon et al. 2005)
HCV NS3	Ser-protease	RNA-dependent RNA polymerase, host p68 RNA helicase, host PTB protein	Replication	(Aizaki et al. 2006; Chang and Luo 2006; Lim et al. 2006)
Phage P4 α-helicase	Primase	Not determined	Replication	(Ziegelin et al. 1993)
SV-40 T-antigen	Not determined	DNA polymerase, SSB	Replication	(Otto et al. 1979)

List of all enzymatic activities present in addition to the helicase activity of various helicases are indicated here. The table also gives a list of proteins that these helicases show physical and/or functional interactions with.

enzyme without change in reaction rates (Levin and Patel 1999; Byrd and Raney 2005; Tackett et al. 2005).

Mechanism of Helicase Action

The unwinding activity of a helicase can be considered an outcome of two fundamental activities of all helicases: (1) unidirectional translocation along single-stranded nucleic acid and, (2) strand-separation activity. To carry out these reactions, a helicase must cycle through a series of energy states driven by NTP binding and/or hydrolysis and subsequently product release. Thus, in order to understand the mechanism of helicase catalyzed unwinding reactions, it is important to understand all the individual steps to it, namely: nucleic-acid binding, NTP binding and hydrolysis, single-stranded translocation, and then finally the strand-separation function. In the following section, each of these aspects of helicase mechanism will be dealt with in detail, with respect to two viral helicases that have been extensively characterized– the Hepatitis C Virus NS3 helicase and the bacteriophage T7 gene4 helicase.

Nucleic-Acid Binding

The binding of the helicase to the nucleic acid forms the first critical step toward unwinding the duplex substrate. Understanding DNA/RNA binding by the enzyme could help answer questions like: does the enzyme require a single-stranded region to initiate unwinding? Does the enzyme interact with only one strand of the nucleic acid or both? Does NTP binding alter the enzyme's affinity for nucleic-acid binding?

Most helicases have been shown to require a short single-stranded tail to load onto the duplex substrate to carry out the unwinding efficiently. The polarity of the single-strand almost always depends on the polarity of the helicase translocation, i.e., a $3' \rightarrow 5'$ helicase uses a short single-stranded $3'$-tail, while a $5' \rightarrow 3'$ helicase uses a $5'$-tail. Many of the ring-helicases like T7 gene4 helicase, DnaB helicase of *E. coli*, etc., require a Y-shaped substrate, having both $3'$ and $5'$-tails. Interestingly, SF3 helicases like E1 helicase of Papillomaviruses and T-antigen from SV40 can initiate unwinding from completely dsDNA by binding to a site-specific region (origin of replication), causing duplex melting and entry of the helicase onto the single-stranded region. The site-specific DNA binding is mediated by the DNA-binding domain, while the unwinding is mediated by the ATPase/helicase domain (Wu et al. 1998; Wu et al. 2001; Enemark and Joshua-Tor 2006).

The directional translocation of the helicase on its nucleic-acid substrate entails that its binding site for the nucleic acid is also polarized with respect to the sugar-phosphate backbone. A direct evidence for this was shown with Rep helicase binding to single-stranded dT_{16} containing the fluorescent base Etheno-adenosine at the $5'$ end or the $3'$ end. The enzyme showed different extents of fluorescence enhancements depending on position of the label (Bjornson et al. 1998). A similar observation was also demonstrated with HCV NS3 helicase domain using duplex substrates with either a $5'$- or a $3'$-overhang. The enzyme bound to the $3'$-overhang with a 45-fold higher affinity than to the 5′-overhang (Levin et al. 2005). Another parameter that can be obtained from nucleic-acid binding experiments is the occlusion site measurement. Occlusion site can be defined as the number of bases/base pairs the enzyme protects when it binds to the nucleic-acid substrate. Nuclease footprinting is often used to assay for the occlusion site. Occlusion site measurements also give an idea about the enzyme's interaction with the duplex region or the displaced strand.

X-ray structures of the oligonucleotide bound HCV NS3 domain (Fig. 20.5) showed that the enzyme bound the oligo nucleotide at a cleft that separated domain 3 (all-helix domain) from domains 1 and 2 (RecA homology domains). The binding polarity of the oligonucleotide was consistent with the biochemical assays – the 3′end positioned away from the enzyme and the 5′-end oriented between domains 2 and 3 of the enzyme. The enzyme sought predominantly backbone interactions with the DNA, with very few base-specific interactions. Trp501 and Val 432, both highly conserved among HCV NS3 sequences, show interactions with the nucleic-acid bases defining the central binding cavity to five nucleotides (Kim et al. 1998).

Nucleic-acid binding has been biochemically characterized using equilibrium-binding experiments, which is the preferred method of studying nucleic-acid binding. These studies have provided valuable information about the binding constants, stoichiometry of binding and at times can also give insights into the oligomeric state of the enzyme. The DNA-binding studies, with fluorimetric titrations and nitrocellulose filter binding assays, on HCV NS3 helicase domain (NS3h) showed that the enzyme bound ssDNA with a very high affinity (K_d ~2–10 nM) in the absence of NTP. The NS3h binding occluded about 8.3 bases and had a stoichiometry of 1:1, enzyme:ssDNA (Levin and Patel 2002). However, the binding affinity dropped 80-fold in the presence of NTP (Levin et al. 2003). Though the enzyme did not bind to blunt-ended duplexes (Levin and Patel 2002; Levin et al. 2005), it showed a high affinity for partial duplexes with a 3′-single-stranded tail (Levin et al. 2005).

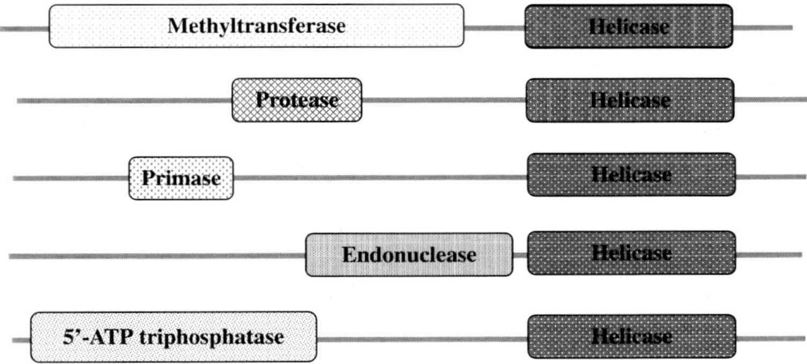

Fig. 20.5 Domain organization of viral helicases. Many viral helicases have been shown to have other enzymatic activities in addition to their helicase function. These associated functions play an important role in replication and/or packaging of the mature virions. The domain organization was recreated based on the data from the Pfam database (Finn et al. 2006)

It is extremely interesting to note that the enzyme's DNA-binding behavior is vastly different from its RNA-binding behavior. It has been shown that NS3h helicase binds ssRNA with a 10-fold lower affinity, at neutral pH, than it binds DNA (Levin and Patel 2002). However, the enzyme seems to have maximal RNA-binding capacity at pH 6.5 (Gwack et al. 1996). Also, the affinity of the ATP bound enzyme for RNA is increased at low pH (Lam et al. 2004). Though NS3 helicase possesses the ability to bind to any single-stranded RNA, it exhibits a preference for its genomic RNA sequences. Banerjee and Dasgupta have shown that NS3 binds to the 3′-UTR of the genomic RNA with a much higher affinity than the 5′-UTR sequence (Banerjee and Dasgupta 2001). They attribute the differential affinities to the secondary structures associated with each of these sequences (Banerjee and Dasgupta 2001). The specificity of the HCV helicase for the 3′-UTR could implicate a role for the helicase in viral replication since the negative strand synthesis would have to initiate at this terminus. Chang et al have demonstrated that the Arginine-rich motif (motif VI) of SF2 helicases is important for RNA binding (Chang et al. 2000). This motif is both structurally and functionally conserved in many flaviviruses including the HGV (Gwack et al. 1999). The conserved Arginine is critical for nucleic-acid binding and helicase activity. The dynamics of the subdomain 2, which contains this conserved motif, revealed that this domain could be responsible for the conformational change associated with ATP-binding and hydrolysis, thereby driving the helicase reaction (Liu et al. 2001). The subdomain 2 has also been implicated in dsDNA binding. Motifs IV and V of subdomain 2 have been shown to undergo local unfolding in order to accommodate the dsDNA in their DNA-binding site (Liu et al. 2003).

Electrostatic analysis of the HCV NS3 helicase by Multi-Conformation Continuum Electrostatics (MCCE) identified two residues crucial for nucleic-acid binding – H369 and E493. H369 and E493 were at 3 and 6 Å distance, respectively, from the reported DNA-binding site. Mutational analysis of the two residues resulted in a drastic decrease in the nucleic-acid-binding affinities, indicating the importance of these two residues in NA binding. Based on these results, Frick et al. propose a model to explain the modulation of nucleic-acid-binding affinity by ATP due to the changes in the intrinsic pKa of these residues that arise from ATP and DNA binding, and the activation of the enzyme at low pH (Frick et al. 2004).

A high-resolution structure of the ring-shaped bacteriophage T7 gp4 helicase bound to DNA is not available as yet. Mutational studies have indicated that the conserved motif H4 is somehow involved in DNA binding. The x-ray of the helicase domain revealed that the residues of this motif lined up near the center of the hexamer, consistent with the enzyme binding the ssDNA in its central channel (Egelman et al. 1995). It has been proposed that nucleotide binding induces a conformational change in the H4 motif causing the region around it to fold into a helical structure. Two residues, R487 and G488, have been implicated in contacting DNA (Washington et al. 1996). In the unliganded state, this region is still disordered implying that nucleotide-binding couples the conformational changes important for DNA binding by the enzyme (Sawaya et al. 1999). Recently, it was also shown that mutation of three lysines to alanines (K467, 471, 473) abolishes DNA binding (Crampton et al. 2006).

The DNA bound structure of the ring-shaped E1 helicase of Papilloma virus has been solved (Fig. 20.5), which shows that all the residues seen to interact with the DNA are within the AAA+ domain. The groups mediate mostly backbone interactions with the DNA – the H507 and K506 forming hydrogen-bonding interactions with the backbone phosphates and all three residues F464, K506, and H507 making Van der Waals interactions with the sugar residue linking the two H-bonding phosphates. 5′ end of the ssDNA is directed toward the N-terminal oligomerization domains, whereas the 3′ end is directed toward the C terminus consistent with its translocation polarity (Enemark and Joshua-Tor 2006).

Unlike HCV NS3 protein, the ring-shaped T7 helicase requires a forked DNA substrate to initiate unwinding. The protein makes contact with both the 3′- and the 5′- strands, and these contacts are important not only for initiating the reaction, but throughout the unwinding reaction (Hingorani and Patel 1993). For optimal unwinding the enzyme requires a 35nt 5′-tail and a 15nt 3′-tail (Ahnert and Patel 1997). The nuclease protection assays indicate a 25–30 base occlusion site on the 5′strand (Hingorani and Patel 1993) which is consistent with the enzyme requiring a 35nt 5′-tail for optimal unwinding. As opposed to HCV helicase, T7 helicase binds to ssDNA tightly only in the presence of dTTP (Hingorani and Patel 1993). Therefore, the NTPase activity of helicases partly serves to modulate interactions with the nucleic acid. Some helicases bind tightly to nucleic acid in their NTP-liganded form, while others in the nucleotide-free or NDP-liganded form and vice versa.

The single-stranded DNA is bound in the central cavity of the hexamer (Egelman et al. 1995). The binding of the helicase to ssDNA is a multistep process that does not utilize NTP (Picha et al. 2000). At a given time only one or two subunits of the hexameric helicase contacts the DNA (Yu et al. 1996). The enzyme binds ssDNA with a K_d of ~10 nM and a stoichiometry of one strand per hexamer (Hingorani and Patel 1993). The T7 gene4 helicase also exhibits dsDNA-binding activity. However, the enzyme has a 50-fold lower affinity for dsDNA as compared to ssDNA (Hingorani and Patel 1993).

The enzyme binds the 5′-strand in its central cavity and excludes the 3′-strand from its active binding site. Replacing the 3′-strand with the biotin–streptavidin complex results in the same outcome, implying that the 3′-strand of the fork provides steric hinderance to the enzyme thereby preventing it from binding the duplex (Hacker and Johnson 1997). At the replication fork of the T7 genome, the enzyme is thought to transiently bind at the primase site, followed by a conformational change accompanied by the ring-opening, ssDNA binding, and ring-closure (Ahnert et al. 2000). In the absence of the 3′-tail, the enzyme can bind and translocate along the duplex DNA (Jeong and Patel, unpublished data).

Experiments involving synthetic substrates also give information about which strands are contacted by the helicase at the unwinding junction. Different helicases show different levels of tolerance to changes in the chemical nature of the loading strand, breaks along the unwinding track, abasic sites, electrostatic disruptions, etc. HCV NS3 helicase is extremely sensitive to the nature of the displaced strand (Tackett et al. 2001a), while the Dda helicase of bacteriophage T4 and NPH-II helicase of *Vaccinia* virus show little or no sensitivity (Tackett et al. 2001b; Kawaoka et al. 2004). The T7 helicase stalls with disruptions on the loading strand (Yong and Romano 1995), while replacing the displaced strand with a morpholino substrate increases the unwinding rate of the enzyme (Jeong and Patel, unpublished data).

Unidirectional Translocation

Translocation of the helicase along the single-stranded nucleic acid is considered to be one of the two key activities of the helicase that is required for its unwinding function. The translocation function is coupled to NTP hydrolysis, and though no one has so far demonstrated this translocation to be strictly unidirectional, the overall movement of the protein is biased to a single direction.

Different approaches have been used to study the translocation of the protein along the single-stranded nucleic acid and the coupling of this action to NTP hydrolysis. One of the earliest approaches was to study the steady-state kinetics of NTP hydrolysis as a function of ssDNA length (Liu and Alberts 1981; Matson and Richardson 1983; Raney and Benkovic 1995). The steady-state kinetics of NTP hydrolysis has also been used to differentiate between the ssDNA translocation activities of PriA protein from its unwinding activity (Lee and Marians 1990). A more recent approach involves biotin labeling the oligonucleotide at either the 3′- or the 5′-end and observing the disruption of the biotin–streptavidin complex by the helicase (Morris et al. 2001). This approach has been used to demonstrate both the polarity and the unidirectional translocation of bacteriophage T4 Dda helicase (Byrd and Raney 2004) HCV NS3 helicase and SV40 T-antigen (Morris et al. 2002). Kim and co-workers not only studied pre-steady state kinetics of dTTP hydrolysis as a function of ssDNA length, of bacteriophage T7 helicase, but also studied the energy coupling of the process using a coupled enzyme assay which measured the amount of inorganic phosphate (Pi) released using phosphate-binding protein (PBP) labeled with MDCC (Kim et al. 2002). This approach originally developed by the Webb lab (Hirshberg et al. 1998) has been used to obtain stepping rates and energy efficiency (coupling constants) of other enzymes including PcrA, UvrD (Raney and Benkovic 1995; Dillingham et al. 1999; Dillingham et al. 2000; Soultanas and Wigley 2000). Extensive modeling of the pre-steady state kinetics of NTP-dependent translocation of the motor proteins have been done by Fischer and Lohman (Fischer and Lohman 2004) and demonstrated on the *E. coli* protein UvrD (Fischer et al. 2004; Tomko et al. 2007).

Models of Unidirectional Translocation

Different mechanisms have been proposed for translocation of helicases along single-stranded nucleic acids. All mechanisms involve NTP hydrolysis, with a coupled conformation change to explain the biased movement.

Stepping mechanism - The stepping mechanism requires that the helicase possesses two DNA-binding sites. The two sites have differential affinities for the

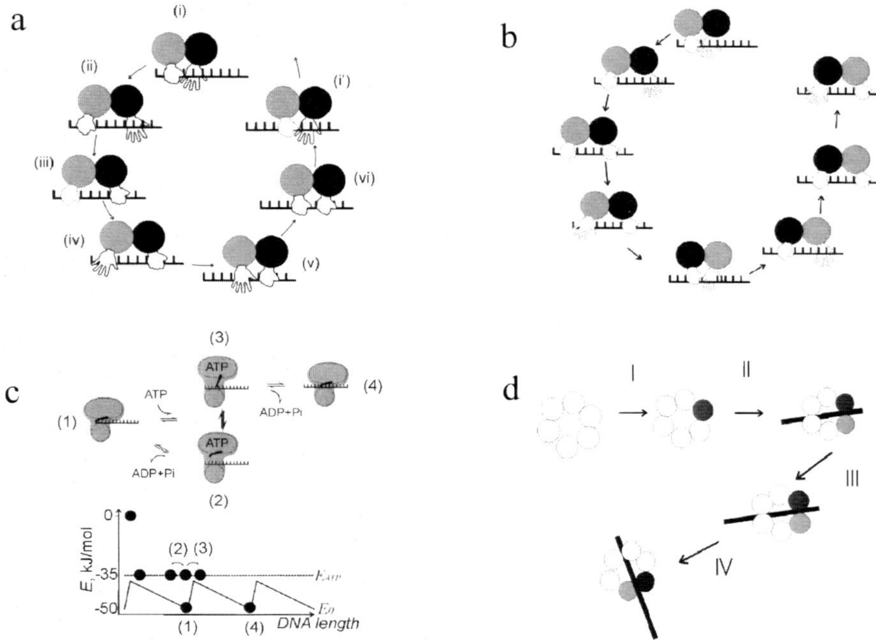

Fig. 20.6 Mechanisms of translocation of helicases on single-stranded nucleic acids. Many mechanisms of single-stranded translocation have been reported, based on structural and biochemical data. Panels (**a**) and (**b**) describe the stepping mechanism of helicase translocation. Amongst the stepping mechanisms, panel (a) represents the "Inchworming" model of helicase translocation. This model is often used to describe the translocation of monomeric helicases with two nucleic-acid-binding sites. The two sites cycle between tight binding and weak moving as dictated by their NTP ligation states, and associated conformational changes. One cycle of inchworming is typically completed in a set of six conformational changes, with the two binding sites (or domains) always retaining their position on the nucleic acid relative to each other constant. Panel (b) describes the "Rolling" mechanism. This model is often used to describe the translocation of dimeric helicases. In this model, the two subunits alternate their positions on the nucleic acid as they change their NTP ligation states. Panel (**c**) describes the "Brownian Ratchet" model reported for the translocation of Hepatitis C virus NS3 helicase domain. In this model, the enzyme's binding affinities for the single-stranded nucleic acid are modulated by its ATP ligation state. ATP binding weakens its affinity while ATP hydrolysis results in tight binding. In its weakly bound state, the enzyme could ratchet back and forth on the single-stranded substrate. Panel (**d**) represents one of the many possible NTP ligation and nucleic-acid occupancy states for the "Sequential hydrolysis" mechanism proposed for T7 gene4 helicase. Here, the enzyme contacts the ssNA two subunits at a time. NTP hydrolysis results in translocation of the helicase and the transfer of the ssNA substrate to the adjacent subunits.

nucleic acid, which are modulated by the different NTP ligation states. In the "inchworm" type stepping model (Fig. 20.6a), one site is bound to the nucleic acid tightly (H), while the other site is weakly bound (T). NTP hydrolysis results in a power stroke, causing the weak site T to dissociate, move away from the tight site H and bind ahead of it. At the new position, the weak site T initiates tight interactions and becomes the new H site, in the process weakening the interactions of the previous H site to generate the new T site. In another power stroke, the sites undergo another round of nucleic-acid affinity changes to obtain their original starting affinities. Thus, one cycle in an inchworm stepping mechanism is completed with six conformational changes (Hill and Tsuchiya 1981; Lohman 1993; Patel and Donmez 2006).

Another stepping model is used to describe the translocation of dimeric helicases. In the inchworming model, both the nucleic-acid-binding sites could be present on the same polypeptide chain. In the "rolling" model (Fig. 20.6b) each monomer contributes one nucleic-acid binding site and that the two subunits of the helicase alternate their binding to the single-stranded nucleic acid depending on the changes in their NTP ligation states (Wong et al. 1992; Lohman 1993).

Brownian motor mechanism – This mechanism was proposed as an alternative mechanism to the stepping mechanism (Fig. 20.6c). This mechanism involves a single nucleic-acid binding site modulated by NTP binding and hydrolysis. In the absence of any bound NTP, the enzyme mediates very tight interactions with the ssNA substrate. In this state, the energy profile of the helicase is deep and saw-tooth shaped. Thus, in the tight state, the helicase is unable to mediate any motion along the ssNA. On NTP binding, its affinity for ssNA drops several fold, resulting in a shallow energy profile. The enzyme is now capable of moving either forward or backward (Brownian motion). In a power stroke coupled with NTP hydrolysis, the enzyme moves forward, going back to its original tight state (Levin et al. 2005; Patel and Donmez 2006).

Sequential "subunit rotation" mechanism of hexameric helicases- This mechanism has been proposed to explain the translocation of hexameric helicases like the bacteriophage T7 helicase on ssDNA (Fig. 20.6d). In this mechanism, three cooperative steps of sequential DNA-binding and release are required for processive translocation along ssDNA. DNA is translocated by power strokes powered by NTP binding to the catalytic site. First, the empty NTP site gets occupied to generate the weak DNA-binding site T*. The DNA-binding step in the next subunit, $T^* \rightarrow N \cdot T^*$, commences when the previous subunit in the sequence has completed its power stroke and is in the $N \cdot T$ state. Geometrically, this is possible if the power stroke of the previous subunit brings the DNA strand into a position where it can quickly fluctuate to the next subunit. Since hydrolysis enables release of the DNA strand, in order to ensure high processivity the unbinding of DNA in one subunit must take place after the binding of DNA to the next subunit. Thus, the transition $N \cdot T \rightarrow DP$ in one catalytic site must follow the binding of nucleotide to the next site, i.e., state $T^* \rightarrow N \cdot T^*$. Finally, the power stroke $N.T^* \rightarrow N \cdot T$, results in translocation (Liao et al. 2005).

Base Pair Separation Mechanisms

Helicases couple the energy of NTP hydrolysis to single-stranded translocation and base-pair separation. Translocation of helicases can take place by any of the above-mentioned mechanisms. Mechanisms of base-pair separation can be in general classified into "active" or "passive" depending on the extent to which the enzyme is involved in the strand-separation function (Lohman 1993; Lohman and Bjornson 1996; von Hippel and Delagoutte 2001; Betterton and Julicher 2005).

In a "passive" mechanism (Fig 20.7a) of strand separation, the enzyme translocates along the single-stranded nucleic acid till it reaches the duplex junction. Now the enzyme waits for the two strands to open due to thermal fraying. Once a base pair opens, the enzyme now moves ahead and this cycle continues till the duplex has completely separated. For a passive helicase, the unwinding step-size is likely to be one, since it is extremely difficult for more than one base pair to open by thermal fluctuations. In the "active" mechanism of helicase action (Fig. 20.7b), the enzyme destabilizes the junction, thereby altering the energy profile of the duplexes at the junction, making them easier to melt. An active mechanism can account for larger step-sizes reported for many of the helicases (Serebrov and Pyle 2004; Spurling et al. 2006; Myong et al. 2007). Force dependence and stability dependence studies have revealed that both T7 gene 4 helicase and the HCV NS3 helicase unwind by an active mechanism (Cheng et al. 2007; Donmez et al. 2007; Johnson et al. 2007).

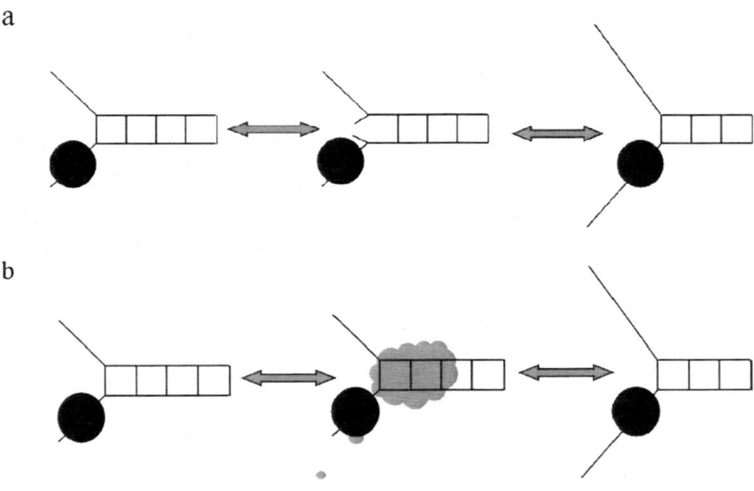

Fig. 20.7 Passive and Active mechanism of nucleic-acid unwinding. Helicases convert the chemical energy of NTP hydrolysis to mechanical work done in-terms of translocation on the single-strand and separation of the two strands of the duplex DNA or RNA. The energy from NTP hydrolysis can be used by the helicase just for single-stranded translocation, and the enzyme relies on thermal fraying to break the base pair. This mechanism is referred to as the Passive mechanism (**A**) On the other hand, the energy from NTP hydrolysis is used for translocation as well as for destabilization of the base pairs (shown as a gray cloud) on the duplex, to enable strand-separation. This mechanism is referred to as the active mechanism (**B**).

A mechanism of strand separation reported for many of the helicases involves excluding the complementary strand, preventing reannealing. The enzyme could use specific residues in the nucleic-acid-binding cleft as a wedge to separate the two strands (Tackett et al. 2001a; Kawaoka et al. 2004), or the entire helicase molecule could assemble in such a way so as to exclude the other strand thereby keeping the two strands separated (Ahnert and Patel 1997; Hacker and Johnson 1997; Donmez and Patel 2006).

Some of the hexameric helicases like the papillomavirus E1 helicase, SV40 T-antigen are involved in viral replication and hence are required to bind sequence specifically to the origin of replication and melt the base pairs to initiate replication. A looping model was suggested for the strand-separation mechanism of T-antigen, where the double hexamer carried out bidirectional unwinding, looping out the separated single strands through the middle (Li et al. 2003). However, this mechanism has now been refined. According to the new model, the separated strands no longer loop out of the double hexamer, but instead an alternative conformation is proposed where the ssDNA exit through an exit channel on the helicase domain of the double helical enzyme (Gai et al. 2004). A few other mechanisms have been used to describe strand separation by hexameric helicases and these include, the torsional model, plough-share model, etc. (Takahashi et al. 2005; Patel and Donmez 2006).

Helicases as Antiviral Drug Targets

Viruses are obligate parasites. They direct the host cellular metabolism for their replication. To date, although a multitude of viral infections can be warded off through vaccinations, there still exist many viral pathogens against which vaccination is not yet available. This includes diseases like hepatitis C and acquired immuno deficiency syndrome (AIDS). The current strategy in handling these conditions have been through chemotherapeutics which include immune system boosters like interferon-α and -γ and a host of antiviral drugs.

Target the Host or the Virus?

While designing antiviral targets one could consider two broad strategies: targeting a cellular factor involved in viral replication or targeting a virus-specific gene product. Targeting the host factors could result in drastic side effects since the targeted protein could also get inhibited in normal non-infected cells. The latter strategy, on the other hand, could confer a higher virus-specific activity and a low toxicity to the host. However, one caveat that could exist is that, if the targeted protein has metabolic functions, then there would a smaller window of specificity since viral and cellular enzymes catalyze similar enzymatic reactions. However, since the viral and cellular proteins are not identical, structure-based drug design can often exploit the differences between the host and the viral enzymes to generate drugs specific for the virus.

Table 20.4 Helicase inhibitors

Antiviral agent	Mode of action	Nature of inhibition	References
RNA aptamers Piperidine derivatives Heterocyclic-substituted carboxamide Benzimidazole and benzoxazole derivatives	Binding to nucleic-acid binding site	Competitive	(Kumar et al. 1997; Phoon et al. 2001; Borowski et al. 2002; Nishikawa et al. 2004)
Doxorubicin Daunorubicin Epirubin Nogalamycin Mitoxantrone	Inhibition of translocation	DNA/RNA binding, intercalating agent	(Bachur et al. 1992; Lun et al. 1998; Borowski et al. 2002)
N7-chloroethyl-guanosine N9-chloroethyl-guanosine	Competitive activation of helicase	Feedback inhibition by ssNA?	(Borowski et al. 2001)
Aminothiozolyl-phenyl derivatives	Inhibition at the replication bubble	Tight binding of the helicase–primase to the ssDNA of the replication fork	(Liuzzi et al. 2004; Kleymann 2005)
Small-molecule aromatics	Weak binding to DNA substrate	Inhibition of cooperative binding between E1 and E2	(White et al. 2005)

List of all small molecules and biologics that have been shown to have anti-helicase activity, along with their modes of action.

Currently, the most targeted virus-specific factors are the polymerases. The polymerases are essentially required for the replication of viruses. The reverse transcriptase (RT) of the retroviruses and the hepadnaviruses is the sole viral enzyme required for the synthesis of DNA from viral RNA. Viral polymerases are therefore an extremely favorable target for the development of antiviral therapy (De Clercq 2004). Another virus-specific target is the viral helicase. Most viral helicases have multiple enzyme activities associated with the unwinding function. Thus drug design against helicases could involve several general strategies.

Helicase Inhibitors: Strategies and Prospects

All helicases are fuelled by NTP hydrolysis for their unwinding function. Thus, small-molecule inhibitors could be used to inhibit the NTPase function in a number of ways. These inhibitor molecules, usually nucleotide analogs, could directly compete for NTP binding, inhibit nucleic-acid binding, inhibit NTP hydrolysis or NDP release, or uncouple NTP hydrolysis and translocation (Borowski et al. 2000; Borowski et al. 2002; Xu et al. 2006).

Another strategy used in helicase inhibition involves disruption of the protein–protein interfaces between the helicase and other proteins of the replication complex. This strategy has been currently deployed for the inhibition of the HPV E1 helicase whereby its interaction with the E2 protein has been disrupted using inhibitors (White et al. 2003). The HSV helicase–primase complex is inhibited by aminothiozolylphenyl-containing drugs and thiozole urea derivatives. These compounds appear to act by enhancing the binding of the complex to ssDNA in the replication bubble preventing DNA polymerization (Crumpacker and Schaffer 2002; Crute et al. 2002; Kleymann 2004; Biswas et al. 2007).

In a more recent approach, Xue et al. have developed a new strategy for the inhibition of HCV replication. They use siRNAs to knock down cellular host factors, which are important for HCV replication (Zhang et al. 2004; Xue et al. 2007). However, this approach cannot be used as a sole approach for anti-HCV therapy since the host factors involved are important not only for HCV replication but a host of other functions related to cellular RNA metabolism. Table 20.4 gives a list of all small-molecule inhibitors against helicases that have been developed so far. (For a more comprehensive study on helicases as antiviral targets see reviews by Yao and Weber 1998; Frick 2003; Kleymann 2004; Kwong et al. 2005; Maga et al. 2005; Frick and Lam 2006; Frick 2007).

References

Abbate, E. A., J. M. Berger, et al. (2004). "The X-ray structure of the papillomavirus helicase in complex with its molecular matchmaker E2." *Genes Dev* **18**(16): 1981–96.

Abrahams, J. P., A. G. Leslie, et al. (1994). "Structure at 2.8 A resolution of F1-ATPase from bovine heart mitochondria." *Nature* **370**(6491): 621–8.

Ackermann, H. W. (2006). Classification of Bacteriophages. *The Bacteriophages*. R. Calender, Oxford University press: 8–16.

Ahnert, P. and S. S. Patel (1997). "Asymmetric interactions of hexameric bacteriophage T7 DNA helicase with the 5′- and 3′-tails of the forked DNA substrate." *J Biol Chem* **272**(51): 32267–73.

Ahnert, P., K. M. Picha, et al. (2000). "A ring-opening mechanism for DNA binding in the central channel of the T7 helicase-primase protein." *Embo J* **19**(13): 3418–27.

Aizaki, H., K. S. Choi, et al. (2006). "Polypyrimidine-tract-binding protein is a component of the HCV RNA replication complex and necessary for RNA synthesis." *J Biomed Sci* **13**(4): 469–80.

Bachur, N. R., F. Yu, et al. (1992). "Helicase inhibition by anthracycline anticancer agents." *Mol Pharmacol* **41**(6): 993–8.

Banerjee, R. and A. Dasgupta (2001). "Specific interaction of hepatitis C virus protease/helicase NS3 with the 3′-terminal sequences of viral positive- and negative-strand RNA." *J Virol* **75**(4): 1708–21.

Barber, G. N. (2001). "Host defense, viruses and apoptosis." *Cell Death Differ* **8**(2): 113–26.

Betterton, M. D. and F. Julicher (2005). "Opening of nucleic-acid double strands by helicases: active versus passive opening." *Phys Rev E Stat Nonlin Soft Matter Phys* **71**(1 Pt 1): 011904.

Bird, L. E., H. S. Subramanya, et al. (1998). "Helicases: a unifying structural theme?" *Curr Opin Struct Biol* **8**(1): 14–8.

Biswas, S., M. Swift, et al. (2007). "High frequency of spontaneous helicase-primase inhibitor (BAY 57-1293) drug-resistant variants in certain laboratory isolates of HSV-1." *Antivir Chem Chemother* **18**(1): 13–23.

Bjornson, K. P., J. Hsieh, et al. (1998). "Kinetic mechanism for the sequential binding of two single-stranded oligodeoxynucleotides to the Escherichia coli Rep helicase dimer." *Biochemistry* **37**(3): 891–9.

Blinov, V. M., E. V. Koonin, et al. (1989). "Two early genes of bacteriophage T5 encode proteins containing an NTP-binding sequence motif and probably involved in DNA replication, recombination and repair." *FEBS Lett* **252**(1–2): 47–52.

Bork, P. and E. V. Koonin (1993). "An expanding family of helicases within the ′DEAD/H′ superfamily." *Nucleic Acids Res* **21**(3): 751–2.

Borowski, P., M. Lang, et al. (2001). "Inhibition of the helicase activity of HCV NTPase/helicase by 1-beta-D-ribofuranosyl-1,2,4-triazole-3-carboxamide-5 ′-triphosphate (ribavirin-TP)." *Acta Biochim Pol* **48**(3): 739–44.

Borowski, P., O. Mueller, et al. (2000). "ATP-binding domain of NTPase/helicase as a target for hepatitis C antiviral therapy." *Acta Biochim Pol* **47**(1): 173–80.

Borowski, P., S. Schalinski, et al. (2002). "Nucleotide triphosphatase/helicase of hepatitis C virus as a target for antiviral therapy." *Antiviral Res* **55**(3): 397–412.

Braedt, G. and G. R. Smith (1989). "Strand specificity of DNA unwinding by RecBCD enzyme." *Proc Natl Acad Sci U S A* **86**(3): 871–5.

Brechot, C. (2004). "Pathogenesis of hepatitis B virus-related hepatocellular carcinoma: old and new paradigms." *Gastroenterology* **127**(5 Suppl 1): S56–61.

Byrd, A. K. and K. D. Raney (2004). "Protein displacement by an assembly of helicase molecules aligned along single-stranded DNA." *Nat Struct Mol Biol* **11**(6): 531–8.

Byrd, A. K. and K. D. Raney (2005). "Increasing the length of the single-stranded overhang enhances unwinding of duplex DNA by bacteriophage T4 Dda helicase." *Biochemistry* **44**(39): 12990–7.

Carrington-Lawrence, S. D. and S. K. Weller (2003). "Recruitment of polymerase to herpes simplex virus type 1 replication foci in cells expressing mutant primase (UL52) proteins." *J Virol* **77**(7): 4237–47.

Chang, K. S. and G. Luo (2006). "The polypyrimidine tract-binding protein (PTB) is required for efficient replication of hepatitis C virus (HCV) RNA." *Virus Res* **115**(1): 1–8.

Chang, S. C., J. C. Cheng, et al. (2000). "Roles of the AX(4)GKS and arginine-rich motifs of hepatitis C virus RNA helicase in ATP- and viral RNA-binding activity." *J Virol* **74**(20): 9732–7.

Chen, J. and P. Ahlquist (2000). "Brome mosaic virus polymerase-like protein 2a is directed to the endoplasmic reticulum by helicase-like viral protein 1a." *J Virol* **74**(9): 4310–8.

Cheng, W., S. Dumont, et al. (2007). "NS3 helicase actively separates RNA strands and senses sequence barriers ahead of the opening fork." *Proc Natl Acad Sci U S A* **104**(35): 13954–9.

Cordin, O., J. Banroques, et al. (2006). "The DEAD-box protein family of RNA helicases." *Gene* **367**: 17–37.

Crampton, D. J., S. Mukherjee, et al. (2006). "DNA-induced switch from independent to sequential dTTP hydrolysis in the bacteriophage T7 DNA helicase." *Mol Cell* **21**(2): 165–74.

Crumpacker, C. S. and P. A. Schaffer (2002). "New anti-HSV therapeutics target the helicase-primase complex." *Nat Med* **8**(4): 327–8.

Crute, J. J., R. C. Bruckner, et al. (1991). "Herpes simplex-1 helicase-primase. Identification of two nucleoside triphosphatase sites that promote DNA helicase action." *J Biol Chem* **266**(31): 21252–6.

Crute, J. J., C. A. Grygon, et al. (2002). "Herpes simplex virus helicase-primase inhibitors are active in animal models of human disease." *Nat Med* **8**(4): 386–91.

De Clercq, E. (2004). "Antivirals and antiviral strategies." *Nat Rev Microbiol* **2**(9): 704–20.

Delagoutte, E. and P. H. von Hippel (2001). "Molecular mechanisms of the functional coupling of the helicase (gp41) and polymerase (gp43) of bacteriophage T4 within the DNA replication fork." *Biochemistry* **40**(14): 4459–77.

Delagoutte, E. and P. H. von Hippel (2002). "Helicase mechanisms and the coupling of helicases within macromolecular machines. Part I: Structures and properties of isolated helicases." *Q Rev Biophys* **35**(4): 431–78.

Dillingham, M. S., P. Soultanas, et al. (1999). "Site-directed mutagenesis of motif III in PcrA helicase reveals a role in coupling ATP hydrolysis to strand separation." *Nucleic Acids Res* **27**(16): 3310–7.

Dillingham, M. S., D. B. Wigley, et al. (2000). "Demonstration of unidirectional single-stranded DNA translocation by PcrA helicase: measurement of step size and translocation speed." *Biochemistry* **39**(1): 205–12.

Dinant, S., M. Janda, et al. (1993). "Bromovirus RNA replication and transcription require compatibility between the polymerase- and helicase-like viral RNA synthesis proteins." *J Virol* **67**(12): 7181–9.

Donmez, I. and S. S. Patel (2006). "Mechanisms of a ring shaped helicase." *Nucleic Acids Res* **34**(15): 4216–24.

Donmez, I., V. Rajagopal, et al. (2007). "Nucleic acid unwinding by hepatitis C virus and bacteriophage t7 helicases is sensitive to base pair stability." *J Biol Chem* **282**(29): 21116–23.

Dreyfus, I. I. a. M. (2006). "DEAD-box RNA helicases in *Escherichia coli*." *Nucleic Acids Research* **34**(15): 4189–4197.

Du, M. X., R. B. Johnson, et al. (2002). "Comparative characterization of two DEAD-box RNA helicases in superfamily II: human translation-initiation factor 4A and hepatitis C virus non-structural protein 3 (NS3) helicase." *Biochem J* **363**(Pt 1): 147–55.

Egelman, E. H. (1998). "Bacterial helicases." *J Struct Biol* **124**(2–3): 123–8.

Egelman, E. H., X. Yu, et al. (1995). "Bacteriophage T7 helicase/primase proteins form rings around single-stranded DNA that suggest a general structure for hexameric helicases." *Proc Natl Acad Sci U S A* **92**(9): 3869–73.

Enemark, E. J. and L. Joshua-Tor (2006). "Mechanism of DNA translocation in a replicative hexameric helicase." *Nature* **442**(7100): 270–5.

Enemark, E. J., A. Stenlund, et al. (2002). "Crystal structures of two intermediates in the assembly of the papillomavirus replication initiation complex." *Embo J* **21**(6): 1487–96.

Fields, B., Knipe D. M., and Howley P. M., (1996). *Fundamental Virology*. Philadelphia, Lippincott-Raven publishers.

Finn, R. D., J. Mistry, et al. (2006). "Pfam: clans, web tools and services." *Nucleic Acids Res* **34**(Database issue): D247–51.

Fischer, C. J. and T. M. Lohman (2004). "ATP-dependent translocation of proteins along single-stranded DNA: models and methods of analysis of pre-steady state kinetics." *J Mol Biol* **344**(5): 1265–86.

Fischer, C. J., N. K. Maluf, et al. (2004). "Mechanism of ATP-dependent translocation of E.coli UvrD monomers along single-stranded DNA." *J Mol Biol* **344**(5): 1287–309.

Frick, D. N. (2003). "Helicases as antiviral drug targets." *Drug News Perspect* **16**(6): 355–62.

Frick, D. N. (2007). "The hepatitis C virus NS3 protein: a model RNA helicase and potential drug target." *Curr Issues Mol Biol* **9**(1): 1–20.

Frick, D. N. and A. M. Lam (2006). "Understanding helicases as a means of virus control." *Curr Pharm Des* **12**(11): 1315–38.

Frick, D. N., R. S. Rypma, et al. (2004). "Electrostatic analysis of the hepatitis C virus NS3 helicase reveals both active and allosteric site locations." *Nucleic Acids Res* **32**(18): 5519–28.

Gai, D., R. Zhao, et al. (2004). "Mechanisms of conformational change for a replicative hexameric helicase of SV40 large tumor antigen." *Cell* **119**(1): 47–60.

Gannon, J. V. and D. P. Lane (1990). "Interactions between SV40 T antigen and DNA polymerase alpha." *New Biol* **2**(1): 84–92.

Gao, Z., A. Krithivas, et al. (1998). "The Epstein-Barr virus lytic transactivator Zta interacts with the helicase-primase replication proteins." *J Virol* **72**(11): 8559–67.

Gillitzer, E., G. Chen, et al. (2000). "Separate domains in E1 and E2 proteins serve architectural and productive roles for cooperative DNA binding." *Embo J* **19**(12): 3069–79.

Goh, P. Y., Y. J. Tan, et al. (2004). "Cellular RNA helicase p68 relocalization and interaction with the hepatitis C virus (HCV) NS5B protein and the potential role of p68 in HCV RNA replication." *J Virol* **78**(10): 5288–98.

Gorbalenya, A. E. and E. V. Koonin (1989). "Viral proteins containing the purine NTP-binding sequence pattern." *Nucleic Acids Res* **17**(21): 8413–40.

Gorbalenya, A. E. and E. V. Koonin (1993). "Helicases – Amino-Acid-Sequence Comparisons and Structure-Function-Relationships." *Curr Opin Struct Biol* **3**(3): 419–429.

Gorbalenya, A. E., E. V. Koonin, et al. (1988a). "A conserved NTP-motif in putative helicases." *Nature* **333**(6168): 22.

Gorbalenya, A. E., E. V. Koonin, et al. (1988b). "A novel superfamily of nucleoside triphosphate-binding motif containing proteins which are probably involved in duplex unwinding in DNA and RNA replication and recombination." *FEBS Lett* **235**(1–2): 16–24.

Gorbalenya, A. E., E. V. Koonin, et al. (1989). "Two related superfamilies of putative helicases involved in replication, recombination, repair and expression of DNA and RNA genomes." *Nucleic Acids Res* **17**(12): 4713–30.

Gorbalenya, A. E., E. V. Koonin, et al. (1990). "A new superfamily of putative NTP-binding domains encoded by genomes of small DNA and RNA viruses." *FEBS Lett* **262**(1): 145–8.

Goregaoker, S. P., D. J. Lewandowski, et al. (2001). "Identification and functional analysis of an interaction between domains of the 126/183-kDa replicase-associated proteins of tobacco mosaic virus." *Virology* **282**(2): 320–8.

Graves-Woodward, K. L., J. Gottlieb, et al. (1997). "Biochemical analyses of mutations in the HSV-1 helicase-primase that alter ATP hydrolysis, DNA unwinding, and coupling between hydrolysis and unwinding." *J Biol Chem* **272**(7): 4623–30.

Guex, N. and M. C. Peitsch (1997). "SWISS-MODEL and the Swiss-PdbViewer: an environment for comparative protein modeling." *Electrophoresis* **18**(15): 2714–23.

Gwack, Y., D. W. Kim, et al. (1996). "Characterization of RNA binding activity and RNA helicase activity of the hepatitis C virus NS3 protein." *Biochem Biophys Res Commun* **225**(2): 654–9.

Gwack, Y., H. Yoo, et al. (1999). "RNA-Stimulated ATPase and RNA helicase activities and RNA binding domain of hepatitis G virus nonstructural protein 3." *J Virol* **73**(4): 2909–15.

Hacker, K. J. and K. A. Johnson (1997). "A hexameric helicase encircles one DNA strand and excludes the other during DNA unwinding." *Biochemistry* **36**(46): 14080–7.

Hall, M. C. and S. W. Matson (1999). "Helicase motifs: the engine that powers DNA unwinding." *Mol Microbiol* **34**(5): 867–77.

Hamatake, R. K., M. Bifano, et al. (1997). "A functional interaction of ICP8, the herpes simplex virus single-stranded DNA-binding protein, and the helicase-primase complex that is dependent on the presence of the UL8 subunit." *J Gen Virol* **78 (Pt 4)**: 857–65.

He, Z. G. and C. C. Richardson (2004). "Effect of single-stranded DNA-binding proteins on the helicase and primase activities of the bacteriophage T7 gene 4 protein." *J Biol Chem* **279**(21): 22190–7.

Hill, T. L. and T. Tsuchiya (1981). "Theoretical aspects of translocation on DNA: adenosine triphosphatases and treadmilling binding proteins." *Proc Natl Acad Sci U S A* **78**(8): 4796–800.

Hingorani, M. M. and S. S. Patel (1993). "Interactions of bacteriophage T7 DNA primase/helicase protein with single-stranded and double-stranded DNAs." *Biochemistry* **32**(46): 12478–87.

Hingorani, M. M. and S. S. Patel (1996). "Cooperative interactions of nucleotide ligands are linked to oligomerization and DNA binding in bacteriophage T7 gene 4 helicases." *Biochemistry* **35**(7): 2218–28.

Hingorani, M. M., M. T. Washington, et al. (1997). "The dTTPase mechanism of T7 DNA helicase resembles the binding change mechanism of the F1-ATPase." *Proc Natl Acad Sci U S A* **94**(10): 5012–7.

Hirshberg, M., K. Henrick, et al. (1998). "Crystal structure of phosphate binding protein labeled with a coumarin fluorophore, a probe for inorganic phosphate." *Biochemistry* **37**(29): 10381–5.

Howe, A. Y., R. Chase, et al. (1999). "A novel recombinant single-chain hepatitis C virus NS3-NS4A protein with improved helicase activity." *Protein Sci* **8**(6): 1332–41.

Hughes, F. J. and M. A. Romanos (1993). "E1 protein of human papillomavirus is a DNA helicase/ATPase." *Nucleic Acids Res* **21**(25): 5817–23.

Ilyina, T. V., A. E. Gorbalenya, et al. (1992). "Organization and evolution of bacterial and bacteriophage primase-helicase systems." *J Mol Evol* **34**(4): 351–7.

Im, D. S. and N. Muzyczka (1990). "The AAV origin binding protein Rep68 is an ATP-dependent site-specific endonuclease with DNA helicase activity." *Cell* **61**(3): 447–57.

Ito, E., D. Sahri, et al. (2004). "Baculovirus proteins IE-1, LEF-3, and P143 interact with DNA in vivo: a formaldehyde cross-linking study." *Virology* **329**(2): 337–47.

Ivanov, K. A., V. Thiel, et al. (2004). "Multiple enzymatic activities associated with severe acute respiratory syndrome coronavirus helicase." *J Virol* **78**(11): 5619–32.

Ivanov, K. A. and J. Ziebuhr (2004). "Human coronavirus 229E nonstructural protein 13: characterization of duplex-unwinding, nucleoside triphosphatase, and RNA 5′-triphosphatase activities." *J Virol* **78**(14): 7833–8.

Iyer, L. M., D. D. Leipe, et al. (2004). "Evolutionary history and higher order classification of AAA+ ATPases." *J Struct Biol* **146**(1–2): 11–31.

James, J. A., A. K. Aggarwal, et al. (2004). "Structure of adeno-associated virus type 2 Rep40-ADP complex: insight into nucleotide recognition and catalysis by superfamily 3 helicases." *Proc Natl Acad Sci U S A* **101**(34): 12455–60.

James, J. A., C. R. Escalante, et al. (2003). "Crystal structure of the SF3 helicase from adeno-associated virus type 2." *Structure* **11**(8): 1025–35.

Jarvis, T. C., J. W. Newport, et al. (1991). "Stimulation of the processivity of the DNA polymerase of bacteriophage T4 by the polymerase accessory proteins. The role of ATP hydrolysis." *J Biol Chem* **266**(3): 1830–40.

Jarvis, T. C., D. M. Ring, et al. (1990). "'Macromolecular crowding': thermodynamic consequences for protein-protein interactions within the T4 DNA replication complex." *J Biol Chem* **265**(25): 15160–7.

Jeang, K. T. and V. Yedavalli (2006). "Role of RNA helicases in HIV-1 replication." *Nucleic Acids Res* **34**(15): 4198–205.

Jeong, Y. J., M. K. Levin, et al. (2004). "The DNA-unwinding mechanism of the ring helicase of bacteriophage T7." *Proc Natl Acad Sci U S A* **101**(19): 7264–9.

Jing, D. H., F. Dong, et al. (1999). "Interactions of bacteriophage T4-coded primase (gp61) with the T4 replication helicase (gp41) and DNA in primosome formation." *J Biol Chem* **274**(38): 27287–98.

Johnson, D. S., L. Bai, et al. (2007). "Single-molecule studies reveal dynamics of DNA unwinding by the ring-shaped T7 helicase." *Cell* **129**(7): 1299–309.

Jongeneel, C. V., T. Formosa, et al. (1984). "Purification and characterization of the bacteriophage T4 dda protein. A DNA helicase that associates with the viral helix-destabilizing protein." *J Biol Chem* **259**(20): 12925–32.

Kadare, G. and A. L. Haenni (1997). "Virus-encoded RNA helicases." *J Virol* **71**(4): 2583–90.

Kato, M., D. N. Frick, et al. (2001). "A complex of the bacteriophage T7 primase-helicase and DNA polymerase directs primer utilization." *J Biol Chem* **276**(24): 21809–20.

Kato, M., T. Ito, et al. (2003). "Modular architecture of the bacteriophage T7 primase couples RNA primer synthesis to DNA synthesis." *Mol Cell* **11**(5): 1349–60.

Kawaoka, J., E. Jankowsky, et al. (2004). "Backbone tracking by the SF2 helicase NPH-II." *Nat Struct Mol Biol* **11**(6): 526–30.

Killoran, M. P. and J. L. Keck (2006). "Sit down, relax and unwind: structural insights into RecQ helicase mechanisms." *Nucleic Acids Res* **34**(15): 4098–105.

Kim, D. E., M. Narayan, et al. (2002). "T7 DNA helicase: a molecular motor that processively and unidirectionally translocates along single-stranded DNA." *J Mol Biol* **321**(5): 807–19.

Kim, D. W., J. Kim, et al. (1997). "Mutational analysis of the hepatitis C virus RNA helicase." *J Virol* **71**(12): 9400–9.

Kim, J. L., K. A. Morgenstern, et al. (1998). "Hepatitis C virus NS3 RNA helicase domain with a bound oligonucleotide: the crystal structure provides insights into the mode of unwinding." *Structure* **6**(1): 89–100.

King, J. A., R. Dubielzig, et al. (2001). "DNA helicase-mediated packaging of adeno-associated virus type 2 genomes into preformed capsids." *Embo J* **20**(12): 3282–91.

Kleymann, G. (2004). "Helicase primase: targeting the Achilles heel of herpes simplex viruses." *Antivir Chem Chemother* **15**(3): 135–40.

Kleymann, G. (2005). "Agents and strategies in development for improved management of herpes simplex virus infection and disease." *Expert Opin Investig Drugs* **14**(2): 135–61.

Kong, D. and C. C. Richardson (1998). "Role of the acidic carboxyl-terminal domain of the single-stranded DNA-binding protein of bacteriophage T7 in specific protein-protein interactions." *J Biol Chem* **273**(11): 6556–64.

Koonin, E. V. (1991). "Similarities in RNA helicases." *Nature* **352**(6333): 290.

Koonin, E. V. (1992). "A new group of putative RNA helicases." *Trends Biochem Sci* **17**(12): 495–7.

Korolev, S., J. Hsieh, et al. (1997). "Major domain swiveling revealed by the crystal structures of complexes of E. coli Rep helicase bound to single-stranded DNA and ADP." *Cell* **90**(4): 635–47.

Korolev, S., N. Yao, et al. (1998). "Comparisons between the structures of HCV and Rep helicases reveal structural similarities between SF1 and SF2 super-families of helicases." *Protein Sci* **7**(3): 605–10.

Kumar, P. K., K. Machida, et al. (1997). "Isolation of RNA aptamers specific to the NS3 protein of hepatitis C virus from a pool of completely random RNA." *Virology* **237**(2): 270–82.

Kwong, A. D., B. G. Rao, et al. (2005). "Viral and cellular RNA helicases as antiviral targets." *Nat Rev Drug Discov* **4**(10): 845–53.

Lam, A. M., R. S. Rypma, et al. (2004). "Enhanced nucleic acid binding to ATP-bound hepatitis C virus NS3 helicase at low pH activates RNA unwinding." *Nucleic Acids Res* **32**(13): 4060–70.

Lavia, P., A. M. Mileo, et al. (2003). "Emerging roles of DNA tumor viruses in cell proliferation: new insights into genomic instability." *Oncogene* **22**(42): 6508–16.

LeBowitz, J. H. and R. McMacken (1986). "The Escherichia coli dnaB replication protein is a DNA helicase." *J Biol Chem* **261**(10): 4738–48.

Ledwaba, T., Z. Dlamini, et al. (2004). "Molecular genetics of human cervical cancer: role of papillomavirus and the apoptotic cascade." *Biol Chem* **385**(8): 671–82.

Lee, J. Y. and W. Yang (2006). "UvrD helicase unwinds DNA one base pair at a time by a two-part power stroke." *Cell* **127**(7): 1349–60.

Lee, M. S. and K. J. Marians (1990). "Differential ATP requirements distinguish the DNA translocation and DNA unwinding activities of the Escherichia coli PRI A protein." *J Biol Chem* **265**(28): 17078–83.

Lefebvre, S. D., M. L. Wong, et al. (1999). "Simultaneous interactions of bacteriophage T4 DNA replication proteins gp59 and gp32 with single-stranded (ss) DNA. Co-modulation of ssDNA binding activities in a DNA helicase assembly intermediate." *J Biol Chem* **274**(32): 22830–8.

Levin, M. (2002). DNA unwinding mechanism of the helicase from Hepatitis C virus. *Biochemistry*. Columbus, Ohio State University, 152.

Levin, M. K., M. Gurjar, et al. (2005). "A Brownian motor mechanism of translocation and strand separation by hepatitis C virus helicase." *Nat Struct Mol Biol* **12**(5): 429–35.

Levin, M. K., M. M. Gurjar, et al. (2003). "ATP binding modulates the nucleic acid affinity of hepatitis C virus helicase." *J Biol Chem* **278**(26): 23311–6.

Levin, M. K. and S. S. Patel (1999). "The helicase from hepatitis C virus is active as an oligomer." *J Biol Chem* **274**(45): 31839–46.

Levin, M. K. and S. S. Patel (2002). "Helicase from hepatitis C virus, energetics of DNA binding." *J Biol Chem* **277**(33): 29377–85.

Levrero, M. (2006). "Viral hepatitis and liver cancer: the case of hepatitis C." *Oncogene* **25**(27): 3834–47.

Li, D., R. Zhao, et al. (2003). "Structure of the replicative helicase of the oncoprotein SV40 large tumour antigen." *Nature* **423**(6939): 512–8.

Liao, J. C., Y. J. Jeong, et al. (2005). "Mechanochemistry of t7 DNA helicase." *J Mol Biol* **350**(3): 452–75.

Lim, S. G., Y. J. Tan, et al. (2006). "Use of an in vitro model and yeast two-hybrid system to investigate the pathogenesis of hepatitis C." *Intervirology* **49**(1–2): 44–50.

Linder, P. (2000). "Quick guide: DEAD-box proteins." *Curr Biol* **10**(24): R887.

Linder, P. and M. C. Daugeron (2000). "Are DEAD-box proteins becoming respectable helicases?" *Nat Struct Biol* **7**(2): 97–9.

Linder, P., P. F. Lasko, et al. (1989). "Birth of the D-E-A-D box." *Nature* **337**(6203): 121–2.

Liu, C. C. and B. M. Alberts (1981). "Characterization of the DNA-dependent GTPase activity of T4 gene 41 protein, an essential component of the T4 bacteriophage DNA replication apparatus." *J Biol Chem* **256**(6): 2813–20.

Liu, D., Y. S. Wang, et al. (2001). "Solution structure and backbone dynamics of an engineered arginine-rich subdomain 2 of the hepatitis C virus NS3 RNA helicase." *J Mol Biol* **314**(3): 543–61.

Liu, D., W. T. Windsor, et al. (2003). "Double-stranded DNA-induced localized unfolding of HCV NS3 helicase subdomain 2." *Protein Sci* **12**(12): 2757–67.

Liuzzi, M., P. Kibler, et al. (2004). "Isolation and characterization of herpes simplex virus type 1 resistant to aminothiazolylphenyl-based inhibitors of the viral helicase-primase." *Antiviral Res* **64**(3): 161–70.

Lohman, T. M. (1993). "Helicase-catalyzed DNA unwinding." *J Biol Chem* **268**(4): 2269–72.

Lohman, T. M. and K. P. Bjornson (1996). "Mechanisms of helicase-catalyzed DNA unwinding." *Annu Rev Biochem* **65**: 169–214.

Luking, A., U. Stahl, et al. (1998). "The protein family of RNA helicases." *Crit Rev Biochem Mol Biol* **33**(4): 259–96.

Lun, L., P. M. Sun, et al. (1998). "Antihelicase action of CI-958, a new drug for prostate cancer." *Cancer Chemother Pharmacol* **42**(6): 447–53.

Maga, G., S. Gemma, et al. (2005). "Specific targeting of hepatitis C virus NS3 RNA helicase. Discovery of the potent and selective competitive nucleotide-mimicking inhibitor QU663." *Biochemistry* **44**(28): 9637–44.

Mahdi, A. A., G. S. Briggs, et al. (2003). "A model for dsDNA translocation revealed by a structural motif common to RecG and Mfd proteins." *Embo J* **22**(3): 724–34.

Mancini, E. J., R. Assenberg, et al. (2007). "Structure of the Murray Valley encephalitis virus RNA helicase at 1.9 Angstrom resolution." *Protein Sci* **16**(10): 2294–300.

Mancini, E. J., D. E. Kainov, et al. (2004). "Atomic snapshots of an RNA packaging motor reveal conformational changes linking ATP hydrolysis to RNA translocation." *Cell* **118**(6): 743–55.

Marintcheva, B. and S. K. Weller (2003). "Helicase motif Ia is involved in single-strand DNA-binding and helicase activities of the herpes simplex virus type 1 origin-binding protein, UL9." *J Virol* **77**(4): 2477–88.

Masterson, P. J., M. A. Stanley, et al. (1998). "A C-terminal helicase domain of the human papillomavirus E1 protein binds E2 and the DNA polymerase alpha-primase p68 subunit." *J Virol* **72**(9): 7407–19.

Mastrangelo, E., M. Bollati, et al. (2006). "Preliminary crystallographic characterization of an RNA helicase from Kunjin virus." *Acta Crystallograph Sect F Struct Biol Cryst Commun* **62**(Pt 9): 876–9.

Matson, S. W. (1986). "Escherichia coli helicase II (urvD gene product) translocates unidirectionally in a 3′ to 5′ direction." *J Biol Chem* **261**(22): 10169–75.

Matson, S. W. (1991). "DNA helicases of Escherichia coli." *Prog Nucleic Acid Res Mol Biol* **40**: 289–326.

Matson, S. W., D. W. Bean, et al. (1994). "DNA helicases: enzymes with essential roles in all aspects of DNA metabolism." *Bioessays* **16**(1): 13–22.

Matson, S. W. and K. A. Kaiser-Rogers (1990). "DNA helicases." *Annu Rev Biochem* **59**: 289–329.

Matson, S. W. and C. C. Richardson (1983). "DNA-dependent nucleoside 5′-triphosphatase activity of the gene 4 protein of bacteriophage T7." *J Biol Chem* **258**(22): 14009–16.

Matson, S. W., S. Tabor, et al. (1983). "The gene 4 protein of bacteriophage T7. Characterization of helicase activity." *J Biol Chem* **258**(22): 14017–24.

Mendonca, V. M., H. D. Klepin, et al. (1995). "DNA helicases in recombination and repair: construction of a delta uvrD delta helD delta recQ mutant deficient in recombination and repair." *J Bacteriol* **177**(5): 1326–35.

Mindell, D. P., J. S. Rest, et al. (2004). Viruses and the tree of life. *Assembling the tree of life*. J. Cracraft. NY, Oxford University Press: 107–118.

Morris, P. D., A. K. Byrd, et al. (2002). "Hepatitis C virus NS3 and simian virus 40 T antigen helicases displace streptavidin from 5′-biotinylated oligonucleotides but not from 3′-biotinylated oligonucleotides: evidence for directional bias in translocation on single-stranded DNA." *Biochemistry* **41**(7): 2372–8.

Morris, P. D., A. J. Tackett, et al. (2001). "Biotin-streptavidin-labeled oligonucleotides as probes of helicase mechanisms." *Methods* **23**(2): 149–59.

Myong, S., M. M. Bruno, et al. (2007). "Spring-loaded mechanism of DNA unwinding by hepatitis C virus NS3 helicase." *Science* **317**(5837): 513–6.

Nakai, H. and C. C. Richardson (1986). "Interactions of the DNA polymerase and gene 4 protein of bacteriophage T7. Protein-protein and protein-DNA interactions involved in RNA-primed DNA synthesis." *J Biol Chem* **261**(32): 15208–16.

Nakai, H. and C. C. Richardson (1988). "The effect of the T7 and Escherichia coli DNA-binding proteins at the replication fork of bacteriophage T7." *J Biol Chem* **263**(20): 9831–9.

Niedenzu, T., D. Roleke, et al. (2001). "Crystal structure of the hexameric replicative helicase RepA of plasmid RSF1010." *J Mol Biol* **306**(3): 479–87.

Nishikawa, F., K. Funaji, et al. (2004). "In vitro selection of RNA aptamers against the HCV NS3 helicase domain." *Oligonucleotides* **14**(2): 114–29.

Notarnicola, S. M., H. L. Mulcahy, et al. (1997). "The acidic carboxyl terminus of the bacteriophage T7 gene 4 helicase/primase interacts with T7 DNA polymerase." *J Biol Chem* **272**(29): 18425–33.

Otto, B., E. Fanning, et al. (1979). "DNA polymerases and a single-strand-specific DNA-binding protein associated with simian virus 40 nucleoprotein complexes." *Cold Spring Harb Symp Quant Biol* **43 Pt 2**: 705–8.

Park, P., W. Copeland, et al. (1994). "The cellular DNA polymerase alpha-primase is required for papillomavirus DNA replication and associates with the viral E1 helicase." *Proc Natl Acad Sci U S A* **91**(18): 8700–4.

Patel, S. S. and I. Donmez (2006). "Mechanisms of helicases." *J Biol Chem* **281**(27): 18265–8.

Patel, S. S. and M. M. Hingorani (1993). "Oligomeric structure of bacteriophage T7 DNA primase/helicase proteins." *J Biol Chem* **268**(14): 10668–75.

Patel, S. S. and M. M. Hingorani (1995). "Nucleotide binding studies of bacteriophage T7 DNA helicase-primase protein." *Biophys J* **68**(4 Suppl): 186S-189S; discussion 189S-190S.

Patel, S. S., M. M. Hingorani, et al. (1994). "The K318A mutant of bacteriophage T7 DNA primase-helicase protein is deficient in helicase but not primase activity and inhibits primase-helicase protein wild-type activities by heterooligomer formation." *Biochemistry* **33**(25): 7857–68.

Patel, S. S. and K. M. Picha (2000). "Structure and function of hexameric helicases." *Annu Rev Biochem* **69**: 651–97.

Pause, A. and N. Sonenberg (1992). "Mutational analysis of a DEAD box RNA helicase: the mammalian translation initiation factor eIF-4A." *Embo J* **11**(7): 2643–54.

Phoon, C. W., P. Y. Ng, et al. (2001). "Biological evaluation of hepatitis C virus helicase inhibitors." *Bioorg Med Chem Lett* **11**(13): 1647–50.

Piccininni, S., A. Varaklioti, et al. (2002). "Modulation of the hepatitis C virus RNA-dependent RNA polymerase activity by the non-structural (NS) 3 helicase and the NS4B membrane protein." *J Biol Chem* **277**(47): 45670–9.

Picha, K. M., P. Ahnert, et al. (2000). "DNA binding in the central channel of bacteriophage T7 helicase-primase is a multistep process. Nucleotide hydrolysis is not required." *Biochemistry* **39**(21): 6401–9.

Picha, K. M. and S. S. Patel (1998). "Bacteriophage T7 DNA helicase binds dTTP, forms hexamers, and binds DNA in the absence of Mg2+. The presence of dTTP is sufficient for hexamer formation and DNA binding." *J Biol Chem* **273**(42): 27315–9.

Raney, K. D. and S. J. Benkovic (1995). "Bacteriophage T4 Dda helicase translocates in a unidirectional fashion on single-stranded DNA." *J Biol Chem* **270**(38): 22236–42.

Rogers, G. W., Jr., A. A. Komar, et al. (2002). "eIF4A: the godfather of the DEAD box helicases." *Prog Nucleic Acid Res Mol Biol* **72**: 307–31.

Rogers, G. W., Jr., N. J. Richter, et al. (2001). "Modulation of the helicase activity of eIF4A by eIF4B, eIF4H, and eIF4F." *J Biol Chem* **276**(33): 30914–22.

Rosenberg, A. H., S. S. Patel, et al. (1992). "Cloning and expression of gene 4 of bacteriophage T7 and creation and analysis of T7 mutants lacking the 4A primase/helicase or the 4B helicase." *J Biol Chem* **267**(21): 15005–12.

Sawaya, M. R., S. Guo, et al. (1999). "Crystal structure of the helicase domain from the replicative helicase-primase of bacteriophage T7." *Cell* **99**(2): 167–77.

Schattner, E. J. (2002). "Apoptosis in lymphocytic leukemias and lymphomas." *Cancer Invest* **20**(5–6): 737–48.

Schmid, S. R. and P. Linder (1992). "D-E-A-D protein family of putative RNA helicases." *Mol Microbiol* **6**(3): 283–91.

Seif, R. (1982). "New properties of simian virus 40 large T antigen." *Mol Cell Biol* **2**(12): 1463–71.

Seo, Y. S., F. Muller, et al. (1993). "Bovine papilloma virus (BPV)-encoded E2 protein enhances binding of E1 protein to the BPV replication origin." *Proc Natl Acad Sci U S A* **90**(7): 2865–9.

Serebrov, V. and A. M. Pyle (2004). "Periodic cycles of RNA unwinding and pausing by hepatitis C virus NS3 helicase." *Nature* **430**(6998): 476–80.

Shiratori, A., T. Shibata, et al. (1999). "Systematic identification, classification, and characterization of the open reading frames which encode novel helicase-related proteins in Saccharomyces cerevisiae by gene disruption and Northern analysis." *Yeast* **15**(3): 219–53.

Shuman, S. (1993). "Vaccinia virus RNA helicase. Directionality and substrate specificity." *J Biol Chem* **268**(16): 11798–802.

Sickmier, E. A., K. N. Kreuzer, et al. (2004). "The crystal structure of the UvsW helicase from bacteriophage T4." *Structure* **12**(4): 583–92.

Singleton, M. R., M. S. Dillingham, et al. (2007). "Structure and mechanism of helicases and nucleic acid translocases." *Annu Rev Biochem* **76**: 23–50.

Singleton, M. R., M. R. Sawaya, et al. (2000). "Crystal structure of T7 gene 4 ring helicase indicates a mechanism for sequential hydrolysis of nucleotides." *Cell* **101**(6): 589–600.

Smale, S. T. and R. Tjian (1986). "T-antigen-DNA polymerase alpha complex implicated in simian virus 40 DNA replication." *Mol Cell Biol* **6**(11): 4077–87.

Soultanas, P. and D. B. Wigley (2000). "DNA helicases: 'inching forward'." *Curr Opin Struct Biol* **10**(1): 124–8.

Spurling, T. L., R. L. Eoff, et al. (2006). "Dda helicase unwinds a DNA-PNA chimeric substrate: evidence for an inchworm mechanism." *Bioorg Med Chem Lett* **16**(7): 1816–20.

Stano, N. M., Y. J. Jeong, et al. (2005). "DNA synthesis provides the driving force to accelerate DNA unwinding by a helicase." *Nature* **435**(7040): 370–3.

Story, R. M., I. T. Weber, et al. (1992). "The structure of the E. coli recA protein monomer and polymer." *Nature* **355**(6358): 318–25.

Strath, J. and G. E. Blair (2006). "Adenovirus subversion of immune surveillance, apoptotic and growth regulatory pathways: a model for tumorigenesis." *Acta Microbiol Immunol Hung* **53**(2): 145–69.

Stukalin, E. B., H. Phillips, 3rd, et al. (2005). "Coupling of two motor proteins: a new motor can move faster." *Phys Rev Lett* **94**(23): 238101.

Subramanya, H. S., L. E. Bird, et al. (1996). "Crystal structure of a DExx box DNA helicase." *Nature* **384**(6607): 379–83.

Tackett, A. J., Y. Chen, et al. (2005). "Multiple full-length NS3 molecules are required for optimal unwinding of oligonucleotide DNA in vitro." *J Biol Chem* **280**(11): 10797–806.

Tackett, A. J., P. D. Morris, et al. (2001a). "Unwinding of unnatural substrates by a DNA helicase." *Biochemistry* **40**(2): 543–8.

Tackett, A. J., L. Wei, et al. (2001b). "Unwinding of nucleic acids by HCV NS3 helicase is sensitive to the structure of the duplex." *Nucleic Acids Res* **29**(2): 565–72.

Takahashi, T. S., D. B. Wigley, et al. (2005). "Pumps, paradoxes and ploughshares: mechanism of the MCM2-7 DNA helicase." *Trends Biochem Sci* **30**(8): 437–44.

Tanner, N. K. (2003). "The newly identified Q motif of DEAD box helicases is involved in adenine recognition." *Cell Cycle* **2**(1): 18–9.

Tanner, N. K., O. Cordin, et al. (2003). "The Q motif: a newly identified motif in DEAD box helicases may regulate ATP binding and hydrolysis." *Mol Cell* **11**(1): 127–38.

Taylor, T. J. and D. M. Knipe (2004). "Proteomics of herpes simplex virus replication compartments: association of cellular DNA replication, repair, recombination, and chromatin remodeling proteins with ICP8." *J Virol* **78**(11): 5856–66.

Tomko, E. J., C. J. Fischer, et al. (2007). "A nonuniform stepping mechanism for E. coli UvrD monomer translocation along single-stranded DNA." *Mol Cell* **26**(3): 335–47.

Toth, E. A., Y. Li, et al. (2003). "The crystal structure of the bifunctional primase-helicase of bacteriophage T7." *Mol Cell* **12**(5): 1113–23.

Trego, K. S. and D. S. Parris (2003). "Functional interaction between the herpes simplex virus type 1 polymerase processivity factor and origin-binding proteins: enhancement of UL9 helicase activity." *J Virol* **77**(23): 12646–59.

Tuteja, N. and R. Tuteja (2004). "Unraveling DNA helicases. Motif, structure, mechanism and function." *Eur J Biochem* **271**(10): 1849–63.

Van Der Heijden, M. W., J. E. Carette, et al. (2001). "Alfalfa mosaic virus replicase proteins P1 and P2 interact and colocalize at the vacuolar membrane." *J Virol* **75**(4): 1879–87.

Velankar, S. S., P. Soultanas, et al. (1999). "Crystal structures of complexes of PcrA DNA helicase with a DNA substrate indicate an inchworm mechanism." *Cell* **97**(1): 75–84.

von Hippel, P. H. and E. Delagoutte (2001). "A general model for nucleic acid helicases and their "coupling" within macromolecular machines." *Cell* **104**(2): 177–90.

Walker, S. L., R. S. Wonderling, et al. (1997). "Mutational analysis of the adeno-associated virus Rep68 protein: identification of critical residues necessary for site-specific endonuclease activity." *J Virol* **71**(4): 2722–30.

Washington, M. T., A. H. Rosenberg, et al. (1996). "Biochemical analysis of mutant T7 primase/helicase proteins defective in DNA binding, nucleotide hydrolysis, and the coupling of hydrolysis with DNA unwinding." *J Biol Chem* **271**(43): 26825–34.

White, P. W., A. M. Faucher, et al. (2005). "Biphenylsulfonacetic acid inhibitors of the human papillomavirus type 6 E1 helicase inhibit ATP hydrolysis by an allosteric mechanism involving tyrosine 486." *Antimicrob Agents Chemother* **49**(12): 4834–42.

White, P. W., S. Titolo, et al. (2003). "Inhibition of human papillomavirus DNA replication by small molecule antagonists of the E1-E2 protein interaction." *J Biol Chem* **278**(29): 26765–72.

Wong, I., K. L. Chao, et al. (1992). "DNA-induced dimerization of the Escherichia coli rep helicase. Allosteric effects of single-stranded and duplex DNA." *J Biol Chem* **267**(11): 7596–610.

Wu, C., D. Edgil, et al. (1998). "The origin DNA-binding and single-stranded DNA-binding domains of simian virus 40 large T antigen are distinct." J Virol **72**(12): 10256–9.

Wu, C., R. Roy, et al. (2001). "Role of single-stranded DNA binding activity of T antigen in simian virus 40 DNA replication." *J Virol* **75**(6): 2839–47.

Wu, J., A. K. Bera, et al. (2005). "Structure of the Flavivirus helicase: implications for catalytic activity, protein interactions, and proteolytic processing." *J Virol* **79**(16): 10268–77.

Xu, H., N. Strater, et al. (2003). "Structure of DNA helicase RepA in complex with sulfate at 1.95 A resolution implicates structural changes to an "open" form." *Acta Crystallogr D Biol Crystallogr* **59**(Pt 5): 815–22.

Xu, T., A. Sampath, et al. (2005). "Structure of the Dengue virus helicase/nucleoside triphosphatase catalytic domain at a resolution of 2.4 A." *J Virol* **79**(16): 10278–88.

Xu, T., A. Sampath, et al. (2006). "Towards the design of flavivirus helicase/NTPase inhibitors: crystallographic and mutagenesis studies of the dengue virus NS3 helicase catalytic domain." *Novartis Found Symp* **277**: 87–97; discussion 97–101, 251–3.

Xue, Q., H. Ding, et al. (2007). "Inhibition of hepatitis C virus replication and expression by small interfering RNA targeting host cellular genes." *Arch Virol*.

Yao, N., T. Hesson, et al. (1997). "Structure of the hepatitis C virus RNA helicase domain." *Nat Struct Biol* **4**(6): 463–7.

Yao, N., P. Reichert, et al. (1999). "Molecular views of viral polyprotein processing revealed by the crystal structure of the hepatitis C virus bifunctional protease-helicase." *Structure* **7**(11): 1353–63.

Yao, N. and P. C. Weber (1998). "Helicase, a target for novel inhibitors of hepatitis C virus." *Antivir Ther* **3**(Suppl 3): 93–7.

Yon, C., T. Teramoto, et al. (2005). "Modulation of the nucleoside triphosphatase/RNA helicase and 5′-RNA triphosphatase activities of Dengue virus type 2 nonstructural protein 3 (NS3) by interaction with NS5, the RNA-dependent RNA polymerase." *J Biol Chem* **280**(29): 27412–9.

Yong, Y. and L. J. Romano (1995). "Nucleotide and DNA-induced conformational changes in the bacteriophage T7 gene 4 protein." *J Biol Chem* **270**(41): 24509–17.

Yoon-Robarts, M., A. G. Blouin, et al. (2004). "Residues within the B′ motif are critical for DNA binding by the superfamily 3 helicase Rep40 of adeno-associated virus type 2." *J Biol Chem* **279**(48): 50472–81.

Yu, X., M. M. Hingorani, et al. (1996). "DNA is bound within the central hole to one or two of the six subunits of the T7 DNA helicase." *Nat Struct Biol* **3**(9): 740–3.

Zhang, J., O. Yamada, et al. (2004). "Down-regulation of viral replication by adenoviral-mediated expression of siRNA against cellular cofactors for hepatitis C virus." *Virology* **320**(1): 135–43.

Ziegelin, G., T. Niedenzu, et al. (2003). "Hexameric RSF1010 helicase RepA: the structural and functional importance of single amino acid residues." *Nucleic Acids Res* **31**(20): 5917–29.

Ziegelin, G., E. Scherzinger, et al. (1993). "Phage P4 alpha protein is multifunctional with origin recognition, helicase and primase activities." *Embo J* **12**(9): 3703–8.

Chapter 21
Integrase: Structure, Function, and Mechanism

James Dolan and Jonathan Leis

Retroviral DNA integration is a multistep process that occurs in defined stages. The viral-encoded integrase (IN) is both necessary and sufficient to catalyze integration (Katz et al. 1990; Li and Craigie 2005; Sinha and Grandgenett 2005). After assembly of a stable complex of IN with specific DNA sequences at the ends of the viral long-terminal repeats (LTRs), terminal dinucleotides are removed from each 3′ end by endonucleolytic processing. The viral DNA 3′ ends are then covalently linked to the host target DNA in a concerted cleavage/ligation reaction. The processing and joining steps from several retroviruses have been analyzed, including avian sarcoma virus (ASV), murine leukemia virus (MLV), and HIV-1 (Kulkosky and Skalka 1994). Both steps require a stable complex composed of IN and at least 16 base pairs of both ends of a linear viral DNA. While INs specifically cleave their respective viral DNA ends, there is little specificity for the target DNA. This results in integration at many sites. By definition then, the binding of viral DNA must be different from the binding of target DNA by IN and must be accounted for in any structural model of the enzyme. In this chapter, we will define in more detail the structural and biochemical properties of the holoenzyme as well as interactions between the enzyme and viral and host DNA substrates. Next we will cover the biochemistry of the 3′ processing and joining followed by host cell proteins that influence the integration reaction. Lastly, we will highlight IN as a chemotherapeutic target and current strategies for development of novel IN inhibitors.

Structure and Oligomerization

All IN proteins share a conserved three-domain structure including a N-terminal, catalytic core, and C-terminal domain that are required for catalytic activity (Engelman et al. 1995). The N-terminal domain (residues 1–49) contains a HHCC zinc-binding site (Cai et al. 1997; van den Ent et al. 1999). A phenylalanine residue

---

J. Leis (✉)
Department of Microbiology and Immunology, Northwestern University
Feinberg School of Medicine, 303 E. Chicago Ave., Chicago, IL 60611, USA
e-mail: j-leis@northwestern.edu

C.E. Cameron et al. (eds.), *Viral Genome Replication*,
DOI 10.1007/b135974_21, © Springer Science+Business Media, LLC 2009

in the first position of a protein enters it into a ubiquitin–proteasome proteolytic pathway called N-end degradation. The IN of HIV-1, HIV-2, and simian immunodeficiency virus (SIV) all have a phenylalanine at position one, and HIV-1 IN has been shown to be subject to degradation in this pathway (Mulder and Muesing 2000). This mechanism has been proposed to maintain host chromosome stability and integrity through IN instability. The C-terminal domain (residues 213–288) is less conserved and forms a SH3 fold through three α-helices. This domain contains multimerization determinants as well as non-specific DNA-binding activity (Andrake and Skalka 1996). The catalytic core domain (residues 50–212) contains the residues involved in catalysis. The conserved DDE motif in the catalytic core domain chelates the required metal cofactor and forms part of the active site (Polard and Chandler 1995). When mutations are introduced at these residues, all enzymatic activities are significantly impaired (Kulkosky and Skalka 1994).

Retroviral IN can be found in monomer, dimer, and tetramer forms in solution. Dimeric IN can catalyze 3′ processing and joining of DNA duplex oligodeoxyribonucleotides, while the homotetramer is required for a concerted DNA integration

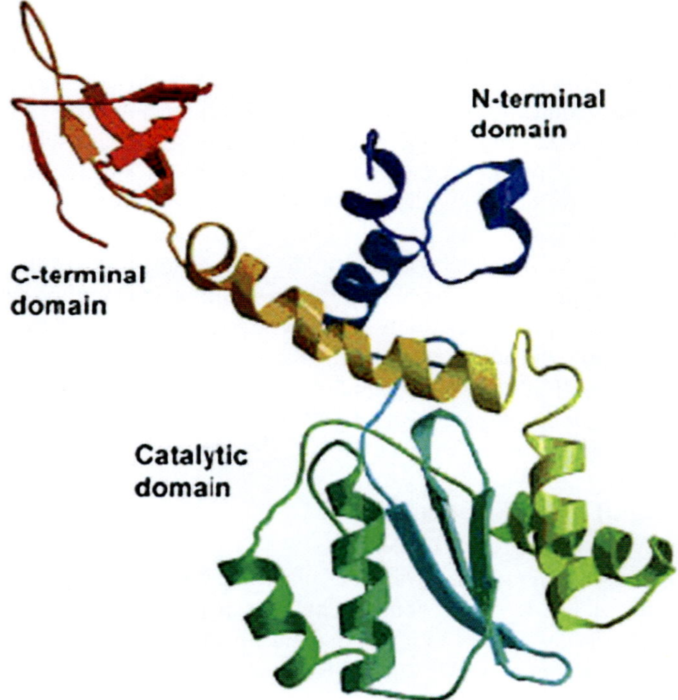

Fig. 21.1 HIV-1 IN full-length monomer. This is a representation of the full-length IN monomer. This structure was assembled based on crystallization data of two-domain structures and homology of the viral integrase to transposases. The formation of a full-length representation was based on super-imposing the catalytic domain from each two-domain structure.

reaction (Bao et al. 2003; Faure et al. 2005; Li et al. 2006). Crystal structures of the individual domains and two-domain fragments have been solved providing information about orientations and interactions of the various domains with each other and with host proteins (Jenkins et al. 1997; Wang et al. 2001; Chiu and Davies 2004; Maroun et al. 2005). Figure 21.1 shows an assembled three-domain structure that is modeled from available crystal data. However, in the absence of a structure for a holoenzyme in complex with DNA substrates, little is known about the positioning of the domains in the active oligomer (Petit et al. 1999; Craigie 2001) or intersubunit interaction, though two residues have been implicated in oligomerization (Jenkins et al. 1996; Kalpana et al. 1999).

Enzymatic Activities

Much of the information we have on the molecular mechanism of integration comes from the use of reconstituted systems employing duplex oligodeoxyribonucleotides (oligos) where the products of the reaction are separated by gel electrophoresis. ASV IN, for example, catalyzes the specific removal of the two bases from the 3′ end of the strands adjacent to a conserved CA dinucleotide using 15 base pair substrates corresponding to either viral DNA ends (called U3 or U5) (Katzman et al. 1989; Kukolj and Skalka 1995). Similar substrates were used to demonstrate the joining (or strand transfer) reaction, where one oligo integrated into another thereby increasing the size of the radiolabeled substrate (Craigie et al. 1990; Katz et al. 1990). For HIV-1 duplex oligo substrates of comparable size, nucleotide substitutions in the U5 and U3 ends were shown to affect one or both of the catalytic functions of HIV-1 IN. Substitutions in the HIV-1 U5 region, for example, inhibit 3′ processing (e.g., positions 1–6 and 9–11) or not (e.g., positions 12–14) (Esposito and Craigie 1998). Other changes at specific nucleic acid positions in the HIV-1 U3 and U5 ends affect each of the catalytic reactions, though changes in one end have a more pronounced effect than in another. The joining reaction can be measured separately from 3′ processing using a "preprocessed" duplex DNA substrate in which the two terminal bases adjacent to the CA dinucleotides are removed. Nevertheless, most do not display the concerted nature of the DNA integration reaction as it was first described (Murphy and Goff 1992). Deletions were placed 5′ to the conserved CA dinucleotides in the MLV U3 LTR region and when transfected into cells the processing of both LTR ends were affected adversely. This finding implies that the two ends of the viral DNA were brought together so that mutations in one affected the processing of the other, though the actual insertion of each DNA end into the target DNA may be sequential (Li et al. 2006). Several assay systems that display concerted DNA integration properties have been described and used to demonstrate changes in retroviral DNA integration are context dependent. For example, base pair substitutions placed into the HIV-1 U5 or ASV U3 ends (referred to as the "dominant" LTR ends) caused decreases in the rate of catalysis. In contrast, comparable substitutions at the same positions in the "non-dominant" ends are associated with changes in mechanism from concerted (two-ended viral DNA insertion) to non-concerted (one-ended viral DNA insertion) integration (Fitzgerald et al. 1992; Lutzke et al. 1994; Vora et al. 1994; Aiyar et al. 1996; Vora et al. 1997; Masuda et al. 1998; McCord et al. 1999; Brin and Leis 2002a,b; Li and Craigie 2005; Li et al. 2006). When the viral DNA is inserted into the target DNA, it occurs in a staggered manner; the size of the stagger is virus specific. The resulting gapped intermediates are repaired by host proteins introducing the 4- to 6-base pair duplications into the target that flank the integrated viral DNA (Brin et al. 2000). The determinants for the duplication size are not known. The sites of integration in the target DNA are for the most part random (Kulkosky and Skalka 1994).

Host Cell Interacting Proteins

The IN complex is comprised of the viral IN, viral and host DNA substrates, and possibly host proteins. Several proteins have been suggested to play a variety of roles in integration, from suppressors of auto-integration to nuclear import of the pre-integration complex and host DNA substrate selection. For HIV-1 these include high-mobility group (HMG) family member HMG-1α, barrier-to-autointegration factor (BAF), Integrase interactor 1 (INi1), lens epithelium-derived growth factor (LEDGF)/p75, hepatoma-derived growth factor related protein 2 (HRP2), and the histone acetyltransferase p300 (Chen and Engelman 1998; Kalpana et al. 1999; Harris and Engelman 2000; Suzuki and Craigie 2002; Cereseto et al. 2005; Ciuffi et al. 2005; Maroun et al. 2006). These proteins can be categorized into two groups based on whether they interact directly with IN or are associated with the IN complex by affinity for DNA or by other protein–protein interactions. It is unclear, however, whether these proteins are absolutely required or there is redundancy among host cell proteins in the integration process. The BAF protein is thought to bind to IN and prevent the autointegration of the viral genome. The INi1 protein binds directly to HIV-1 IN and is thought to function in both early and late stages of viral replication (Landau 2002; Maroun et al. 2006). High-mobility group family member HMG-1α stimulates integration reactions in vitro (Aiyar et al. 1996; Farnet and Bushman 1997; Hindmarsh et al. 1999), though a genetic knockout of the HMG gene family in DT40 cells does not appear to affect viral replication (Beitzel and Bushman 2003; Maroun et al. 2006). The acetylation of IN by p300 has been shown to occur at target lysine residues in the C-terminus of HIV-1 IN (Cereseto et al. 2005). These modifications are thought to influence affinity of IN for target DNA. It has been postulated that this modification, in concert with acetylation of the nucleosomes for uncoiling of chromatin, promotes the integration of *Lentivirus* DNA into actively transcribed regions of the host genome (Cereseto et al. 2005). LEDGF and HRP2 share domain structure homology and both contain the IN-binding domain (IBD) motif. The main difference between LEDGF and HRP2 is that the chromatin-tethering activity is lacking in the HRP2 protein, suggesting that LEDGF is biologically relevant within the host cell nucleus (Vanegas et al. 2005). A crystal structure of LEDGF/p75 bound to the catalytic core domain of HIV-1 IN was solved and the critical residues for this interaction were within the 165–173 region of IN (Cherepanov et al. 2005). LEDGF has been shown to stimulate IN activity in vitro, but this effect is specific to half-site products (Raghavendra and Engelman 2007). Another group has demonstrated that when LEDGF is added after 3′ processing (and presumably after stable IN complex formation), there is no stimulation (Yu et al. 2007). Studies have also implicated that regions of active transcription are preferential sites for integration, and proteins that interact with these regions can influence integration (Schroder et al. 2002). These results highlight the differences that are intrinsic in the various assay systems as well as demonstrate the need for in vivo experimental data. Further characterization of these host proteins has proven difficult because they are not required for 3′ processing or joining in vitro.

Integrase Holoenzyme Properties

Recognition of viral DNA by integrase is specific; however, there is cross-recognition within respective virus families. For example, HIV-1 IN will 3′-process LTR end sequences derived from simian immunodeficiency virus (SIV), another *Lentivirus* family member. Several techniques including NMR spectroscopy, use of cross-linking agents, and nucleotide analog-modified substrates have been used to probe DNA–IN interactions. The most informative have been cross-linking studies that identified contacts on the enzyme surface with both viral and target DNAs, including residues 67, 117, 143, 148, 152, 153, 159, 230, 263, and 264 (Esposito and Craigie 1998; Heuer and Brown 1998; Johnson et al. 2006). These studies used a variety of DNA substrates and cross-linking agents, and many of these residues have been confirmed in secondary studies to interact with DNA (either viral or target) through additional mutational and/or chemical investigation. Analysis of nucleotide analog-modified substrates to identify contacts between the viral DNA and the enzyme surface has also been informative (Agapkina et al. 2004). In this case, a map was assembled of putative contacts between the enzyme surface and either the phosphate-backbone or heterocyclic bases. The processing reaction was also determined to require local destabilization of the third A-T base pair (in the CA dinucleotide) for efficient activity (Agapkina et al. 2004). Specific IN residues that were identified in this study include K159 and E152, which interact with the N-7 position of the third adenosine on the processed strand. Residues nearer the active site, such as Q148 and Y143, interact with the terminal nucleotides of the processed strand.

The specificity of the viral DNA interaction is not shared with the target DNA. The IN tetramer binds the target DNA in a different trench than the viral DNA and both substrates are bound at the same time during the joining reaction. As the specific DNA sequences for integration are not conserved, it is possible that conserved residues among all IN proteins are involved in the non-specific DNA-binding activity. The use of cross-linking reagents during the incubation of IN with a non-viral DNA substrate oligo has implicated several peptides representing residues 49–69, 139–152, 213–246, 247–270, and 271–288 to be close enough to the target DNA to cross-link (Heuer and Brown 1998). A few other residues have been identified to interact with the target DNA and influence the joining reaction (Harper et al. 2001; Harper et al. 2003; Li and Craigie 2005; Al-Mawsawi et al. 2006 and 2008; Dolan et al., 2009) including residues 119, 130, 132, 181, 185, and 219.

Mutations introduced at these residues result in a loss of the joining reaction with little affect on 3′ processing. These residues align along a trench on the IN surface approximately perpendicular to the LTR. DNA building sites (see black arrow in Fig. 21.2.)

A homotetramer model for HIV-1 IN with DNA representing 20 base pairs of the U3 and U5 termini was assembled using structural and biochemical data and molecular dynamics simulations (Chen et al. 2006). Construction of this model used separate two-domain crystal structures of the N-terminal domain and catalytic core (PDB code 1K6Y) and the catalytic core with the C-terminal domain (PDB code 1EX4). Superimposition of the two crystals at the catalytic core led to the assembly

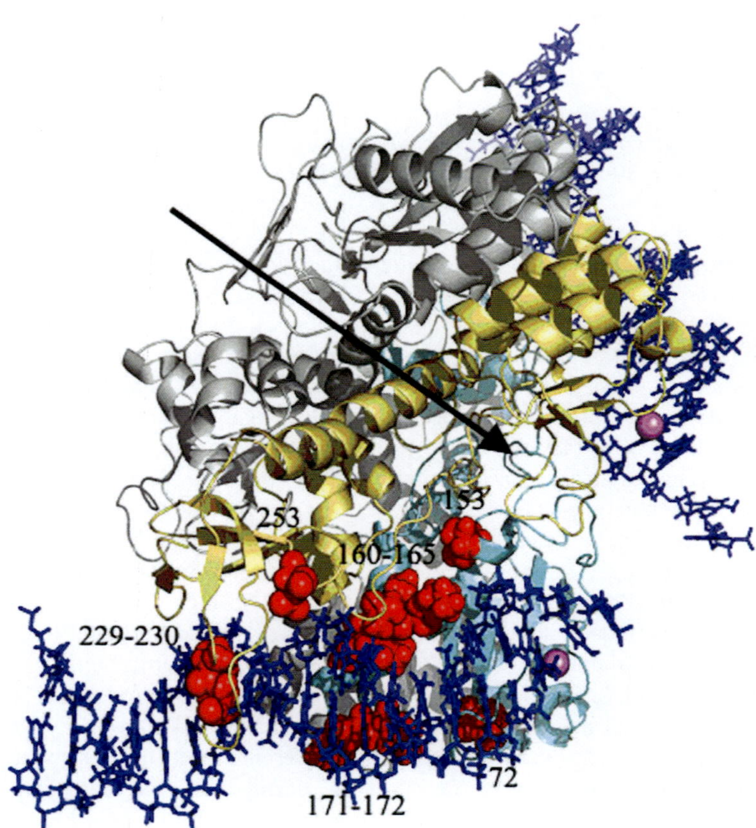

Fig. 21.2 HIV-1 IN homotetramer model with bound viral DNA ends. IN tetramer model with subunits colored *cyan* and *yellow* that interact with the two viral DNA ends and two other subunits colored in *gray* which do not. The DNAs are in a stick representation (*blue*) and Mg^{+2} in the two catalytic sites are indicated by spheres (*magenta*). The *black arrow* is oriented to highlight the proposed target DNA-binding site. The amino acids that specifically recognize the viral DNA ends are shown in space fill model (*red*) for only one of the two viral DNA ends. Amino acid substitutions at these positions affect the specificity of the enzyme for the DNA substrate.

of the three-domain structure (See Fig. 21.1). The tetramer was assembled using crystal lattice contacts. The viral DNA was placed in the tetramer using critical positions derived from Transposase structural data and use of a molecular dynamics simulation. The viral DNAs face one another with their 3′ ends positioned adjacent to the active site Asp residues. This positioning is in agreement with cross-linking of viral DNA to peptides 49–69, 139–152, 247–270, 271–288, and 153–167 (Heuer and Brown 1997; Drake et al. 1998; Esposito and Craigie 1998; Heuer and Brown 1998; Johnson et al. 2006) and mutagenesis studies (Lutzke et al. 1994; Sayasith et al. 2000; Agapkina et al. 2004; Chen et al. 2006; Puglia et al. 2006). To validate the model, Chen et al. identified amino acids that were spatially within 10 Å from the viral DNA ends. Using a structural alignment of HIV-1, SIV, and ASV INs, unique residues were identified. Construction and purification of 16 HIV-1 enzymes in which the amino acids from ASV were placed into the structurally related positions of HIV-1 IN was carried out, and the chimeric enzymes were tested for changes in specificity for 3′ processing of the viral DNA ends using duplex oligo substrates. This analysis demonstrated that there are multiple HIV-1 IN amino acid contacts with the viral DNA and that substitution of ASV IN amino acids at many positions conferred the partial ability to cleave ASV substrates with a concomitant loss in cleavage of the homologous HIV-1 substrate. Substitutions at HIV-1 IN residues that changed specificity include V72, S153, K160, I161, G163, Q164, V165, H171, L172 (Chen et al. 2006), Q44, L68, E69, D229, S230, and D253 (Dolan et al. 2009). Figure 21.2 highlights the viral DNA bound in one trench of the IN tetramer, and the labeled residues represent those that were shown to change the specificity of the enzyme for substrate DNA. Several other models of IN coupled to DNA have been reported (Gao et al. 2001; Podtelezhnikov et al. 2003; Karki et al. 2004; Wielens et al. 2005). These models do not explain all of the cross-linking, mutagenesis, and kinetics data which are accounted for in the Chen et al. model.

Chemotherapeutics: Design and Strategy

Inhibitors of integrase are currently a topic of investigation for antiretroviral therapy. Current anti-retroviral therapies use a combination of inhibitors targeted against the viral enzymes, protease (PR), and reverse transcriptase (RT). Molecules that inhibit viral fusion are also being used in drug regimens. Resistance mutations to the enzyme inhibitors for PR and RT occur frequently enough that escape mutants after selection are common. Therefore, molecules that inhibit integrase are being sought. Several classes of compounds show promise of efficacy against IN at physiologically relevant levels. Diketo acids and naphthyridine carboxamides are two examples of molecules currently under investigation and development as IN inhibitors.

The diketo acids are selective inhibitors of IN and act at the level of strand transfer (Hazuda et al. 2000). The diketo acid moiety showed positive inhibition against IN at nanomolar concentrations, suggesting that these molecules could be developed to function at physiologically relevant levels. Two molecules of this class were used for selection of resistant mutants. Three positions changed in a majority of the viruses that were sequenced. Mutations at residues 66, 153, and 154 were commonly selected. These residues are located near the active site, suggesting that the molecule binds near the catalytic center of the enzyme and presumably does not interfere with 3′ processing but affects the joining reaction in a currently undefined manner. Since the initial discovery of this class of compounds, numerous derivatives have been synthesized and examined at their inhibitory potential (Fikkert et al. 2003; Lee and Robinson 2004; Svarovskaia et al. 2004; Brigo et al. 2005).

Naphthyridine carboxamides are also small molecules that were selected to inhibit IN at the mechanism of strand transfer (Hazuda et al. 2004). These molecules are nanomolar inhibitors, and many derivatives have yielded good pharmacokinetics (Embrey et al. 2005; Guare et al. 2006). Similar to the diketo acid selection, one molecule of this set demonstrated good pharmacokinetics and was used for further study. Selection of resistant mutants in the presence of the inhibitor highlighted the putative binding site to be near residues 72, 121, and 125. Although naphthyridine carboxamides are derivatives of diketo acids, there is no cross resistance for the selected mutations. Naphthyridine carboxamides and diketo acids therefore have distinct binding sites for their inhibitory mechanism.

Several of the mutations that show up in drug-resistant enzymes affect 3′ processing and not just the joining reaction. This would result in partially defective INs that would place a pressure for the selection of second site mutations that compensate for lost 3′ processing activity by the initial drug-resistance mutation. For example, position 153 is important in the recognition of the viral DNA, and mutation at this position only mildly affects the activity of the enzyme. As such, this model would predict that this mutation would be found by itself when subjected to drug selection with little replication defect (Lee and Robinson 2004; Lu et al. 2005). In contrast, mutation at position 72 affects the catalytic activity of the 3′ processing as well as strand-transfer reactions in vitro (Chen et al. 2006). Not surprisingly, a mutation at this residue is found in combination with mutations at positions 121 and 125 (Hazuda et al. 2004). It has been demonstrated that the T125S substitution increases the 3′ processing of HIV-1 duplex (Chen et al. 2006) as well as the joining of a HIV-1 preprocessed substrate (Dolan et al. 2009) so that substitutions at this position could compensate for the low activity caused by the mutation at Val 72 (Chen et al. 2006). In identifying critical residues involved in the particular enzymatic reactions, investigation with small molecules has also proven useful in understanding the biology of the enzyme. Studies involving the binding of high-affinity drugs that target the enzymatic activities of IN have demonstrated that binding of viral DNA promotes a distinct active conformation (Espeseth et al. 2000; Grobler et al. 2002; Pommier et al. 2005). Continued research in this area will yield more insight into the properties of integrase in addition to providing new targets for inhibitors.

Conclusion

A great wealth of knowledge and information has been gathered about the biochemistry and structure of retroviral IN recently. Identification of residues that are responsible for interacting with a specific DNA sequence is one recent development, as well as the assembly of a predictive computer model of the IN holoenzyme. These tools advance the understanding of the enzyme system in addition to open new areas of research. Gene therapy using targeted DNA vectors may one day be feasible using an enzyme that has been engineered to interact with specific DNA sequences for insertion in a specific DNA target. It is clear that this area of research is poised for significant discovery in the near future.

References

Agapkina, J., M. Smolov, et al. (2004). "HIV-1 integrase can process a 3′-end crosslinked substrate." *Eur J Biochem* **271**(1): 205–11.

Aiyar, A., P. Hindmarsh, et al. (1996). "Concerted integration of linear retroviral DNA by the avian sarcoma virus integrase in vitro: dependence on both long terminal repeat termini." *J Virol* **70**(6): 3571–80.

Al-Mawsawi, L. Q., V. Fikkert, et al. (2006). "Discovery of a small-molecule HIV-1 integrase inhibitor-binding site." *Proc Natl Acad Sci USA* **103**(26): 10080–5.

Al-Mawsawi, L. Q., A. Hombrouck, et al. (2008). "Four-tiered pi interaction at the dimeric interface of HIV-1 integrase critical for DNA integration and viral infectivity." *Virology* **377**: 355–63.

Andrake, M. D. and A. M. Skalka (1996). "Retroviral integrase, putting the pieces together." *J Biol Chem* **271**(33): 19633–6.

Bao, K. K., H. Wang, et al. (2003). "Functional oligomeric state of avian sarcoma virus integrase." *J Biol Chem* **278**(2): 1323–7.

Beitzel, B. and F. Bushman (2003). "Construction and analysis of cells lacking the HMGA gene family." *Nucleic Acids Res* **31**(17): 5025–32.

Brigo, A., K. W. Lee, et al. (2005). "Comparative molecular dynamics simulations of HIV-1 integrase and the T66I/M154I mutant: binding modes and drug resistance to a diketo acid inhibitor." *Proteins* **59**(4): 723–41.

Brin, E. and J. Leis (2002a). "Changes in the mechanism of DNA integration in vitro induced by base substitutions in the HIV-1 U5 and U3 terminal sequences." *J Biol Chem* **277**(13): 10938–48.

Brin, E. and J. Leis (2002b). "HIV-1 integrase interaction with U3 and U5 terminal sequences in vitro defined using substrates with random sequences." *J Biol Chem* **277**(21): 18357–64.

Brin, E., J. Yi, et al. (2000). "Modeling the late steps in HIV-1 retroviral integrase-catalyzed DNA integration." *J Biol Chem* **275**(50): 39287–95.

Cai, M., R. Zheng, et al. (1997). "Solution structure of the N-terminal zinc binding domain of HIV-1 integrase." *Nat Struct Biol* **4**(7): 567–77.

Cereseto, A., L. Manganaro, et al. (2005). "Acetylation of HIV-1 integrase by p300 regulates viral integration." *Embo J* **24**(17): 3070–81.

Chen, A., I. T. Weber, et al. (2006). "Identification of amino acids in HIV-1 and avian sarcoma virus integrase subsites required for specific recognition of the long terminal repeat Ends." *J Biol Chem* **281**(7): 4173–82.

Chen, H. and A. Engelman (1998). "The barrier-to-autointegration protein is a host factor for HIV type 1 integration." *Proc Natl Acad Sci USA* **95**(26): 15270–4.

Cherepanov, P., Z. Y. Sun, et al. (2005). "Solution structure of the HIV-1 integrase-binding domain in LEDGF/p75." *Nat Struct Mol Biol* **12**(6): 526–32.

Chiu, T. K. and D. R. Davies (2004). "Structure and function of HIV-1 integrase." *Curr Top Med Chem* **4**(9): 965–77.

Ciuffi, A., M. Llano, et al. (2005). "A role for LEDGF/p75 in targeting HIV DNA integration." *Nat Med* **11**(12): 1287–9.

Craigie, R. (2001). "HIV integrase, a brief overview from chemistry to therapeutics." *J Biol Chem* **276**(26): 23213–6.

Craigie, R., T. Fujiwara, et al. (1990). "The IN protein of Moloney murine leukemia virus processes the viral DNA ends and accomplishes their integration in vitro." *Cell* **62**(4): 829–37.

Dolan, J., A. Chen, et al. (2009). "Defining the DNA substrate binding sites on HIV-1 integrase." *J. Mol. Biol.* **385**: 568–79.

Drake, R. R., N. Neamati, et al. (1998). "Identification of a nucleotide binding site in HIV-1 integrase." *Proc Natl Acad Sci USA* **95**(8): 4170–5.

Embrey, M. W., J. S. Wai, et al. (2005). "A series of 5-(5,6)-dihydrouracil substituted 8-hydroxy-[1,6]naphthyridine-7-carboxylic acid 4-fluorobenzylamide inhibitors of HIV-1 integrase and viral replication in cells." *Bioorg Med Chem Lett* **15**(20): 4550–4.

Engelman, A., G. Englund, et al. (1995). "Multiple effects of mutations in human immunodeficiency virus type 1 integrase on viral replication." *J Virol* **69**(5): 2729–36.

Espeseth, A. S., P. Felock, et al. (2000). "HIV-1 integrase inhibitors that compete with the target DNA substrate define a unique strand transfer conformation for integrase." *Proc Natl Acad Sci USA* **97**(21): 11244–9.

Esposito, D. and R. Craigie (1998). "Sequence specificity of viral end DNA binding by HIV-1 integrase reveals critical regions for protein-DNA interaction." *Embo J* **17**(19): 5832–43.

Farnet, C. M. and F. D. Bushman (1997). "HIV-1 cDNA integration: requirement of HMG I(Y) protein for function of preintegration complexes in vitro." *Cell* **88**(4): 483–92.

Faure, A., C. Calmels, et al. (2005). "HIV-1 integrase crosslinked oligomers are active in vitro." *Nucleic Acids Res* **33**(3): 977–86.

Fikkert, V., B. Van Maele, et al. (2003). "Development of resistance against diketo derivatives of human immunodeficiency virus type 1 by progressive accumulation of integrase mutations." *J Virol* **77**(21): 11459–70.

Fitzgerald, M. L., A. C. Vora, et al. (1992). "Concerted integration of viral DNA termini by purified avian myeloblastosis virus integrase." *J Virol* **66**(11): 6257–63.

Gao, K., S. L. Butler, et al. (2001). "Human immunodeficiency virus type 1 integrase: arrangement of protein domains in active cDNA complexes." *Embo J* **20**(13): 3565–76.

Grobler, J. A., K. Stillmock, et al. (2002). "Diketo acid inhibitor mechanism and HIV-1 integrase: implications for metal binding in the active site of phosphotransferase enzymes." *Proc Natl Acad Sci USA* **99**(10): 6661–6.

Guare, J. P., J. S. Wai, et al. (2006). "A series of 5-aminosubstituted 4-fluorobenzyl-8-hydroxy-[1,6]naphthyridine-7-carboxamide HIV-1 integrase inhibitors." *Bioorg Med Chem Lett* **16**(11): 2900–4.

Harper, A. L., L. M. Skinner, et al. (2001). "Use of patient-derived human immunodeficiency virus type 1 integrases to identify a protein residue that affects target site selection." *J Virol* **75**(16): 7756–62.

Harper, A. L., M. Sudol, et al. (2003). "An amino acid in the central catalytic domain of three retroviral integrases that affects target site selection in nonviral DNA." *J Virol* **77**(6): 3838–45.

Harris, D. and A. Engelman (2000). "Both the structure and DNA binding function of the barrier-to-autointegration factor contribute to reconstitution of HIV type 1 integration in vitro." *J Biol Chem* **275**(50): 39671–7.

Hazuda, D. J., N. J. Anthony, et al. (2004). "A naphthyridine carboxamide provides evidence for discordant resistance between mechanistically identical inhibitors of HIV-1 integrase." *Proc Natl Acad Sci USA* **101**(31): 11233–8.

Hazuda, D. J., P. Felock, et al. (2000). "Inhibitors of strand transfer that prevent integration and inhibit HIV-1 replication in cells." *Science* **287**(5453): 646–50.

Heuer, T. S. and P. O. Brown (1997). "Mapping features of HIV-1 integrase near selected sites on viral and target DNA molecules in an active enzyme-DNA complex by photo-cross-linking." *Biochemistry* **36**(35): 10655–65.

Heuer, T. S. and P. O. Brown (1998). "Photo-cross-linking studies suggest a model for the architecture of an active human immunodeficiency virus type 1 integrase-DNA complex." *Biochemistry* **37**(19): 6667–78.

Hindmarsh, P., T. Ridky, et al. (1999). "HMG protein family members stimulate human immunodeficiency virus type 1 and avian sarcoma virus concerted DNA integration in vitro." *J Virol* **73**(4): 2994–3003.

Jenkins, T. M., A. Engelman, et al. (1996). "A soluble active mutant of HIV-1 integrase: involvement of both the core and carboxyl-terminal domains in multimerization." *J Biol Chem* **271**(13): 7712–8.

Jenkins, T. M., D. Esposito, et al. (1997). "Critical contacts between HIV-1 integrase and viral DNA identified by structure-based analysis and photo-crosslinking." *Embo J* **16**(22): 6849–59.

Johnson, A. A., W. Santos, et al. (2006). "Integration requires a specific interaction of the donor DNA terminal 5′-cytosine with glutamine 148 of the HIV-1 integrase flexible loop." *J Biol Chem* **281**(1): 461–7.

Kalpana, G. V., A. Reicin, et al. (1999). "Isolation and characterization of an oligomerization-negative mutant of HIV-1 integrase." *Virology* **259**(2): 274–85.

Karki, R. G., Y. Tang, et al. (2004). "Model of full-length HIV-1 integrase complexed with viral DNA as template for anti-HIV drug design." *J Comput Aided Mol Des* **18**(12): 739–60.

Katz, R. A., G. Merkel, et al. (1990). "The avian retroviral IN protein is both necessary and sufficient for integrative recombination in vitro." *Cell* **63**(1): 87–95.

Katzman, M., R. A. Katz, et al. (1989). "The avian retroviral integration protein cleaves the terminal sequences of linear viral DNA at the in vivo sites of integration." *J Virol* **63**(12): 5319–27.

Kukolj, G. and A. M. Skalka (1995). "Enhanced and coordinated processing of synapsed viral DNA ends by retroviral integrases in vitro." *Genes Dev* **9**(20): 2556–67.

Kulkosky, J. and A. M. Skalka (1994). "Molecular mechanism of retroviral DNA integration." *Pharmacol Ther* **61**(1–2): 185–203.

Landau, N. R. (2002). "A cellular cofactor in HIV-1 assembly: INI1 is also an "outtie"." *Trends Pharmacol Sci* **23**(6): 252–3.

Lee, D. J. and W. E. Robinson, Jr. (2004). "Human immunodeficiency virus type 1 (HIV-1) integrase: resistance to diketo acid integrase inhibitors impairs HIV-1 replication and integration and confers cross-resistance to L-chicoric acid." *J Virol* **78**(11): 5835–47.

Li, M. and R. Craigie (2005). "Processing of viral DNA ends channels the HIV-1 integration reaction to concerted integration." *J Biol Chem* **280**(32): 29334–9.

Li, M., M. Mizuuchi, et al. (2006). "Retroviral DNA integration: reaction pathway and critical intermediates." *Embo J* **25**(6): 1295–304.

Lu, R., A. Limon, et al. (2005). "Genetic analyses of DNA-binding mutants in the catalytic core domain of human immunodeficiency virus type 1 integrase." *J Virol* **79**(4): 2493–505.

Lutzke, R. A., C. Vink, et al. (1994). "Characterization of the minimal DNA-binding domain of the HIV integrase protein." *Nucleic Acids Res* **22**(20): 4125–31.

Maroun, M., O. Delelis, et al. (2006). "Inhibition of early steps of HIV-1 replication by SNF5/Ini1." *J Biol Chem* **281**(32): 22736–43.

Maroun, R. G., L. Zargarian, et al. (2005). "A structural study of model peptides derived from HIV-1 integrase central domain." *Rapid Commun Mass Spectrom* **19**(18): 2539–48.

Masuda, T., M. J. Kuroda, et al. (1998). "Specific and independent recognition of U3 and U5 att sites by human immunodeficiency virus type 1 integrase in vivo." *J Virol* **72**(10): 8396–402.

McCord, M., R. Chiu, et al. (1999). "Retrovirus DNA termini bound by integrase communicate in trans for full-site integration in vitro." *Virology* **259**(2): 392–401.

Mulder, L. C. and M. A. Muesing (2000). "Degradation of HIV-1 integrase by the N-end rule pathway." *J Biol Chem* **275**(38): 29749–53.

Murphy, J. E. and S. P. Goff (1992). "A mutation at one end of Moloney murine leukemia virus DNA blocks cleavage of both ends by the viral integrase in vivo." *J Virol* **66**(8): 5092–5.

Petit, C., O. Schwartz, et al. (1999). "Oligomerization within virions and subcellular localization of human immunodeficiency virus type 1 integrase." *J Virol* **73**(6): 5079–88.

Podtelezhnikov, A. A., K. Gao, et al. (2003). "Modeling HIV-1 integrase complexes based on their hydrodynamic properties." *Biopolymers* **68**(1): 110–20.

Polard, P. and M. Chandler (1995). "Bacterial transposases and retroviral integrases." *Mol Microbiol* **15**(1): 13–23.

Pommier, Y., A. A. Johnson, et al. (2005). "Integrase inhibitors to treat HIV/AIDS." *Nat Rev Drug Discov* **4**(3): 236–48.

Puglia, J., T. Wang, et al. (2006). "Revealing domain structure through linker-scanning analysis of the murine leukemia virus (MuLV) RNase H and MuLV and human immunodeficiency virus type 1 integrase proteins." *J Virol* **80**(19): 9497–510.

Raghavendra, N. K. and A. Engelman (2007). "LEDGF/p75 interferes with the formation of synaptic nucleoprotein complexes that catalyze full-site HIV-1 DNA integration in vitro: implications for the mechanism of viral cDNA integration." *Virology* **360**(1): 1–5.

Sayasith, K., G. Sauve, et al. (2000). "Characterization of mutant HIV-1 integrase carrying amino acid changes in the catalytic domain." *Mol Cells* **10**(5): 525–32.

Schroder, A. R., P. Shinn, et al. (2002). "HIV-1 integration in the human genome favors active genes and local hotspots." *Cell* **110**(4): 521–9.

Sinha, S. and D. P. Grandgenett (2005). "Recombinant human immunodeficiency virus type 1 integrase exhibits a capacity for full-site integration in vitro that is comparable to that of purified preintegration complexes from virus-infected cells." *J Virol* **79**(13): 8208–16.

Suzuki, Y. and R. Craigie (2002). "Regulatory mechanisms by which barrier-to-autointegration factor blocks autointegration and stimulates intermolecular integration of Moloney murine leukemia virus preintegration complexes." *J Virol* **76**(23): 12376–80.

Svarovskaia, E. S., R. Barr, et al. (2004). "Azido-containing diketo acid derivatives inhibit human immunodeficiency virus type 1 integrase in vivo and influence the frequency of deletions at two-long-terminal-repeat-circle junctions." *J Virol* **78**(7): 3210–22.

van den Ent, F. M., A. Vos, et al. (1999). "Dissecting the role of the N-terminal domain of human immunodeficiency virus integrase by trans-complementation analysis." *J Virol* **73**(4): 3176–83.

Vanegas, M., M. Llano, et al. (2005). "Identification of the LEDGF/p75 HIV-1 integrase-interaction domain and NLS reveals NLS-independent chromatin tethering." *J Cell Sci* **118**(Pt 8): 1733–43.

Vora, A. C., R. Chiu, et al. (1997). "Avian retrovirus U3 and U5 DNA inverted repeats. Role Of nonsymmetrical nucleotides in promoting full-site integration by purified virion and bacterial recombinant integrases." *J Biol Chem* **272**(38): 23938–45.

Vora, A. C., M. McCord, et al. (1994). "Efficient concerted integration of retrovirus-like DNA in vitro by avian myeloblastosis virus integrase." *Nucleic Acids Res* **22**(21): 4454–61.

Wang, J. Y., H. Ling, et al. (2001). "Structure of a two-domain fragment of HIV-1 integrase: implications for domain organization in the intact protein." *Embo J* **20**(24): 7333–43.

Wielens, J., I. T. Crosby, et al. (2005). "A three-dimensional model of the human immunodeficiency virus type 1 integration complex." *J Comput Aided Mol Des* **19**(5): 301–17.

Yu, F., G. S. Jones, et al. (2007). "HIV-1 Integrase Preassembled on Donor DNA Is Refractory to Activity Stimulation by LEDGF/p75." *Biochemistry* **16**(10): 2899–2908.

Part III
Antivirals: Targets, Mechanisms and Resistance

Chapter 22
Viral DNA Polymerase Inhibitors

Graciela Andrei, Erik De Clercq, and Robert Snoeck

Introduction

DNA viruses, as well as their host cells, require a DNA-dependent DNA polymerase to faithfully replicate their genomes. Viruses with small DNA genomes, such as papillomaviruses and polyomaviruses, have a limited coding capacity and utilize mainly the host replication machinery for their genome amplification. In contrast, large DNA viruses encode a specific polymerase equipped with a proofreading 3′–5′-exonuclease activity and other replication proteins that assure the replication of their genomic information. As a critical component of the viral replication machinery, viral DNA polymerases are the specific target of a number of antiviral drugs currently used to inhibit viral replication. Most antiviral drugs approved by the US Food and Drug Administration (FDA) inhibit viral genome replication, nearly all of these inhibit a DNA polymerase and most of these drugs are nucleoside analogs. FDA-approved inhibitors of viral polymerases target certain human herpesviruses, the retrovirus HIV (human immunodeficiency virus) and the hepadnavirus HBV (hepatitis B virus). This chapter focuses on the description of viral DNA polymerase inhibitors, whether currently approved or candidate drugs, that are particularly active against herpesviruses.

Four types of polymerase inhibitors are recognized: substrate analogs (nucleoside and nucleotide analogs), product analogs (pyrophosphate analogs), allosteric inhibitors (non-nucleoside inhibitors), and inhibitors that intercalate or directly interact with nucleic acids.

The development of antiviral drugs against large DNA viruses (i.e., herpesviruses and poxviruses) was initially focused on viral DNA polymerase inhibitors and, as a consequence, most of the antiviral drugs currently on the market target viral DNA polymerases (Table 22.1). However, advances in the molecular biology of DNA virus replication allowed the identification of novel targets for drug development

G. Andrei (✉)
Laboratory of Virology, Rega Institute for Medical Research, Katholieke Universiteit Leuven, B-3000 Leuven, Belgium,
e-mail: graciela.andrei@rega.kuleuven.ac.be

C.E. Cameron et al. (eds.), *Viral Genome Replication*,
DOI 10.1007/b135974_22,

Table 22.1 Viral DNA polymerase inhibitors that have been marketed

Type of inhibitor	Spectrum of activity	Principal indications	Route of administration
A. Nucleoside analogs			
- Acyclovir (ACV) Zovirax®	- Herpesviruses (HSV-1, HSV-2, VZV, HCMV, EBV, HHV-6)	- Mucosal, cutaneous, and systemic HSV-1 and HSV-2 infections (including herpetic keratitis, herpetic encephalitis, genital herpes, neonatal herpes, and herpes labialis) - VZV infections (including varicella and herpes zoster)	- Orally (genital herpes, herpes zoster) - Topically (herpetic keratitis, herpes labialis) - Intravenously (herpetic encephalitis, and other severe HSV and VZV infections)
- Valaciclovir (VALC) Zelitrex®, Valtrex®	- Idem as acyclovir	- HSV and VZV infections that can be approached by oral therapy (i.e., genital herpes, herpes zoster) - Also used in the prophylaxis of HCMV in transplant patients	- Orally
- Penciclovir (PCV) Denavir®, Vectavir®	- Herpesviruses (HSV-1, HSV-2, and VZV)	- Mucocutaneous HSV-1 and HSV-2 infections, particularly recurrent herpes labialis	- Topically
- Famciclovir (FCV) Famvir®	- Idem as penciclovir	- Mucocutaneous HSV-1 and HSV-2 infections and VZV infections	- Orally
- Bromovinyldeoxiuridine (BVDU) Zostex®, Brivirac®, Zerpex®	- Herpesviruses (HSV-1 and VZV)	- HSV-1 and VZV infections, particularly herpes zoster, but also herpes labialis and herpetic keratitis	- Orally (herpes zoster) - Topically (herpes labialis, herpetic keratitis)
- Ganciclovir (GCV) Cymevene®, Cytovene®	- Herpesviruses (HSV-1, HSV-2, and HCMV)	- HCMV infections in immunocompromised patients (prevention and treatment)	- Intravenously - Intraocular (intravitreal) implant (Vitrasert®) for HCMV retinitis

Table 22.1 (continued)

Type of inhibitor	Spectrum of activity	Principal indications	Route of administration
- Valganciclovir (VGCV) Valcyte®	- Idem as ganciclovir	- HCMV infections (prevention and treatment in immunocompromised patients	- Orally
B. Nucleotide analogs			
- Cidofovir (CDV, HPMPC) Vistide®	- Herpesviruses (HSV, VZV, HCMV, EBV, etc.), papilloma-, polyoma-, adeno-, and poxviruses	- Officially HCMV retinitis in AIDS patients - Off-labeled used in the treatment of acyclovir and/or foscarnet-resistant HSV infections, recurrent genital herpes, genital warts, cutaneous papillomatous lesions, cervical intraepithelial neoplasia, recurrent respiratory papillomatosis, molluscum contagiosum lesions, adenovirus infections, and polyomavirus-associated diseases [i.e., progressive multifocal leukoencephalopathy (PML), hemorrhagic cystitis) - Potential use in the case of a bioterroristic attack with smallpox	- Intravenously - Topically
C. Pyrophosphate analogs			
- Foscarnet (PFA) Foscavir®	- Herpesviruses (HSV, VZV, HCMV, etc. and also HIV)	- HCMV retinitis in AIDS patients - Mucocutaneous acyclovir-resistant HSV and VZV infections and GCV-resistant CMV infections in immunocompromised patients	- Intravenously

(Kleymann 2005; Biron 2006). Progress in different areas of research, such as gene expression, protein purification, proteomics, bioinformatics, and efficient high-throughput screening, has facilitated the characterization and functional assays of these new targets (DeFilippis et al. 2003; Coen & Schaffer 2003).

Currently Approved DNA Pol Inhibitors

Nucleoside Analogs

Acyclic Nucleoside Analogues

Nucleoside analogs represent the most productive source of antiviral agents. These agents need to be phosphorylated to their active form, the triphosphate form, to be able to target the viral DNA polymerase. The active forms inhibit polymerases by competing with natural dNTP substrates and/or incorporation into the growing DNA chain, where they can often terminate DNA elongation. Nucleoside analogs, acting as competitive inhibitors or substrates for viral polymerases, afford a reduction in viral DNA synthesis in infected cells. The selectivity of a nucleoside analog as inhibitor of viral replication depends on two parameters: (i) the efficiency by which viral enzymes phosphorylate the drug compared to cellular enzymes and (ii) the potency and efficiency by which viral genome replication is inhibited in comparison with cellular functions.

The first nucleoside analogs were synthesized as part of anti-metabolite cancer research programs in the late 1970s. Retrospectively, it can be stated that antiviral chemotherapy came of age with the discovery of acyclovir (ACV) in 1977 (Elion et al. 1977). This compound is a potent and selective inhibitor of herpes simplex virus (HSV) and varicella-zoster virus (VZV) replication and has demonstrated an excellent safety profile in clinical practice. Notably, in 1988 Dr. G. Elion received the Nobel Prize award for her work on the mechanism of action of nucleoside analogs including acyclovir.

Prior to ACV, nucleoside analogs such as vidarabine (adenine arabinoside, Ara-A), idoxuridine (IDU), and trifluridine (TFT) were associated with variable antiviral activity in humans and significant toxicity when systemically administered. The use of IDU and TFT was limited to topical therapy of herpetic keratitis (Kaufman et al. 1962; Kaufman 1963; Kaufman & Heidelberger 1964; Kaufman 1962). Neither IDU nor TFT can be used systemically because they are too toxic, especially to bone marrow. Vidarabine can be administered systemically, and it was the first antiviral drug used to treat herpetic encephalitis (Whitley et al. 1977). The therapeutic window for vidarabine is very narrow since this drug is phosphorylated by cellular adenosine kinase, resulting in significant side effects (mostly megaloblastic anemia) in patients. Shortly after the introduction of ACV as a highly specific anti-HSV agent, vidarabine was replaced by ACV in the treatment of herpetic encephalitis (Whitley et al. 1986; Whitley et al. 1981; Skoldenberg et al. 1984).

ACV represents the first generation of effective antiherpesvirus drugs. Despite its safety profile and potency against HSV and VZV, this compound had two major limitations: modest activity against other herpesviruses and poor oral bioavailability (only 15–30%). Three different approaches have been followed to obtain better oral bioavailability, intracellular pharmacokinetics, and antiviral activity: (i) synthesis of oral prodrugs of ACV, (ii) improvement nucleoside structure, and (iii) development of entirely novel structures. Soon after the discovery of ACV, other purine analogs, including penciclovir (PCV) and ganciclovir (GCV) and pyrimidine analogs, such as BVDU (brivudin) were described.

Acyclovir and Valacyclovir

ACV [9-(2-hydroxyethoxymethyl)guanine, Zovirax®] (Fig. 22.1), a structural analog of the natural 2′-deoxyguanosine, can be considered the first truly specific antiviral agent (Elion et al. 1977; Schaeffer et al. 1978) with potent activity against herpes simplex type 1 (HSV-1) and type 2 (HSV-2), VZV, and Epstein–Barr virus (EBV) and modest activity against human cytomegalovirus (HCMV). It consists of a guanine base attached to an acyclic sugar-like molecule. Shortly after its discovery, ACV became the drug of choice for the treatment of HSV and VZV infections, particularly primary and recurrent genital herpes, and mucocutaneous HSV lesions and VZV infections in immunosuppressed patients (Table 22.1). Nowadays, ACV and its prodrug, the L-valyl ester valacyclovir, have become the gold standard for prophylaxis and treatment of diseases caused by HSV and VZV, and both compounds have shown benefit in the management of HCMV diseases in transplant recipients.

The mechanism of action of ACV is shown in Fig. 22.2. The critical determinant of the selective activity of ACV against HSV and VZV is its preferential phosphorylation by the virus-encoded thymidine kinase (TK) (Fyfe et al. 1978; Keller et al. 1981). This enzyme converts ACV to ACV monophosphate (ACV-MP) which is then phosphorylated by the cellular GMP kinase to ACV diphosphate (ACV-DP) and further to ACV triphosphate (ACV-TP) by nucleoside diphosphate kinase or other cellular enzymes (Miller & Miller 1980; Miller & Miller 1982). ACV-TP, the active form of ACV, is more inhibitory to HSV DNA polymerase than to cellular DNA polymerases (Furman et al. 1979; St Clair et al. 1980) and this inhibition is competitive with respect to the natural substrate dGTP. High concentrations of dGTP can reverse the antiviral activity of ACV. In addition, ACV-TP can also serve as a substrate of the DNA polymerase reaction and hence be incorporated into DNA at its 3′-terminus. As 3′-terminal ACV-MP residues cannot be excised by the DNA polymerase-associated 3′–5′exonuclease (Derse et al. 1981), they prevent further chain elongation and thus act as DNA chain terminators because ACV does not contain the 3′-hydroxyl group required for DNA elongation. This explains the occurrence of short DNA fragments in HSV-infected cells exposed to ACV (McGuirt et al. 1984). It has also been demonstrated that following incorporation of ACV-TP, the

A. Nucleoside (substrate) analogues		
Natural nucleoside	**Nucleoside analog**	**Prodrug**
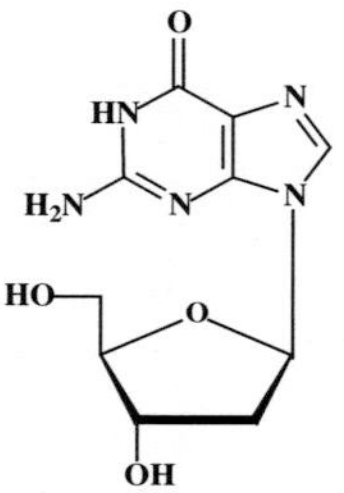 **2’-Deoxyguanosine**	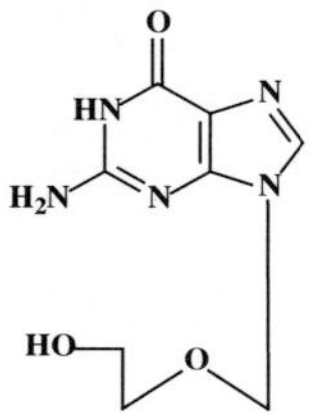 **Acyclovir (ACV)** 9-(2-hydroxyethoxymethyl)guanine **Zovirax®**	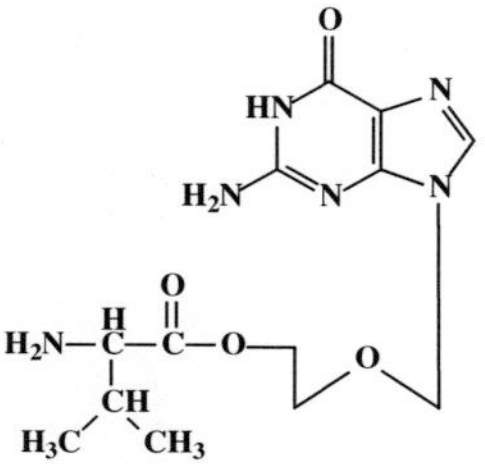 **Valacyclovir (VACV)** L-valine ester of acyclovir **Valtrex®, Zelitrex®**
	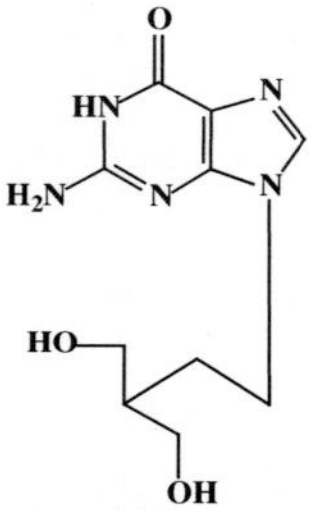 **Penciclovir (PCV)** 9-(4-hydroxy-3-hydroxymethyl-but- 1-yl)guanine **Denavir®, Vectavir®**	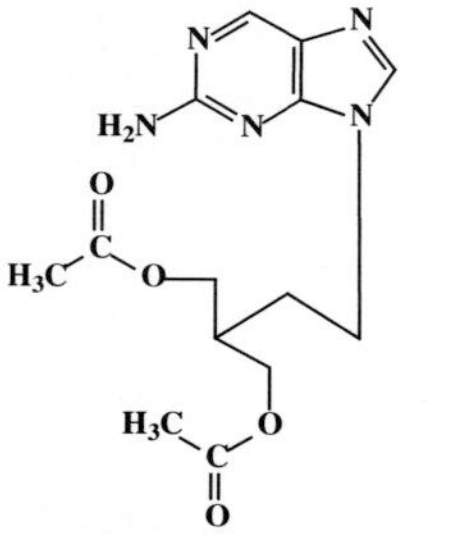 **Famciclovir (FCV)** Diacetyl ester of 9-(4-hydroxy-3-hydroxymethyl-but- 1-yl)-6-deoxyguanine **Famvir®**
	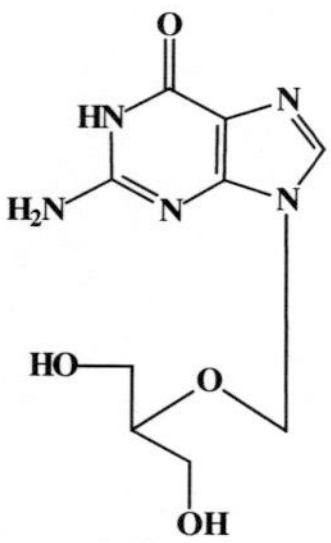 **Ganciclovir (GCV)** 9-(1,3-dihydroxy-2-propoxymethyl)guanine (DHPG) **Cymevene®, Cytovene®**	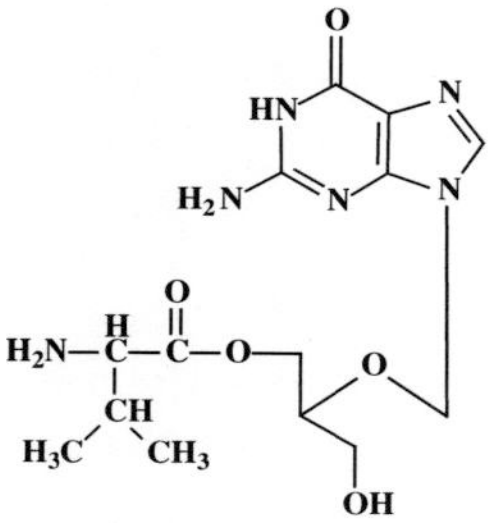 **Valganciclovir (VGCV)** L-valyl ester of ganciclovir **Valcyte®**

Fig. 22.1 Structural formulae of viral DNA polymerase inhibitors that have been marketed.

2'deoxythymidine	**Brivudin** (*E*)-5-(2-bromovynil)-2'-deoxyuridine (BVDU), bromovynildeoxyuridine **Zostex®, Zonavir®, Zerpex®**	

B. Nucleotide (substrate) analogues		
Natural nucleotide	**Nucleotide analog**	
2'-deoxycytosine 5'-monophosphate	**Cidofovir (CDV)** (S)-1-(3-hydroxy-2-phosphonylmethoxypropyl)cytosine (HPMPC) **Vistide®**	

C. Pyrophosphate (product) analogue

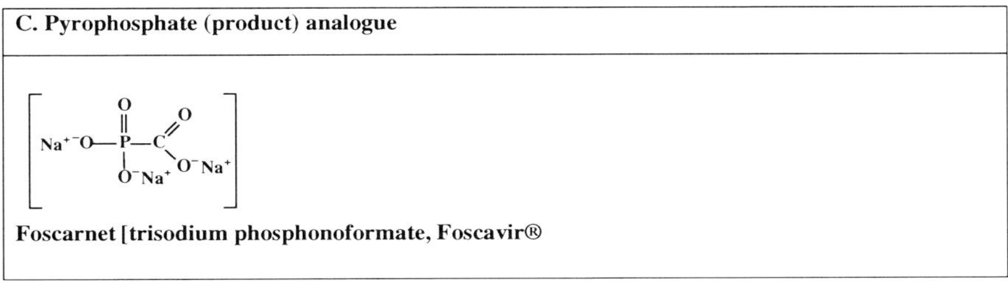

Foscarnet [trisodium phosphonoformate, Foscavir®

Fig. 22.1 (continued)

viral polymerase becomes trapped on the ACV-terminated DNA chain when the next deoxynucleoside triphosphate binds (Martin et al. 1994; Ilsley et al. 1995; Reardon 1989). ACV has been described as a "suicide" inhibitor, because it inactivates the HSV DNA polymerase as well as being a substrate for it (Furman et al. 1984).

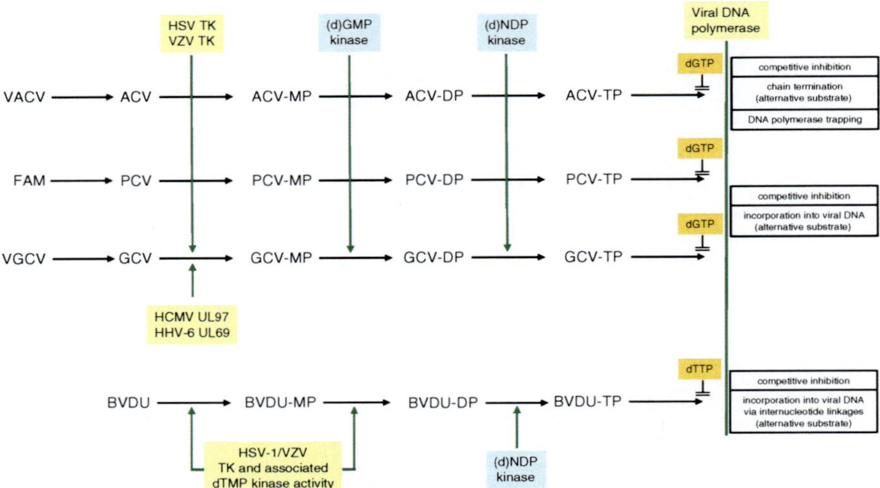

Fig. 22.2 Intracellular metabolism and mechanism of action of nucleoside analogs. Acyclic nucleoside analogs such as ACV, PCV, and GCV need to be selectively phosphorylated intracellularly in three steps, to the triphosphate (TP) active forms. The first phosphorylation step to the monophosphate forms (MP) is carried out by the HSV- or VZV-encoded thymidine kinase (TK), the HCMV UL97 open reading frame that encodes for a protein kinase or its HHV-6 homolog UL69. Therefore, the first phosphorylation is limited to virus-infected cells. Further phosphorylation to the diphosphate (DP) and triphosphate (TP) forms is carried out by cellular enzymes [i.e., dGMP kinase and nucleoside 5′-diphosphate (NDP) kinase] The triphosphate forms inhibit viral DNA polymerases acting as competitive inhibitors of dGTP binding and/or as alternative substrates if incorporated into the growing DNA chain. In the case of ACV, the incorporation of ACV-TP into the viral DNA leads to termination of chain elongation and trapping of the viral polymerase on the terminated DNA chain when the next deoxynucleoside triphosphate binds. For the pyrimidine nucleoside analog BVDU, following uptake by the (virus-infected) cells the compound is phosphorylated by the HSV-1- or VZV-encoded TK to the 5′-monophosphate (BVDU-MP) and 5′-diphosphate (BVDU-DP), and further to the 5′-triphosphate (BVDU-TP) by cellular kinases, i.e., NDP kinase. BVDU-TP can act as competitive inhibitor/alternative substrate of the viral DNA polymerase and as a substrate it can be incorporated internally (via nucleotide linkages) into the (growing) DNA chain

As ACV has limited oral bioavailability and limited solubility in water (~0.2%, 25°C) relatively large doses and frequent administration were thus required to maintain plasma levels high enough to achieve viral inhibition. Much research effort was directed to improve the solubility and oral bioavailability of ACV, and several water-soluble esters of ACV were investigated. The valine ester of ACV, valacyclovir (VACV, Valtrex®, Zelitrex®) (Fig. 22.1), proved to be a safe and efficacious drug (Perry & Faulds 1996; Darby 1994). The absolute oral bioavailability of ACV following oral administration of VACV is 54% (Weller et al. 1993). The increased oral bioavailability of VACV is due to a carrier-mediated intestinal absorption, via the human intestinal peptide transporter hPEPT1 (Guo et al. 1999), followed by the rapid conversion to ACV by ester hydrolysis in the small intestine (Perry & Faulds 1996; De Clercq & Field 2006). Following oral administration, VACV is rapidly metabolized to yield ACV and the essential amino acid L-valine (Perry & Faulds 1996).

Several clinical studies have demonstrated that VACV has a safety profile comparable to ACV in patients with genital herpes, herpes labialis, and herpes zoster (Warren et al. 2004; Corey et al. 2004; DeJesus et al. 2003; Corey et al. 2004; Beutner 1995; Beutner et al. 1995; Gupta et al. 2004). VACV appears a more attractive option in the treatment of HSV and VZV infections due to a less frequent dosing regimen, which may contribute to increased patient adherence to therapy. Although VACV is not potent enough for the treatment of established HCMV disease, it has been approved in several countries for prophylaxis of HCMV infection in transplant recipients (Biron 2006). The safety and efficacy of VACV for prevention of HCMV have been documented in several studies (Hodson et al. 2005; Lowance et al. 1999).

Penciclovir and Famciclovir

Penciclovir [PCV, 9-(4-hydroxy-3-hydroxymethyl-but-1-yl)guanine Denavir®, Vectavir®] (Fig. 22.1) is also a 2′-deoxyguanosine analog resembling ACV in chemical structure, mechanism of action, and spectrum of antiviral activity (Perry & Wagstaff 1995). Like ACV, PCV depends on the HSV- and VZV-encoded TK for activation to the monophosphate form (Fig. 22.2). Cellular enzymes are responsible for further phosphorylation to the triphosphate form. PCV-TP inhibits the viral DNA polymerase through competition with 2′-deoxyguanosine triphosphate and is incorporated in the viral DNA. Unlike ACV-TP, PCV-TP is not an obligate chain terminator as PCV has two hydroxyl groups on the acyclic chain and can be incorporated into the growing DNA chain. Intracellular concentrations of PCV-TP are at least 30-fold higher than those of ACV-TP (Vere Hodge et al. 1989). However, HSV and VZV DNA polymerases have higher affinity for ACV-TP than for PCV-TP. As a result, the relative activities ACV and PCV against HSV and VZV in cell culture are similar. PCV-TP is more stable within infected cells than ACV-TP and, therefore, it has longer antiviral action (Boyd et al. 1987).

Like ACV, PCV is very poorly absorbed when given orally. Famciclovir (FCV) (Fig. 22.1), the diacetylester of 6-deoxypenciclovir, was developed as an oral prodrug. After oral administration, FCV is rapidly and extensively absorbed and efficiently converted to PCV in two steps: (i) removal of the two acetyl groups (the first acetyl side chain of FCV is cleaved by esterases found in the intestinal wall, and the second is removed on first pass through the liver), and (ii) oxidation at the six position catalyzed by aldehyde oxidase that account for the conversion of 6-deoxypenciclovir to PCV (Rashidi et al. 1997; Perry & Wagstaff 1995). The total oral bioavailability of PCV from FCV is 77%. Data collected from several clinical studies demonstrated that FCV is well tolerated in patients and is effective against HSV-1 and HSV-2 for both therapy and long-term suppression of recurrences and is also efficacious in the treatment of herpes zoster (Table 22.1) (Simpson & Lyseng-Williamson 2006; Mertz et al. 1997; Sacks et al. 2005; Wald et al. 2006).

Ganciclovir and Valganciclovir

ACV has poor activity against HCMV due to the much lower accumulation of phosphorylated ACV in HCMV-infected cells than in HSV- or VZV-infected cells (Furman et al. 1981). The lack of activity of ACV against HCMV prompted the development of other acyclic guanosine analogs, including ganciclovir [GCV, 9-(1,3-dihydroxy-2-propoxymethyl)guanine, DHPG, Cymevene®, Cytovene®] (Fig. 22.1), which is more potent against HCMV than ACV. GCV was the first antiviral agent approved for the treatment of infections caused by HCMV. It has become the treatment of choice for the management of HCMV diseases and remains the first-line treatment of HCMV disease in transplant recipients (Razonable & Emery 2004; Biron 2006).

GCV is an acyclic guanosine analog of 2′-deoxyguanosine and unlike ACV, but similar to PCV, has the equivalent of a 3′-hydroxyl group in the acyclic chain (Fig. 22.1). Like ACV and PCV, GCV is converted to GCV-TP (the active form against HCMV) in a multi-step process involving viral and cellular enzymes (Fig. 22.2). Because HCMV does not encode for a TK, in HCMV-infected cells the initial phosphorylation of GCV is catalyzed by a protein kinase encoded by the HCMV UL97 open reading frame (Sullivan et al. 1992; Littler et al. 1992). Cellular kinases further convert GCV-MP to GCV-TP, which is both a competitive inhibitor and a substrate for the viral DNA polymerase. GCV-TP is a better inhibitor for the HCMV DNA polymerase than for cellular DNA polymerases (Martin et al. 1994; Biron et al. 1985), and it is a better substrate for HCMV DNA polymerase than for cellular DNA polymerases (Reid et al. 1988). GCV-TP is not an obligate chain terminator; however, after incorporating GCV-MP, HCMV DNA polymerase stalls after incorporating one additional nucleotide (Reid et al. 1988). The preferential phosphorylation of GCV in HCMV-infected cells and the higher inhibition of viral DNA polymerization than cellular DNA synthesis account for the selectivity of GCV against HCMV. However, it appears that the selectivity of GCV at each of these steps is lower than the selectivity of ACV against HSV or VZV. This correlates with a higher toxicity of GCV in the clinic. The side effects of GCV are mostly hematologic abnormalities (neutropenia, anemia and thrombocytopenia) and probable long-term reproductive toxicity. The HCMV-encoded protein kinase UL97 also catalyses the initial phosphorylation step of ACV (Talarico et al. 1999); however, ACV is a less efficient substrate than GCV, which explains in part the lower activity of ACV compared to GCV against HCMV. Another difference between these two compounds is the four- to five-fold shorter half-life of ACV-TP compared to GCV-TP in infected cells, resulting in lower intracellular levels of the active ACV-TP.

GCV can be given intravenously (IV), orally (Cytovene®), or as an ocular implant (Vitraset®, Chiron) for the treatment of HCMV retinitis (Table 22.1). The low bioavailability of GCV (~6%) prompted the development of the prodrug, valganciclovir (VGCV, Valcyte®), the L-valyl ester of GCV (Fig. 22.1). Following oral administration, it is metabolized to the active form (GCV) in the intestinal wall and liver. VGCV exhibits oral bioavailability of about 60%. VGCV has now replaced oral GCV, and there is still a debate whether prophylaxis therapy or pre-emptive therapy should be used for HCMV infections in asymptomatic high-risk transplant recipients (Biron 2006).

Human herpesvirus 6 (HHV-6) encodes UL69 phosphotransferase, a homolog of HCMV UL97, which has been shown to be able to phosphorylate GCV (Michel & Mertens 2004; Ansari & Emery 1999). This compound shows reasonable activity against HHV-6 in vitro, although 20-fold lower levels of GCV-MP are attained by pUL69 phosphotransferase than by its HCMV homolog pUL97 (De Bolle et al. 2002).

Pyrimidine Nucleoside Analogs

Brivudin

In the late 1970 s, brivudin [BVDU, (*E*)-5-(2-bromovinyl)-2′-deoxyuridine, bromovinyldeoxyuridine Zostex®, Brivirac®, Zerpex®] (Fig. 22.1) was described as a highly selective antiviral agent (De Clercq et al. 1979; De Clercq 2004), which proved specifically active against HSV-1 and VZV (De Clercq 2004). Several congeners of BVDU have been synthesized, including BVaraU (sorivudine), the arabinofuranosyl counterpart of BVDU. The 5-2-(*E*)-bromovinyl group, with the bromine in the *trans* configuration, is crucial for the antiviral selectivity of all BVDU derivatives.

BVDU and BVaraU are among the most potent inhibitors of VZV. The selective activity of BVDU against HSV-1 and VZV is dependent on specific phosphorylation by HSV-1 or VZV TK which converts BVDU subsequently to its monophosphate (BVDU-MP) and diphosphate (BVDU-DP) (Fig. 22.2). The latter is then converted to the triphosphate BVDU-TP by a nucleoside diphosphate (NDP) kinase or other cellular kinase, whereupon BVDU-TP enters in competition with the natural substrate dTTP for the viral DNA polymerase. It can inhibit the incorporation of dTTP into viral DNA, or, as an alternate substrate, it can be itself incorporated, thus leading to the formation of structurally and functionally disabled viral DNA (De Clercq 2004). The conversion of BVDU-MP to BVDU-DP does not occur in HSV-2-infected cells due to the lack of thymidylate kinase activity of the HSV-2 TK (Fyfe 1982). This explains the poor activity of BVDU against HSV-2. Thus, the predominant determinant in the antiviral activity of BVDU, as well as of the different BVDU derivatives, is the virus-encoded TK and its associated thymidylate (dTMP) kinase activity. As seen in HSV-2, in some clinical isolates and in in vitro-selected BVDU-resistant mutants, thymidylate kinase activity can be regulated independently from TK activity (Docherty et al. 1991; Andrei et al. 2005a).

Following oral administration of BVDU, approximately 90% is absorbed and about 70% of the oral dose is rapidly transformed to bromovinyl–uracil (BVU) during first passage through the liver (Wutzler 1997). Clinical studies have confirmed that BVDU is effective in treatment of herpes zoster, both in short-term (formation of new lesions) and long-term effects (prevention of post-herpetic neuralgia), BVDU being more efficient and/or convenient than the other anti-VZV drugs acyclovir, valacyclovir, and famciclovir (De Clercq 2004). BVDU has been marketed in several European countries for the treatment of herpes zoster (shingles, zona) (Table 22.1).

There is one limitation for the use of BVDU: it should not be given to patients under 5-fluorouracil therapy since BVU, the degradation product of BVDU, is a potent inhibitor of dihydropyrimidine dehydrogenase (DHP), the enzyme responsible for the first step in the catabolic pathway of pyrimidines. DHP is also needed for the degradation of 5-fluorouracil. Therefore, concomitant administration of this drug together with BVDU results in increased exposure to 5-fluorouracil since BVU protects 5-fluorouracil against breakdown by DHP and significantly increases the half-life of BVU (Desgranges et al. 1986). Like BVDU, sorivudine is metabolized to bromovinyl–uracil and, therefore, its administration with 5-fluorouracil is contraindicated. Sorivudine was licensed in Japan in 1993 for the treatment of herpes zoster, but the product was withdrawn following several deaths related to co-administration with 5-fluorouracil (Okuda et al. 1997; Okuda et al. 1998)

Nucleotide Analogs

Acyclic Nucleoside Phosphonates (ANPs)

The discovery of acyclic nucleoside phosphonates (ANPs) represented a breakthrough in the treatment of DNA virus and retrovirus infections (De Clercq & Holý 2005). According to their activity spectrum, the first generation of ANPs can be classified in two categories: (i) the "HPMP" (i.e., 3-hydroxy-2-phosphonylmethoxypropyl) derivatives, represented by HPMPC (cidofovir, CDV) (Fig. 22.1), which displays activity against a broad variety of DNA viruses, and (ii) the "PME" (i.e., 2-phosphonylmethoxyethyl) and "PMP" (i.e., 2-phosphonylmethoxypropyl) derivatives, represented by PMEA (adefovir) and PMPA (tenofovir), respectively. These three representative compounds have been licensed for treatment of HCMV retinitis in AIDS patients (cidofovir, Vistide®), chronic hepatitis B virus (HBV) infections (adefovir dipivoxil, Hepsera®), and HIV infections (tenofovir disoproxil fumarate, TDF, Viread®). CDV can also be used "off-label" in treatment of herpesvirus (other than HCMV) infections, as well as polyoma-, papilloma-, adeno-, and poxvirus infections (Safrin et al. 1997; Snoeck & De Clercq 2002; De Clercq 2003). TDF is also available in a fixed-dose combination form with emtricitabine (Truvada®), with emtricitabine, and efavirenz (Atripla®) for the treatment of AIDS and has recently been approved for treatment of HBV infections as well.

In natural nucleotides (or nucleoside phosphates), the phosphate group is attached through an ester bound (–P–O–C–) to the nucleoside, as, for example, when a phosphate group has been linked to ACV (ACV-MP) during the first step of phosphorylation carried out by HSV- or VZV-encoded TK. In ANPs, the phosphate group, in the form of a phosphonate group, has already been attached to the (acyclic) nucleoside analog, thus resulting in formation of a phosphonomethyl ether (–P–C–O–), which unlike the phosphate ester linkage should resis attack by esterases.

The first ANP, identified in 1986, was (*S*)-9-(3-hydroxy-2-phosphonylmethoxypropyl)adenine (HPMPA), emerging as a broad-spectrum anti-DNA agent (De Clercq et al. 1986). This compound can be considered a hybrid between acyclic nucleoside analogs, such as (*S*)-9-(2,3-dihydroxypropyl)adenine (DHPA), which was previously described (De Clercq et al. 1978) as an acyclic nucleoside analog with broad-spectrum antiviral activity and a phosphonate analog such as phosphonoformic acic (PFA) or phosphonoacetic acid (PAA). PMEA was developed in parallel with HPMPA, whereas CDV was derived from HPMPA by simply substituting a pyrimidine (cytosine) for the purine (adenine) moiety. Further modifications of the acyclic side chain of HPMPA and PMEA led to PMPA and PMPDAP. The transition of HPMPA to PMEA allowed the activity spectrum to be extended to retroviruses (while maintaining activity against herpes- and hepadnaviruses); further modification to PMPA restricted the activity spectrum to retro- and hepadnaviruses.

ANPs can be taken up by cells and need only two phosphorylation steps to be converted to their active phosphoryl derivatives (Fig. 22.3) (Ho et al. 1992). These two phosphorylation steps are carried out by cellular enzymes. In this way, ANPs are independent of the first phosphorylation step which in the nucleoside analogs is catalyzed by the HSV- or VZV-encoded TK or the HCMV-encoded protein kinase UL97. Therefore, ANPs are active against TK-deficient HSV and VZV mutants. The diphosphoryl derivatives of the ANPs (i.e., CDVpp and PMEApp) interact with the viral DNA polymerase as either competitive inhibitors [with respect to the natural substrates (i.e., dCTP, dATP)] or alternate substrates (thus leading to incorporation of ANP into DNA). For PMEA, this incorporation inevitably leads to DNA chain termination, but CDV and HPMPA contain a hydroxyl function in the acyclic side chain that would allow further chain elongation. Incorporation of CDVpp into viral DNA slows down elongation and results in chain termination when two consecutive CDVpp residues are incorporated as has been demonstrated for HCMV (Xiong et al. 1997; Xiong et al. 1996). The diphosphorylated forms of ANPs inhibit viral DNA polymerases more potently than the cellular DNA polymerases.

The metabolites of the ANPs (i.e., CDV) show an unusually long intracellular half-life; this may account for the long-lasting antiviral activity of the compounds. This prolonged antiviral action may be attributed to the formation of the CDVp-choline adduct, which could serve as an intracellular reservoir for the mono- and diphosphoryl derivatives of CDV.

CDV administered by the intravenous route has been approved for the treatment of HCMV retinitis in AIDS patients, but has also been used successfully in the treatment of HSV-1, HSV-2, and VZV infections (i.e., those that are resistant to first-line therapy), EBV, human herpesvirus 6 (HHV-6), HHV-7, and HHV-8 infections, polyomavirus infections [i.e., progressive multifocal leukoencephalopathy (PML) due to JC virus and hemorrhagic cystitis due to BK virus], papillomavirus infections (i.e., disseminated respiratory papillomatosis), adenovirus infections, and poxvirus infections (Table 22.1). When given topically,

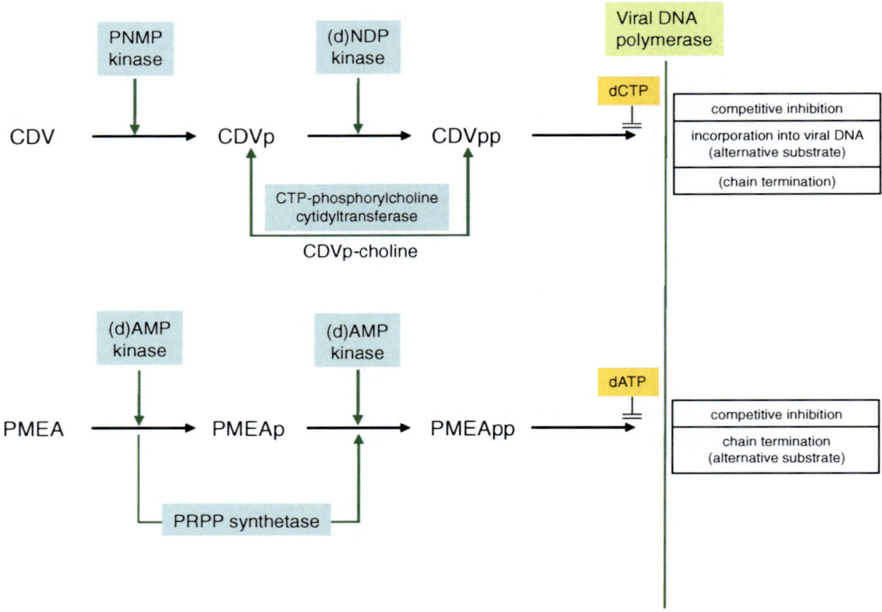

Fig. 22.3 Intracellular metabolism and mechanism of action of CDV (cidofovir) and PMEA (adefovir) against DNA viruses. Once inside the cells, ANPs need to be activated by cellular enzymes, which are different for the pyrimidine (i.e., CDV) and for the purine (i.e., PMEA) series. Pyrimidine nucleoside monophosphate (PNMP) kinase catalyses conversion of CDV to CDV-monophosphoryl (CDVp), which is then further phosphorylated to the active form, CDV-diphosphoryl (CDVpp) by nucleoside 5′-diphosphate (NDP) kinase. CDVp-choline is considered to serve as an intracellular reservoir for the mono- and diphosphoryl derivatives of CDV. Two different pathways have been suggested for the phosphorylation of PMEA and related purines. PMEA can be converted directly to PMEA-diphosphoryl (PMEApp) by 5′-phosphoribosyl-1-pyrophosphate (PRPP) synthetase. Alternatively, AMP kinase may be involved in the two consecutive steps of phosphorylation of PMEA to PMEAp and PMEApp. The diphosphoryl derivatives of the ANPs (i.e., CDVpp and PMEApp) interact with the viral DNA polymerase as either competitive inhibitors [with respect to natural substrates (i.e., dCTP, dATP)] or alternative substrates (thus leading to incorporation of ANPs into DNA). For PMEA, this incorporation inevitably leads to DNA chain termination, but CDV has a hydroxyl function in the acyclic side chain that might allow further chain elongation. Chain termination occurs when two consecutive CDVpp are incorporated into the growing DNA chain.

CDV has also proven beneficial in the treatment of mucocutaneous HSV-1 and HSV-2 infections, HPV-associated papillomatosous lesions [i.e., anogenital warts, plantar warts, recurrent laryngeal papillomatosis, cervical intraepithelial neoplasia (CIN) grade III], and poxvirus infections [i.e., molluscum contagiosum and orf virus] (Safrin et al. 1997; Snoeck & De Clercq 2002; De Clercq 2003). CDV present two major disadvantages that have restrained its use: (i) low oral bioavailability (<5%) requiring IV administration, usually once a week (or every other week), and (ii) dose-dependent nephrotoxicity that can be limited by pre-hydration and co-administration of probenecid.

Pyrophosphate Analogs

Foscarnet

Foscarnet (PFA, Foscavir®, FOS) is the trisodium salt of phosphonoformic acid (Fig. 22.1), a pyrophosphate analogue, which is a product of polymerization of nucleic acids. PFA, which does not require phosphorylation by viral or cellular kinases, inhibits directly the activity of the viral DNA polymerase. PFA acts as a product analog by binding to the pyrophosphate binding site and blocking the release of pyrophosphate from the terminal nucleoside triphosphate when added onto the growing DNA chain (Ostrander & Cheng 1980; Eriksson et al. 1982; Crumpacker 1992).

PFA can be considered as a second-line therapy and its use is reserved to patients that have failed ACV or GCV therapy due to viral resistance or that cannot be treated with GCV due to side effects of the drug (Table 22.1). Because PFA does not require a phosphorylation step to inhibit the viral DNA polymerase, it remains active against ACV^r HSV and VZV strains due to mutations in the viral TK or against GCV^r HCMV strains harboring mutations in the UL97 gene. Indeed, several reports have documented the efficacy of PFA in management of infections caused by these types of drug-resistant viruses (Morfin & Thouvenot 2003; Gilbert et al. 2002; Baldanti & Gerna 2003).

Mechanisms of Resistance

Since the introduction of ACV, now 25 years ago, for treatment of herpesvirus infections (Bacon et al. 2003), several studies have characterized drug-resistant herpesvirus mutants isolated either in vitro or in vivo. Different mechanisms exist by which HSV can acquire resistance to ACV. Three of these mechanisms involve the viral TK: (i) alteration of viral TK ($TK^{altered}$) resulting in less efficient phosphorylation of the drug, (ii) deletion of the viral TK gene ($TK^{negative}$, TK^{-}), and (iii) reduction of the expression level of TK ($TK^{partial}$) (Coen 1991; Hill et al. 1991; Harris et al. 2003; Chibo et al. 2004). Alternatively, alterations at the level of the viral DNA polymerase gene have also been observed (Collins et al. 1989; Sacks et al. 1989; Hwang et al. 1992; Gaudreau et al. 1998). Although viruses of all four phenotypes have been isolated from patients, the predominant drug-resistant phenotype recovered in vivo (similar to the in vitro situation) exhibited TK deficiency ($TK^{negative}$ or $TK^{partial}$, albeit rather low activity), as shown by several investigators (Gaudreau et al. 1998; Pottage, Jr. & Kessler 1995; Sacks et al. 1989; Coen et al. 1982).

Surveys among immunocompetent individuals have shown that ACV-resistant (ACV^r) HSV is extremely rare in this group of patients, the prevalence varying between 0 and 0.6% (Englund et al. 1990; Nugier et al. 1992; Christophers et al. 1998; Danve-Szatanek et al. 2004). In contrast to usually self-limiting infections in healthy individuals, in immunocompromised individuals (i.e., patients with HIV infection and recipients of solid organ or bone marrow transplants) HSV infection can be severe and persistent. In these cases, prolonged antiviral therapy is required for management of the infection, resulting in emergence of drug-resistant mutants in approximately 4–7% of immunocompromised patients (Christophers et al. 1998; Chakrabarti et al. 2000; Chen et al. 2000). The probability of developing unresponsive lesions appears to be related to degree of immunosuppression.

The TK gene is not essential for virus replication in cell culture, although in vivo it is involved in HSV virulence, pathogenicity, and reactivation from latency (Coen et al. 1989; Efstathiou et al. 1989; Jacobson et al. 1993). Nevertheless, about 95% of clinical HSV isolates resistant to ACV contain mutations in the viral TK and not in the viral DNA polymerase (Pottage, Jr. & Kessler 1995; Christophers et al. 1998). The lower frequency of DNA polymerase mutations could be due to fewer mutations that can result in a drug-resistant phenotype relative to TK, where there are several mutations that can confer resistance. Mutations in the TK gene that are associated with ACV resistance are mostly due to the addition or deletion of nucleotides in long homopolymer runs of G's and C's resulting in frame shift mutations and consequently in a truncated enzyme (Hwang & Chen 1995; Sasadeusz et al. 1997; Morfin et al. 2000; Sarisky et al. 2001). In fact, two studies have demonstrated that about 50% of the clinical ACV-resistant (ACV^r) HSV strains contain such type of mutations and the other half harbor single nucleotide substitutions in conserved and/or non-conserved regions of the TK gene (Gilbert et al. 2002). Mutations identified in PCV^r mutant herpesviruses isolated in vitro were generally not found within homopolymeric G and C nucleotide stretches (Sarisky et al. 2001). In a subsequent study (Suzutani et al. 2003), it was found that mutations in the TK genes from ACV^r HSV-1 mutants consisted of 50% single nucleotide substitutions and 50% frameshift mutations, while the corresponding figures for the PCV^r mutants were 4 and 96%, respectively. Recently, it was described that mutations in the TK genes of mutant viruses produced under single-round high-dose selection with BVDU consisted of 41.7% frameshift mutations within homopolymer repeats of Gs and Cs and single nucleotide substitutions (58.3 %) (Andrei et al. 2005a). The A168T change, which proved to be associated with an altered TK phenotype, appeared to be the most common substitution.

HSV-1 DNA polymerase, encoded by the UL30 gene, has 1,235 amino acid residues. It exhibits all the enzymatic functions of a polymerase and belongs to the pol α family, which includes human polymerase α and DNA polymerases from animals and other viruses. HSV-1 DNA polymerase exhibits in addition to a deoxyribonucleotide polymerizing (catalytic) function, $5' \rightarrow 3'$exonuclease/RNase H function and $3' \rightarrow 5'$ exonuclease editing activity (Knopf 1979; Boehmer & Lehman 1997). The COOH terminus of the polymerase interacts with an accessory factor, UL42 that serves to increase processivity of the polymerase (Digard et al. 1993; Zuccola et al. 2000). The overall architecture of HSV-1 DNA polymerase closely resembles that of other polymerases belonging to the polymerase α family despite being at least 300 amino acids longer and exhibiting low sequence similarity (range 16–50%). The crystal structure of the HSV-1 DNA polymerase has recently been reported (Liu et al. 2006) showing that the 3′–5′-exonuclease domain and the polymerase palm, fingers, and thumb domains of HSV polymerase can be individually superimposed with the equivalent domain structures from other polymerase α structures. HSV-1 polymerase is comprised of six structural domains: a pre-NH_2 domain, an NH_2 domain terminal, a 3′–5′-exonuclease domain, and polymerase palm, fingers, and thumb domains. Based on sequence conservation in the pol α polymerase family, the exonuclease domain contains conserved regions exo I, exo II (region IV), and exo III (δ- region C). Regions III and VI are located in the fingers subdomain, regions I, II, and VII belong to the palm subdomain, and the thumb subdomain contains conserved region V (Fig. 22.4). These regions appear to flank the catalytic site in the palm subdomain and may play a role in positioning the template and primer strands.

Most of the work with regard to mutations in the HSV DNA polymerase has been performed with ACV and PFA, the majority of the mutations being mapped to regions I, II, and III and δ-region C, indicating that these regions are important for the binding of dNTP and pyrophosphate. Although PFA inhibits the viral DNA polymerase by a different mechanism than that of nucleoside analogs, mutations in the HSV DNA polymerase that confer resistance to ACV also confer resistance to PFA. Mutations associated with resistance to HPMP derivatives such as CDV and HPMPA have been linked to changes in non-conserved regions of HSV DNA polymerase. Interestingly, these mutations did not confer resistance to PME derivatives such as PMEA (Andrei et al. 2000). These findings suggest that these two subclasses of ANPs differ in their mode of interaction with the viral DNA polymerase. Moreover, mutants resistant to HPMP derivatives remained sensitive to PFA and ACV, while different degrees of cross-resistance between PME derivatives, PFA, and ACV were noted (Andrei et al. 2007a; Andrei et al. 2007b; Bestman-Smith & Boivin 2002). Most DNA polymerase mutations conferring resistance to PFA have been associated with cross-resistance to PME derivatives, but some HSV polymerase mutations have been linked to resistance to PFA and sensitivity to PME derivatives (Andrei et al. 2007a). These data support possible use of CDV and not PMEA in treatment of PFA^r HSV infections. Indeed, CDV has proven efficacious in treatment of ACV^r and/or PFA^r HSV-associated diseases (Snoeck et al. 1994a; LoPresti et al. 1998; Safrin et al. 1997).

Pathogenicity of drug-resistant HSV mutants has been mostly studied in mouse models (Coen 1994). One assay of pathogenesis in mice entails infection of the central nervous system via intracerebral inoculation (i.c.), which leads to encephalitis and death (neurovirulence). In a second assay of pathogenesis, virus is inoculated at a peripheral site, where it replicates and, following axonal transport, reaches the trigeminal ganglia (secondary site of replication). There, HSV establishes and maintains a latent infection. The assay that can be considered as most sensitive to drug-resistance mutation is ability of virus to kill mice after i.c. inoculation (Coen 1994). In this assay, mutants described as $TK^{negative}$ and $TK^{partial}$ have been shown to be significantly less virulent than wild-type viruses (Field & Wildy 1978; Field

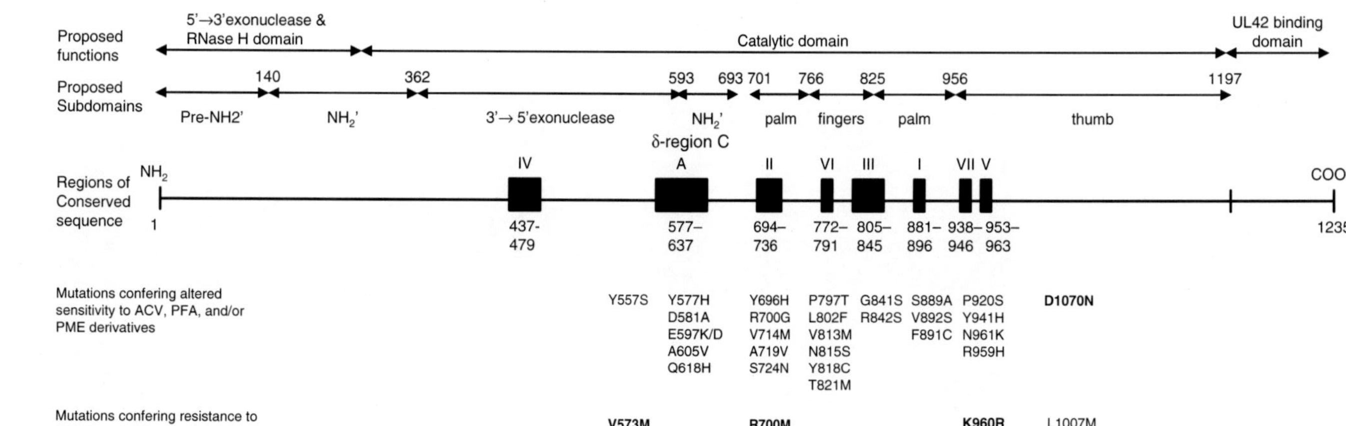

Fig. 22.4 Organization of HSV-1 DNA polymerase and amino acid substitutions leading to drug resistance. Locations of functional sites on HSV-1 DNA polymerase and linear N terminus to C terminus configuration of the polypeptide are shown. Solid boxes indicate regions numbered I to VII based on conservation among the DNA polymerases and region A, a region showing slight conservation. Locations of point mutations that result in altered drug sensitivity are shown in the lower part of the figure.

& Darby 1980; Darby et al. 1981; Tenser et al. 1983; Chrisp et al. 1989; Suzutani et al. 1995; Andrei et al. 2005a; Andrei et al. 2007a). However, Pelosi and colleagues have reported a TK mutant with a large deletion mutation which was only slightly impaired in neurovirulence and one $TK^{partial}$ mutant which was fully neurovirulent following i.c. inoculation (Pelosi et al. 1998a). Degrees of attenuation of $TK^{altered}$ mutants and some DNA polymerase mutants may vary substantially (Pelosi et al. 1998a; Pelosi et al. 1998b; Darby et al. 1981; Darby et al. 1984; Field & Darby 1980; Field & Coen 1986). Among $TK^{altered}$ and DNA polymerase mutants the most pathogenic drug-resistant mutants can be found, although it appears that drug-resistant mutants arising under pressure of the HPMP derivatives have the lowest levels of neurovirulence (Andrei et al. 2007a) Although it is generally accepted that $TK^{deficient}$ viruses are unable to be reactivated from latency (Coen et al. 1989; Efstathiou et al. 1989; Jacobson et al. 1993), some reports have emerged that describe individual $TK^{deficient}$ isolates (in particular, mutants harboring mutations within the 7G homopolymer repeat of the HSV-1 TK gene) that can be recovered, albeit inefficiently, from latently infected animals and can cause recurrent infections in patients (Hwang et al. 1994; Sasadeusz & Sacks 1996; Horsburgh et al. 1998; Morfin et al. 2000; Harris et al. 2003). Three different mechanisms can be exploited by HSV-1 to compensate for a mutation that would otherwise inactivate TK and prevent reactivation from latency: (i) ribosomal frameshifting during which low levels of TK are expressed (Hwang et al. 1994; Griffiths et al. 2003), (ii) replication errors due to reduced fidelity of the DNA polymerase when replicating homopolymeric runs that create subpopulations of wild-type virus (Sasadeusz & Sacks 1996; Griffiths & Coen 2003; Grey et al. 2003), and (iii) alleles in loci other than TK that have potential to complement TK function and influence the ability of HSV to replicate in the nervous system and to reactivate from latency (Horsburgh et al. 1998). An important factor that influences drug-resistance and pathogenicity is heterogeneity of the viral population. Mixtures of different drug-resistant mutants or mixtures of drug-resistant virus and wild-type virus can complement for both drug resistance and pathogenicity. This can be seen in the case of TK frameshift mutations in homopolymeric sequences, the most common drug-resistant mutations, which can easily revert resulting in mixed populations that are reactivated from latency (Sasadeusz & Sacks 1996; Griffiths & Coen 2003; Grey et al. 2003).

Similarly to HSV, ACV treatment for VZV infections does not generate ACV^r viruses in immunocompetent hosts. However, in immunocompromised hosts, VZV infection tends to be severe and prolonged and ACV^r mutants have been isolated after long-term treatment with ACV (Pahwa et al. 1988; Snoeck et al. 1994b; Talarico et al. 1993; Boivin et al. 1994). Resistance to ACV in VZV appears as a consequence of mutations either in TK or DNA polymerase genes. These are the most frequent TK mutants isolated both in cell culture and in the clinic (Gilbert et al. 2002). Mutations conferring resistance to nucleoside analogs have been found all along the VZV TK gene, although specific regions including ATP- and nucleoside-binding sites are recognized as mutagenic hot spots as well as amino acid 231 (Boivin et al. 1994; Morfin et al. 1999; Talarico et al. 1993). Resistances to PFA associated with mutations in the VZV DNA polymerase gene have also been described in immunocompromised patients (Visse et al. 1998; Visse et al. 1999). Amino acid substitutions in VZV DNA polymerase described in ACV^r and PFA^r mutants corresponded to changes described in the HSV DNA polymerase, although some mutants exhibited a discrepancy in their sensitivity to ACV or aphidicolin in comparison with corresponding HSV-1 mutants (Kamiyama et al. 2001; Visse et al. 1998; Visse et al. 1999). These findings indicate that identical or similar amino acid substitutions may create different conformations in HSV and VZV DNA polymerase that account for a discrepancy in drug susceptibility of the ACV^r mutants (Kamiyama et al. 2001). In contrast, VZV mutants selected in vitro under pressure with PME derivatives harbored mutations in viral DNA not corresponding to those described in HSV. Interestingly, it appears that ACV and PCV select in vitro for different drug-resistant VZV genotypes: ACV selects for TK mutants, while PCV selects for DNA polymerase mutants (Andrei et al. 2004). Furthermore, alterations in the viral DNA polymerase that confer resistance to PCV were shown to be also responsible for cross-resistance to PFA (Andrei et al. 2004). Several reports have indicated that PCV remains active against some HSV-1 and VZV TK and DNA polymerase mutants that are resistant to ACV (Boyd et al. 1987; Pelosi et al. 1998a; Hasegawa et al. 1995; Andrei et al. 2004). These findings indicate that interactions between HSV or VZV TK and PCV or ACV, and likewise between viral DNA polymerases and triphosphates of PCV or ACV, are distinct and may account for the differences observed between ACV^r and PCV^r VZV strains. Furthermore, emergence frequency of resistant VZV mutants proved to be significantly higher following ACV exposure than following PCV exposure (Ida et al. 1999)

HCMV resistance to GCV may arise from mutations in either the UL97 (phosphotransferase) or UL54 (DNA polymerase) genes (Baldanti & Gerna 2003; Gilbert et al. 2002; Chou 1999; Chou & Meichsner 2000). In contrast to ACV, the selection of GCV^r HCMV mutants requires a longer time in cell culture than for HSV. This is probably due to higher fidelity of HCMV DNA polymerase compared to HSV DNA polymerase (Sullivan & Coen 1991). Also different from HSV and VZV TK gene mutations, HCMV UL97 mutations have limited distribution in the gene: changes clustered in codons 460–520 (proposed ATP-binding site) or codons 590–607 (function in substrate recognition) (Wolf et al. 1995; Chou et al. 1995; Gilbert et al. 1998; Abraham et al. 1999). This is probably due to the fact that HCMV UL97 is very important for viral replication, presenting various UL97 mutants a fitness loss compared to wild-type strains. The HCMV pUL97 has been characterized as a protein which is autophosphorylated and is capable of phosphorylating GCV. The role of pUL97 protein kinase in HCMV replication and pathogenesis is still under investigation. Wolf and colleagues reported that UL97 kinase has an impact on at least two distinct phases of viral replication: DNA synthesis as well as capsid assembly and nuclear egress, indicating that protein phosphorylation mediated by this kinase increases efficiency of these two phases of virus replication (Wolf et al. 2001). The fact that HCMV UL97 protein kinase mutations associated with antiviral resistance are localized at specific codons simplifies identification of mutations by targeted PCR sequencing or restriction length polymorphism analysis (Scott et al. 2004; Chou 1999; Lurain et al. 2001). As PFA and CDV are independent of pUL97 protein kinase for their antiviral action, these two drugs are recommended for treatment of HCMV infections resistant to GCV due to alterations in UL97. It is worth mentioning that valganciclovir does not appear to select for an increased number of GCV^r strains (Boivin et al. 2004).

A number of different mutations in the HCMV UL54 gene have been associated with resistance to GCV (Jabs et al. 1998; Smith et al. 1997). However, mutations in the UL54 gene are less common than in the UL97 gene. Most mutations conferring resistance to GCV cluster in specific regions of the HCMV DNA polymerase that are conserved among α-like DNA polymerases and show simultaneous cross-resistance to CDV (Baldanti & Gerna 2003; Gilbert et al. 2002; Chou 1999). Most of the GCV^r/CDV^r UL54 mutants retained sensitivity to PFA, while mutations in domains II, III, and IV of HCMV DNA polymerase responsible for PFA resistance do not show cross-resistance to GCV or CDV (Cihlar et al. 1998a; Cihlar et al. 1998b; Baldanti et al. 1996). However, multi-drug-resistant HCMV strains have been recently isolated in immunocompromised patients emphasizing the necessity of developing new treatment options. Furthermore, previously unrecognized mutations in the HCMV DNA polymerase gene continue to be isolated from patients receiving antiviral therapy (Weinberg et al. 2003; Scott et al. 2004). CDV^r strains isolated in vitro were shown to be cross-resistant to GCV. Because CDV is not the first-line therapy for HCMV infections due to its renal toxicity, selection of CDV^r strains is an extremely rare event and it appears to be mostly driven by long-term administration of GCV (Smith et al. 1997).

Similar to HCMV, resistance to GCV in HHV-6 has been mapped in pUL69 and DNA polymerase (encoded by the U38 gene). GCV^r HHV-6 have been detected in clinical specimens and generated in the laboratory (Manichanh et al. 2001). Amino acid changes in the HHV-6 UL69 protein kinase, homologous to those in the HCMV UL97 phosphotransferase, were shown to cause resistance to GCV (Safronetz et al. 2003). Mutations in the UL38 gene associated with resistance to GCV and CDV were shown to be different from those conferring resistance to PFA (Manichanh et al. 2001; Bonnafous et al. 2007).

Amino acid substitutions in vaccinia virus DNA polymerase (encoded by the E9L gene) which are linked to CDV resistance have been recently reported (Andrei et al. 2006). These mutations are located within the 3′–5′ exonuclease (A314T) and polymerase (A684V) catalytic domain. By marker transfer experiments it could be demonstrated that either mutation alone could confer a drug-resistant phenotype although degree of resistance was significantly lower than in virus encoding both mutations. A314T recombinant virus was shown to be associated with hypersensitivity to the pyrophosphate analogue phosphonoacetic acid (PAA). A684V appeared to increase resistance to PAA. Presence of both mutations resulted in no change in susceptibility to PAA. All CDV^r viruses exhibited reduced virulence in mice, demonstrating that these E9L mutations are inextricably linked to reduced fitness in vivo. Interestingly, it was observed that treatment for 5 days with CDV still protected mice against a lethal intranasal challenge with drug-resistant virus bearing both mutations. Sequence analysis of adenovirus DNA polymerase of in vitro-selected CDV^r strains allowed identification of amino acid changes associated with drug resistance (Kinchington et al. 2002). These substitutions are located in conserved regions of adenovirus DNA polymerase predicted to be involved in nucleotide binding.

Different reports based on crystallographic studies have provided details on how mutations in different viral DNA polymerases domains affect binding and catalysis (Liu et al. 2006; Zuccola et al. 2000; Loregian et al. 2004; Appleton et al. 2006; Huang et al. 1999; Shi et al. 2006; Tchesnokov et al. 2006).

Candidate Viral DNA Polymerase Inhibitors

In Table 22.2 are summarized the spectrum of antiviral activity and mechanism of action of nucleoside analogs, nucleotide analogs, and non-nucleoside analogs that target viral DNA polymerases which have been developed or are under development.

Nucleoside Analogs

Lobucavir

Lobucavir (LBV) is a cyclobutyl analog of guanine (Fig. 22.5) that has activity against most herpesviruses and also against HIV and HBV. LBV is a potent inhibitor of HCMV DNA polymerase in vitro. However, this nucleoside analogue is phosphorylated intracellularly to its triphosphate form both in infected and in uninfected cells, being the phosphorylated metabolite levels in HCMV-infected cells being only two- to three-fold higher compared to uninfected cells (Tenney et al. 1997). Due to lack of selective metabolism of LBV in virus-infected cells, the compound can be used as a substrate by host cell polymerases thus increasing toxicity and safety risks. Although promising results were obtained in early clinical trials against HCMV and HBV, development of LBV was halted due to safety concerns. Toxicologic studies in rodents showed an increase in number of different cancers with long-term administration of the drug.

H2G

H2G, (*R*)-9-[4-hydroxy-2-(hydroxymethyl)butyl]guanine (omaciclovir), is an acyclic nucleoside analog (Fig. 22.5) that has shown potent activity against different herpesviruses, especially against VZV (Abele et al. 1988). This compound has a mode of action similar to that of ACV, but with less selectivity as a substrate for TK. Also, resistance to H2G has been mapped in TK (Ng et al. 2001). In contrast to ACV, H2G is not an obligate chain terminator, although incorporation of the triphosphate form (H2G-TP) results in limited chain elongation. Another difference with ACV-TP is the longer intracellular half-life of H2G-TP. A prodrug of H2G, MIV-606, the L-valine ester of H2G, with higher oral bioavailability has been synthesized and phase I/II clinical trails have been initiated with this compound.

Table 22.2 New viral DNA polymerase inhibitors under development

Type of inhibitor	Spectrum of activity	Mechanism of action	Comments
A. Nucleoside analogs			
Lobucavir (LBV) (cyclobutyl guanosine analog)	- HSV, VZV, HCMV, HHV-6, HHV-7, HBV	- Phosphorylated by cellular enzymes - The active form, LBV-TP, is a potent inhibitor of HCMV DNA polymerase	- Anti-HCMV activity shown in HIV-infected patients - Development stopped in 2002
Omaciclovir (H2G) (carbocyclic guanosine analog)	- HSV, VZV, HSV-1, HSV-2, EBV, HHV-8	- Phosphorylated by VZV- and HSV-encoded TK - Resistance is associated with mutations in the viral TK gene. Cross-resistance with ACV and other TK-dependent drugs - Competitive inhibitor of viral DNA polymerase. It is not an obligate chain terminator, but reduces chain extension - H2G-TP has a longer intracellular half-life than ACV-TP; therefore it has a more prolonged antiviral activity	- Development terminated for VZV - Possible development for EBV (infectious mononucleosis)
A-5021 (guanosine analog)	- HSV, VZV	- Mode of action similar to ACV . Substrate for HSV and VZV TK . Competitive inhibitor of viral polymerase . A-5A21-TP has a longer intracellular half-life than ACV-TP . It is not a chain terminator but limits chain elongation - Resistance profile similar as for ACV mutants	- Development no actively pursued
S2242 (guanosine analog)	- Broad-spectrum herpesviruses (i.e., HSV, VZV, HCMV, EBV, HHV-6, HHV-7, and HHV-8) and poxviruses	- Active against $TK^{deficient}$ HSV and VZV strains - Not a substrate for HSV TK or HCMV protein kinase	Development stopped due to safety concerns

Table 22.2 (continued)

Type of inhibitor	Spectrum of activity	Mechanism of action	Comments
		- The diacetate ester prodrug (HOE961) is orally bioavailable - More potent than ACV against HSV-1 and more efficacious than GCV against murine HCMV in different animal models	
BCNAs (bicyclic pyrimidine nucleoside analogs	- VZV	- Mechanism of action not fully elucidated . Substrate for VZV TK . BCNAs resistant mutants show mutations in the VZV TK gene. . Cross-resistance with ACV and BVDU . No triphosphate detected in VZV-infected cells	- Oral prodrug (FV-100) under development for the treatment of VZV infections
B. Nucleotide analogs			
HDP-CDV (alkoxyalkyl ester of CDV)	- Broad-spectrum DNA viruses	- Idem as for CDV after removal of alkoxyalkyl group	- Currently under development - Orally bioavailable - Diminished accumulation of the drug in the kidney compared to CDV - Higher intracellular levels in comparison with CDV
HPMPO-DAPy	- Spectrum of activity similar to that of CDV, except for weak activity against HCMV	- Idem as for HPMPC	- Not (yet) in development
HPMP-5-azaC	- Spectrum of activity similar to that of CDV	- Idem as for HPMPC	- Not (yet) in development
C. Non-nucleoside analogs			-
PNU-183792	- VZV, HSV, HCMV, HHV-8, EBV	- Inhibition of the binding of the natural substrate (dTTP) to the DNA polymerase	- Limited information on the current state of development of the compound

A-5021

In a search for novel nucleoside analogs with antiherpesvirus activity, (1′*S*,2′*R*)-9-[[1′,2′-bis(hydroxymethyl)cycloprop-1′-yl]methyl]-guanine (A-5021) (Fig. 22.5)

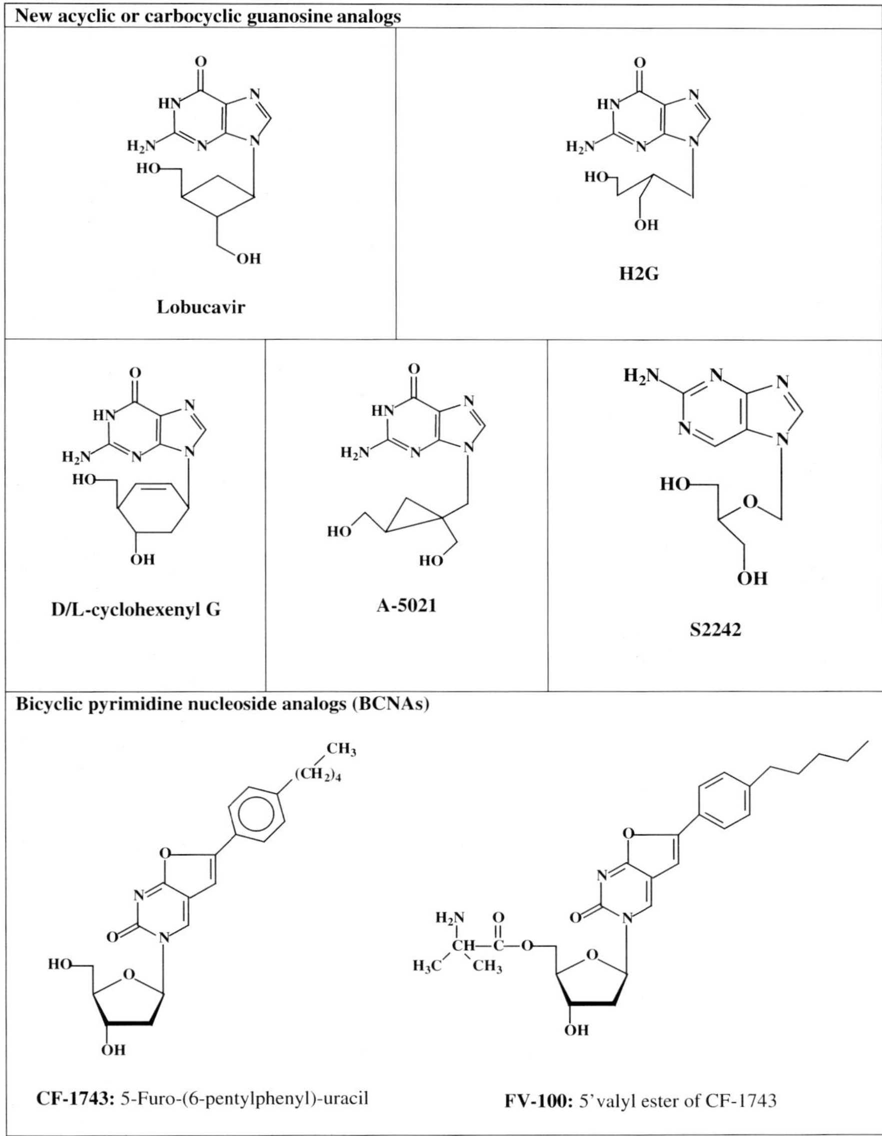

Fig. 22.5 Structural formulae of candidate viral DNA polymerase inhibitors that are candidates for further development.

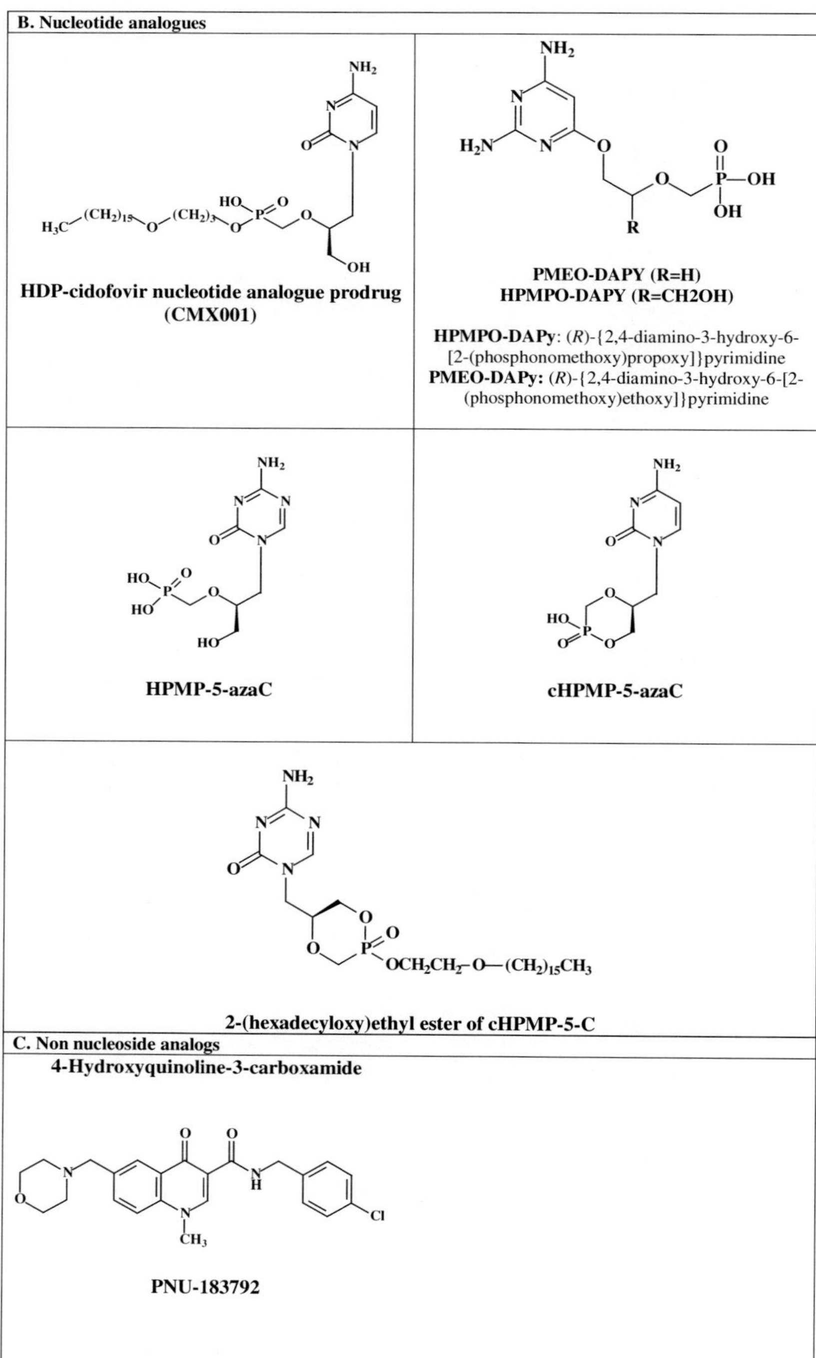

Fig. 22.5 (continued)

emerged as a potent inhibitor of HSV-1, HSV-2, VZV, and HCMV (Iwayama et al. 1998). In vitro antiviral activity of this guanosine analog was higher than that of ACV or PCV against HSV-1 and VZV; however, activity against HSV-2 was comparable to that of the gold standards ACV and PCV (Iwayama et al. 1998). A-5021 also proved active against EBV and HHV-6 but not HHV-8 (De Clercq et al. 2001). The mechanism of action of this compound was shown to be similar to that of ACV and PCV (Ono et al. 1998). A-5021 is monophosphorylated by viral TK and to the di- and triphosphate forms by cellular kinases. A-5021-TP appeared to accumulate more than ACV-TP but less than PCV-TP in MRC-5 cells infected with HSV-1 or VZV, whereas HSV-2-infected MRC-5 cells had comparable levels of A-5021 and ACV triphosphates. The intracellular half-life of A-5021-TP was shown to be considerably longer than that of ACV-TP and shorter than that of PCV-TP. A-5021-TP is a competitive inhibitor of herpes DNA polymerases with respect to dGTP and it can be incorporated into DNA instead of dGTP and terminate elongation, although it permits limited chain extension. Thus, the strong antiviral activity of A-5021 appears to depend on a more rapid and stable accumulation of its triphosphate in infected cells; however, a disadvantage for the compound is its cross-resistance with ACV and PCV. In vivo efficacy of A-5021 against HSV-1 has been demonstrated in different animal models, its activity being superior to that of ACV in all HSV-1 models of infection (Iwayama et al. 1999). However, no advantage over PCV was seen against a model of systemic HSV-2 infection. The compound has entered clinical development, albeit transiently, and it remains to be demonstrated whether the stronger potency and prolonged antiviral activity compared to ACV observed in vitro and in animal models will be observed in the clinical studies.

S2242

The acyclic purine nucleoside analog, 2-amino-7-[(1,3-dihydroxy-2-propoxy) methyl]purine (S2242) (Fig. 22.5), the only known antivirally active acyclic nucleoside analogue with the side chain substituted at the N7 position of the purine ring, has been reported to possess potent activity against several herpesviruses including HSV, VZV, HCMV, EBV, HHV-6, HHV-7, and HHV-8 and also against poxviruses (Neyts & De Clercq 1997; Meerbach et al. 1998; Neyts et al. 1994; Zhang et al. 1999; De Clercq et al. 2001). Of special interest is the potent activity of the compound against HCMV and against $TK^{deficient}$ mutants of HSV and VZV. Indeed, it was demonstrated that S2242 is not phosphorylated by either HSV TK or HCMV-encoded UL-97 kinase (Zimmermann et al. 1997). S2242 was found to be a substrate for deoxycytidine (dCK) kinase and for deoxyguanosine (dGK) kinase. S2242 was shown to be phosphorylated in a time- and concentration-dependent manner to its monophosphate, diphosphate, and triphosphate (Neyts et al. 1998). This nucleoside analog was not preferentially phosphorylated in HSV-1-, VZV-, or HHV-6-infected cells. In HCMV-infected human embryonic lung cells, a 5- to 25-fold increase in S2242 metabolite formation was observed compared with noninfected cells, suggesting that an HCMV-encoded or -induced enzyme causes specific phosphorylation of S2242.

S2242 proved to be more effective than ACV in different mouse models of HSV-1 infection and it was also far more effective than GCV in preventing or delaying murine cytomegalovirus-induced mortality in immunocompetent and severe combined immune deficiency (SCID) mice (Neyts et al. 1995; Neyts & De Clercq 2001). In addition, S2242 and its orally active diacetate ester prodrug (HOE961) were reported to be potent inhibitors of vaccinia virus and cowpox virus replication in cell culture and in infected mice (Smee et al. 2002). However, development of the compound was stopped due to safety concerns.

BCNAs

In 1999, some unusual bicyclic nucleoside analogues (BCNAs) with significant and selective anti-VZV activity were reported (McGuigan et al. 1999). The early drug leads had a long alkyl side chain on the bicyclic base part, with optimum length of ca. C8–C10. In vitro potencies of these compounds against VZV were ca. 300-fold more potent than the clinically established anti-herpetic agent ACV. Replacement of the alkyl chain by a p-alkylphenyl unit led to a significant boost in potency against VZV (McGuigan & Balzarini 2006). The most potent analogue was the p-pentylphenyl BCNA analogue Cf1743 (Fig. 22.5), which exhibited activity against a broad range of VZV clinical isolates at subnanomolar concentrations. After extensive studies of structure–activity relationship and pharmacology of the BCNAs, Cf1743 was identified as a potential clinical candidate. Pharmocokinetic studies indicated the need to improve its bioavailability. The HCl salt of the 5'-valyl ester of Cf1743 (i.e., FV-100 designed in analogy with valaciclovir with respect to acyclovir) emerged as the most promising prodrug (McGuigan et al. 2007). Phase 1 studies have been planned. Unlike ACV, which is a broad-spectrum anti-herpetic agent, the BCNAs are highly specific for VZV with no significant activity against any other virus, including other members of the herpesvirus family and the closely related simian varicella virus (SVV).

Although the precise mechanism of action of these compounds remains to be elucidated, it is clear that for their antiviral activity they depend on phosphorylation by the VZV-encoded thymidine kinase, since Cf1743 completely lost its anti-VZV efficacy against both laboratory and clinical virus isolates with a deficient TK (Andrei et al. 2005b). Furthermore, mutant viruses emerging under selective pressure with the BCNAs present amino acid changes in the viral TK. The BCNAs owe at least part of their antiviral selectivity to a specific activation/phosphorylation by the VZV-encoded TK and associated thymidylate kinase (dTMP-K) activity, while not being recognized by the closely related HSV-1-encoded TK/dTMP-K enzyme (Sienaert et al. 2002). Strikingly, there was no close correlation between affinity for VZV TK and antiviral activity, pointing to a different structure–activity relationship for the eventual antiviral target for the BCNAs. It should be noted that the closely related SVV TK is also able to recognize the BCNAs as a substrate, but SVV replication is not affected by the BCNAs (Sienaert et al. 2004). Notably, this class of compounds is not recognized by cellular kinases that participate in the anabolism of other pyrimidine analogs. Among the cellular kinases that do not recognize the

BCNAs as substrate is the nucleoside diphosphate (NDK) kinase, which converts BVDU-DP to the active triphosphate form (Sienaert et al. 2003). Consequently, no 5′-triphosphate of BCNAs could be detected in VZV-infected cells. Further studies are highly warranted to decipher the mode of action of the BCNAs and to determine the effects of the BCNAs anabolites on the VZV DNA polymerase.

BCNAs are not susceptible to degradation by human or bacterial thymidine phosphorylase, and thus are not cleaved to their free (inactive) base (Balzarini et al. 2002). Also, the latter is not inhibitory to dihydropyrimidine dehydrogenase, an enzyme involved in the degradation of thymine, uracil, and the anticancer agent 5-fluorouracil. This means that BCNAs do not interfere with degradation of 5-fluorouracil and, in contrast to BVDU, they could eventually be used in therapy of VZV infections in patients under 5-fluorouracil treatment. It is evident that BCNAs represent highly promising anti-VZV compounds that are not susceptible to breakdown by nucleoside/nucleobase catabolic enzymes and are not expected to interfere with cellular catabolic processes such as those involved in 5-fluorouracil catabolism.

Nucleotide Analogs

Cidofovir Esters

The clinical use of cidofovir is limited by its poor oral bioavailability and renal toxicity. To overcome these restrictions, Hostetler's group has synthesized alkoxyalkyl esters of CDV and its cyclic form, i.e., cCDV (Beadle et al. 2002). In these prodrugs, a fatty acid been linked to the parent molecule to facilitate drug absorption in the gastrointestinal tract. These alkoxyalkyl esters of CDV and cCDV were much more active in vitro than the parent compounds against several herpesviruses, including HSV, VZV, CMV, EBV, HHV-6, and HHV-8 and poxviruses. A 2.5- to 4-log increase in antiviral activity against HCMV replication in vitro was observed. In addition, these derivatives showed improved uptake and absorption and had oral bioavailabilities in mice of 88–97%, compared to less than 5% for CDV. Studies with radiolabeled compound confirmed increased cell penetration (10 to 20-fold) and higher intracellular levels (100-fold) of CDV-PP (the active form of the compound) than those measured following treatment of the cells with CDV (Aldern et al. 2003). In vivo, oral administration of hexadecyloxypropyl-CDV (HDP-CDV) (Fig. 22.5) proved as effective as parental CDV in treatment of herpes- and poxvirus infection in several mouse models (Kern et al. 2004; Bidanset et al. 2004; Wan et al. 2005; Keith et al. 2004; Keith et al. 2004; Quenelle et al. 2004). Importantly, diminished accumulation of the drug in the kidney was reported according to studies evaluating tissue distribution of radiolabeled HDP-CDV and other alkoxyalkyl esters of CDV in mice (Ciesla et al. 2003; Kern et al. 2004).

HDP-CDV (CMX001) in an oral formulation is presently under development by Chimerix. A Phase I clinical study to evaluate the safety and pharmacokinetics of orally administered CMX001 in healthy volunteers has been announced (Painter & Hostetler, 2004). CMX001 will be developed for potential use in treatment of smallpox or vaccine-related side effects. This drug could also provide a safer therapy for ACV- and GCV-resistant herpesviruses in the immunocompromised host.

In addition to herpes- and poxviruses, increased activity of alkoxyalkyl esters of CDV compared to parent compound CDV was also shown against adenovirus, polyomavirus, and papillomavirus (Hartline et al. 2005; Hostetler et al., 2006). A similar prodrug strategy was applied to other ANPs, such as HPMPA, the enhancement of antiviral potency being similar to that of CDV (Ruiz et al. 2007; Lebeau et al. 2006; Choo et al. 2007; Dal Pozzo et al. 2007; Ruiz et al. 2006; Beadle et al. 2006).

New Generations of Acyclic Nucleoside Phosphonates

Following the success of the first ANPs, two new generations of ANPs have been recently described. The "second generation" ANPs include the "open ring" or "O-linked" ANP analogues or 6-[2-phosphonomethoxyalkoxy]-2,4-diaminopyrimidines (DAPys), which showed substantial potential for the treatment of a broad range of DNA virus and retrovirus infections (Balzarini et al. 2004; Hockova et al. 2003; Hockova et al. 2004; De Clercq et al. 2005). HPMPO-DAPy (Fig. 22.5) offers an activity similar to that of CDV except for HCMV that is poorly inhibited, while PMEO-DAPy (Fig. 22.5) displays an analogous activity spectrum as that of PMEA. The "third generation" of ANPs encompasses HPMP derivatives with a 5-azacytosine moiety such as HPMP-5-azaC (Fig. 22.3) and its cyclic form (i.e., cHPMP-5-azaC) (Krecmerova et al. 2007a). These compounds were at least as potent as CDV against several DNA viruses, including HCMV. Furthermore, CDV and HPMP-5-azaC proved equally potent in pathogenic models of HSV and poxvirus infections in mice. Among several prodrugs of cHPMP-5-azaC synthesized, the hexadecyloxyethyl ester proved to be about 250-fold more active than the parent compound (Krecmerova et al. 2007b). Further studies are needed to determine the clinical potential of these compounds.

Non-nucleoside Analogs

The use of an in vitro HCMV DNA polymerase assay in high-throughput screening allowed identification of a novel class of non-nucleoside herpesvirus polymerase inhibitors, the naphthalene–carboxamides, PNU-26370 being the lead compound of this series of non-nucleoside DNA polymerase inhibitors (Vaillancourt et al. 2000; Tucker et al. 2000). Structure–activity relationship (SAR) studies demonstrated that a quinoline ring could be substituted for naphthalene, leading to the discovery of the 4-oxo-dihydroquinoline-3-carboxamides (4-oxo-DHQ), represented by PNU-181128, PNU-181465, and PNU-183792 (Fig. 22.5), that demonstrated inhibition of HCMV, HSV, and VZV polymerases (Brideau et al. 2002; Oien et al. 2002). High specificity for viral DNA polymerases compared to human alpha (α), gamma (γ), delta (δ) polymerases was observed. PNU-183792 displays broad spectrum of activity in cell culture against different herpesviruses, including HSV-1, HSV-2,

VZV, SVV, HCMV, murine and rat cytomegaloviruses, EBV, and HHV-8 (Kaposi-associated herpesvirus), exceptions being HHV-6 and HHV-7 (Wathen 2002). PNU-183792 was inactive against unrelated DNA or RNA viruses indicating specificity for herpesviruses. A strong correlation between inhibition of the viral DNA polymerases and antiviral activity for this class of compounds supports inhibition of the viral DNA polymerase as the mechanism of antiviral activity. The 4-oxo-DHQs were found to be competitive inhibitors of nucleoside binding; however, no cross-resistance could be detected with GCV-resistant HCMV or ACV-resistant HSV mutants. In vitro antiviral activity of the 4-oxo-DHQs was comparable or superior to existing antiherpesvirus drugs and drug resistance to these compounds correlated with point mutations in conserved domain III of the HCMV DNA polymerase (V823A + V824L) (Thomsen et al. 2003). V823 is conserved in the DNA polymerases of human herpesviruses, except for HHV-6 and HHV-7 that contain an alanine at this position. Mutations associated with resistance to 4-oxo-DHQs did not confer resistance to existing anti-herpesvirus nucleoside analogs. Based on the crystal structure of the HSV DNA polymerase, Liu et al. proposed a HSV DNA polymerase model, suggesting that the 4-oxo-DHQs bind at the polymerase active site interacting non-covalently with the DNA duplex–DNA polymerase complex and not with enzyme or DNA duplex alone (Liu et al. 2006). PNU-183792 is orally bioavailable, and activity was demonstrated in a model of lethal murine cytomegalovirus (MCMV) infection (Brideau et al. 2002). So far, these non-nucleoside DNA polymerase inhibitors have not been evaluated in any clinical trials.

Further SAR studies led to the discovery of 4-oxo-4,7-dihydrothienopyridines (DHTPs) (Schnute et al. 2005; Schnute et al. 2007) and 7-oxo-4,7-dihydrothieno [3,2-b]pyridine-6-carboxamides (Larsen et al. 2007). Some of these compounds demonstrated broad-spectrum inhibition of the herpesvirus polymerases HCMV, HSV-1, EBV, and VZV with high specificity compared to human DNA polymerases. DHTPs, in contrast to the kinetics determined for the 4-oxo-DHQs, proved to be competitive inhibitors of dTTP incorporation into primer template by HCMV DNA polymerase (Schnute et al. 2005).

Inhibitors of Protein–Protein Interactions

Protein–protein interactions among proteins implicated in herpesvirus DNA replication are essential for viral genome replication and are considered as attractive potential drug targets (Coen & Schaffer 2003). Similar to other herpesviruses, HCMV DNA polymerase contains a catalytic subunit (UL54) and an accessory protein (UL44) that is thought to increase the processivity of the enzyme. Loregian and collaborators (Loregian et al. 2003) have identified peptides from the C terminus of UL54 which could efficiently disrupt the physical interaction between UL54 and UL44 and specifically inhibit the stimulation of UL54 by UL44. These findings provide the basis for developing new classes of anti-HCMV inhibitors that act by disrupting the UL54/UL44 interaction. Indeed, small molecules that disrupt the in vitro interaction between the HSV DNA polymerase (UL30) and the accessory protein (UL42) and exhibit anti-HSV activity have been identified (Coen & Schaffer 2003). Recently, mutations that decrease DNA binding of the processivity factor of HSV DNA polymerase were shown to reduce viral yield, to alter kinetics of viral replication, and to decrease fidelity of DNA replication (Jiang et al. 2007b; Jiang et al. 2007a).

Small Interfering RNAs

RNA interference (RNAi) is a natural mechanism of post-transcriptional gene silencing, widely conserved in multicellular organisms. This pathway is thought to be an ancient mechanism for protecting the host and its genome against viruses and transposable genetic elements (Hannon 2002). The molecular mediators of RNAi are double-stranded RNAs of 21–23 nucleotides in length that induce the sequence-specific degradation of homologous RNAs. RNAi has been used as a means to manipulate gene expression experimentally and to probe gene function. It has also been proposed that this biological response might be exploited therapeutically as an antiviral defense mechanism. The siRNA approaches have been shown to be effective against a variety of viruses in cell culture (Leonard & Schaffer 2006; Silva et al. 2002; Dykxhoorn & Lieberman 2006). Recently, it has been described that synthetic siRNA against essential gene products of HCMV such as UL54 (DNA polymerase) and UL97 (protein kinase) can trigger RNAi in infected cells leading to effective inhibition of viral replication (Wiebusch et al. 2004; Shin et al. 2006). These results demonstrated the effectiveness of siRNAs against experimental HCMV infection and open new possibilities for antiviral strategies.

References

Abele, G., B. Eriksson, J. Harmenberg & B. Wahren, 1988. Inhibition of varicella-zoster virus-induced DNA polymerase by a new guanosine analog, 9-[4-hydroxy-2-(hydroxymethyl) butyl]guanine triphosphate, Antimicrob. Agents Chemother. 32: 1137–1142.

Abraham, B., S. Lastere, J. Reynes, F. Bibollet-Ruche, N. Vidal & M. Segondy, 1999. Ganciclovir resistance and UL97 gene mutations in cytomegalovirus blood isolates from patients with AIDS treated with ganciclovir, J. Clin. Virol. 13: 141–148.

Aldern, K. A., S. L. Ciesla, K. L. Winegarden & K. Y. Hostetler, 2003. Increased antiviral activity of 1-O-hexadecyloxypropyl-[2-(14)C]cidofovir in MRC-5 human lung fibroblasts is explained by unique cellular uptake and metabolism, Mol. Pharmacol. 63: 678–681.

Andrei, G., J. Balzarini, P. Fiten, E. De Clercq, G. Opdenakker & R. Snoeck, 2005a. Characterization of herpes simplex virus type 1 thymidine kinase mutants selected under a single round of high-dose brivudin, J. Virol. 79: 5863–5869.

Andrei, G., E. De Clercq & R. Snoeck, 2004. In vitro selection of drug-resistant varicella-zoster virus (VZV) mutants (OKA strain): differences between acyclovir and penciclovir ? Antiviral Res. 61: 181–187.

Andrei, G., P. Fiten, M. Froeyen, E. De Clercq, G. Opdenakker & R. Snoeck, 2007a. DNA polymerase mutations in drug-resistant herpes simplex virus mutants determine in vivo neurovirulence and drug-enzyme interactions, Antiviral. Ther. 12: 719–732.

Andrei, G., P. Fiten, P. Goubau, H. van Landuyt, B. Gordts, D. Selleslag, E. De Clercq, G. Opdenakker & R. Snoeck, 2007b. Dual infection with polyomavirus BK and acyclovir-resistant herpes simplex virus successfully treated with cidofovir in a bone marrow transplant recipient, Transpl. Infect. Dis. 9: 126–131.

Andrei, G., D. B. Gammon, P. Fiten, E. De Clercq, G. Opdenakker, R. Snoeck & D. H. Evans, 2006. Cidofovir resistance in vaccinia virus is linked to diminished virulence in mice, J. Virol. 80: 9391–9401.

Andrei, G., R. Sienaert, C. McGuigan, E. De Clercq, J. Balzarini & R. Snoeck, 2005b. Susceptibilities of several clinical varicella-zoster virus (VZV) isolates and drug-resistant VZV strains to bicyclic furano pyrimidine nucleosides, Antimicrob. Agents Chemother. 49: 1081–1086.

Andrei, G., R. Snoeck, E. De Clercq, R. Esnouf, P. Fiten & G. Opdenakker, 2000. Resistance of herpes simplex virus type 1 against different phosphonylmethoxyalkyl derivatives of purines and pyrimidines due to specific mutations in the viral DNA polymerase gene, J. Gen. Virol. 81: 639–648.

Ansari, A.& V. C. Emery, 1999. The U69 gene of human herpesvirus 6 encodes a protein kinase which can confer ganciclovir sensitivity to baculoviruses, J. Virol. 73: 3284–3291.

Appleton, B. A., J. Brooks, A. Loregian, D. J. Filman, D. M. Coen & J. M. Hogle, 2006. Crystal structure of the cytomegalovirus DNA polymerase subunit UL44 in complex with the C terminus from the catalytic subunit. Differences in structure and function relative to unliganded UL44, J. Biol. Chem. 281: 5224–5232.

Bacon, T. H., M. J. Levin, J. J. Leary, R. T. Sarisky & D. Sutton, 2003. Herpes simplex virus resistance to acyclovir and penciclovir after two decades of antiviral therapy, Clin. Microbiol. Rev. 16: 114–128.

Baldanti, F. & G. Gerna, 2003. Human cytomegalovirus resistance to antiviral drugs: diagnosis, monitoring and clinical impact, J. Antimicrob. Chemother. 52: 324–330.

Baldanti, F., M. R. Underwood, S. C. Stanat, K. K. Biron, S. Chou, A. Sarasini, E. Silini & G. Gerna, 1996. Single amino acid changes in the DNA polymerase confer foscarnet resistance and slow-growth phenotype, while mutations in the UL97-encoded phosphotransferase confer ganciclovir resistance in three double-resistant human cytomegalovirus strains recovered from patients with AIDS, J. Virol. 70: 1390–1395.

Balzarini, J., C. Pannecouque, L. Naesens, G. Andrei, R. Snoeck, E. De Clercq, D. Hockova & A. Holy, 2004. 6-[2-phosphonomethoxy)alkoxy]-2,4-diaminopyrimidines: a new class of acyclic pyrimidine nucleoside phosphonates with antiviral activity, Nucleosides Nucleotides Nucleic Acids 23: 1321–1327.

Balzarini, J., R. Sienaert, S. Liekens, A. Van Kuilenburg, A. Carangio, R. Esnouf, E. De Clercq & C. McGuigan, 2002. Lack of susceptibility of bicyclic nucleoside analogs, highly potent inhibitors of varicella-zoster virus, to the catabolic action of thymidine phosphorylase and dihydropyrimidine dehydrogenase, Mol. Pharmacol. 61: 1140–1145.

Beadle, J. R., C. Hartline, K. A. Aldern, N. Rodriguez, E. Harden, E. R. Kern & K. Y. Hostetler, 2002. Alkoxyalkyl esters of cidofovir and cyclic cidofovir exhibit multiple-log enhancement of antiviral activity against cytomegalovirus and herpesvirus replication in vitro, Antimicrob. Agents Chemother. 46: 2381–2386.

Beadle, J. R., W. B. Wan, S. L. Ciesla, K. A. Keith, C. Hartline, E. R. Kern & K. Y. Hostetler, 2006. Synthesis and antiviral evaluation of alkoxyalkyl derivatives of 9-(S)-(3-hydroxy-2-phosphonomethoxypropyl)adenine against cytomegalovirus and orthopoxviruses, J. Med. Chem. 49: 2010–2015.

Bestman-Smith, J.& G. Boivin, 2002. Herpes simplex virus isolates with reduced adefovir susceptibility selected in vivo by foscarnet therapy, J. Med. Virol. 67: 88–91.

Beutner, K. R., 1995. Valacyclovir: a review of its antiviral activity, pharmacokinetic properties, and clinical efficacy, Antiviral Res. 28: 281–290.

Beutner, K. R., D. J. Friedman, C. Forszpaniak, P. L. Andersen & M. J. Wood, 1995. Valacyclovir compared with acyclovir for improved therapy for herpes zoster in immunocompetent adults, Antimicrob. Agents Chemother. 39: 1546–1553.

Bidanset, D. J., J. R. Beadle, W. B. Wan, K. Y. Hostetler & E. R. Kern, 2004. Oral activity of ether lipid ester prodrugs of cidofovir against experimental human cytomegalovirus infection, J. Infect. Dis. 190: 499–503.

Biron, K. K., 2006. Antiviral drugs for cytomegalovirus diseases, Antiviral Res. 71: 154–163.

Biron, K. K., S. C. Stanat, J. B. Sorrell, J. A. Fyfe, P. M. Keller, C. U. Lambe & D. J. Nelson, 1985. Metabolic activation of the nucleoside analog 9-[(2-hydroxy-1-(hydroxymethyl)ethoxy]methyl)guanine in human diploid fibroblasts infected with human cytomegalovirus, Proc. Natl. Acad. Sci. U. S. A 82: 2473–2477.

Boehmer, P. E. & I. R. Lehman, 1997. Herpes simplex virus DNA replication, Annu. Rev. Biochem. 66: 347–384.

Boivin, G., C. K. Edelman, L. Pedneault, C. L. Talarico, K. K. Biron & H. H. Balfour, Jr., 1994. Phenotypic and genotypic characterization of acyclovir-resistant varicella-zoster viruses isolated from persons with AIDS, J. Infect. Dis. 170: 68–75.

Boivin, G., N. Goyette, C. Gilbert, N. Roberts, K. Macey, C. Paya, M. D. Pescovitz, A. Humar, E. Dominguez, K. Washburn, E. Blumberg, B. Alexander, R. Freeman, N. Heaton & E. Covington, 2004. Absence of cytomegalovirus-resistance mutations after valganciclovir prophylaxis, in a prospective multicenter study of solid-organ transplant recipients, J. Infect. Dis. 189: 1615–1618.

Bonnafous, P., L. Naesens, S. Petrella, A. Gautheret-Dejean, D. Boutolleau, W. Sougakoff & H. Agut, 2007. Different mutations in the HHV-6 DNA polymerase gene accounting for resistance to foscarnet, Antivir. Ther. 12: 877–888.

Boyd, M. R., T. H. Bacon, D. Sutton & M. Cole, 1987. Antiherpesvirus activity of 9-(4-hydroxy-3-hydroxy-methylbut-1-yl)guanine (BRL 39123) in cell culture, Antimicrob. Agents Chemother. 31: 1238–1242.

Brideau, R. J., M. L. Knechtel, A. Huang, V. A. Vaillancourt, E. E. Vera, N. L. Oien, T. A. Hopkins, J. L. Wieber, K. F. Wilkinson, B. D. Rush, F. J. Schwende & M. W. Wathen, 2002. Broad-spectrum antiviral activity of PNU-183792, a 4-oxo-dihydroquinoline, against human and animal herpesviruses, Antiviral Res. 54: 19–28.

Chakrabarti, S., D. Pillay, D. Ratcliffe, P. A. Cane, K. E. Collingham & D. W. Milligan, 2000. Resistance to antiviral drugs in herpes simplex virus infections among allogeneic stem cell transplant recipients: risk factors and prognostic significance, J. Infect. Dis. 181: 2055–2058.

Chen, Y., C. Scieux, V. Garrait, G. Socie, V. Rocha, J. M. Molina, D. Thouvenot, F. Morfin, L. Hocqueloux, L. Garderet, H. Esperou, F. Selimi, A. Devergie, G. Leleu, M. Aymard, F. Morinet, E. Gluckman & P. Ribaud, 2000. Resistant herpes simplex virus type 1 infection: an emerging concern after allogeneic stem cell transplantation, Clin. Infect. Dis. 31: 927–935.

Chibo, D., J. Druce, J. Sasadeusz & C. Birch, 2004. Molecular analysis of clinical isolates of acyclovir resistant herpes simplex virus, Antiviral Res. 61: 83–91.

Choo, H., J. R. Beadle, Y. Chong, J. Trahan & K. Y. Hostetler, 2007. Synthesis of the 5-phosphonopent-2-en-1-yl nucleosides: a new class of antiviral acyclic nucleoside phosphonates, Bioorg. Med. Chem. 15: 1771–1779.

Chou, S., 1999. Antiviral drug resistance in human cytomegalovirus, Transpl. Infect. Dis. 1: 105–114.

Chou, S., S. Guentzel, K. R. Michels, R. C. Miner & W. L. Drew, 1995. Frequency of UL97 phosphotransferase mutations related to ganciclovir resistance in clinical cytomegalovirus isolates, J. Infect. Dis. 172: 239–242.

Chou, S.& C. L. Meichsner, 2000. A nine-codon deletion mutation in the cytomegalovirus UL97 phosphotransferase gene confers resistance to ganciclovir, Antimicrob. Agents Chemother. 44: 183–185.

Chrisp, C. E., J. C. Sunstrum, D. R. Averill, Jr., M. Levine & J. C. Glorioso, 1989. Characterization of encephalitis in adult mice induced by intracerebral inoculation of herpes simplex virus type 1 (KOS) and comparison with mutants showing decreased virulence, Lab Invest 60: 822–830.

Christophers, J., J. Clayton, J. Craske, R. Ward, P. Collins, M. Trowbridge & G. Darby, 1998. Survey of resistance of herpes simplex virus to acyclovir in northwest England, Antimicrob. Agents Chemother. 42: 868–872.

Ciesla, S. L., J. Trahan, W. B. Wan, J. R. Beadle, K. A. Aldern, G. R. Painter & K. Y. Hostetler, 2003. Esterification of cidofovir with alkoxyalkanols increases oral bioavailability and diminishes drug accumulation in kidney, Antiviral Res. 59: 163–171.

Cihlar, T., M. D. Fuller & J. M. Cherrington, 1998a. Characterization of drug resistance-associated mutations in the human cytomegalovirus DNA polymerase gene by using recombinant mutant viruses generated from overlapping DNA fragments, J. Virol. 72: 5927–5936.

Cihlar, T., M. D. Fuller, A. S. Mulato & J. M. Cherrington, 1998b. A point mutation in the human cytomegalovirus DNA polymerase gene selected in vitro by cidofovir confers a slow replication phenotype in cell culture, Virology 248: 382–393.

Coen, D. M., 1991. The implications of resistance to antiviral agents for herpesvirus drug targets and drug therapy, Antiviral Res. 15: 287–300.

Coen, D. M., 1994. Acyclovir-resistant, pathogenic herpesviruses, Trends Microbiol. 2: 481–485.

Coen, D. M., M. Kosz-Vnenchak, J. G. Jacobson, D. A. Leib, C. L. Bogard, P. A. Schaffer, K. L. Tyler & D. M. Knipe, 1989. Thymidine kinase-negative herpes simplex virus mutants establish latency in mouse trigeminal ganglia but do not reactivate, Proc. Natl. Acad. Sci. U. S. A 86: 4736–4740.

Coen, D. M.& P. A. Schaffer, 2003. Antiherpesvirus drugs: a promising spectrum of new drugs and drug targets, Nat. Rev. Drug Discov. 2: 278–288.

Coen, D. M., P. A. Schaffer, P. A. Furman, P. M. Keller & M. H. St Clair, 1982. Biochemical and genetic analysis of acyclovir-resistant mutants of herpes simplex virus type 1, Am. J. Med. 73: 351–360.

Collins, P., B. A. Larder, N. M. Oliver, S. Kemp, I. W. Smith & G. Darby, 1989. Characterization of a DNA polymerase mutant of herpes simplex virus from a severely immunocompromised patient receiving acyclovir, J. Gen. Virol. 70 (Pt 2): 375–382.

Corey, L., A. Wald, R. Patel, S. L. Sacks, S. K. Tyring, T. Warren, J. M. Douglas, Jr., J. Paavonen, R. A. Morrow, K. R. Beutner, L. S. Stratchounsky, G. Mertz, O. N. Keene, H. A. Watson, D. Tait & M. Vargas-Cortes, 2004. Once-daily valacyclovir to reduce the risk of transmission of genital herpes, N. Engl. J. Med. 350: 11–20.

Crumpacker, C. S., 1992. Mechanism of action of foscarnet against viral polymerases, Am. J. Med. 92: 3S-7S.

Dal Pozzo, F., G. Andrei, I. Lebeau, J. R. Beadle, K. Y. Hostetler, E. De Clercq & R. Snoeck, 2007. In vitro evaluation of the anti-orf virus activity of alkoxyalkyl esters of CDV, cCDV and (S)-HPMPA, Antiviral Res. 75: 52–57.

Danve-Szatanek, C., M. Aymard, D. Thouvenot, F. Morfin, G. Agius, I. Bertin, S. Billaudel, B. Chanzy, M. Coste-Burel, L. Finkielsztejn, H. Fleury, T. Hadou, C. Henquell, H. Lafeuille, M. E. Lafon, A. Le Faou, M. C. Legrand, L. Maille, C. Mengelle, P. Morand, F. Morinet, E. Nicand, S. Omar, B. Picard, B. Pozzetto, J. Puel, D. Raoult, C. Scieux, M. Segondy, J. M. Seigneurin, R. Teyssou & C. Zandotti, 2004. Surveillance network for herpes simplex virus resistance to antiviral drugs: 3-year follow-up, J. Clin. Microbiol. 42: 242–249.

Darby, G., 1994. Acyclovir–and beyond, J. Int. Med. Res. 22 Suppl 1: 33A–42A.

Darby, G., M. J. Churcher & B. A. Larder, 1984. Cooperative effects between two acyclovir resistance loci in herpes simplex virus, J. Virol. 50: 838–846.

Darby, G., H. J. Field & S. A. Salisbury, 1981. Altered substrate specificity of herpes simplex virus thymidine kinase confers acyclovir-resistance, Nature 289: 81–83.

De Bolle, L., D. Michel, T. Mertens, C. Manichanh, H. Agut, E. De Clercq & L. Naesens, 2002. Role of the human herpesvirus 6 u69-encoded kinase in the phosphorylation of ganciclovir, Mol. Pharmacol. 62: 714–721.

De Clercq, E., 2003. Clinical potential of the acyclic nucleoside phosphonates cidofovir, adefovir, and tenofovir in treatment of DNA virus and retrovirus infections, Clin. Microbiol. Rev. 16: 569–596.

De Clercq, E., 2004. Discovery and development of BVDU (brivudin) as a therapeutic for the treatment of herpes zoster, Biochem. Pharmacol. 68: 2301–2315.

De Clercq, E., J. Descamps, P. De Somer & A. Holý, 1978. (S)-9-(2,3-Dihydroxypropyl)adenine: an aliphatic nucleoside analog with broad spectrum antiviral activity. Science 200: 563–565.

De Clercq, E., G. Andrei, J. Balzarini, P. Leyssen, L. Naesens, J. Neyts, C. Pannecouque, R. Snoeck, C. Ying, D. Hockova & A. Holý, 2005. Antiviral potential of a new generation of acyclic nucleoside phosphonates, the 6-[2-(phosphonomethoxy)alkoxy]-2,4-diaminopyrimidines, Nucleosides Nucleotides Nucleic Acids 24: 331–341.

De Clercq, E., J. Descamps, P. de Somer, P. J. Barr, A. S. Jones & R. T. Walker, 1979. (E)-5-(2-Bromovinyl)-2′-deoxyuridine: a potent and selective anti-herpes agent, Proc. Natl. Acad. Sci. U. S. A 76: 2947–2951.

De Clercq, E.& H. J. Field, 2006. Antiviral prodrugs - the development of successful prodrug strategies for antiviral chemotherapy, Br. J. Pharmacol. 147: 1–11.

De Clercq, E.& A. Holý, 2005. Acyclic nucleoside phosphonates: a key class of antiviral drugs, Nat. Rev. Drug Discov. 4: 928–940.

De Clercq, E., A. Holy, I. Rosenberg, T. Sakuma, J. Balzarini & P. C. Maudgal, 1986. A novel selective broad-spectrum anti-DNA virus agent, Nature 323: 464–467.

De Clercq, E., L. Naesens, L. De Bolle, D. Schols, Y. Zhang & J. Neyts, 2001. Antiviral agents active against human herpesviruses HHV-6, HHV-7 and HHV-8, Rev. Med. Virol. 11: 381–395.

DeFilippis, V., C. Raggo, A. Moses & K. Fruh, 2003. Functional genomics in virology and antiviral drug discovery, Trends Biotechnol. 21: 452–457.

DeJesus, E., A. Wald, T. Warren, T. W. Schacker, S. Trottier, M. Shahmanesh, J. L. Hill & C. A. Brennan, 2003. Valacyclovir for the suppression of recurrent genital herpes in human immunodeficiency virus-infected subjects, J. Infect. Dis. 188: 1009–1016.

Derse, D., Y. C. Cheng, P. A. Furman, M. H. St Clair & G. B. Elion, 1981. Inhibition of purified human and herpes simplex virus-induced DNA polymerases by 9-(2-hydroxyethoxymethyl)guanine triphosphate. Effects on primer-template function, J. Biol. Chem. 256: 11447–11451.

Desgranges, C., G. Razaka, E. De Clercq, P. Herdewijn, J. Balzarini, F. Drouillet & H. Bricaud, 1986. Effect of (E)-5-(2-bromovinyl)uracil on the catabolism and antitumor activity of 5-fluorouracil in rats and leukemic mice, Cancer Res. 46: 1094–1101.

Digard, P., W. R. Bebrin, K. Weisshart & D. M. Coen, 1993. The extreme C terminus of herpes simplex virus DNA polymerase is crucial for functional interaction with processivity factor UL42 and for viral replication, J. Virol. 67: 398–406.

Docherty, J. J., A. T. Dobson, J. J. Trimble & B. A. Jennings, 1991. Herpes simplex virus type 1 that exhibits herpes simplex virus type 2 sensitivity to (E)-5-(2-bromovinyl)-2′-deoxyuridine, Intervirology 32: 308–315.

Dykxhoorn, D. M.& J. Lieberman, 2006. Silencing viral infection, PLoS. Med. 3: e242.

Efstathiou, S., S. Kemp, G. Darby & A. C. Minson, 1989. The role of herpes simplex virus type 1 thymidine kinase in pathogenesis, J. Gen. Virol. 70 (Pt 4): 869–879.

Elion, G. B., P. A. Furman, J. A. Fyfe, P. de Miranda, L. Beauchamp & H. J. Schaeffer, 1977. Selectivity of action of an antiherpetic agent, 9-(2-hydroxyethoxymethyl) guanine, Proc. Natl. Acad. Sci. U. S. A 74: 5716–5720.

Englund, J. A., M. E. Zimmerman, E. M. Swierkosz, J. L. Goodman, D. R. Scholl & H. H. Balfour, Jr., 1990. Herpes simplex virus resistant to acyclovir. A study in a tertiary care center, Ann. Intern. Med. 112: 416–422.

Eriksson, B., B. Oberg & B. Wahren, 1982. Pyrophosphate analogues as inhibitors of DNA polymerases of cytomegalovirus, herpes simplex virus and cellular origin, Biochim. Biophys. Acta 696: 115–123.

Field, H. J.& D. M. Coen, 1986. Pathogenicity of herpes simplex virus mutants containing drug resistance mutations in the viral DNA polymerase gene, J. Virol. 60: 286–289.

Field, H. J. & G. Darby, 1980. Pathogenicity in mice of strains of herpes simplex virus which are resistant to acyclovir in vitro and in vivo, Antimicrob. Agents Chemother. 17:209–216.

Field, H. J.& P. Wildy, 1978. The pathogenicity of thymidine kinase-deficient mutants of herpes simplex virus in mice, J. Hyg. (Lond) 81: 267–277.

Furman, P. A., P. de Miranda, M. H. St Clair & G. B. Elion, 1981. Metabolism of acyclovir in virus-infected and uninfected cells, Antimicrob. Agents Chemother. 20: 518–524.

Furman, P. A., M. H. St Clair, J. A. Fyfe, J. L. Rideout, P. M. Keller & G. B. Elion, 1979. Inhibition of herpes simplex virus-induced DNA polymerase activity and viral DNA replication by 9-(2-hydroxyethoxymethyl) guanine and its triphosphate, J. Virol. 32: 72–77.

Furman, P. A., M. H. St Clair & T. Spector, 1984. Acyclovir triphosphate is a suicide inactivator of the herpes simplex virus DNA polymerase, J. Biol. Chem. 259: 9575–9579.

Fyfe, J. A., 1982. Differential phosphorylation of (E)-5-(2-bromovinyl)-2′-deoxyuridine monophosphate by thymidylate kinases from herpes simplex viruses types 1 and 2 and varicella zoster virus, Mol. Pharmacol. 21: 432–437.

Fyfe, J. A., P. M. Keller, P. A. Furman, R. L. Miller & G. B. Elion, 1978. Thymidine kinase from herpes simplex virus phosphorylates the new antiviral compound, 9-(2-hydroxyethoxymethyl)guanine, J. Biol. Chem. 253: 8721–8727.

Gaudreau, A., E. Hill, H. H. Balfour, Jr., A. Erice & G. Boivin, 1998. Phenotypic and genotypic characterization of acyclovir-resistant herpes simplex viruses from immunocompromised patients, J. Infect. Dis. 178: 297–303.

Gilbert, C., J. Bestman-Smith & G. Boivin, 2002. Resistance of herpesviruses to antiviral drugs: clinical impacts and molecular mechanisms, Drug Resist. Updat. 5: 88–114.

Gilbert, C., J. Handfield, E. Toma, R. Lalonde, M. G. Bergeron & G. Boivin, 1998. Emergence and prevalence of cytomegalovirus UL97 mutations associated with ganciclovir resistance in AIDS patients, AIDS 12: 125–129.

Grey, F., M. Sowa, P. Collins, R. J. Fenton, W. Harris, W. Snowden, S. Efstathiou & G. Darby, 2003. Characterization of a neurovirulent acyclovir-resistant variant of herpes simplex virus, J. Gen. Virol. 84: 1403–1410.

Griffiths, A., S. H. Chen, B. C. Horsburgh & D. M. Coen, 2003. Translational compensation of a frameshift mutation affecting herpes simplex virus thymidine kinase is sufficient to permit reactivation from latency, J. Virol. 77: 4703–4709.

Griffiths, A.& D. M. Coen, 2003. High-frequency phenotypic reversion and pathogenicity of an acyclovir-resistant herpes simplex virus mutant, J. Virol. 77: 2282–2286.

Guo, A., P. Hu, P. V. Balimane, F. H. Leibach & P. J. Sinko, 1999. Interactions of a nonpeptidic drug, valacyclovir, with the human intestinal peptide transporter (hPEPT1) expressed in a mammalian cell line, J. Pharmacol. Exp. Ther. 289: 448–454.

Gupta, R., A. Wald, E. Krantz, S. Selke, T. Warren, M. Vargas-Cortes, G. Miller & L. Corey, 2004. Valacyclovir and acyclovir for suppression of shedding of herpes simplex virus in the genital tract, J. Infect. Dis. 190: 1374–1381.

Hannon, G. J., 2002. RNA interference, Nature 418: 244–251.

Harris, W., P. Collins, R. J. Fenton, W. Snowden, M. Sowa & G. Darby, 2003. Phenotypic and genotypic characterization of clinical isolates of herpes simplex virus resistant to aciclovir, J. Gen. Virol. 84: 1393–1401.

Hartline, C. B., K. M. Gustin, W. B. Wan, S. L. Ciesla, J. R. Beadle, K. Y. Hostetler & E. R. Kern, 2005. Ether lipid-ester prodrugs of acyclic nucleoside phosphonates: activity against adenovirus replication in vitro, J. Infect. Dis. 191: 396–399.

Hasegawa, T., M. Kurokawa, T. A. Yukawa, M. Horii & K. Shiraki, 1995. Inhibitory action of acyclovir (ACV) and penciclovir (PCV) on plaque formation and partial cross-resistance of ACV-resistant varicella-zoster virus to PCV, Antiviral Res. 27: 271–279.

Hill, E. L., G. A. Hunter & M. N. Ellis, 1991. In vitro and in vivo characterization of herpes simplex virus clinical isolates recovered from patients infected with human immunodeficiency virus, Antimicrob. Agents Chemother. 35: 2322–2328.

Ho, H. T., K. L. Woods, J. J. Bronson, H. De Boeck, J. C. Martin & M. J. Hitchcock, 1992. Intracellular metabolism of the antiherpes agent (S)-1-[3-hydroxy-2-(phosphonylmethoxy) propyl]cytosine, Mol. Pharmacol. 41: 197–202.

Hockova, D., A. Holý, M. Masojidkova, G. Andrei, R. Snoeck, E. De Clercq & J. Balzarini, 2003. 5-Substituted-2,4-diamino-6-[2-(phosphonomethoxy)ethoxy]pyrimidines-acycli c nucleoside phosphonate analogues with antiviral activity, J. Med. Chem. 46: 5064–5073.

Hockova, D., A. Holý, M. Masojidkova, G. Andrei, R. Snoeck, E. De Clercq & J. Balzarini, 2004. Synthesis and antiviral activity of 2,4-diamino-5-cyano-6-[2-(phosphonomethoxy) ethoxy]pyrimidine and related compounds, Bioorg. Med. Chem. 12: 3197–3202.

Hodson, E. M., C. A. Jones, A. C. Webster, G. F. Strippoli, P. G. Barclay, K. Kable, D. Vimalachandra & J. C. Craig, 2005. Antiviral medications to prevent cytomegalovirus disease and early death in recipients of solid-organ transplants: a systematic review of randomised controlled trials, Lancet 365: 2105–2115.

Horsburgh, B. C., S. H. Chen, A. Hu, G. B. Mulamba, W. H. Burns & D. M. Coen, 1998. Recurrent acyclovir-resistant herpes simplex in an immunocompromised patient: can strain differences compensate for loss of thymidine kinase in pathogenesis?, J. Infect. Dis. 178: 618–625.

Hostetler, K. Y., S. Rought, K. A. Aldern, J. Trahan, J. R. Beadle & J. Corbeil, 2006. Enhanced antiproliferative effects of alkoxyalkyl esters of cidofovir in human cervical cancer cells in vitro, Mol. Cancer Ther. 5: 156–159.

Huang, L., K. K. Ishii, H. Zuccola, A. M. Gehring, C. B. Hwang, J. Hogle & D. M. Coen, 1999. The enzymological basis for resistance of herpesvirus DNA polymerase mutants to acyclovir: relationship to the structure of alpha-like DNA polymerases, Proc. Natl. Acad. Sci. U. S. A 96: 447–452.

Hwang, C. B.& H. J. Chen, 1995. An altered spectrum of herpes simplex virus mutations mediated by an antimutator DNA polymerase, Gene 152: 191–193.

Hwang, C. B., B. Horsburgh, E. Pelosi, S. Roberts, P. Digard & D. M. Coen, 1994. A net +1 frameshift permits synthesis of thymidine kinase from a drug-resistant herpes simplex virus mutant, Proc. Natl. Acad. Sci. U. S. A 91: 5461–5465.

Hwang, C. B., K. L. Ruffner & D. M. Coen, 1992. A point mutation within a distinct conserved region of the herpes simplex virus DNA polymerase gene confers drug resistance, J. Virol. 66: 1774–1776.

Ida, M., S. Kageyama, H. Sato, T. Kamiyama, J. Yamamura, M. Kurokawa, M. Morohashi & K. Shiraki, 1999. Emergence of resistance to acyclovir and penciclovir in varicella-zoster virus and genetic analysis of acyclovir-resistant variants, Antiviral Res. 40: 155–166.

Ilsley, D. D., S. H. Lee, W. H. Miller & R. D. Kuchta, 1995. Acyclic guanosine analogs inhibit DNA polymerases alpha, delta, and epsilon with very different potencies and have unique mechanisms of action, Biochemistry 34: 2504–2510.

Iwayama, S., Y. Ohmura, K. Suzuki, N. Ono, H. Nakazawa, M. Aoki, I. Tanabe, T. Sekiyama, T. Tsuji, M. Okunishi, K. Yamanishi & Y. Nishiyama, 1999. Evaluation of anti-herpesvirus activity of (1′S,2′R)-9-[[1′,2′-bis(hydroxymethyl)cycloprop-1′-yl]methyl]- guanine (A-5021) in mice, Antiviral Res. 42: 139–148.

Iwayama, S., N. Ono, Y. Ohmura, K. Suzuki, M. Aoki, H. Nakazawa, M. Oikawa, T. Kato, M. Okunishi, Y. Nishiyama & K. Yamanishi, 1998. Antiherpesvirus activities of (1′S,2′R)-9-[[1′,2′-bis(hydroxymethyl)cycloprop-1′-yl]methyl]guanine (A-5021) in cell culture, Antimicrob. Agents Chemother. 42: 1666–1670.

Jabs, D. A., C. Enger, M. Forman & J. P. Dunn, 1998. Incidence of foscarnet resistance and cidofovir resistance in patients treated for cytomegalovirus retinitis. The cytomegalovirus retinitis and viral resistance study group, Antimicrob. Agents Chemother. 42: 2240–2244.

Jacobson, J. G., K. L. Ruffner, M. Kosz-Vnenchak, C. B. Hwang, K. K. Wobbe, D. M. Knipe & D. M. Coen, 1993. Herpes simplex virus thymidine kinase and specific stages of latency in murine trigeminal ganglia, J. Virol. 67: 6903–6908.

Jiang, C., Y. T. Hwang, J. C. Randell, D. M. Coen & C. B. Hwang, 2007a. Mutations that decrease DNA binding of the processivity factor of the herpes simplex virus DNA polymerase reduce viral yield, alter the kinetics of viral DNA replication, and decrease the fidelity of DNA replication, J. Virol. 81: 3495–3502.

Jiang, C., Y. T. Hwang, G. Wang, J. C. Randell, D. M. Coen & C. B. Hwang, 2007b. Herpes simplex virus mutants with multiple substitutions affecting DNA binding of UL42 are impaired for viral replication and DNA synthesis, J. Virol. 81: 12077–12079.

Kamiyama, T., M. Kurokawa & K. Shiraki, 2001. Characterization of the DNA polymerase gene of varicella-zoster viruses resistant to acyclovir, J. Gen. Virol. 82: 2761–2765.

Kaufman, H., E. L. Martola & C. Dohlman, 1962. Use of 5-iodo-2′-deoxyuridine (IDU) in treatment of herpes simplex keratitis, Arch. Ophthalmol. 68: 235–239.

Kaufman, H. E., 1962. Clinical cure of herpes simplex keratitis by 5-iodo-2-deoxyuridine, Proc. Soc. Exp. Biol. Med. 109: 251–252.

Kaufman, H. E., 1963. Chemotherapy of Herpes Keratitis, Invest Ophthalmol. 2: 504–518.

Kaufman, H. E.& C. Heidelberger, 1964. Therapeutic Antiviral Action of 5-trifluoromethyl-2′-deoxyuridine in herpes simplex keratitis, Science 145: 585–586.

Keith, K. A., W. B. Wan, S. L. Ciesla, J. R. Beadle, K. Y. Hostetler & E. R. Kern, 2004. Inhibitory activity of alkoxyalkyl and alkyl esters of cidofovir and cyclic cidofovir against orthopoxvirus replication in vitro, Antimicrob. Agents Chemother. 48: 1869–1871.

Keller, P. M., J. A. Fyfe, L. Beauchamp, C. M. Lubbers, P. A. Furman, H. J. Schaeffer & G. B. Elion, 1981. Enzymatic phosphorylation of acyclic nucleoside analogs and correlations with antiherpetic activities, Biochem. Pharmacol. 30: 3071–3077.

Kern, E. R., D. J. Collins, W. B. Wan, J. R. Beadle, K. Y. Hostetler & D. C. Quenelle, 2004. Oral treatment of murine cytomegalovirus infections with ether lipid esters of cidofovir, Antimicrob. Agents Chemother. 48: 3516–3522.

Kinchington, P. R., T. Araullo-Cruz, J. P. Vergnes, K. Yates & Y. J. Gordon, 2002. Sequence changes in the human adenovirus type 5 DNA polymerase associated with resistance to the broad spectrum antiviral cidofovir, Antiviral Res. 56: 73–84.

Kleymann, G., 2005. Agents and strategies in development for improved management of herpes simplex virus infection and disease, Expert. Opin. Investig. Drugs 14: 135–161.

Knopf, K. W., 1979. Properties of herpes simplex virus DNA polymerase and characterization of its associated exonuclease activity, Eur. J. Biochem. 98: 231–244.

Krecmerova, M., A. Holý, A. Piskala, M. Masojidkova, G. Andrei, L. Naesens, J. Neyts, J. Balzarini, E. De Clercq & R. Snoeck, 2007a. Antiviral activity of triazine analogues of 1-(S)-[3-hydroxy-2-(phosphonomethoxy)propyl]cytosine (cidofovir) and related compounds, J. Med. Chem. 50: 1069–1077.

Krecmerova, M., A. Holý, R. Pohl, M. Masojidkova, G. Andrei, L. Naesens, J. Neyts, J. Balzarini, E. De Clercq & R. Snoeck, 2007b. Ester Prodrugs of Cyclic 1-(S)- [3-Hydroxy-2-(phosphonomethoxy)propyl]-5-azacytosine: Synthesis and antiviral activity, J. Med. Chem. 50(23): 5765–5772.

Larsen, S. D., Z. Zhang, B. A. DiPaolo, P. R. Manninen, D. C. Rohrer, M. J. Hageman, T. A. Hopkins, M. L. Knechtel, N. L. Oien, B. D. Rush, F. J. Schwende, K. J. Stefanski, J. L. Wieber, K. F. Wilkinson, K. M. Zamora, M. W. Wathen & R. J. Brideau, 2007. 7-Oxo-4,7-dihydrothieno[3,2-b]pyridine-6-carboxamides: synthesis and biological activity of a new class of highly potent inhibitors of human cytomegalovirus DNA polymerase, Bioorg. Med. Chem. Lett. 17:3840–3844.

Lebeau, I., G. Andrei, F. Dal Pozzo, J. R. Beadle, K. Y. Hostetler, E. De Clercq, O. J. van den Oord & R. Snoeck, 2006. Activities of alkoxyalkyl esters of cidofovir (CDV), cyclic CDV, and (S)-9-(3-hydroxy-2-phosphonylmethoxypropyl)adenine against orthopoxviruses in cell monolayers and in organotypic cultures, Antimicrob. Agents Chemother. 50: 2525–2529.

Leonard, J. N.& D. V. Schaffer, 2006. Antiviral RNAi therapy: emerging approaches for hitting a moving target, Gene Ther. 13: 532–540.

Littler, E., A. D. Stuart & M. S. Chee, 1992. Human cytomegalovirus UL97 open reading frame encodes a protein that phosphorylates the antiviral nucleoside analogue ganciclovir, Nature 358: 160–162.

Liu, S., J. D. Knafels, J. S. Chang, G. A. Waszak, E. T. Baldwin, M. R. Deibel, Jr., D. R. Thomsen, F. L. Homa, P. A. Wells, M. C. Tory, R. A. Poorman, H. Gao, X. Qiu & A. P. Seddon, 2006. Crystal structure of the herpes simplex virus 1 DNA polymerase, J. Biol. Chem. 281: 18193–18200.

LoPresti, A. E., J. F. Levine, G. B. Munk, C. Y. Tai & D. B. Mendel, 1998. Successful treatment of an acyclovir- and foscarnet-resistant herpes simplex virus type 1 lesion with intravenous cidofovir, Clin. Infect. Dis. 26: 512–513.

Loregian, A., B. A. Appleton, J. M. Hogle & D. M. Coen, 2004. Specific residues in the connector loop of the human cytomegalovirus DNA polymerase accessory protein UL44 are crucial for interaction with the UL54 catalytic subunit, J. Virol. 78: 9084–9092.

Loregian, A., R. Rigatti, M. Murphy, E. Schievano, G. Palu & H. S. Marsden, 2003. Inhibition of human cytomegalovirus DNA polymerase by C-terminal peptides from the UL54 subunit, J. Virol. 77: 8336–8344.

Lowance, D., H. H. Neumayer, C. M. Legendre, J. P. Squifflet, J. Kovarik, P. J. Brennan, D. Norman, R. Mendez, M. R. Keating, G. L. Coggon, A. Crisp & I. C. Lee, 1999. Valacyclovir for the prevention of cytomegalovirus disease after renal transplantation. International Valacyclovir Cytomegalovirus Prophylaxis Transplantation Study Group, N. Engl. J. Med. 340: 1462–1470.

Lurain, N. S., A. Weinberg, C. S. Crumpacker & S. Chou, 2001. Sequencing of cytomegalovirus UL97 gene for genotypic antiviral resistance testing, 1, Antimicrob. Agents Chemother. 45: 2775–2780.

Manichanh, C., C. Olivier-Aubron, J. P. Lagarde, J. T. Aubin, P. Bossi, A. Gautheret-Dejean, J. M. Huraux & H. Agut, 2001. Selection of the same mutation in the U69 protein kinase gene of human herpesvirus-6 after prolonged exposure to ganciclovir in vitro and in vivo, J. Gen. Virol. 82: 2767–2776.

Martin, J. L., C. E. Brown, N. Matthews-Davis & J. E. Reardon, 1994. Effects of antiviral nucleoside analogs on human DNA polymerases and mitochondrial DNA synthesis, Antimicrob. Agents Chemother. 38: 2743–2749.

McGuigan, C. & J. Balzarini, 2006. Aryl furano pyrimidines: the most potent and selective anti-VZV agents reported to date, Antiviral Res. 71: 149–153.

McGuigan, C., R. N. Pathirana, M. Migliore, R. Adak, G. Luoni, A. T. Jones, A. Diez-Torrubia, M. J. Camarasa, S. Velazquez, G. Henson, E. Verbeken, R. Sienaert, L. Naesens, R. Snoeck, G. Andrei & J. Balzarini, 2007. Preclinical development of bicyclic nucleoside analogues as potent and selective inhibitors of varicella zoster virus, J. Antimicrob. Chemother. 60(6): 1316–1330.

McGuigan, C., C. J. Yarnold, G. Jones, S. Velazquez, H. Barucki, A. Brancale, G. Andrei, R. Snoeck, E. De Clercq & J. Balzarini, 1999. Potent and selective inhibition of varicella-zoster virus (VZV) by nucleoside analogues with an unusual bicyclic base, J. Med. Chem. 42: 4479–4484.

McGuirt, P. V., J. E. Shaw, G. B. Elion & P. A. Furman, 1984. Identification of small DNA fragments synthesized in herpes simplex virus-infected cells in the presence of acyclovir, Antimicrob. Agents Chemother. 25: 507–509.

Meerbach, A., A. Holý, P. Wutzler, E. De Clercq & J. Neyts, 1998. Inhibitory effects of novel nucleoside and nucleotide analogues on Epstein-Barr virus replication, Antivir. Chem. Chemother. 9: 275–282.

Mertz, G. J., M. O. Loveless, M. J. Levin, S. J. Kraus, S. L. Fowler, D. Goade & S. K. Tyring, 1997. Oral famciclovir for suppression of recurrent genital herpes simplex virus infection in women. A multicenter, double-blind, placebo-controlled trial. Collaborative Famciclovir Genital Herpes Research Group, Arch. Intern. Med. 157: 343–349.

Michel, D. & T. Mertens, 2004. The UL97 protein kinase of human cytomegalovirus and homologues in other herpesviruses: impact on virus and host, Biochim. Biophys. Acta 1697: 169–180.

Miller, W. H. & R. L. Miller, 1980. Phosphorylation of acyclovir (acycloguanosine) monophosphate by GMP kinase, J. Biol. Chem. 255: 7204–7207.

Miller, W. H. & R. L. Miller, 1982. Phosphorylation of acyclovir diphosphate by cellular enzymes, Biochem. Pharmacol. 31: 3879–3884.

Morfin, F. & D. Thouvenot, 2003. Herpes simplex virus resistance to antiviral drugs, J. Clin. Virol. 26: 29–37.

Morfin, F., D. Thouvenot, M. Aymard & G. Souillet, 2000. Reactivation of acyclovir-resistant thymidine kinase-deficient herpes simplex virus harbouring single base insertion within a 7 Gs homopolymer repeat of the thymidine kinase gene, J. Med. Virol. 62: 247–250.

Morfin, F., D. Thouvenot, M. Turenne-Tessier, B. Lina, M. Aymard & T. Ooka, 1999. Phenotypic and genetic characterization of thymidine kinase from clinical strains of varicella-zoster virus resistant to acyclovir, Antimicrob. Agents Chemother. 43: 2412–2416.

Neyts, J., G. Andrei, R. Snoeck, G. Jahne, I. Winkler, M. Helsberg, J. Balzarini & E. De Clercq, 1994. The N-7-substituted acyclic nucleoside analog 2-amino-7-[(1,3-dihydroxy-2-propoxy)methyl]purine is a potent and selective inhibitor of herpesvirus replication, Antimicrob. Agents Chemother. 38: 2710–2716.

Neyts, J., J. Balzarini, G. Andrei, Z. Chaoyong, R. Snoeck, A. Zimmermann, T. Mertens, A. Karlsson & E. De Clercq, 1998. Intracellular metabolism of the N7-substituted acyclic nucleoside analog 2-amino-7-(1,3-dihydroxy-2-propoxymethyl)purine, a potent inhibitor of herpesvirus replication, Mol. Pharmacol. 53: 157–165.

Neyts, J. & E. De Clercq, 1997. Antiviral drug susceptibility of human herpesvirus 8, Antimicrob. Agents Chemother. 41: 2754–2756.

Neyts, J. & E. De Clercq, 2001. Efficacy of 2-amino-7-(1,3-dihydroxy-2-propoxymethyl)purine for treatment of vaccinia virus (orthopoxvirus) infections in mice, Antimicrob. Agents Chemother. 45: 84–87.

Neyts, J., G. Jahne, G. Andrei, R. Snoeck, I. Winkler & E. De Clercq, 1995. In vivo antiherpesvirus activity of N-7-substituted acyclic nucleoside analog 2-amino-7-[(1,3-dihydroxy-2-propoxy)methyl]purine, Antimicrob. Agents Chemother. 39: 56–60.

Ng, T. I., Y. Shi, H. J. Huffaker, W. Kati, Y. Liu, C. M. Chen, Z. Lin, C. Maring, W. E. Kohlbrenner & A. Molla, 2001. Selection and characterization of varicella-zoster virus variants resistant to (R)-9-[4-hydroxy-2-(hydroxymethy)butyl]guanine, Antimicrob. Agents Chemother. 45: 1629–1636.

Nugier, F., J. N. Colin, M. Aymard & M. Langlois, 1992. Occurrence and characterization of acyclovir-resistant herpes simplex virus isolates: report on a two-year sensitivity screening survey, J. Med. Virol. 36: 1–12.

Oien, N. L., R. J. Brideau, T. A. Hopkins, J. L. Wieber, M. L. Knechtel, J. A. Shelly, R. A. Anstadt, P. A. Wells, R. A. Poorman, A. Huang, V. A. Vaillancourt, T. L. Clayton, J. A. Tucker & M. W. Wathen, 2002. Broad-spectrum antiherpes activities of 4-hydroxyquinoline carboxamides, a novel class of herpesvirus polymerase inhibitors, Antimicrob. Agents Chemother. 46: 724–730.

Okuda, H., T. Nishiyama, K. Ogura, S. Nagayama, K. Ikeda, S. Yamaguchi, Y. Nakamura, K. Kawaguchi, T. Watabe & Y. Ogura, 1997. Lethal drug interactions of sorivudine, a new antiviral drug, with oral 5-fluorouracil prodrugs, Drug Metab. Dispos. 25: 270–273.

Okuda, H., K. Ogura, A. Kato, H. Takubo & T. Watabe, 1998. A possible mechanism of eighteen patient deaths caused by interactions of sorivudine, a new antiviral drug, with oral 5-fluorouracil prodrugs, J. Pharmacol. Exp. Ther. 287: 791–799.

Ono, N., S. Iwayama, K. Suzuki, T. Sekiyama, H. Nakazawa, T. Tsuji, M. Okunishi, T. Daikoku & Y. Nishiyama, 1998. Mode of action of (1′S,2′R)-9-[[1′,2′-bis(hydroxymethyl) cycloprop-1′-yl]methyl]guanine (A-5021) against herpes simplex virus type 1 and type 2 and varicella-zoster virus, Antimicrob. Agents Chemother. 42: 2095–2102.

Ostrander, M.& Y. C. Cheng, 1980. Properties of herpes simplex virus type 1 and type 2 DNA polymerase, Biochim. Biophys. Acta 609: 232–245.

Pahwa, S., K. Biron, W. Lim, P. Swenson, M. H. Kaplan, N. Sadick & R. Pahwa, 1988. Continuous varicella-zoster infection associated with acyclovir resistance in a child with AIDS, JAMA 260: 2879–2882.

Painter, G. R.& K. Y. Hostetler, 2004. Design and development of oral drugs for the prophylaxis and treatment of smallpox infection, Trends Biotechnol. 22: 423–427.

Pelosi, E., G. B. Mulamba & D. M. Coen, 1998a. Penciclovir and pathogenesis phenotypes of drug-resistant Herpes simplex virus mutants, Antiviral Res. 37: 17–28.

Pelosi, E., F. Rozenberg, D. M. Coen & K. L. Tyler, 1998b. A herpes simplex virus DNA polymerase mutation that specifically attenuates neurovirulence in mice, Virology 252: 364–372.

Perry, C. M. & D. Faulds, 1996. Valaciclovir. A review of its antiviral activity, pharmacokinetic properties and therapeutic efficacy in herpesvirus infections, Drugs 52: 754–772.

Perry, C. M. & A. J. Wagstaff, 1995. Famciclovir. A review of its pharmacological properties and therapeutic efficacy in herpesvirus infections, Drugs 50: 396–415.

Pottage, J. C., Jr. & H. A. Kessler, 1995. Herpes simplex virus resistance to acyclovir: clinical relevance, Infect. Agents Dis. 4: 115–124.

Quenelle, D. C., D. J. Collins, W. B. Wan, J. R. Beadle, K. Y. Hostetler & E. R. Kern, 2004. Oral treatment of cowpox and vaccinia virus infections in mice with ether lipid esters of cidofovir, Antimicrob. Agents Chemother. 48: 404–412.

Rashidi, M. R., J. A. Smith, S. E. Clarke & C. Beedham, 1997. In vitro oxidation of famciclovir and 6-deoxypenciclovir by aldehyde oxidase from human, guinea pig, rabbit, and rat liver, Drug Metab Dispos. 25: 805–813.

Razonable, R. R. & V. C. Emery, 2004. Management of CMV infection and disease in transplant patients. 27–29 February 2004, Herpes. 11: 77–86.

Reardon, J. E., 1989. Herpes simplex virus type 1 and human DNA polymerase interactions with 2′-deoxyguanosine 5′-triphosphate analogues. Kinetics of incorporation into DNA and induction of inhibition, J. Biol. Chem. 264: 19039–19044.

Reid, R., E. C. Mar, E. S. Huang & M. D. Topal, 1988. Insertion and extension of acyclic, dideoxy, and ara nucleotides by herpesviridae, human alpha and human beta polymerases. A unique inhibition mechanism for 9-(1,3-dihydroxy-2-propoxymethyl)guanine triphosphate, J. Biol. Chem. 263: 3898–3904.

Ruiz, J. C., K. A. Aldern, J. R. Beadle, C. B. Hartline, E. R. Kern & K. Y. Hostetler, 2006. Synthesis and antiviral evaluation of alkoxyalkyl esters of phosphonopropoxymethyl-guanine and phosphonopropoxymethyl-diaminopurine, Antivir. Chem. Chemother. 17: 89–95.

Ruiz, J. C., J. R. Beadle, K. A. Aldern, K. A. Keith, C. B. Hartline, E. R. Kern & K. Y. Hostetler, 2007. Synthesis and antiviral evaluation of alkoxyalkyl-phosphate conjugates of cidofovir and adefovir, Antiviral Res. 75: 87–90.

Sacks, S. L., F. Y. Aoki, A. Y. Martel, S. D. Shafran & M. Lassonde, 2005. Clinic-initiated, twice-daily oral famciclovir for treatment of recurrent genital herpes: a randomized, double-blind, controlled trial, Clin. Infect. Dis. 41: 1097–1104.

Sacks, S. L., R. J. Wanklin, D. E. Reece, K. A. Hicks, K. L. Tyler & D. M. Coen, 1989. Progressive esophagitis from acyclovir-resistant herpes simplex. Clinical roles for DNA polymerase mutants and viral heterogeneity? Ann. Intern. Med. 111: 893–899.

Safrin, S., J. Cherrington & H. S. Jaffe, 1997. Clinical uses of cidofovir, Rev. Med. Virol. 7: 145–156.

Safronetz, D., M. Petric, R. Tellier, B. Parvez & G. A. Tipples, 2003. Mapping ganciclovir resistance in the human herpesvirus-6 U69 protein kinase, J. Med. Virol. 71: 434–439.

Sarisky, R. T., M. R. Quail, P. E. Clark, T. T. Nguyen, W. S. Halsey, R. J. Wittrock, B. J. O′Leary, M. M. Van Horn, G. M. Sathe, S. Van Horn, M. D. Kelly, T. H. Bacon & J. J. Leary, 2001. Characterization of herpes simplex viruses selected in culture for resistance to penciclovir or acyclovir, J. Virol. 75: 1761–1769.

Sasadeusz, J. J. & S. L. Sacks, 1996. Spontaneous reactivation of thymidine kinase-deficient, acyclovir-resistant type-2 herpes simplex virus: masked heterogeneity or reversion?, J. Infect. Dis. 174: 476–482.

Sasadeusz, J. J., F. Tufaro, S. Safrin, K. Schubert, M. M. Hubinette, P. K. Cheung & S. L. Sacks, 1997. Homopolymer mutational hot spots mediate herpes simplex virus resistance to acyclovir, J. Virol. 71: 3872–3878.

Schaeffer, H. J., L. Beauchamp, P. de Miranda, G. B. Elion, D. J. Bauer & P. Collins, 1978. 9-(2-hydroxyethoxymethyl) guanine activity against viruses of the herpes group, Nature 272: 583–585.

Schnute, M. E., D. J. Anderson, R. J. Brideau, F. L. Ciske, S. A. Collier, M. M. Cudahy, M. Eggen, M. J. Genin, T. A. Hopkins, T. M. Judge, E. J. Kim, M. L. Knechtel, S. K. Nair, J. A. Nieman, N. L. Oien, A. Scott, S. P. Tanis, V. A. Vaillancourt, M. W. Wathen & J. L. Wieber, 2007. 2-Aryl-2-hydroxyethylamine substituted 4-oxo-4,7-dihydrothieno[2,3-b]pyridines as broad-spectrum inhibitors of human herpesvirus polymerases, Bioorg. Med. Chem. Lett. 17: 3349–3353.

Schnute, M. E., M. M. Cudahy, R. J. Brideau, F. L. Homa, T. A. Hopkins, M. L. Knechtel, N. L. Oien, T. W. Pitts, R. A. Poorman, M. W. Wathen & J. L. Wieber, 2005. 4-Oxo-4,7-dihydrothieno[2,3-b]pyridines as non-nucleoside inhibitors of human cytomegalovirus and related herpesvirus polymerases, J. Med. Chem. 48: 5794–5804.

Scott, G. M., M. A. Isaacs, F. Zeng, A. M. Kesson & W. D. Rawlinson, 2004. Cytomegalovirus antiviral resistance associated with treatment induced UL97 (protein kinase) and UL54 (DNA polymerase) mutations, J. Med. Virol. 74: 85–93.

Shi, R., A. Azzi, C. Gilbert, G. Boivin & S. X. Lin, 2006. Three-dimensional modeling of cytomegalovirus DNA polymerase and preliminary analysis of drug resistance, Proteins 64: 301–307.

Shin, M. C., S. K. Hong, J. S. Yoon, S. S. Park, S. G. Lee, D. G. Lee, W. S. Min, W. S. Shin & S. Y. Paik, 2006. Inhibition of UL54 and UL97 genes of human cytomegalovirus by RNA interference, Acta Virol. 50: 263–268.

Sienaert, R., G. Andrei, R. Snoeck, E. De Clercq, C. McGuigan & J. Balzarini, 2004. Inactivity of the bicyclic pyrimidine nucleoside analogues against simian varicella virus (SVV) does not correlate with their substrate activity for SVV-encoded thymidine kinase, Biochem. Biophys. Res. Commun. 315: 877–883.

Sienaert, R., L. Naesens, A. Brancale, A. Carangio, G. Andrei, R. Snoeck, A. Van Kuilenburg, E. De Clercq, C. McGuigan & J. Balzarini, 2003. Metabolic and pharmacological characteristics of the bicyclic nucleoside analogues (BCNAs) as highly selective inhibitors of varicella-zoster virus (VZV), Nucleosides Nucleotides Nucleic Acids 22: 995–997.

Sienaert, R., L. Naesens, A. Brancale, E. De Clercq, C. McGuigan & J. Balzarini, 2002. Specific recognition of the bicyclic pyrimidine nucleoside analogs, a new class of highly potent and selective inhibitors of varicella-zoster virus (VZV), by the VZV-encoded thymidine kinase, Mol. Pharmacol. 61: 249–254.

Silva, J. M., S. M. Hammond & G. J. Hannon, 2002. RNA interference: a promising approach to antiviral therapy?, Trends Mol. Med. 8: 505–508.

Simpson, D. & K. A. Lyseng-Williamson, 2006. Famciclovir: a review of its use in herpes zoster and genital and orolabial herpes, Drugs 66: 2397–2416.

Skoldenberg, B., M. Forsgren, K. Alestig, T. Bergstrom, L. Burman, E. Dahlqvist, A. Forkman, A. Fryden, K. Lovgren & K. Norlin, 1984. Acyclovir versus vidarabine in herpes simplex encephalitis. Randomised multicentre study in consecutive Swedish patients, Lancet 2: 707–711.

Smee, D. F., K. W. Bailey & R. W. Sidwell, 2002. Treatment of lethal cowpox virus respiratory infections in mice with 2-amino-7-[(1,3-dihydroxy-2-propoxy)methyl]purine and its orally active diacetate ester prodrug, Antiviral Res. 54: 113–120.

Smith, I. L., J. M. Cherrington, R. E. Jiles, M. D. Fuller, W. R. Freeman & S. A. Spector, 1997. High-level resistance of cytomegalovirus to ganciclovir is associated with alterations in both the UL97 and DNA polymerase genes, J. Infect. Dis. 176: 69–77.

Snoeck, R., G. Andrei, M. Gerard, A. Silverman, A. Hedderman, J. Balzarini, C. Sadzot-Delvaux, G. Tricot, N. Clumeck & E. De Clercq, 1994a. Successful treatment of progressive mucocutaneous infection due to acyclovir- and foscarnet-resistant herpes simplex virus with (S)-1-(3-hydroxy-2-phosphonylmethoxypropyl)cytosine (HPMPC), Clin. Infect. Dis. 18: 570–578.

Snoeck, R. & E. De Clercq, 2002. Role of cidofovir in the treatment of DNA virus infections, other than CMV infections, in immunocompromised patients, Curr. Opin. Investig. Drugs 3: 1561–1566.

Snoeck, R., M. Gerard, C. Sadzot-Delvaux, G. Andrei, J. Balzarini, D. Reymen, N. Ahadi, J. M. De Bruyn, J. Piette & B. Rentier, 1994b. Meningoradiculoneuritis due to acyclovir-resistant varicella zoster virus in an acquired immune deficiency syndrome patient, J. Med. Virol. 42: 338–347.

St Clair, M. H., P. A. Furman, C. M. Lubbers & G. B. Elion, 1980. Inhibition of cellular alpha and virally induced deoxyribonucleic acid polymerases by the triphosphate of acyclovir, Antimicrob. Agents Chemother. 18: 741–745.

Sullivan, V. & D. M. Coen, 1991. Isolation of foscarnet-resistant human cytomegalovirus patterns of resistance and sensitivity to other antiviral drugs, J. Infect. Dis. 164: 781–784.

Sullivan, V., C. L. Talarico, S. C. Stanat, M. Davis, D. M. Coen & K. K. Biron, 1992. A protein kinase homologue controls phosphorylation of ganciclovir in human cytomegalovirus-infected cells, Nature 359: 85.

Suzutani, T., K. Ishioka, E. De Clercq, K. Ishibashi, H. Kaneko, T. Kira, K. Hashimoto, M. Ogasawara, K. Ohtani, N. Wakamiya & M. Saijo, 2003. Differential mutation patterns in thymidine kinase and DNA polymerase genes of herpes simplex virus type 1 clones passaged in the presence of acyclovir or penciclovir, Antimicrob. Agents Chemother. 47: 1707–1713.

Suzutani, T., S. Koyano, M. Takada, I. Yoshida & M. Azuma, 1995. Analysis of the relationship between cellular thymidine kinase activity and virulence of thymidine kinase-negative herpes simplex virus types 1 and 2, Microbiol. Immunol. 39: 787–794.

Talarico, C. L., T. C. Burnette, W. H. Miller, S. L. Smith, M. G. Davis, S. C. Stanat, T. I. Ng, Z. He, D. M. Coen, B. Roizman & K. K. Biron, 1999. Acyclovir is phosphorylated by the human cytomegalovirus UL97 protein, Antimicrob. Agents Chemother. 43: 1941–1946.

Talarico, C. L., W. C. Phelps & K. K. Biron, 1993. Analysis of the thymidine kinase genes from acyclovir-resistant mutants of varicella-zoster virus isolated from patients with AIDS, J. Virol. 67: 1024–1033.

Tchesnokov, E. P., C. Gilbert, G. Boivin & M. Gotte, 2006. Role of helix P of the human cytomegalovirus DNA polymerase in resistance and hypersusceptibility to the antiviral drug foscarnet, J. Virol. 80: 1440–1450.

Tenney, D. J., G. Yamanaka, S. M. Voss, C. W. Cianci, A. V. Tuomari, A. K. Sheaffer, M. Alam & R. J. Colonno, 1997. Lobucavir is phosphorylated in human cytomegalovirus-infected and -uninfected cells and inhibits the viral DNA polymerase, Antimicrob. Agents Chemother. 41: 2680–2685.

Tenser, R. B., J. C. Jones, S. J. Ressel & F. A. Fralish, 1983. Thymidine plaque autoradiography of thymidine kinase-positive and thymidine kinase-negative herpesviruses, J. Clin. Microbiol. 17: 122–127.

Thomsen, D. R., N. L. Oien, T. A. Hopkins, M. L. Knechtel, R. J. Brideau, M. W. Wathen & F. L. Homa, 2003. Amino acid changes within conserved region III of the herpes simplex virus and human cytomegalovirus DNA polymerases confer resistance to 4-oxo-dihydroquinolines, a novel class of herpesvirus antiviral agents, J. Virol. 77: 1868–1876.

Tucker, J. A., T. L. Clayton, C. G. Chidester, M. W. Schulz, L. E. Harrington, S. J. Conrad, Y. Yagi, N. L. Oien, D. Yurek & M. S. Kuo, 2000. Structure-activity relationships of acyloxyamidine cytomegalovirus DNA polymerase inhibitors, Bioorg. Med. Chem. 8: 601–615.

Vaillancourt, V. A., M. M. Cudahy, S. A. Staley, R. J. Brideau, S. J. Conrad, M. L. Knechtel, N. L. Oien, J. L. Wieber, Y. Yagi & M. W. Wathen, 2000. Naphthalene carboxamides as inhibitors of human cytomegalovirus DNA polymerase, Bioorg. Med. Chem. Lett. 10: 2079–2081.

Vere Hodge, R. A., D. Sutton, M. R. Boyd, M. R. Harnden & R. L. Jarvest, 1989. Selection of an oral prodrug (BRL 42810; famciclovir) for the antiherpesvirus agent BRL 39123 [9-(4-hydroxy-3-hydroxymethylbut-l-yl)guanine; penciclovir], Antimicrob. Agents Chemother. 33: 1765–1773.

Visse, B., B. Dumont, J. M. Huraux & A. M. Fillet, 1998. Single amino acid change in DNA polymerase is associated with foscarnet resistance in a varicella-zoster virus strain recovered from a patient with AIDS, J. Infect. Dis. 178 Suppl 1: S55–S57.

Visse, B., J. M. Huraux & A. M. Fillet, 1999. Point mutations in the varicella-zoster virus DNA polymerase gene confers resistance to foscarnet and slow growth phenotype, J. Med. Virol. 59: 84–90.

Wald, A., S. Selke, T. Warren, F. Y. Aoki, S. Sacks, F. Diaz-Mitoma & L. Corey, 2006. Comparative efficacy of famciclovir and valacyclovir for suppression of recurrent genital herpes and viral shedding, Sex Transm. Dis. 33: 529–533.

Wan, W. B., J. R. Beadle, C. Hartline, E. R. Kern, S. L. Ciesla, N. Valiaeva & K. Y. Hostetler, 2005. Comparison of the antiviral activities of alkoxyalkyl and alkyl esters of cidofovir against human and murine cytomegalovirus replication in vitro, Antimicrob. Agents Chemother. 49: 656–662.

Warren, T., J. Harris & C. A. Brennan, 2004. Efficacy and safety of valacyclovir for the suppression and episodic treatment of herpes simplex virus in patients with HIV, Clin. Infect. Dis. 39 Suppl 5: S258-S266.

Wathen, M. W., 2002. Non-nucleoside inhibitors of herpesviruses, Rev. Med. Virol. 12: 167–178.

Weinberg, A., D. A. Jabs, S. Chou, B. K. Martin, N. S. Lurain, M. S. Forman & C. Crumpacker, 2003. Mutations conferring foscarnet resistance in a cohort of patients with acquired immunodeficiency syndrome and cytomegalovirus retinitis, J. Infect. Dis. 187: 777–784.

Weller, S., M. R. Blum, M. Doucette, T. Burnette, D. M. Cederberg, P. de Miranda & M. L. Smiley, 1993. Pharmacokinetics of the acyclovir pro-drug valaciclovir after escalating single- and multiple-dose administration to normal volunteers, Clin. Pharmacol. Ther. 54: 595–605.

Whitley, R. J., C. A. Alford, M. S. Hirsch, R. T. Schooley, J. P. Luby, F. Y. Aoki, D. Hanley, A. J. Nahmias & S. J. Soong, 1986. Vidarabine versus acyclovir therapy in herpes simplex encephalitis, N. Engl. J. Med. 314: 144–149.

Whitley, R. J., S. J. Soong, R. Dolin, G. J. Galasso, L. T. Ch'ien & C. A. Alford, 1977. Adenine arabinoside therapy of biopsy-proved herpes simplex encephalitis. National Institute of Allergy and Infectious Diseases collaborative antiviral study, N. Engl. J. Med. 297: 289–294.

Whitley, R. J., S. J. Soong, M. S. Hirsch, A. W. Karchmer, R. Dolin, G. Galasso, J. K. Dunnick & C. A. Alford, 1981. Herpes simplex encephalitis: vidarabine therapy and diagnostic problems, N. Engl. J. Med. 304: 313–318.

Wiebusch, L., M. Truss & C. Hagemeier, 2004. Inhibition of human cytomegalovirus replication by small interfering RNAs, J. Gen. Virol. 85: 179–184.

Wolf, D. G., C. T. Courcelle, M. N. Prichard & E. S. Mocarski, 2001. Distinct and separate roles for herpesvirus-conserved UL97 kinase in cytomegalovirus DNA synthesis and encapsidation, Proc. Natl. Acad. Sci. U. S. A 98: 1895–1900.

Wolf, D. G., D. J. Lee & S. A. Spector, 1995. Detection of human cytomegalovirus mutations associated with ganciclovir resistance in cerebrospinal fluid of AIDS patients with central nervous system disease, Antimicrob. Agents Chemother. 39: 2552–2554.

Wutzler, P., 1997. Antiviral therapy of herpes simplex and varicella-zoster virus infections, Intervirology 40: 343–356.

Xiong, X., J. L. Smith & M. S. Chen, 1997. Effect of incorporation of cidofovir into DNA by human cytomegalovirus DNA polymerase on DNA elongation, Antimicrob. Agents Chemother. 41: 594–599.

Xiong, X., J. L. Smith, C. Kim, E. S. Huang & M. S. Chen, 1996. Kinetic analysis of the interaction of cidofovir diphosphate with human cytomegalovirus DNA polymerase, Biochem. Pharmacol. 51: 1563–1567.

Zhang, Y., D. Schols & E. De Clercq, 1999. Selective activity of various antiviral compounds against HHV-7 infection, Antiviral Res. 43: 23–35.

Zimmermann, A., D. Michel, I. Pavic, W. Hampl, A. Luske, J. Neyts, E. De Clercq & T. Mertens, 1997. Phosphorylation of aciclovir, ganciclovir, penciclovir and S2242 by the cytomegalovirus UL97 protein: a quantitative analysis using recombinant vaccinia viruses, Antiviral Res. 36: 35–42.

Zuccola, H. J., D. J. Filman, D. M. Coen & J. M. Hogle, 2000. The crystal structure of an unusual processivity factor, herpes simplex virus UL42, bound to the C terminus of its cognate polymerase, Mol. Cell 5: 267–278.

Chapter 23
Viral RNA Polymerase Inhibitors

Todd Appleby, I-hung Shih, and Weidong Zhong

Infections by RNA viruses continue to exist as significant public health problems worldwide. In response to the urgent need for safer and more efficacious treatment options against infections caused by RNA viruses, the pharmaceutical and biotechnology industries have devoted significant efforts over the last two decades to discovering and developing new antiviral agents. As the primary viral enzyme responsible for genome replication and transcription, RNA-dependent RNA polymerases (RdRps) emerged early and remained as one of the most promising targets for therapeutic intervention of RNA virus infections. Advances in both basic research and drug discovery technology have resulted in the identification of a significant number of nucleoside (NIs) and non-nucleoside inhibitors (NNIs) of viral RdRps. In this chapter, we will focus our attention on various classes of viral RdRp inhibitors, with main emphases on those of hepatitis C virus (HCV) due to its significant unmet medical need. Recent progress in understanding their mechanism of action, antiviral activity profiles, and emergence of drug resistance mutations will be discussed.

Hepatitis C virus (HCV), a positive-stranded RNA virus belonging to the *Flaviviridae* family, is the main causative agent of non-A, non-B post-transfusion hepatitis which affects approximately 170–200 million individuals worldwide. Current standard therapies consist of pegylated interferon α (PEG-IFNα) and ribavirin (RBV) with treatment durations ranging from 6 to 12 months depending on the infected HCV genotypes. The sustained viral response (SVR), defined as no detectable viral RNA in patient plasma 6 months after cessation of therapy, typically occurred in 40–60% of treated patients. The sustained response rate is significantly lower in patients infected with genotype 1 HCV, and in patients associated with characteristics unfavorable to current therapies. Because of lack of broad efficacy, long treatment duration, and significant side effects associated with current therapies, development of more efficacious and safer treatment options are urgently needed in order to effectively combat this debilitating viral disease. New therapies

W. Zhong (✉)
Gilead Sciences, Inc., 333 Lakeside Drive, Foster City, CA 94404, USA
e-mail: weidong.zhong@gilead.com

C.E. Cameron et al. (eds.), *Viral Genome Replication*,
DOI 10.1007/b135974_23, © Springer Science+Business Media, LLC 2009

are especially critical for individuals who have either failed or are contraindicated to current therapies.

Nucleoside Viral RNA Polymerase Inhibitors (NIs)

Nucleoside and nucleotide analogues constitute a major class of clinically used antiviral agents. Currently they count for more than half of marketed therapeutics in treatment of viral infections such as HIV, HBV, and herpes viruses. These drugs are generally dideoxynucleoside/tide analogues that inhibit viral replication by targeting viral polymerases whose primary function is DNA synthesis. The success of nucleoside/tide analogues in treating viral infections provides a compelling rationale for exploring specific nucleoside inhibitors for RNA-dependent RNA polymerases encoded by RNA viruses of significant medical needs, such as picornaviruses, respiratory syncytial virus (RSV), and several members of *Flaviviridae* family including HCV. In the case of antiviral development for HCV NS5B RdRp, significant progress has been made in the past several years, attributed mainly to availability of various biochemical and cell-based assays as well as crystal structures of NS5B.

Four nucleoside inhibitors have been studied clinically against HCV infection (NM283, R7128, MK0608, and R1626) (Table 23.1). Valopicitabine (NM283) is the 3′-O-valinyl ester of 2′-β-C-methylcytidine; MK0608 is 2′-β-C-methyl-7-deazaadenosine; R7128 is 2′-deoxy-2′-F-2′-C-methylcytidine, and R1626 is the 2′, 3′, 5′ tri-isobutyl ester prodrug of 4′-azidocytidine. The first three candidates share the common feature of 2′-C-methyl substitution on the ribose, while R1626 contains a novel 4′ modification. Triphosphates of all four nucleoside analogues are able to serve as substrate for HCV NS5B and result in chain termination once incorporated into the nascent RNA strand (Migliaccio et al. 2003; Klumpp et al. 2006).

Drug resistance selection identified a single mutation in NS5B, Ser282Thr (S282T), which conferred high levels of resistance to 2′-C-methylnucleoside inhibitors (Migliaccio et al. 2003). Cross-resistance of this mutation against all 2′-C-methylpurines/pyrimidines suggests that serine residue is directed specifically toward the 2′-methyl moiety in the ribose. Conversion of serine to bulkier threonine may impose steric hindrance to the 2′-C-methyl group, and therefore blocks 2′-C-methylnucleosides from binding to the enzyme. Replicon RNA carrying the S282T mutation has proved to be extremely less fit in vitro. Nonetheless, it was identified in one of two chimpanzees treated with MK-0608 (Conf. abstract reference). In addition, sequence analysis revealed the emergence of the S282T mutation in one of 15 HCV patients receiving monotherapy of NM283 (Conf. abstract reference). The significance of the S282T mutation in 2′-C-methylnucleoside therapies remains to be determined in future clinical studies.

Table 23.1 Structures of HCV NS5B nucleoside inhibitors (NIs)

Compound	Structure
NM283	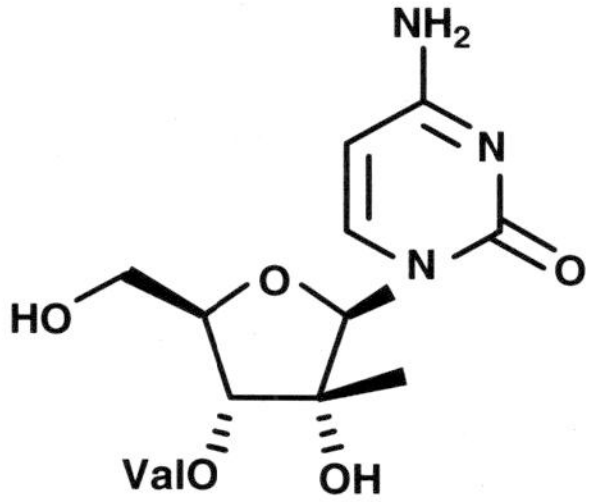
R1626	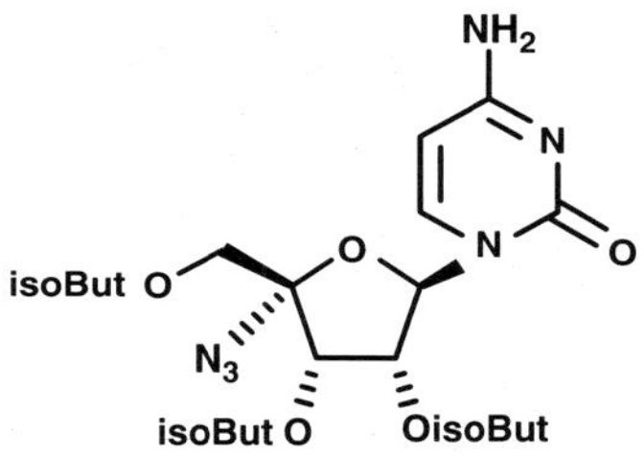
MK0608	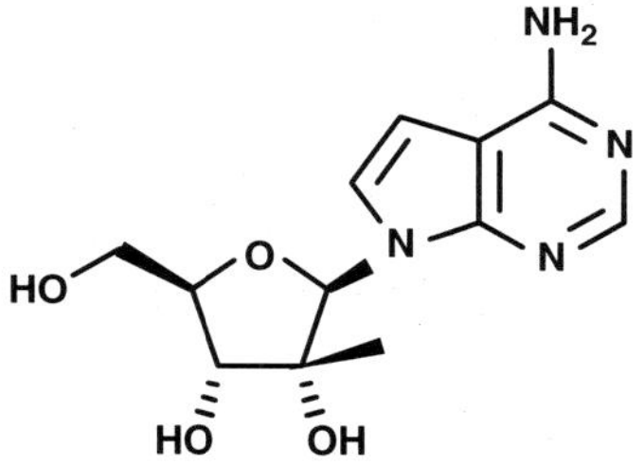
R7128 (prodrug of PSI-6130, chemical structure not revealed)	PSI-6130 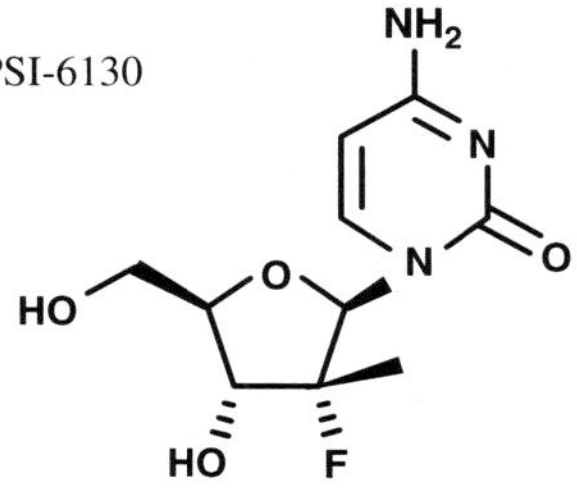

2′-C-Methylnucleoside Inhibitors

Valopicitabine, NM283 (3′-O-valinyl-2′-C-methylcytidine)

An antiviral effect of 2′-C-methylcytidine (NM107), the parent molecule of NM283, was first demonstrated against bovine viral diarrhea virus (BVDV), a surrogate virus for HCV, and the 5′-triphosphate of NM107 was also shown to possess good potency against HCV NS5B in vitro (Table 23.2). However, NM107 showed poor oral bioavailability, partly due to metabolic conversion by cytidine or cytidylic deaminase. This limitation was alleviated by addition of a prodrug moiety (3′ valinyl ester) to NM107 to boost active absorption via a peptide transport mechanism. In clinical studies, valopicitabine alone resulted in moderate viral load reduction (1.2 log10 after 15 days, 800 mg daily dosing) (Afdhal 2004). The antiviral efficacy was enhanced when combined with Peg-IFN and ribavirin. However, significant side effects associated with gastrointestinal tract were observed and further clinical development of NM283 was recently suspended.

MK0608 (7-deaza-2′-C-Methyladenosine)

MK0608 possesses the most potent antiviral activity in replicon assays (EC50 = 0.25 μM, Table 23.2) among the four nucleoside candidates (Olsen et al. 2004; Carroll and Olsen 2006). Replacing the nitrogen with carbon at the 7-position of adenosine greatly improved its binding to the NS5B active site and therefore reduced IC50 for MK0608 triphosphate by 10-fold versus the adenosine analogue. 7-Deaza modification also reduced deamination by adenosine deaminase. MK0608 showed superior pharmacokinetic properties in animals over 2′-C-methylcytidine (NM107), with oral bioavailability ranging from 50 to 90% among different preclinical animal species.

Dosing of MK0608 in chimpanzees chronically infected with HCV showed promising antiviral efficacy after both oral and IV administration of the compound (Olsen et al. 2006; Olsen 2006). In the best case, a chimpanzee treated for 7 days with MK0608 at 2 mg/kg produced a 5.7 log10 reduction in viral load. These encouraging results suggested that NI could achieve significant antiviral efficacy against HCV replication, comparable to that observed with HCV NS3 protease inhibitors.

PSI-6130/R7128

PSI-6130 (2′-deoxy-2′fluoro-2′-C-methylcytidine) possesses good antiviral activity against HCV replicons in vitro. R7128, a prodrug form of PSI-6130 (chemical structure not released), is currently under clinical development for the treatment of chronic HCV infection. The pyrimidine analogue combined ribose modifications of two HCV NS5B nucleoside inhibitors, 2′β-C-methyl and 2′-α-fluoro, to achieve enhanced antiviral activity and reduced activity against cellular RNA polymerases. Interestingly, neither 2′-deoxy-2′-fluorocytidine (FdC) nor PSI-6130

Table 23.2 Activity profiles of HCV NS5B nucleoside inhibitors

	NM283	R1626	MK0608	R7128
Active entity	2′-C-Methylcytidine	4′-Azidocytidine	7-Deaza-2′-C-methyladenosine	2′-Deoxy-2′-fluoro-2′-C-methylcytidine
Triphosphate IC_{50}	0.2 μM*	0.3 μM¶	0.07 μM*	ND
Replicon EC_{50}	5 μM*	1.1 μM¶	0.25 μM*	4.6 μM (EC_{90})¥
Clinical status	Terminated	Terminated	Unknown	Phase II
Prodrug moiety	2′-O-Val-prodrug	2′,3′, 5′-O-isobutyl-prodrug	Nucleoside	Prodrug structure not released

*(Olsen et al. 2006)
¶(Klumpp et al. 2006)
¥ (Stuyver et al. 2006)

showed antiviral activity against cytopathic BVDV in MDBK cells, suggesting unique recognition of the 2′-fluoro substitution by HCV NS5B (Stuyver et al. 2006).The inhibitory potency of PSI-6130 is contributed by the chain-terminating effect of 2′-C-methyl moiety and a specific interaction with the 2′-F modification. These features enhanced the EC90 of PSI-6130 to 5.4 μM, fourfold lower than that of 2′-C-methylcytidine (NM107) in replicon assays. Like other 2′-C-methylnucleoside inhibitors, PS-6130 showed reduced potency against HCV replicon that carried the S282T mutation, but likely to a lesser extent.

4′-Azidonucleoside Inhibitor

R1626, the 2′, 3′, 5′ tri-isobutyl ester prodrug of R1479 (4′-azidocytidine), is the only HCV NS5B NI under clinical development without the 2′-C-methyl modification. Previously, several 4′-substituted 2′-deoxynucleosides were identified to be potent inhibitors against HIV RT. In particular, 4′-ethynyldeoxynucleosides showed antiviral activity against various laboratory and clinical HIV-1 strains (Kodama et al. 2001; Hayakawa et al. 2004). In HCV replicon assays, R1479 was shown to effectively inhibit HCV RNA replication with a good selectivity index (EC50 ~1 μM and CC50 > 2000 μM). In contrast, 4′-ethynylribosenucleoside exhibited minimal inhibitory effects (Smith et al. 2007). Subsequently it was demonstrated that the 5′-triphosphate of R1479 inhibited NS5B in vitro with an IC50 of 0.3 μM and caused chain termination in single-nucleotide incorporation assays (Klumpp et al. 2006). It is noteworthy that 4′-azidouridine triphosphate showed similar potency as its cytidine counterpart against HCV NS5B in vitro, but 4′-azidouridine showed weak inhibitory effects in replicon assays due to limited intracellular phosphorylation. A recent report demonstrated that use of a phosphoramidate prodrug could bypass the rate-limiting phosphorylation step for 4′-azidouridine and enhance its replicon EC50 to 0.22 μM (Perrone et al. 2007).

In vitro resistance studies showed that an HCV replicon carrying the S96T mutation in NS5B conferred moderate resistance to R1479 (Le Pogam et al. 2006a). However, replication fitness of the S96T replicon is significantly reduced in vitro. Furthermore, sequence analysis of 64 clinical HCV isolates failed to identify a pre-existing S96T variant, consistent with its low replication fitness.

In a phase I study, HCV-1-infected patients treated with R1626 showed dose-dependent viral load reduction. At the highest dosing of 4500 mg twice daily, a 3.7 log10 drop was observed (Roberts et al. 2006). However, moderate hematological changes were observed in a dose-dependent manner. Unfortunately, development of R1626 was terminated recently due to additional safety concerns in combination studies (Roche Press Release, 2008).

Other Nucleoside Inhibitors

3′-Modified Inhibitors

Nucleoside analogues that lack the 3′-OH group are obligatory chain terminators. With the simplest modifications, 3′-deoxynucleosides are potent inhibitors of HCV NS5B in vitro. Other modifications at the 3′-position either did not improve the potency against NS5B (e.g., 3′-F) or abolished its substrate capacity (e.g., 3′-methoxy) (Shih et al. 2007). Of the four natural bases, 3′-deoxy-GTP and 3′-deoxy-CTP showed superior potency against HCV NS5B with K_i values in the sub-μM range (Shim et al. 2003). Modification of the 7-position on the guanosine base further enhanced potency by eightfold. However, none of these 3′-deoxynucleosides exhibited significant activity in replicon assays due to inefficient intracellular activation. Antiviral activity (EC50) of 3′-deoxynucleosides against HCV replicon was improved to low-μM range after addition of a phosphate prodrug moiety to increase intracellular phosphorylation (Prakash et al. 2005).

2′-Modified Inhibitors

In addition to the 2′-β-methyl modification described above, 2′-O-methyl derivatives also proved to be potent inhibitors of HCV NS5B in vitro with IC50 values ranging from sub- to low-μM (Tomassini et al. 2005). However, due to limited intracellular phosphorylation, significant inhibition in HCV replicon assays was only observed with phosphate prodrug derivatives (e.g., SATE-2′-O-methyl-cytosine and SATE-2′-O-methyl-cytosine) (Prakash et al. 2005). Another 2′-modified nucleoside that showed inhibitory activity in HCV replicon assays is 2′-deoxy-2-fluorocytidine (FdC). However, the weak activity of FdC-TP against NS5B in vitro suggests that the observed replicon activity might be partly contributed by its cytostatic effect (Stuyver et al. 2004).

Activity Against Other Viral RdRps

Several HCV NS5B nucleoside inhibitors mentioned above also showed antiviral activity against other RNA viruses. In fact, many of the nucleoside inhibitors were initially discovered in the surrogate BVDV assay, prior to the discovery of the HCV replicon cell culture model. For 2′-C-methylnucleosides, antiviral activity is specific to (+) strand RNA viruses and is inactive against (–) strand RNA viruses or dsDNA viruses (Olsen et al. 2004). In particular, 2′-C-methylpurines (A, G, and 7-deazaG) showed potent antiviral activity against several human pathogens in the *Flaviviridae* family, including West Nile, Dengue, and Yellow fever viruses. 2′-C-Methylnucleoside (including NM107) also showed modest antiviral activity against several members of *Picornaviridae* family (Poliovirus, Foot-and-mouth disease virus, and Rhinoviruses), but not against the more distant

Togaviridae family. These results may suggest a broad spectrum of application for 2′-C-methylnucleoside in antiviral therapies.

Non-nucleoside Viral RNA Polymerase Inhibitors (NNIs)

In recent years, antiviral drug discovery platforms utilizing high throughput screening technology (HTS) have afforded discovery of many initial lead series of non-nucleoside viral RNA polymerase inhibitors (NNIs), especially against HCV NS5B polymerase. The use of modern structure-based drug design techniques has further enabled detailed molecular and mechanistic characterizations of some of the NNIs. In particular, our understanding of the structure, function, and inhibition of viral RdRps has been greatly enhanced by the use of X-ray crystallography.

Non-nucleoside Inhibitor of Poliovirus 3D^{pol}

In 1992, it was shown that the fungal metabolite, gliotoxin (Fig. 23.1), was capable of blocking viral RNA synthesis in cells that were infected with poliovirus while cellular RNA synthesis was only mildly affected (Rodriguez and Carrasco 1992). Similar results were obtained in an in vitro enzyme assay utilizing purified recombinant poliovirus RdRp (3D^{pol}). In this experiment, poliovirus 3D^{pol}-catalyzed RNA synthesis was significantly inhibited in the presence of gliotoxin (94% inhibition at 600 μM gliotoxin). This evidence indicates that the inhibition of viral replication observed in cell culture is due to gliotoxin acting directly on poliovirus 3D^{pol}. Consistent with observations from other biological processes that are disrupted by gliotoxin (Kroll et al. 1999), the effect of this fungal metabolite was slightly abrogated by addition of 5 mM of the reducing agent, dithiothreitol, to the enzyme assay mixture, suggesting that the disulfide bridge in gliotoxin must be intact for optimal inhibitory activity. Due to its modest antiviral activity and limited selectivity index,

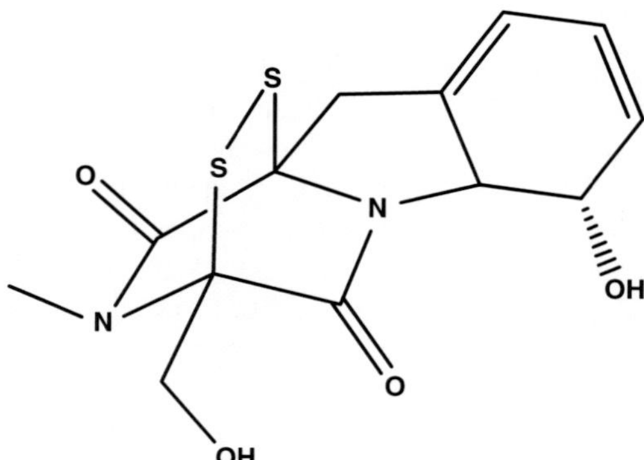

Fig. 23.1 Gliotoxin is a natural product inhibitor of poliovirus 3D^{pol}.

gliotoxin has not been further developed as an antiviral agent. However, this study did demonstrate the therapeutic potential of directly targeting viral RdRps.

Non-nucleoside Inhibitors of Bovine Viral Diarrhea Virus RdRp (NS5B)

Prior to the establishment of robust in vitro cell culture systems for HCV, bovine viral diarrhea virus (BVDV) has frequently been used as a surrogate model for evaluation of anti-HCV inhibitors because it is physiologically close to HCV. Several classes of potent non-nucleoside inhibitors have been discovered against BVDV which were subsequently determined to target the NS5B RdRp of BVDV (Fig. 23.2) (Baginski et al. 2000; Sun et al. 2003; Paeshuyse et al. 2006).

VP32947, a triazinoindole analogue, was shown to inhibit BVDV replication in cell culture with an EC_{50} of 20 nM (Baginski et al. 2000). Time addition experiments showed that VP32947 affects BVDV replication at early steps of the viral life cycle. Drug resistance studies further mapped the resistance mutation to the NS5B polymerase gene, in which a single mutation (F224S) conferred high levels of resistance to this compound. These results, coupled with potent inhibitory activity against a recombinant BVDV NS5B in biochemical assays, established VP32947 as a BVDV NS5B polymerase inhibitor.

BPIP, 5-[(4-bromophenyl)methyl]-2-phenyl-*5H*-inidazol[4,5-c]pyridine (Fig. 23.2), is another potent BVDV inhibitor discovered in antiviral assays, with an EC_{50} of approximately 40 nM. This compound showed cross-resistance with VP32947 and shared the same resistance mutation (F224S) in the BVDV NS5B polymerase gene, suggesting that it is a viral polymerase inhibitor. Interestingly, unlike VP32947, BPIP was not active in polymerization assays using a recombinant BVDV NS5B enzyme but showed activity against the replicase complex isolated from BVDV-infected cells (Paeshuyse et al. 2006).

Similar approach by scientists at Bristol–Myers Squibb also identified a novel class of specific BVDV inhibitors (Fig. 23.2). Compound-1453, as the example of the series, has an EC_{50} of $\sim$2.2 μM against BVDV (Sun et al. 2003). In order to further dissect the mechanism of action of compound-1453, resistant virus was

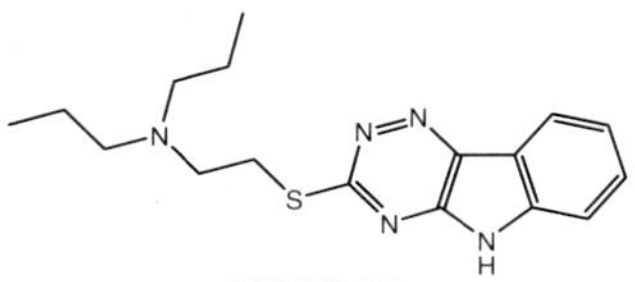

VP32947

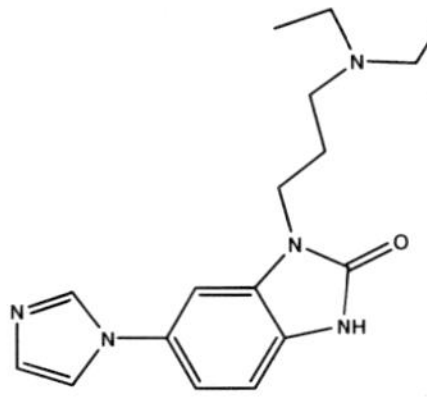

Compound-1453

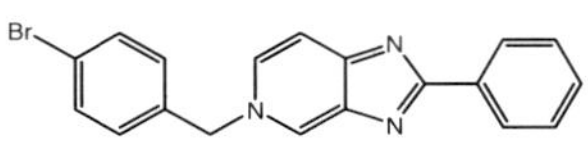

BPIP

Fig. 23.2 Non-nucleoside inhibitors of BVDV NS5B polymerase.

generated by propagating BVDV for multiple passages in the presence of inhibitor concentrations up to 33 μM. Subsequent sequencing analysis identified one common mutation among all drug resistant isolates. This mutation results in a glutamate acid to glycine residue change in the NS5B polymerase of BVDV at position 291. Although this evidence strongly suggests that NS5B is the target for inhibition by the compound, a corresponding in vitro enzyme assay utilizing recombinant BVDV NS5B failed to show inhibition (IC_{50} >300 μM). Interestingly, the activity of compound-1453 was mostly recovered in a replicase complex assay (IC_{50} ~17 μM) which utilizes the isolated membrane fraction from cells that were infected by BVDV. The results from the replicase complex assay suggest that the compound acts on BVDV NS5B in the context of intact replicase complexes. One explanation of such an observation is that physiologically relevant viral RNA replication may require additional host or viral factors which interact with the RNA-dependent RNA polymerase. Assays utilizing only the isolated enzyme may fail to identify inhibitors that act by disrupting critical interactions between the polymerase and these additional regulatory factors. The alternative explanation is that conformation(s) critical for binding with this class of compounds may have been altered slightly in the recombinant NS5B enzyme which renders it unable to bind with the inhibitor, again suggesting the importance of cell-based antiviral assays.

Non-nucleoside Inhibitors of HCV NS5B

In the past decade, HCV NS5B RdRp has been pursued aggressively in drug discovery by various groups and, as a result, a relatively large number of inhibitor molecules have been identified. To date, extensive structural information has become available for HCV NS5B including complex structures between NS5B and various non-nucleoside inhibitors. Those non-nucleoside inhibitors bind to different allosteric pockets in NS5B. We will discuss the various classes of inhibitors according to their binding site, assuming that they adopt a similar mechanism of action in inhibiting NS5B enzymatic activity.

Extensive work in this area has led to the identification of several classes of NNIs which act directly against the RdRp from HCV (NS5B) and show potency in cell-based HCV replicon assays. Of even greater importance, it has recently been demonstrated that HCV NS5B NNIs do show anti-HCV activity in vivo. In addition to extensive biological and kinetic characterization of HCV NS5B inhibitors, X-ray crystallography has allowed for the determination of numerous structures of C-terminally truncated NS5B alone (Fig. 23.3) (Ago et al. 1999; Bressanelli et al. 1999; Lesburg et al. 1999; Biswal et al. 2005) and in combination with various compounds of interest providing insight into the detailed molecular mechanisms of inhibitor action. Currently, non-nucleoside HCV NS5B inhibitors comprise a diverse set of chemical scaffolds that bind in either the active site or to multiple allosteric sites in the enzyme. These inhibitor-binding sites can be roughly divided into four regions of NS5B (Fig. 23.4). NNI site I is located directly above the metal

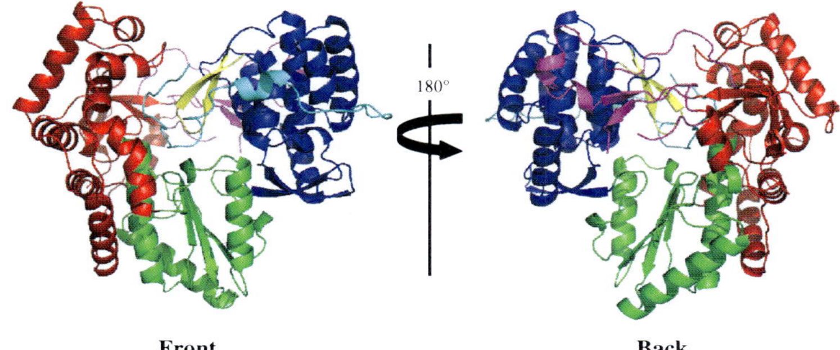

Fig. 23.3 Two views of the domain structure of a 21-residue C-terminally truncated HCV NS5B. The figure shows the fingers (*red*), palm (*green*), thumb (*blue*), β-hairpin (*yellow*), 'fingertips' connection domain (*magenta*), and C-terminal region (*cyan*). The two views are related by a 180° rotation about a vertical axis. The standard 'front' and 'back' views are labeled accordingly. Coordinates from the Protein Data Bank entry 1C2P.

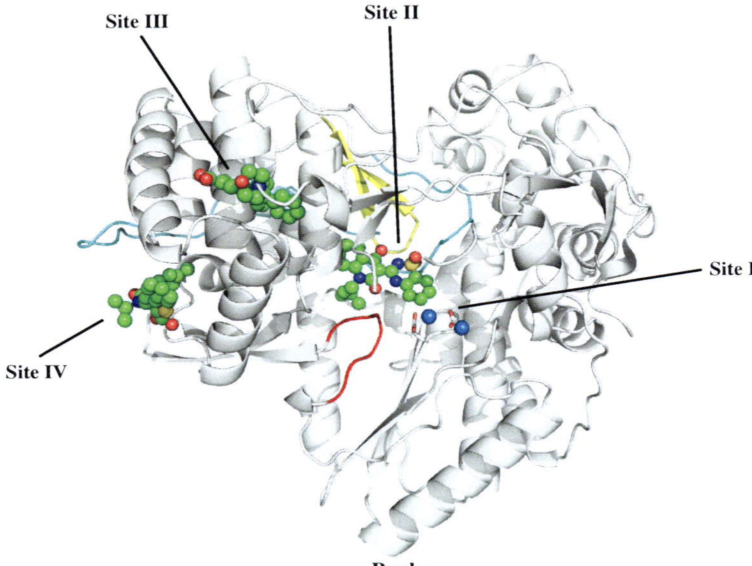

Fig. 23.4 Location of the four HCV NS5B NNI-binding sites. NS5B is viewed from the back with the β-hairpin (*yellow*), C-terminus (*cyan*), and conserved structural motif E (*red*) colored accordingly. The active-site, metal-binding region (NNI Site I) is shown as blue spheres coordinated by the catalytic aspartic acid residues 318 and 319. The three allosteric sites (NNI Site II, NNI Site III, and NNI Site IV) are labeled accordingly.

ion-binding site at the catalytic center of the polymerase. NNI site II is located adjacent to the active site near the putative RNA-binding groove which comprises the largest cavity in the apo form of the enzyme and contains the conserved structural motif E, also known as the primer grip. NNI site III is located in a region of the polymerase where the fingers subdomain makes contact with the back of the thumb subdomain through a series of loop structures that form an interdomain linkage also known as the fingertips. Finally, NNI site IV is located on the outer perimeter of the thumb subdomain near the site where the C-terminal residues of NS5B exit the thumb subdomain and wrap around the front of the polymerase before folding back into the putative RNA-binding groove.

NNI-Binding Site I

Small molecule library screenings against HCV NS5B have led to the discovery of a series of diketo acid and dicarboxylic acid compounds (Pace et al. 2004; Summa et al. 2004). Based on the chemical structure of these reversible inhibitors, it is believed that the compounds act as pyrophosphate analogues and chelate the divalent magnesium ions coordinated by the catalytic aspartate residues in the polymerase active site. Although the early diketo and dicarboxylic acid compounds showed sub-μM activity against isolated NS5B in in vitro enzyme assays, they demonstrated poor inhibition in cell-based HCV replicon assays. The drop in cell-based potency is most likely the result of poor membrane permeability due to the polar anionic character of the compounds. In order to alleviate complications due to the general reactivity and poor physiochemical properties of the earlier diketo acids, several compounds based on a 5,6,-dihydroxypyrimidine-4-carboxylic acid scaffold (Fig. 23.5a) were chosen as improved metal chelators (Koch et al. 2006). One of the more potent compounds discovered following intensive structure–activity relationship (SAR) studies, achieved an enzymatic IC_{50} of 150 nM and demonstrated improved activity in cell-based replicon assays (EC_{50} ~9 μM). The mechanism of action of these inhibitors has been validated by kinetic experiments that show they are competitive with respect to elongating nucleotide substrates and that they inhibit both initiation and elongation phases of HCV RNA replication (Liu et al. 2006). Active-site inhibitors remain highly desirable for antiviral therapy due to the strong conservation of active-site residues and the reduced likelihood for the emergence of viral resistance.

NNI-Binding Site II

Numerous X-ray crystal structures are available which show a wide variety of chemical scaffolds binding in the general region of NNI-binding site II (Pfefferkorn et al. 2005a,b; Gopalsamy, Chopra et al. 2006; Powers, Piper et al. 2006; Tedesco, Shaw et al. 2006; Slater et al. 2007; Yan et al. 2007). Together with biochemical data, this structural information has provided insight into how these inhibitors may disrupt the replication of HCV RNA leading to an antiviral effect. NNI-binding site II lies at the interface between the palm and thumb subdomains adjacent to the catalytic

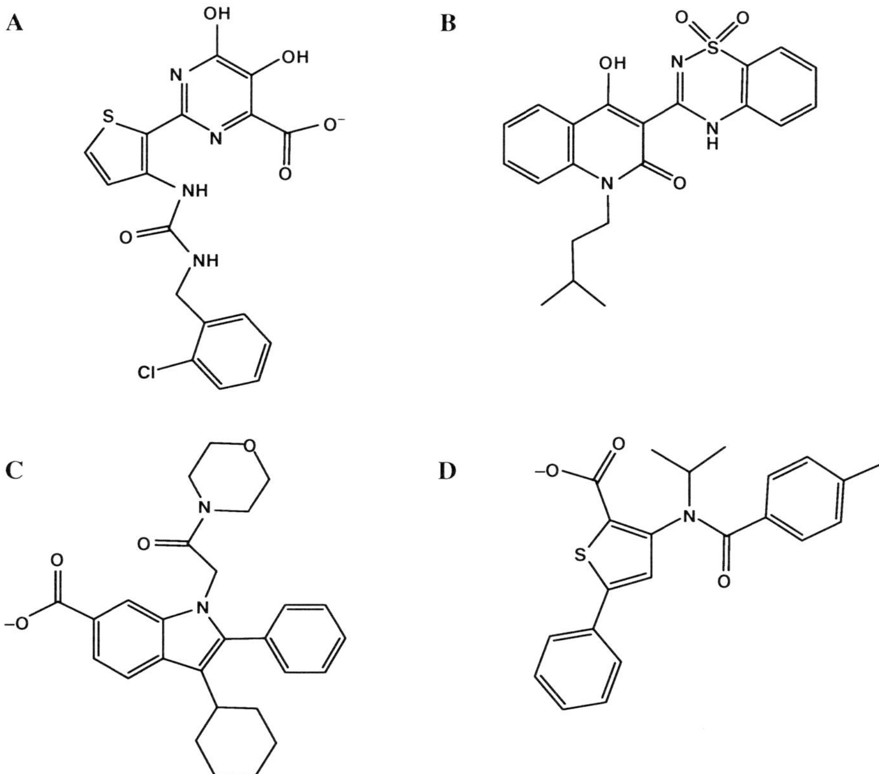

Fig. 23.5 Representative scaffolds from the four NNI-binding sites. (**A**) NNI site I, a 2-(2-thienyl)-5,6-dihydroxy-4-carboxypyrimidine; (**B**) NNI site II, a benzothiadiazine; (**C**) NNI site III, an indole-N-acetamide; (**D**) NNI site IV, a thiophene-2-carboxylic acid.

center of the enzyme and includes a short loop of residues (362–376) that form the conserved structural motif E. This structural feature is conserved among RNA-dependent RNA polymerases and is believed to play a role in positioning the 3′-end of the growing RNA strand and has therefore been termed the primer-grip. Residues on the tip of the β-hairpin also form a portion of the packing interface with a number of compounds that bind to NNI site II. The β-hairpin is an important structural feature unique to HCV NS5B and is believed to play a critical role in selecting the 3′-terminal end of single-stranded HCV RNA templates during de novo initiation (Hong et al. 2001). As one class of NNI site II inhibitors, the benzothiadiazines (Fig. 23.5b) display potency in both enzymatic NS5B assays and cell-based HCV replicon assays (Tedesco et al. 2006). Kinetic experiments suggest that these compounds are non-competitive inhibitors of RNA replication initiation (Tomei et al. 2004; Liu et al. 2006; Yang et al. 2007). A crystal structure of a benzothiadiazine bound to a 21-residue C-terminal truncated form of NS5B supports such a mecha-

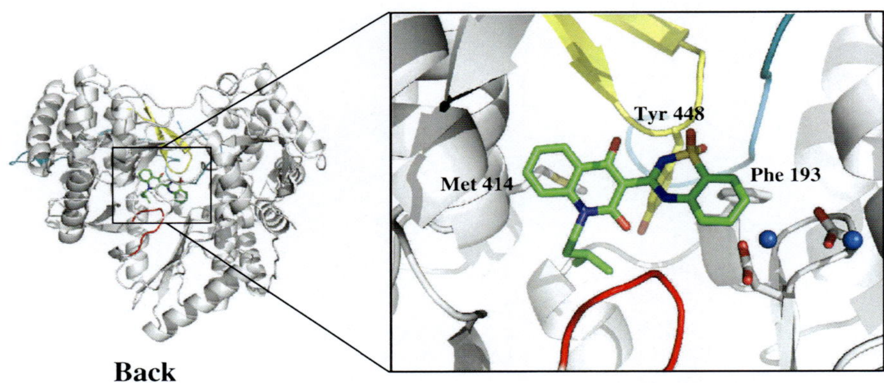

Fig. 23.6 A benzothiadiazine inhibitor bound to NNI-binding site II. NS5B is viewed from the back with the β-hairpin (*yellow*), C-terminus (*cyan*), and conserved structural motif E (*red*) colored accordingly. The active-site, metal-binding region (NNI Site I) is shown as blue spheres coordinated by the catalytic aspartic acid residues 318 and 319. Side chains for three of the key residues which form a packing interface with the compound are shown as sticks and are labeled accordingly.

nism of inhibition (Fig. 23.6) (Tedesco et al. 2006). The benzothiadiazine and quinolinone rings of the inhibitor are nearly coplanar and stack against the side chains of residues from the thumb subdomain (Met 414), palm subdomain (Phe193), and β-hairpin (Tyr 448). The branched alkyl chain of the compound tucks into a deep pocket formed at the interface of the palm and thumb subdomains that includes residues from the primer-grip. It has also been determined that HCV replicon cell lines cultured in the presence of benzothiadiazine compounds give rise to resistance mutations that map to NS5B (Nguyen et al. 2003; Mo et al. 2005; Le Pogam et al. 2006b). Interestingly, one of the most significant sites of resistance occurs at Met 414. Generally, the long, unbranched Met residue is mutated to a residue that contains a side chain with branching at the beta or gamma position. It is clear from the structure that such mutations would lead to clashing between the compound and the mutant enzyme. Due to the fact that benzothiadiazines bind in contact with critical structural elements of HCV NS5B, it is possible that these compounds act by imposing conformational constraints on the enzyme, subsequently disrupting the ability of the polymerase to form a productive initiation complex.

Recently, ViroPharma Incorporated (Exton, PA), in collaboration with Wyeth Research (Collegeville, PA), has disclosed a compound, HCV-796 (Fig. 23.7), a potent and specific inhibitor of HCV NS5B activity (IC_{50} ~40 nM) (Poster Presentations from the 13th International Meeting on Hepatitis C Virus and Related Viruses, 2006; Cairnes, Australia, August 27–31, 2006). HCV-796 is non-competitive with respect to NTPs and RNA substrates and is reported to also bind in NNI site II at the interface of the thumb and palm subdomains adjacent to the NS5B active site. The structure of HCV-796 bound to NS5B reveals slight rearrangements of several protein residues in the primer-grip region of the polymerase which results in the for-

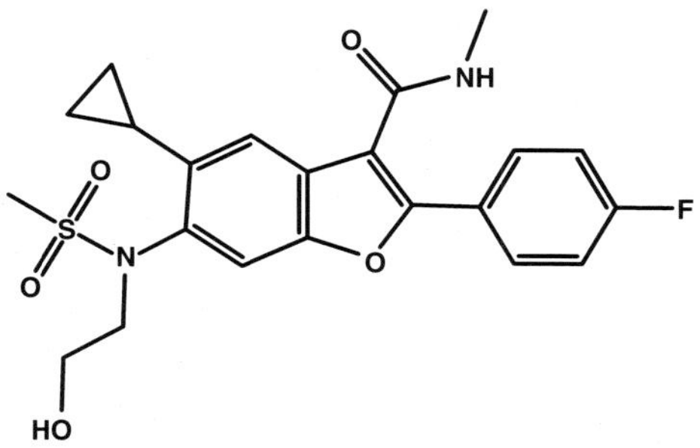

Fig. 23.7 HCV-796 is a potent HCV NS5B inhibitor that has shown anti-viral activity in clinical trials.

mation of a deep binding pocket for this compound compared to what is normally observed in the apoenzyme and other inhibitor-bound crystal structures. Consistent with other NNI site II inhibitors, biochemical analysis has shown that HCV-796 also acts at the initiation phase of HCV RNA synthesis. In addition to displaying potent activity in the cell-based HCV replicon assay (EC_{50} ~9 nM vs. HCV-1b replicon), HCV-796 has shown effectiveness in vivo. After oral dosing with the compound, uPA-SCID mice carrying chimeric human livers that are infected with HCV genotype 1a demonstrate a marked reduction in viral titer. In chronically infected HCV patients, HCV-796 also demonstrate clinical efficacy either alone or in combination with pegylated interferon. However, further clinical development of HCV-796 was halted recently as a result of observation of potential hepatotoxicity in some of the treated patients (ViroPharma Press Release, August 10, 2007).

In addition to the growing number of non-covalent inhibitors which bind to the HCV NS5B NNI site II, there have also been scaffolds identified which are now structurally characterized as covalently binding to a conserved cysteine residue (Cys 366) on the tip of motif E in a reversible fashion. The two scaffolds reported to date include a class of substituted aminorhodanine derivatives (Powers et al. 2006) and a series of isothiazole-based inhibitors (Yan et al. 2007), both of which contain examples of compounds that attain submicromolar activity in in vitro NS5B enzyme assays (Fig. 23.8).In addition to containing reactive centers which are attacked by the thiolate of Cys 366, these compounds contain additional aromatic substitutions that overlap at least partially with other NNI site II inhibitors. Surprisingly, although these compounds contain reactive centers, the isothiazoles display comparable slightly lowered activity in cell-based HCV replicon assays. This result suggests that the reactivity of the compounds is relatively specific toward Cys 366 in NS5B. In the case of aminorhodanine derivatives, a compound that inhibited HCV NS5B with a potency of 200 nM, showed no inhibition against a panel of cysteine proteases that included cathepsin B, calpain, human caspases 1, 3, 6, 7, and 8, or aldose reductase, again demonstrating the relative specificity of this class of inhibitors toward NS5B. VX-950, an HCV NS3 protease inhibitor with potent antiviral activity currently in late-stage clinical trials, has been shown to act as a

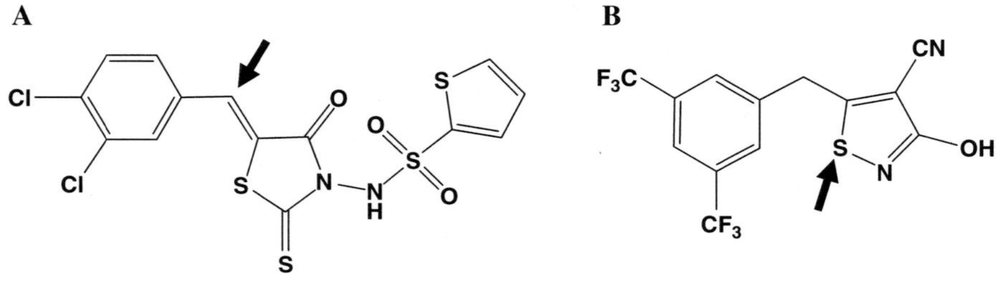

A **B**

Fig. 23.8 Example of compounds known to modify Cys 366 of NS5B and inhibit polymerase activity. (**A**) Substituted aminorhodanine derivative. (**B**) Substituted isothiazole. Reactive centers are denoted by *arrows*. The disulfide bond formation between the thiolate of Cys366 and the sulfur of the isothiazole occurs through ring opening.

reaction product analog that binds in the active site of the enzyme and forms a covalent adduct with the catalytic serine residue in reversible fashion (Lin et al. 2006). By analogy, if potency can be increased while still retaining tight specificity, NS5B inhibitors that react with Cys 366 may prove to be safe, effective anti-HCV agents.

NNI-Binding Site III

Recently, a class of benzimidazole-based (Beaulieu et al. 2004a,b; Ishida et al. 2006) and indole-N-acetamide (Harper et al. 2005) compounds has been identified and characterized as NNI of HCV NS5B which bind in a region of the protein distinct from NNI sites I and II (Kukolj et al. 2005). Furthermore, two of the indole-N-acetamide inhibitors have now been characterized by X-ray crystallography and in fact, confirm the existence of a third NNI-binding site on the polymerase (Di Marco et al. 2005). The crystal structures of a 55-residue C-terminally truncated NS5B alone and in complex with NNI site III compounds (Fig. 23.9) reveal that the inhibitors have displaced a short α-helical loop segment which, in the apo form of the enzyme, forms a portion of the fingertips which extends outward from the fingers subdomain and forms contacts on the back of the thumb subdomain.Disruption of this interdomain contact appears to be stabilized by the phenyl and cyclohexyl substituents of the compound binding into predominantly hydrophobic pockets while the acid moiety forms a salt bridge with Arg 503 on the surface of the enzyme. Consistent with this structural result, kinetic experiments on these and related compounds suggest that NS5B NNI site III inhibitors are non-competitive with respect to nucleotide substrates and act before the enzyme transitions into RNA elongation mode (Tomei et al. 2003; McKercher et al. 2004; Liu et al. 2006). Based on the fact that binding of this class of compounds slightly alters the structure observed in the apo form of the crystallized enzyme, the mechanism of inhibition likely proceeds by disrupting interdomain contacts or interfering with dynamic interdomain communication that takes place prior to the RNA elongation phase of HCV RNA replication.

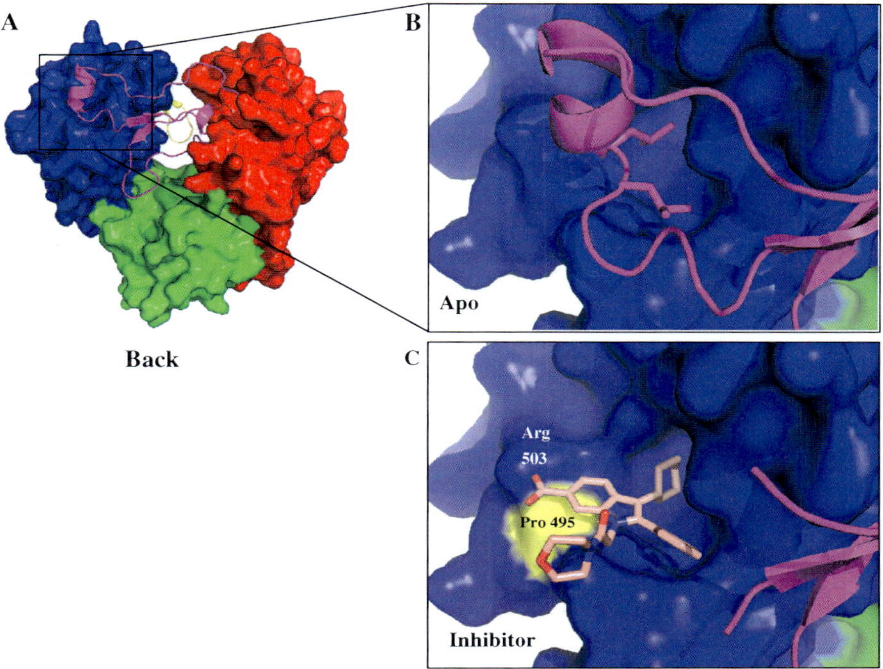

Fig. 23.9 NNI-binding site III. Crystal structures of a 55-residue C-terminally truncated form of NS5B alone and bound to an indole-N-acetamide inhibitor. (**A**) NS5B is viewed from the back to highlight the fingertips (*magenta*) which tether the fingers (*red*) and thumb (*blue*) subdomains in the apo enzyme to form a fully encircled active site. (**B**) An enlarged view of the fingertips interaction with the thumb subdomain. Leu 30 and Leu 31 are shown as stick models. (**C**) The inhibitor (*orange*) displaces a stretch of 14 residues (22–35) of the fingertips disrupting the interdomain linkage. The cyclohexyl and phenyl substitutions of the inhibitor appear to mimic the pair of leucine residues which anchor the fingertips to the thumb subdomain by binding in deep hydrophobic pockets. The carboxylate moiety of the inhibitor interacts with the positively charged Arg 503 side chain on the surface of NS5B. A significant site of resistance, Pro 495, is labeled and colored *yellow*.

Compounds derived from both the benzimidazole and the indole-N-acetamide class of inhibitors display potency in the single-digit nanomolar range in the in vitro enzymatic assays and slightly lower activity in cell-based HCV replicon assays (>100 nM). Not surprisingly, culturing HCV replicon cell lines in the presence of NNI site III inhibitors gives rise to resistance mutations (Tomei et al. 2003; Kukolj et al. 2005). One of the more significant mutations maps to Pro 495, which lies in direct contact with a portion of the indole ring and carboxylate substituent of the indole-N-acetamide class of inhibitors. At least two of the benzimidazole-based inhibitors, JTK-003 and JTK-009 (Japan Tabacco Inc., Yokohama, Japan), have advanced into clinical trials.

NNI-Binding Site IV

Several enzyme-inhibitor co-crystal structures are currently available which reveal a fourth, distinct NNI-binding site on the outer perimeter of the thumb subdomain of HCV NS5B (Love et al. 2003; Wang et al. 2003; Biswal et al. 2005; Biswal et al. 2006; Li et al. 2006; Yan et al. 2006). An example of one such inhibitor is shown in Figs. 23.5d and 23.10. NNI site IV inhibitors bind in a shallow groove that runs along the outer surface of the thumb subdomain nearly 35 Å from the catalytic center of the enzyme. These compounds typically share a general feature of having a large, hydrophobic moiety, often a substituted aromatic ring, which penetrates into a predominantly hydrophobic pocket formed by the side chains from residues Leu 419, Arg 422, Met 423, Leu 474, Tyr 477, and Trp 528. These compounds demonstrate a wide range of chemical diversity, but generally also include one or two hydrogen bond acceptors that interact with one or both of the backbone amide hydrogens from residues Ser 476 and Tyr 477. Despite the large number of crystal structures, it remains unclear what the precise molecular mechanism of action for this class of compounds is. In addition to rearrangements observed for residues, Met 423 and Leu 497, which are in direct van der Waals contact with the compounds, it has been speculated that slight perturbations in a nearby α-helix upon inhibitor binding may interfere with the allosteric regulation, conformational flexibility, or possible oligermization of NS5B (Biswal et al. 2006). As expected from the structural analysis, these compounds are non-competitive with respect to nucleotide substrate (Howe et al. 2006). Although, certain classes of NNI site IV inhibitors can show low nanomolar potency in NS5B enzyme assays (Li et al. 2006), this activity

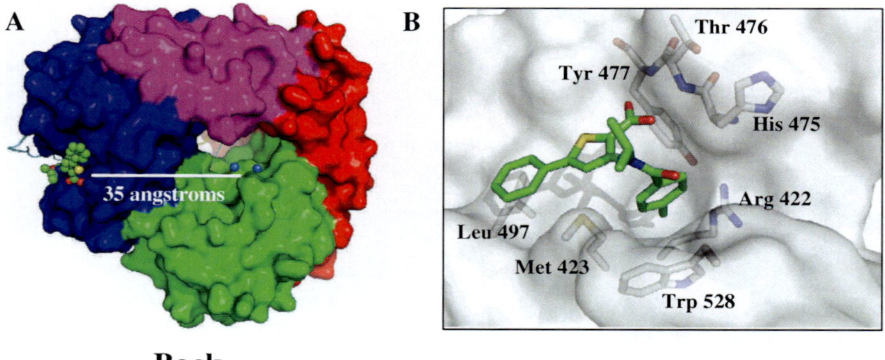

Fig. 23.10 NNI-binding site IV. (**A**) NS5B is viewed from the back and shown as a surface with the fingers (*red*), thumb (*blue*), and palm (*green*) subdomains colored accordingly. The thiophene-2-carboxylic acid inhibitor is bound on the perimeter of the thumb subdomain in a pocket ~35 Å from the active site (catalytic metal ions denoted by *blue spheres*). (**B**) Enlarged view of the inhibitor-binding site. The para-substituted aromatic ring penetrates into the deepest portion of the binding pocket, while the carboxylate moiety interacts with the backbone nitrogen of residues Thr 476 and Tyr 477. Met 423 is a key site of mutational resistance against NNI site IV inhibitors.

does not translate to the cell-based HCV replicon assay with EC_{50} values generally in the micromolar range. Despite the lower cell-based activity, cell lines harboring resistant HCV replicons can be selected in the presence of these allosteric inhibitors (Howe et al. 2006). In agreement with structural data, two different single-site mutations, Leu 419 and Met 423, map to the NNI-binding site IV and most likely disrupt ability of the compounds to bind by altering the local structure of the deep hydrophobic pocket. To date, several NNI site IV inhibitors described in the literature have progressed into clinical stages of development.

Conclusions

RNA viruses completely rely on the activity of virally encoded RNA-dependent RNA polymerases for replication and transcription of their genomes. For this reason, these enzymes remain attractive targets for therapeutic intervention of viral infections. Considering the large number and diversity of compounds in the discovery and preclinical stages of development, it is likely that more and more of these RdRp inhibitors will advance to studies in humans. With compounds already in the clinic for the treatment of HCV, it will not be long before we realize the true potential of RNA-dependent RNA polymerase inhibitors as antiviral drugs.

References

Afdhal, N. (2004). *Annual Meeting of the American Association for the Study of Liver Diseases*. Boston.

Ago, H., T. Adachi, et al. (1999). "Crystal structure of the RNA-dependent RNA polymerase of hepatitis C virus." *Structure* **7**(11): 1417–26.

Baginski, S. G., D. C. Pevear, et al. (2000). "Mechanism of action of a pestivirus antiviral compound." *Proc Natl Acad Sci USA* **97**(14): 7981–6.

Beaulieu, P. L., M. Bos, et al. (2004a). "Non-nucleoside inhibitors of the hepatitis C virus NS5B polymerase: discovery of benzimidazole 5-carboxylic amide derivatives with low-nanomolar potency." *Bioorg Med Chem Lett* **14**(4): 967–71.

Beaulieu, P. L., M. Bos, et al. (2004b). "Non-nucleoside inhibitors of the hepatitis C virus NS5B polymerase: discovery and preliminary SAR of benzimidazole derivatives." *Bioorg Med Chem Lett* **14**(1): 119–24.

Biswal, B. K., M. M. Cherney, et al. (2005). "Crystal structures of the RNA-dependent RNA polymerase genotype 2a of hepatitis C virus reveal two conformations and suggest mechanisms of inhibition by non-nucleoside inhibitors." *J Biol Chem* **280**(18): 18202–10.

Biswal, B. K., M. Wang, et al. (2006). "Non-nucleoside inhibitors binding to hepatitis C virus NS5B polymerase reveal a novel mechanism of inhibition." *J Mol Biol* **361**(1): 33–45.

Bressanelli, S., L. Tomei, et al. (1999). "Crystal structure of the RNA-dependent RNA polymerase of hepatitis C virus." *Proc Natl Acad Sci USA* **96**(23): 13034–9.

Carroll, S. S. and D. B. Olsen (2006). "Nucleoside analog inhibitors of hepatitis C virus replication." *Infect Disord Drug Targets* **6**(1): 17–29.

Di Marco, S., C. Volpari, et al. (2005). "Interdomain communication in hepatitis C virus polymerase abolished by small molecule inhibitors bound to a novel allosteric site." *J Biol Chem* **280**(33): 29765–70.

Gopalsamy, A., R. Chopra, et al. (2006). "Discovery of proline sulfonamides as potent and selective hepatitis C virus NS5b polymerase inhibitors. Evidence for a new NS5b polymerase binding site." *J Med Chem* **49**(11): 3052–5.

Harper, S., B. Pacini, et al. (2005). "Development and preliminary optimization of indole-N-acetamide inhibitors of hepatitis C virus NS5B polymerase." *J Med Chem* **48**(5): 1314–7.

Hayakawa, H., S. Kohgo, et al. (2004). "Potential of 4′-C-substituted nucleosides for the treatment of HIV-1." *Antivir Chem Chemother* **15**(4): 169–87.

Hong, Z., C. E. Cameron, et al. (2001). "A novel mechanism to ensure terminal initiation by hepatitis C virus NS5B polymerase." *Virology* **285**(1): 6–11.

Howe, A. Y., H. Cheng, et al. (2006). "Molecular mechanism of a thumb domain hepatitis C virus nonnucleoside RNA-dependent RNA polymerase inhibitor." *Antimicrob Agents Chemother* **50**(12): 4103–13.

Ishida, T., T. Suzuki, et al. (2006). "Benzimidazole inhibitors of hepatitis C virus NS5B polymerase: identification of 2-[(4-diarylmethoxy)phenyl]-benzimidazole." *Bioorg Med Chem Lett* **16**(7): 1859–63.

Klumpp, K., V. Leveque, et al. (2006). "The novel nucleoside analog R1479 (4′-azidocytidine) is a potent inhibitor of NS5B-dependent RNA synthesis and hepatitis C virus replication in cell culture." *J Biol Chem* **281**(7): 3793–9.

Koch, U., B. Attenni, et al. (2006). "2-(2-Thienyl)-5,6-dihydroxy-4-carboxypyrimidines as inhibitors of the hepatitis C virus NS5B polymerase: discovery, SAR, modeling, and mutagenesis." *J Med Chem* **49**(5): 1693–705.

Kodama, E. I., S. Kohgo, et al. (2001). "4′-Ethynyl nucleoside analogs: potent inhibitors of multidrug-resistant human immunodeficiency virus variants in vitro." *Antimicrob Agents Chemother* **45**(5): 1539–46.

Kroll, M., F. Arenzana-Seisdedos, et al. (1999). "The secondary fungal metabolite gliotoxin targets proteolytic activities of the proteasome." *Chem Biol* **6**(10): 689–98.

Kukolj, G., G. A. McGibbon, et al. (2005). "Binding site characterization and resistance to a class of non-nucleoside inhibitors of the hepatitis C virus NS5B polymerase." *J Biol Chem* **280**(47): 39260–7.

Le Pogam, S., W. R. Jiang, et al. (2006a). "In vitro selected Con1 subgenomic replicons resistant to 2′-C-methyl-cytidine or to R1479 show lack of cross resistance." *Virology* **351**(2): 349–59.

Le Pogam, S., H. Kang, et al. (2006b). "Selection and characterization of replicon variants dually resistant to thumb- and palm-binding nonnucleoside polymerase inhibitors of the hepatitis C virus." *J Virol* **80**(12): 6146–54.

Lesburg, C. A., M. B. Cable, et al. (1999). "Crystal structure of the RNA-dependent RNA polymerase from hepatitis C virus reveals a fully encircled active site." *Nat Struct Biol* **6**(10): 937–43.

Li, H., J. Tatlock, et al. (2006). "Identification and structure-based optimization of novel dihydropyrones as potent HCV RNA polymerase inhibitors." *Bioorg Med Chem Lett* **16**(18): 4834–8.

Lin, C., A. D. Kwong, et al. (2006). "Discovery and development of VX-950, a novel, covalent, and reversible inhibitor of hepatitis C virus NS3.4A serine protease." *Infect Disord Drug Targets* **6**(1): 3–16.

Liu, Y., W. W. Jiang, et al. (2006). "Mechanistic study of HCV polymerase inhibitors at individual steps of the polymerization reaction." *Biochemistry* **45**(38): 11312–23.

Love, R. A., H. E. Parge, et al. (2003). "Crystallographic identification of a noncompetitive inhibitor binding site on the hepatitis C virus NS5B RNA polymerase enzyme." *J Virol* **77**(13): 7575–81.

McKercher, G., P. L. Beaulieu, et al. (2004). "Specific inhibitors of HCV polymerase identified using an NS5B with lower affinity for template/primer substrate." *Nucleic Acids Res* **32**(2): 422–31.

Migliaccio, G., J. E. Tomassini, et al. (2003). "Characterization of resistance to non-obligate chain-terminating ribonucleoside analogs that inhibit hepatitis C virus replication in vitro." *J Biol Chem* **278**(49): 49164–70.

Mo, H., L. Lu, et al. (2005). "Mutations conferring resistance to a hepatitis C virus (HCV) RNA-dependent RNA polymerase inhibitor alone or in combination with an HCV serine protease inhibitor in vitro." *Antimicrob Agents Chemother* **49**(10): 4305–14.

Nguyen, T. T., A. T. Gates, et al. (2003). "Resistance profile of a hepatitis C virus RNA-dependent RNA polymerase benzothiadiazine inhibitor." *Antimicrob Agents Chemother* **47**(11): 3525–30.

Olsen, D., M. Davies, et al. (2006). The Nucleoside Inhibitor MK-0608 Mediates Suppression of HCV Replication for >30 Days in Chronically Infected Chimpanzees *46th Interscience Conference on Antimicrobial Agents and Chemotherapy*. San Francisco.

Olsen, D. B. (2006). Hepatitis C virus resistance to new antivirals *15th International HIV Drug Resistance Workshop*. Sitges, Spain.

Olsen, D. B., A. B. Eldrup, et al. (2004). "A 7-deaza-adenosine analog is a potent and selective inhibitor of hepatitis C virus replication with excellent pharmacokinetic properties." *Antimicrob Agents Chemother* **48**(10): 3944–53.

Pace, P., E. Nizi, et al. (2004). "The monoethyl ester of meconic acid is an active site inhibitor of HCV NS5B RNA-dependent RNA polymerase." *Bioorg Med Chem Lett* **14**(12): 3257–61.

Paeshuyse, J., P. Leyssen, et al. (2006). "A novel, highly selective inhibitor of pestivirus replication that targets the viral RNA-dependent RNA polymerase." *J Virol* **80**(1): 149–60.

Perrone, P., G. M. Luoni, et al. (2007). "Application of the phosphoramidate ProTide approach to 4′-azidouridine confers sub-micromolar potency versus hepatitis C virus on an inactive nucleoside." *J Med Chem* **50**(8): 1840–9.

Pfefferkorn, J. A., M. L. Greene, et al. (2005a). "Inhibitors of HCV NS5B polymerase. Part 1: Evaluation of the southern region of (2Z)-2-(benzoylamino)-3-(5-phenyl-2-furyl)acrylic acid." *Bioorg Med Chem Lett* **15**(10): 2481–6.

Pfefferkorn, J. A., R. Nugent, et al. (2005b). "Inhibitors of HCV NS5B polymerase. Part 2: Evaluation of the northern region of (2Z)-2-benzoylamino-3-(4-phenoxy-phenyl)-acrylic acid." *Bioorg Med Chem Lett* **15**(11): 2812–8.

Powers, J. P., D. E. Piper, et al. (2006). "SAR and mode of action of novel non-nucleoside inhibitors of hepatitis C NS5b RNA polymerase." *J Med Chem* **49**(3): 1034–46.

Prakash, T. P., M. Prhavc, et al. (2005). "Synthesis and evaluation of S-acyl-2-thioethyl esters of modified nucleoside 5′-monophosphates as inhibitors of hepatitis C virus RNA replication." *J Med Chem* **48**(4): 1199–210.

Roberts, S., G. Cooksley, et al. (2006). Results of a Phase 1B, Multiple Dose Study of R1626, a Novel Nucleoside Analog Targeting HCV Polymerase in Chronic HCV Genotype 1 Patients. *57th Annual Meeting of the American Association* Boston, MA.

Rodriguez, P. L. and L. Carrasco (1992). "Gliotoxin: inhibitor of poliovirus RNA synthesis that blocks the viral RNA polymerase 3Dpol." *J Virol* **66**(4): 1971–6.

Shih, I., A. Huang, et al. (2007). 14th International Symposium On Hepatitis C Virus and Related Viruses, Glasgow, UK.

Shim, J., G. Larson, et al. (2003). "Canonical 3′-deoxyribonucleotides as a chain terminator for HCV NS5B RNA-dependent RNA polymerase." *Antiviral Res* **58**(3): 243–51.

Slater, M. J., E. M. Amphlett, et al. (2007). "Optimization of Novel Acyl Pyrrolidine Inhibitors of Hepatitis C Virus RNA-Dependent RNA Polymerase Leading to a Development Candidate." *J Med Chem* **50**(5): 897–900.

Smith, D. B., J. A. Martin, et al. (2007). "Design, synthesis, and antiviral properties of 4′-substituted ribonucleosides as inhibitors of hepatitis C virus replication: the discovery of R1479." *Bioorg Med Chem Lett* **17**(9): 2570–6.

Stuyver, L. J., T. R. McBrayer, et al. (2006). "Inhibition of hepatitis C replicon RNA synthesis by beta-D-2′-deoxy-2′-fluoro-2′-C-methylcytidine: a specific inhibitor of hepatitis C virus replication." *Antivir Chem Chemother* **17**(2): 79–87.

Stuyver, L. J., T. R. McBrayer, et al. (2004). "Inhibition of the subgenomic hepatitis C virus replicon in huh-7 cells by 2′-deoxy-2′-fluorocytidine." *Antimicrob Agents Chemother* **48**(2): 651–4.

Summa, V., A. Petrocchi, et al. (2004). "Discovery of alpha, gamma-diketo acids as potent selective and reversible inhibitors of hepatitis C virus NS5b RNA-dependent RNA polymerase." *J Med Chem* **47**(1): 14–7.

Sun, J. H., J. A. Lemm, et al. (2003). "Specific inhibition of bovine viral diarrhea virus replicase." *J Virol* **77**(12): 6753–60.

Tedesco, R., A. N. Shaw, et al. (2006). "3-(1,1-dioxo-2H-(1,2,4)-benzothiadiazin-3-yl)-4-hydroxy-2(1H)-quinolinones, potent inhibitors of hepatitis C virus RNA-dependent RNA polymerase." *J Med Chem* **49**(3): 971–83.

Tomassini, J. E., K. Getty, et al. (2005). "Inhibitory effect of 2′-substituted nucleosides on hepatitis C virus replication correlates with metabolic properties in replicon cells." *Antimicrob Agents Chemother* **49**(5): 2050–8.

Tomei, L., S. Altamura, et al. (2003). "Mechanism of action and antiviral activity of benzimidazole-based allosteric inhibitors of the hepatitis C virus RNA-dependent RNA polymerase." *J Virol* **77**(24): 13225–31.

Tomei, L., S. Altamura, et al. (2004). "Characterization of the inhibition of hepatitis C virus RNA replication by nonnucleosides." *J Virol* **78**(2): 938–46.

Wang, M., K. K. Ng, et al. (2003). "Non-nucleoside analogue inhibitors bind to an allosteric site on HCV NS5B polymerase. Crystal structures and mechanism of inhibition." *J Biol Chem* **278**(11): 9489–95.

Yan, S., T. Appleby, et al. (2007). "Isothiazoles as active-site inhibitors of HCV NS5B polymerase." *Bioorg Med Chem Lett* **17**(1): 28–33.

Yan, S., T. Appleby, et al. (2006). "Structure-based design of a novel thiazolone scaffold as HCV NS5B polymerase allosteric inhibitors." *Bioorg Med Chem Lett* **16**(22): 5888–91.

Yang, W., Y. Sun, et al. (2007). "Hepatitis C Virus (HCV) NS5B Nonnucleoside Inhibitors Specifically Block Single-Stranded Viral RNA Synthesis Catalyzed by HCV Replication Complexes In Vitro." *Antimicrob Agents Chemother* **51**(1): 338–42.

Chapter 24
HIV-1 Reverse Transcriptase Inhibitors and Mechanisms of Resistance

Bruno Marchand and Stefan G. Sarafianos

Due to its vital role in the viral life cycle and to the lack of a mammalian equivalent, the reverse transcriptase (RT) is one of the main targets in antiretroviral therapy. More than half of the drugs approved for the treatment of HIV infections target the RT of the virus (http://www.fda.gov/oashi/aids/virals.html). Inhibitors targeting RT are divided in two classes: nucleoside analog RT inhibitors (NRTIs) and non-nucleoside analogs RT inhibitors (NNRTIs). Both classes target the polymerase activity of RT, but differ significantly in their mechanisms of inhibition and resistance.

Nucleoside Analog RT Inhibitors

Nucleoside analog RT inhibitors (NRTIs) resemble nucleosides, the building blocks that are used by DNA polymerases for the synthesis of DNA. NRTIs may have modifications present in both the sugar and the base moieties. All NRTIs used in the clinic lack a 3′-OH group on the sugar moiety, necessary for phosphodiester bond formation. The 3′-OH group may be either absent or replaced by another chemical group. There are eight nucleoside analogs currently approved for the treatment of HIV infections: 3′-azido-2′,3′-dideoxythymidine (AZT, zidovudine), 2′,3′-dideoxyinosine (ddI, didanosine), 2′,3′-dideoxycytidine (ddC, zalcitabine), 2′,3′-didehydro-2′,3′-dideoxythymidine (d4T, stavudine), (−)-β-L-3′-thia-2′,3′-dideoxycytidine (3TC, lamivudine), (1*S*, 4*R*)-4-[2-amino-6-(cyclopropyl-amino)-9*H*-purin-9-yl]-2-cyclopentene-1-methanol succinate (ABC, abacavir), (-)-β-L-3′-thia-2′,3′-dideoxy-5-fluorocytidine (FTC, emtricitabine), and (*R*)-9-(2-phosphonylmethoxypropyl)adenine (TDF, tenofovir) (Fig. 24.1). Nucleoside analogs are administered as unphosphorylated prodrugs and are phopshorylated to their active triphosphate form by cellular kinases. The only exception is tenofovir (Fig. 24.1), which already contains a phosphonate group

S.G. Sarafianos (✉)
Department of Molecular Microbiology and Immunology, University of Missouri School of Medicine, 471d Life Sciences Center, 1201 E. Rollins Drive, Columbia, MO 65211-7310, USA
e-mail: sarafianoss@missouri.edu

C.E. Cameron et al. (eds.), *Viral Genome Replication*,
DOI 10.1007/b135974_24, © Springer Science+Business Media, LLC 2009

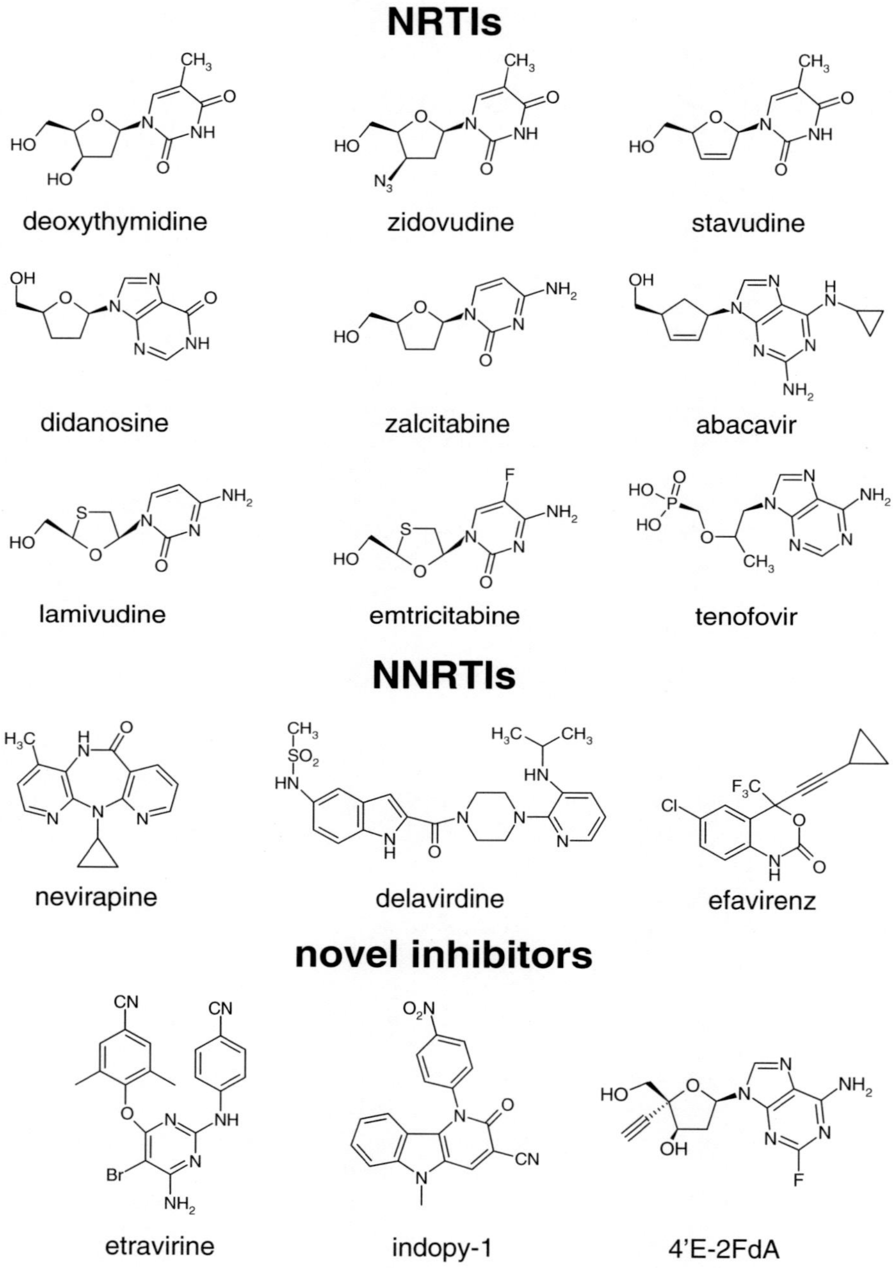

NRTIs

deoxythymidine zidovudine stavudine

didanosine zalcitabine abacavir

lamivudine emtricitabine tenofovir

NNRTIs

nevirapine delavirdine efavirenz

novel inhibitors

etravirine indopy-1 4'E-2FdA

Fig. 24.1 Chemical structures of HIV-1 reverse transcriptase inhibitors currently approved for clinical use or under development. Deoxythymidine, a natural nucleoside, was included for comparison purpose.

replacing the α-phosphate present on natural nucleotides. Hence, tenofovir requires the addition of only two phosphate groups for activation. The active form of all NRTIs (triphosphate or TP) competes with natural nucleotides for incorporation in the growing DNA chain. Once incorporated, they act as chain-terminators due to the lack of the 3′-OH group and they arrest further DNA synthesis (Mitsuya et al. 1987). Abacavir and didanosine (Fig. 24.1) must undergo additional modifications by cellular enzymes on their base moiety to create the active forms: carbovir-triphosphate and dideoxyadenosine-triphosphate (ddATP), respectively. There are significant differences in the rates by which RT incorporates the various analogs with respect to the corresponding natural nucleotides. For example, the thymidine analogs AZT-TP and d4T-TP are barely discriminated against when competing with dTTP for incorporation (Kerr and Anderson 1997). Conversely, the incorporation of ddCTP is carried out less efficiently than dCTP (Feng and Anderson 1999). 3TC-TP, a cytidine analog, has a lower level of incorporation than dCTP and ddCTP (Krebs et al. 1997). Finally, the adenosine analog tenofovir–diphosphate is not as good a substrate for HIV RT as is dATP. The basis for this discrepancy lies mainly on differences in the interactions of RT active site residues with the incoming dNTPs (deoxynucleotide triphosphates) or the activated NRTIs.

While incorporation of NRTIs efficiently terminates DNA elongation, RT is able to remove an incorporated chain-terminator in the presence of pyrophosphate, or a pyrophosphate donor such as ATP (Arion et al. 1998, Meyer et al. 1998). The removal reaction resembles the nucleotide incorporation in reverse. The enzyme uses a pyrophosphate donor to attack the phosphate junction between the chain-terminator and the penultimate nucleotide and reforms the triphosphate form of the analog or a dinucleotide tetraphosphate, respectively. The rate of removal varies significantly between different nucleotide analogs, with AZT-MP being removed at a relatively fast rate compared to other chain-terminators, and 3TC-MP being cleaved at a very slow rate (Isel et al. 2001, Mas et al. 2002, Naeger et al. 2002). It should be noted, however, that the template sequence used affects the rate of removal of a chain-terminator (Meyer et al. 2004) and that no systematic analysis of the rate of removal of the different NRTIs was conducted on an identical sequence.

Mechanisms of Resistance to Nucleoside Analog RT Inhibitors

Highly active antiretroviral therapy (HAART) has resulted in dramatic reductions in the morbidity and mortality associated with HIV infections (Palella et al. 1998). HAART regimens are most often composed of a backbone of two NRTIs and either a protease inhibitor or an NNRTI. The relatively low fidelity of HIV RT (approximately one mutation per transcribed genome) (Menendez-Arias 2002, Svarovskaia et al. 2003) results in quasi-species that have errors in the viral genomic sequences that code for HIV proteins, including RT (Domingo 2002). Under the pressure of inhibitors, HIV variants are selected harboring mutations that confer drug resistance to specific inhibitors or an entire class of drugs. Resistance to NRTIs can occur with a single mutation, as is the case with 3TC (Tisdale et al. 1993), FTC (Schinazi et al.

1993), or tenofovir (Wainberg et al. 1999), or may be the result of the accumulation of several mutations, as is the case with the AZT resistance mutations (Table 24.1).

At least two distinct mechanisms of resistance to NRTIs have been described. In the first one, the mutant enzyme discriminates against the nucleoside analog at the level of incorporation into DNA. This results in lower levels of chain-termination. The second mechanism involves an excision reaction that selectively unblocks chain-terminated primers using pyrophosphate or ATP as a pyrophosphate donor. In this case, the mutant enzyme is more efficient at removing the nucleotide analog, increasing the possibility of completing viral replication.

Mechanism of Discrimination

The best known example of resistance through discrimination at the level of incorporation is the mechanism of HIV resistance to 3TC and FTC. A single mutation at Met184 of HIV-1 RT (M184V/I) causes high-level resistance to these NRTIs. Met184 is part of the conserved Tyr183-Met184-Asp185-Asp186 motif (YMDD) (Fig. 24.2A and discussed in Wendeler et al. in this volume). The biochemical basis of this mechanism of resistance has been studied extensively. In steady-state kinetic assays, K_m and V_{max} values for dCTP incorporation are the same for wild-type and the 3TC-resistant M184V RT. However, incorporation of 3TC-MP by the resistant enzyme was decreased significantly (Quan et al. 1996). Pre-steady-state studies with M184V RT demonstrated that the decreased incorporation was due to a decrease in catalytic turnover (k_{pol}) rather than a decrease in the affinity for 3TC-TP (Krebs et al. 1997). The M184V enzyme also confers low level of resistance (4- to 8-fold) to ddI and ddC in cell culture (Gu et al. 1992) in accordance with the low-level resistance observed in vivo.

The structural description of the mechanism of resistance to 3TC has been based on the crystal structures of the ternary complex of HIV RT with DNA and an incoming dNTP (Huang et al. 1998) and of the 3TC-resistant M184I RT bound to a double-stranded DNA substrate (Sarafianos et al. 1999). When dCTP and 3TC-TP are modeled in the structure of the 3TC-resistant enzyme (Sarafianos et al. 1999) the biggest difference is observed in the positioning of the sugar moiety of the nucleotides. 3TC is an L enantiomer of the normal nucleotide, and as such, it projects its oxathiolane ring toward residue 184. This results in steric hindrance between the β-branched side chain of Ile184 and the oxathiolane ring of 3TC-TP (or FTC-TP) and a repositioning of the inhibitor that decreases its rate of incorporation. This model is consistent with the observation that other RT mutants with β-branched residues at position 184 (valine or threonine) would have increased resistance to 3TC and FTC and predicts that inhibitors that are designed to have reduced steric conflict should have an improved resistance profile with the M184V/I mutants.

The L74V RT mutation develops during the treatment of HIV-infected patients treated with ddI (St Clair et al. 1991). It confers moderate resistance to ddI (6- to 26-fold) and low-level resistance to ddC (~15-fold) (Winters et al. 1997). L74V in combination with M184V also confers low-level resistance to ABC (Harrigan et al. 2000). Enzymatic assays using wild-type and L74V RT show no significant

Table 24.1 Reverse transcriptase mutations associated to resistance to NRTIs and NNRTIs

Drug		Resistance mutations
	Nucleoside Analogue Reverse Transcriptase Inhibitors (NRTIs)	
Zidovudine	Excision	M41L, D67N, K70R, L210W, T215F/Y, K219Q/E, 69ins[1]
(AZT)	Discrimination	MDR[2]
Stavudine	Excision	M41L, D67N, K70R, V75T, L210W, T215F/Y, K219Q/E, 69ins
(d4T)	Discrimination	MDR, V75T
Didanosine	Excision	69ins
(ddI)	Discrimination	K65R, L74V, MDR
Zalcitabine	Excision	69ins
(ddC)	Discrimination	K65R, L74V, MDR
Abacavir	Excision	69ins
(ABC)	Discrimination	K65R, L74V, Y115F, MDR
Lamivudine	Excision	69ins, E44A/D, V118I[3]
(3TC)	Discrimination	K65R, M184V/I
Emtricitabine	Excision	69ins
(FTC)	Discrimination	K65R, M184V/I
Tenofovir	Excision	M41L, L210W, T215Y, 69ins
(TDF)	Discrimination	K65R
	Non-nucleoside Analogue Reverse Transcriptase Inhibitors (NNRTIs)	
Nevirapine	L100I, L101E/P, K103N, V106M/A, V108I, Y181C/I, Y188L/H/C, G190A/S/E, F227L, M230L, Y318F	
Delavirdine	L100I, L101E/P, K103N, V106A/M, V108I, Y181C/I, Y188L/H/C, G190E, M230L, P236L, Y318F	
Efavirenz	L100I, L101E/P, K103N, V106M/A, V108I, Y181C/I, Y188L, G190S/E, M230L	

The information on drug-resistance mutations was obtained from the Stanford HIV Drug Resistance Database; http://hivdb.stanford.edu/cgi-bin/NRTIResiNote.cgi

[1]69ins consist of the mutation T69S and the insertion of two or more amino acids between positions 69 and 70 (generally SS, SA or SG) in combination with mutations M41L, A62V, K70R, L210W, T215Y/F, and K219Q

[2]Multi-drug-resistant (MDR) RT includes mutations Q151M, A62V, V75I, F77L, and F116Y

[3]Not confirmed by biochemical analysis

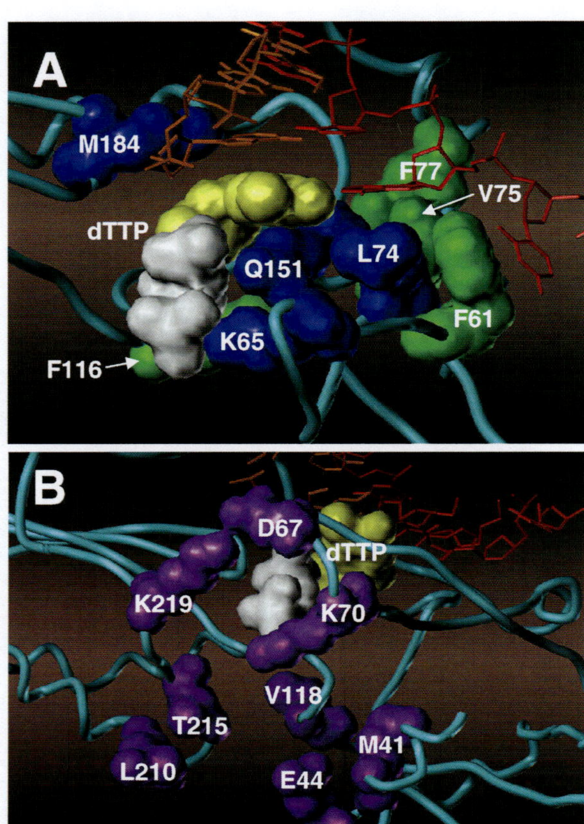

Fig. 24.2 Structure of the active site of HIV-1 RT bound to a DNA/DNA primer–template and an incoming nucleotide (dTTP) showing the main amino acids involved in NRTI resistance (pdb code 1RTD). Amino acids directly involved in the discrimination mechanism (**A**) are shown in *blue*, while amino acids generally found in the Q151M complex are shown in *green*. In (**B**), the residues involved in resistance through the excision mechanism are shown in *purple*. In both cases, the DNA primer is shown in *orange* and the DNA template in *red*. The bound-incoming nucleotide (Van Der Waals volume) is shown in *yellow* with the triphosphate moiety in *white*. The main chain of p66 is shown in *cyan*.

difference in the kinetic constants for dNTP incorporation. However, the K_i values for ddATP and ddCTP increased for L74V by 4.5- and 3.3-fold, respectively, suggesting that the mechanism of resistance is based on discrimination at the level of incorporation (Martin et al. 1993). The same conclusion can be drawn with pre-steady-state incorporation assays of dGTP and its analog abacavir (ABC) by wild-type and L74V RT. While no major difference is observed between the two enzymes for the incorporation of dGTP, a decrease in the k_{pol}/K_d ratio of 4.2-fold is observed with the L74V mutant compared to the wild-type enzyme (Ray et al. 2002).

Leu74 contacts the template base that is paired to the incoming dNTP (Fig. 24.2A). The crystal structure of the unliganded L74V enzyme shows no significant changes compared to the wild-type enzyme (Ren et al. 1998). It has been hypothesized that the diminished incorporation of ddNTPs by the L74V enzyme is due to a repositioning of the template–primer at the active site, which would yield an unstable complex for ddNTP incorporation (Boyer et al. 1994).

The K65R mutation in RT is a relatively infrequent mutation associated with didanosine, abacavir (ABC), and tenofovir DF (TDF) therapy. K65R causes reduced susceptibility to those NRTIs and to a lesser extent to 3TC and ddC (Wainberg et al. 1999, Winters et al. 1997, Harrigan et al. 2000, Miller et al. 2000). Steady-state and pre-steady-state kinetic studies have shown that the mechanism of resistance conferred by the K65R mutation is based on reduced incorporation of the inhibitor, mainly because of a decrease in the catalytic efficiency (k_{pol}) (Deval et al. 2004, White et al. 2002). The ability of K65R to escape inhibition comes at a cost, as the mutant enzyme has a reduced replicative capacity compared to the wild-type enzyme, caused by a decreased ability to use natural nucleotides (Deval et al. 2004).

In the structure of the ternary complex of wild-type RT with DNA and dNTP, the Lys65 side chain forms a hydrogen bond with the γ-phosphate of the incoming dNTP (Huang et al. 1998, Tuske et al. 2004) (Fig. 24.2A). In the structure of a ternary complex of the wild-type RT with tenofovir-diphosphate (DP) as the incoming substrate, Lys65 is in a position to make contact with the α-phosphate of tenofovir-DP as well (Tuske et al. 2004). Preliminary structural data of the ternary complex of K65R RT with DNA and tenofovir-diphosphate (DP) suggest that the mutated residue (Arg65) interacts with Arg72 of RT, a highly conserved residue known to affect the chemistry of polymerization (Sarafianos et al. 1995). The new position of Arg65 may help explain the observed decreased catalytic efficiency that has been associated with the K65R mutation (Sarafianos et al., CSH 2006). Furthermore, it has been shown that both the Lys65 and the 3′-OH of the incoming nucleotide are important for the conformation of the ternary complex. Hence, the loss of both of these contacts may contribute to the poor use of dideoxynucleotides and of other analogs lacking a 3′-OH group compared to natural deoxynucleotides.

The Q151M mutation confers resistance to a wide variety of NRTIs. This mutation is observed relatively rarely, especially after treatment of infected individuals with thymidine analogs (AZT or d4T) along with ddI or ddC (Shirasaka et al. 1995). Although the Q151M mutation confers NRTI resistance, it is at the expense of the viral replicative capacity. The appearance of other mutations (A62V, V75I, F77L and F116Y) along with Q151M corrects the replication deficiencies as well as increases the level of resistance (Ueno et al. 1995). An enzyme harboring all five of these mutations was resistant to ddATP, AZT-TP, d4T-TP, and ddCTP, but was still sensitive to adefovir-diphosphate (an acyclic NRTI related to tenofovir-diphosphate) and to 3TC-TP (Ueno and Mitsuya 1997).

In the structure of RT/DNA/dNTP complex, the side chain of Gln151 forms a hydrogen bond with the 3′OH group of the incoming nucleotide (Fig. 24.2A). The loss of this contact in the Q151M mutant has an effect on incorporation of the canonical dNTP substrates and their NRTI analogs. However, the change is more detrimental to the inhibitors that lack the 3′OH, because they are already affected negatively by the loss of the hydrogen bond between the 3′OH and the α-phosphate of the dNTP. This intramolecular interaction is expected to activate the α-phosphate for the nucleophilic attack and by providing torsional restraints that facilitate correct alignment of the reactants in the transition state (Huang et al. 1998).

Mechanism of Excision

Resistance to the thymidine analogs AZT and d4T occurs with the appearance of a specific set of mutations including M41L, D67N, K70R, L210W, T215Y/F, and K219Q (AZT-R) (Kellam et al. 1992, Larder and Kemp, 1989). These residues are not located at the dNTP-binding site where the incoming NRTI-triphosphate is expected to bind (Fig. 24.2B). Furthermore, they do not seem to interact with the dNTP in the crystal structure of RT in complex with DNA and dNTP (Huang et al. 1998). Biochemical studies have shown that the presence of these mutations does not affect the incorporation of nucleotide analogs in the elongating DNA chain. Instead, they accelerate the ATP-dependent removal of the incorporated NRTI that occurs at the same active site as the polymerization reaction and that allows resumption of the elongation of the formerly terminated DNA strand (Meyer et al. 1999). Detailed kinetic analysis of this reaction has revealed that the increased rate of chain-terminator removal conferred by the AZT-R set of mutations is mainly due to an increase in the maximum rate of the removal reaction while less pronounced difference in ATP binding was also observed (Ray et al. 2003). This observation suggests a favorable modification in the orientation of the bound ATP at the catalytic site of RT. A basic structure–activity relationship study on the substrate requirements for efficient excision showed that an aromatic base enhances the removal reaction, suggesting that there is a π–π interaction between the ATP and the aromatic side chain of one of the residues involved in excision (Meyer et al. 2002).

A structural model has been proposed to address the biochemical data on the excision reaction (Boyer et al. 2001). In this model, the ATP-binding cleft is surrounded by residues involved in the excision-based resistance to NRTIs (41, 44, 67, 70, 210, 215, 219) (Fig. 24.3). The aromatic ring of ATP interacts with the aromatic ring of Tyr215 through π–π interactions. This interaction changes the interactions of ATP and affects its orientation. The γ-phosphate of ATP is located near the phosphate that joins the last two nucleotides of the primer. It is likely that changes in the interactions of ATP at its binding site by introducing new mutations may affect its orientation and alter the specificity of the excision reaction and the ability of the enzyme to unblock more efficiently other NRTIs.

It was hypothesized that the NRTI-terminated primer is susceptible to excision when positioned at the pre-translocation site surrounded by residues of the nucleotide-binding site (N site). This hypothesis was supported by the inhibitory effect of the next complementary nucleotide on the removal reaction, which suggests that the chain-terminator is moved to the priming site (P site) in order to bind the incoming nucleotide, forming a dead-end complex (Meyer et al. 1998). The hypothesis was later confirmed using site-specific footprinting and cross-linking assays. It was shown with site-specific footprinting assays that complexes which have a preference for the pre-translocation state are more efficient at removing the 3′ nucleotide of the primer than complexes showing a preference for the post-translocation state (Marchand and Gotte 2003). Furthermore, using a cross-linking reaction, stable covalent RT–DNA complexes of the pre- and post-translocation states were prepared. Using these trapped complexes, it was demonstrated that only

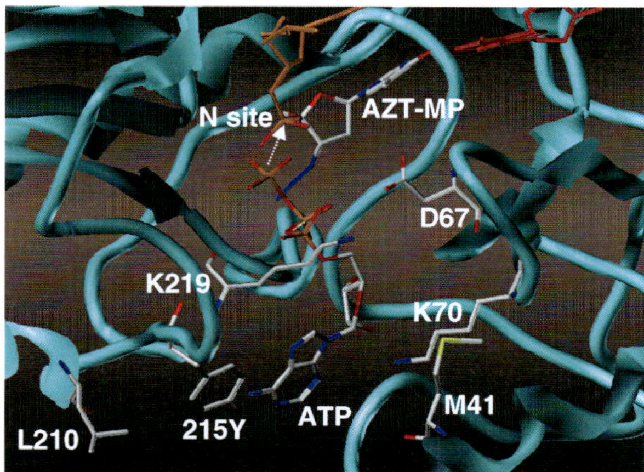

Fig. 24.3 Structural model of the active site of HIV-1 RT bound to an AZT-terminated DNA/DNA primer–template and an ATP molecule. AZT-MP, forming the 3′ end of the bound primer, is located at the nucleotide-binding site (N site), in a pre-translocation conformation. ATP forms π–π interactions with the mutated 215Y, and has its triphosphate positioned toward the α-phosphate of the AZT-MP, poised to attack the phosphodiester bond linking AZT-MP to the penultimate nucleotide of the primer. Other residues commonly associated with the excision resistance mechanism are also shown. The main chain of p66 is shown in *cyan*, and the primer and template molecules are shown in orange and red, respectively. The model was constructed using protocols described in (Larder and Kemp 1989).

the pre-translocation complex is able to conduct the removal reaction (Sarafianos et al. 2003).

As mentioned earlier, different chain-terminators are removed at different rates. However, other factors are also important in determining the extent of the excision reaction. For example, it has been shown that the excision of NRTIs is susceptible to the presence of the next incoming substrate. This susceptibility varies among different NRTIs. Specifically, the RT with mutations D67N, K70R, T215F, and K219Q were shown to remove efficiently in in vitro assays both AZT-MP and ddAMP. However, HIV clinical samples carrying these mutations are resistant in vivo only to AZT, and not to ddI, the prodrug of ddATP. This discrepancy was reconciled when it was shown that the removal of ddAMP at the 3′ end of the primer is much more susceptible to inhibition by the next complementary nucleotide (IC_{50} of 12 μM) compared to the removal of AZT (IC_{50} of 230 μM) (Meyer et al. 1999). At physiologically relevant concentrations of dNTPs, the excision of ddAMP would most likely be inhibited while the excision of AZT-MP would not. The effect of the presence of the next incoming dNTP to the excision of chain-terminated complexes can be assessed by following the stability of "dead-end complexes" in band mobility shift assays (Tong et al. 1997). These complexes contain RT with chain-terminated nucleic acid and the next complementary dNTP. They can be stable enough to be resolved on non-denaturing gels and appear as shifted bands compared to a less stable complex in the absence of dNTP. Using this assay, it was shown that ddAMP-terminated complex is more easily shifted compared to an AZT-MP terminated complex, suggesting a more efficient binding of the next complementary nucleotide when ddAMP is the chain-terminator. It was proposed that this difference is due to the bulky azido group at the 3′ end of AZT (Boyer et al. 2001). This is further supported by the cross-resistant phenotype to all azido-containing nucleoside analogs and the observation using site-specific footprinting that azido-containing analogs tend to reside in the nucleotide-binding site after incorporation at higher concentration of the next complementary nucleotide when compared to dideoxynuleotides (Marchand and Gotte 2003). Two complexes of HIV RT were crystallized with a double-stranded DNA substrate containing an AZT-MP molecule at the 3′ end of its primer ($DNA_{AZT\text{-}MP}$) (Sarafianos et al. 2002). One of the complexes had the AZT-MP molecule in its nucleotide-binding site (N site, $RT/DNA_{AZT\text{-}MP\,(N)}$), prior to translocation (pre-translocation complex), while the other had the AZT-MP at the priming site (P site, $RT/DNA_{AZT\text{-}MP\,(P)}$), after translocation (post-translocation complex). The structure of the post-translocation complex $RT/DNA_{AZT\text{-}MP\,(P)}$ shows that the azido group of AZT-MP at the P site is flexible in this binary complex. Modeling of an incoming nucleotide in this complex ($RT/DNA_{AZT\text{-}MP\,(P)}/dNTP$) predicts steric hindrance between the azido group of $DNA_{AZT\text{-}MP\,(P)}$ and the incoming nucleotide. This strongly suggests that the azido group prevents the next complementary nucleotide from inhibiting the excision reaction.

AZT resistance mutations are also selected by the NRTI d4T (Coakley et al. 2000), and the presence of these mutations decreases the efficacy of d4T in antiretroviral treatment (Izopet et al. 1999). In addition to d4T, mutations M41L and L210W in combination with other AZT-R mutations have also been associated with reduced sensitivity to the nucleoside analog tenofovir (Miller et al. 2004). Resistance to these analogs has been related to the excision reaction (Meyer et al. 2000, White et al. 2004). The level of resistance observed with d4T and tenofovir is lower when compared to AZT, which can be explained by the higher susceptibility to nucleotide concentration for inhibition of the removal reaction (Meyer et al. 1999, Marchand et al. 2007).

Another factor that seems to affect the excision reaction is the presence of specific mutations. In that respect, some mutational patterns appear to be incompatible. For instance, when the 3TC-resistance mutation M184V is added to an enzyme that has the AZT-R-resistance mutations, it abrogates the ability of the enzyme to unblock AZT-terminated primers, thus causing resensitization to AZT (Larder et al. 1995). Similar phenotypes have been observed with the mutations K65R and L74V (St Clair et al. 1991), as well as the NNRTI-resistance mutations L100I and Y181C (Larder 1992, Byrnes et al. 1994), and the foscarnet-resistance mutations W88G and E89K (Tachedjian et al. 1996). It has been demonstrated that many of these mutations decrease the rate of AZT removal from the 3′ end of the primer, thus decreasing the resistance level of AZT resistant enzymes (Boyer et al. 2002b, Gotte et al. 2000, Miranda et al. 2005, Selmi et al. 2003, Sluis-Cremer et al. 2000, White et al. 2005). In the case of the M184V mutation, it has been proposed that its decreased rate of excision is related to a repositioning of the nucleic acid in the mutant enzyme which may affect the alignment of the excision reaction components. Unfortunately, the resensitization to AZT conferred by the M184V mutation can be compensated for by the appearance of other mutations (H208Y, R211K and L214F) in addition to the classical AZT-R-resistance mutations (Sturmer et al. 2003). These residues are in close proximity to the proposed ATP-binding site and may alter the orientation of ATP in a favorable manner for the excision reaction to occur in the presence of the M184V mutation.

Other RT mutations affect the specificity of the excision reaction in a different way. Such mutations are usually selected during HIV-1 treatment with multiple NRTIs. For example, while the AZT-R mutations (M41L, D67N, K70R, L210W, T215Y/F and K219Q) cause resistance to AZT, addition of the E44D/A mutation in the AZT-R background confers additional moderate resistance to 3TC (Hertogs et al. 2000). It was shown that the 3TC resistance due to mutation E44D/A is associated with the excision mechanism (Girouard et al. 2003). In the presence of the latter mutation, 3TC is more efficiently cleaved from the 3′ end of the primer.

Another mutation that appears in the background of the AZT-R resistance mutations and broadens the specific of the excision reaction is the "fingers insertion complex". This complex consists of an AZT-R mutation backbone expanded by a mutational pattern at the fingers subdomain of RT. Specifically, it has an additional mutation at residue 69 of Thr to Ser and an insertion of two or more amino acids (usually SS, SA or SG) between residues 69 and 70. This mutational pattern confers resistance to all NRTIs used in the treatment of HIV infections (De Antoni et al. 1997, Winters et al. 1998). It has been shown that the fingers insertion complex confers multi-NRTI resistance by increasing the rate of chain-terminator removal (White et al. 2004, Mas et al. 2000). Site-specific footprinting assays showed that the increase in chain-terminator removal is due to an increased access to the pre-translocational state (Marchand and Gotte 2003) associated with a decreased binding for NRTI and dNTP (Mas et al. 2000). This observation is also supported by gel mobility shift assays (Boyer et al. 2002a).

It was demonstrated that the selection of a mechanism of resistance is based on the initial properties of the inhibitor. For instance, nucleoside analogs that are poorly incorporated, such as 3TC, will favor a discrimination mechanism of resistance as opposed to AZT, which is efficiently incorporated, but also easily removed, which will favor the removal mechanism (Isel et al. 2001).

Non-nucleoside Analog RT Inhibitors

Non-nucleoside analog RT inhibitors are part of a chemically diverse class of inhibitors (Fig. 24.1). They all bind at the same hydrophobic pocket at the base of the thumb of the p66 subunit and inhibit the nucleotide incorporation activity of RT. There are three NNRTIs approved for the treatment of HIV infections; nevirapine, delavirdine, and efavirenz. Crystal structures of HIV RT in complex with these compounds showed that all three of these inhibitors bind to the NNRTI-binding pocket (NNIBP) which is close to (~10 Å), but distinct from the polymerase active site of RT (Esnouf et al. 1997, Kohlstaedt et al. 1992, Lindberg et al. 2002), which is defined by three aspartic acid residues (D185, D186, and D110 of the β6-β9-β10 sheet). The NNIBP is formed mainly by components of two β-sheets from the p66 subunit. First, by the β9-β10 hairpin of the polymerase active site and the β6-β9-β10 sheet. Second, by the β12-β13 hairpin or "primer grip", a structure that is known to position the primer strand at the polymerase active site. In addition to the p66 residues, the NNIBP also consists of the β7-β8-connecting loop of p51. Specific residues that form the NNIBP are Y181, Y188, F227, W229, Y318, P95, L100, V106, V179, L234, and P236 from p66 and E138 from p51.

Mechanism of Action

Comparison of the structures of many RT/NNRTI complexes with those of RT in complex with substrates (DNA or DNA and dNTP) or in the unliganded form reveals that binding of NNRTI causes a number of conformational changes in HIV RT (Sarafianos et al. 2004, Ren et al. 1995, Ding et al. 1995). First, the β12-β13 hairpin (primer grip) which is part of the β12-β13-β14 sheet is substantially displaced to make room for the incoming NNRTI. Second, because the β14 strand of the β12-β13-β14 sheet is part of the p66 thumb, the whole p66 thumb subdomain moves from a folded-down position in the unliganded form to an upright configuration which is even more extended than that seen in the RT/DNA or RT/DNA/dNTP complexes (Rodgers et al. 1995). Interestingly, the NNIBP is not present in the unliganded form of the enzyme. The specific space is occupied by the side chains of Y181, Y188, and W229 in the unliganded structure. Binding of NNRTIs results in the reorientation of these chains, creating space to accommodate the incoming NNRTI. Despite the availability of multiple RT/NNRTI crystal structures from the research groups of Dr. E. Arnold, Dr. D. Stammers, Dr. T. Steitz, and Dr. T. Unge (reviewed in Sarafianos et al. 2004), the structures of the mechanistically important RT/DNA/NNRTI and/or RT/DNA/dNTP/NNRTI complexes are still missing.

Several kinetic studies have contributed to our understanding of the mechanism of action of NNRTIs. Steady-state kinetic studies have shown that inhibition is non-competitive with respect to both template–primer and dNTP. Pre-steady-state kinetic studies have shown that NNRTIs do not affect significantly the binding constants for nucleic acid or dNTP. Instead, they appear to inhibit the chemical step of phosphodiester bond formation and render it the rate-limiting step of the polymerazation reaction (Rittinger et al. 1995, Spence et al. 1995). According to available structures, this decrease in the catalytic rate may be due to structural changes close to the active site as well as a repositioning of the primer grip, which may affect the orientation of the primer and the incoming nucleotide. Also, the thumb repositioning is likely to restrict the mobility of the thumb, a requisite for efficient polymerization. Furthermore, photoaffinity cross-linking experiments also showed that the fingers mobility is reduced in the presence of NNRTIs (Peletskaya et al. 2004). These observations suggest a more rigid structure of the polymerase domain in the presence of NNRTIs which can yield unfavorable complexes for nucleotide incorporation. Importantly, structural analysis of the intermediates of the polymerization reaction (Sarafianos et al. 2002) as well as recent crystallographic evidence suggests that NNRTIs restrict the flexibility of the YMDD loop and prevent the catalytic aspartate residues from adopting their metal-binding conformations (Das et al. 2007).

Mechanisms of NNRTI Resistance

Unlike the NRTI-resistance mutations that are dispersed throughout the polymerase subdomain of RT, mutations conferring resistance to NNRTIs are all located in or around the NNIBP (Fig. 24.4). The most frequent mutations observed during NNRTI treatment are Y181C, Y188C/L, and K103N (Mellors et al. 1992, Nunberg et al. 1991, Richman et al. 1991). Other mutations include L100I, V106A, and G190A .

The aromatic rings of Tyr181 and Tyr188 make important contributions to the binding of NNRTIs (Kohlstaedt et al. 1992, Ren et al. 1995). Mutation to non-aromatic residues, usually Y181C and Y188C/L reduces the affinity for NNRTIs due a loss of contacts at these two residues. These residues are less important in efavirenz binding, therefore mutations at 181 or 188 do not confer a high level of resistance to this inhibitor (Ren et al. 2000).

K103N confers resistance to all NNRTIs. Crystal structures showed a hydrogen bond between Tyr188 and Asn103 in the absence of NNRTI, which closes the entrance to the NNIBP, efficiently reducing its access to multiple NNRTIs (Lindberg et al. 2002, Ren et al. 2000, Hsiou et al. 2001).

The L100I mutation results in resistance to delavirdine. Crystal structures showed that the presence of an Ile at position 100 causes steric hindrance either with the NNRTI or in the NNIBP (Ren et al. 2004). The NNIBP is not rigid and can adapt to the conformation of different NNRTIs. NNRTIs can also adapt to different conformation of the NNIBP, but delavirdine cannot easily reposition itself in the presence of L100I rendering it vulnerable to this mutation.

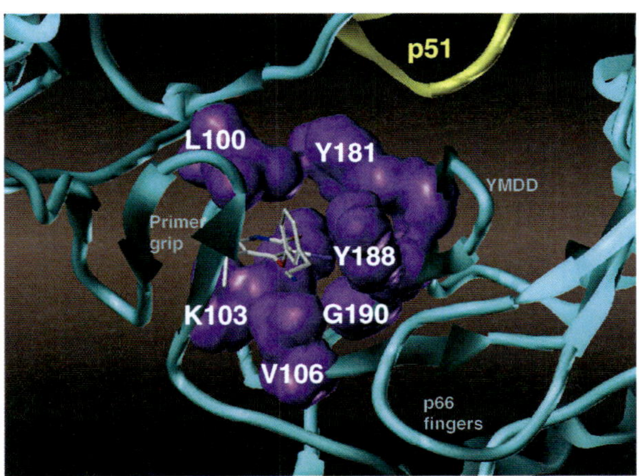

Fig. 24.4 Structure of HIV-1 RT bound to the NNRTI TMC125 (pdb code 2B5J). The main chain of p66 is shown in *cyan*, while the main chain of p51 is shown in *yellow*. The amino acids commonly associated with resistance to NNRTIs are displayed in *violet*. TMC125 is colored by atoms, C being *grey*, O *red*, and N *blue*.

Mutation V106A confers resistance mainly to nevirapine. Crystal structures of wild-type RT and the V106A mutant show little structural differences (Ren et al. 2004). One difference is the loss of a Van der Waals interaction with the smaller Ala at 106 compared to the wild-type Val. A perturbation of the nevirapine's ring stacking with Tyr181 and Tyr188 can also be observed. It seems that those small changes are sufficient to decrease the susceptibility of RT to nevirapine.

The mutation G190A is selected by a NNRTI not used in the clinic, HBY 097, and causes cross-resistance to nevirapine and efavirenz. A crystal structure of RT with HBY 097 shows a contact between the inhibitor and the Gly190 (Hsiou et al. 1998). Modeling of an Ala at position 190 shows that steric hindrance between the Ala190 and the inhibitor would occur (Sarafianos et al. 2004). Similar conflicts could be expected with other NNRTIs.

Connection and RNase H Subdomain Resistance Mutations

Recently, it has been reported that a number of mutations at the connection, or RNase H subdomains of RT, affect resistance to RT inhibitors (Nikolenko et al. 2007, Delviks-Frankenberry et al. 2007). The biochemical mechanism of this effect is still the subject of investigation. One of the connection subdomain mutations, N348I, was shown to be the first mutation conferring cross-resistance to both NRTIs and NNRTIs (Yap et al., *Retroviruses* 2007; Hachiya et al., *CROI* 2007; Yap et al., *CROI* 2007). This mutation is present in a significant number of clinical isolates in the presence of other NRTI-resistance mutations. The involvement of residues from the connection and RNase H subdomains in drug resistance highlights the need for expanding the current genotypic assays that currently do not characterize the entire pol region.

Novel RT Inhibitors

Although a number of RT inhibitors are available for the treatment of HIV-infected patients, the development of resistance to these drugs highlights the need for novel inhibitors that are active against drug-resistant viruses. New molecules that inhibit RT by novel mechanisms are under development.

A multidisciplinary effort has led to the discovery of the diaryltriazine (DATA) (Das et al. 2004) and diarylpyrimidine (DAPY) classes of NNRTIs that are highly potent against a wide range of HIV-1-resistant viral strains when tested in recombinant virus assays and in clinical trials (Fig. 24.1). TMC125 (etravirine) decreased the HIV viral load as efficiently as a five-drug combination in naïve patients and retained potency in patients infected with NNRTI-resistant HIV-1 variants (Andries et al. 2004). TMC125 and related anti-AIDS drug candidates can bind the RT target in multiple conformations and escape the effects of drug-resistant mutations. Crystallographic studies showed that DAPY inhibitors can adapt to changes in the NNIBP that cause resistance to other NNRTIs (Das et al. 2004). A related inhibitor

(TMC278) from the same family (Janssen et al. 2005) has a much higher bioavailability and is also currently in clinical trials. It has high bioavailability (Frenkel et al. 2005) and displays outstanding potency against wild-type and viruses resistant to other NNRTIs.

Indopy is an RT inhibitor with an indolopyridone core (Fig. 24.1). It inhibits RT by a novel mechanism (Jochmans et al. 2006). The in vivo EC_{50} is 30 nM, whereas the IC_{50} in cell-free assays is 290 nM. Indopy inhibition is competitive with regard to dNTP, suggesting an overlapping binding site. Gel mobility shift assays showed that indopy stabilizes a pre-formed RT-DNA/DNA complex, as it was earlier shown for dNTPs (Tong et al. 1997), suggesting again a common binding site. Site-specific footprinting experiments confirmed that both dNTP and indopy bind to a post-translocation complex of RT-DNA/DNA. A preference for a pyrimidine, particularly T, at the 3′ end of the primer was observed. Indopy retained inhibitory activity against AZT- and NNRTI-resistant viruses. The Y115F and M184V mutations conferred low-level resistance by themselves and high-level resistance when combined. K65R conferred RT hypersensitivity to indopy. All the mutations affecting the sensitivity to indopy are located proximal to the nucleotide-binding site, suggesting that the inhibitor binds at this site.

A class of NRTIs that retains the 3′OH and also have a 4′ substituent on the ribose ring of an adenosine are unlike the classical chain-terminators and have been reported to have potent antiviral activity (Sugimoto et al. 1999). In designing NRTIs as antiviral agents, the accepted dogma has been that the 3′ OH group must be missing in order to effectively terminate the elongating DNA chain. This is highlighted by the fact that all NRTIs used in the clinic today and those in clinical trials lack a 3′OH. In contrast to this dogma, 4′-substituted NRTIs contain a 3′ OH group and are orders of magnitude more potent than all approved NRTIs (Kodama et al. 2001) (Fig. 24.1). Early compounds had 4′ substitution such as azido, cyano, methyl, fluoromethyl, ethyl, hydroxyethyl, vinyl, ethynyl, and propynyl groups (Kodama et al. 2001, Maag et al. 1994). Clearly, the 4′ analogs displaying the best inhibitory activity against HIV-1 and the best selectivity indexes are the 4′ethynyl-containing adenine analogs (Kodama et al. 2001).

The role of the 3′ OH in 4′ substituted nucleosides was analyzed by synthesizing 4′ ethynyl 2′,3′-dideoxycytosine (Siddiqui et al. 2004). The nucleoside analog had no detectable activity against HIV-1 growth in vitro at concentrations up to 10 μM. The triphosphate form of 4′ ethynyl 2′,3′-dideoxycytosine was slightly more potent than AZT in blocking DNA elongation by HIV-1 RT. This suggests that the 3′OH plays a role in the activation step of the nucleoside analog by cellular kinases. Modifications in the base moiety were also found to influence the antiviral activity. A novel analog, 4′-ethynyl 2′-deoxy-2-fluoroadenosine (Ohrui et al. 2006), was reported to be the most active antiretroviral known, inhibiting HIV with an EC_{50} of 70 picomolar in cell-based assays, and with a selectivity index up to 150,000. This inhibitor retained significant activity against multi-drug-resistant viruses isolated from patients. In DNA elongation assays, it showed very low inhibition of DNA polymerase α and β and mitochondrial DNA polymerase γ and high stability to catabolic enzymes such as adenosine deaminase (Frenkel et al. 2005). It was shown to be efficiently phosphorylated to its triphosphate form and to be able to protect cells from HIV-1 infection after being removed from the supernatant by 75% and 47% after 24 and 48 hours, respectively (Nakata et al. 2007). Hence, this compound holds promise as a next generation therapeutic for the treatment of HIV infections.

In conclusion, despite being the first antiretroviral drugs used in the clinic, RT inhibitors still constitute the backbone of highly active antiretroviral therapies (HAART). The development of resistance to RT inhibitors impairs the efficiency of HAART, which highlights the need for novel inhibitors that are efficient against RT inhibitors resistant viruses. In that respect, important progress is being made in the development of new NRTIs and NNRTIs that hold promise for the future of improved treatments of HIV infections.

References

Andries K, Azijn H, Thielemans T, Ludovici D, Kukla M, Heeres J, Janssen P, De Corte B, Vingerhoets J, Pauwels R, de Bethune MP: TMC125, a novel next-generation nonnucleoside reverse transcriptase inhibitor active against nonnucleoside reverse transcriptase inhibitor-resistant human immunodeficiency virus type 1. *Antimicrob Agents Chemother* 2004, **48:** 4680–4686.

Arion D, Kaushik N, McCormick S, Borkow G, Parniak MA: Phenotypic mechanism of HIV-1 resistance to 3′-azido-3′-deoxythymidine (AZT): increased polymerization processivity and enhanced sensitivity to pyrophosphate of the mutant viral reverse transcriptase. *Biochemistry* 1998, **37:**15908–15917.

Boyer PL, Sarafianos SG, Arnold E, Hughes SH: Nucleoside analog resistance caused by insertions in the fingers of human immunodeficiency virus type 1 reverse transcriptase involves ATP-mediated excision. *J Virol* 2002a, **76:**9143–9151.

Boyer PL, Sarafianos SG, Arnold E, Hughes SH: Selective excision of AZTMP by drug-resistant human immunodeficiency virus reverse transcriptase. *J Virol* 2001, **75:**4832–4842.

Boyer PL, Sarafianos SG, Arnold E, Hughes SH: The M184V mutation reduces the selective excision of zidovudine 5′-monophosphate (AZTMP) by the reverse transcriptase of human immunodeficiency virus type 1. *J Virol* 2002b, **76:**3248–3256.

Boyer PL, Tantillo C, Jacobo-Molina A, Nanni RG, Ding J, Arnold E, Hughes SH: Sensitivity of wild-type human immunodeficiency virus type 1 reverse transcriptase to dideoxynucleotides depends on template length; the sensitivity of drug-resistant mutants does not. *Proc Natl Acad Sci USA* 1994, **91:**4882–4886.

Byrnes VW, Emini EA, Schleif WA, Condra JH, Schneider CL, Long WJ, Wolfgang JA, Graham DJ, Gotlib L, Schlabach AJ, et al.: Susceptibilities of human immunodeficiency virus type 1 enzyme and viral variants expressing multiple resistance-engendering amino acid substitutions to reserve transcriptase inhibitors. *Antimicrob Agents Chemother* 1994, **38:**1404–1407.

Coakley EP, Gillis JM, Hammer SM: Phenotypic and genotypic resistance patterns of HIV-1 isolates derived from individuals treated with didanosine and stavudine. *Aids* 2000, **14:**F9–15.

Das K, Clark AD, Jr., Lewi PJ, Heeres J, De Jonge MR, Koymans LM, Vinkers HM, Daeyaert F, Ludovici DW, Kukla MJ, et al: Roles of conformational and positional adaptability in structure-based design of TMC125-R165335 (etravirine) and related non-nucleoside reverse transcriptase inhibitors that are highly potent and effective against wild-type and drug-resistant HIV-1 variants. *J Med Chem* 2004, **47:**2550–2560.

Das K, Sarafianos SG, Clark AD, Jr., Boyer PL, Hughes SH, Arnold E: Crystal structures of clinically relevant Lys103Asn/Tyr181Cys double mutant HIV-1 reverse transcriptase in complexes with ATP and non-nucleoside inhibitor HBY 097. *J Mol Biol* 2007, **365:**77–89.

De Antoni A, Foli A, Lisziewicz J, Lori F: Mutations in the pol gene of human immunodeficiency virus type 1 in infected patients receiving didanosine and hydroxyurea combination therapy. *J Infect Dis* 1997, **176:**899–903.

Delviks-Frankenberry KA, Nikolenko GN, Barr R, Pathak VK: Mutations in human immunodeficiency virus type 1 RNase H primer grip enhance 3′-azido-3′-deoxythymidine resistance. *J Virol* 2007, **81:**6837–6845.

Deval J, White KL, Miller MD, Parkin NT, Courcambeck J, Halfon P, Selmi B, Boretto J, Canard B: Mechanistic basis for reduced viral and enzymatic fitness of HIV-1 reverse transcriptase containing both K65R and M184V mutations. *J Biol Chem* 2004, **279:**509–516.

Ding J, Das K, Moereels H, Koymans L, Andries K, Janssen PA, Hughes SH, Arnold E: Structure of HIV-1 RT/TIBO R 86183 complex reveals similarity in the binding of diverse nonnucleoside inhibitors. *Nat Struct Biol* 1995, **2:**407–415.

Domingo E: Quasispecies theory in Virology. *J Virol* 2002, **76:**463–465.

Esnouf RM, Ren J, Hopkins AL, Ross CK, Jones EY, Stammers DK, Stuart DI: Unique features in the structure of the complex between HIV-1 reverse transcriptase and the bis(heteroaryl)piperazine (BHAP) U-90152 explain resistance mutations for this nonnucleoside inhibitor. *Proc Natl Acad Sci USA* 1997, **94:**3984–3989.

Feng JY, Anderson KS: Mechanistic studies comparing the incorporation of (+) and (−) isomers of 3TCTP by HIV-1 reverse transcriptase. *Biochemistry* 1999, **38:**55–63.

Frenkel YV, Clark AD, Jr., Das K, Wang YH, Lewi PJ, Janssen PA, Arnold E: Concentration and pH dependent aggregation of hydrophobic drug molecules and relevance to oral bioavailability. *J Med Chem* 2005, **48:**1974–1983.

Girouard M, Diallo K, Marchand B, McCormick S, Gotte M: Mutations E44D and V118I in the reverse transcriptase of HIV-1 play distinct mechanistic roles in dual resistance to AZT and 3TC. *J Biol Chem* 2003, **278:**34403–34410.

Gotte M, Arion D, Parniak MA, Wainberg MA: The M184V mutation in the reverse transcriptase of human immunodeficiency virus type 1 impairs rescue of chain-terminated DNA synthesis. *J Virol* 2000, **74:**3579–3585.

Gu Z, Gao Q, Li X, Parniak MA, Wainberg MA: Novel mutation in the human immunodeficiency virus type 1 reverse transcriptase gene that encodes cross-resistance to 2′,3′-dideoxyinosine and 2′,3′-dideoxycytidine. *J Virol* 1992, **66:**7128–7135.

Harrigan PR, Stone C, Griffin P, Najera I, Bloor S, Kemp S, Tisdale M, Larder B: Resistance profile of the human immunodeficiency virus type 1 reverse transcriptase inhibitor abacavir (1592U89) after monotherapy and combination therapy. CNA2001 Investigative Group. *J Infect Dis* 2000, **181:**912–920.

Hertogs K, Bloor S, De Vroey V, van Den Eynde C, Dehertogh P, van Cauwenberge A, Sturmer M, Alcorn T, Wegner S, van Houtte M, et al: A novel human immunodeficiency virus type 1 reverse transcriptase mutational pattern confers phenotypic lamivudine resistance in the absence of mutation 184V. *Antimicrob Agents Chemother* 2000, **44:**568–573.

Hsiou Y, Das K, Ding J, Clark AD, Jr., Kleim JP, Rosner M, Winkler I, Riess G, Hughes SH, Arnold E: Structures of Tyr188Leu mutant and wild-type HIV-1 reverse transcriptase complexed with the non-nucleoside inhibitor HBY 097: inhibitor flexibility is a useful design feature for reducing drug resistance. *J Mol Biol* 1998, **284:**313–323.

Hsiou Y, Ding J, Das K, Clark AD, Jr., Boyer PL, Lewi P, Janssen PA, Kleim JP, Rosner M, Hughes SH, Arnold E: The Lys103Asn mutation of HIV-1 RT: a novel mechanism of drug resistance. *J Mol Biol* 2001, **309:**437–445.

Huang H, Chopra R, Verdine GL, Harrison SC: Structure of a covalently trapped catalytic complex of HIV-1 reverse transcriptase: implications for drug resistance. *Science* 1998, **282:** 1669–1675.

Isel C, Ehresmann C, Walter P, Ehresmann B, Marquet R: The emergence of different resistance mechanisms toward nucleoside inhibitors is explained by the properties of the wild type HIV-1 reverse transcriptase. *J Biol Chem* 2001, **276:**48725–48732.

Izopet J, Bicart-See A, Pasquier C, Sandres K, Bonnet E, Marchou B, Puel J, Massip P: Mutations conferring resistance to zidovudine diminish the antiviral effect of stavudine plus didanosine. *J Med Virol* 1999, **59:**507–511.

Janssen PA, Lewi PJ, Arnold E, Daeyaert F, de Jonge M, Heeres J, Koymans L, Vinkers M, Guillemont J, Pasquier E, et al: In search of a novel anti-HIV drug: multidisciplinary coordination in the discovery of 4-[[4-[[4-[(1E)-2-cyanoethenyl]-2,6-dimethylphenyl]amino]-2- pyrimidinyl]amino]benzonitrile (R278474, rilpivirine). *J Med Chem* 2005, **48:** 1901–1909.

Jochmans D, Deval J, Kesteleyn B, Van Marck H, Bettens E, De Baere I, Dehertogh P, Ivens T, Van Ginderen M, Van Schoubroeck B, et al: Indolopyridones inhibit human immunodeficiency virus reverse transcriptase with a novel mechanism of action. *J Virol* 2006, **80:** 12283–12292.

Kellam P, Boucher CA, Larder BA: Fifth mutation in human immunodeficiency virus type 1 reverse transcriptase contributes to the development of high-level resistance to zidovudine. *Proc Natl Acad Sci USA* 1992, **89:**1934–1938.

Kerr SG, Anderson KS: Pre-steady-state kinetic characterization of wild type and 3′-azido-3′-deoxythymidine (AZT) resistant human immunodeficiency virus type 1 reverse transcriptase: implication of RNA directed DNA polymerization in the mechanism of AZT resistance. *Biochemistry* 1997, **36:**14064–14070.

Kodama EI, Kohgo S, Kitano K, Machida H, Gatanaga H, Shigeta S, Matsuoka M, Ohrui H, Mitsuya H: 4′-Ethynyl nucleoside analogs: potent inhibitors of multidrug-resistant human immunodeficiency virus variants in vitro. *Antimicrob Agents Chemother* 2001, **45:**1539–1546.

Kohlstaedt LA, Wang J, Friedman JM, Rice PA, Steitz TA: Crystal structure at 3.5 A resolution of HIV-1 reverse transcriptase complexed with an inhibitor. *Science* 1992, **256:**1783–1790.

Krebs R, Immendorfer U, Thrall SH, Wohrl BM, Goody RS: Single-step kinetics of HIV-1 reverse transcriptase mutants responsible for virus resistance to nucleoside inhibitors zidovudine and 3-TC. *Biochemistry* 1997, **36:**10292–10300.

Larder BA, Kemp SD, Harrigan PR: Potential mechanism for sustained antiretroviral efficacy of AZT-3TC combination therapy. *Science* 1995, **269:**696–699.

Larder BA, Kemp SD: Multiple mutations in HIV-1 reverse transcriptase confer high-level resistance to zidovudine (AZT). *Science* 1989, **246:**1155–1158.

Larder BA: 3′-Azido-3′-deoxythymidine resistance suppressed by a mutation conferring human immunodeficiency virus type 1 resistance to nonnucleoside reverse transcriptase inhibitors. *Antimicrob Agents Chemother* 1992, **36:**2664–2669.

Lindberg J, Sigurdsson S, Lowgren S, Andersson HO, Sahlberg C, Noreen R, Fridborg K, Zhang H, Unge T: Structural basis for the inhibitory efficacy of efavirenz (DMP-266), MSC194 and PNU142721 towards the HIV-1 RT K103N mutant. *Eur J Biochem* 2002, **269:** 1670–1677.

Maag H, Nelson JT, Steiner JL, Prisbe EJ: Solid-state and solution conformations of the potent HIV inhibitor, 4′-azidothymidine. *J Med Chem* 1994, **37:**431–438.

Marchand B, Gotte M: Site-specific footprinting reveals differences in the translocation status of HIV-1 reverse transcriptase. Implications for polymerase translocation and drug resistance. *J Biol Chem* 2003, **278:**35362–35372.

Marchand B, White KL, Ly JK, Margot NA, Wang R, McDermott M, Miller MD, Gotte M: Effects of the translocation status of human immunodeficiency virus type 1 reverse transcriptase on the efficiency of excision of tenofovir. *Antimicrob Agents Chemother* 2007, **51:**2911–2919.

Martin JL, Wilson JE, Haynes RL, Furman PA: Mechanism of resistance of human immunodeficiency virus type 1 to 2′,3′-dideoxyinosine. *Proc Natl Acad Sci USA* 1993, **90:** 6135–6139.

Mas A, Parera M, Briones C, Soriano V, Martinez MA, Domingo E, Menendez-Arias L: Role of a dipeptide insertion between codons 69 and 70 of HIV-1 reverse transcriptase in the mechanism of AZT resistance. *Embo J* 2000, **19:**5752–5761.

Mas A, Vazquez-Alvarez BM, Domingo E, Menendez-Arias L: Multidrug-resistant HIV-1 reverse transcriptase: involvement of ribonucleotide-dependent phosphorolysis in cross-resistance to nucleoside analogue inhibitors. *J Mol Biol* 2002, **323:**181–197.

Mellors JW, Dutschman GE, Im GJ, Tramontano E, Winkler SR, Cheng YC: In vitro selection and molecular characterization of human immunodeficiency virus-1 resistant to non-nucleoside inhibitors of reverse transcriptase. *Mol Pharmacol* 1992, **41:**446–451.

Menendez-Arias L: Molecular basis of fidelity of DNA synthesis and nucleotide specificity of retroviral reverse transcriptases. *Prog Nucleic Acid Res Mol Biol* 2002, **71:**91–147.

Meyer PR, Matsuura SE, Mian AM, So AG, Scott WA: A mechanism of AZT resistance: an increase in nucleotide-dependent primer unblocking by mutant HIV-1 reverse transcriptase. *Mol Cell* 1999, **4:**35–43.

Meyer PR, Matsuura SE, Schinazi RF, So AG, Scott WA: Differential removal of thymidine nucleotide analogues from blocked DNA chains by human immunodeficiency virus reverse transcriptase in the presence of physiological concentrations of 2′-deoxynucleoside triphosphates. *Antimicrob Agents Chemother* 2000, **44:**3465–3472.

Meyer PR, Matsuura SE, So AG, Scott WA: Unblocking of chain-terminated primer by HIV-1 reverse transcriptase through a nucleotide-dependent mechanism. *Proc Natl Acad Sci USA* 1998, **95:**13471–13476.

Meyer PR, Matsuura SE, Tolun AA, Pfeifer I, So AG, Mellors JW, Scott WA: Effects of specific zidovudine resistance mutations and substrate structure on nucleotide-dependent primer unblocking by human immunodeficiency virus type 1 reverse transcriptase. *Antimicrob Agents Chemother* 2002, **46:**1540–1545.

Meyer PR, Smith AJ, Matsuura SE, Scott WA: Effects of primer-template sequence on ATP-dependent removal of chain-terminating nucleotide analogues by HIV-1 reverse transcriptase. *J Biol Chem* 2004, **279:**45389–45398.

Miller MD, Margot N, Lu B, Zhong L, Chen SS, Cheng A, Wulfsohn M: Genotypic and phenotypic predictors of the magnitude of response to tenofovir disoproxil fumarate treatment in antiretroviral-experienced patients. *J Infect Dis* 2004, **189:**837–846.

Miller V, Ait-Khaled M, Stone C, Griffin P, Mesogiti D, Cutrell A, Harrigan R, Staszewski S, Katlama C, Pearce G, Tisdale M: HIV-1 reverse transcriptase (RT) genotype and susceptibility to RT inhibitors during abacavir monotherapy and combination therapy. *Aids* 2000, **14:** 163–171.

Miranda LR, Gotte M, Liang F, Kuritzkes DR: The L74V mutation in human immunodeficiency virus type 1 reverse transcriptase counteracts enhanced excision of zidovudine monophosphate associated with thymidine analog resistance mutations. *Antimicrob Agents Chemother* 2005, **49:**2648–2656.

Mitsuya H, Jarrett RF, Matsukura M, Di Marzo Veronese F, DeVico AL, Sarngadharan MG, Johns DG, Reitz MS, Broder S: Long-term inhibition of human T-lymphotropic virus type III/lymphadenopathy-associated virus (human immunodeficiency virus) DNA synthesis and RNA expression in T cells protected by 2′,3′-dideoxynucleosides in vitro. *Proc Natl Acad Sci USA* 1987, **84:**2033–2037.

Naeger LK, Margot NA, Miller MD: ATP-dependent removal of nucleoside reverse transcriptase inhibitors by human immunodeficiency virus type 1 reverse transcriptase. *Antimicrob Agents Chemother* 2002, **46:**2179–2184.

Nakata H, Amano M, Koh Y, Kodama E, Yang G, Bailey CM, Kohgo S, Hayakawa H, Matsuoka M, Anderson KS, et al: Activity against human immunodeficiency virus type 1, intracellular metabolism, and effects on human DNA polymerases of 4′-ethynyl-2-fluoro-2′-deoxyadenosine. *Antimicrob Agents Chemother* 2007, **51:**2701–2708.

Nikolenko GN, Delviks-Frankenberry KA, Palmer S, Maldarelli F, Fivash MJ, Jr., Coffin JM, Pathak VK: Mutations in the connection domain of HIV-1 reverse transcriptase increase 3′-azido-3′-deoxythymidine resistance. *Proc Natl Acad Sci USA* 2007, **104:**317–322.

Nunberg JH, Schleif WA, Boots EJ, O′Brien JA, Quintero JC, Hoffman JM, Emini EA, Goldman ME: Viral resistance to human immunodeficiency virus type 1-specific pyridinone reverse transcriptase inhibitors. *J Virol* 1991, **65:**4887–4892.

Ohrui H, Kohgo S, Hayakawa H, Kodama E, Matsuoka M, Nakata T, Mitsuya H: 2′-Deoxy-4′-C-ethynyl-2-fluoroadenosine: a nucleoside reverse transcriptase inhibitor with highly potent activity against all HIV-1 strains, favorable toxic profiles and stability in plasma. *Nucleic Acids Symp Ser (Oxf)* 2006:1–2.

Palella FJ, Jr., Delaney KM, Moorman AC, Loveless MO, Fuhrer J, Satten GA, Aschman DJ, Holmberg SD: Declining morbidity and mortality among patients with advanced human immunodeficiency virus infection. HIV Outpatient Study Investigators. *N Engl J Med* 1998, **338:**853–860.

Peletskaya EN, Kogon AA, Tuske S, Arnold E, Hughes SH: Nonnucleoside inhibitor binding affects the interactions of the fingers subdomain of human immunodeficiency virus type 1 reverse transcriptase with DNA. *J Virol* 2004, **78:**3387–3397.

Quan Y, Gu Z, Li X, Li Z, Morrow CD, Wainberg MA: Endogenous reverse transcription assays reveal high-level resistance to the triphosphate of (−)2′-dideoxy-3′-thiacytidine by mutated M184V human immunodeficiency virus type 1. *J Virol* 1996, **70:**5642–5645.

Ray AS, Basavapathruni A, Anderson KS: Mechanistic studies to understand the progressive development of resistance in human immunodeficiency virus type 1 reverse transcriptase to abacavir. *J Biol Chem* 2002, **277:**40479–40490.

Ray AS, Murakami E, Basavapathruni A, Vaccaro JA, Ulrich D, Chu CK, Schinazi RF, Anderson KS: Probing the molecular mechanisms of AZT drug resistance mediated by HIV-1 reverse transcriptase using a transient kinetic analysis. *Biochemistry* 2003, **42:**8831–8841.

Ren J, Esnouf R, Garman E, Somers D, Ross C, Kirby I, Keeling J, Darby G, Jones Y, Stuart D, et al.: High resolution structures of HIV-1 RT from four RT-inhibitor complexes. *Nat Struct Biol* 1995, **2:**293–302.

Ren J, Esnouf RM, Hopkins AL, Jones EY, Kirby I, Keeling J, Ross CK, Larder BA, Stuart DI, Stammers DK: 3′-Azido-3′-deoxythymidine drug resistance mutations in HIV-1 reverse transcriptase can induce long range conformational changes. *Proc Natl Acad Sci USA* 1998, **95:**9518–9523.

Ren J, Milton J, Weaver KL, Short SA, Stuart DI, Stammers DK: Structural basis for the resilience of efavirenz (DMP-266) to drug resistance mutations in HIV-1 reverse transcriptase. *Structure* 2000, **8:**1089–1094.

Ren J, Nichols CE, Chamberlain PP, Weaver KL, Short SA, Stammers DK: Crystal structures of HIV-1 reverse transcriptases mutated at codons 100, 106 and 108 and mechanisms of resistance to non-nucleoside inhibitors. *J Mol Biol* 2004, **336:**569–578.

Richman D, Shih CK, Lowy I, Rose J, Prodanovich P, Goff S, Griffin J: Human immunodeficiency virus type 1 mutants resistant to nonnucleoside inhibitors of reverse transcriptase arise in tissue culture. *Proc Natl Acad Sci USA* 1991, **88:**11241–11245.

Rittinger K, Divita G, Goody RS: Human immunodeficiency virus reverse transcriptase substrate-induced conformational changes and the mechanism of inhibition by nonnucleoside inhibitors. *Proc Natl Acad Sci USA* 1995, **92:**8046–8049.

Rodgers DW, Gamblin SJ, Harris BA, Ray S, Culp JS, Hellmig B, Woolf DJ, Debouck C, Harrison SC: The structure of unliganded reverse transcriptase from the human immunodeficiency virus type 1. *Proc Natl Acad Sci USA* 1995, **92:**1222–1226.

Sarafianos SG, Clark AD, Jr., Das K, Tuske S, Birktoft JJ, Ilankumaran P, Ramesha AR, Sayer JM, Jerina DM, Boyer PL, et al: Structures of HIV-1 reverse transcriptase with pre- and post-translocation AZTMP-terminated DNA. *Embo J* 2002, **21:**6614–6624.

Sarafianos SG, Clark AD, Jr., Tuske S, Squire CJ, Das K, Sheng D, Ilankumaran P, Ramesha AR, Kroth H, Sayer JM, et al: Trapping HIV-1 reverse transcriptase before and after translocation on DNA. *J Biol Chem* 2003, **278:**16280–16288.

Sarafianos SG, Das K, Clark AD, Jr., Ding J, Boyer PL, Hughes SH, Arnold E: Lamivudine (3TC) resistance in HIV-1 reverse transcriptase involves steric hindrance with beta-branched amino acids. *Proc Natl Acad Sci USA* 1999, **96:**10027–10032.

Sarafianos SG, Das K, Hughes SH, Arnold E: Taking aim at a moving target: designing drugs to inhibit drug-resistant HIV-1 reverse transcriptases. *Curr Opin Struct Biol* 2004, **14:** 716–730.

Sarafianos SG, Pandey VN, Kaushik N, Modak MJ: Site-directed mutagenesis of arginine 72 of HIV-1 reverse transcriptase. Catalytic role and inhibitor sensitivity. *J Biol Chem* 1995, **270:**19729–19735.

Schinazi RF, Lloyd RM, Jr., Nguyen MH, Cannon DL, McMillan A, Ilksoy N, Chu CK, Liotta DC, Bazmi HZ, Mellors JW: Characterization of human immunodeficiency viruses resistant to oxathiolane-cytosine nucleosides. *Antimicrob Agents Chemother* 1993, **37:** 875–881.

Selmi B, Deval J, Alvarez K, Boretto J, Sarfati S, Guerreiro C, Canard B: The Y181C substitution in 3′-azido-3′-deoxythymidine-resistant human immunodeficiency virus, type 1, reverse transcriptase suppresses the ATP-mediated repair of the 3′-azido-3′-deoxythymidine 5′-monophosphate-terminated primer. *J Biol Chem* 2003, **278:**40464–40472.

Shirasaka T, Kavlick MF, Ueno T, Gao WY, Kojima E, Alcaide ML, Chokekijchai S, Roy BM, Arnold E, Yarchoan R, et al.: Emergence of human immunodeficiency virus type 1 variants with resistance to multiple dideoxynucleosides in patients receiving therapy with dideoxynucleosides. *Proc Natl Acad Sci USA* 1995, **92:**2398–2402.

Siddiqui MA, Hughes SH, Boyer PL, Mitsuya H, Van QN, George C, Sarafinanos SG, Marquez VE: A 4′-C-ethynyl-2′,3′-dideoxynucleoside analogue highlights the role of the 3′-OH in anti-HIV active 4′-C-ethynyl-2′-deoxy nucleosides. *J Med Chem* 2004, **47:**5041–5048.

Sluis-Cremer N, Arion D, Kaushik N, Lim H, Parniak MA: Mutational analysis of Lys65 of HIV-1 reverse transcriptase. *Biochem J* 2000, **348 Pt 1:**77–82.

Spence RA, Kati WM, Anderson KS, Johnson KA: Mechanism of inhibition of HIV-1 reverse transcriptase by nonnucleoside inhibitors. *Science* 1995, **267:**988–993.

St Clair MH, Martin JL, Tudor-Williams G, Bach MC, Vavro CL, King DM, Kellam P, Kemp SD, Larder BA: Resistance to ddI and sensitivity to AZT induced by a mutation in HIV-1 reverse transcriptase. *Science* 1991, **253:**1557–1559.

Sturmer M, Staszewski S, Doerr HW, Larder B, Bloor S, Hertogs K: Correlation of phenotypic zidovudine resistance with mutational patterns in the reverse transcriptase of human immunodeficiency virus type 1: interpretation of established mutations and characterization of new polymorphisms at codons 208, 211, and 214. *Antimicrob Agents Chemother* 2003, **47:** 54–61.

Sugimoto I, Shuto S, Mori S, Shigeta S, Matsuda A: Nucleosides and nucleotides. 183. Synthesis of 4′alpha-branched thymidines as a new type of antiviral agent. *Bioorg Med Chem Lett* 1999, **9:**385–388.

Svarovskaia ES, Cheslock SR, Zhang WH, Hu WS, Pathak VK: Retroviral mutation rates and reverse transcriptase fidelity. *Front Biosci* 2003, **8:**d117–134.

Tachedjian G, Mellors J, Bazmi H, Birch C, Mills J: Zidovudine resistance is suppressed by mutations conferring resistance of human immunodeficiency virus type 1 to foscarnet. *J Virol* 1996, **70:**7171–7181.

Tisdale M, Kemp SD, Parry NR, Larder BA: Rapid in vitro selection of human immunodeficiency virus type 1 resistant to 3′-thiacytidine inhibitors due to a mutation in the YMDD region of reverse transcriptase. *Proc Natl Acad Sci USA* 1993, **90:**5653–5656.

Tong W, Lu CD, Sharma SK, Matsuura S, So AG, Scott WA: Nucleotide-induced stable complex formation by HIV-1 reverse transcriptase. *Biochemistry* 1997, **36:**5749–5757.

Tuske S, Sarafianos SG, Clark AD, Jr., Ding J, Naeger LK, White KL, Miller MD, Gibbs CS, Boyer PL, Clark P, et al: Structures of HIV-1 RT-DNA complexes before and after incorporation of the anti-AIDS drug tenofovir. *Nat Struct Mol Biol* 2004, **11:**469–474.

Ueno T, Mitsuya H: Comparative enzymatic study of HIV-1 reverse transcriptase resistant to 2′,3′-dideoxynucleotide analogs using the single-nucleotide incorporation assay. *Biochemistry* 1997, **36:**1092–1099.

Ueno T, Shirasaka T, Mitsuya H: Enzymatic characterization of human immunodeficiency virus type 1 reverse transcriptase resistant to multiple 2′,3′-dideoxynucleoside 5′-triphosphates. *J Biol Chem* 1995, **270:**23605–23611.

Wainberg MA, Miller MD, Quan Y, Salomon H, Mulato AS, Lamy PD, Margot NA, Anton KE, Cherrington JM: In vitro selection and characterization of HIV-1 with reduced susceptibility to PMPA. *Antivir Ther* 1999, **4:**87–94.

White KL, Chen JM, Margot NA, Wrin T, Petropoulos CJ, Naeger LK, Swaminathan S, Miller MD: Molecular mechanisms of tenofovir resistance conferred by human immunodeficiency virus type 1 reverse transcriptase containing a diserine insertion after residue 69 and multiple thymidine analog-associated mutations. *Antimicrob Agents Chemother* 2004, **48:**992–1003.

White KL, Margot NA, Ly JK, Chen JM, Ray AS, Pavelko M, Wang R, McDermott M, Swaminathan S, Miller MD: A combination of decreased NRTI incorporation and decreased excision determines the resistance profile of HIV-1 K65R RT. *Aids* 2005, **19:**1751–1760.

White KL, Margot NA, Wrin T, Petropoulos CJ, Miller MD, Naeger LK: Molecular mechanisms of resistance to human immunodeficiency virus type 1 with reverse transcriptase mutations K65R and K65R+M184V and their effects on enzyme function and viral replication capacity. *Antimicrob Agents Chemother* 2002, **46:**3437–3446.

Winters MA, Coolley KL, Girard YA, Levee DJ, Hamdan H, Shafer RW, Katzenstein DA, Merigan TC: A 6-basepair insert in the reverse transcriptase gene of human immunodeficiency virus type 1 confers resistance to multiple nucleoside inhibitors. *J Clin Invest* 1998, **102:**1769–1775.

Winters MA, Shafer RW, Jellinger RA, Mamtora G, Gingeras T, Merigan TC: Human immunodeficiency virus type 1 reverse transcriptase genotype and drug susceptibility changes in infected individuals receiving dideoxyinosine monotherapy for 1 to 2 years. *Antimicrob Agents Chemother* 1997, **41:**757–762.

Chapter 25
Lethal Mutagenesis

Kathleen Too and David Loakes

The concept of error catastrophe or lethal mutagenesis has become a significant area of research. Predominantly it is being used as a method of treating RNA viral targets, though it has also been used as a model to explain aging (Edelmann and Gallant 1977). This review will consider only the role of lethal mutagenesis as it pertains to viral diseases. The process of evolution requires genetic variation, and this derives primarily from random mutations in a genome population. It has long been known that RNA polymerases are more error prone than DNA polymerases and that there are essentially no mechanisms for the correction of errors. It stands to reason then that RNA viruses will generate more random mutations than their DNA counterparts. In fact RNA viruses exhibit 10 times higher mutation rates than retroviruses and 300 times higher than in DNA viruses (Drake et al. 1998).

Much has been written on the theory of lethal mutagenesis, and a review of this literature would constitute a volume of its own. Due to the low fidelity of the viral RNA-dependent RNA polymerase (RdRp), viruses have high mutation rates and exist as a pool of quasispecies (Manrubia et al. 2005, Vignuzzi et al. 2006). This heterogeneity allows for rapid adaptation by the virus to changing environments and selection pressures. However, as viruses live at the edge of maximum variation, the error threshold (Biebricher and Eigen 2005), any small change in mutation rate by external sources, forces the virus beyond tolerable mutation and into error catastrophe. The cost of replication fidelity can be measured in terms of viral fitness, and an increase in mutation rate then becomes a choice between replication rate and replication fidelity. This review aims to look beyond the theory of error catastrophe and presents evidence supporting it. We examine the use of mutagenic nucleoside analogues to force viral species into extinction, where the target is a polymerase, or for retroviruses a reverse transcriptase. There are few cases where this strategy has been applied clinically and where it has we examine the consequences of lethal mutagenesis.

Mutagenic nucleoside analogues will base pair with more than one of the natural bases. There are essentially three different methods for altering the base pairing

K. Too (✉)
Medical Research Council, Laboratory of Molecular Biology, Hills Road, Cambridge, CB2 2QH, United Kingdom
e-mail: kathleentoo@gmail.com

C.E. Cameron et al. (eds.), *Viral Genome Replication*,
DOI 10.1007/b135974_25,

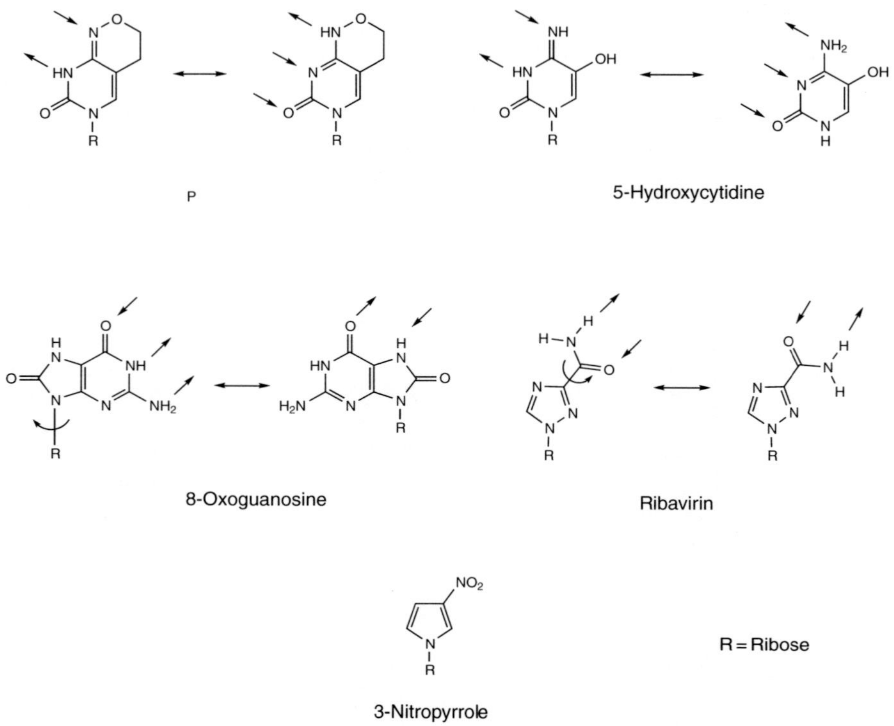

Fig. 25.1 Three different classes of nucleoside analogues that can behave as viral mutagens: Addition of electronegative atoms (O or N) alters tautomeric constant, rotation around a bond presents different hydrogen bonding faces, or non-hydrogen bonding analogues.

properties of a nucleoside (Fig. 25.1). (I) Introduction of an electronegative element (O or N) onto the exocyclic amino group or to position 5 of pyrimidines alters the tautomeric constant, bringing it closer to unity. Examples include the nucleoside P and 5-hydroxycytidine. (II) Rotation about a bond can introduce different hydrogen bonding faces. Examples of this are rotation around the glycosidic bond of 8-oxoguanosine, presenting either Watson–Crick or Hoogsteen hydrogen bonding faces, or rotation around an amide bond as in the case of ribavirin. (III) The third method is to use non-hydrogen bonding analogues, for example, the nucleoside derivative of 3-nitropyrrole. Each of these groups of analogue has been assessed as lethal mutagens and will be discussed below.

Ribavirin as an Inducer of Lethal Mutagenesis

The nucleoside ribavirin (RBV) is by far the most studied mutagen that will induce lethal mutagenesis in RNA virus populations (Fig. 25.1). RBV, introduced in 1972, has been licensed for use in combination with interferon-α (IFNalpha) for treatment against hepatitis C virus (HCV). It was long believed that RBV decreased GTP pools by inhibition of the enzyme inosine monophosphate dehydrogenase (IMPDH) because the 5′-monophosphate derivative of RBV resembles GMP. This, however, does not account completely for its observed antiviral effect. Crotty et al. demonstrated that RBV acts as an RNA virus mutagen and that its antiviral effect can be attributed to lethal mutagenesis, and a single dose of RBV at concentrations that induces >9-fold increase in mutagenesis can result in >99% loss in viral infectivity against HCV (Crotty, Cameron and Andino 2001). Since then there has been discussion in the literature as to whether or not the effect of RBV against HCV is due to lethal mutagenesis or as an IMPDH inhibitor. The field remains open to some debate and evidence has been presented that RBV activity against flaviviruses and paramyxoviruses (Leyssen et al. 2005) is similar to that observed in the presence of mycophenolic acid (an IMPDH inhibitor) and is therefore independent of error-prone replication. Addition of guanosine can suppress the effects of RBV against HCV, but is insufficient to account for the increased replicon error rate. Thus, the antiviral effect of RBV is probably not due exclusively to lethal mutagenesis for all viruses that it is active against. The purpose of this part of the review, however, is not to delve too deeply into this argument, rather to concentrate on the evidence that has accumulated subsequently on the broadening antiviral spectrum of RBV, and the increasing evidence that supports lethal mutagenesis (Crotty, Cameron and Andino 2002; Graci and Cameron 2006).

Ribavirin has the potential to elicit mutagenic events through rotation around the amide bond, presenting two different hydrogen bonding faces (Fig. 25.1). In principle it is therefore able to form two-hydrogen-bonded base pairs with each of the native nucleobases. Cameron and co-workers have shown that template RBV directs the incorporation of both UTP and CTP with almost equal efficiency with poliovirus $3D^{pol}$ and HCV RNA-dependent RNA polymerase (RdRp) (NS5B) (Maag et al. 2001), and it has been shown that HCV RNA polymerase incorporated only opposite pyrimidines (Vo, Young and Lai 2003). These data demonstrate that RBV must be capable of using both of these two hydrogen bonding faces and therefore be mutagenic. RBV triphosphate is itself a much poorer substrate than native NTPs, being incorporated at 10^3–10^4 times lower efficiency. Nevertheless, the activity of RBV correlates directly with its incorporation frequency by these viral polymerases, and this is strong evidence that its activity is due to an enhanced mutation rate.

Does RBV increase the mutation frequency of viral RNA? Treatment of subgenomic HCV virus RNA in Huh-7 cells with IFN, RBV, or a combination, followed by RT-PCR analysis of the RNA-activated protein kinase binding domain revealed that the number of mutations developed were 5 (control, 6), 36, and 57, respectively (Kanda et al. 2004), which corresponded with the observed inhibition of HCV replication. The most dominant mutations were G→A and A→G, both with RBV alone and in combination. While RNA viruses rely on error-prone replication to ensure high rates of genetic variation, they are also required to maintain highly conserved genomic segments. Contreras studied the mutation rate in five genomic regions of HCV under the influence of RBV or IFN (Contreras et al. 2002). In the absence of RBV, conserved regions (5' UTR, core, NS5A) exhibited no mutations, but a significant number of mutations arose on treatment with RBV in these regions. In addition, treatment with high doses of IFN to inhibit HCV RNA synthesis showed no mutations on treatment with RBV. Taken together this data support the view that RBV acts by promoting non-viable mutation rates in HCV in variable as well conserved genomic regions.

Other Antiviral Targets

The use of RBV to target a number of other RNA viruses has also been reported. Passage experiments using RBV to treat the foot-and-mouth disease virus (FMDV) showed that infectious virus could not be detected after three rounds, and after four passages intracellular FMDV RNA, viral protein, and minus-stranded FMDV RNA could not be detected (Gu et al. 2006). A single treatment of 1000 μM RBV caused a high mutation rate in the 3D and P1 regions and caused a 99.7% reduction in viral infectivity. Similar rates of mutation in 3D and P1 regions were observed by Domingo (Sierra et al. 2007), who report that the mutagenic and hence antiviral, effect of RBV was independent of depletion of the intracellular GTP concentration. The mutation spectrum was examined in some detail and included a mutant that had reduced sensitivity to RBV.

The flaviviridae comprises three genera: hepacivirus which includes HCV, pestivirus including bovine viral diarrhea virus (BVDV), and Flavivirus including Yellow Fever virus (YFV). Dengue virus (DEN) is another pathogenic virus within this family, and it is therefore perhaps not surprising that these viruses have been subject to treatment with RBV, given the compound's already known activity against HCV. BVDV, YFV, and DEN respond poorly to monotherapy treatment with RBV, but BVDV and YFV respond to combination therapy with IFN (Buckwold et al. 2003), while DEN responds to treatment in combination with mycophenolic acid (Takhampunya et al. 2006). In the case of Dengue virus therapy, the effect is reversed on treatment with guanosine, suggesting that inhibition of IMPDH is also involved in viral extinction.

Ribavirin has also been used to treat the West Nile virus (WNV) by serial passages in various cell lines. In HeLa cells, the rate of mutation of WNV was found to be very low for a single passage and did not decrease the viral population, but led to extinction after accumulation of mutations over four passages (Day et al. 2005). Mutations were assayed in the viral NS5 and UTR regions. The mutation spectrum gave rise to transition mutations, with RBV being incorporated preferentially as GTP rather than ATP, and the mutation frequency could also be reduced by addition of guanosine. Thus error-prone replication of WNV may only account for one of the modes of action of RBV.

Hantavirus (HTNV) is a growing disease which has no approved drug treatment. Treatment of HTNV with RBV in Vero E6 cells has been shown to inhibit the production of infectious virus, yielding high mutation rates in viral RNA (9.5/1000 nucleotides). The antiviral effect of RBV was shown to be due to error catastrophe and not inhibition of IMPDH (Jonsson, Milligan and Arterburn 2005). Other nucleoside analogues that have been shown to reduce viral titer include the

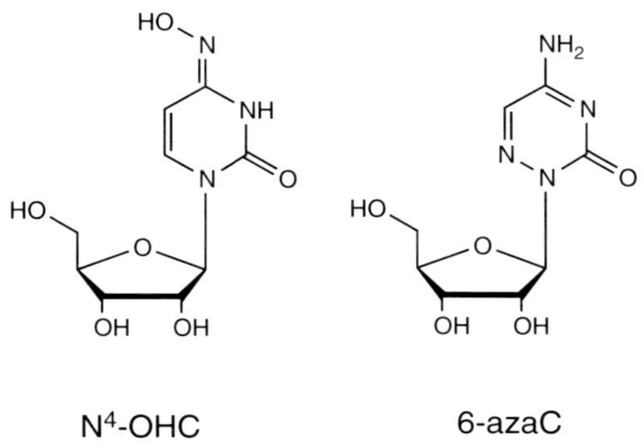

Fig. 25.2 Antiviral cytidine analogues: N^4-hydroxycytidine (N^4-OHC) and 6-azacytidine (6-azaC).

nucleoside analogues N^4-hydroxycytidine (N^4-OHC) (Fig. 25.2) and 6-azacytidine (6-azaC) (Pyrc et al. 2006). These compounds were found to inhibit Coronovirus NL63 infection (IC_{50} 400 nM and 32 nM respectively), and it was suggested that they may interfere with UMP synthesis. However, both compounds are known mutagens, and therefore may be acting through error catastrophe.

Another class of nucleosides that could be used to induce error catastrophe are non-hydrogen base analogues. Harki et al. assayed the universal base analogue 3-nitropyrrole (3-NP) (Fig. 25.1) against poliovirus and showed that it was preferentially incorporated opposite A and U by the RdRp (Harki et al. 2002). However, the rate of incorporation of 3-NP was lower by a factor of 100 and the analogue exhibited no antiviral effect. Other non-hydrogen bonding analogues include dichlorobenzimdazoles, but these were found to be viral polymerase inhibitors rather than inducers of lethal mutagenesis (Chang et al. 2006). Therefore this particular class of nucleoside has yet to be shown to be of use to induce error catastrophe. The fact that they lack hydrogen bonding functionality reduces the potential to be incorporated into viral RNA, despite the reduced fidelity of these polymerases. Nevertheless, the examples described are for only two such compounds, and this is therefore chemical space that could be explored further.

Lethal Mutagenesis Targeting HIV

Given that HIV replicates its genome at exceptionally high mutation rates, analogues that can increase viral mutation should lead the virus into error catastrophe. An obvious target for lethal mutagenesis in HIV is its reverse transcriptase, and to this end various deoxynucleoside analogues have been examined. Key compounds in this field are the Koronis compounds KP-1212/1461 (Fig. 25.3) (Harris et al. 2005). KP-1461, the prodrug of KP-1212, is hydrolyzed in the liver prior to phosphorylation and incorporation into DNA. One major disadvantage of this approach is that the analogue may also be incorporated into host DNA, whereon it will cause mutations,

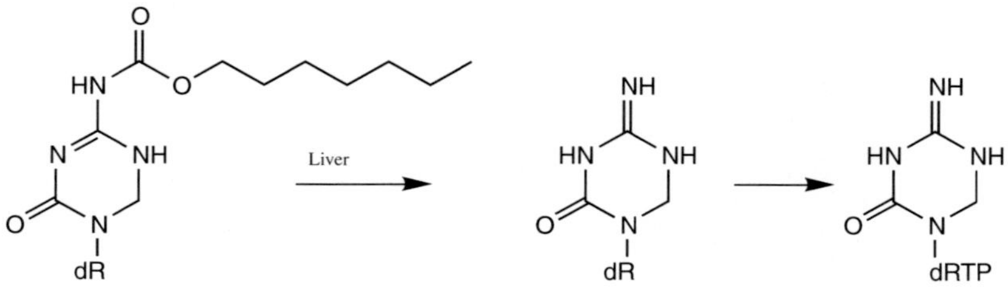

KP-1461 KP-1212 KP-1212 triphosphate

Fig. 25.3 Lethal mutagens with activity against HIV developed by Koronis Pharmaceuticals.

thus it is important to have a high degree of selectivity for the RT over the host DNA polymerases. KP-1212/1461 is reached phase 2a clinical trial. It has been shown to be active against HIV-1 and -2, and drug-resistant strains of HIV, and shows signs of synergy with Zidovudine. KP-1212 exhibits low genotoxicity profile, without any toxic effects to mitochondrial DNA synthesis.

Another nucleoside analogue that has been used to examine lethal mutagenesis of HIV is 5-hydroxy-dC (5-OHdC). The analogue 5-OHdC induces transition mutations by serial passages in CEM cells, which after 9–24 passages resulted in loss of infective viral particles (Loeb et al. 1999). As its 5′-triphosphate it is a good substrate for HIV-RT being incorporated as both dC and dT, with only slightly reduced efficiency compared to native dNTPs. The analogue 5-OHdC has been shown to have low toxicity in MT-4 cells (>2 mM). When used in combination with AZT to assess its ability to induce error catastrophe it has been reported that either AZT or 5-OHdC alone do not reduce viral fitness after serial passaging, but a combination of the two drugs led to extinction of the virus (Tapia et al. 2005).

As yet, work on lethal mutagenesis of HIV is still too early to establish whether the virus is capable of developing resistance. It is most likely that if it does occur that mutations will arise in the RT in the same way that this has already occurred with Zidovudine. One potential strategy is the use of ribonucleoside analogues. The incorporation of ribonucleotides requires the host cell RNA polymerase rather than RT. Incorporation of analogues into viral RNA should lead to lethal mutagenesis, while incorporation into the host should be relatively minor. mRNAs have short half-lives, and translation into enzymes/proteins may not be deleterious as enzymes tolerate a range of mutations even within the catalytic site. A number of ribonucleoside analogues have been studied in cell-free systems to probe their use as lethal mutagens, and the requirements of such analogues to induce error catastrophe have been reviewed (Graci and Cameron 2004). Most of these analogues have been cytosine analogues capable of behaving as either cytosine or uracil (Suzuki et al. 2006). Thus a number of nucleoside analogues have been identified as inducing mutations during transcription/reverse transcription, but as yet none of them have been shown to exhibit antiviral properties.

Foot-and-Mouth Virus as a Target for Lethal Mutagenesis

Domingo and co-workers have been investigating for some time the use of lethal mutagenesis, in particular against the Picornaviridae Foot-and-Mouth Disease virus (FMDV). FMDV is heavily influenced by high mutation rates and quasispecies dynamics, and viral adaptation can be measured in terms of fitness; large populations lead to fitness gain, while population bottlenecks lead to fitness loss, and low fitness favors viral extinction. They have shown that the nucleoside 5-azacytidine and the nucleobase 5-fluorouracil (5FU) (Fig. 25.4), which is converted to 5-fluorouridine-5′-monophosphate by the nucleoside phosphoribosyltransferase enzymes, can push FMDV into error catastrophe, leading to a 2- to 6.4-fold increase in mutation frequency (Sierra et al. 2000). As was found with RBV, the majority of mutations occurred in the 3D (polymerase) coding region, though a number of mutations were identified that did not lead to viral extinction (Pariente, Airaksinen and Domingo 2003). Viral extinction using mutagenic nucleoside analogues with or without other antiviral inhibitors (guanidine hydrochloride, heparin) occurred most frequently with low fitness viral populations, and high fitness virus could only be achieved using mutagenic nucleosides. Persistent infections also responded to RBV, where mutations in 3D were close to the maximal mutation frequency, whereas in the capsid protein VP1 mutation levels were four times higher, most of which were lethal (Pariente, Sierra and Airaksinen 2005). The analogue 5FU can also be used for the treatment of viral hemorrhagic fevers (lymphocytic choriomeningitis virus [LCMV]) where it induced modest increase in virus mutations and prevented the establishment of a persistent infection in (RAG2-KO) mice infected with LCMV.

Fig. 25.4 Lethal mutagens with activity against FMDV: 5-azacytidine (5-azaC) and 5-fluorouracil (5FU).

Emerging Resistance to Lethal Mutagens

Mutations which increase polymerase fidelity will decrease the quasispecies diversity and decrease virus fitness, and hence resistance to lethal mutagens should be under a negative selection pressure. The use of mutagenic agents, in particular nucleoside analogues, to eliminate RNA viral infections has been seen by some as an ideal solution. One of the reasons is because it was anticipated that mutagenic nucleoside analogues, as their 5′-triphosphate derivatives, would be incorporated infrequently and randomly into the viral RNA genome and that therefore the emergence of drug resistance would be limited. However, there are already reports of resistance arising in response to mutagenic nucleoside analogues. Serial passage of poliovirus with 400 μM RBV led to extinction of the virus after five passages. Under less stringent conditions in which the virus was exposed to 100 μM RBV for four rounds followed by 400 μM led to pools of virus that were able to withstand treatment with the drug. Analysis of these resistant strains identified a mutation in the RdRp which conferred enhanced fidelity (Pfeiffer and Kirkegard 2003). The mutation (G64S) occurs in the finger domain of the polymerase, and these mutants show a 3-fold increase in fidelity in the guanidine-resistance assay. In addition, the G64S mutant was also less susceptible to error catastrophe in the presence of other mutagens (5-azacytidine). It is suggested that this mutation both reduces the incorporation of incorrect nucleotides and an increase in fidelity in the presence of mutagenic analogues.

Resistance to RBV treatment has also been observed in clinical samples. Samples derived from patients with HCV on RBV monotherapy showed a resistant mutation in NS5B (F415Y), located in the P helix of the thumb domain (Young et al. 2003). The F415Y mutation reverted back to F415 when treatment was discontinued. In another study, patients on RBV and IFN combination therapy were also found to develop resistance, with mutations found on NS5B around the substrate entry site and the NTP tunnel (Hamano et al. 2005). Resistance to RBV has also been observed in FMDV, resulting in M296I mutants in the viral (3D) polymerase (Sierra et al. 2007). This mutation occurs in the conserved region for binding of the primer–template complex and shows a reduced capacity to accept RBV triphosphate in place of GTP or ATP.

Lethal mutagenesis is still in its infancy as a method to treat viral infections, but there is growing evidence to support it as a viable therapeutic tool. Many will object to the idea of a mutagenic compound used to treat a viral infection, yet many of the viral targets are life threatening. Lethal mutagenesis is not a method to treat a common cold, but may prove highly efficient at eliminating pathogenic viruses. It is perhaps inevitable that viruses will become resistant to lethal mutagens, as they have for all other antiviral agents. Thus this should be seen as another weapon in the armory for tackling deadly viral diseases.

Lethal Mutagenesis of Retroviruses via APOBEC Proteins

A recent perspective entitled "Weapons of mutational destruction" describe an alternative method to induce error catastrophe in virus species (KewalRamani and Coffin 2003). Proteins belonging to the family of cytidine deaminases, the APOlipoprotein B mRNA-editing enzyme catalytic polypeptide-like complex (APOBEC) family have been identified, which are capable of blocking the replication of retroviruses, such as HIV, in the absence of virion infectivity factor (Vif) protein. These include APOBEC3G (Sheehy et al. 2002), APOBEC3F (Wiegand et al. 2004), and APOBEC3B (Doehle, Schäfer and Cullen 2005). APOBEC3G is the most studied member of the APOBEC group of proteins and was found to reduce the infectivity of Vif-deficient HIV-1 virions by 10- to >20-fold. Similar results were observed for APOBEC3F, while APOBEC3B showed an inhibition of Vif-deficient HIV-1 infectivity of only about 5- and 2-fold, respectively. However, of all the APOBEC proteins, APOBEC3B is the only member which is resistant to Vif (Doehle, Schäfer and Cullen 2005). But APOBEC3B is not expressed in many of the natural cellular targets of HIV-1 and hence does not normally affect HIV-1 replication in vivo. The mechanism of action of the APOBEC proteins has been intensively studied and is still being debated.

In the absence of Vif, APOBEC3G (and probably 3F) are incorporated into virions. They are then transferred from producer cells to target cells by the virus. Following viral entry and uncoating, the first step of reverse transcription is initiated whereby APOBEC3F and 3G target the cytosines on the nascent proviral minus strand DNA and convert them to uracil. It is speculated that such dC to dU transitions cause inhibition of viral replication via several mechanisms. First, because uracil is not well tolerated in DNA; it is degraded by uracil *N*-glycosidases (UNG) leading to the formation of nicked DNA, which are further cleaved by the host DNA repair enzymes such as APE1 (Schröfelbauer et al. 2005). Second should the mutated proviruses survive and integrate, the introduced dU residues would induce the incorporation of dA residues into the proviral plus strand DNA, resulting in a dG-to-dA hypermutation of the HIV-1 provirus (Harris et al. 2003, Lecossier et al. 2003). It is speculated that such elevated levels of dG-to-dA mutations leads to lethal mutagenesis due to the reduced capacity for the virus to encode functional proteins (Fig. 25.5). Moreover, based on the population level analysis of HIV-1 hypermutation and its relationship with APOBEC3G and Vif genetic variation, Pace et al. (2006) have showed that APOBEC3G-induced HIV hypermutation represents a potent host antiviral factor in vivo. Further biochemical experiments have confirmed the correlation between deaminase and antiviral activity (Iwantani et al. 2006).

However, recent reports by several independent research groups have identified APOBEC3G and 3F mutant proteins which have lost their cytidine deaminase ability but yet retain their anti-HIV-1 activity in transfection-based assays (Holmes et al. 2007 and references therein). Similar results were also observed for mutants of APOBEC3G against the human T-cell leukaemia virus type 1 (HTLV-1) (Sasada et al. 2005). These findings suggest that the antiviral activity of the APOBEC proteins might be due to several mechanisms of action in addition to lethal mutagenesis. Moreover, Greene and co-workers found that when APOBEC3G levels in $CD4^+$ T-cells were suppressed using RNA interference (RNAi) methods, HIV-1 infection of T-cells occurred. Surprisingly, sequencing of the reverse transcripts slowly formed in untreated $CD4^+$ T cells reveals only low levels of dGdA hypermutation, raising the possibility that the APOBEC3G-restricting activity may not

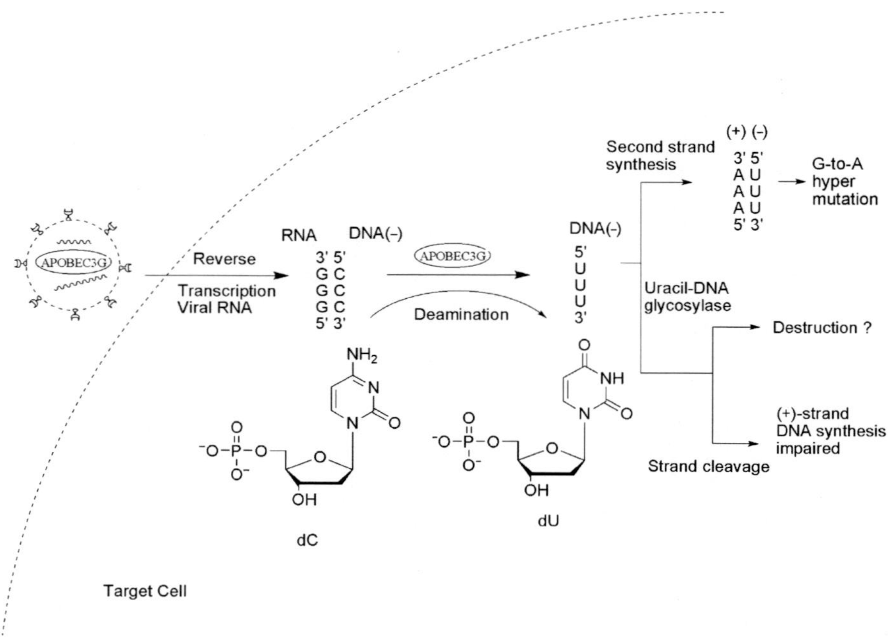

Fig. 25.5 Effects of APOBEC3G on Vif-deficient HIV-1 particles. APOBEC3G deaminates the deoxycytidine to deoxyuracil on minus-strand DNA causing G-to-A hypermutation during the reverse transcription step. Alternatively, the uracil containing DNA can be potentially degraded by uracil DNA-glycosylase and apurinic–apyrimidinic endonuclease.

be strictly dependent on deamination (Chiu et al. 2005). Recent evidence suggests that APOBEC3G associates with ribonucleoprotein (RNP), which is present in both the cytosol and the discrete microdomains and plays an important role as a possible regulator of cellular RNA function (Gallois-Montbrun et al. 2007).

Mechanisms of Action of Vif and APOBEC Proteins

HIV is a complex retrovirus which encodes six accessory proteins namely, Tat, Rev, Nef, Vpr, Vpu, and Vif. Recent attention has centered on the function of virion infectivity factor (Vif), mainly because of its interaction with the cellular protein APOBEC3 proteins. Vif was demonstrated to bind to APOBEC3G and 3F and generates their rapid degradation in producer cells, thus preventing their incorporation into virions (Conticello, Harris and Neuberger 2003, Marin et al. 2003). APOBEC3G exists in two forms: an enzymatically active low molecular mass (LMM) and an inactive high molecular mass (HMM) ribonucleoprotein complex. In cells prone to HIV infection, such as H9 T cells, APOBEC3G exists as an inactive HMM complex that does not have any anti-HIV activity. Vif induces the polyubiquitination of HMM APOBEC3G by interacting with an E3 ubiquitin ligase complex consisting of the cellular proteins Cullin5, Elongin B, Elongin C, and Rbx1 (Yu et al. 2003). The latter interaction is mediated via a highly conserved suppressor of cytokine signalling (SOCS) box-like motif (SQL(Y/F)LA) that binds to Elongin C which in turn recruits the cellular proteins Cullin5, Elongin B, and Rbx1 to form the E3 ubiquitin ligase complex ElonginB/C-Cul5-SOCS-box (Mehle et al. 2004). As a result of these interactions, polyubiquitination of APOBEC3G occurs leading to its degradation.

Potential Therapeutic Targets and Agents: Inhibition of Vif-APOBEC Interaction

Ways to target the Vif-APOBEC protein interactions have been extensively reviewed and discussed (Carr et al. 2006, Chiu and Greene 2006, Cullen 2006). Table 25.1 provides a summary of potential anti-HIV therapeutic strategies, including inhibition of APOBEC3G-Vif, increased APOBEC3G synthesis, inhibition of APOBEC degradation, promoting APOBEC3G-Gag/RNA binding, inhibition of Vif synthesis, and inhibition of Vif proteolysis. Vif appears to be highly species specific. For example, it was found that human (h) APOBEC3G is non-responsive to the Vif protein of African green monkey's (agm) SIV and, similarly, agmAPOBEC3G is non-responsive to HIV-1 Vif. This is due to a single amino acid difference between hAPOBEC3G and agmAPOBEC3G: D128 (hAPOBEC3G) to K128 (agmAPOBEC3G) (Bogerd et al. 2004). Switching these two charged residues reversed the observed specificity, i.e., agmAPOBEC3G (D128) was fully responsive to HIV-1 Vif. Hence small charged molecules or peptide inhibitors targeting this region of hAPOBEC3G might interfere with its interaction with HIV-1 Vif. More importantly, it is predicted that small molecules targeting APOBEC3G at position 128 will not affect APOBEC3G enzymatic activity, which is conferred by other domains of the protein.

Increase in Cellular Levels of APOBEC3G

An alternative approach is to suppress virus replication by interfering with the levels of Vif or APOBEC proteins in cells. Rose et al. (2004) recently reported that APOBEC3G mRNA levels are enhanced up to 20-fold by phorbol myristate acetate (PMA) via a mechanism involving activation of a cascade of cellular kinases including protein kinase C (PKC), mitogen-activated protein kinase kinase (MEK), and extracellular signal regulated kinase (ERK). Therefore a potential antiviral strategy is to develop chemical inducers such as PMA in order to stimulate the different kinases: PKC, MEK, and ERK lead to an increase in cellular levels of APOBEC3G which is a natural inhibitor of HIV.

Table 25.1 Summary of the ways to target Vif-APOBEC protein interactions- potential therapeutic drug targets

Strategy	Outcomes	Mechanism	Agents	Reference
Inhibit APOBEC3G-Vif	Decrease APOBEC3G degradation Increase APOBEC3G in Gag/RNA/virions	Target amino acid 128 of APOBEC3G	peptide inhibitor charged molecule	(Fang and Landau 2007)
Increase APOBEC3G synthesis	Increase APOBEC3G transcription	PKC, MEK, ERK activation	chemical inducers	(Rose et al. 2004)
Inhibit APOBEC degradation	Increase APOBEC3G levels	Proteasome inhibitors	chemical & peptide inhibitors (MG132)	(Liu et al. 2004, Yu et al. 2003).
		E3 Ub-ligase inhibitors		(Xiao et al. 2007)
Enhance APOBEC3G-Gag/RNA binding	Increase APOBEC3G virion incorporation	Increase monoubiquitination		(Dussart et al. 2005)
		Increase phosphorylation of APOBEC3G		(Douaisi et al. 2005)
Inhibit Vif synthesis	Decrease RNA levels	Ribozymes/antisense RNA	Gene therapy vectors	(Barnor et al. 2004, Lorentzen et al. 1991)
Enhance Vif proteolysis	Decrease Vif levels	Structural changes to increase proteolytic susceptibility Increase binding to E3 Ub-ligase Aa63-70, 88-89 of Vif		(Alexander et al. 2002, Fujita et al. 2004, Fujita et al. 2003)

Inhibition of APOBEC3G Degradation by Vif

Vif forms a complex with Elongin B/C, Cullin-5, and Rbx-1 to induce the polyubiquitination and proteosome-mediated degradation of hAPOBEC3G. These interactions serve as potential targets for anti-HIV-1 drug development. It was reported that the addition of the proteosome inhibitor MG132 (Fig. 25.6) to cells co-expressing APOBEC3G and Vif resulted in inhibition of APOBEC3G degradation (Yu et al. 2003). Further, in vitro experiments showed that MG132 caused increased cellular and intravirion levels of APOBEC3G and a decrease in viral infectivity (Liu et al. 2004). More recently, Xiao et al. (2007) used a zinc-chelating reagent to inhibit Cullin-5 E3 ubiquitin ligase. Inhibition of the zinc-dependent E3 ubiquitin ligase resulted in the inhibition of HIV Vif activity and the liberation of natural HIV-inhibitor APOBEC3G. Furthermore, (Fang and Landau 2007) have developed a Vif α-complementation assay for the rapid identification of small molecule Vif inhibitors specific for the E3 Ubiquitin-ligase enzymes such as Elongin B/C, Cullin-5, and Rbx-1.

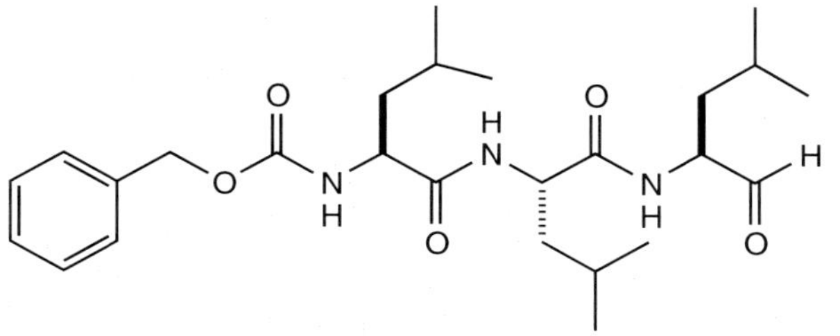

MG132

Fig. 25.6 Proteasome inhibitor MG132.

Enhancing APOBEC3G-Gag/RNA Binding

APOBEC3G packaging into HIV-1 virions involves assembly with the nucleocapsid region of the Gag viral protein and possible binding of RNA (Schäfer, Bogerd and Cullen 2004). It has been demonstrated that both phosphorylation by tyrosine kinases Fyn and Hck (Douaisi et al. 2005) and monoubiquitination of APOBEC3G by Nedd4-1 (Dussart et al. 2005) are necessary for the APOBEC3G-Gag-RNA binding resulting in the incorporation of APOBEC3G into virions. Ways to enhance such phosphorylation and monoubiquitation could lead to a potential antiviral strategy.

Gene Therapy

Antisense RNAs targeting either Vif sequences have been shown to possess anti-HIV activity (Barnor et al. 2004). Moreover, hairpin self-cleaving ribozymes have been designed against Vif mRNA (Lorentzen et al. 1991). In both cases the presence of these RNAs in HIV-infected cells leads to a decrease in p24 antigen production, showing that reducing Vif RNA levels suppresses HIV production. This kind of gene therapy approach offers another avenue for the development of antiviral agents.

Enhance Vif Proteolysis

Recently, a naturally occurring mutant Vif containing a two amino acids, DS, insertion at position 62, has been shown to rapidly cleave in viral producer cells and is not active in enhancing HIV replication in vitro. The two patients infected with this HIV strain are long-term non-progressors (Alexander et al. 2002). Furthermore, it has been revealed that amino acid residues 63–70 and 88–89 are important for maintaining intracellular levels of Vif (Fujita et al. 2003; Fujita et al. 2004). Further understanding of the mechanisms of ubiquitin-mediated Vif degradation, the structure of the proteolysis susceptible DS mutant Vif, and the determinants and effects of intravirion Vif proteolysis may yield a novel drug target for anti-HIV therapy by reducing intracellular levels of Vif.

In summary, hAPOBEC3 proteins form part of a natural anti-retroviral defense mechanism against HIV-1 causing its lethal mutagenesis by virion incorporated hAPOBEC3G cytidine deaminase. Unfortunately, the HIV-1 Vif protein circumvents this potent antiviral action. By inhibiting Vif action, the APOBEC3 proteins are free to induce its antiviral activity in all natural target cells in vivo. Future development of chemical agents to target actions of Vif and APOBEC3 proteins would be greatly assisted by further understanding of the mechanisms of action of Vif and APOBEC3 interactions.

References

Alexander, L., Aquino-DeJesus, M.J., Chan, M. and Andiman, W.A. (2002) Inhibition of human immunodeficiency virus type 1 (HIV-1) replication by a two-amino-acid insertion in HIV-1 Vif from a nonprogressing mother and child. J. Virol. 76, 10533–10539.

Barnor, J.S., Miyano-Kurosaki, N., Yamaguchi, K., Sakamoto, A., Ishikawa, K., Inagaki, Y., Yamamoto, N., Osei-Kwasi, M., Ofori-Adjei, D. and Takaku, H. (2004) Intracellular expression of antisense RNA. transcripts complementary to the human immunodeficiency virus type-1 vif gene inhibits viral replication in infected T-lymphoblastoid cells. Biochem. Biophys. Res. Commun. 320, 544–550.

Biebricher, C.K. and Eigen, M. (2005) The error threshold. Virus Res. 107, 117–127.

Bogerd, H.P., Doehle, B.P., Wiegand, H.L. and Cullen, B.R. (2004) A single amino acid difference in the host APOBEC3G protein controls the primate species specificity of HIV-1 Vif. Proc. Natl. Acad. Sci. USA 101, 3770–3774.

Buckwold, V.E., Wei, J., Wenzel-Mathers, M. and Russell, J. (2003) Synergistic in vitro interactions between alpha interferon and ribavirin against viral diarrhea virus and yellow fever virus as surrogate models of hepatitis C replication. Antimicrob. Agents Chemother. 47, 2293–2298.

Carr, J.M., Davis, A.J., Feng, F., Burrell, C.J. and Li, P. (2006) Cellular interactions of virion infectivity factor (Vif) as potential therapeutic targets: APOBEC3G and more? Curr. Drug Targets 7, 1583–1593.

Chang, J., Nie, X., Gudima, S. and Taylor, J. (2006) Action of inhibitors of accumulation of processed Hepatitis Delta virus RNAs. J. Virol. 80, 3205–3214.

Chiu, Y.L. and Greene, W.C. (2006) APOBEC3 cytidine deaminase distinct antiviral actions along the retroviral life cycle. J. Biol. Chem. 281, 8309–8312.

Chiu, Y.L., Soros, V.B., Kreisberg, J.F., Stopak, K., Yonemoto, W. and Greene, W.C. (2005) Cellular APOBEC3G restricts HIV-1 infection in resting CD4+ T cells. Nature 435, 108–114.

Conticello, S.G., Harris, R.S. and Neuberger, M.S. (2003) The Vif protein of HIV triggers degradation of the human antiretroviral DNA deaminase APOBEC3G. Curr. Biol. 13, 2009–2013.

Contreras, A.M., Hiasa, Y., He, W., Terella, A., Schmidt, E.V. and Chung, R.T. (2002) Viral RNA mutations are region specific and increased by ribavirin in a full-length hepatitic C virus replication system. J. Virol. 76, 8505–8517.

Crotty, S., Cameron, C.E. and Andino, R. (2001) RNA virus error catastrophe: direct molecular test by using ribavirin. Proc. Natl. Acad. Sci. USA 98, 6895–6900.

Crotty, S., Cameron, C.E. and Andino, R. (2002) Ribavirin's antiviral mechanism of action: lethal mutagenesis? J. Mol. Med. 80, 86–95.

Cullen, B.R. (2006) Role and mechanism of action of the APOBEC3 family of antiretroviral resistance factors. J. Virol. 80, 1067–1076.

Day, C.W., Smee, D.F., Julander, J.G., Yamshchikov, V.F., Sidwell, R.W. and Morrey, J.D. (2005) Error-prone replication of West Nile virus caused by ribavirin. Antiviral Res. 67, 38–45.

Doehle, B.P., Schäfer, A. and Cullen, B.R. (2005) Human APOBEC3B is a potent inhibitor of HIV-1 infectivity and is resistant to HIV-1 Vif. Virology 339, 281–288.

Douaisi, M., Dussart, S., Courcoul, M., Bessou, G., Lerner, E.C., Decroly, E. and Vigne, R. (2005) The tyrosine kinases Fyn and Hck favor the recruitment of tyrosine-phosphorylated APOBEC3G into vif-defective HIV-1 particles. Biochem. Biophys. Res. Commun. 329(3), 917–924.

Drake, J.W., Charlesworth, B., Charlesworth, D. and Crow, J.E. (1998) Rates of spontaneous mutation. Genetics 148, 1667–1686.

Dussart, S., Douaisi, M., Courcoul, M., Bessou, G., Vigne, R. and Decroly, E. (2005) APOBEC3G ubiquitination by Nedd4-1 favors its packaging into HIV-1 particles. J. Mol. Biol. 345, 547–558.

Edelmann, P. and Gallant, J. (1977) On the translational error theory of aging. Proc. Natl. Acad. Sci. USA 74, 3396–3398.

Fang, L. and Landau, N.R. (2007) Analysis of Vif-induced APOBEC3G degradation using an alpha-complementation assay. Virology 359, 162–169.

Fujita, M., Akari, H., Sakurai, A., Yoshida, A., Chiba, T., Tanaka, K., Strebel, K. and Adachi, A. (2004) Expression of HIV-1 accessory protein Vif is controlled uniquely to be low and optimal by proteasome degradation. Microbes Infect. 6, 791–798.

Fujita, M., Sakurai, A., Yoshida, A., Miyaura, M., Koyama, A.H., Sakai, K. and Adachi, A. (2003) Amino acid residues 88 and 89 in the central hydrophilic region of human immunodeficiency virus type 1 Vif are critical for viral infectivity by enhancing the steady-state expression of Vif. J. Virol. 77, 1626–1932.

Gallois-Montbrun, S., Kramer, B., Swanson, C.M., Byers, H., Lynham, S., Ward, M. and Malim, M.H. (2007) Antiviral protein APOBEC3G localizes to ribonucleoprotein complexes found in P bodies and stress granules. J. Virol. 81, 2165–2178.

Graci, J.D. and Cameron, C.E. (2004) Challenges for the development of ribonucleoside analogues as inducers of error catastrophe. Antivir. Chem. Chemother. 15, 1–13.

Graci, J.D. and Cameron, C.E. (2006) Mechanisms of action of ribavirin against distinct diseases. Rev. Med. Virol. 16, 37–48.

Gu, C.-j., Zheng, C.-y., Zhang, Q., Shi, L.-l. and Qu, S.-f. (2006) An antiviral mechanism investigated with ribavirin as an RNA virus mutagen for foot-and-mouth disease virus. J. Biochem. Mol. Biol. 39, 9–15.

Hamano, K., Sakamoto, N., Enomoto, N., Izumi, N., Asahina, Y., Kurosaki, M., Ueda, E., Tanabe, Y., Maekawa, S., Itakura, J., Watanabe, H., Kakinuma, S. and Watanabe, M. (2005)

Mutations in the NS5B region of the hepatitis C virus genome correlate with clinical outcomes of interferon-alpha plus ribavirin combination therapy. J. Gastroenterol. Hepatol. 20, 1401–1409.

Harki, D.A., Graci, J.D., Korneeva, V.S., Ghosh, S.K.B., Hong, Z., Cameron, C.E. and Peterson, B.R. (2002) Synthesis and evaluation of a mutagenic and non-hydrogen bonding ribonucleoside analogue: 1-β-D-ribofuranosyl-3-nitropyrrole. Biochemistry 41, 9026–9033.

Harris, K.S., Brabant, W., Styrchak, S., Gall, A. and Daifuku, R. (2005) KP-1212/1461, a nucleoside designed for the treatment of HIV by viral mutagenesis. Antiviral Res. 67, 1–9.

Harris, R.S., Bishop, K.N., Sheehy, A.M., Craig, H.M., Petersen-Mahrt, S.K., Watt, I.N., Neuberger, M.S. and Malim, M.H. (2003) DNA deamination mediates innate immunity to retroviral infection. Cell 113, 803–809.

Holmes, R.K., Koning, F.A., Bishop, K.N. and Malim, M.H. (2007) APOBEC3F can inhibit the accumulation of HIV-1 reverse transcription products in the absence of hypermutation. Comparisons with APOBEC3G. J. Biol. Chem. 282, 2587–95.

Iwantani, Y., Takeuchi, H., Strebel, K. and Levin, J.G. (2006) Biochemical activities of highly purified, catalytically active human APOBEC3G: correlation with antiviral effect. J. Virol. 80, 5992–6002.

Jonsson, C.B., Milligan, B.G. and Arterburn, J.B. (2005) Potential importance of error catastrophe to the development of antiviral strategies for hantaviruses. Virus Res. 107, 195–205.

Kanda, T., Yokosuka, O., Imazeki, F., Tanaka, M., Shino, Y., Shimada, H., Tomonaga, T., Nomura, F., Nagao, K., Ochiai, T. and Saisho, H. (2004) Inhibition of subgenomic hepatitis C virus RNA in Huh-7 cells: ribavirin induces mutagenesis in HCV RNA. J. Viral Hepatitis 11, 479–487.

KewalRamani, V.N. and Coffin, J.M. (2003) Weapons of mutational destruction. Science 301, 923–925.

Lecossier, D., Bouchonnet, F., Clavel, F. and Hance, A.J. (2003) Hypermutation of HIV-1 DNA in the absence of the Vif protein. Science 300, 1112.

Leyssen, P., Balzarini, J., De Clercq, E. and Neyts, J. (2005) The predominant mechanism by which ribavirin exerts its antiviral activity in vitro against flaviviruses and paramyxoviruses is mediated by inhibition of IMP dehydrogenase. J. Virol. 79, 1943–1947.

Liu, B.D., Yu, X.H., Luo, K., Yu, Y.K. and Yu, X.F. (2004) Influence of primate lentiviral vif and proteasome inhibitors on human immunodeficiency virus type 1 virion packaging of APOBEC3G. J. Virol. 78, 2072–2081.

Loeb, L.A., Essigmann, J.M., Kazazi, F., Zhang, J., Rose, K.D. and Mullins, J.I. (1999) Lethal mutagenesis of HIV with mutagenic nucleoside analogs. Proc. Natl. Acad. Sci. USA 96, 1492–1497.

Lorentzen, E.U., Wieland, U., Kuhn, J.E. and Braun, R.W. (1991) In vitro cleavage of HIV-1 Vif RNA by a synthetic Ribozyme. Virus Genes 5, 17–23.

Maag, D., Castro, C., Hong, Z. and Cameron, C.E. (2001) HepatitisC virus RNA-dependent RNA polymerase (NS5B) as a mediator of the antiviral activity of ribavirin. J. Biol. Chem. 276, 46094–46098.

Manrubia, S.C., Escarmis, C., Domingo, E. and Lazaro, E. (2005) High mutation rates, bottlenecks and robustness of RNA viral quasiespecies. Gene 347, 273–282.

Marin, M., Rose, K.M., Kozak, S.L. and Kabat, D. (2003) HIV-1 Vif protein binds the editing enzyme APOBEC3G and induces its degradation. Nature Med. 9, 1398–1403.

Mehle, A., Goncalves, J., Santa-Marta, M., McPike, M. and Gabuzda, D. (2004) Phosphorylation of a novel SOCS-box regulates assembly of the HIV-1 Vif-Cul5 complex that promotes APOBEC3G degradation. Genes & Development 18, 2861–2866.

Pace, C., Keller, J., Nolan, D., James, I., Gaudieri, S., Moore, C. and Mallal, S. (2006) Population level analysis of Human Immunodeficiency Virus type 1 hypermutation and its relationship with APOBEC3G and Vif genetic variation. J. Virol. 80, 9259–9269.

Pariente, N., Airaksinen, A. and Domingo, E. (2003) Mutagenesis versus inhibition in the efficiency of extinction of foot-and-mouth disease virus. J. Virol. 77, 7131–7138.

Pariente, N., Sierra, S. and Airaksinen, A. (2005) Action of mutagenic agents and antiviral inhibitors on foot-and-mouth disease virus. Virus Res. 107, 183–193.

Pfeiffer, J.K. and Kirkegard, K. (2003) A single mutation in poliovirus RNA-dependent RNA polymerase confers resistance to mutagenic nucleotide analogs via increased fidelity. Proc. Natl. Acad. Sci. USA 100, 7289–7294.

Pyrc, K., Bosch, B.J., Berkhout, B., Jebbink, M.F., Dijkman, R., Rottier, P. and van der Hoek, L. (2006) Inhibition of human coronavirus NL63 infection at early stages of the replication cycle. Antimicrob. Agents Chemother. 50, 2000–2008.

Rose, K.M., Marin, M., Kozak, S.L. and Kabat, D. (2004) Transcriptional regulation of APOBEC3G, a cytidine deaminase that hypermutates human immunodeficiency virus. J. Biol. Chem. 279, 41744–41749.

Sasada, A., Takaori-Kondo, A., Shirakawa, K., Kobayashi, M., Abudu, A., Hishizawa, M., Imada, K., Tanaka, Y. and Uchiyama, T. (2005) APOBEC3G targets human T-cell leukemia virus type 1. Retrovirology 2, 32.

Schäfer, A., Bogerd, H.P. and Cullen, B.R. (2004) Specific packaging of APOBEC3G into HIV-1 virions is mediated by the nucleocapsid domain of the gag polyprotein precursor. Virology 328, 163–168.

Schröfelbauer, B., Yu, Q., Zeitlin, S.G. and Landau, N.R. (2005) Human immunodeficiency virus type 1 Vpr induces the degradation of the UNG and SMUG uracil-DNA glycosidases. J. Virol. 79, 10978–10987.

Sheehy, A.M., Gaddis, N.C., Choi, J.D. and Malim, M.H. (2002) Isolation of a human gene that inhibits HIV-1 infection and is suppressed by the viral Vif protein. Nature 418, 646–650.

Sierra, M., Airaksinen, A., Gonzalez-Lopez, C., Agudo, R., Arias, A. and Domingo, E. (2007) Foot-and-mouth disease virus mutant with decreased sensitivity to ribavirin: implications for error catastrophe. J. Virol. 81, 2012–2024.

Sierra, S., Davila, M., Lowenstein, P.R. and Domingo, E. (2000) Response of foot-and-mouth disease to increased mutagenesis: influence of viral load and fitness in loss of infectivity. J. Virol. 74, 8316–8323.

Suzuki, T., Moriyama, K., Otsuka, C., Loakes, D. and Negishi, K. (2006) Template properties of mutagenic cytosine analogues in reverse transcription. Nucleic Acids Res. 34, 6438–6449.

Takhampunya, R., Ubol, S., Houng, H.-S., Cameron, C.E. and Padmanabhan, R. (2006) Inhibition of dengue virus replication by mycophenolic acid and ribavirin. J. Gen. Virol. 87, 1947–1952.

Tapia, N., Fernandez, G., Parera, M., Gomez-Mariano, G., Clotet, B., Quinones-Mateu, M., Domingo, E. and Martinez, M.A. (2005) Combination of a mutagenic agent with reverse transcriptase inhibitor results in a systematic inhibition of HIV-1 infection. Virology 338, 1–8.

Vignuzzi, M., Stone, J.K., Arnold, J.J., Cameron, C.E. and Andino, R. (2006) Quasiespecies diversity determines pathogenesis through cooperative interactions in a viral population. Nature 439, 344–348.

Vo, N.V., Young, K.-C. and Lai, M.M.C. (2003) Mutagenic and inhibitory effects of ribavirin on hepatitis C virus RNA polymerase. Biochemistry 42, 10462–10471.

Wiegand, H.L., Doehle, B.P., Bogerd, H.P. and Cullen, B.R. (2004) A second human antiretroviral factor, APOBEC3F, is suppressed by th HIV-1 and HIV-2 Vif proteins. EMBO J. 23, 2451–2458.

Xiao, Z., Ehrlich, E., Luo, K., Xiong, Y. and Yu, X.F. (2007) Zinc chelation inhibits HIV Vif activity and liberates antiviral function of the cytidine deaminase APOBEC3G. FASEB J. 21, 217–222.

Young, K.-C., Lindsay, K.L., Lee, K.-J., Liu, W.-C., He, J.-W., Milstein, S.L. and Lai, M.M.C. (2003) Identification of a ribavirin-resistant NS5B mutation of hepatitis C virus during ribavirin monotherapy. Hepatology 38, 869–878.

Yu, X.H., Yu, Y.K., Liu, B.D., Luo, K., Kong, W., Mao, P.Y. and Yu, X.F. (2003) Induction of APOBEC3G ubiquitination and degradation by an HIV-1 Vif-Cul5-SCF complex. Science 302, 1056rr–1060.

Chapter 26
Clinical Implications of Reverse Transcriptase Inhibitor Resistance

Kristel Van Laethem and Anne-Mieke Vandamme

Introduction

Currently, several drugs are approved by the US Food and Drug Association or available through expanded access programs for the treatment of HIV-1 patients. They target four distinct steps within the HIV-1 replication cycle, i.e., the entry of the virus particle within the target cell (entry inhibitors [EI] subdivided into fusion inhibitors and CCR5 antagonists), the conversion of viral RNA into DNA (nucleoside and non-nucleoside reverse transcriptase inhibitors [NRTI and NNRTI]), the integration of the viral DNA into the host genome (integrase inhibitors [INI]), and the maturation of the virus particle (protease inhibitors [PI]). At the moment, NRTI remain the backbone components for antiretroviral drug combinations that generally comprise 2 NRTI and 1 NNRTI or 2 NRTI and 1 PI (DHHS 2006).

The introduction of combination therapy in the mid-1990 s resulted in a decrease of the morbidity and mortality in the HIV-1 patient population that has access to therapy (CASCADE Collaboration 2003). Nevertheless, therapy failure can occur due to severe adverse events (toxicity failure), due to rising viral load (virological failure), or due to decreasing CD4 cell count (immunological failure) which ultimately can result into clinical progression (clinical failure). As changes in viral load mostly precede changes in CD4 count, the viral load is the most relevant laboratory marker to monitor the short-term in vivo activity of a therapy given to an individual patient (Mellors et al. 1997). Incomplete adherence to therapy, suboptimal therapy potency, suboptimal pharmacokinetics, and pre-existing drug resistance can be factors that result into rising viral loads and therapy failure. Residual replication under the continuous selective pressure of drugs will result into the gradual overgrowth of the wild-type variants by variants with some level of resistance and into the accumulation of additional mutations (Vandamme et al. 1998). Thus, the main goal of HAART is to reduce the viral load as much as possible to prevent further clinical progression, but its immediate goal is to reduce the viral load below the detection

K. Van Laethem (✉)
Katholieke Universiteit Leuven, Rega Institute for Medical Research, Clinical and Epidemiological Virology, Minderbroedersstraat 10, 3000 Leuven, Belgium
e-mail: kristel.vanlaethem@uz.kuleuven.ac.be

C.E. Cameron et al. (eds.), *Viral Genome Replication*,
DOI 10.1007/b135974_26, © Springer Science+Business Media, LLC 2009

limit of current assays (<50 RNA copies/ml) to prevent the development of mutations and the formation of variants that result in antiviral resistance to the current therapy.

Increasing numbers of drug-resistant mutations have been identified within gag, pol, and env, reflecting the genetic flexibility of HIV-1. These mutations can directly boost the ability of the virus to specifically replicate in presence of the drug (major mutations) or indirectly by increasing the replication capacity of the virus in general, whether in presence or absence of the drug (minor, compensatory or accessory mutations). In general, the major mutations are found at the drug-binding sites, whereas the others can be found at distinct sites within the target protein or even other proteins. In this respect, PI mutations have not only been observed within the protease but also at several sites within its gag substrate (Doyon et al., 1996; Nijhuis et al., 2007).

Resistance development against anti-HIV inhibitors is a fascinating field, resistance has been reported for all currently available drugs. However, for the scope of this book, this review will focus on resistance against RTI. For reviews on resistance against other classes of drugs, see Clark et al. (2007), Marcelin et al. (2005), Poveda et al. (2005), and Weber et al. (2006).

Key Drug-Resistant Mutations Within Reverse Transcriptase

Nucleoside Reverse Transcriptase Inhibitor (NRTI) Mutations

Resistance Mechanisms

After conversion into its triphosphate form, NRTI compete with the natural deoxynucleotide-triphosphate (dNTPs) substrates of RT by binding to the active site of the enzyme and by incorporation into the newly synthesized DNA chain. Incorporation of a NRTI will result into chain termination as it lacks a 3′-hydroxyl group which prevents the formation of a phosphodiester bond between the NRTI and the next incoming dNTP. Resistance toward NRTI can involve at least two resistance mechanisms. Some mutations confer resistance through an enhanced discrimination between the NRTI and the natural dNTP substrate, i.e., by a decreased affinity of the NRTI-triphosphate for the RT active site (e.g., V75T, M184V) or a decreased rate of incorporation of the NRTI-triphosphate (e.g., K65R, K70E, L74V, Q151M) (Feng et al. 1999; 2006; Lennerstrand et al. 2001; Selmi et al. 2003a; Sluis-Cremer et al. 2007). The second NRTI resistance mechanism is mediated by thymidine analogue mutations (TAMs, M41L, D67N, K70R, L210W, T215FY, K219EQ) that increase the rate of phosphorolytic removal of the chain terminating NRTI and enable again continued DNA synthesis (Arion et al. 1998; Meyer et al. 2000).

Mutations Associated with Discrimination Against NRTI

The mutation M184IV occurs rapidly under the selective pressure of lamivudine and emtricitabine. In a background of a limited number of TAMs, this mutation resensitizes the virus to zidovudine, stavudine, and tenofovir (Larder et al. 1995; Whitcomb et al. 2003). Similar findings have been observed in vitro for the interactions between, respectively, K65R, L74V, L100I or Y181C, and TAMs (Eron et al. 1993; Miranda et al. 2005; Selmi et al. 2003b; White et al. 2006).

The most common mutation observed during didanosine monotherapy is L74V (St. Clair et al. 1991). In addition to L74V, in vitro didanosine resistance occurs due to mutations at codons 65 and 184 (Winters et al. 1997). The K65R has been isolated from several patients receiving long-term treatment with didanosine monotherapy. M184V also decreases susceptibility to didanosine. However, it is only rarely observed in HIV-1-infected patients receiving didanosine.

Although the V75T mutation has been observed after in vitro resistance selection experiments with stavudine and it is associated with reduced susceptibility to the drug (Lacey et al. 1994), this mutation is rarely detected in patients failing stavudine therapy. Instead, other substitutions at this position, such as V75AMS, are sometimes selected in vivo.

Combinations of the mutations K65R, L74V, Y115F, and M184V are necessary to confer significant resistance to abacavir in vitro (Tisdale et al. 1997). Single, double, and triple combinations of these mutations have been observed in therapy-naïve patients receiving abacavir monotherapy. The most common mutational pattern was L74V + M184V (Harrigan et al. 2000; Miller et al. 2000). Recently, L74I has also been linked to the use of abacavir or efavirenz (Wirden et al. 2006).

In vitro and in vivo resistance development to tenofovir has been associated with the RT mutations K65R and K70E (Margot et al. 2002; Delaugerre et al. 2005). The increasing prevalence of the previously rare K65R during the last years has been associated with the clinical use of tenofovir, but also with exposure to didanosine, abacavir, stavudine, and lamivudine whether or not in combination with NNRT nevirapine or efavirenz (Kagan et al. 2004; Segondy et al. 2005; Sungkanuparph et al. 2008; Trotta et al. 2006).

During combination therapy complex resistance patterns may emerge. It has been reported that combinations of NRTIs can select for a rare multi-nucleoside resistance pathway, characterized by the Q151M marker mutation. This set of mutations, A62V, S68G, V75I, F77L, F116Y, and Q151M confers cross-resistance to all NRTIs, with levels of resistance to didanosine, zalcitabine, and stavudine that are significantly higher than the resistance levels observed for virus mutants developed under monotherapy (Shirasaka et al. 1993; Schmit et al. 1996). The presence of other mutations in combination with the multi-nucleoside resistance set, such as M184V or K65R, are necessary to confer high-level resistance toward lamivudine, abacavir, or tenofovir (Schmit et al. 1998; Van Laethem et al. 2000, 2007).

Mutations Associated with Excision of NRTI

The reverse transcriptase mutations M41L, D67N, K70R, T210W, T215FY, and K219EQ were first ascribed to the selective pressure of zidovudine and appear traditionally step by step (Boucher et al. 1992). Afterward it was shown that stavudine could also select for this resistance pathway (Lin et al. 1994) and therefore this particular set of mutations was assigned as thymidine-associated mutations (TAMs). Genotypic analysis of samples obtained from patients receiving lamivudine + zidovudine or lamivudine + stavudine revealed similar levels of TAMs (50 vs 45%, p = 0.79) after 72 weeks of follow-up (Kutizkes et al. 2004).

A second multi-nucleoside resistance pattern has been described with an estimated prevalence of 1% in a population of patients failing antiretroviral therapy (Winters et al. 1998). It is marked by insertions between RT codons 68 and 70 (Eggink et al. 2007). Various combinations of amino acids are observed in the insert complex, most commonly T69S-SA, T69S-SG, and T69S-SS. In addition to the insert, TAMs are usually observed in these strains, with T215Y appearing in all strains. Phenotypic resistance tests of the patient isolates showed reduced susceptibility toward all NRTIs which was confirmed by site-directed mutagenesis experiments (Eggink et al. 2007; Prado et al. 2004; Winters et al. 1998).

Non-nucleoside Reverse Transcriptase Inhibitor (NNRTI) Mutations

Although the NNRTI drug class is extremely potent in reducing the viral load, it is characterized by a low genetic barrier toward drug resistance and by broad cross-resistance which limits the consecutive use of an NNRTI after virological failure (Richman et al. 1994; Antinori et al. 2002). In most instances, single mutations at RT positions 98–108, 138, 179–190, 221–238, and 318 are observed at virological failure and they are sufficient for class resistance (Table 26.1). They confer resistance through a reduced association and/or an increased dissociation rate between the inhibitor and the enzyme. The K103N mutation is the most commonly observed mutation in vivo. It is suggested that the observed cross-resistance of the K103N mutant is due to the stabilizing effect of the asparagine side chain at 103 on the closed-pocket conformation of the unliganded RT structure. This could provide resistance to a wide range of NNRTIs by giving a reduced rate of association between inhibitor and enzyme (Ren et al. 2000).

Table 26.1 Mutations associated with resistance toward non-nucleoside reverse transcriptase inhibitors

Region 98–108	Region 179–190	Region 221–238	Single positions
A98G	V179DEFGI	H221Y	E138GKQ
L100I	Y181CIV	P225H	Y318FT
K101ENPQR	Y188CHFL	F227CL	
K103HNRST	G190ACEQSTV	M230IL	
V106AIM		L234I	
V108I		P236L	
		K238NT	

Adapted from Rimsky et al. (2007) and Rega v7.1.1 (www.kuleuven.ac.be/rega/cev).

Compared to the FDA-approved NNRTI (nevirapine, delavirdine, and efavirenz), etravirine that currently is available through open access programs shows a higher genetic barrier and greater resilience to NNRTI resistance mutations (Andries et al. 2004; Vingerhoets et al. 2005). This is probably due to the torsion flexibility and the repositioning and reorientation capacities of etravirine in the NNRTI-binding pocket of the mutant RT and to conformational rearrangements within the enzyme (Das et al. 2004; Rodriguez-Barrios et al. 2005). This (partial) activity of etravirine against NNRTI-resistant virus was recently also confirmed in clinical trials (TMC125 Writing Group 2007; Madruga et al. 2007; Lazzarin et al. 2007). These data suggest that sequential use of etravirine after NNRTI failure is possible. Nevertheless, preliminary analyses have shown that in presence of an increasing number of mutations among V90I, A98G, L100I, K101EP, V106I, V179DF, Y181CIV, and G190AS, decreasing virological response was observed with the largest impact in patients carrying three or more of these mutations (Vingerhoets et al. 2007).

Accessory Mutations

Based upon statistical analyses of large data sets, additional mutations within RT have been identified (Svicher et al. 2006; Rhee et al. 2007; Nebbia et al. 2007). These accessory mutations almost always follow primary NRTI mutations. T39A, K43EQ, E44D, V118I, K122E, E203K, H208Y, D218E, respectively V35I, R83K, F214L have been positively and negatively associated with TAM mutations, whereas I50V has been negatively correlated with M184V. The clinical impact of these mutations, some of which also occur as a natural polymorphism, remains currently uncertain, but preliminary analyses have shown that the presence of, e.g., R83K at baseline correlates with a favorable virological response to zidovudine or stavudine-containing regimens and with less development of TAMs (Ceccherini-Silberstein et al. 2007).

A few studies have identified mutations within the connection and RNaseH domains of RT. They increase resistance toward thymidine analogues in the presence of TAMs by reducing the rate of RNA degradation. This reduction leads to an increase of time available for the excision of the already incorporated NRTI by the TAM-mutated enzyme (Nikolenko et al. 2005; 2007; Brehm et al. 2007; Delviks-Frankenberry et al. 2007). The majority of resistance assays do not cover this region. However, further data concerning the additional value of these mutations in guiding therapy choices are still required before changing present technology.

Assessing Drug Resistance

Clinical Indications for Resistance Testing

Resistance testing was at first solely performed to diagnose antiviral resistance at therapy failure. Gradually the idea arose to use this information for subsequently selecting new therapy combinations. A number of retrospective and prospective studies showed a short-term modest beneficial effect of resistance testing on response to the following therapy (Durant et al. 1999; Baxter et al. 2000; Meynard et al. 2002; Cohen et al. 2002; Tural et al. 2002; Cingolani et al. 2002) (Table 26.2). These data resulted into the recommendation of resistance testing at therapy failure by all expert panels. Although an optimal first-line therapy could be critical for long-term success, the use of resistance testing for designing first-line therapies remained more controversial. In a retrospective study, Little et al. (2002) could show that first-line therapy was more likely to fail in patients recently infected with drug-resistant virus. In a prospective multi-center study, first-line combination therapies guided by resistance testing had similar efficacy in chronically infected patients with baseline drug resistance as compared with patients carrying wild-type virus (Oette et al. 2006). Additionally, Sax et al. (2005) suggested that resistance testing for chronically infected patients and not only patients with acute infection could be cost-effective as soon as baseline drug resistance levels exceed 1%. As baseline resistance levels vary around 10% within the United States and Europe (Little et al. 2002; Wensing et al. 2005), resistance testing for therapy naïve patients is currently also recommended in these regions (The European HIV Drug Resistance Guidelines Panel 2006; The DHHS Panel on Antiretroviral Guidelines for Adults and Adolescents 2006; Hammer et al. 2006; Gazzard et al. 2006).

Table 26.2 Trials testing the prospective use of drug resistance testing

Study	Design	Viral response
VIRADAPT	genotype vs SOC	–1.04 vs –0.46 RNA log copies/ml at week 12 –1.15 vs –0.67 RNA log copies/ml at week 24
GART	genotype + expert advice vs SOC	–1.19 vs –0.61 RNA log copies/ml averaged at week 4 and 8
NARVAL	phenotype vs genotype vs SOC	35 vs 44 vs 36% below 200 RNA copies/ml at week 12
HAVANA	genotype vs SOC	49 vs 36 below 400 RNA copies/ml at week 24
	expert advice vs no expert advice	47 vs 37% below 400 RNA copies/ml at week 24
ARGENTA	genotype + expert advice vs SOC + expert advice	27 vs 12% below 500 RNA copies/ml at week 12
VIRA3001	phenotype vs SOC	46 vs 34% below 400 RNA copies/ml at week 16

Drug Resistance Assays

Phenotypic Drug Resistance Assays

Phenotypic assays measure the in vitro viral replication in the presence of increasing drug concentrations and determine the concentration of drug required to inhibit the replication of the patient's virus by 50% (IC_{50}). Although the IC_{90} is a better measurement for the requested in vivo effect, the IC_{50} is more reproducible as it is detected at the steepest part of the dose–response curve. A small change in response has therefore a less dramatic impact on the IC_{50} than on the IC_{90}. In addition, it is not the IC_{50} itself that is clinically used, but a factor that compares the IC_{50} of the patient's virus to the IC_{50} measured for a wild-type reference strain (fold-resistance).

Genotypic Drug Resistance Assays

Genotypic assays determine the nucleotide sequence of the viral gene that is targeted by the drugs within the therapy combination. The nucleotide sequence is subsequently translated into an amino acid sequence and compared to a reference wild-type strain to generate a mutation list. These mutations are then compared with a list of "resistance-associated" mutations. Mutations are defined as resistance associated (1) when they are known to be selected in vitro or in vivo in the presence of a certain drug, (2) when they have been associated with a reduction in phenotypic susceptibility, or (3) when they have been associated with a reduced clinical response (Johnson et al. 2006). This implies that the result of a genotypic assay always depends on pre-existing knowledge.

Recommendations for Drug Resistance Testing

Although improvements within molecular biology technologies currently allow the amplification of samples with very low viral loads, most of the expert panels do not recommend the testing of samples with viral loads less than 500–1000 RNA copies/ml as it might lead to unrepresentative selection of particular quasi-species from the total viral population which could not fully explain the observed therapy failure.

Additionally, it is highly important to monitor the resistance pattern of the active replicating plasma viral population under the selective pressure of the failing therapy as it is this population that is responsible for the failure. Therefore, resistance testing has to be performed before stopping or changing therapy.

At an initial virological failure, patients do not necessarily display resistance mutations against all drugs in that failing regimen. The loss in activity of one single component within a triple regimen is often sufficient for suboptimal suppression of the viral replication and the initial rising viral load. Drugs within the regimen that display a low genetic barrier, such as the NRTI lamivudine, the NNRTI, and the EI enfuvirtide, often select for drug-resistant virus at early viral breakthrough as one single mutation can be sufficient for reducing or loosing their antiviral activity, whereas still no resistance against the higher genetic barrier drugs (NRTI and boosted PI) is detected (Kempf et al., 2004). A quick intervention is required as soon as a rising viral load is observed to prevent the accumulation of other mutations. This could maintain the option for recycling some of the components within the failing regimen. Concomitantly, the chance for the development of cross-resistance against drugs within the same classes can be reduced which should broaden the future treatment options. Therefore, the viral load should be monitored at regular intervals in patients. A detectable viral load is then an indication for the performance of a drug resistance test.

Results of resistance assays represent only the genotypic or phenotypic profile of the majority of viral variants present in vivo. They have difficulties to detect minor variants that reflect less than 25% of the total viral population (Van Laethem et al. 1999). If the sample is collected without the presence of selective pressure, the chances are high that wild type will be detected instead of the mutant (Devereux et al. 1999; Venturi et al. 2002). In this respect, Palmer et al. (2006) could show that minor nevirapine resistant variants persisted above predose levels for more than 1 year in more than 23% of women after receiving a single-dose nevirapine for prevention of mother-to-child transmission. These mutant variants were detected by allele-specific PCR analysis at levels ranging from 0.1 to 20% 1 year after therapy in samples that were negative for nevirapine resistance by standard population sequencing. Mutant viruses display often a reduced replication capacity in absence of drugs, in comparison toward wild-type, and therefore they are rapidly overgrown by the wild-type. However, they remain as minor variants and can rapidly re-emergence in a subsequent therapy causing therapy failure. When no selective pressure is present, reversal of resistance (defined as back-mutation to the wild-type codon at a particular position) can also occur, although at a much lower frequency and much slower rate than the re-emergence of wild-type.

Resistance assays are most accurate in determining resistance to the current therapy combination. Not detecting resistance to any previously used drug does not guarantee complete drug susceptibility. If a previous isolate from a patient has been scored resistant to a particular drug, these resistant variants can still exist as minor variants or can be archived in latently infected cells and the response might be limited when that drug is re-used (Ghosn et al. 2006). Higher rates of virological failure were for instance observed in women who started nevirapine-containing therapy within 6 months after receipt of single, peripartum dose of nevirapine (Lockman et al. 2007). In conclusion, the therapy and viral load histories combined with the results from historical resistance tests could add valuable information to the resistance results at the observed therapy failure for designing a new potent combination. This historical information is not available at baseline and population sequencing could result into an underestimation of the present resistance levels. The impact of minority drug-resistant variants is still unclear in the setting of primary drug resistance, however, there are some indications that they can matter (Van Laethem et al. 2007a). A recent study showed that the risk of virological failure was significantly higher for patients starting a first-line therapy consisting of NRTI and NNRTI with minority NNRTI-resistant variants detected by ultra-deep sequencing than for patients without NNRTI resistance (Simen et al. 2007).

Genotypic and Phenotypic Drug Resistance Assays Provide Complementary Information for Clinical Decision Making

Genotypic resistance assays are more widely spread than phenotypic assays. They can deliver results within a few days, whereas more time is required to obtain phenotypic resistance results. This turn-around time can be an important factor in some clinical situations where results are needed as soon as possible, e.g., post-exposure prophylaxis. Additionally, they are less costly and they can be more easily implemented in regular molecular biology facilities which is not the case for phenotypic assays. Therefore, genotyping is mostly performed locally, whereas phenotyping is commonly put out to commercial companies (Monogram Biosciences, CA, US and Virco, Mechelen, Belgium). Nevertheless, these and other companies can also supply genotypic results and interpretations.

Phenotypic resistance assays give a direct measurement of susceptibility toward the tested inhibitors that includes the effect of all mutations and their interactions. In this manner, they are ideal assays for testing the susceptibility toward new antiviral drugs from which hardly any genotypic data are yet available (e.g., etravirine, darunavir and tipranavir), for testing samples from patients who failed multiple therapies, and for testing samples with atypical or less prevalent mutations.

Before the mechanism clarifying the antagonistic effect of M184V/I within the background of TAMs was known and thus before its inclusion into genotypic drug resistance rules, it was already measured by phenotypic assays (Schmit et al. 1998). This reversal in phenotypic resistance is reflected by a partial virological response (Masquelier et al. 1999). The interaction between TAMs and K103N results into the reversal of phenotypic resistance to efavirenz. However, it is not reflected by a change in virological response. Patients displaying this pattern of mutations still fail efavirenz-containing therapy despite the phenotypic susceptible score (Shulman et al. 2001). Genotypic assays can make a distinction based upon the detection of the mutations, whereas phenotypic assays cannot.

The current phenotypic assays have difficulties in detecting resistance to some NRTIs, especially when associated with TAMs (Zhang et al. 2005). These drugs compete with the natural dNTPs for binding to the reverse transcriptase and incorporation into the growing DNA chain. The phenotypic assays use activated cells with high intracellular dNTP pools which leads to the generation of dead-end complexes at low NRTI concentrations and prevents the excision of the incorporated NRTI (Meyer et al. 2000). As a consequence, the fold resistance values often fall within the reproducibility range of the assay. However, these in vitro conditions do not always reflect in vivo cell conditions with low intracellular dNTP pools and therefore, such genotypic resistance patterns are associated with a relevant in vivo reduced clinical response despite the low in vitro resistance levels.

Sequencing has the advantage of detecting substitutions that are considered to be associated with a lower genetic barrier to resistance development, i.e., mutations that are considered as markers for developing resistance patterns (e.g., K70R for the TAM or Q151M for the multi-nucleoside resistance pathway) or reversal mutations (215 revertants) (de Ronde et al. 2001). Variants displaying these substitutions require less additional nucleic acid changes than wild-type to develop a phenotypic resistant mutant or to obtain a reduced clinical response. Although these mutations are associated with reduced clinical response, they are not always linked with reduced phenotypic susceptibility.

Interpretation of Drug Resistance

Virologists who are responsible for genotypic drug resistance testing ought to convert the generated mutation list into an advice for the clinician. However, due to the complex nature of resistance patterns, it is not always possible to keep informed of the latest discoveries. Therefore, most of them rely on genotypic drug resistance interpretation systems that are updated at regular intervals. The commercial genotypic assays are concerted with an interpretation system (TRUGENE® HIV-1 Genotyping System (Siemens Medical Solutions Diagnostics, NY, US) and ViroSeq™ HIV-1 Genotyping System (Abbott Molecular Inc., IL, US). However, sequences can also be submitted to commercial interpretation systems (virco®TYPE HIV-1 [Virco]) and ViroScore (ABL, Luxembourg) or publicly available systems (e.g., geno2pheno, ANRS, HIVdb, and Rega). Their goal is to translate complex patterns of resistance-associated mutations into a categorical (e.g., susceptible and resistant) or continuous (fold change) variable that more easily can be implemented into clinical practice. These systems can be based on rules (e.g., ANRS and Rega) on mutation tables with additive and/or subtracting scoring derived from literature (e.g., HIVdb and Rega), on bio-informatics' analysis (e.g., geno2pheno and Rega), or combinations thereof (virco®TYPE HIV-1 and GeneSeqHIV™).

Phenotypic drug resistance assays deliver a continuous variable and this is in agreement with the drug resistance continuum in vivo as resistance is not an all-or-none phenomenon. However, a clinician has to decide whether or not to include a certain drug and therefore this variable is currently categorized into a more workable final output (e.g., sensitive/partially sensitive/resistant (PhenoSense™, Monogram Biosciences) or within/above normal susceptibility range (Antivirogram®, Virco)). Initially, arbitrary cut-offs were used, reflecting the reproducibility range of the test. This was recognized to be inadequate and therefore biological cut-offs were determined (Harrigan et al. 2001). These cut-offs reflect the range in susceptibilities toward each individual drug observed in therapy-naïve patients. Later on, clinical cut-offs have been determined for some drugs. They are based on the distinction between fold-resistance levels for which a reduced virologic response and fold-resistance levels for which no virologic response is observed. The clinical cut-offs for the two commercial assays can be found at their respective web sites (http://www.monogramhiv.com and http://www.vircolab.com). Bearing in mind the drug resistance continuum, it is hypothesized that a continuous resistance score might even improve the prediction of therapy outcome. However, a recent retrospective study showed that incremental phenotypic drug susceptibility scores more accurately predicted the virologic and immunologic outcome in patients starting salvage therapy compared to dichotomous or continuous scores (De Luca et al. 2006).

A few studies have compared genotypic drug resistance testing, combined with interpretation systems, and phenotypic drug resistance testing for the determination of drug susceptibilities. In one limited study, concordance ranged between 81 and 91% with the lowest value obtained for NRTI (Dunne et al., 2001). Complete concordance for all tested drugs was obtained for only 17% of the samples. Ross et al. (2005) calculated the concordance between two commercially available phenotypic tests and several interpretation systems and determined the kappa coefficients. Overall agreement between the phenotypic drug resistance tests was 86.9% with kappa value of 0.621. The lowest concordance was again observed for NRTI. The overall concordance between each individual interpretation system and phenotypic drug resistance test was greater than 80%, with kappa values ranging between 0.474 and 0.675. The highest discordances were observed for abacavir, didanosine, zalcitabine, and saquinavir.

Epidemiology of Drug Resistance RTI Mutations in Therapy-Naïve and Therapy-Experienced Patients

In the last years, several epidemiological studies have investigated the prevalence of drug resistance in therapy-experienced patients failing antiretroviral therapy. This could provide insight into the extent of the need for new active antiretrovirals but also into the potential of spreading drug resistance.

The UK HIV Drug Resistance Database was used to estimate the prevalence of drug resistance between 1998 and 2002 (UK Collaborative Group on HIV Drug Resistance, 2005). When resistance levels per number of performed resistance tests were investigated, the prevalence of resistance to more than one drug class varied between 75 and 82%, which is of the same magnitude as a prevalence of 76% in the United States (Richman et al. 2004). Resistance to two or more drug classes remained constant at 17% in the United Kingdom during the investigated period. Resistance toward NRTIs was the most prevalent (~75%). Over time, the most frequent dual class resistance changed from NRTI and PI to NRTI and NNRTI which reflected the change in prescription behavior. When estimates were based on the number of patients on therapy, from whom the majority is successfully suppressed, the prevalence of resistance to any drug class was 4.5% in 1998 and increased to 17% in 2002. In France, resistance assays, performed between 1997 and 2002, displayed genotypic resistance to at least one NRTI in 78.3%, to at least one NNRTI in 38.9%, and to at least one PI in 47.0%. Triple-class resistance peaked at 25.9% in 2000 and stabilized at 25.5% in 2002 (Tamalet et al. 2003). In a Portuguese study more stringent definitions of resistance were used (Vercauteren et al., 2007). Multi-drug resistance was defined as no more than one drug susceptible and class resistance as no drug susceptible within a certain drug class. Here, the incidence of multi-drug resistance decreased from 5.7 to 2.7% in the investigated period 2001–2006. The incidence of NRTI class resistance also declined from 13.5% in 2001 to 6.1% in 2006. NNRTI class resistance initially increased and subsequently decreased (35.6% [2001–2002] – 47.7% [2003–2004] – 42.0% [2005–2006]). The opposite was observed for PI class resistance (10.5% [2001–2002] – 5.6% [2003–2004] – 7.3% [2005–2006]).

In patients with primary infection followed at 10 US cities, the frequency of resistance mutations increased from 8% in 1995–1998 to 22.7% in 1999–2000 (Little et al. 2002). In Europe, the percentage of therapy-naïve patients with acute and chronic infection who carried at least one resistance mutation was 10.4% during 1996–2002 (Wensing et al. 2005). More than 7% showed resistance mutations against NRTI, 2.9% against NNRTI, and 2.5% against PI. Among a Canadian study population of newly diagnosed patients from 2000 to 2001, 8.1% displayed resistance mutations: 4.1% against NRTI, 1.4% against NNRTI, and 1.5% against PI (Jayaraman et al. 2006). More than 2000 patients in the United Kingdom were tested for drug resistance before receiving antiretroviral therapy between 1996 and 2003 and 14.2% of them had mutations that conferred resistance against at least one antiretroviral drug (Cane et al. 2005). In total, 9.9% had NRTI, 4.5% NNRTI, and 4.6% PI mutations. Most samples (10.9%) were resistant to one drug class only (mostly NRTI), 1.9% showed evidence of resistance to two drug classes, and 1.4% showed evidence of resistance to three classes. Masquelier et al. (2005) reported the results from a European and Canadian seroconverters cohort. In this cohort, 9.1, 0.5, and 0.7% carried drug-resistant mutations against NRTI, NNRTI, or PI, respectively. In conclusion, as for therapy-experienced patients, NRTI resistance was also the most prevalent form of class resistance in therapy-naïve patients across all studies.

Based upon a summary of manuscripts describing the prevalence of drug-resistant mutations in therapy-experienced from Europe, America, and Asia (Fig. 26.1), the most common single RT mutation in therapy-experienced patients is M184IV (mean 44%), followed by T215FY (34%), M41L (30%), D67N (24%), K103NRST (19%), L210W (18%), K70R (16%), K219EQR (14%), and Y181CIV (14%). The most common mutations in therapy-naïve patients from studies from around the world are K103NRST (1.9%), 215 revertants (1.9%), M41L (1.7%), K219EQR (0.9%), M184IV (0.9%), and T215FY (0.8%) (Fig. 26.1). The observed difference in hierarchy of drug-resistant mutations within experienced and naïve patients suggests a differential selection of drug-resistant mutations during transmission (Corvasce et al. 2006). In this respect, it has been claimed that M184IV and T215FY display a low relative efficiency of transmission. In contrast, the 215 revertants are found at the same proportion in both patient populations which could suggest an efficient transmissibility. However, the so-called disproportional transmission of some mutants could also be a reflection of resistance reversion in the therapy-experienced patient before the transmission event, or the reversion after transmission but before detection efforts (in chronically infected drug naïve patients). A recent study in which sequences from primary HIV infections and from therapy-naïve and therapy-experienced chronically infections were investigated, delivered an additional explanation (Brenner et al. 2007). In that particular study, early infection accounted for approximately half of the onward transmissions.

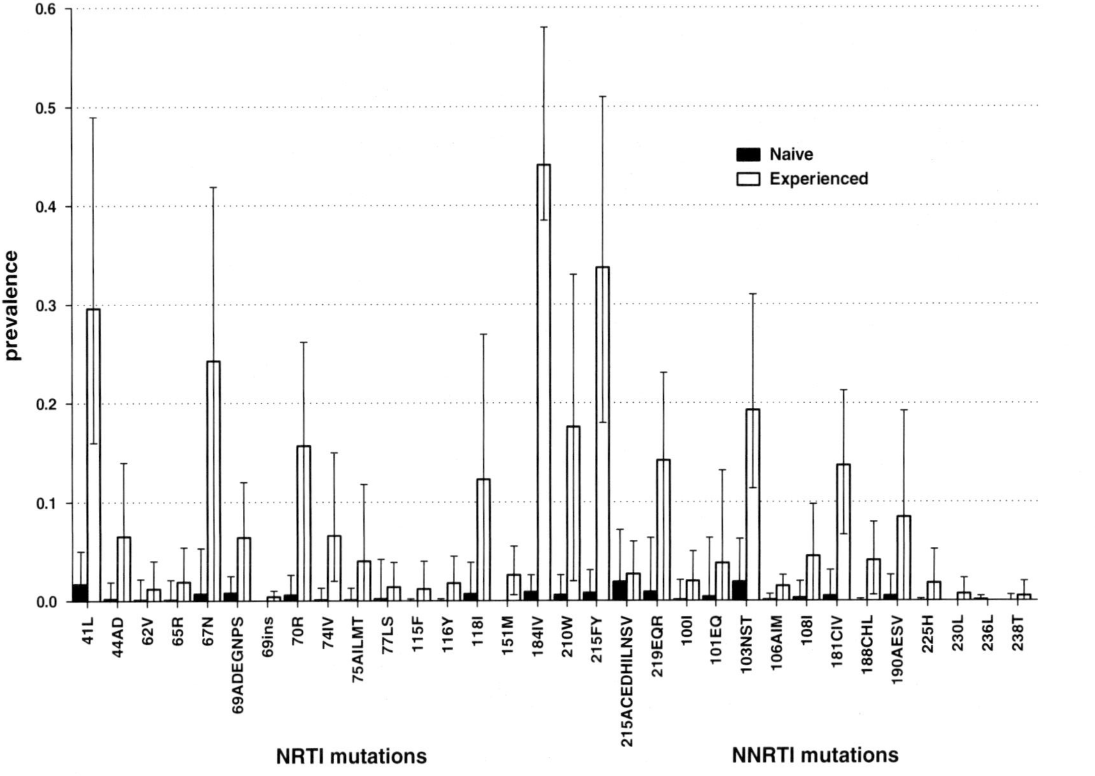

Fig. 26.1 Prevalence of drug-resistant mutations in HIV-1 reverse transcriptase of therapy-experienced and therapy-naïve patients. The mean proportion of the prevalence of a particular drug resistance mutation within therapy-experienced (*right bar*) and therapy-naïve (*left bar*) patients is displayed. The *line* denotes the minimum–maximum range. NRTI mutations are displayed at the *left side* of the graphic and the NNRTI mutations at the *right side*. The values are based on published studies with available numeric data on prevalence of individual drug-resistant mutations in Europe (Corvasce et al. 2006; Palma et al. 2007; Cane et al. 2005; Masquelier et al. 2005; Wensing et al. 2005; Descamps et al. 2005; Booth et al. 2007; Costagliola et al. 2007), America (Ríos et al. 2007; Pérez et al. 2007; Lama et al. 2006; Jayaraman et al. 2005; Shet et al. 2006; Rodrigues et al. 2006; Escoto-Delgadillo et al. 2005; Little et al. 2002), Africa (Vidal et al. 2007; Maréchal et al. 2006; Maslin et al. 2005; Vergne et al. 2006) and Asia (Sen et al. 2007; Han et al. 2007; Sukasem et al. 2007).

Half of the primary infections (293 out of 593 infections) grouped into 75 different transmission clusters. About 49% of the clusters contained 2.7 ± 0.8 infections and the remaining 8.8 ± 3.5 infections (mean ± SD). Forward transmission intervals ranged from 1 to 37 months with overall transmission intervals of 15.2 ± 9.5 months and thus reversion of some mutations in untreated patients before the onward transmission might explain the different drug resistance patterns in therapy-naïve and therapy-experienced patients.

Clinical Impact of RT Mutations

As M184IV and TAMs are the most prevalent NRTI mutations in therapy-experienced patients and TAMs confer cross-resistance to some or all NRTIs, studies investigating the clinical impact of drug resistance within reverse transcriptase are often confined to these mutations and their interactions. This high prevalence in the current population of therapy-experienced patients is a reflection of the extensive use of lamivudine, zidovudine, and stavudine often as part of suboptimal mono- and bi-therapy combinations in pre-HAART era.

Recent studies have revealed the occurrence of at least two distinct TAM pathways. De Luca et al. (2006) defined three TAM profiles: TAM1 (1 of M41L, L210W, and T215Y or ≥2 TAMs with K70 or ≥1 TAM of M41L and L210W), TAM2 (1 of D67N, K70R, T215F, K219EQ or ≥2 TAMs with M41 and L210), and TAM3 (all others) (Fig. 26.2). When using 41 and 210 as markers for TAM1 and 70R as marker for TAM2, 88% of all TAM profiles within the time period 1991–2004 could be classified. About 51% of the profiles belonged to TAM1, 37% to TAM2, and 12% to TAM3. The six most frequent TAM profiles of the 19 profiles that were observed were M41L, L210W, T215Y; M41L, D67N, L210W, T215Y; M41L, T215Y; K70R; D67N, K70R, K219Q; and D67N, K70R, T215F, K219Q. About 51% of the TAM profiles belonged to one of these six. Before 1997, TAM2 was more prevalent than TAM1. After 1997, TAM2 was less prevalent. TAM3 remained stable. In the majority of patients for whom longitudinal data was available, the initial TAM profile was retained, especially TAM1. TAM3 tended to evolve into TAM1. In multivariate analysis, previous treatment with NNRTI and previous experience with stavudine

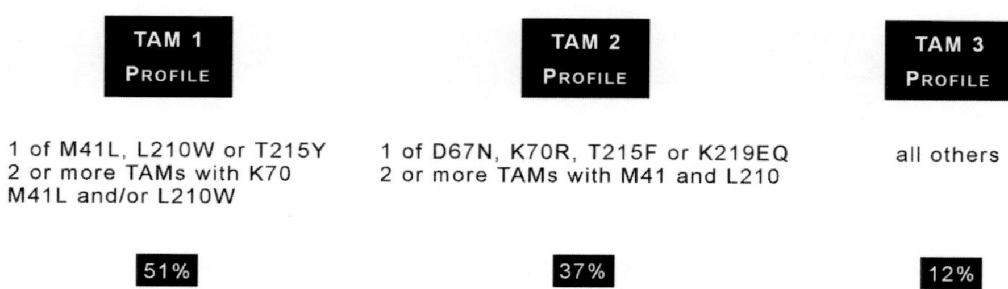

Fig. 26.2 Thymidine-associated mutational profiles in sequences obtained for drug resistance testing in the time period 1991–2004. Adapted from De Luca et al. (2006).

+ lamivudine were associated with higher chances to display TAM1 profile. Time on monotherapy (most often zidovudine) was associated with less chance to display TAM1 and more chance to have TAM2. In a smaller subset of patients, patients with TAM1 who were treated with stavudine containing HAART showed a trend toward a higher probability of failing than did those with TAM1 who were receiving zidovudine containing HAART.

The study of Cozzi-Lepri et al. (2005) confirmed the presence of two distinct pathways: TAM1 (M41L+L210W+T215Y) and TAM2 (D67N+K70R+K219EQ). Patients with longer zidovudine experience tended to display TAM2 profile more often. However, TAM2 occurred less frequently than TAM1. E44AD and V118I tended to cluster with TAM1. M184V occurred as often with both profiles. TAM1 had a similar impact on the virological response toward stavudine and zidovudine at 6 months. However, in the presence of TAM2 the adjusted mean difference in the reduction of viral load at 6 months was 0.96 $\log_{10}$ RNA copies/ml (95% CI: 0.20–1.73) higher in patients who started stavudine-containing therapy instead of zidovudine-containing therapy.

In vivo studies from the last years suggest that the in vivo relevance of M184V for didanosine might be limited or that the mutation might even be advantage for the activity of didanosine-containing therapy.

Among patients on didanosine-containing therapy, median-fold changes in phenotypic susceptibility to didanosine were greater in patients with the M184V mutation (fold changes of 2.2 vs 1.2, $p < 0.001$). Nonetheless, the median change in viral load and percentage of patients attaining an undetectable viral load were similar in those taking didanosine, irrespective of whether the M184V was present at baseline (Sproat et al. 2005).

In the randomized, placebo-controlled Jaguar trial, the impact of RT mutations on the virological response to didanosine or placebo, added to the failing regimen, was investigated at a follow-up of 4 weeks (Marcelin et al. 2005). Eight mutations, with a prevalence of more than 5%, were associated with a reduced response to didanosine, M41L, D67N, T69D, L74V, V118I, L210W, T215FY, and K219EQ, and two mutations were associated with a better response, K70R and M184IV. The best prediction of the virological response at week 4 was obtained by the rule M41L + T69D + L74V + T215FY + K219EQ - K70R - M184IV. Patients had a viral load reduction of −1.24, −0.84, −0.61, −0.4, and −0.07 $\log_{10}$ RNA copies/ml when they were ranked as having a genotypic score of −2, −1, 0, 1, and 2 or more, respectively. Isolates with a score of −2 displayed all M184V and K70R with no other mutations. Isolates with a score of −1 displayed or only M184V (22 isolates) or M184V + K70R + K219EQ/M41L (8 isolates). Patients carrying TAM2 profile, defined as D67N, K70R, T215F, and K219Q (25 isolates), had a viral load reduction of −0.95 $\log_{10}$ RNA copies/ml. Patients carrying TAM1 profile, defined as M41L, L210W, and T215Y (24 isolates), showed a viral load reduction of −0.47 $\log_{10}$ RNA copies/ml. Patients with wild type at 184 displayed a worse viral load response (−0.15 $\log_{10}$ RNA copies/ml) in comparison to those with 184 V (−0.61 $\log_{10}$ RNA copies/ml) ($p = 0.042$). However, only eight patients were WT at 184. More than 90% of the population carried the mutant M184V.

Due to the conflicting data regarding the effect of M184IV on the phenotypic susceptibility and in vivo activity of didanosine, Eron et al. (2007) retrospectively investigated data from patients receiving didanosine as single antiviral drug. They hoped to enhance their ability to interpret the results as, in contrast to the other studies, no other drugs were administered and the viruses contained besides M184V (or not) no or a modest number of other NRTI mutations. At week 8, patients naïve to lamivudine had an approximately 0.4 $\log_{10}$ RNA copies/ml greater decline in viral load than lamivudine-experienced patients. The decline in viral load was less for patients with M184V irrespective of lamivudine experience, although not significantly. It confirms the results of others that didanosine retains partly its in vivo activity in the presence of M184V and after exposure to lamivudine but that these factors diminish somewhat its antiviral effect.

The antiviral efficacy of abacavir was investigated in therapy-experienced patients where abacavir was added to a background antiretroviral therapy and the virological response 4 weeks after addition of abacavir was investigated. It revealed that patients with only M184V (−0.74 $\log_{10}$ RNA copies/ml) and patients with M184V + 1 TAM (−0.95 $\log_{10}$ RNA copies/ml) showed similar responses as patients with wild-type virus (−0.96 $\log_{10}$ RNA copies/ml). Patients with at least 2 TAMs or M184IV + 2 or 3 TAMs showed a reduced response (−0.38 until −0.18 $\log_{10}$ RNA copies/ml), whereas patients with M184V + at least 4 TAMs showed no response at all (−0 $\log_{10}$ RNA copies/ml) (Lanier et al. 2004). A clinical interpretation rule for abacavir was deduced from the Narval trial (Brun-Vézinet et al. 2003). Within this trial, patients with less than four mutations of M41L, D67N, L210W, T215YF, M184IV, and L74V showed a decrease in viral load of −1.64 $\log_{10}$ RNA copies/ml at week 12, while a reduction of −0.69 $\log_{10}$ RNA copies/ml was obtained in patients with four of these mutations and −0.19 $\log_{10}$ RNA copies/ml in patients with five or six mutations.

The phase 2 trials GS-99-907 and GS-98-902 were used to determine the genotypic marker mutations of the virological response toward tenofovir in therapy-experienced patients (Miller et al. 2004). Patients with K65R at baseline did not show a treatment response. Significantly reduced responses were observed for patients with at least 3 TAMs, relative to patients without TAMs (−0.40 vs −0.80 $\log_{10}$ RNA copies/ml at week 24). The response, relative to the placebo group (+0.03 $\log_{10}$ RNA copies/ml), was however still significant. In the absence of M41L and L210W, patients with at least three TAMs showed significant response relative to placebo group, but the response did not differ significantly from the patients without TAMs anymore (−0.67, +0.07, and −0.80 $\log_{10}$ RNA copies/ml at week 24, respectively). Patients with T215FY but without M41L or L210W showed a response not significantly different from patients without TAMs (−0.70 $\log_{10}$ RNA copies/ml at week 24). A multivariate linear regression model of viral load response to tenofovir in therapy-experienced patients was developed by Waters et al. (2006): −0.66 $\log_{10}$ RNA copies/ml for tenofovir treatment, −0.2 $\log_{10}$ per $\log_{10}$ RNA copies/ml increase baseline viral load, +0.14 $\log_{10}$ RNA copies/ml per mutation of M41L, D67N, L210W, T215Y, +0.61 $\log_{10}$ RNA copies/ml for K65R, +0.29 $\log_{10}$ RNA copies/ml for L74V and −0.13 $\log_{10}$ RNA copies/ml for M184V. This confirmed the resensitization effect of M184V on TAM towards tenofovir. A negative effect of L74V on the virological response towards tenofovir was observed despite the limited effect of this single mutation on the phenotypic resistance levels towards the drug and its resensitization effect in the presence of TAMs. It is hypothesized that L74V could be a marker for previous experience with abacavir or didanosine. These drugs are known to select also for K65R and it is possible that its concentration within the total viral population returned below the detection limit of present genotypic technology.

The prevalence of K65R increased from 0.4% in 1998 to 3.6% in 2003 (Parikh et al. 2006a). Among samples with K65R, a strong negative association was evident with TAMs, but not with other NRTI mutations. K65R reduced zidovudine resistance in a M41L, L210W, T215Y and in a D67N, K70R, T215F, and K219Q background. TAMs antagonized the effect of K65R, reducing resistance to tenofovir, abacavir, zalcitabine, didanosine, and stavudine (Parikh et al. 2006b). K65R and TAMs exhibit bidirectional phenotypic antagonism which likely explains the negative association of these mutations, the rare emergence of K65R with therapy-containing zidovudine and its more frequent emergence with combinations that exclude zidovudine. Similar negative associations, although not to the same extent, have been observed for TAMs and other resensitization mutations such as L74V and Y181C, resulting in a lower frequency of both resistance pathways and even the development of alternative pathways (Eron et al. 1993; Richman et al. 1994; Shafer et al. 1994). In addition, K65R was never found on the same genome with T215FY and at least two other TAMs, except in the presence of Q151M mutation complex (Parikh et al. 2006a). Of the 45 single genomes that had both K65R and T215FY, 22 also had Q151M and/or at least 2 other mutations of this complex. Of the other 23 single genomes, all had T215FY alone or with only 1 other TAM. The antagonism between K65R and TAMs might have a clinical benefit when simultaneous therapy with NRTIs that select for either mutations is prescribed which might prolong therapy response and delay the emergence of NRTI resistance (Gianotti et al. 2004).

Also the opposite can occur, early failure was indeed observed in a clinical trial comparing the efficacy of the combination abacavir + lamivudine + tenofovir with abacavir + lamivudine + efavirenz despite the fact that it was anticipated as a potent combination (49 vs 5% non-responders, $p < 0.001$) (Gallant et al. 2005). This study made clear that this combination should not be used in clinical practice. The most likely explanation for the early failure was the low genetic barrier for resistance to the whole regimen. In almost all non-responders, the M184IV was observed and in the majority of them K65R was also present. K65R is selected by tenofovir and abacavir and causes cross-resistance toward lamivudine, whereas lamivudine and tenofovir both select for M184V. The rising prevalence of K65R within genotypes from a clinic in the United Kingdom in the period 2000–2002 (Winston et al. 2004) was not associated with the use of one single drug, but with the dual use of tenofovir + didanosine and the triple combination tenofovir + didanosine + abacavir. Similar results have been found in a Portuguese database containing genotypes from patients failing antiretroviral therapy between 2002 and 2005 (Theys et al. 2009). In that study, a significant association of K65R selection was observed with the backbone tenofovir + didanosine and the regimens tenofovir + abacavir + NNRTI or lamivudine/emtricitabine, didanosine + stavudine + NNRTI, and tenofovir + lamivudine/emtricitabine + NNRTI. After 2004 the incidence of K65R at therapy failure decreased again which correlated with the sharp decrease of the previous mentioned regimens that currently are not recommended anymore (DHHS, 2006).

Strategies for Managing Drug Resistance

During the last years many potent antiretrovirals have become available which enables the successful treatment of HIV-1-infected patients. Several clinical trials have compared different combination therapies of which some showed equivalent virological and immunological efficacies (DHHS, 2006). Although potent first-line therapies nevertheless can fail and treatment of HIV-1 is a life-long commitment, the choice of this first regimen in relation to its resistance costs becomes more important as this affect future therapy options. Barlett et al. (2006) compared retrospectively 15 randomized clinical trials with 18 different triple-drug combinations. They found that the virological success rates of NNRTI-containing and PI/r-containing combinations overlap, but that failure on a PI/r-containing combination preserves more therapy options. The results of a randomized, open-label, prospective study comparing class-sparing combinations (lopinavir/r + efavirenz vs lopinavir/r + 2 NRTI vs efavirenz + 2 NRTI) have recently been presented (Haubrich et al. 2007). At a median of 112 weeks, virological failure occurred in 29, 37, and 24%, respectively, with significant differences between lopinavir/r + 2 NRTI and efavirenz + 2 NRTI. However, resistance mutations in two drug classes were more frequent in the efavirenz + 2 NRTI arm in comparison to lopinavir/r + 2 NRTI (26 vs 1%). NNRTI mutations were more often detected at failure on lopinavir/r + efavirenz than on efavirenz + 2 NRTI (66 vs 44%). These findings confirm the complexity of selecting first-line therapies.

With the introduction of new potent drugs in the last years, some against new targets such as integrase and envelope, the efficacy of combination therapy in highly treatment-experienced patients with multi-drug-resistant virus has improved impressively. A substantial proportion of these patients reached a sustainable undetectable viral load in the TORO, RESIST, POWER, DUET, MOTIVATE, and BENCHMARK trials which was previously not always feasible with the standard of care regimens at that time (Hill et al., 2007). The RESIST trial was first to acknowledge the long-standing virological impact of one additional active drug within the backbone regimen, next to the investigated drug (Hicks et al. 2006). The time to viral failure in the tipranavir/r treatment arm increased by almost threefold when enfuvirtide was also included within the regimen. However, the inclusion of enfuvirtide did not prolong the time to viral failure in the comparator arms. This observation highlighted the importance of the total susceptibility score of the combination therapy to obtain a durable response and it led to the recommendation to include at least two new active drugs within the subsequent regimen, supported by other drugs with possibly some partial activity (DHHS 2006).

However, in cases of multi-drug resistance the reduction of the viral load to undetectable levels cannot always be achieved with the available antiretroviral drugs. Treatment interruption in this subset of patients was associated with a reversal to the wild-type virus and a decrease of the CD4 cell count (Deeks et al. 2001). For these instances a new treatment strategy, therapy simplification, was designed and tested. The aim of this strategy was to maintain as much as possible the immunological status of the patient by retaining the less fit, and presumed less virulent, mutant virus with a simplified regimen (Deeks et al. 2005). Interruption of PIs or NNRTIs was associated with stable viral loads. However, interrupting NRTIs resulted into an immediate increase in viral load (median of 0.66 $\log_{10}$ RNA copies/ml during the first 16 weeks) suggesting that NRTIs often maintained some antiviral activity even in the presence of many NRTI mutations. Ongoing administration of a failing regimen could however lead to further accumulation of resistance mutations that later could limit the efficacy of new drug candidates within existing drug classes. Therefore, the Experienced-184 V study was set up to compare the effects of lamivudine monotherapy against complete therapy interruption in patients harbouring a M184V mutant virus as it is presumed that lamivudine does not select for any other mutations (Castagna et al. 2006). By week 48, 20 of 29 patients in the treatment-interruption arm in comparison to 12 of 29 in the lamivudine arm had discontinued the study because of immunological or clinical failure which was statistically significant ($p = 0.018$). Lamivudine monotherapy may thus lead to a better outcome than complete therapy interruption. Recently, retrospective studies have investigated the accumulation of mutations in patients left on virologically suboptimal NRTI therapies and found that these combinations maintain immunological benefit with no or minimal accumulation of NRTI mutations (Llibre et al. 2007; Cozzi-Lepri et al. 2007).

Planned or unplanned therapy interruption can occur for many reasons and there are indications that stopping antiretroviral therapy combinations containing drugs with different half-lives may lead to the development and selection of drug resistance. The risk for drug resistance depends on the genetic barrier of the drug with the longest half-life, the time on suboptimal drug concentrations sufficient for selecting resistance, and the level of viral replication. Taylor et al. (2007) has classified several stopping strategies such as simultaneous stop, staggered stop, switch or exchange stop, and protected stop, which should preferentially be applied in different clinical situations. The protected stop in which all drugs are stopped concomitantly and replaced by lopinavir/r monotherapy during 4 weeks is considered to be a good candidate for a universal stopping strategy. Indeed, lopinavir/r monotherapy has already proven to be able to maintain viral suppression with the minimal development of resistance (Arribas et al. 2005). Of course, the potential of lopinavir/r in this respect will probably depend on pre-existing PI resistance.

Conclusion and Future Directions

Improvements in the potency and tolerability of prescribed therapy combinations have resulted into more durable suppression of viral replication. Together with a stricter viral load follow-up and more swift therapeutic interventions, it resulted into changes of drug resistance patterns and their prevalence within, not only therapy-experienced but also therapy-naïve patients. During the last years, drug resistance mechanisms have been more clarified and the impact of mutations or fold-resistance levels on the virological response toward a particular drug have been better understood. This has improved the clinical usefulness of genotypic and phenotypic resistance testing and in the meantime resistance testing has become a cornerstone in the management of HIV-1 therapy.

Despite our progress, continuous basic research remains crucial. New mutations are still being identified, linked to particular known resistance mutations or to experience with a certain therapy. Each time, their mechanism of action and clinical impact remains to be investigated. Ongoing efforts also illustrate a possible negative impact of minority drug resistant variants on the virological response, however, these studies are still anecdotic and have not revealed any clinical cut-off in sensitivity yet. More clinically oriented research also remains a high priority, since despite the success of current antiretroviral combinations, therapy failure can still occur resulting into the accumulation of resistance mutations. Interpretation systems need to be improved, especially toward HIV-1 variants of different subtypes. Also, as antiretroviral therapy of HIV-1 infection remains a long-term effort and with the rationale to maintain potent therapy options for as long as possible, further research is required to deliver insights on how successive regimens should best strategically be started and stopped with long term planning, taking into account their efficacy, susceptibility score and potential for development of particular drug-resistant mutations to a potential next regimen. Designing optimal 'waiting' regimen is often problematic and needs further research attention, trading between too quickly burning new drug in suboptimal combinations versus risking the ultimate deterioration of the patient's immune system with death as result.

References

Andries, K., Azijn, H., Thielemans, T., Ludovici, D., Kukla, M., Heeres, J., Janssen, P., De Corte, B., Vingerhoets, J., Pauwels, R. and de Béthune, M.-P. 2004. TMC125, a novel next-generation nonnucleoside reverse transcriptase inhibitor active against nonnucleoside reverse transcriptase inhibitor-resistant human immunodeficiency virus type 1. Antimicrob. Agents Chemother. 48: 4680–4686.

Antinori, A., Zaccarelli, M., Cingolani, A., Forbici, F., Rizzo, M.G., Trotta, M.P., Di Giambenedetto, S., Narciso, P., Ammassari, A., Girardi, E., De Luva, A. and Perno C.F. 2002. Cross-resistance among nonnucleoside reverse transcriptase inhibitors limits recycling efavirezn after nevirapine failure. AIDS Res. Hum. Retroviruses 18: 835–838.

Arion, D., Kaushik, N., McCormick, S., Borkow, G. and Parniak, M.A. 1998. Phenotypic mechanism of HIV-1 resistance to 3'-azido-3'-deoxythymidine (AZT): increased polymerization processivity and enhanced sensitivity to pyrophosphate of the mutant viral reverse transcriptase. Biochemistry 37: 15908–15917.

Arribas, J.R., Pulido, F., Delgado, R., Lorenzo, A., Miralles, P., Arranz, A., González-García, J.J., Cepeda, C., Hervás, R., Paño, J.R., Gaya, F., Carcas, A., Montes, M.L., Costa, J.R. and Peña, J.M. 2005. Lopinavir/ritonavir as single-drug therapy for maintenance of HIV-1 viral suppression: 48-weeks results of a randomized, controlled, open-label, proof-of-concept pilot clinical trial (OK Study). J. Acquir. Immune Defic. Syndr. 40: 280–287.

Bartett, J.A., Fath, M.J., DeMasi, R., Hermes, A., Quinn, J., Mondou, E. and Rousseau, F. 2006. An updated systematic overview of triple combination therapy in antiretroviral-naïve HIV-infected adults. AIDS 20: 2051–2064.

Baxter, J.D., Mayers, D.L., Wentworth, D.N., Neaton, J.D., Hoover, M.L., Winters, M.A., Mannheimer, S.B., Thompson, M.A., Abrams, D.I., Brizz, B.J., Ioannidis, J.P. and Merigan, T.C. 2000. A randomized study of antiretroviral management based on plasma genotypic antiretroviral resistance testing in patients failing therapy. AIDS 14: F83–F93.

Booth, C.L., Garcia-Diaz, A.M., Youle, M.S., Johnson, M.A., Phillips, A. and Geretti, A.M. 2007. Prevalence and predictors of antiretroviral drug resistance in newly diagnosed HIV-1 infection. J. Antimicrob. Chemother. 59: 517–524.

Boucher, C.A.B., O'Sullivan, E., Mulder, J.W., Ramautarsing, C., Kellam, P., Darby, G., Lange, J.M.A., Goudsmit, J. and Larder B.A. 1992. Ordered appearance of zidovudine resistance mutations during treatment of 18 human immunodeficiency virus-positive subjects. J. Infect. Dis. 165: 105–110.

Brehm, J., Koontz, D., Sluis-Cremer, N. and Mellors J. 2007. HIV-1 reverse transcriptase mutations A371V and Q509L decrease DNA-dependent RNase H cleavage and increase the rate of AZT-MP excision. Antivir. Ther. 12: S124.

Brenner, B.G., Roger, M., Routy, J.-P., Moisi, D., Ntemgwa, M., Matte, C., Baril, J.-G., Thomas, R., Rouleau, D., Bruneau, J., Leblanc, R., Legault, M., Tremblay, C., Charest, H., Wainberg, M.A. and the Quebec Primary HIV Infection Study Group. 2007. High rates of forward transmission events after acute/early HIV-1 infection. J. Infect. Dis. 195: 951–959.

Brun-Vézinet, F., Descamps, D., Ruffault, A., Masquelier, B., Calvez V., Peytavin, G., Telles, F., Morand-Joubert, L., Meynard, J.-L., Vray, M., Costagliola, D. and the Narval (ANRS 088) Study Group. 2003. Clinically relevant interpretation of genotype for resistance to abacavir. AIDS 17: 1795–1802.

Cane, P., Chrystie, I., Dunn, D., Evans, B., Geretti, A.-M., Green, H., Phillips, A., Pillay, D., Porter, K., Pozniak, A., Sabin, C., Smit, E., Weber, J., Zuckerman, M. and UK Group on Transmitted HIV Drug Resistance. 2005. Time trends in primary resistance to HIV drugs in the United Kingdom: multicentre observational study. BMJ 331: 1368.

CASCADE Collaboration. 2003. Determinants of survival following HIV-1 seroconversion after the introduction of HAART. Lancet 362: 1267–1274.

Castagna, A., Danise, A., Menzo, S., Galli, L., Gianotti, N., Carini, E., Boeri, E., Galli, A., Cernuschi, M., Hasson, H., Clementi, M. and Lazzarin, A. 2006. Lamivudine monotherapy in HIV-1-infected patients harbouring a lamivudine-resistant virus: a randomized pilot study (E-184 V study). AIDS 20: 795–803.

Ceccherini-Silberstein, F., Svicher, V., Santoro, M.M., Prosperi, M., Forbici, F., Bertolli, A., Mussini, C., De Luca, A., Palamara, G., Narciso, P., Antinori, A., Balotta, C., d'Arminio Monforte, A. and Perno C. – IcoNA Study Group for Clinical use of HIV Genotype Resistance Test. 2007. Impact of HIV-1 reverse transcriptase polymorphism R83K on virological response in drug-naïve patients starting thymidine-analogue-containing HAART. Antivir. Ther. 12: S77.

Cingolani, A., Antinori, A., Rizzo, M.G., Murri, R., Ammassari, A., Baldini, F., Di Giambenedetto, S., Cauda, R. and De Luca, A. 2002. Usefulness of monitoring HIV drug resistance and adherence in individuals failing highly active antiretroviral therapy: a randomized trial (ARGENTA). AIDS 16: 369–379.

Clark, S., Calef, C. and Mellors, J. 2007. Mutations in retroviral genes associated with drug resistance, pp. 58–158 in HIV Sequence Compendium 2006/2007. Edited by: Thomas Leitner, T., Foley, B., Hahn, B., Marx, P., McCutchan, F., Mellors, J., Wolinsky, S. and Korber, B. Published by: Theoretical Biology and Biophysics Group, Los Alamos National Laboratory, Los Alamos, NM. LA-UR 07–4826.

Cohen, C., Hunt, S., Sension, M., Farthing, C., Conant, M., Jacobson, S., Nadler, J., Verbiest, W., Hertogs, K., Ames, M., Rinehart, A.R., Graham, N.M. and VIRA3001 Study Team. 2002. A randomized trial assessing the impact of phenotypic resistance testing on antiretroviral therapy. AIDS 16: 579–588.

Corvasce, S., Violin, M., Romano, L., Razzoloni, F., Vicenti, I., Galli, A., Duca, P., Caramma, I., Balotta, C. and Zazzi, M. 2006. Evidence of differential selection of HIV-1 variants carrying drug-resistant mutations in seroconverters. Antivir. Ther. 11: 329–334.

Costagliola, D., Descamps, D., Assoumou, L., Morand-Joubert, L., Marcelin, A.-G., Brodard, V., Delaugerre, C., Mackiewicz, V., Ruffault, A., Izopet, J., Plantier, J.-C., Tamalet, C., Yerly, S., Saidi, S., Brun-Vézinet, F., Masquelier, B. and the Agence Nationale de Recherches sur le SIDA et les Hepatites Virales (ANRS) AC11 Resistance Study Group. 2007. Prevalence of HIV-1 drug resistance in treated patients. A French nationwide study. J Acquir. Immune Defic. Syndr., 46: 12–18.

Cozzi-Lepri, A., Ruiz, L., Loveday, C., Phillips, A.N., Clotet, B., Reiss, P., Ledergerber, B., Holkmann, C., Staszewski, S. and Lundgren, J.D. for the EuroSIDA Study Group. 2005. Thymidine analogue mutation profiles: factors associated with acquiring specific profiles and their impact on the virological response to therapy. Antivir. Ther. 10: 791–802.

Cozzi-Lepri, A., Phillips, A.N., Martínez-Picado, J., d'Arminio Monforte A., Katlama, C., Eg Hansen, A.-B., Horban, A., Bruun, J., Clotet, B. and Lundgren, J. for the EuroSIDA study group. 2007. Rate of accumulation of thymidine analogue mutations (TAM) in patients left on virologically failing regimens containing zidovudine (ZDV) or stavudine (d4T). Antivir. Ther. 12: S81.

Das, K., Clark, A.D., Lewi, P.J., Heeres, J., De Jonge, M.R., Koymans, L.M., Vinkers, H.M., Daeyaert, F., Ludovici, D.W., Kukla, M.J., De Corte, B., Kavash, R.W., Ho, C.Y., Ye, H., Lichtenstein, M.A., Andries, K., Pauwels, R., de Béthune, M.-P., Boyer, P.L., Clark, P., Hughes, S.H., Janssen, P.A. and Arnold, E. 2004. Roles of conformational and positional adaptability in structure-based design of TMC125-R165335 (etravirine) and related non-nucleoside reverse transcriptase inhibitors that are highly potent and effective against wild-type and drug-resistant HIV-1 variants. J. Med. Chem. 47: 2550–2560.

Deeks, S.G., Wrin, T., Liegler, T., Hoh, R., Hayden, M., Barbour, J.D., Hellmann, N.S., Petropoulos, C.J., McCune, J.M., Hellerstein, M.K. and Grant, R.M. 2001. Virologic and immunogic consequences of discontinuing combination antiretroviral-drug therapy in HIV-infected patients with detectable viremia. N. Engl. J. Med. 344: 472–480.

Deeks, S.G., Hoh, R., Neilands, T.B., Liegler, T., Aweeka, F., Petropoulos, C.J., Grant, R.M. and Martin, J.N. 2005. Interruption of treatment with individual therapeutic drug classes in adults with multidrug-resistant HIV-1 infection. J. Infect. Dis. 192: 1537–1544.

De Luca, A., Weidler, J., Di Giambenedetto, S., Coakley, E., Bacarelli, A., Bates, M., Lie, Y., Pesano, R., Cauda, R. and Schapiro, J. 2006. Incremental and continuous phenotypic drug susceptibility scores more accurately predict virologic and immunologic treatment outcomes in HIV+ patients starting salvage therapy: finding from the Argenta trial. 13th Conference on Retroviruses and Opportunistic Infections, Denver, February 5–8 2006, abstract 650.

De Luca, A., Di Giambenedetto, S., Romano, L., Gonnelli, A., Corsi, P., Baldari, M., Di Pietro, M., Menzo, S., Francisci, D., Almi, P., Zazzi, M. and the Antiretroviral Resistance Cohort Analysis Study Group. 2006. Frequency and treatment-related predictors of thymidine-analogue mutation patterns in HIV-1 isolates after unsuccessful antiretroviral therapy. J. Infect. Dis. 193: 1219–1222.

Delaugerre, C., Roudiere, L., Peytavin, G., Rouzioux, C., Viard, J.-P. and Chaix, M.-L. 2005. Selection of a rare resistance profile in an HIV-1-infected patient exhibiting a failure to an antiretroviral regimen including tenofovir DF. J. Clin. Virol. 32: 241–244.

Delviks-Frankenberry, K.A., Nikolenko, G.N., Barr, R. and Pathak, V.K. 2007. Probing the mechanism by which connection domain mutations enhance AZT resistance: mutational analysis of the RNase H primer grip. Antivir. Ther. 12: S125.

de Ronde, A., van Dooren, M., van Der Hoek, L., Bouwhuis, D., de Rooij, E., van Gemen, B., de Boer, R. and Goudsmit, J. 2001. Establishment of new transmissible and drug-sensitive human immunodeficiency virus type 1 wild types due to transmission of nucleoside analogue-resistant virus. J. Virol. 75: 595–602.

Descamps, D., Chaix, M.L., Andre, P., Brodard, V., Cottalorda, J., Deveau, C., Harzic, M., Ingrand, D., Izopet, J., Kohli, E., Masquelier, B., Mouajjah, S., Palmer, P., Pellegrin, I., Plantier, J.C., Poggi, C., Rogez, S., Ruffault, A., Schneider, V., Signori-Schmuck, A., Tamalet, C., Wirden, M., Rouzioux, C., Brun-Vézinet, F., Meyer, L. and Costagliola, D. 2005. French national sentinel survey of antiretroviral drug resistance in patients with HIV-1 primary infection and in antiretroviral-naïve chronically infected patients in 2001–2002. J Acquir. Immune Defic. Syndr. 38: 545–552.

Devereux, H., Youle, M., Johnson, M. and Loveday, C. 1999. Rapid decline in detectability of HIV-1 drug resistance mutations after stopping therapy. AIDS 13: F123–F127.

Doyon, L., Croteau, G., Thibeault, D., Poulin, F., Pilote, L. and Lamarre, D. 1996. Second locus involved in human immunodeficiency virus type 1 resistance to protease inhibitors. J. Virol. 70: 3763–3769.

Dunne, A.L., Mitchell, F.M., Coberly, S.K., Hellmann, N.S., Hoy, J., Mijch, A., Petropoulos, C.J., Mills, J. and Crowe, S.M. 2001. Comparison of genotyping and phenotyping methods for determining susceptibility of HIV-1 to antiretroviral drugs. AIDS 15: 1471–1475.

Durant, J., Clevenbergh, P., Halfon, P., Delgiudice, P., Porsin, P., Simonet, P., Montagne, N., Boucher, C.A., Schapiro, J.M. and Dellamonica, P. 1999. Drug-resistance genotyping in HIV-1 therapy: the VIRADAPT randomized controlled trial. Lancet 353: 2195–2199.

Eggink, D., Huigen, M.C.D.G., Boucher, C.A.B., Götte, M. and Nijhuis, M. 2007. Insertions in the β3-β4 loop of reverse transcriptase of human immunodeficiency virus type 1 and their mechanism of action, influence on drug susceptibility and viral replication. Antivir. Res. 75: 93–103.

Eron, J.J., Chow, Y.-K., Caliendo, A.M., Videler, J., Devore, K.M., Cooley, T.P., Liebman, H.A., Kaplan, J.C., Hirsch, M.S. and D'Aquila, R.T. 1993. Pol mutations conferring zidovudine and didanosine resistance with different effects in vitro yield multiply resistant human immunodeficiency virus type 1 isolates in vivo. Antimicrob. Agents Chemother. 37: 1480–1487.

Eron, J.J., Bosch, R.J., Bettendorf, D., Petch, L., Fiscus, S., Frank, I. and the Adult Clinical Trials Group 307 Protocol Team. 2007. The effect of lamivudine therapy and M184V on the antiretroviral activity of didanosine. J. Acquir. Immune Defic. Syndr. 45: 249–251.

Escoto-Delgadillo, M., Vázquez-Valls, E., Ramírez-Rodríguez, M., Corona-Nakamura, A., Amaya-Tapia, G., Quintero-Pérez, N., Panduro-Cerda, A. and Torres-Mendoza, B.M. 2005. Drug-resistance mutations in antiretroviral-naive patients with established HIV-1 infection in Mexico. HIV Med. 6: 403–409.

Feng, J.Y. and Anderson K.S. 1999. Mechanistic studies examining the efficiency and fidelity of DNA synthesis by the 3TC-resistant mutant (184 V) of HIV-1 reverse transcriptase. Biochemistry 38: 9440–9448.

Feng, J.Y., Myrick, F.T., Margot, N.A., Mulamba, G.B., Rimsky, L., Borroto-Esoda, K., Selmi, B. and Canard B. 2006. Virologic and enzymatic studies revealing the mechanism of K65R- and Q151M-associated HIV-1 drug resistance towards emtricitabine and lamivudine. Nucleosides Nucleotides Nucleic Acids 25: 89–107.

Gallant, J.E., Rodriguez A.E., Weinberg, W.G., Young, B., Berger, D.S., Lim, M.L., Liao, Q., Ross, L., Johnson, J. and Shaefer, M.S. for the ESS30009 Study. 2005 Early virologic nonresponse to tenofovir, abacavir, and lamivudine in HIV-infected antiretroviral-naïve subjects. J. Infect. Dis. 192: 1921–1930.

Gazzard, B. on behalf of the BHIVA Writing Committee. 2006. British HIV Association (BHIVA) guidelines for the treatment of HIV-infected adults with antiretroviral therapy (2006). HIV Med. 7: 487–503.

Ghosn, J., Pellegrin, I., Goujard, C., Deveau, C., Viard, J.-P., Galimand, J., Harzic, M., Tamalet, C., Meyer, L., Rouzioux, C. and Chaix, M.-L. for the French PRIMO Cohort Study Group (ANRS CO 06). 2006. HIV-1 resistant strains acquired at the time of primary infection massively fuel the cellular reservoir and persist for lengthy periods of time. AIDS 20: 159–170.

Gianotti, N., Seminari, E., Fusetti, G., Salpietro, S., Boeri, E., Galli, A., Lazzarin, A., Clementi, M. and Castagna, A. 2004. Impact of a treatment including tenofovir plus didanosine on the selection of the 65R mutation in highly drug-experienced HIV-infected patients. AIDS 18: 2205–2208.

Hammer, S.M., Saag, M.S., Schechter, M., Montaner, J.S.G., Schooley, R.T., Jacobsen, D.M., Thompson, M.A., Carpenter, C.C.J., Fischl, M.A., Gazzard, B.G., Gatell, J.M., Hirsch, M.S., Katzenstein, D.A., Richman, D.D., Vella, S., Yeni, P.G. and Volberding, P.A. 2006. Treatment for adult HIV infection. 2006 recommendations of the International AIDS Society-USA Panel. JAMA 296: 827–843.

Han, X., Zhang, M., Dai, D., Wang, Y., Zhang, Z., Liu, J., Geng, W., Jiang, Y., Takebe, Y. and Shang, H. 2007. Genotypic resistance mutations to antiretroviral drugs in treatment-naïve HIV/AIDS patients living in Liaoning province, China: baseline prevalence and subtype-specific difference. AIDS Res. Hum. Retroviruses 23: 357–364.

Harrigan, P.R., Stone, C., Griffin, P., Nájera, I., Bloor, S., Kemp, S., Tisdale, M., Larder, B. and the CNA2001 Investigative Group. 2000. Resistance profile of the human immunodeficiency virus type 1 reverse transcriptase inhibitor abacavir (1592U89) after monotherapy and combination therapy. J. Infect. Dis. 181: 912–920.

Harrigan, P.R., Montaner, J.S.G., Wegner, S.A., Verbiest, W., Miller, V., Wood, R. and Larder, B.A. 2001. World-wide variation in HIV-1 phenotypic susceptibility in untreated individuals: biologically relevant values for resistance testing. AIDS 15: 1671–1677.

Haubrich, R.H., Riddler, S.A., DiRienzo, A.G., Peeples, L., Klingman, K.L., Garren, K.W., George, T., Rooney, J.F., Brizz, B., Havlir, D., Mellors, J.W. and the AIDS Clinical Trials Group 5142 Study Team. 2007. Drug resistance at virological failure in a randomized, phase III trial of NRTI-, PI- and NNRTI-sparing regimens for initial treatment of HIV-1 infection (ACTG 5142). Antivir. Ther. 12: S66.

Hicks, C.B., Cahn, P., Cooper, D.A., Walmsley, S.L., Katlama, C., Clotet, B., Lazzarin, A., Johnson, M.A., Neubacher, D., Mayers, D., Valdez, H. on behalf of the RESIST investigator group. 2006. Durable efficacy of tipranavir-ritonavir in combination with an optimized background regimen of antiretroviral drugs for treatment-experienced HIV-1-infected patients at 48 weeks in the Randomized Evaluation of Strategic Intervention in multi-drug resistant patients with Tipranavir (RESIST) studies: an analysis of combined data from two randomized open-label trials. Lancet 368: 466–475.

Hill, A., Miralles, D., Vangeneugden, T. and Lefebvre, E. 2007. Should we now adopt the HIV-RNA <50 copy endpoint for clinical trials of antiretroviral-experienced as well as naïve patients? AIDS 21: 1651–1653.

Jayaraman, G.C., Archibald, C.P., Kim, J., Rekart, M.L., Singh, A.E., Harmen, S., Wood, M. and Sandstrom, P. 2006. A population-based approach to determine the prevalence of transmitted drug-resistant HIV among recent versus established HIV infections. Results from the Canadian HIV strain and drug resistance surveillance program. J. Acquir. Immune Defic. Syndr. 42: 86–90.

Johnson, V.A., Brun-Vézinet, F., Clotet, B., Kuritzkes, D.R., Pillay, D., Schapiro, J.M. and Richman DD. 2006. Update of the drug resistance mutations in HIV-1: Fall 2006. Top. HIV Med. 14: 125–130.

Kagan, R.M., Merigan, T.C., Winters, M.A. and Heseltine, P.N.R. 2004. Increasing prevalence of HIV-1 reverse transcriptase mutation K65R correlates with tenofovir utilization. Antivir. Ther. 9: 827–828.

Kempf D.J., King M.S., Bernstein B., Cernohous P., Bauer E., Moseley J., Gu K., Hsu A., Brun S. and Sun E. 2004. Incidence of resistance in a double-blind study comparing lopinavir/ritonavir plus stavudine and lamivudine to nelfinavir plus stavudine and lamivudine. J. Infect. Dis. 189: 51–60.

Kuritzkes, D.R., Bassett, R.L., Hazelwood, J.D., Barrett, H., Rhodes, R.A., Young, R.K. and Johnson, V.A. for the Adult ACTG Protocol 306 and 370 Teams. 2004. Rate of thymidine analogue resistance mutation accumulation with zidovudine- or stavudine-based regimens. J. Acquir. Immune Defic. Syndr. 36: 600–603.

Lacey, S.F. and Larder, B.A. 1994. Novel mutation (V75T) in human immunodeficiency virus type 1 reverse transcriptase confers resistance to 2',3'-didehydro-2',3'-dideoxythymidine in cell culture. Antimicrob. Agents Chemother. 38: 1428–1432.

Lama, J.R., Sanchez, J., Suarez, L., Caballero, P., Laguna, A., Sanchez, J.L., Whittington, W.L.H., Celum, C. and Grant, R.M. for the Peruvian HIV Sentinel Surveillance Working Group. 2006. Linking HIV and antiretroviral drug resistance surveillance in Peru: a model for a third-generation HIV sentinel surveillance. J. Acquir. Immune Defic. Syndr. 42: 501–505.

Lanier, E.R., Ait-Khaled, M., Scott, J., Stone, C., Melby, T., Sturge, G., St. Clair, M., Steel, H., Hetherington, S., Pearce, G., Spreen, W. and Lafon, S. 2004. Antiviral efficacy of abacavir in antiretroviral therapy-experienced adults harbouring HIV-1 with specific patterns of resistance to nucleoside reverse transcriptase inhibitors. Antivir. Ther. 9: 37–45.

Larder, B.A., Kemp, S.D. and Harrigan, P.R. 1995. Potential mechanism for sustained antiretroviral efficacy of AZT-3TC combination therapy. Science 269: 696–699.

Lazzarin, A., Campbell, T., Clotet, B., Johnson, M., Katlama, C., Moll, A., Twoner, W., Trottier, B., Peeters, M., Vingerhoets, J., de Smedt, G., Baeten, B., Beets, G., Sinha, R., Woodfall, B., on behalf of the DUET-2 study group. 2007. Efficacy and safety of TMC125 (etravirine) in treatment-experienced HIV-1-infected patients in DUET-2: 24-week results from a randomised, double-blind, placebo-controlled trial. Lancet 370: 39–48.

Lennerstrand, J., Stammers, D.K. and Larder B.A. 2001. Biochemical mechanism of human immunodeficiency virus type 1 reverse transcriptase resistance to stavudine. Antimicrob. Agents Chemother. 45: 2144–2146.

Llibre, J.M., Bonjoch, A., Iribarren, J.A., Galindo, M.J., Martínez-Picado, X., Arazo, P., Santos, J.R., Domingo, P. and Clotet, B. 2007. Viral evolution in HIV-1 reverse transcriptase in patients receiving a holding regimen with Trizivir and tenofovir with persistent viral replication. Antivir. Ther. 12: S80.

Lin, P.F., Samanta, H., Rose, R.E., Patick, A.K., Trimble, J., Bechtold, C.M., Revie, D.R., Khan, N.C., Federici, M.E., Li, H., et al. 1994. Genotypic and phenotypic analysis of human immunodeficiency virus type 1 isolates from patients on prolonged stavudine therapy. J. Infect. Dis. 170: 1157–1164.

Little, S.J., Holte, S., Routy, J.-P., Daar, E.S., Markowitz, M., Collier, A.C., Koup, R.A., Mellors, J.W., Connick, E., Conway, B., Kilby, M., Wang, L., Whitcomb, J.M., Hellmann, N.S. nad Richman, D.D. 2002. Antiretroviral-drug resistance among patients recently infected with HIV. N. Engl. J. Med. 347: 385–394.

Lockman, S., Shapiro, R.L., Smeaton, L.M., Wester, C., Thior, I., Stevens, L., Chand, F., Makhema, J., Moffat, C., Asmelash, A., Ndase, P., Arimi, P., van Widenfelt, E., Mazhani, L., Novitsky, V., Lagakos, S. and Essex, M. 2007. Response to antiretroviral therapy after a single, peripartum dose of nevirapine. N. Engl. J. Med. 356: 135–147.

Madruga, J.V., Cahn, P., Grinsztejn, B., Haubrich, R., Lalezari, J., Mills, A., Pialoux, G., Wilkin, T., Peeters, M., Vingerhoets, J., de Smedt, G., Leopold, L., Trefiglio, R., Woodfall, B., on behalf of the DUET-1 study group. 2007. Efficacy and safety of TMC125 (etravirine) in treatment-experienced HIV-1-infected patients in DUET-1: 24-weeks results from a randomised, double-blind, placebo-controlled trial. Lancet 370: 29–38.

Marcelin, A.G., Flandre, P., Peytavin, G. and Calvez, V. 2005. Predictors of virologic response to ritonavir-boosted protease inhibitors. AIDS Rev. 7: 225–232.

Marcelin, A.G., Flandre, P., Pavie, J., Schmidely, N., Wirden, M., Lada, O., Chiche, D., Molina, J.M., Calvez, V. and the AI454-176 Jaguar Study Team. 2005. Clinically relevant genotype interpretation of resistance to didanosine. Antimicrob. Agents Chemother. 49: 1739–1744.

Maréchal, V., Jauvin, V., Selekon, B., Leal, J., Pelembi, P., Fikouma, V., Gabrie, P., Heredeibona, L.S., Goumba, C., Serdouma, E., Ayouba, A. and Fleury, H. 2006. Increasing HIV type 1 polymorphic diversity but no resistance to antiretroviral drugs in untreated patients from Central African Republic: a 2005 study. AIDS Res. Hum. Retroviruses 22: 1036–1044.

Margot, N.A., Isaacson, E., McGowan, I., Cheng, A.K., Schooley, R.T. and Miller, M.D. 2002. Genotypic and phenotypic analyses of HIV-1 in antiretroviral-experienced patients with tenofovir DF. AIDS 16: 1227–1235.

Maslin, J., Rogier, C., Caron, M., Grandadam, M., Koeck, J.-L. and Nicand, E. 2005. Antiretroviral drug resistance among drug-naïve HIV-1-infected individuals in Djibouti (Horn of Africa). Antivir. Ther. 10: 855–859.

Masquelier, B., Descamps, D., Carriere, I., Ferchal, F., Collin, G., Denayrolles, M., Ruffault, A., Chanzy, B., Izopet, J., Buffet-Janvresse, C., Schmitt, M.P., Race, E., Fleury, H.J., Aboulker, J.P., Yeni, P. and Brun-Vézinet, F. 1999. Zidovudine resensitization and dual HIV-1 resistance to zidovudine and lamivudine in the delta lamivudine roll-over study. Antivir. Ther. 4: 69–77.

Masquelier, B., Bhaskaran, K., Pillay, D., Gifford, R., Balestre, E., Jorgensen, L.B., Pedersen, C., van der Hoek, L., Prins, M., Balotta, C., Longo, B., Kücherer, C., Poggensee, G., Ortiz, M., de Mendoza, C., Gill, J., Fleury, H. and Porter, K. on behalf of the CASCADE Collaboration. 2005. Prevalence of transmitted HIV-1 drug resistance and the role of resistance algorithms. Data from seroconverters in the CASCADE Collaboration from 1987 to 2003. J. Acquir. Immune Defic. Syndr. 40: 505–511.

Mellors, J.W., Muñoz, A., Giorgi, J.V., Margolick, J.B., Tassoni, C.J., Gupta, P., Kingsley, L.A., Todd, J.A., Saah, A.J., Detels, R., Phair, J.P. and Rinaldo, C.R. 1997. Plasma viral load and $CD4^+$ lymphocytes as prognostic markers of HIV-1 infection. Ann. Intern. Med. 126: 946–954.

Meyer, P.R., Matsuura, S.E., Schinazi, R.F., So A.G. and Scott W.A. 2000. Differential removal of thymidine nucleotide analogues from blocked DNA chains by human immunodeficiency virus reverse transcriptase in the presence of physiological concentrations of 2'-deoxynucleoside triphosphates. Antimicrob. Agents Chemother. 44: 3465–3472.

Meynard, J.L., Vray, M., Morand-Joubert, L., Race, E., Descamps, D., Peytavin, G., Matheron, S., Lamotte, C., Guiramand, S., Costagliola, D., Brun-Vézinet, F., Clavel, F. and Girard, P.M. Narval Trial Group. 2002. Phenotypic or genotypic resistance testing for choosing antiretroviral therapy after treatment failure: a randomized trial. AIDS 16: 727–736.

Miller, V., Ait-Khaled, M., Stone, C., Griffin, P., Mesogiti, D., Cutrell, A., Harrigan, R., Staszewski, S., Katlama, C., Pearce, G. and Tisdale, M. 2000. HIV-1 reverse transcriptase (RT) genotype and susceptibility to RT inhibitors during abacavir monotherapy and combination therapy. AIDS 14: 163–171.

Miller, M.D., Margot, N., Lu, B., Zhong, L., Chen, S.-S., Cheng, A. and Wulfsohn, M. 2004. Genotypic and phenotypic predictors of the magnitude of response to tenofovir disproxil fumarate treatment in antiretroviral-experienced patients. J. Infect. Dis. 189: 837–846.

Miranda, L.R., Götte, M., Liang, F. and Kuritzkes, D.R. 2005. The L74V mutation in human immunodeficiency virus type 1 reverse transcriptase counteracts enhanced excision of zidovudine monophosphate associated with thymidine analog resistance mutations. Antimicrob. Agents Chemother. 49: 2648–2656.

Nebbia, G., Sabin, C.A., Dunn, D.T. and Geretti A.-M. on behalf of the UK Collaborative Group on HIV Drug Resistance and the UK Collaborative HIV Cohort (CHIC) Study Group. 2007. J. Antimicrob. Chemother. 59: 1013–1016.

Nijhuis, M., van Maarseveen, N.M., Lastere, S., Schipper, P., Coakley, E., Glass, B., Rovenska, M., de Jong, D., Chappey, C., Goedegebuure, I.W., Heilek-Snyder, G., Dulude, D., Cammack, N., Brakier-Gingras, L., Konvalinka, J., Parkin, N., Kräusslich, H.-G., Brun-Vézinet, F., and Boucher, C.A.B. 2007. A novel substrate-based HIV-1 protease inhibitor drug resistance mechanism. PLoS Medicine 4(1): e36.

Nikolenko, G.N., Palmer, S., Maldarelli, F., Mellors, J.W., Coffin, J.M. and Pathak, V.K. 2005. Mechanism for nucleoside analog-mediated abrogation of HIV-1 replication: balance between RNase H activity and nucleotide excision. Proc. Natl. Acad. Sci. USA 102: 2093–2098.

Nikolenko, G.N., Delviks-Frankenberry, K.A., Palmer, S., Maldarelli, F., Fivash, M.J. Jr., Coffin, J.M. and Pathak, V.K. 2007. Mutations in the connection domain of HIV-1 reverse transcriptase increase 3'-azido-3'-deoxythymidine resistance. Proc. Natl. Acad. Sci. USA 104: 317–322.

Oette, M., Kaiser, R., Däumer, M., Petch, R., Fätkenheuer, G., Carls, H., Rockstroh, J.K., Schmalöer, D., Stechel, J., Feldt, T., Pfister, H., Häussinger, D. and the RESINA Study Team. 2006. Primary HIV drug resistance and efficacy of first-line antiretroviral therapy guided by resistance testing. J. Acquir. Immune Defic. Syndr. 41: 573–581.

Palma, A.C., Araújo, F., Duque, V., Borges, F., Paixão, M.T., Camacho, R. On behalf of the Portuguese SPREAD Network. 2007. Molecular epidemiology and prevalence of drug resistance-associated mutations in newly diagnosed HIV-1 patients in Portugal. Infect. Genet. Evol. 7: 391–398.

Palmer, S., Boltz, V., Martinson, N., Maldarelli, F., Gray, G., McIntyre, J., Mellors, J., Morris, L. and Coffin, J. 2006. Persistence of nevirapine-resistant HIV-1 in women after single-dose nevirapine therapy for prevention of maternal-to-fetal HIV-1 transmission. Proc. Natl. Acad. Sci. USA 103: 7094–7099.

Parikh, U.M., Barnas, D.C., Faruki, H. and Mellors, J.W. 2006a. Antagonism between the HIV-1 reverse-transcriptase mutation K65R and thymidine-analogue mutations at the genomic level. J. Infect. Dis. 194: 651–660.

Parikh, U.M., Bachelor, L., Koontz, D., and Mellors, J.W. 2006b. The K65R mutation in human immunodeficiency virus type 1 reverse transcriptase exhibits bidirectional phenotypic antagonism with thymidine analog mutations. J. Virol. 80: 4971–4977.

Pérez, L., Pérez Álvarez, L., Carmona, R., Aragonés, C., Delgado, E., Thomson, M.M., González, Z., Contreras, G., Pérez J. and Nájera R. 2007. Genotypic resistance to antiretroviral drugs in patients infected with several HIV type 1 genetic forms in Cuba. AIDS Res. Hum. Retroviruses 23: 407–414.

Poveda, E., Briz, V. and Soriano, V. 2005. Enfuvirtide, the first fusion inhibitor to treat HIV infection. AIDS Reviews 7: 139–147.

Prado, J.G., Franco, S., Matamoros, T., Ruiz, L., Clotet, B., Menéndez-Arias, L., Martínez, M.A. and Martinez-Picado J. 2004. Relative replication fitness of multi-nucleoside analogue-resistant HIV-1 strains bearing a dipeptide insertion in the fingers subdomain of the reverse transcriptase and mutations at codons 67 and 215. Virology 326: 103–112.

Ren, J., Milton, J., Weaver, K.L., Short, S.A., Stuart, D.I. and Stammers D.K. 2000. Structural basis for the resilience of efavirenz (DMP-266) to drug resistance mutations in HIV-1 reverse transcriptase. Structure 8: 1089–1094.

Rhee, S.-Y., Liu, T.F., Holmes, S.P. and Shafer, R.W. 2007. HIV-1 subtype B protease and reverse transcriptase amino acid covariation. PLoS Comput. Biol. 3: e87.

Richman, D.D., Havlir, D., Corbeil, J., Looney, D., Ignacio, C., Spector, S.A., Sullivan, J., Cheeseman, S., Barringer, K., Pauletti, D., Shih, C.-K., Myers, M. and Griffin, J. 1994. Nevirapine resistance mutations of human immunodeficiency virus type 1 selected during therapy. J. Virol. 68: 1660–1666.

Richman, D.D., Morton, S.C., Wrin, T., Hellmann, N., Berry, S., Shapiro, M.F. and Bozzette, S.A. 2004. The prevalence of antiretroviral drug resistance in the United States. AIDS 18: 1393–1401.

Rimsky, L.T., Tambuyzer, L., Vingerhoets, J., Azijn, H., Staes, M., Picchio, G., Kraus, G. and de Béthune, M.-P. 2007. Compilation of mutations associated with resistance to NNRTIs deduced from clinical samples, in vitro analyses and bibliographical studies. Antivir. Ther. 12: S72.

Ríos, M., Delgado, E., Péréz-Álvarez, L., Fernández, J., Gálvez, P., Vázquez de Parga, E., Yung, V., Thomson, M.M. and Nájera, R. 2007. Antiretroviral drug resistance and phylogenetic diversity of HIV-1 in Chile. J. Med. Virol. 79: 647–656.

Rodrigues, R., Scherer, L.C., Oliveira, C.M., Franco, H.M., Sperhacke, R.D., Ferreira, J.L., Castro, S.M., Stella, I.M. and Brigido, L.F. 2006. Low prevalence of primary antiretroviral resistance mutations and predominance of HIV-1 clade C at polymerase gene in newly diagnosed individuals from south Brazil. Virus Res. 116: 201–207.

Rodriguez-Barrios, F., Balzarini, J. and Gago, F. 2005. The molecular basis of resilience to the effect of the Lys103Asn mutation in non-nucleoside HIV-1 reverse transcriptase inhibitors studied by targeted molecular dynamics simulations. J. Am. Chem. Soc. 127: 7570–7578.

Ross, L., Boulmé, R., Fisher, R., Hernandez, J., Florance, A., Schmit, J.-C. and Williams, V. 2005. A direct comparison of drug susceptibility to HIV type 1 from antiretroviral experienced subjects as assessed by the antivirogram and phenosense assays and by seven resistance algorithms. AIDS Res. Hum. Retroviruses 21: 933–939.

Sax, P.E., Islam, R., Walensky, R.P., Losina, E., Weinstein, M.C., Goldie, S.J., Sadownik, S.N. and Freedberg, K.A. 2005. Should resistance testing be performed for treatment-naïve HIV-infected patients? A cost-effectiveness analysis. Clin. Infect. Dis. 41: 1316–1323.

Schmit, J.C., Cogniaux, J., Hermans, P., Van Vaeck, C., Sprecher, S., Van Remoortel, B., Witvrouw, M., Balzarini, J., Desmyter, J., De Clercq, E. and Vandamme A.-M. 1996. Multiple drug resistance to nucleoside analogues and nonnucleoside reverse transcriptase inhibitors in an efficiently replicating human immunodeficiency virus type 1 patient strain. J. Infect. Dis. 174: 962–968.

Schmit, J.-C., Van Laethem, K., Ruiz, L., Hermans, P., Sprecher, S., Sönnerborg, A., Leal, M., Harrer, T., Clotet, B., Arendt, V., Lissen, E., Witvrouw, M., Desmyter, J., De Clercq, E. and Vandamme, A.-M. 1998. Multiple dideoxynucleoside analogue-resistant (MddNR) HIV-1 strains isolated from patients from different European countries. AIDS 12: 2007–2015.

Schmit, J.-C., Martinez-Picado, J., Ruiz, L., Tural, C., Van Laethem, K., Cabrera, C., Ibanez, A., Puig, T., Witvrouw, M., Desmyter, J., De Clercq, E., Clotet, B. and Vandamme, A.-M. 1998. Evolution of HIV drug resistance in zidovudine/zalcitabine- and zidovudine/didanosine-experienced patients receiving lamivudine-containing combination therapy. Antivir. Ther. 3: 81–88.

Segondy M. and Montes, B. 2005. Prevalence and conditions of selection of the K65R mutation in the reverse transcriptase gene of HIV-1. J. Acquir. Immune Defic. Syndr. 38: 110–111.

Selmi, B., Deval, J., Boretto, J. and Canard, B. 2003a. Nucleotide analogue binding, catalysis and primer unblocking in the mechanisms of HIV-1 reverse transcriptase-mediated resistance to nucleoside analogues. Antivir. Ther. 8: 143–154.

Selmi, B., Deval, J., Alvarez, K., Boretto, J., Sarfati, S., Guerreiro, C., and Canard, B. 2003b. The Y181C substitution in 3'-azido-3'-deoxythymidine-resistant human immunodeficiency virus type 1 reverse transcriptase suppresses the ATP-mediated repair of the 3'-azido-3'-deoxythymidine 5'-monophosphate-terminated primer. J. Biol. Chem. 278: 40464–40472.

Sen, S., Tripathy, S.P., Chimanpure, V.M., Patil, A.A., Bagul, R.D. and Paranjape, R.S. 2007. Human immunodeficiency virus type 1 drug resistance mutations in peripheral blood mononuclear cell proviral DNA among antiretroviral treatment-naïve and treatment-experienced patients from Pune, India. AIDS Res. Hum. Retroviruses 23: 489–497.

Shafer, R.W., Kozal, M.J., Winters, M.A., Iversen, A.K., Katzenstein, D.A., Ragni, M.V., Meyer, W.A. III, Gupta, P., Rasheed, S., Coombs, R., Katzman, M., Fiscus, S. and Merigan T.C. 1994. Combination therapy with zidovudine and didanosine selects for drug-resistant human immunodeficiency virus type 1 strains with unique patterns of pol gene mutations. J. Infect. Dis. 169: 722–729.

Shet, A., Berry, L., Mohri, H., Mehandru, S., Chung, C., Kim, A., Jean-Pierre, P., Hogan, C., Simon, V., Boden, D. and Markowitz, M. 2006. Tracking the prevalence of transmitted antiretroviral drug-resistant HIV-1. A decade of experience. J. Acquir. Immune Defic. Syndr. 41: 439–446.

Shirasaka, T., Yarchoan, R., O'Brien, M.C., Husson, R.N., Anderson, B.D., Kojima, E., Shimada, T., Broder, S. and Mitsuya, H. 1993. Changes in drug sensitivity of human immunodeficiency virus type 1 during therapy with azidothymidine, dideoxycytidine, and dideoxyinosine: an in vitro comparative study. Proc. Natl. Acad. Sci. USA 90: 562–566.

Shulman, N., Zolopa, A.R., Passaro, D., Shafer, R.W., Huang, W., Katzenstein, D., Israelski, D.M., Hellmann, N., Petropoulos, C. and Whitcomb, J. 2001. Phenotypic hypersusceptibility to non-nucleoside reverse transcriptase inhibitors in treatment-experienced HIV-infected patients: impact on virological response to efavirenz-based therapy. AIDS 15: 1125–1132.

Simen, B.B., Huppler Hullsiek, K., Novak, R.M., MacArthur, R.D., Baxter, J.D., Huang, C., Lubeski, C., Turenchalk, G.S., Braverman, M.S., Desany, B., Simons, J.F., Rothberg, J.M.,

Egholm, M. and Kozal, M.J. 2007. Prevalence of low abundant drug-resistant variants by ultra-deep sequencing in chronically HIV-infected antiretroviral (ARV)-naïve patients and the impact on virological outcomes. Antivir. Ther. 12: S149.

Sluis-Cremer, N., Sheen, C.-W., Zelina, S., Argoti Torres P.S., Parikh U.M. and Mellors J.W. 2007. Molecular mechanism by which the K70E mutation in human immunodeficiency virus type 1 reverse transcriptase confers resistance to nucleoside reverse transcriptase inhibitors. Antimicrob. Agents Chemother. 51: 48–53.

Sproat, M., Pozniak, A.L., Peeters, M., Winters, B., Hoetelmans, R., Graham, N.M. and Gazzard, B.G. 2005. The influence of the M184V mutation in the HIV-1 reverse transcriptase on the virological outcome of highly active antiretroviral therapy regimens with or without didanosine. Antivir. Ther. 10: 357–361.

St. Clair, M.H., Martin, J.L., Tudor-Williams, G., Bach, M.C., Vavro, C.L., King, D.M., Kellam, P., Kemp, S.D. and Larder, B.A. 1991. Resistance to ddI and sensitivity to AZT induced by a mutation in HIV-1 reverse transcriptase. Science 253: 1557–1559.

Sukasem, C., Churdboonchart, V., Chasombat, S., Kohreanudom, S., Watitpun, C., Pasomsub, E., Piroj, W., Tiensuwan, M. and Chantratita, W. 2007. Surveillance of genotypic resistance mutations in chronic HIV-1 treated individuals after completion of the national access to antiretroviral program in Thailand. Infection 35: 81–88.

Sungkanuparph, S., Manosuthi, W., Kiertiburanakul, S., Saekang, N., Pairoj, W. and Chantratita, W. 2008. Prevalence and risk factors for developing K65R mutations among HIV-1 infected patients who fail an initial regimen of fixed-dose combination of stavudine, lamivudine, and nevirapine. J. Clin. Virol. 41: 310–313.

Svicher, V., Sing, T., Santoro, M.M., Forbici, F., Rodríguez-Barrios, F., Bertoli, A., Beerenwinkel, N., Bellocchi, M.C., Gago, F., d'Arminio Monforte, A., Antinori, A., Lengauer, T., Ceccherini-Silberstein, F. and Perno, C.F. 2006. Involvement of novel human immunodeficiency virus type 1 reverse transcriptase mutations in the regulation of resistance to nucleoside inhibitors. J. Virol. 80: 7186–7198.

Tamalet, C., Fantini, J., Tourres, C. and Yahi, N. 2003. Resistance of HIV-1 to multiple antiretroviral drugs in France : a 6-year survey (1997–2002) based on an analysis of over 7000 genotypes. AIDS 17: 2383–2388.

Taylor, S., Boffito, M., Khoo, S., Smit, E. and Back, D. 2007. Stopping antiretroviral therapy. AIDS 21: 1673–1682.

The DHHS Panel on Antiretroviral Guidelines for Adults and Adolescents 2006.

The European HIV Drug Resistance Guidelines Panel 2006.

The TMC125-C223 Writing Group. 2007. Efficacy and safety of etravirine (TMC125) in patients with highly resistant HIV-1: primary 24-week analysis. AIDS 21: F1–F10.

Theys, K., Vercauteren, J., Abecasis, A.B., Libin, P., Deforche, K., Vandamme, A.-M. and Camacho, R. 2009. The rise and fall of K65R in a Portuguese HIV-1 Drug Resistance database, despite continuously increasing use of tenofovir. Infect. Genet. Evol. in press.

Tisdale, M., Alnadaf, T. and Cousens, D. 1997. Combination of mutations in human immunodeficiency virus type 1 reverse transcriptase required for resistance to the carbocyclic nucleoside 1592U89. Antimicrob. Agents Chemother. 41: 1094–1098.

Trotta, M.P., Bonfigli, S., Ceccherini-Silberstein, F., Bellagamba, R., D'Arrigo, R., Soldani, F., Zaccarelli, M., Bellocchi, M.C., Lorenzini, P., Marconi, P., Boumis, E., Forbici, F., Comandini, U.V., Tozzi, V., Narciso, P., Perno, C.F. and Antinori, A. 2006. Clinical and genotypic correlates of mutation K65R in HIV-infected patients failing regimens not including tenofovir. J. Med. Virol. 78: 535–541.

Tural, C., Ruiz, L., Holtzer, C., Schapiro, J., Viciana, P., Gonzalez, J., Domingo, P., Boucher, C., Rey-Joly, C., Clotet, B. and the Havana Study Group. 2002. Clinical utility of HIV-1 genotyping and expert advice: the Havana trial. AIDS 16: 209–218.

UK Collaborative Group on HIV Drug Resistance. 2005. Estimating HIV-1 drug resistance in antiretroviral-treated individuals in the United Kingdom. J. Infect. Dis. 192: 967–973.

Vandamme, A.-M., Van Vaerenbergh, K. and De Clercq, E. 1998. Anti-human immunodeficiency virus drug combination strategies. Antivir. Chem. Chemother. 9: 187–203.

Van Laethem, K., Van Vaerenbergh, K., Schmit, J.-C., Sprecher, S., Hermans, P., De Vroey, V., Schuurman, R., Harrer, T., Witvrouw, M., Van Wijngaerden, E., Stuyver, L., Van Ranst, M., Desmyter, J., De Clercq, E. and Vandamme, A.-M. 1999. Phenotypic assays and sequencing are less sensitive than point mutation assays for the detection of resistance in mixed HIV-1 genotypic populations. J. Acquir. Immune Defic. Syndr. 22: 107–118.

Van Laethem, K., Witvrouw, M., Balzarini, J., Schmit, J.-C., Sprecher, S., Hermans, P., Leal, M., Harrer, T., Ruiz, L., Clotet, B., Van Ranst, M., Desmyter, J., De Clercq, E. and Vandamme A.-M. 2000. Patient HIV-1 strains carrying the multiple nucleoside resistance mutations are cross-resistant to abacavir. AIDS 10: 469–471.

Van Laethem, K., De Munter, P., Schrooten, Y., Verbesselt, R., Van Ranst, M., Van Wijngaerden, E. and Vandamme, A.-M. 2007a. No response to first-line tenofovir + lamivudine + efavirenz despite optimization according to baseline resistance testing: impact of resistant minority variants on efficacy of low genetic barrier drugs. J. Clin. Virol. 39: 43–47.

Van Laethem, K., Pannecouque, C. and Vandamme A.-M. 2007b. Mutations at 65 and 70 within the context of a Q151M cluster in human immunodeficiency virus type 1 reverse transcriptase impact the susceptibility to the different nucleoside reverse transcriptase inhibitors in distinct ways. Infect. Genet. Evol. 7: 600–603.

Venturi, G., Romano, L., Carli, T., Corsi, P., Pippi, L., Valensin, P.E. and Zazzi, M. 2002. Divergent distribution of HIV-1 drug-resistant variants on and off antiretroviral therapy. Antivir. Ther. 7: 245–250.

Vercauteren, J., Theys, K., Debruyne, M., Deforche, K., Duque, J.L., Peres, S., Carvalho, A.P., Mansinho, K., Vandamme, A.-M. and Camacho, R. 2007. The incidence of multidrug and class resistance in HIV-1 infected patients is decreasing over time (2001–2006). Fifth European HIV Drug Resistance Workshop, Cascais, Portugal. Abstract 1.

Vergne, L., Diagbouga, S., Kouanfack, C., Aghokeng, A., Butel, C., Laurent, C., Noumssi, N., Tardy, M., Sawadogo, A., Drabo, J., Hien, H., Zekeng, L., Delaporte, E. and Peeters, M. 2006. HIV-1 drug-resistance mutations among newly diagnosed patients before scaling-up programmes in Burkina Faso and Cameroon. Antivir. Ther. 11: 575–579.

Vidal, N., Niyongabo, T., Nduwimana, J., Butel, C., Ndayiragije, A., Wakana, J., Nduwimana, M., Delaporte, E. and Peeters, M. 2007. HIV type 1 diversity and antiretroviral drug resistance mutations in Burundi. AIDS Res. Hum. Retroviruses 23: 175–180.

Vingerhoets, J., Azijn, H., Fransen, E., De Baere, I., Smeulders, L., Jochmans, D., Andries, K., Pauwels, R. and de Béthune M.-P. 2005. TMC125 displays a high genetic barrier to the development of resistance: evidence from in vitro selection experiments. J. Virol. 79: 12773–12782.

Vingerhoets, J., Buelens, A, Peeters, M, Picchio, G., Tambuyzer, L., Van Marck, H., De Smedt, G., Woodfall, B. and de Béthune, M.-P. 2007. Impact of baseline NNRTI mutations on the virological response to TMC125 in the phase III clinical trials DUET-1 and DUET-2. Antivir. Ther. 12: S34.

Waters et al. 2006. K65R, L74V and thymidine analog mutations (TAMs) in HIV-1 RT associated with reduced response to tenofovir DF in antiretroviral-experienced patients. CROI, Abstract 633.

Weber, J., Piontkivska, H. and Quiñones-Mateu, M.E. 2006. HIV type 1 tropism and inhibitors of viral entry: clinical implications. AIDS Rev. 8: 60–77.

Wensing, A.M., van de Vijver, D.A., Angarano, G., Asjõ, B., Balotta, C., Boeri, E., Camacho, R., Chaix, M.-L., Costagliola, D., De Luca, A., Derdelinckx, I., Grossman, Z., Hamouda, O., Hatzakis, A., Hemmer, R., Hoepelman, A., Horban, A., Korn, K., Kücherer, C., Leitner, T., Loveday, C., MacRae, E., Maljkovic, I., de Mendoza, C., Meyer, L., Nielsen, C., Op de Coul, E.L., Ormaasen, V., Paraskevis, D., Perrin, L., Puchhammer-Stockl, E., Ruiz, L., Salminen, M., Schmit, J.-C., Schneider, F., Schuurman, R., Soriano, V., Stanczak, G., Stanojevic, M., Vandamme, A.-M., Van Laethem, K., Violin, M., Wilbe, K., Yerly, S., Zazzi, M., Boucher, C.A. for the SPREAD Programme. 2005. Prevalence of drug-resistant HIV-1 variants in untreated individuals in Europe: implications for clinical management. J. Infect. Dis. 192: 958–966.

Whitcomb, J.M., Parkin, N.T., Chappey, C., Hellmann, N.S. and Petropoulos, C.J. 2003. Broad nucleoside reverse-transcriptase inhibitor cross-resistance in human immunodeficiency virus type 1 isolates. J. Infect. Dis. 188: 992–1000.

White, K.L., Chen, J.M., Feng, J.Y., Margot, N.A., Ly, J.K., Ray, A.S., MacArthur, H.L., McDermott, M.J., Swaminathan, S. and Miller, M.D. 2006. The K65R reverse transcriptase mutation in HIV-1 reverses the excision phenotype of zidovudine resistance mutations. Antivir. Ther. 11: 155–163.

Winston, A., Pozniak, A., Mandalia, S., Gazzard, B., Pillay, D. and Nelson, M. 2004. Which nucleoside and nucleotide backbone combinations select for the K65R mutation in HIV-1 reverse transcriptase. AIDS 18: 949–951.

Winters, M.A., Shafer, R.W., Jellinger, R.A., Mamtora, G., Gingeras, T. and Merigan, T.C. 1997. Human immunodeficiency virus type 1 reverse transcriptase genotype and drug susceptibility changes in infected individuals receiving dideoxyinosine monotherapy for 1 to 2 years. Antimicrob. Agents Chemother. 41: 757–762.

Winters, M.A., Coolley, K.L., Girard, Y.A., Levee, D.J., Hamdan, H., Shafer, R.W., Katzenstein, D.A. and Merigan, T.C. 1998. A 6-basepair insert in the reverse transcriptase gene of human immunodeficiency virus type 1 confers resistance to multiple nucleoside inhibitors. J. Clin. Invest. 102: 1769–1775.

Wirden, M., Roquebert, B., Derache, A., Simon, A., Duvivier, C., Ghosn, J., Dominguez, S., Boutonnet, V., Ait-Arkoub, Z., Katlama, C., Calvez, V. and Marcelin, A.-G. 2006. Risk factors for selection of the L74I reverse transcriptase mutation in human immunodeficiency virus type 1-infected patients. Antimicrob. Agents Chemother. 50: 2553–2556.

Zhang, J., Rhee, S.Y., Taylor, J. and Shafer, R.W. 2005. Comparison of the precision and sensitivity of the Antivirogram and PhenoSense HIV drug susceptibility assays. J. Acquir. Immune Defic. Syndr. 38: 439–444.

Index

L

M

N